Non-Noble Metal Catalysis

Non-Noble Metal Catalysis

Molecular Approaches and Reactions

Edited by
Robertus J. M. Klein Gebbink
Marc-Etienne Moret

Editors

Prof. Dr. Robertus J. M. Klein Gebbink
Organic Chemistry & Catalysis
Debye Institute for Nanomaterials
Science
Utrecht University
Universiteitsweg 99
3584 CG Utrecht
Netherlands

Dr. Marc-Etienne Moret
Organic Chemistry & Catalysis
Debye Institute for Nanomaterials
Science
Utrecht University
Universiteitsweg 99
3584 CG Utrecht
Netherlands

Cover Image: © julie deshaies/
Shutterstock

Library of Congress Card No.:
applied for

British Library Cataloguing-in-Publication Data
A catalogue record for this book is available from the British Library.

Bibliographic information published by the Deutsche Nationalbibliothek
The Deutsche Nationalbibliothek lists this publication in the Deutsche Nationalbibliografie; detailed bibliographic data are available on the Internet at <http://dnb.d-nb.de>.

Print ISBN: 978-3-527-34061-3
ePDF ISBN: 978-3-527-69911-7
ePub ISBN: 978-3-527-69910-0
oBook ISBN: 978-3-527-69908-7

Cover Design: Wiley
Typesetting SPi Global, Chennai, India
Printing and Binding C.O.S. Printers Pte Ltd, Singapore

Printed on acid-free paper

10 9 8 7 6 5 4 3 2 1

Contents

Preface

Since its early development in the 1960s, the field of homogeneous catalysis has led to a plethora of industrially applied organometallic catalysts and, not the least, to an in-depth fundamental understanding of the reactivity of transition metal complexes. The threefold awarding of the Nobel Prize to the field of homogeneous catalysis in the very beginning of the twenty-first century highlights the impact the homogeneous catalysis field has made on chemistry and synthesis in general [1–3]. Remarkably, the reactions for which these awards have been given predominantly make use of noble, platinum group metals. This illustrates the historical importance and dominance of the use of noble metals in the field of homogeneous catalysis at large, from gram-scale, exploratory organic synthesis in pharmaceutical labs to large-scale industrial processes.

Although non-noble metals such as iron have been investigated from the early days of catalysis on, their noble counterparts have quickly and durably come to occupy the center of the stage. However, many recent endeavors in the field shift the focus back to non-noble metals, sometimes referred to as "base metals," in the development of new homogeneous catalysts. This move is largely driven by economic and environmental considerations. Not only are market prices of noble metals generally high, which is largely due to their relatively low abundancy in the earth crust, but these prices are often rather volatile as well. In addition, many of the noble metals are associated with toxicity issues for humans and the environment. As a consequence, the use of noble metal catalysts in, e.g., later stages of active pharmaceutical ingredient synthesis requires stringent purification procedures with the associated energetic and financial costs.

Motivated by many of these considerations, the scientific community has become interested in the study and development of homogeneous catalysts that are based on non-noble metals. The practical use of metals such as manganese, iron, and cobalt promises to alleviate, at least partly, some of these issues. A recent analysis by the EU on the criticality of raw materials furthermore shows that the late first-row transition metals are all above the economic importance threshold, whereas all except cobalt are below the supply risk threshold [4]. This is in contrast with many other raw materials, including the platinum group metals, where geopolitical issues come in to play as well.

One should not forget, though, that the current blossoming of the field of non-noble metal catalysis is for a large part simply born out of scientific curiosity. The availability of multiple oxidation states, often spaced by one-electron differences, and the strong tendency to adopt high-spin electron configurations lead to markedly different chemistry for non-noble metals with respect to noble metals, e.g. in terms of kinetic lability and lifetimes of intermediates. The investigation of non-noble metals in homogeneous catalysis is therefore expected to unravel fundamentally new reactivity patterns, leading to new catalysts, and, not unimportantly, to new applications. In contrast to the early days of catalysis, the current availability of advanced spectroscopic and analytical tools, including density functional theory and other computational methods, now allows for a detailed characterization and understanding of non-noble metal complexes, catalysts, and reactive intermediates. This situation is clearly different from the times when Kochi was exploring iron-mediated C—C coupling chemistry in the 1940s (see Chapter 11 by Neidig et al.).

Although the terms "non-noble metals" and "base metals" are broadly defined, we opted to focus this book on the late, first-row transition metals Mn, Fe, Co, Ni, and Cu, given the volume of recent interest in and the development of the catalytic chemistry of these metals. Only in selected cases will examples using other metals be discussed, and if so mainly to put recent developments in perspective. In this sense, the book adds on and complements earlier books on related topics, such as the book edited by Bullock on "catalysis without precious metals" [5].

The first four chapters of the book deal with conceptual aspects of non-noble metal catalysis in order to provide the reader with some further background. These chapters include discussions on non-innocent ligands (de Bruin, Chapter 1), computational methods (Ye, Neese, Chapter 2), multinuclear complexes (Mankad, Chapter 3), and industrial applications (Alsters, Le Fort, Chapter 4). Subsequent chapters discuss typical reaction classes, such as additions to C=C , C=N, and C=O double bonds (Chapters 5–10), the formation of C—C and C—hetero atom bonds through cross-coupling (Chapters 11–13), (formal) oxidation reactions (Chapters 14–16), and small-molecule activation (Chapters 16–20). These reaction classes are chosen to be representative of the broad range of reactions for which non-noble metal catalysts are being investigated. These chapters are presented from the point of view of synthetic method development or of catalyst development and may focus on the use of a single metal for a particular reaction or on a particular reaction itself. Accordingly, a particular reaction or catalyst may appear in more than one chapter.

We hope this book provides the more experienced reader with a contemporary overview of the current standing in the field of homogeneous non-noble metal catalysis and appeals to the less experienced reader in raising further interest in the field.

A big "thank you" not only goes out to all the contributors to this book, who have kept up with us as editors, but also the support staff at Wiley for their help and patience. We would also like to thank our collaborators within the European training network NoNoMeCat on homogeneous "non-noble metal catalysis" for the joint and stimulating efforts in further developing the field and training the next general generation of researchers in the field [6]. Not surprisingly, many of these collaborators are contributors to this book.

Utrecht, July 2018

Robertus J. M. Klein Gebbink
Marc-Etienne Moret

References

1 The Nobel Prize in Chemistry 2001 was awarded to William S. Knowles and Ryoji Noyori “for their work on chirally catalyzed hydrogenation reactions” and to K. Barry Sharpless “for his work on chirally catalyzed oxidation reactions.” See: www.nobelprize.org/nobel_prizes/chemistry/laureates/2001/.
2 The Nobel Prize in Chemistry 2005 was awarded to Yves Chauvin, Robert H. Grubbs, and Richard R. Schrock “for the development of the metathesis method in organic synthesis.” See: www.nobelprize.org/nobel_prizes/chemistry/laureates/2005/.
3 The Nobel Prize in Chemistry 2010 was awarded jointly to Richard F. Heck, Ei-ichi Negishi, and Akira Suzuki “for palladium-catalyzed cross couplings in organic synthesis.” See: www.nobelprize.org/nobel_prizes/chemistry/laureates/2010/.
4 European Commission. Study on the review of the list of Critical Raw Materials. https://publications.europa.eu/en/publication-detail/-/publication/08fdab5f-9766-11e7-b92d-01aa75ed71a1/language-en (accessed 19 July 2018).
5 Bullock, R.M. (ed.) (2010). *Catalysis without Precious Metals*. Wiley.
6 For information on the NoNoMeCat network see: www.nonomecat.eu (accessed 17 July 2018).

1

Application of Stimuli-Responsive and "Non-innocent" Ligands in Base Metal Catalysis

Andrei Chirila, Braja Gopal Das, Petrus F. Kuijpers, Vivek Sinha, and Bas de Bruin

University of Amsterdam (UvA), Van 't Hof Institute for Molecular Sciences (HIMS), Homogeneous, Supramolecular and Bio-Inspired Catalysis (HomKat), Science Park 904, 1098 XH Amsterdam, The Netherlands

1.1 Introduction

The development of efficient and selective catalysts is an important goal of modern research in chemistry – the science of matter and its transformations. Our society needs new catalysts to become more sustainable, and a desire for selectivity and efficiency in the preparation of medicines and materials has boosted our interest in developing new methods based on homogeneous catalysis, particularly on the development of new ligands that can be fine-tuned to specific needs. The properties of a metal complex as a whole are the result of the interaction between the metal center and its surrounding ligands. In traditional approaches, the steric and electronic properties of the spectator ligand are used to control the performance of the catalyst, but most of the reactivity takes place at the metal. Recent new approaches deviate from this concept and make use of ligands that play a more prominent role in the elementary bond activation steps in a catalytic cycle [1, 2]. The central idea is that the metal and the ligand can act in a synergistic manner to facilitate a chemical process. In this light, complexes based on the so-called "non-innocent" ligands offer interesting prospects and have attracted quite some attention.

The term "non-innocent" is broadly used, and diverse authors give different interpretations to the term. It was originally introduced by Jørgensen [3] to indicate that assigning metal oxidation states can be ambiguous when complexes contain redox-active ligands. As such, ligands that get reduced or oxidized in a redox process of a transition metal complex are often referred to as "redox non-innocent." [4, 5] With modern spectroscopic techniques, combined with computational studies, assigning metal and ligand oxidations states has become less ambiguous, and hence, many authors started to use the term "redox-active ligands" instead. Gradually, many authors also started to use the term "non-innocent" for ligands that are more than just an ancillary ligand, frequently involving ligands that have reactive moieties that can act in cooperative (catalytic) chemical transformations, act as temporary electron reservoirs, or respond to external triggers to modify the properties or reactivity of a complex.

Non-Noble Metal Catalysis: Molecular Approaches and Reactions,
First Edition. Edited by Robertus J. M. Klein Gebbink and Marc-Etienne Moret.

A common objective of many of these investigations is to achieve better control over the catalytic reactivity of first-row transition metal complexes, with the ultimate goal to replace the scarce, expensive noble metals currently used in a variety of catalytic processes by cheap and abundant first-row transition metals. Instead of providing a comprehensive overview of redox non-innocent [6, 7] and cooperative ligands [1, 8, 9], this chapter is intended to provide a conceptual introduction into the topic of achieving control over the catalytic reactivity of non-noble metals using non-innocent ligands on the basis of recent examples.

Noble metals are frequently used in several catalytic synthetic methodologies and many industrial processes [10]. Their catalytic reactivity is most frequently based on their well-established "two-electron reactivity," involving typical elementary steps such as reductive elimination and oxidative addition. These elementary steps easily occur for late (mostly second and third rows) transition metals having two stable oxidation states differing by two electrons. However, most noble metals are scarce and are therefore expensive (and sometimes toxic [11]). Therefore, it is necessary to reinvestigate the use of cheaper, abundant, and benign metals to arrive at cost-effective alternatives. This is not an easy task, as base metals (Fe, Co, Cu, Ni, etc.) often favor one-electron redox processes, and typical elementary steps commonly observed in noble metal catalysis are only scarcely observed for base metals. As such, the unique properties of non-innocent ligands are advantageous to gain better control over the reactivity of base metals. In some cases, this leads to reactivity comparable to that of noble metal complexes (but more cost-effective and benign), whereas in other cases, the combination of a base metal with a "non-innocent" ligand can actually give access to unique new types of reactivity.

This chapter has four parts. In Section 1.2, the concept of responsive ligands is discussed, giving examples of a series of ligands that can be tuned using external stimuli such as light, pH, or ligand-based redox reactions. These can trigger a change in the properties of the ligand, thereby modifying the reactivity of the metal. Section 1.3 deals with redox-active ligands that behave as electron reservoirs. In the examples provided, this feature enables oxidative addition and reductive elimination steps for first-row transition metal complexes that, without the aid of redox-active ligands, are less inclined to undergo these catalytically relevant elementary steps. Section 1.4 focuses on recent examples of cooperative catalysis, in which non-noble metal reactivity is combined with ligand-based reactivity in key substrate activation steps. The last part (Section 1.5) deals with examples in which the coordinated substrate itself acts as a redox-active moiety in key elementary steps of a catalytic reaction. More specifically, these coordinated substrates get oxidized or reduced by the metal by a single electron, thus creating "substrate radicals," which play an important role in catalytic radical-type transformations.

1.2 Stimuli-Responsive Ligands

Common ancillary (innocent) ligands in homogeneous catalysis typically control the activity and selectivity of the catalyst by affecting the steric and electronic properties around the reactive metal center. As such, changing the reactivity of the active metal center usually requires the synthesis of new ligands, which is often associated with elaborate synthetic procedures [6]. However, the electronic and steric properties of ligands can sometimes be influenced in an easier manner by using external stimuli, involving, for example, ligand protonation/deprotonation, ligand oxidation/reduction, or (reversible) light-induced ligand transformations (Scheme 1.1) [12].

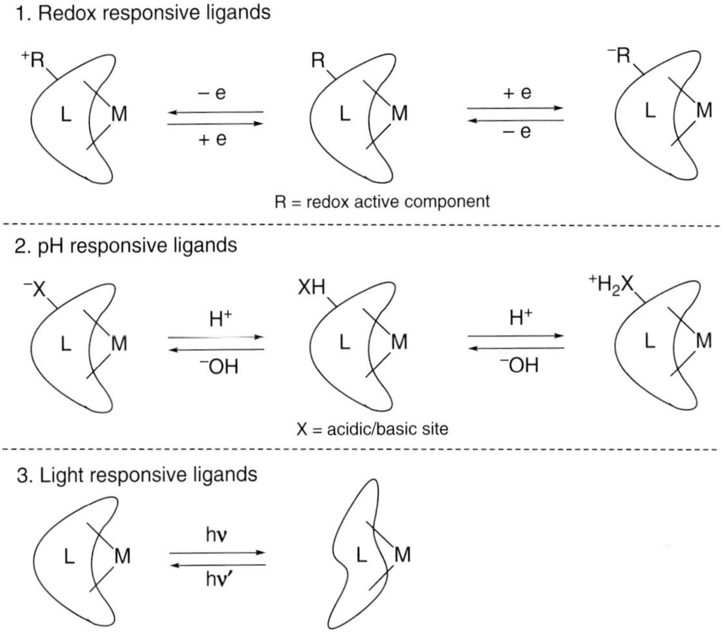

Scheme 1.1 Switching catalytic properties of a catalyst using external stimuli.

When using such responsive ligands, the metal oxidation state is typically unaffected, but its reactivity is nonetheless influenced by the new electronic and steric properties of the ligand. Furthermore, the solubility of the metal complex can sometimes be significantly influenced by such external stimuli. In most current literature, these ligands are nevertheless considered to be "innocent" ligands as they are not directly involved in substrate bond making/breaking processes nor lead to ambiguities in assigning the metal oxidation state. Stimuli-responsive ligands are particularly useful to influence the catalyst *during* a catalytic reaction and are therefore mainly applied to develop switchable catalytic systems.

1.2.1 Redox-Responsive Ligands

Oxidation or reduction of a complex containing one or more redox-active ligands can lead to oxidation or reduction of the ligand rather than the metal. As such, the ligand can switch between one or multiple oxidized and reduced states, by which the electronic properties of the ligand (and thereby the metal) change.

These redox processes can be triggered either chemically or electrochemically [13]. Often metallocenes such as ferrocene or cobaltocene are used because of their reversible oxidation and reduction cycles [14]. In other cases, the redox-active part of the ligand of interest is actually a metallocene moiety [15]. Upon oxidation of a ferrocenyl to a ferrocenylium group attached to the ligand, the electron density of the donor ligand decreases and thereby also that of the metal bound to this ligand, as can be observed in a shift of the CO stretch frequency to higher wavenumbers for carbonyl complexes [16]. Recently, a review appeared reporting a variety of chemical oxidants and reductants that allow the design of new catalysts with switchable ligands at a specific desired potential [17]. Examples of the use of redox-active ligands in catalysis frequently involve redox processes that partly occur at the redox-active ligand and partly at the catalytic metal center (see Section 1.3). Examples of redox-responsive ligands in catalysis wherein ligand-based redox processes affect the metal center and its catalytic properties indirectly are rare, especially for base metals. The main application of such reported examples is in the field of switchable catalysis. Furthermore, the solubility of the ligand can change significantly because of charge buildup, thus enabling separation of the catalyst from the reaction mixture after a catalytic reaction [18].

By oxidation or reduction of the ligand, the overall charge of the complex changes, which affects the catalytic activity of the central metal, and in some cases, this can be used to switch a catalyst ON and OFF. Most of the recently reported examples of such switchable catalysts involve systems based on noble metals [18–20], but a few examples of base metals are known as described below. One of the first redox-responsive base metal catalysts reported involves a titanium-based salen-type ligand substituted with two ferrocene (Fc) moieties (Figure 1.1a) [21]. The catalyst was used in the ring-opening polymerization of

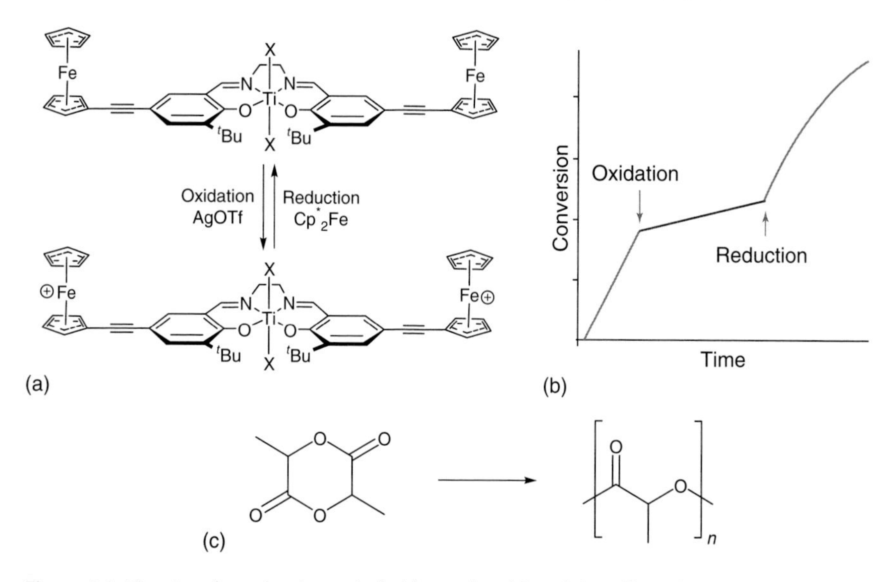

Figure 1.1 Titanium-based redox-switchable catalyst (a) and the effect of switching on the catalysis (b) on the polymerization reaction (c).

lactides, during which the neutral catalyst showed a 30-fold enhanced rate with respect to the oxidized complex. Oxidation of the ferrocenyl moieties of the catalyst does not completely shut down the catalytic activity, but by addition of small amounts of oxidant or reductant, the catalyst can nonetheless be switched between a more active (ON) and less active (OFF) state *during catalysis* (Figure 1.1b).

More recently, new titanium and zirconium catalysts were developed based on salfan (Y = NMe) and thiolfan (Y = S) ligands (Figure 1.2a) containing a ferrocene moiety closer to the metal center [22]. The reduced and oxidized catalysts showed opposing rates for the ring-opening polymerization of L-lactide and ϵ-caprolactone, respectively (Figure 1.2b). By switching between the two states during the polymerization reaction, the catalyst can be used to generate block copolymers with a high degree of regularity. In particular, this last example elegantly shows the power of switchable catalysts for application in polymerization reactions. Given the potential of such systems, we expect that many more examples of redox-switchable catalysts used for a variety of other catalytic reactions are likely to be disclosed in the next couple of years.

1.2.2 pH-Responsive Ligands

Ligands that can be easily protonated or deprotonated by applying relatively mild pH changes are commonly used to affect the solubility of catalysts. With this method, homogeneous catalysts can be easily recycled, thus saving cost and avoiding metal contamination in the products. Reversible protonation of amine groups to obtain water-soluble complexes has been applied to noble-metal-catalyzed reactions such as olefin metathesis [23] and cross-coupling reactions [24]. The selectivity of rhodium metathesis catalysts can be further altered upon protonation of the ligand [25]. By using similar

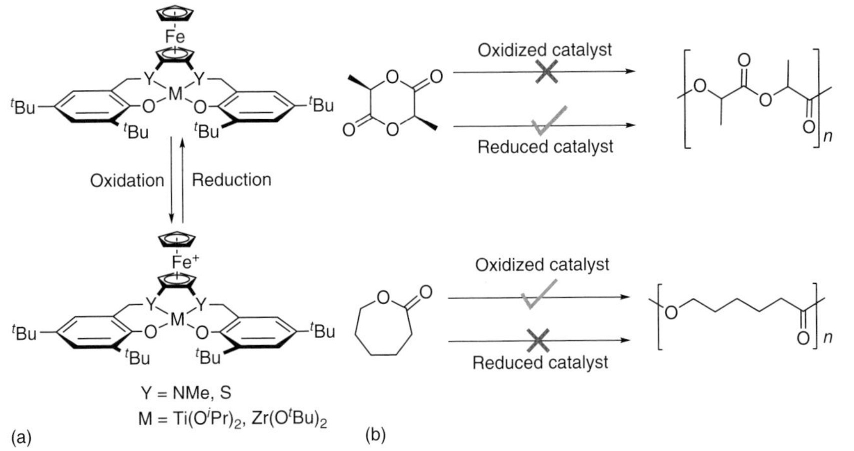

Figure 1.2 Ferrocene containing redox-switchable catalysts (a) and inverted reactivity for the resulting oxidized and reduced complexes (b). Source: Wang et al. 2014 [22]. Reproduced with permission of ACS.

ammonium-tagged NHC ligands, a copper-catalyzed click reaction in water was developed by Li and coworkers [26]. The products could simply be extracted in order to recycle the catalyst several times with a small loss of overall yield, but the catalyst was not switchable. In 2012, the same group reported a similar copper complex for the carbonylation of boronic acids, benzoxazoles, and terminal alkynes [27]. In this case, the catalyst precipitates upon protonation and could be separated by centrifugation (Figure 1.3). The catalyst can be recycled up to four times with only moderate loss in activity. Related copper-catalyzed reactions based on NHC complexes with pendant bases have also been reported [28], but the effects of deprotonation on the catalysis or recyclability of the complex were not discussed in detail for these systems.

The second type of proton-switchable ligands is composed of bipyridine and phenanthroline ligands equipped with moieties that can be (de)protonated. Many late transition metal catalysts based on iridium [29–31], rhodium [32], and rhenium [33] have been reported to use this class of ligands. Recent base metal examples include a switchable copper catalyst for the Ullmann reaction of aryl bromides. The catalyst can be deprotonated in basic water to obtain a highly active catalyst, which could be recycled by acidification (Figure 1.4) [34].

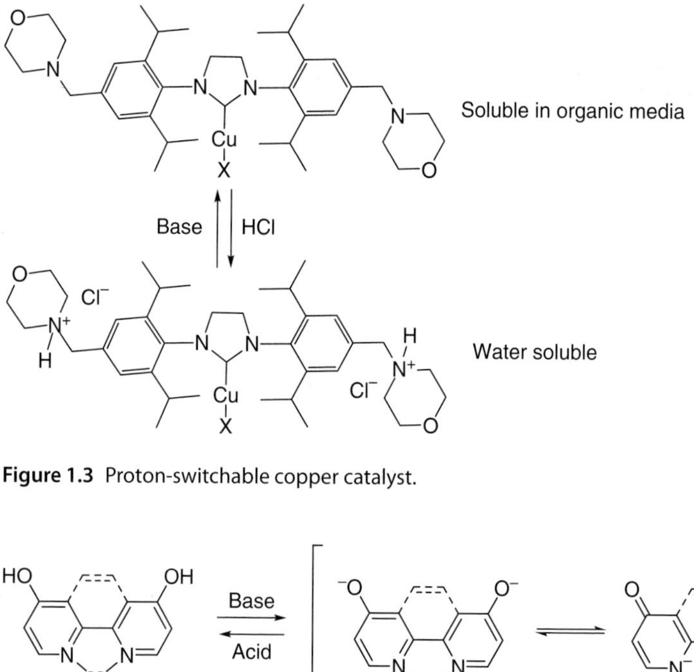

Figure 1.3 Proton-switchable copper catalyst.

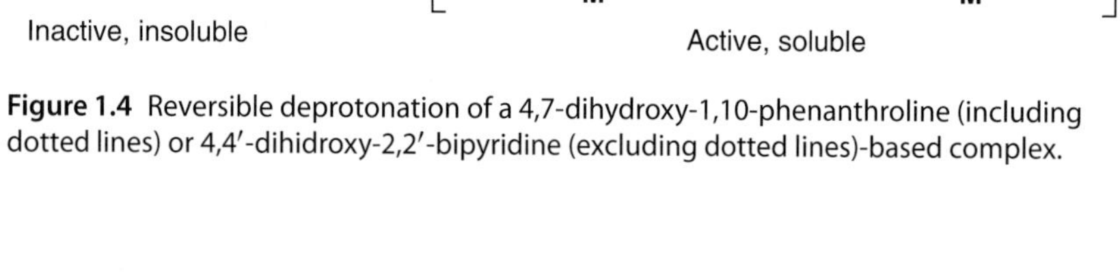

Figure 1.4 Reversible deprotonation of a 4,7-dihydroxy-1,10-phenanthroline (including dotted lines) or 4,4′-dihidroxy-2,2′-bipyridine (excluding dotted lines)-based complex.

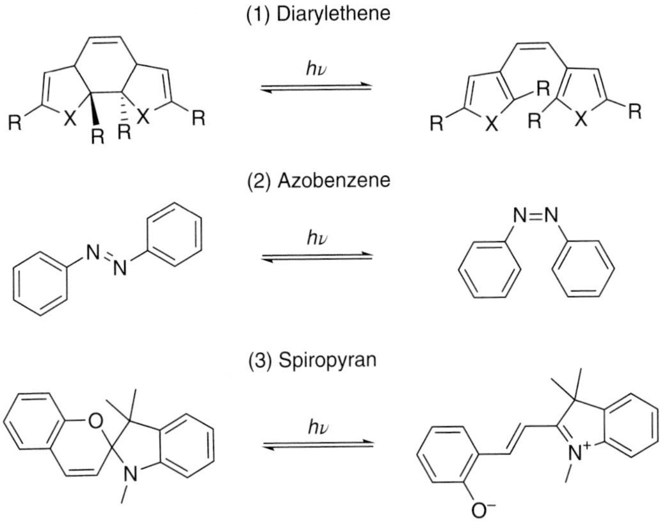

(1) Diarylethene

(2) Azobenzene

(3) Spiropyran

Figure 1.5 Light-active scaffolds that undergo structural changes upon irradiation.

Another example of a proton-switchable catalyst involves a cobalt complex based on bipyridine for the hydrogenation of carbon dioxide to formate [35]. The alcohol substituents were introduced either at the 4,4′- or the 6,6′-positions. The obtained complexes show a large dependence on the concentration of base as the deprotonated complex is active and more stable under the reaction conditions. Recyclability data were not reported for these systems, but the complexes do, however, show a significantly higher activity after deprotonation of the ligand.

1.2.3 Light-Responsive Ligands

Light, being rather non-invasive, is perhaps the most interesting external trigger to switch a bistable catalyst. Upon irradiation with light, many molecules such as diarylethenes, azobenzenes, or spiropyrans can undergo structural rearrangement (Figure 1.5). Incorporation of these switchable moieties in a catalyst could result in easy control of its catalytic activity [36, 37], and use of different wavelengths typically allows two-way switching of these scaffolds.

An elegant example of this type of responsive catalyst was reported by the group of Branda for a copper-catalyzed cyclopropanation reaction (Figure 1.6) [38].

Upon reversible isomerization of the open ligand (Figure 1.6, right complex) to the cyclized complex (Figure 1.6, left complex), almost all stereoselectivity was lost. Although switching the ligand was more difficult after copper coordination, it was still feasible after addition of a small amount of a coordinating solvent to the reaction mixture.

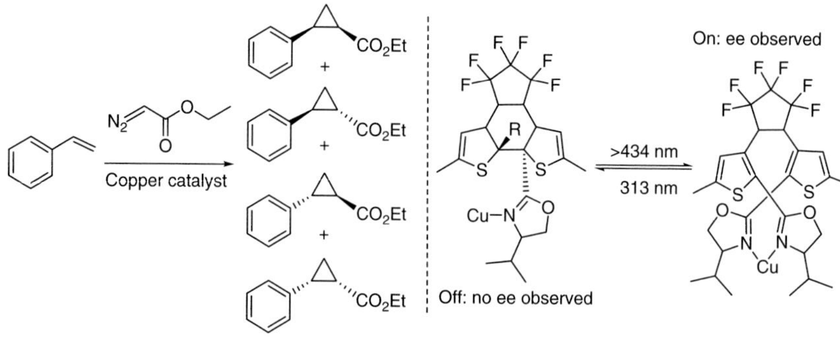

Figure 1.6 Light-induced enantioselective cyclopropanation.

1.3 Redox-Active Ligands as Electron Reservoirs

The most straightforward application of redox-active ligands is as electron reservoir, to facilitate redox processes for base metals that would otherwise be difficult or impossible. As such, redox-active ligands can participate in key redox processes of a catalytic cycle, such as oxidative addition or reductive elimination steps (Scheme 1.2). The ligand can temporarily store or release additional electrons allowing the metal complex to perform multielectron steps, avoiding formation of high-energy oxidation states of the metal if the energy levels of redox-active ligands are more accessible [39]. In this way, even purely ligand-centered redox processes become possible leaving the metal in the same oxidation state throughout an entire catalytic cycle. As such, by making use of redox-active ligands, the reactivity of first-row transition metals can be tuned toward catalytic properties more typically observed for noble metals [40].

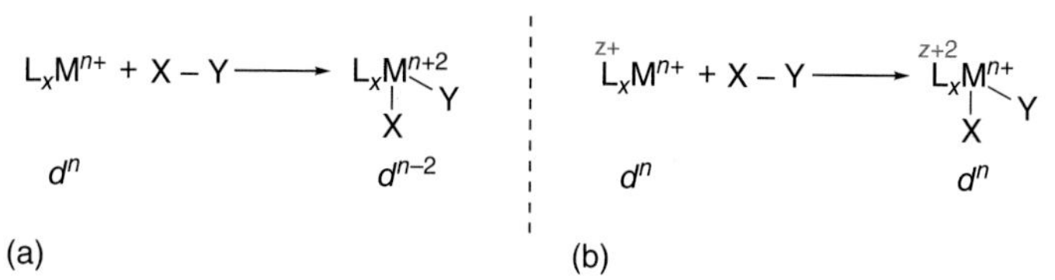

Scheme 1.2 (a) Classic oxidative addition and (b) oxidative addition in metal complexes with redox-active ligands.

1.3.1 Bis(imino)pyridine (BIP)

The bis(imino)pyridine (BIP) ligand (Scheme 1.3) has perhaps been most frequently used as an electron reservoir. This class of ligands consists of pyridine derivatives with imine functionalities at the 2,6-positions and stabilizes metals in low (formal)-oxidation states. The three nitrogen centers of the ligand bind to a metal in a tridentate manner, forming pincer complexes (Scheme 1.3, left). The obtained non-innocent ligand can have more than one oxidation state, as the ligand π^*-orbitals can accept several electrons. The ligand can easily be synthesized

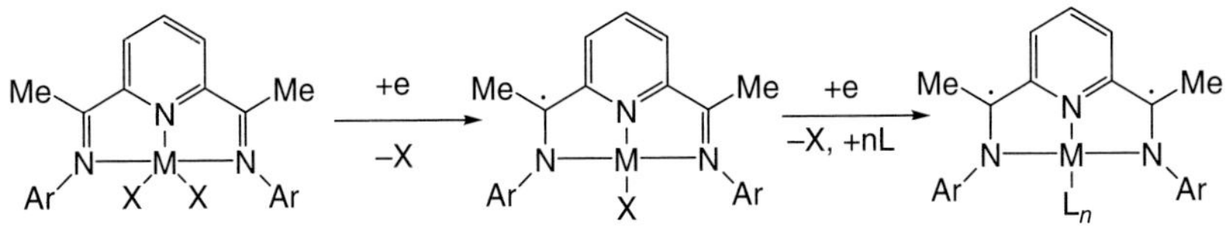

Scheme 1.3 Bis(imino)pyridine complex (left), mono-reduced (middle), and bis-reduced complexes (right).

by Schiff base condensation of commercially available 2,6-diacetylpyridine with 2 equiv. of an aniline derivative. Most commonly, variations in the ligand are made by changing the anilines in the condensation reaction. The highly conjugated ligand framework of bis(imino) pyridine stabilizes unusual formal oxidation states of the metal. A neutral complex is able to accept up to three electrons, leading to ambiguity about the oxidation states of the metal center [41–45].

A variety of coordination complexes with different transition metals have been prepared. Extensive studies by Chirik and coworkers [46–48], Wieghardt [42, 46–48], Budzelaar and coworkers [42, 43], de Bruin [41, 42], deBeer [48], and others have established unusual electronic structures of first-row transition metal complexes containing the BIP ligand. In many cases, the studies revealed the presence of unpaired electrons at the ligand, coupled antiferromagnetically to unpaired electrons at the metal. For example, the four-coordinated compound, (BIP)Fe(N_2) is best described as an intermediate spin ferrous derivative ($S_{Fe} = 1$) antiferromagnetically coupled to a bis(imino)pyridine triplet dianion (Scheme 1.3, right) [46–48]. BIP complexes of first-row transition metals have been used for various multiple electron transfer processes. The obtained complexes occasionally even outperform noble metal complexes.

1.3.1.1 Ethylene Polymerization with BIP

In 1998, Brookhart and coworkers [49] and Gibson and coworkers [50] introduced BIP complexes of mid-to-late first-row transition metals for ethylene polymerization [51]. This was a major breakthrough in the field of olefin polymerization catalysis, as most catalysts explored until then were based on early d^0 transition metals. The abundance of high-valent Ti^{IV}, Zr^{IV}, and Hf^{IV} complexes in polymerization reactions is readily understood from the fact that β-hydrogen elimination is a suppressed chain transfer/chain termination process for these metals, as it requires not only a vacant site but also the presence of (at least two) *d*-electrons. Some palladium catalysts equipped with bulky ligands shielding the axial positions are known to produce polymers by slowing down chain transfer. This is because direct olefin dissociation (after β-hydrogen elimination) is a thermodynamically uphill process for these systems, and the bulky ligand prevents/suppresses olefin substitution and chain transfer to monomer. However, β-hydrogen elimination is still rapid, leading to chain-walking and production of highly branched polymers. As such, it is quite remarkable that (^{i}PrBIP)FeX_2 complexes (Figure 1.7**A**) show a high activity to produce linear, high-density polyethylene in the presence of MMAO (a modified methylaluminoxane activator). The activity is even higher than many of the typical metallocene-based

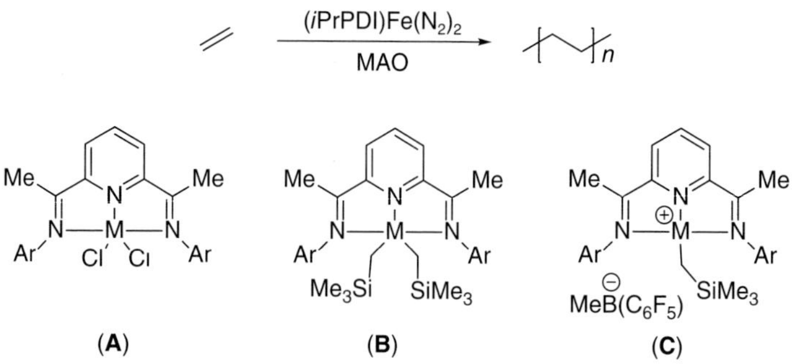

Figure 1.7 Bis(imino)pyridyl complexes used in the polymerization of ethylene.

catalysts [52–54]. The bulky aryl substituents are crucial for the production of high-molecular-weight polymers, presumably because they slow down the rate of chain transfer to the monomer (like Pd). Following these seminal works, Chirik and coworkers developed a new family of related mono- and dialkyl complexes (Figure 1.7**B**) [55]. The corresponding cationic complex (Figure 1.7**C**) was obtained by addition of $[PhMe_2NH][BPh_4]$. The cationic complex proved to be even more active and produced polymers with higher molecular weight (MW) and smaller polydispersity than with MMAO-activated catalysts. These results are consistent with chain termination by β-H elimination, which is, however, much slower than olefin insertion into the Fe—C bond of the growing chain.

Interestingly, Gambarotta and Budzelaar re-examined the alkylation process and found that ligand alkylation as well as ligand reduction occurred under the catalytic conditions, at least during the activation process of the bis-halide precursor to the active catalysts with (M)MAO [56, 57]. The newly obtained complexes also proved highly active in olefin polymerization with (M)MAO activators. Hence, the nature of the "real" active species was unclear for a long time. Despite these confusing findings, Chirik was able to show that the "active Brookhart catalyst" involved in the polymerization reaction is a cationic $[(BIP)Fe^{II}$-alkyl$]$ with an unmodified and non-reduced BIP ligand [50]. As such, it seems that the redox activity of the BIP ligand scaffold is not directly involved in the chain growth process (which is not a redox process anyway). It has been suggested in some reports that an Fe^{III} complex can also be an effective catalyst. From DFT-calculated energy barriers, the Fe^{III} catalyst was found to be more effective during the propagation steps (10.8 kcal mol^{-1} for Fe^{III} vs. 14.2 kcal mol^{-1} for Fe^{II}) [58]. However, the termination/propagation ratio and the experimental polymer MW favor an Fe^{II} catalyst as the active species.

1.3.1.2 Cycloaddition Reactions

Although the redox activity of the BIP ligand does not seem to play a direct role in chain propagation by the Brookhart/Gibson catalysts described above (although it does seem to play a role in the catalyst activation steps), the Chirik group recently reported a number of catalytic reactions in which metal–ligand redox cooperation does seem to play a direct role in some of the key steps of the catalytic mechanism. This seems to be particularly relevant in a series of [2+2] cycloaddition reactions reported by the Chirik group (see below). Although the redox activity of the BIP ligand is difficult to study under the catalytic conditions, mechanistic model studies clearly revealed the importance of the redox-active BIP ligand. To determine where the electrons end up after oxidative addition, a C—C bond cleavage of biphenylene was explored. The reaction is relatively easy because of the thermodynamic driving force of ring-opening of the trained four-membered ring and formation of two strong metal–aryl bonds in the metallocyclic product. The electronic structure of the iron metallocycle **D** (Figure 1.8) was studied by a combination of X-ray diffraction, SQUID magnetometry, NMR spectroscopy, X-ray absorption and emission spectroscopies, and DFT. The combined experimental and computational data established an Fe^{III} product with a bis(imino)pyridine radical anion. The net two-electron process occurs with one electron oxidation at the supporting ligand and one electron oxidation at the iron center [59].

Chirik and coworkers applied similar concepts in intramolecular [2+2] cycloaddition reactions (Scheme 1.4, top) [60]. According to the proposed mechanism, initial reaction of the $(PDI)FeN_2$ complex **E** with the diene substrate forms the corresponding π-complex **F**. Here, both complexes have the BIP ligand in the two-electron reduced form. Complex (**F**) is proposed to undergo a subsequent two-electron oxidative addition process to generate complex **G**. Similar to the above model studies, the electrons required for this transformation are proposed to derive from the reduced ligand, in this case both electrons. Therefore, the iron center can maintain the energetically favorable Fe^{II} oxidation state (instead of the less favorable Fe^{IV} oxidation state). Subsequently, intermediate **G** participates in a two-electron ligand-based reductive elimination reaction to release the product and regenerate the catalyst (**E**). Here, the electron storage capacity of the ligand allows the metal to maintain its stable Fe^{II} oxidation state instead of a high energy Fe^0 oxidation state. These complexes have also been applied successfully in related enyne cyclizations [61].

The same catalysts are also active in the intermolecular reaction between ethylene and various 1,3-butadienes to form the corresponding derivatives (Scheme 1.4, bottom) [62]. In these reactions, a β-H elimination step follows the initial cycloaddition step. An equimolar mixture of ethylene and butadiene in the presence of 5 mol% iron catalyst at 23 °C afforded the expected vinyl cyclobutane in a good yield. When a methyl group was introduced into the diene, a 1,4-addition of ethylene to the 1,3-diene occurred, as described previously by Ritter and coworkers [63]. The sterically more hindered isoprene

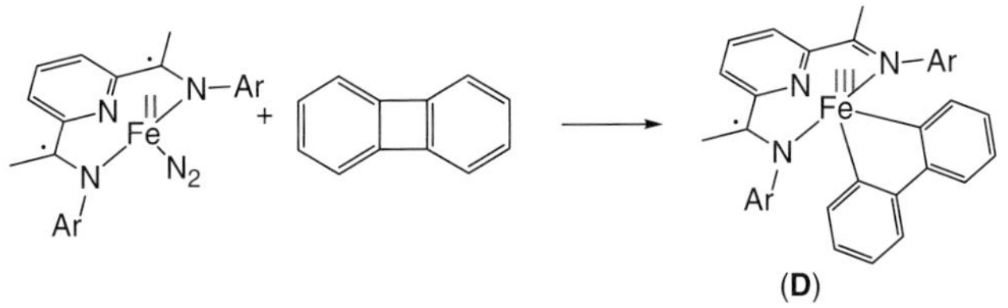

Figure 1.8 Ligand-mediated oxidative addition.

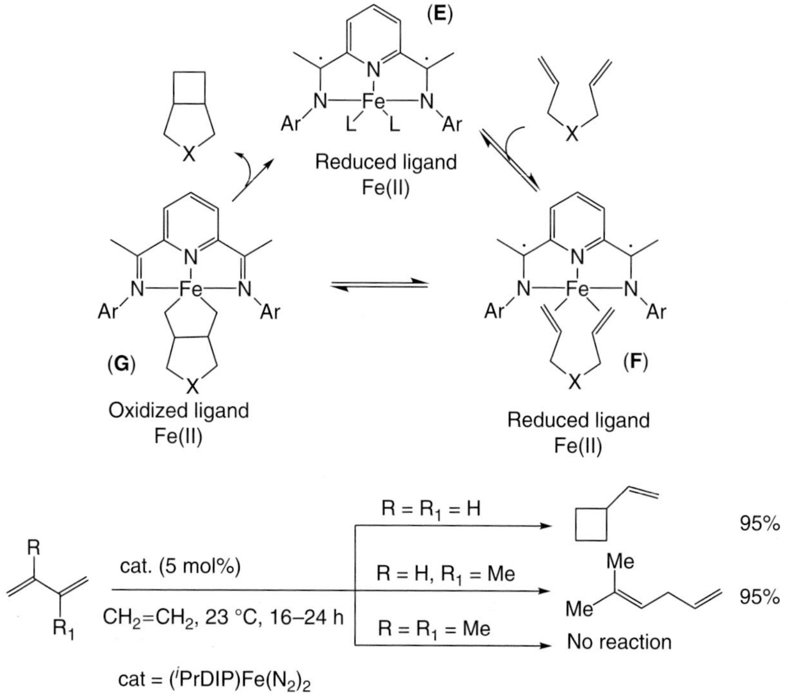

Scheme 1.4 Ligand-mediated cycloaddition reactions.

is a weaker ligand, disfavoring ligand-induced reductive elimination over β-H elimination.

1.3.1.3 Hydrogenation and Hydro-addition Reactions

Addition of H—X (X = H, Si, B, etc.) to alkenes has long relied on precious metal catalysts supported by strong field ligands to enable highly predictable two-electron redox chemistry that constitutes key bond-breaking and bond-forming steps during catalytic turnover. Recent advancements in the field, making efficient use of redox-active ligands, have, however, made it possible to also use base metals for these transformations. Electron transfer from and to the ligand framework in the oxidative addition of H—X bonds and reductive elimination of C—H bonds seems to play an important role in these base metal-catalyzed reactions. Substituted (BIP)Fe$(N_2)_2$ catalysts exhibit high turnover frequencies at both low catalyst loadings and hydrogen pressures for the hydrogenation of α,β-unsaturated alkenes. Exploration of structure–reactivity relationships established smaller aryl substituents (**I** over **H**) and more electron-donating ligands (**J** over **H**, **I**) resulted in an improved performance [64] (Figure 1.9). Synthesis of enantiopure, C_1 symmetric complex **K** has led to the development of highly enantioselective hydrogenation reactions of substituted styrene derivatives [65].

The observation of improved hydrogenation activity upon introduction of more electron-donating chelates inspired the synthesis of NHC pincer complexes

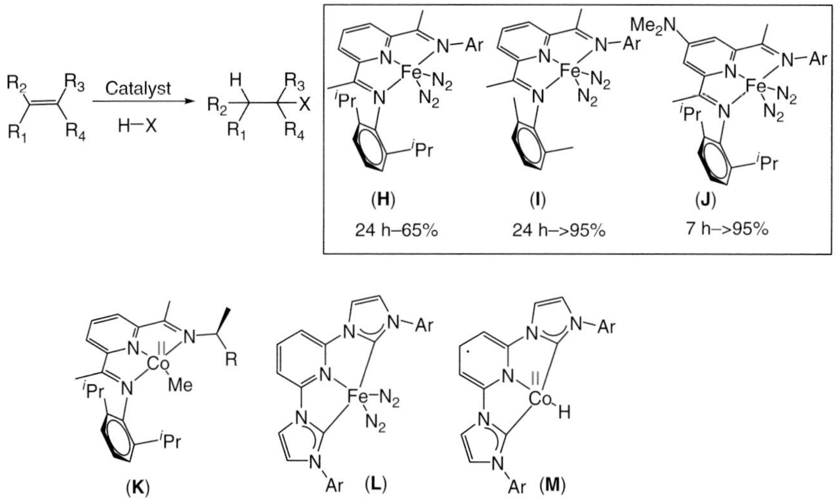

Figure 1.9 Family of BIP-related complexes (**H–M**) for the hydrogenation of alkenes.

L and **M** [66, 67], which also show high activity for unactivated di-, tri-, and tetra-substituted alkenes [68]. However, in contrast to the BIP ligand complexes, detailed spectroscopic studies indicate that the carbon–nitrogen–carbon (CNC) pincer acts as a classical ancillary ligand without involvement of ligand redox activity [69]. As such, one can conclude that application of strong field ligands, forcing low spin configurations, is a valuable alternative strategy to enable two-electron oxidative addition/reductive elimination reactions at iron and cobalt.

Substituted (BIP)Fe$(N_2)_2$ complexes have also been successfully applied for hydroboration and isomerization of alkenes with pinacolboranes [70]. An analogous cobalt catalyst has been found to be even more reactive and was applied for hindered alkenes and alkynes as well [71–74]. The mechanism involves selective insertion of an alkynyl boronate ester into a Co—H bond (the oxidative addition product), which was also proven spectroscopically.

Redox non-innocent bis(imino)pyridine complexes of iron have also been successfully applied for hydrosilylation of alkenes. Both $PhSiH_3$ and Ph_2SiH_2 were found to be effective in silylation and give anti-Markovnikov addition products within minutes [75]. The mechanism is the same as described for hydrogenation and hydroboration. The carbon–silicon bond formation reaction was also studied by the Ritter group using bidentate imino-pyridine complex **N** (Figure 1.10) [76]. The X-ray crystal structure indicates that the C—N bond lengths in the imino functionalities (1.343 ± 0.015 Å) are clearly intermediate between a C—N double bond (c. 1.28 Å) and a single bond (c. 1.46 Å). Similarly, the C—C bond length in the pyridine group is 1.382 ± 0.015 Å, which is the intermediate between a single bond (1.47 Å) and a double bond (1.35 Å). These parameters are clearly indicating a radical anion state of the ligand. The hydrosilylation of carbonyls has also been investigated using manganese complexes **O**, **P** [77]. However, the

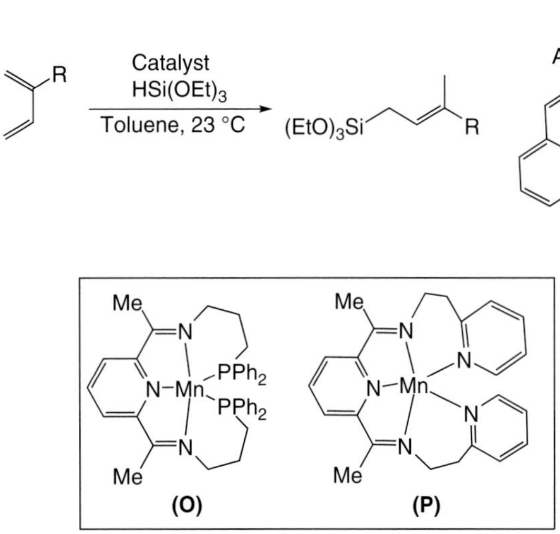

Figure 1.10 N,P Catalysts for hydrosilylation of alkenes.

redox involvement of the non-innocent macrocyclic BIP analog was not detailed or investigated in all these cases.

1.3.2 Other Ligands as Electron Reservoirs

Dithion, catechol, o-aminophenol, and o-phenylenediamine-type bidentate ligands have also been reported to show non-innocent behavior in combination with base metals (Scheme 1.5, left). The coordination behavior of these ligands in their different oxidation states has been studied in great detail [78, 79]. The ligands have three oxidation states (for 1,2-diol, it is catecholato in the fully reduced form, semiquinonato in the one-electron oxidized form, and quinone in the fully oxidized form). The phenyl backbone of these ligands is often substituted to tune the electronic properties, prevent unwanted radical–radical coupling reactions, and stabilize different oxidation states. Very recently, Pinter, de Proft, and coworkers reported a DFT study, which revealed that the reduced ligands strengthen the metal–ligand bonds, resulting in stabilized $M{-}L^{-1/2}$ configurations [80]. This strongly contributes to the overall thermodynamic driving force for ligand-centered electron transfer.

A key development in the field of C—C coupling involving redox-active ligands coordinated to a Co center came from the work of Soper and coworkers [81]. The unusual square planar nucleophilic triplet ground state of the Co^{III} bis-iminophenolate (Scheme 1.5, **Q**) is able to accommodate the formal oxidative addition of an alkyl fragment to yield a five-coordinate square pyramidal Co^{III} species (Scheme 1.5, **R**) with anti-ferromagnetically coupled ligand diradicals. Subsequently, the complex can transfer a formal R^+ group to either aryl or alkyl zinc bromides to yield the corresponding C—C coupled products. This sets the stage for further development of catalytic cross-coupling methodologies involving first-row metals, exploiting the role of redox-active ligands.

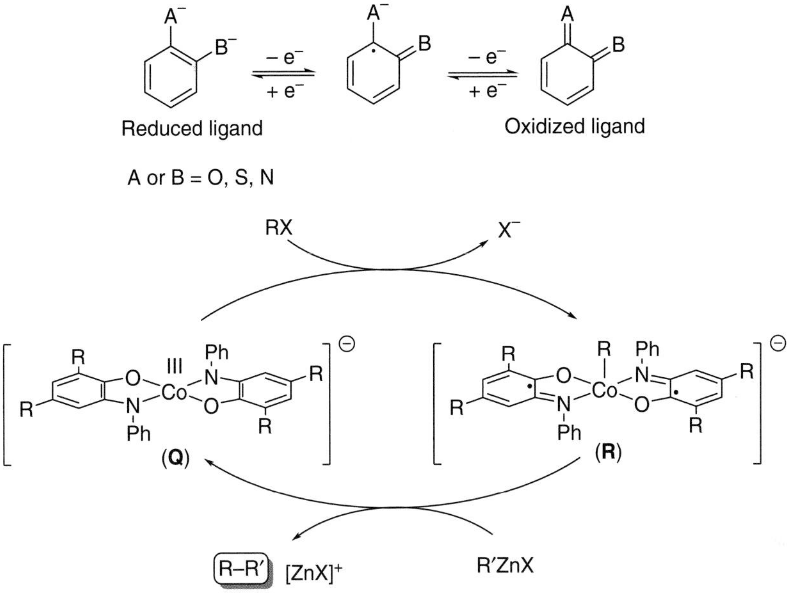

Scheme 1.5 Catechol-derived redox non-innocent ligands reported in literature (left) and corresponding cobalt reactivity (right).

Other types of coupling reactions have also been reported with base metals such as canonical Ni^0-catalyzed Kumada coupling between an aryl bromide and an aryl Grignard reagent [82] and homodimerization of benzyl halide [83]. The coupling reaction proceeds via a similar mechanism as the corresponding noble-metal-catalyzed reaction. The ligand-assisted oxidative addition product has been successfully isolated and characterized by Chaudhuri, Fensterbank, and coworkers [84, 85].

1.4 Cooperative Ligands

In cooperative catalysis, the metal and the ligand act together to activate the substrate. This is a useful approach to enhance and control the reactivity of (first-row) transition metals in catalytic reactions. The first and most well-known examples are catalysts containing ligands that function as internal bases or acids, as pioneered by Noyori, Beller and Milstein for noble metal catalysis [86, 87]. However, catalysts containing other reactive ligand moieties such as ligand radicals are gradually being explored as well (Scheme 1.6).

In the cooperative mode of action, the substrate may initially bind to the metal [88–91] or directly interact with the reactive part of the ligand [92]. These initial interactions are key to bringing the substrate geometrically close and physically accessible to the main reactive center. Scheme 1.6 illustrates the general substrate activation in cooperative non-innocent ligand catalysis. The substrate activation usually involves abstraction of a hydrogen atom or a proton from the substrate.

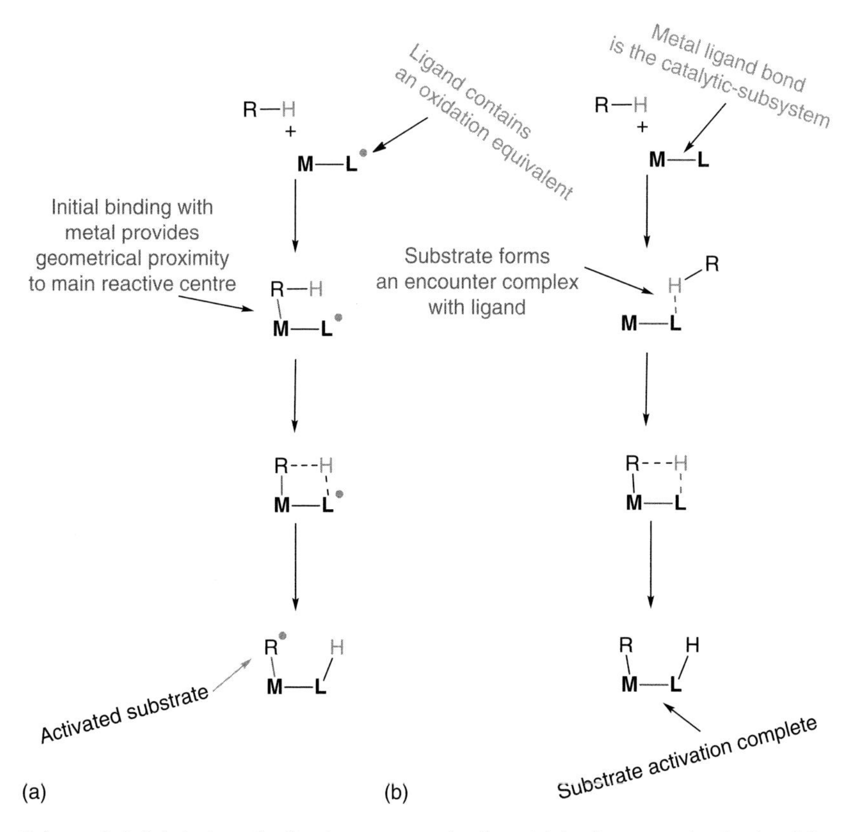

Scheme 1.6 Substrate activation by a cooperative ligand: (a) substrate activation involving ligand radicals and (b) substrate activation by the ligand acting as an internal base.

1.4.1 Cooperative Reactivity with Ligand Radicals

1.4.1.1 Galactose Oxidase (GOase) and its Models

Perhaps the most studied example of cooperative reactivity involving the reactivity of a ligand radical is the alcohol oxidation reaction catalyzed by the enzyme galactose oxidase (GOase). The first step in galactose oxidation by this enzyme is activation via one-electron oxidation of the sulfur-modified tyrosine-272 moiety to form an oxygen-centered (tyrosyl) radical (Scheme 1.7, **S**). The CH_2OH group on the galactose binds over the Cu—O-Tyr-495 bond to form the Cu(II) alkoxide complex **T** with the release of TyrOH (Scheme 1.7). Subsequent proton-coupled electronic transfer (PCET) shifts the radical to the galactose-alkoxide moiety, which, in turn, reduces the Cu(II) center of the enzyme to Cu(I) with the formation of the oxidized product. The reduced enzyme then reacts with dioxygen via a PCET pathway to form H_2O_2, hence completing the catalytic cycle.

In this mechanism, the metal and the ligand cooperate to facilitate the reaction. The initial enzyme activation produces a chemically active oxygen-centered radical. However, this radical alone is incapable of performing the selective reaction. Binding of the substrate to the metal center is also essential to bring

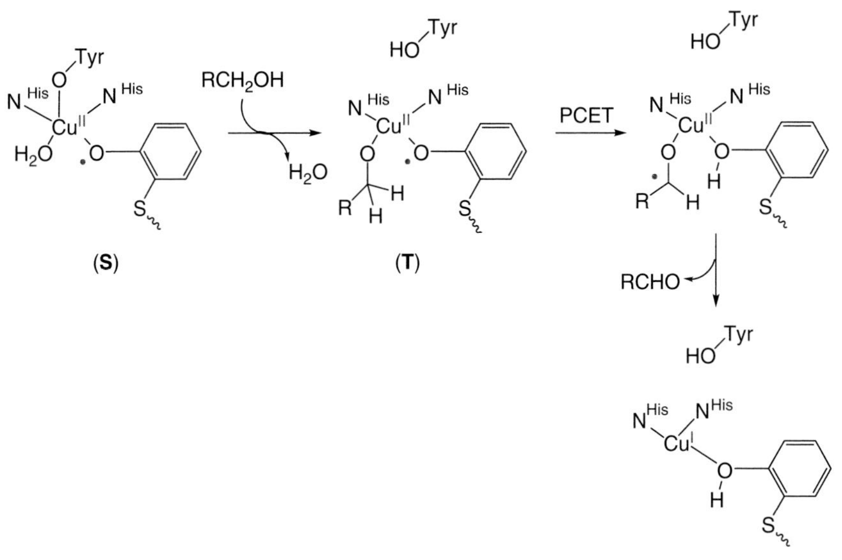

Scheme 1.7 Key steps in substrate activation and catalysis by the enzyme galactose oxidase.

the substrate and the ligand-centered radical close together. This geometrical arrangement enables the actual bond activation process. Subsequent electron transfer from the activated substrate to the metal is also important, hence the need of the redox-active Cu metal in the enzyme.

Analogous to the GOase system, Wieghardt and coworkers [90] reported a bioinspired Cu^{II}–thiophenol catalytic system (Figure 1.11). The initial catalyst activation step occurs by cooperative activation of the catalyst and the ligand to form a diradical system. In contrast to the GOase enzyme, this system has biradical characteristics. Therefore, it can carry out oxidation of two primary alcohols in a single catalytic turnover, enabling alcoholate-derived radical C—C coupling reactions with the formation of secondary diols.

Figure 1.11 (a) Cu(II)–thiophenol-based catalyst described by Wieghardt and Chaudhary. (b) Activation of two alcohol molecules.

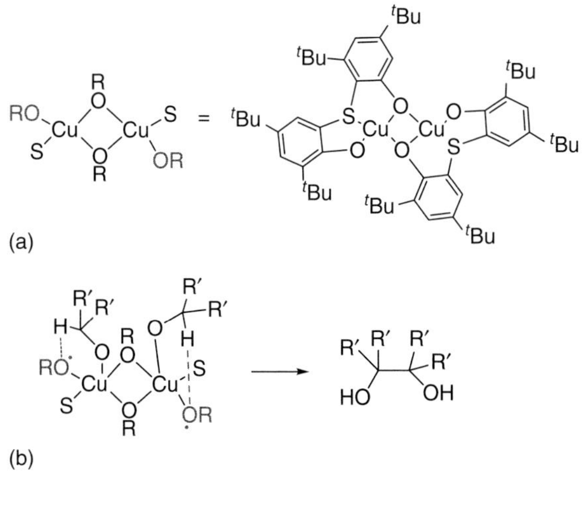

1.4.1.2 Alcohol Oxidation by Salen Complexes

The Zn–Salen catalyst (Scheme 1.8) reported by Wieghardt and coworkers [91] is another good example of catalysis carried out by a ligand radical. Remarkably, this system works even with the redox inert Zn^{2+} metal ion (having a completely filled *d*-shell). The highly conjugated ligand framework presents the possibility to store an oxidizing equivalent on the ligand, which can be used to drive alcohol oxidation catalysis. The substrate first gets deprotonated over the metal–oxygen bond and the resulting alkoxide binds to the metal to form complex **V** (Scheme 1.8). Zn^{2+} is needed to bring the substrates together, but the bond-breaking processes are entirely based on ligand in this case.

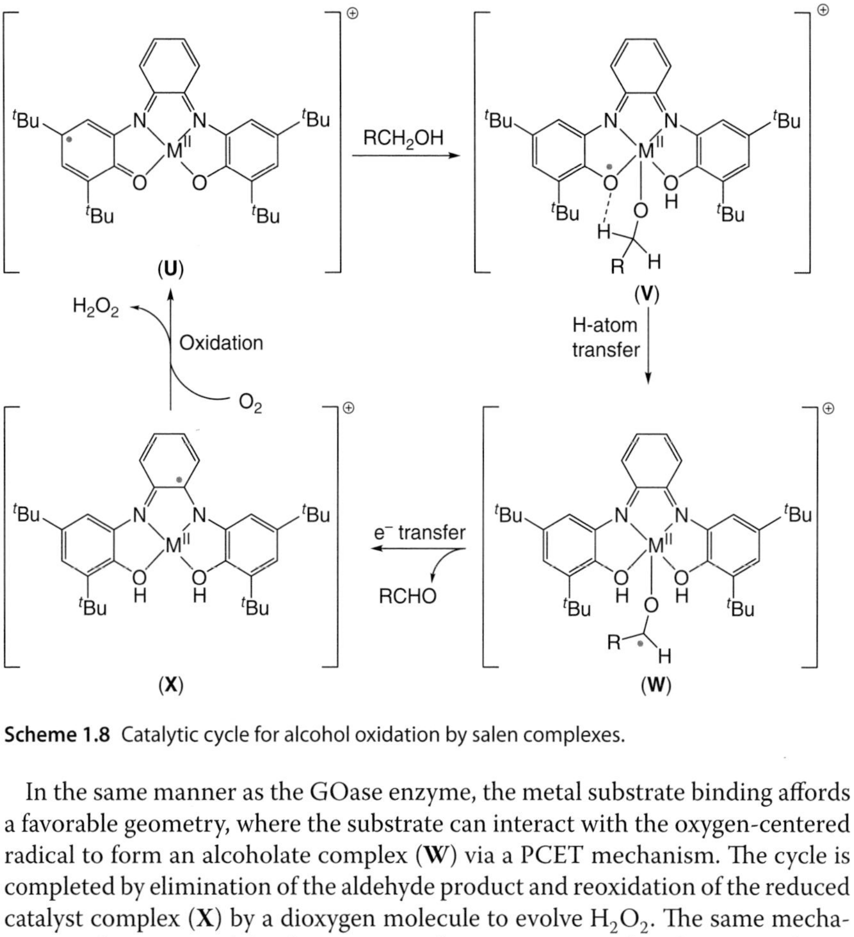

Scheme 1.8 Catalytic cycle for alcohol oxidation by salen complexes.

In the same manner as the GOase enzyme, the metal substrate binding affords a favorable geometry, where the substrate can interact with the oxygen-centered radical to form an alcoholate complex (**W**) via a PCET mechanism. The cycle is completed by elimination of the aldehyde product and reoxidation of the reduced catalyst complex (**X**) by a dioxygen molecule to evolve H_2O_2. The same mechanism is also proposed for the corresponding copper complex, despite Cu being redox active.

1.4.2 Base Metal Cooperative Catalysis with Ligands Acting as an Internal Base

Several well-described catalysts containing ligands that function as an internal base or acid were pioneered by Noyori, Beller and Milstein, initially using primarily noble metals [86, 87]. Application of these types of ligands to use base metals in catalysis is widely setting the stage though, and in several cases, the use of cooperative ligands to shift part of the reactivity from the metal to the ligand is taken to advantage. Some illustrative recent examples are discussed below.

1.4.2.1 Fe–Pincer Complexes

The Fe–pincer system reported by Holthausen and coworkers [92] catalyzes oxidation of methanol, methanediol, and formic acid to CO_2 with the release of H_2. The Fe—N bond is the active catalytic subsystem in this case over which the whole catalytic cycle is carried out cooperatively. In contrast to the above examples of GOase and Cu–thiophenol systems (Section 1.4.1), the substrate first interacts with the ligand (Scheme 1.9). This brings the substrate in proximity to the metal to drive the cooperative double oxidation of the substrate over the Fe—N bond. The catalyst releases formaldehyde, which is thought to convert to methanediol for further dehydrogenation to CO_2. Dihydrogen is believed to be released from the FeH–NH moiety, aided by approach of another alcohol substrate molecule (Scheme 1.9).

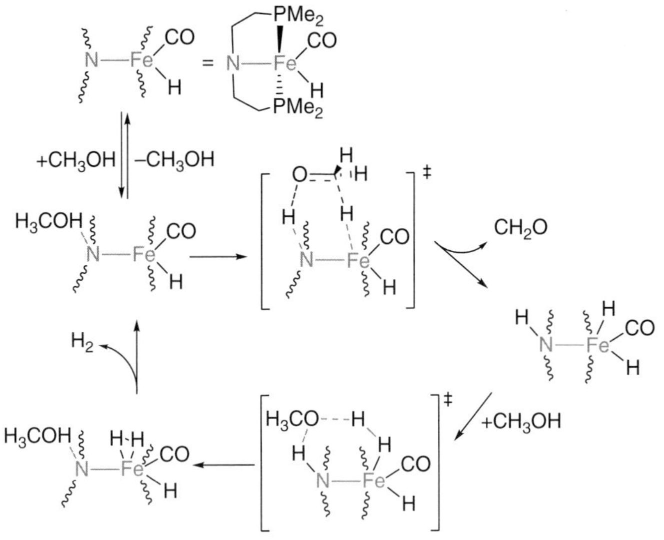

PMe2
CO
N—Fe
H
PMe2
+CH3OH
−CH3OH
H3COH
CO
N—Fe
H
O
H
H
H
H
CO
N—Fe
H
‡
CH2O
H
H
CO
N—Fe
H
H2
H3COH
H-H
CO
N—Fe
H
H3CO
H
H
H
CO
N—Fe
H
‡
+CH3OH

Scheme 1.9 Cooperative activation and oxidation of methanol over an iron–pincer complex.

The catalyst is also believed to catalyze the proposed hydrolysis of formaldehyde to methanediol, as is required for further dehydrogenation (Figure 1.12). The carbonyl carbon in formaldehyde is susceptible to attack by a nucleophile. However, splitting a single water molecule over the C=O bond is energetically unfavorable. This process can be accelerated by another water molecule, which leads to a more relaxed transition state (TS) geometry (Figure 1.12a). The second water molecule assists in the polarization of the water molecule to generate the nucleophile–electrophile (OH^-–H^+) pair. The Fe—N bond in the catalyst further stabilizes this process by allowing for spontaneous splitting of a water molecule (Figure 1.12b). This generates the nucleophile–electrophile pair in a relatively

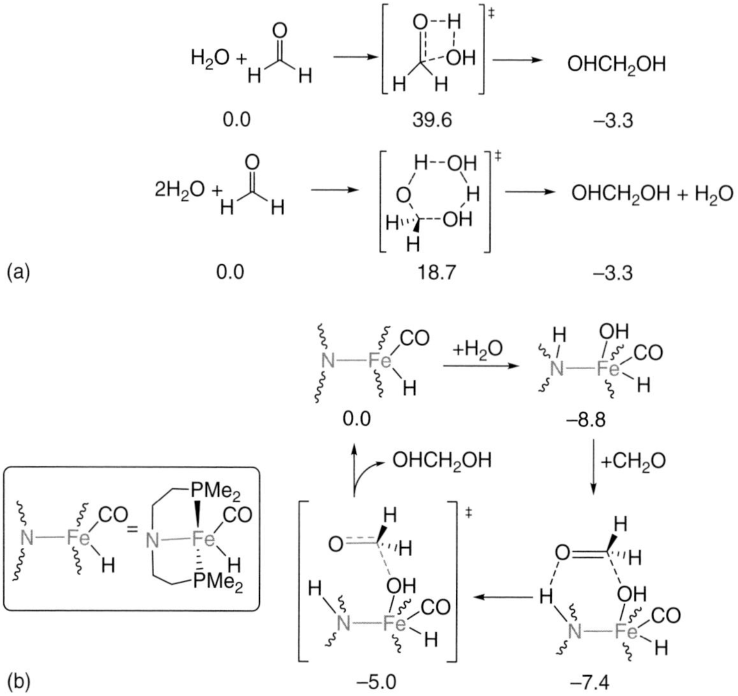

Figure 1.12 Cooperative activation of formaldehyde by an iron–pincer complex.

easy manner. The formaldehyde molecule can now easily be attacked by the hydroxide ligand at the metal to produce methanediol in a formaldehyde/water mixture. Once formed, methanediol is believed to undergo similar "alcohol" activation steps as described in Scheme 1.9.

1.4.2.2 Ligands Containing a Pendant Base

Activation of dihydrogen by base metals is still a challenging reaction in homogeneous catalysis. Catalytic systems that can bind and cleave molecular hydrogen are of particular interest in this regard. Inspired by the Fe–hydrogenase enzyme, DuBois and coworkers [93] proposed a mononuclear nickel complex that contains cyclic diphosphine ligands (Figure 1.13). Nitrogen bases were also incorporated in the ligand backbone. Because of the close proximity of the base around the metal, these are typically known as pendant bases. The system reported by Dubois and coworkers is able to reversibly bind and cleave dihydrogen by cooperative activation of the metal center and the pendant base. The molecular hydrogen molecule initially forms a sigma complex with the metal, which acidifies the molecule for cooperative proton abstraction by the nitrogen base, so to catalyze cleavage of the dihydrogen bond. Further improvements in the nickel-based catalytic system were also reported by varying the substituents on the ligand [94]. Chen and Yang [95] recently demonstrated the potential for applications of pendant bases with an iron center to catalyze the production of methanol from CO_2 and H_2 mixtures. In principle, the dihydrogen oxidation

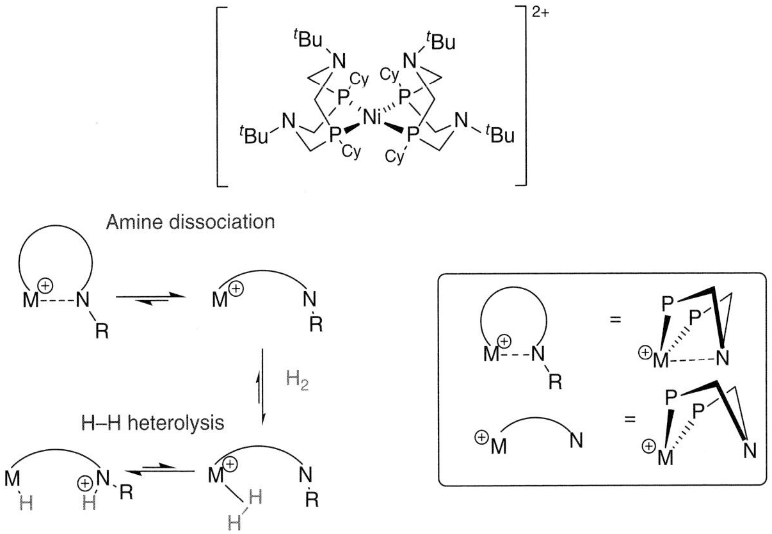

Figure 1.13 Dihydrogen cleavage catalyzed by a metal-pendant nitrogen-based system (DuBois system).

involves the same steps as described for the DuBois system in Figure 1.13. Further details on the DuBois system are described in Chapter 18.

Over the years, pendant catalysts for hydrogen oxidation have also been reported for other iron [96–99] and manganese [100] complexes. The catalytic activity in these systems is largely determined by the geometry of the ligand and the N-metal distance. In general, the metal center is responsible for electronic control of the catalysis and the pendant base controls the protonation step [99]. This cooperative activation thereby enables substrate activation, which is inaccessible without the functionalized ligand.

1.5 Substrate Radicals in Catalysis

Quite recently, several examples of catalytic reactions were disclosed in which formation and detection of discrete metal-bound substrate radicals was reported. These substrate-derived ligand radicals play a key role in a variety of synthetically useful C—C, C—N, and C—O bond formation reactions. These reactions proceed almost without exception via one-electron substrate activation and subsequent controlled radical steps (Figure 1.14). The carbene–radical and nitrene–radical examples discussed in this section provide perhaps the most clear-cut examples of the usefulness of ligand/substrate "non-innocence" involvement in catalysis.

Transition metal carbenes (M=CR$_2$) and nitrenes (M=NR) are the most clear-cut examples for which one-electron activation of the substrate has been well documented in the chemistry of non-noble metals [7, 101]. They are usually formed by addition of a high-energy carbene or nitrene precursor,

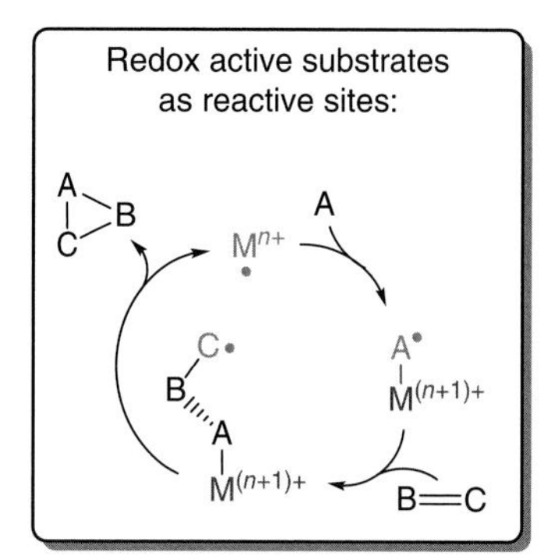

Figure 1.14 One-electron substrate activation and subsequent controlled radical steps.

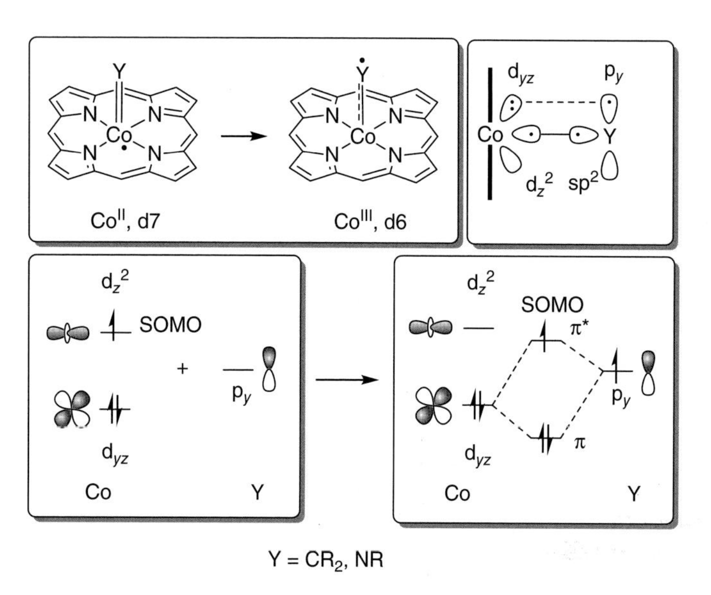

Figure 1.15 Electron transfer from metal to substrate (transformation of a metalloradical into a substrate radical).

such as diazo compounds (to generate carbenes) or iminoiodanes/azides (to generate nitrenes). By choosing a specific combination of a first-row transition metal and spectator ligands, one-electron transfer can occur from the metal to the metal-bound carbene or nitrene moiety, thus forming a carbon- or nitrogen-centered radical. The initial metalloradical is transferred to an organic radical, bound to the metal, and the formed species are typically called "carbene radicals" or "nitrene radicals." This specific situation occurs only when the energy level of the p_y orbital of the carbene or nitrene is lower than the d_{z^2} orbital of the metal (Figure 1.15).

1.5.1 Carbene Radicals

Carbene radicals are perhaps one of the most useful "non-innocent" substrates to react via well-defined and controlled radical-type reactions in the coordination sphere of base metals. The first carbene radicals bound to non-noble transition metals were reported by the group of Casey in the 1970s [102, 103]. The radical was obtained by one-electron reduction of Fischer-type carbenes using external reducing agents (Scheme 1.10). Fischer-type carbene complexes behave as electrophilic species with their LUMO centered on the carbene carbon atom, and hence, the reduction occurs at the carbene carbon rather than the metal. Several examples have been reported involving early transition metal complexes of group 6 (Cr, Mo, and W), which, in most cases, were reduced by sodium/potassium alloy. Formation of persistent carbon-centered radical anions at −50 °C has been confirmed using electron paramagnetic resonance (EPR) spectroscopic measurements. However, none of these early examples were used in catalysis, and they were long considered to be just chemical curiosities.

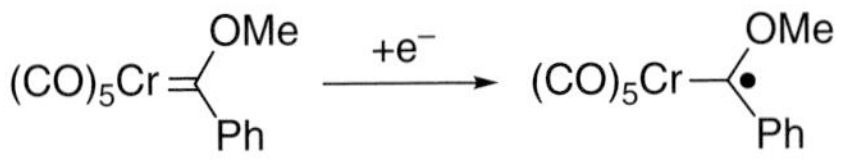

$$(CO)_5Cr{=}C(OMe)Ph \xrightarrow{+e^-} (CO)_5Cr{-}\dot{C}(OMe)Ph$$

Scheme 1.10 First example of a carbene radical complex by Casey and coworkers.

More recently, however, a series of base metal-catalyzed reactions were developed, in which carbene radicals are generated directly upon reaction of a carbene precursor with a metalloradical catalyst. In other words, the carbene radical formation involves a direct 1e⁻ reduction of the carbene by the same metal complex that facilitates its formation [104, 105]. As a direct result of the redox process being intramolecular, the carbene radical is formed without the need of an external reducing agent, in a catalytic manner. It was determined that low spin Group 9 transition metal complexes with metals in the +II oxidation state such as Co^{II} are suitable. The groups of Zhang and De Bruin have detected formation of carbene radicals upon metalloradical activation of diazo compounds (or their tosylhydrazone precursors) by cobalt(II) porphyrin ([Co(por)]) complexes, using complementary techniques such as DFT and EPR [106]. Conclusive evidence of the existence of carbene radicals bound to metal complexes has been brought forward. Subsequently, several catalytic reactions have been developed in which C—C, C—O, and C—H bonds are formed by the involvement of carbene radicals (Scheme 1.11).

These examples include cyclopropanation [107, 108] C—H activation [109], cyclo-propenation [110], as well as ketene [111], alkene [112], 2*H*-chromene [113], furane [114], and indene [115] formation (Scheme 1.11). They all have in common the use of a substituted cobalt(II) porphyrin as the catalyst and a diazo or tosylhydrazone as a high-energy substrate to generate the carbene radical intermediate. After formation of the intermediate radical species, trapping it with different substrates such as alkenes, alkynes, carbon monoxide, or ketones yields an entire series of substituted organic molecules (Scheme 1.11). The reactivity difference between the carbene radical and that of a Fischer carbene is attributed to the more nucleophilic character of the radical. The radical can easily react with, for example, electron-deficient alkenes during cyclopropanation, making this method complementary to the more classical approaches toward cyclopropanation [116–118].

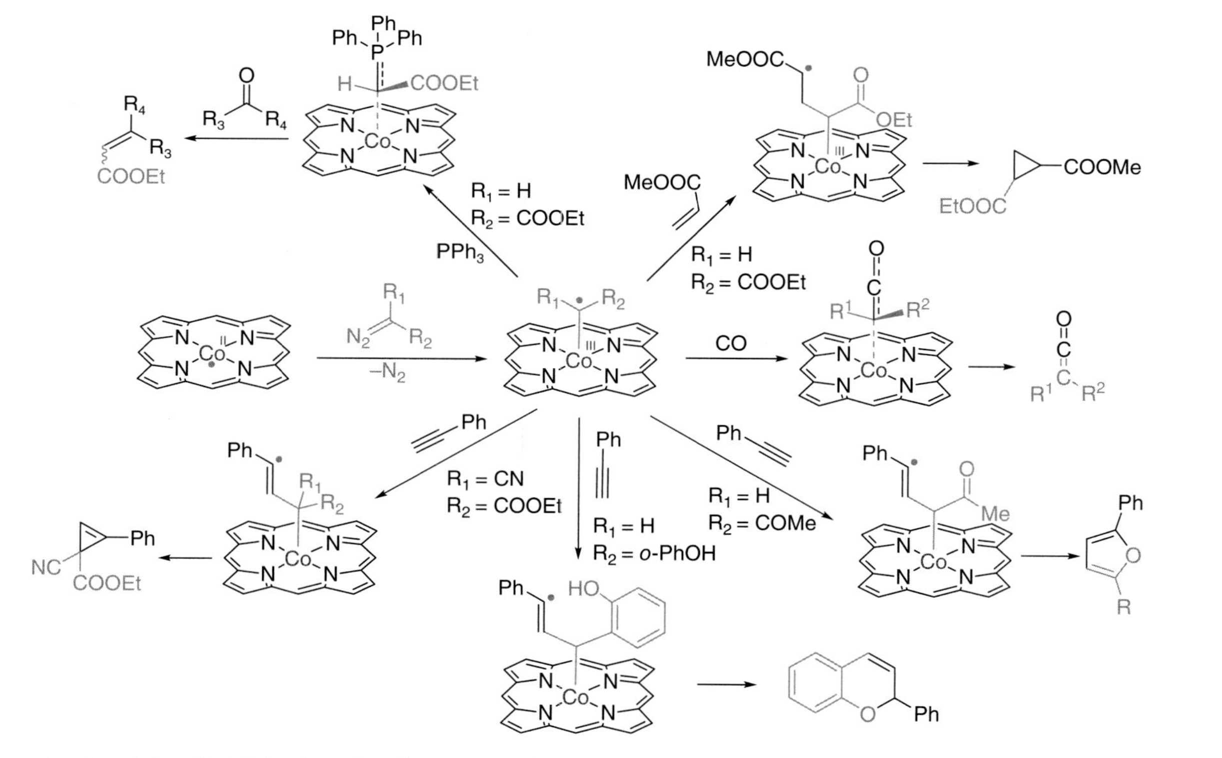

Scheme 1.11 Examples of metalloradical Co(por)-catalyzed reactions with carbene radicals as intermediates.

1.5.2 Nitrene Radicals

Similar to the formation of carbene radicals, using azides or iminoiodanes as substrates instead of diazo compounds results in nitrene formation [119]. In the presence of [Co^{II}(por)] complexes, reduction by one electron of the nitrene is favored, thus generating a nitrogen-based organic radical. Depending on the source of the nitrene transfer reagent, either a mono-nitrene radical or bis-nitrene radicals can be formed, giving rise to interesting reactivity in catalysis (Scheme 1.12). Compared to their carbene counterparts, nitrene radicals are more persistent in solution, thus allowing for detection at room temperature using a variety of spectroscopic techniques [120].

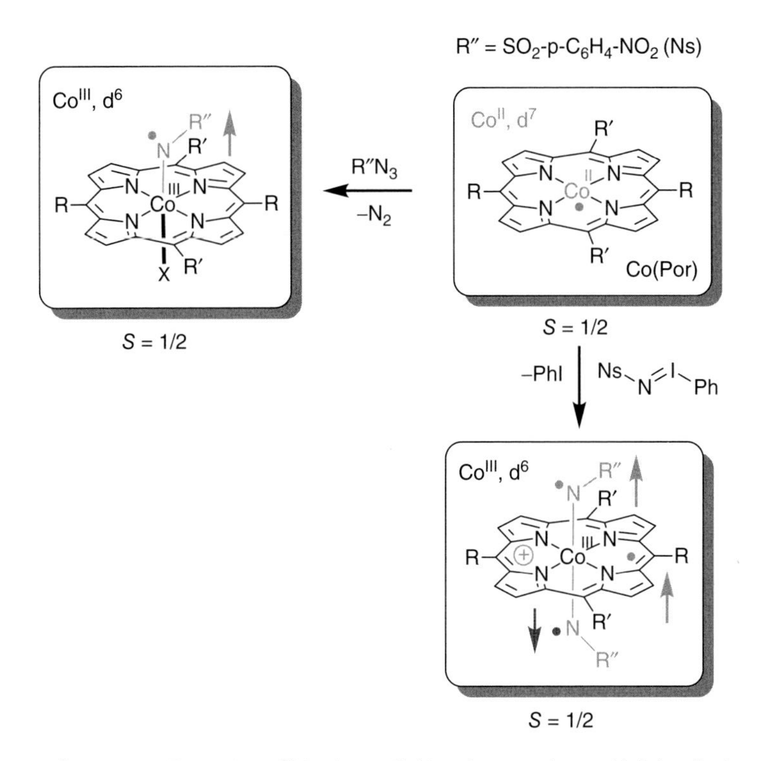

Scheme 1.12 Formation of bis-nitrene (left) and mono-nitrene (right) radicals.

Several examples of catalytic reactions in which nitrene radicals have been proposed and detected as intermediates are shown in Scheme 1.13. Addition to double bonds gives rise to aziridines [121], and activation of benzylic [122] or aldehydic [123] C—H bonds produces secondary amines or amides, respectively. Nitrene radical intermediates are more prone to C—H activations than their carbene equivalents, which are more susceptible to additions. Cobalt complexes are not the only species that can give rise to metalloradical catalysis involving nitrene radicals. Betley and coworkers proposed an Fe^{II} complex than can react with organic azides forming formally one-electron reduced nitrenes and catalytically activating benzylic C—H bonds to form secondary amines [124].

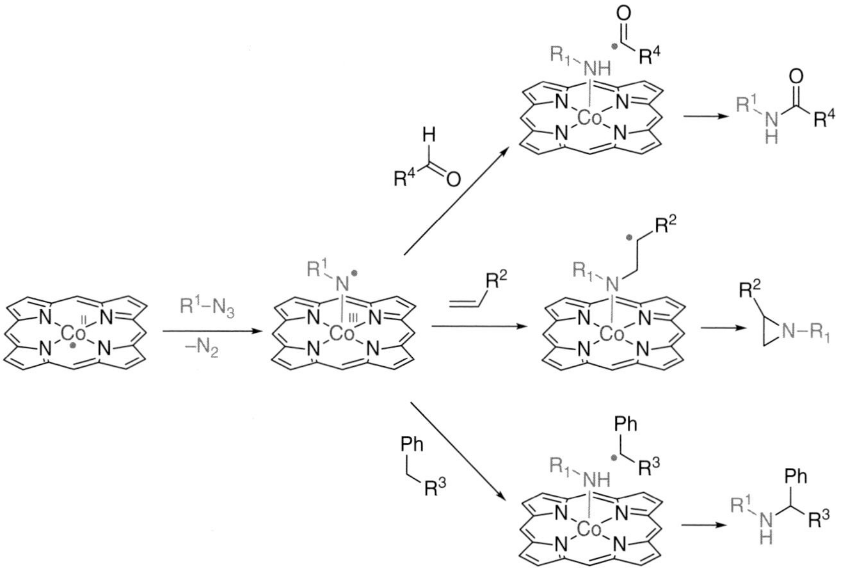

Scheme 1.13 Examples of metalloradical Co(Por)-catalyzed reactions with nitrene radicals as intermediates.

1.6 Summary and Conclusions

Material scarcity and environmental issues emerge an increasing demand on the development of new, cheap, and selective catalysts for sustainable synthesis in a variety of processes. As such, replacing noble metals by cheaper base metals in homogeneous catalysis is tremendously desirable. "Non-innocent" ligands offer several opportunities to achieve this goal. At its core, homogeneous catalysis is based on the properties of a metal complex and its surrounding ligands. Therefore, choosing the right combination of the metal and its surrounding ligands is key to the development of new catalysts. The use of "non-innocent" ligands goes beyond that of classical ancillary ligands, and a "non-innocent" ligand is typically directly involved in one of the key elementary steps of a catalytic reaction. In a broad description, "non-innocent" ligands act synergistically with the metal to enhance the selectivity and activity of the catalyst. In some cases, they facilitate reactions at base metals that are normally reserved to noble metals. In other cases, they enable entirely new reaction pathways.

Besides the classical ancillary ligands, four classes of "non-innocent" ligands can be distinguished in the field of base metal catalysis: (i) Stimuli-responsive ligands are mainly used in the development of switchable catalysts, in which external stimuli such as pH, light, or ligand-based redox reactions modify the properties of the ligand, and thereby the catalyst. (ii) Redox-active ligands are ligands that act as electron reservoirs, which are useful to facilitate two-electron elementary steps such as oxidative addition and reductive elimination at first-row transition metals, which more typically prefer one-electron transformations. (iii) Cooperative ligands participate actively in substrate bond-breaking and bond-making processes, most typically in a synergistic manner with metal participation. Hydrogen atom or proton abstraction from the substrate by the ligand is most typically observed for this class of ligands. (iv) The last class of "non-innocent" ligands are coordinated substrates that behave as redox-active ligands. One-electron transfer from the first-row transition metal to the coordinated substrate leads to formation of discrete "substrate radicals," which actively participate in a variety of catalytic radical-type transformations, giving access to a wide variety of ring-closing and C—H bond functionalization reactions.

Further developments in the field, taking advantage of the intrinsic reactivity of the ligand acting in synergy with the metal, will likely lead to many exiting new discoveries in the near future. This is expected not only to enable the replacement of noble metals in several important processes in homogeneous catalysis but also to uncover new reactivity with various synthetic possibilities. Controlled catalytic radical-type reactions, especially those in which all open-shell elementary steps occur in the coordination sphere of the metal without the formation of "free radicals," provide exciting possibilities for future development of base metal catalysis taking advantage of the "non-innocent" nature of ligands and substrates.

References

1 Grützmacher, H. (2008). *Angew. Chem. Int. Ed.* 47: 1814–1818.
2 van der Vlugt, J.I. and Reek, J.N.H. (2009). *Angew. Chem. Int. Ed.* 48: 8832–8846.
3 Jørgensen, C.K. (1966). *Coord. Chem. Rev.* 1: 164–178.
4 Kaim, W. and Schwederski, B. (2010). *Coord. Chem. Rev.* 254: 1580–1588.
5 Kaim, W. (1987). *Coord. Chem. Rev.* 76: 187–235.
6 Lyaskovskyy, V. and de Bruin, B. (2012). *ACS Catal.* 2: 270–279.
7 Luca, O.R. and Crabtree, R.H. (2013). *Chem. Soc. Rev.* 42: 1440–1459.
8 Trincado, M. and Grützmacher, H. (2015). Cooperating ligands in catalysis. In: *Cooperative Catalysis: Designing Efficient Catalysts for Synthesis* (ed. R. Peters). Weinheim: Wiley-VCH Verlag GmbH & Co. KGaA. doi: 10.1002/9783527681020.ch3.
9 van de Vlugt, J.I. (2012). *Eur. J. Inorg. Chem.* 3: 363–375.
10 Crabtree, R.H. (2009). *The Organometallic Chemistry of the Transition Metals*, 5e. Hoboken, NJ: John Wiley & Sons.
11 Egorova, K.S. and Ananikov, V.P. (2016). *Angew. Chem. Int. Ed.* 40: 12150–12162.
12 Crabtree, R.H. (2011). *New J. Chem.* 35: 18–23.
13 Allgeier, A.M. and Mirkin, C.A. (1998). *Angew. Chem. Int. Ed.* 37: 894–908.
14 Kotz, J.C. and Nivert, C.L. (1973). *J. Organomet. Chem.* 52: 387–406.
15 Lorkovic, I.M., Duff, R.R. Jr., and Wrighton, M.S. (1995). *J. Am. Chem. Soc.* 117: 3617–3618.
16 Yeung, L.K., Kim, J.E., Chung, Y.K. et al. (1996). *Organometallics* 15: 3891–3897.
17 Luca, O.R., Gustafson, J.L., Maddox, S.M. et al. (2015). *Org. Chem. Front.* 2: 823–848.

18 Süßner, M. and Plenio, H. (2005). *Angew. Chem. Int. Ed.* 44: 6885–6888.
19 Savka, R., Foro, S., Gallei, M. et al. (2013). *Chem. Eur. J.* 19: 10655–10662.
20 Arumugam, K., Varnado, C.D., Sproules, S. et al. (2013). *Chem. Eur. J.* 19: 10866–10875.
21 Gregson, C.K.A., Gibson, V.C., Long, N.J. et al. (2006). *J. Am. Chem. Soc.* 128: 7410–7411.
22 Wang, X., Thevenon, A., Brosmer, J.L. et al. (2014). *J. Am. Chem. Soc.* 136: 11264–11267.
23 Balof, S.L., P'Pool, S.J., Berger, N.J. et al. (2008). *Dalton Trans.* 42: 5791–5799.
24 Li, L., Wang, J., Zhou, C. et al. (2011). *Green Chem.* 13: 2071–2077.
25 Peeck, L.H., Leuthäusser, S., and Plenio, H. (2010). *Organometallics* 29: 4339–4345.
26 Wang, W., Wu, J., Xia, C., and Li, F. (2011). *Green Chem.* 13: 3440–3445.
27 Wang, W., Zhang, G., Lang, R. et al. (2013). *Green Chem.* 15: 635–640.
28 Gaulier, C., Hospital, A., Legeret, B. et al. (2012). *Chem. Commun.* 48: 4005–4007.
29 Onishi, N., Xu, S., Manaka, Y. et al. (2015). *Inorg. Chem.* 54: 5114–5123.
30 Himeda, Y., Onozawa-Komatsuzaki, N., Miyazawa, S. et al. (2008). *Chem. Eur. J.* 14: 11076–11081.
31 Suna, Y., Ertem, M.Z., Wang, W.H. et al. (2014). *Organometallics* 33: 6519–6530.
32 Himeda, Y., Miyazawa, S., and Hirose, T. (2011). *ChemSusChem* 4: 487–493.
33 Manbeck, G.F., Muckerman, J.T., Szalda, D.J. et al. (2015). *J. Phys. Chem. B* 119: 7457–7466.
34 Lv, R., Wang, Y., Zhou, C. et al. (2013). *ChemCatChem* 5: 2978–2982.
35 Badiei, Y.M., Wang, W.H., Hull, J.F. et al. (2013). *Inorg. Chem.* 52: 12576–12586.
36 Akita, M. (2011). *Organometallics* 30: 43–51.
37 Neilson, B.M. and Bielawski, C.W. (2013). *ACS Catal.* 3: 1874–1885.
38 Sud, D., Norsten, T.B., and Branda, N.R. (2005). *Angew. Chem. Int. Ed.* 44: 2019–2021.
39 Chirik, P.J. and Wieghardt, K. (2010). *Science* 327: 794–795.
40 Chirik, P.J. (2011). *Inorg. Chem.* 50: 9737–9914.
41 de Bruin, B., Bill, E., Bothe, E. et al. (2000). *Inorg. Chem.* 39: 2936–2947.
42 Budzelaar, P.H.M., de Bruin, B., Gal, A.W. et al. (2001). *Inorg. Chem.* 40: 4649–4655.
43 Enright, D., Gambarotta, S., Yap, G.P.A., and Budzelaar, P.H.M. (2002). *Angew. Chem. Int. Ed.* 41: 3873–3876.
44 Gibson, V.C. and Spitzmesser, S.K. (2003). *Chem. Rev.* 103: 283–315.
45 Gibson, V.C. and Solan, G.A. (2011). *Top. Organomet. Chem.* 1–52.
46 Bart, S.C., Chlopek, K., Bill, E. et al. (2006). *J. Am. Chem. Soc.* 128: 13901–13912.
47 Bart, S.C., Lobkovsky, E., Bill, E. et al. (2007). *Inorg. Chem.* 46: 7055–7063.
48 Stieber, S.C.E., Milsmann, C., Hoyt, J.M. et al. (2012). *Inorg. Chem.* 51: 3770–3785.

49 Small, B.L., Brookhart, M., and Bennett, A.M.A. (1998). *J. Am. Chem. Soc.* 120: 4049–4050.
50 Britovsek, G.J.P., Gibson, V.C., Kimberley, B.S. et al. (1998). *ChemCommun.* 849–850.
51 Gibson, V.C., Redshaw, C., and Solan, G.A. (2007). *Chem. Rev.* 107: 1745–1776.
52 Bennett, A.M.A. (1998). *EI Du Pont de Nemours and Co*, USA, PCT Int. Appl., WO 9827124.
53 Britovsek, G.J.P., Bruce, M., Gibson, V.C. et al. (1999). *J. Am. Chem. Soc.* 121: 8728–8740.
54 Britovsek, G.J.P., Mastroianni, S., Solan, G.A. et al. (2000). *Chem. Eur. J.* 6: 2221–2231.
55 Bouwkamp, M.W., Lobkovsky, E., and Chirik, P.J. (2005). *J. Am. Chem. Soc.* 127: 9660–9661.
56 Scott, J., Gambarotta, S., Korobkov, I., and Budzelaar, P.H.M. (2005). *J. Am. Chem. Soc.* 127: 13019–13029.
57 Poli, R. (2011). *Eur. J. Inorg. Chem.* 1513–1530.
58 Cruz, V.L., Ramos, J., Martinez-Salazar, J. et al. (2009). *Organometallics* 28: 5889–5895.
59 Darmon, J.M., Stieber, S.C.E., Sylvester, K.T. et al. (2012). *J. Am. Chem. Soc.* 134: 17125–17137.
60 Bouwkamp, M.W., Bowman, A.C., Lobkovsky, E., and Chirik, P.J. (2006). *J. Am. Chem. Soc.* 128: 13340–13341.
61 Sylvester, K.T. and Chirik, P.J. (2009). *J. Am. Chem. Soc.* 131: 8772–8774.
62 Russell, S.K., Lobkovsky, E., and Chirik, P.J. (2011). *J. Am. Chem. Soc.* 133: 8858–8861.
63 Moreau, B., Wu, J.Y., and Ritter, T. (2009). *Org. Lett.* 11: 337–339.
64 Russell, S.K., Darmon, J.M., Lobkovsky, E., and Chirik, P.J. (2010). *Inorg. Chem.* 49: 2782–2792.
65 Monfette, S., Turner, Z.R., Semproni, S.P., and Chirik, P.J. (2012). *J. Am. Chem. Soc.* 134: 4561–4564.
66 Danopoulos, A.A., Wright, J.A., Motherwell, W.B., and Ellwood, S. (2004). *Organometallics* 23: 4807–4810.
67 Yu, R.P., Darmon, J.M., Milsmann, C. et al. (2013). *J. Am. Chem. Soc.* 135: 13168–13184.
68 Gibson, V.C., Tellmann, K.P., Humphries, M.J., and Wass, D.F. (2002). *Chem. Commun.* 2316–2317.
69 Darmon, J.M., Yu, R.P., Semproni, S.P. et al. (2014). *Organometallics* 33: 5423–5433.
70 Obligacion, J.V. and Chirik, P.J. (2013). *Org. Lett.* 15: 2680–2268.
71 Obligacion, J.V., Neely, J.M., Yazdani, A.N. et al. (2015). *J. Am. Chem. Soc.* 137: 5855–5858.
72 Obligacion, J.V. and Chirik, P.J. (2013). *J. Am. Chem. Soc.* 135: 19107–19110.
73 Obligacion, J.V., Semproni, S.P., and Chirik, P.J. (2014). *J. Am. Chem. Soc.* 136: 4133–4136.
74 Palmer, W.N., Diao, T., Pappas, L., and Chirik, P.J. (2015). *ACS Catal.* 5: 622–626.

75 Tondreau, A.M., Atienza, C.C.H., Weller, K.J. et al. *Science* 2012 (6068): 335, 576–570.

76 Wu, J.Y., Stanzl, B.N., and Ritter, T. (2010). *J. Am. Chem. Soc.* 132: 13214–13216.

77 Mukhopadhyay, T.K., Flores, M., Groy, T.L., and Trovitch, R.J. (2014). *J. Am. Chem. Soc.* 136: 882–885.

78 Pierpont, C.G. (2011). *Inorg. Chem.* 50: 9766–9772.

79 Poddelsky, A.I., Cherkasov, V.K., and Abakumov, G.A. (2009). *Coord. Chem. Rev.* 253: 291–324.

80 Skara, G., Pinter, B., Geerlings, P., and De Proft, F. (2015). *Chem. Sci.* 6: 4109–4117.

81 Smith, A.L., Hardcastle, K.I., and Soper, J.D. (2010). *J. Am. Chem. Soc.* 132: 14358–14360.

82 Tennyson, A.G., Lynch, V.M., and Bielawski, C.W. (2010). *J. Am. Chem. Soc.* 132: 9420–9429.

83 van der Meer, M., Rechkemmer, Y., Peremykin, I. et al. (2014). *Chem. Commun.* 50: 11104–11106.

84 Mukherjee, C., Weyhermüller, T., Bothe, E., and Chaudhuri, P. (2008). *Inorg. Chem.* 47: 2740–2746.

85 Jacquet, J., Salanouve, E., Orio, M. et al. (2014). *Chem. Commun.* 50: 10394–10397.

86 Werkmeister, S., Junge, K., and Beller, M. (2014). *Org. Process Res. Dev.* 18: 289–302.

87 Gunanathan, C. and Milstein, D. (2014). *Chem. Rev.* 114: 12024–12087.

88 Que, L. and Tolman, W.B. (2008). *Nature* 455: 333–340.

89 Whittaker, J.W. (2003). *Chem. Rev.* 103: 2347–2364.

90 Chaudhuri, P., Hess, M., Flörke, U., and Wieghardt, K. (1998). *Angew. Chem. Int. Ed.* 37: 2217–2220.

91 Chaudhari, P., Hess, M., Müller, J. et al. (1999). *J. Am. Chem. Soc.* 121: 9599–9610.

92 Bielinski, E., Förster, M., Zhang, Y. et al. (2015). *ACS Catal.* 5: 2404–2415.

93 Henry, R.M., Shoemaker, R.K., DuBois, D.L., and DuBois, M.L. (2006). *J. Am. Chem. Soc.* 128: 3002–3010.

94 Yang, J.Y., Chen, S., Dougherty, W.G. et al. (2010). *Chem. Commun.* 46: 8618–8620.

95 Chen, X. and Yang, X. (2016). *J. Phys. Chem. Lett.* 7: 1035–1041.

96 Liu, T., Chen, S., O'Hagan, M.J. et al. (2012). *J. Am. Chem. Soc.* 134: 6257–6272.

97 Liu, T., DuBois, D.L., and Bullock, R.M. (2013). *Nat. Chem.* 5: 228–233.

98 Darmon, J.M., Raugei, S., Liu, T. et al. (2014). *ACS Catal.* 4: 1246–1260.

99 Liu, T., Wang, X., Hoffmann, C. et al. (2014). *Angew. Chem. Int. Ed.* 53: 5300–5304.

100 Hulley, E.B., Helm, M.L., and Bullock, R.M. (2014). *Chem. Sci.* 5: 4729–4741.

101 Dzik, W.I., Zhang, X.P., and de Bruin, B. (2011). *Inorg. Chem.* 50: 9896–9903.

102 Krusic, P.J., Klabunde, U., Casey, C.P., and Block, T.F. (1976). *J. Am. Chem. Soc.* 98: 2015–2018.

103 Block, T.F., Fenske, R.F., and Casey, C.P. (1976). *J. Am. Chem. Soc.* 98: 441–443.

104 Fuchibe, K. and Iwasawa, N. (2003). *Chem. Eur. J.* 9: 905–914.

105 Dzik, W.I., Xu, X., Zhang, X.P. et al. (2010). *J. Am. Chem. Soc.* 132: 10891–10902.

106 Lu, H., Dzik, W.I., Xu, X. et al. (2011). *J. Am. Chem. Soc.* 133: 8518–8521.

107 Xu, X., Zhu, S.F., Cui, X. et al. (2013). *Angew. Chem. Int. Ed.* 52: 11857–11861.

108 Zhu, S.F., Xu, X., Perman, J.A., and Zhang, X.P. (2010). *J. Am. Chem. Soc.* 132: 12796–12799.

109 Cui, X., Xu, X., Jin, L.M. et al. (2015). *Chem. Sci.* 6: 1219–1224.

110 Cui, X., Xu, X., Lu, H.J. et al. (2011). *J. Am. Chem. Soc.* 133: 3304–3307.

111 Paul, N.D., Chirila, A., Lu, H.J. et al. (2013). *Chem. Eur. J.* 19: 12953–12958.

112 Lee, M.Y., Chen, Y., and Zhang, X.P. (2003). *Organometallics* 22: 4905–4909.

113 Paul, N.D., Mandal, S., Otte, M. et al. (2014). *J. Am. Chem. Soc.* 136: 1090–1096.

114 Cui, X., Xu, X., Wojtas, L. et al. (2012). *J. Am. Chem. Soc.* 134: 19981–19984.

115 Das, B.G., Chirila, A., Tromp, M. et al. (2016). *J. Am. Chem. Soc.* 138: 8968–8975.

116 Lebel, H., Marcoux, J.F., Molinaro, C., and Charette, A.B. (2003). *Chem. Rev.* 103: 977–1050.

117 Davies, H.M.L. and Beckwith, R.E.J. (2003). *Chem. Rev.* 103: 2861–2903.

118 Li, A.H., Dai, L.X., and Aggarwal, V.K. (1997). *Chem. Rev.* 97: 2341–2372.

119 Olivos Suarez, A.I., Lyaskovskyy, V., Reek, J.N.H. et al. (2013). *Angew. Chem. Int. Ed. Engl.* 52: 12510–12529.

120 Goswami, M., Lyaskovskyy, V., Domingos, S.R. et al. (2015). *J. Am. Chem. Soc.* 137: 5468–5479.

121 Suarez, A.I.O., Jiang, H.L., Zhang, X.P., and de Bruin, B. (2011). *Dalton Trans.* 40: 5697–5705.

122 Lyaskovskyy, V., Suarez, A.I.O., Lu, H.J. et al. (2011). *J. Am. Chem. Soc.* 133: 12264–12273.

123 Jin, L.M., Lu, H.J., Cui, Y. et al. (2014). *Chem. Sci.* 5: 2422–2427.

124 King, E.R., Hennessy, E.T., and Betley, T.A. (2011). *J. Am. Chem. Soc.* 133: 4917–4923.

2

Computational Insights into Chemical Reactivity and Road to Catalyst Design: The Paradigm of CO_2 Hydrogenation

Bhaskar Mondal, Frank Neese, and Shengfa Ye

Department of Molecular Theory and Spectroscopy, Max-Planck Institut für Chemische Energiekonversion, Stiftstraße 34–36, D45470, Mülheim an der Ruhr, Germany

2.1 Introduction

In the past few decades, computational chemistry together with the development of fast computers and accurate quantum chemical methods has emerged to be a bridge between experimental observations and theoretical predictions. Calculations have proven to be very useful because they can provide atomic level mechanistic details of a given reaction and predict the product selectivity at the level that chemists seek. The traditional catalyst design and development typically employ trial-and-error experimental approaches, which mainly rely on chemical intuitions and extensive empirical observations. In contrast, the rational design, especially based on a deep understanding of the reaction mechanism, is considered to be the most efficient because it may entail an exquisite control over each elementary step in the catalytic cycle [1].

CO_2 functionalization to liquid fuels and useful chemicals is of fundamental concern in recent research because of increasing global energy demand and environmental issues. Among several CO_2 conversion processes, CO_2 hydrogenation producing formic acid or formate has captivated the most interest of chemists as it is the first step to methanol and hydrocarbon synthesis [2]. The use of non-noble metal complexes as catalysts acquires more incentives owing to their high abundance and low cost [3]. In this chapter, we show how computational chemistry helps to derive a rational design strategy for developing base metal catalysts for CO_2 hydrogenation.

2.1.1 Chemical Reactions: Conceptual Thoughts

Before delving into the detailed discussion about CO_2 hydrogenation, it is helpful to recapitulate some general thoughts of chemical reactions that serve as a basis to understand the reaction mechanism and rationalize reaction energetics and kinetics for a given chemical conversion. In a consecutive

Non-Noble Metal Catalysis: Molecular Approaches and Reactions,
First Edition. Edited by Robertus J. M. Klein Gebbink and Marc-Etienne Moret.

chemical reaction involving several elementary steps, most catalytic reactions belonging to this category, the overall rate is usually dictated by the slowest transformation of the reaction, called the rate-determining step (RDS) [4]. Evidently, the RDS is the transformation that possesses the highest barrier in the catalytic cycle. The barrier of an elementary step can be determined by transition state theory (TST) [5]. The Born–Oppenheimer (BO) approximation assumes that in a molecule, the motions of the nuclei and electrons can be separated; thus, one can compute the electronic energy for a given nuclear arrangement and construct a potential energy surface (PES) that defines the energy of the molecule as a function of the nuclear coordinates. For a diatomic molecule, the PES depicts how its energy varies with respect to the different internuclear distance. Within the BO approximation, molecules at their equilibrium geometries correspond to local minima on the PES, and a chemical reaction can therefore be described as nuclei move from one minimum to another along the lowest energy path. According to TST, for a reaction $A + B \rightarrow C + D$, the reactants have to pass through a special geometric arrangement, called the transition state (TS, $[AB]^{\ddagger}$ in Figure 2.1a) before decaying to the product. The reaction barrier ($\Delta G^{\ddagger}$) is simply the free energy difference between the reactant ($A + B$) and the transition state ($[AB]^{\ddagger}$). If one assumes that the molecules in the TS ($[AB]^{\ddagger}$) are in (*quasi*)-equilibrium with reactants A and B, the famous Eyring equation can be obtained, $k = (k_B T/h) \times \exp(-\Delta G^{\ddagger}/RT)$, which relates the reaction rate of an elementary step with its activation barrier (where k is the rate constant and $\Delta G^{\ddagger}$ is the activation free energy of the reaction). For a multistep reaction, the reaction barrier for all elementary steps can be calculated theoretically and the RDS can be readily identified. Clearly, the unambiguous determination of the RDS entirely relies on the differential barrier heights for all elementary steps. Therefore, it is very crucial to compute them accurately. The most popular density functional theory (DFT) [6] often breaks down to calculate reaction barriers within the chemical accuracy (~2 kcal mol^{-1}) [7]. To this end, highly correlated *ab initio* methods are in great demand. For instance, the recently developed local coupled-cluster (L-CC) method that utilizes pair natural orbitals (PNOs) has been proven to be very efficient at a comparable cost to DFT [8]. Specifically, the domain-based version of the local coupled-cluster methods up to single double and perturbative triple excitations (DLPNO-CCSD(T)) [8b] is a method of choice for equivocally determining the RDS.

For catalyst design, one often needs to move from a given reaction to a range of similar chemical transformations. Their thermodynamic and kinetic properties can be correlated through the Bell–Evans–Polanyi (BEP) [9] principle, correlating the reaction enthalpy with the activation barrier by a linear equation $\Delta H^{\ddagger} = \Delta H_o^{\ddagger} + \alpha \Delta H$ (Figure 2.1b, inset). In other words, the more exothermic the reaction is, the lower the activation barrier is, or vice versa as illustrated in Figure 2.1b. Note that not all reactions follow the BEP principle, for which O_2 activation represents a prototypical example. Oxygenation of organic substrates is highly exothermic, yet encounters tremendous barrier, because of the spin-forbidden nature of this conversion [10].

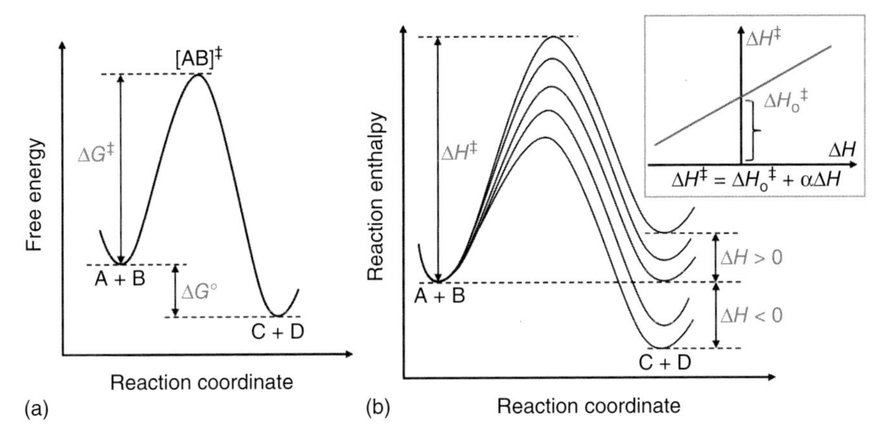

Figure 2.1 (a) Schematic free energy profile of a reaction. (b) Illustration of the Bell–Evans–Polanyi (BEP) principle.

2.1.2 Motivation Behind Studying CO_2 Hydrogenation

The transformation of CO_2 into value-added products and fuels is tantalizing because of the exponentially increasing global energy demand [11]. More importantly, CO_2 is a major contributor to the greenhouse gases, and its ever-growing anthropogenic release poses an environmental threat [12]. Nature, since its existence, has designed an ingenious way through photosynthesis to convert CO_2 into carbohydrates [13]. This motivates researchers to use CO_2 as C1 carbon source in many organic syntheses with an attractive advantage that CO_2 is nontoxic and highly abundant in nature [14]. However, the thermodynamic stability and kinetic inertness of CO_2 demand high-energy input and proper catalysts to achieve CO_2 conversion [15]. Over decades, countless efforts have been devoted to preparing catalysts for efficiently functionalizing CO_2 chemically or electrochemically [16]. Homogeneous CO_2 hydrogenation is one of the most well-studied pathways as it leads to useful products, such as formic acid or formate [16b, 17]. Most precious transition metals, for instance, rhodium, iridium, and ruthenium, have been tested in this direction, and impressive reactivity has been reported. On the other hand, earth-abundant metals, such as iron, cobalt, and nickel, remain under-explored, presumably because of their low reactivity. Only recently, some remarkable developments of iron and cobalt-based catalysts were published in the literature [18]. Despite exhibiting promising reactivity, base metal catalysts still cannot compete with noble metals. Therefore, understanding the reaction mechanism and providing new ideas for designing non-noble metal catalysts are highly desired.

2.1.3 Challenges of CO_2 Reduction

CO_2 is a linear molecule featuring very short C—O bond lengths (1.17 Å). Moreover, the central carbon possesses its highest oxidation state (+4) and hence is often the ultimate product in many chemical and biological oxidation processes.

Therefore, CO_2 is thermodynamically very stable. In CO_2 reduction processes, additional electrons are usually shifted to one of the doubly degenerate π^*-orbitals ($2\pi_u$ in Figure 2.2a), the lowest energy unoccupied molecular orbital (LUMO), which not only causes the lengthening of C—O bonds but also the bending of linear CO_2 molecule. Such geometric distortions entail a significant energy penalty as demonstrated in Figure 2.2b; as a consequence, the reduction potential measured for the one-electron reduction of CO_2 is exceedingly negative (−1.90 V vs NHE). These structural changes, which manifest themselves at least partly in the transition state, induce a high barrier for CO_2 reduction; thus, CO_2 is kinetically rather inert. Therefore, CO_2 functionalization is thermodynamically and kinetically very challenging and demands high-energy input. On the other hand, upon CO_2 bending, the energy of one of the $2\pi_u$ orbitals decreases (Figure 2.2c) and simultaneously the C-atom contribution to this orbital increases (from 61% to 66%, Figure 2.2a) [15b], thereby rendering the C-center more susceptible toward nucleophilic attack.

The high kinetic barrier arising from CO_2 reduction can be circumvented by carefully controlling the reaction trajectory. Given the fact that one-electron reduction of CO_2 is energetically unfavorable, CO_2 usually undergoes two- or multi-electron reduction. Proton-coupled electron transfers (PCETs, Eqs. (2.2)–(2.4)) are reasonable choices and can occur at modest potentials. The

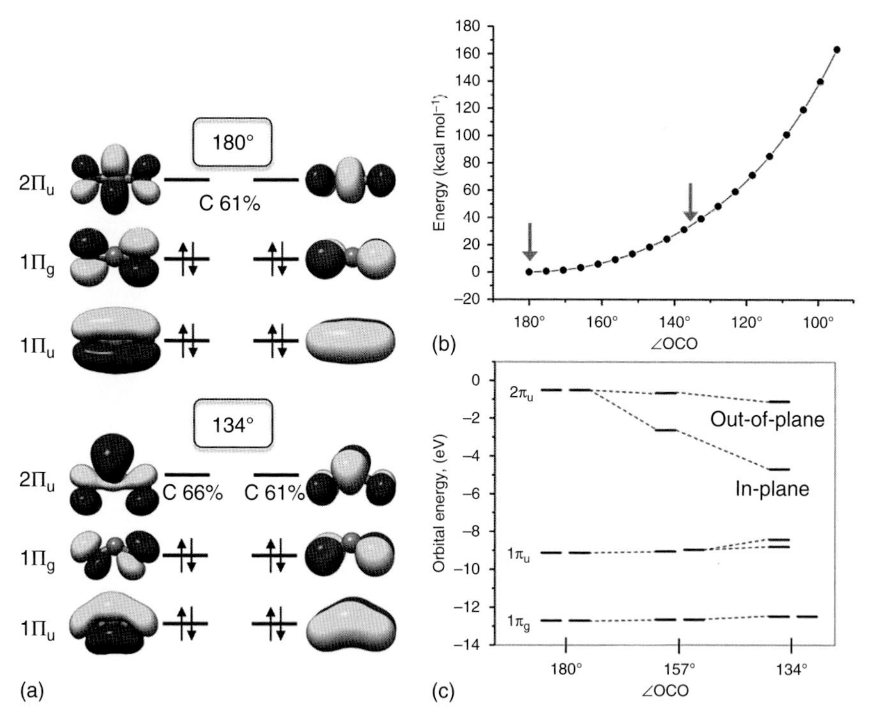

Figure 2.2 (a) π-Type molecular orbitals of CO_2 at different OCO angles. (b) and (c) The variation of total energy (in kcal mol^{-1}) and orbital energy (eV), respectively, as a function of the OCO angle.

details of the PCET process are beyond the scope of this chapter. The $E^{\circ\prime}$ listed in the equations are the formal potentials with respect to the normal hydrogen electrode (NHE) at standard conditions in aqueous solution. The principles of PCET can be visualized in a more general picture as a combined nucleophilic and electrophilic attack, thereby signifying that bifunctional activation is a favorable pathway for CO_2 conversion [11]. In this chapter, we focus on CO_2 hydrogenation to produce formic acid or formate (Eq. 2.3). The entire reaction involves two key elementary steps, viz, H_2 splitting and hydride transfer to CO_2 (vide supra), and the later process can be interpreted as a two-electron transfer along with a proton shift, $H^- = H^+ + 2e^-$.

$$CO_2\ (aq) + e^- \rightarrow CO_2^{\bullet -}\ (aq) \quad E^{\circ\prime} = -1.90\ V \tag{2.1}$$

$$CO_2\ (g) + 2H^+ + 2e^- \rightarrow CO\ (g) + H_2O \quad E^{\circ\prime} = -0.52\ V \tag{2.2}$$

$$CO_2\ (g) + H^+ + 2e^- \rightarrow HCO_2^-\ (aq) \quad E^{\circ\prime} = -0.43\ V \tag{2.3}$$

$$CO_2\ (g) + 4H^+ + 4e^- \rightarrow HCHO\ (aq) + H_2O \quad E^{\circ\prime} = -0.51\ V \tag{2.4}$$

2.1.4 CO_2 Hydrogenation

Catalytic reaction of CO_2 and H_2 in the presence or absence of reactive species, such as base, is the most studied reaction for CO_2 functionalization [18] with the immediate product being formic acid or formate. Hydrogenation of CO_2 beyond formic acid is more challenging and often needs an oxygen sink. Herein, we only discuss CO_2 hydrogenation up to the formic acid level. The first homogenous CO_2 hydrogenation process was reported in 1976 by Inoue et al. [19] It was generally accepted that the addition of a base is often required as the standalone reaction exhibits a significantly high positive Gibbs free energy change (Eq. (2.5)).

$$CO_2\ (g) + H_2\ (g) \rightarrow HCOOH\ (l) \tag{2.5}$$

$$\Delta G^\circ = 7.8\ kcal\ mol^{-1};\ \Delta H^\circ = -7.5\ kcal\ mol^{-1};\ \Delta S^\circ = -51.2\ cal\ (mol\ K)^{-1}$$

$$CO_2\ (g) + H_2\ (g) + NH_3\ (aq) \rightarrow HCO_2^-\ (aq) + NH_4^+\ (aq) \tag{2.6}$$

$$\Delta G^\circ = -2.3\ kcal\ mol^{-1};\ \Delta H^\circ = -20.1\ kcal\ mol^{-1};\ \Delta S^\circ = -59.5\ cal\ (mol\ K)^{-1}$$

$$CO_2\ (aq) + H_2\ (aq) + NH_3\ (aq) \rightarrow HCO_2^-\ (aq) + NH_4^+\ (aq) \tag{2.7}$$

$$\Delta G^\circ = -8.4\ kcal\ mol^{-1};\ \Delta H^\circ = -14.2\ kcal\ mol^{-1};\ \Delta S^\circ = -19.3\ cal\ (mol\ K)^{-1}$$

The above equations show that adding a base (NH_3) makes the conversion much more exothermic, and dissolving the gaseous reactants lowers the unfavorable entropy change. Clearly, the transformation of formic acid as an acid–base complex increases the driving force and is the underlying thermodynamic reason behind the requirement of a base for CO_2 hydrogenation.

2.1.5 Noble vs Non-noble Metal Catalysis

A plethora of efficient noble metal catalysts, mostly based on ruthenium, rhodium, and iridium, can be found in the literature. Excellent reviews covering all precious metal catalysts for CO_2 hydrogenation have been published [11, 18]; herein, we only give several remarkable examples. Among Rh complexes, $[RhCl(TPPTS)_3]$ (TPPTS = tris(*m*-sulfonatophenyl)phosphine) synthesized by Leitner and coworker in 1993 [20] catalyzes CO_2 hydrogenation with a turnover number (TON) of 3439 in the presence of water-soluble amine $HNMe_2$ under mild reaction conditions. The reaction mediated by an iridium pincer catalyst, $[IrH_3(PNP^{iPr})]$ ($PNP^{iPr} = 2,6\text{-}(CH_2PiPr_2)_2C_5H_3N$) prepared by Nozaki and coworkers in 2009 [21], can achieve a turnover frequency (TOF) of 150 000 h^{-1} at 200 °C in aqueous KOH solution, and a maximum TON of 3 500 000 at 120 °C after 48 h. Very recently, a similar Ru pincer complex, $[RuCl(H)(CO)(PNP^{tBu})]$ ($PNP^{tBu} = 2,6\text{-}(CH_2PtBu_2)_2C_5H_3N$), has been shown to exhibit a TOF of 1 100 000 h^{-1} at 120 °C and 40 bar, the highest TOF value reported to date [22].

In comparison to the highly impressive catalytic activity of precious metal catalysts, non-noble metal catalysts are far behind in the race. Only a handful of catalysts involving nickel, iron, and cobalt have been reported. In 2003, Jessop and coworkers developed a high-pressure method and showed that a combination of $NiCl_2$/dcpe (dcpe = $Cy_2PCH_2CH_2PCy_2$) can yield a TON of 4400 and a TOF of 20 h^{-1} in the presence of the base DBU (DBU = 1,8-diazabicyclo[5.4.0]undec-7-ene) at 50 °C and 200 bar [23]. For an *in situ*-generated tetra-dentate iron–phosphine complex, $[Fe(PP_3^{Ph})(H_2)(H)]$ (PP_3^{Ph} = tris(2-(diphenylphosphino)phenyl)phosphine) [24], a TON of 1897 and a TOF of 95 h^{-1} after 20 h at 100 °C for the formation of formic acid have been found. For all base metal complexes tested so far, a phosphine-coordinated low-valent cobalt(I) species, $[Co^I(dmpe)H]$ (dmpe = 1,2-bis(dimethylphosphino)ethane), exhibits the highest catalytic efficacy with a TOF of 74 000 h^{-1} [25]. The reaction proceeds at room temperature and 20 bar pressure. However, a very strong and expensive base, Verkade's base, had to be applied in order to achieve such an outstanding activity. Despite operating under extreme conditions, non-noble metal catalysts still show much lower reactivity than their noble metal congeners. Undoubtedly, mechanistic insights into the reactivity can provide the clue to improving the catalytic efficiency of base metal complexes or leading to new design for more efficient and robust catalysts.

2.1.6 CO_2 Hydrogenation: Basic Mechanistic Considerations

Most noble or non-noble metal catalysts for CO_2 hydrogenation follow a common mechanism as presented in Figure 2.3. In the reaction, transition metal hydrides functioning as hydride donors transfer hydride to CO_2 [17], and the formation of metal hydrides is often accomplished through base-promoted heterolytic H_2 splitting. As such, hydride transfer and H_2 splitting represent the two crucial steps of the entire catalytic cycle. In fact, earlier mechanistic studies showed that either of these two transformations could be the RDS. For instance, H_2 splitting is computationally predicted to be the RDS for CO_2 reduction with $[(PNP)Ir^{III}(H)_3]$, $[(PNP)Co^{III}(H)_3]$, and $[(PNP)Fe^{II}(H)_2(CO)]$ (PNP = 2,6-bis(dialkylphosphinomethyl)-pyridine) [26]. A similar prediction

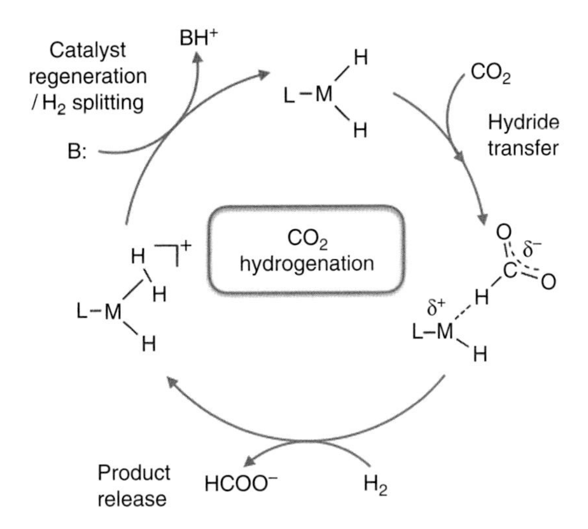

Figure 2.3 CO_2 hydrogenation via hydride transfer by transition metal catalysts.

was also reported for half-sandwich complexes $[Cp^*M^{III}(6,6'\text{-}O^-\text{-bpy})(H_2O)]$ (M = Co, Rh, and Ir; $Cp^* = \eta^5\text{-}C_5Me_5$, bpy = 2,2′-bipyridine) [27]. Interestingly, an experimental kinetic investigation on $[(\eta^6\text{-}C_6Me_6)Ru^{II}(bpy)(OH_2)]^{2+}$ and $[Cp^*Ir^{III}(bpy)(OH_2)]^{2+}$ by Ogo et al. demonstrated different RDSs for the reactions mediated by these two isoelectronic complexes, viz H_2-splitting RDS for the former and hydride transfer RDS for the latter [28]. The RDS was determined by monitoring the change in the TONs as a function of the H_2 and CO_2 pressures. In the case of the Ru(II) complex, the TONs showed a saturation behavior with respect to the CO_2 pressure, whereas the TONs increased linearly with the H_2 pressure, thereby indicating that H_2 splitting is the RDS for the reaction with the Ru(II) complex. As a consequence, the catalytically active Ru hydride species, the intermediate generated in the RDS, was not detected. In contrast, for the Ir(III) complex, the TONs leveled off at a certain pressure of H_2, and the variation of the TONs relied only on the CO_2 pressure. Both observations evidenced a hydride transfer RDS for the transformation catalyzed by the Ir(III) complexes. A recent experimental study on $[Co^I(dmpe)_2H]$ (dmpe = 1,2-bis(dimethylphosphino)ethane) suggested hydride transfer to be the RDS in the presence of Verkade's base [25]. In this case, the catalytic rate was found not to depend on the base concentration and the H_2 pressure but to have a first-order dependence on the CO_2 pressure. The fickle nature of the RDS for CO_2 hydrogenation raises an intriguing question about what factors govern the chemical identity of the RDS. To address this issue, we first undertook a theoretical mechanistic investigation on an existing non-noble metal catalyst.

2.2 Reaction Energetics and Governing Factor

A phosphine-coordinated Fe(II) complex, $[Fe^{II}(H)(\eta^2\text{-}H_2)(PP_3^{Ph})]^+$ ($\mathbf{R}_{Fe}$, PP_3^{Ph} = tris(2-(diphenylphosphino)phenyl)phosphine), developed by Beller and coworkers, serves as a representative example of base metal catalysts for CO_2 hydrogenation. This species has been shown to exhibit high catalytic activity for formate production [24]. To obtain reliable PES, we used a highly correlated *ab initio* DLPNO-CCSD(T) method in conjunction with a large and flexible def2-TZVPP basis set to compute accurate energies of key intermediates and transition states along the reaction pathway. Our theoretical results proposed a catalytic cycle presented in Figure 2.4. Experimentally, it was found that only upon adding a base (e.g. NEt_3) to complex $\mathbf{R}_{Fe}$, the reaction could turn over. This observation suggested that the base-promoted heterolytic H_2 splitting is necessary to generate the catalytically active intermediate $\mathbf{I}_{Fe}$ from $\mathbf{R}_{Fe}$. Complex $\mathbf{I}_{Fe}$ carries out direct hydride transfer to CO_2, leading to the formation of formate-bound complex $\mathbf{P}_{Fe}$, which then undergoes formate dissociation through H_2 binding to regenerate complex $\mathbf{R}_{Fe}$. Our calculations revealed that the subsequent hydride transfer step ($\mathbf{I}_{Fe} \rightarrow \mathbf{P}_{Fe}$) transverses a slightly lower barrier, and hence, heterolytic H_2 splitting ($\mathbf{R}_{Fe} \rightarrow \mathbf{I}_{Fe}$) is the RDS of the overall catalytic cycle [29]. The conjugate acid $HNEt_3^+$ was shown to assist the final product release through H-bonding effects. For comparison, we have investigated the isoelectronic Co(III) complex, $[Co^{III}(H)(\eta^2\text{-}H_2)(PP_3^{Ph})]^{2+}$ ($\mathbf{R}_{Co}$), in the same ligand environment. According to the computed free energies depicted in Figure 2.5, the H_2-splitting process mediated by $\mathbf{R}_{Co}$ via $\mathbf{TS}^{Co}_{H_2}$ involves a much lower barrier than that by $\mathbf{R}_{Fe}$, whereas hydride transfer ($\mathbf{TS}^{Co}_{H^-}$) starting from $\mathbf{I}_{Co}$ appeared unlikely to occur. Thus, even if the reaction with $\mathbf{R}_{Co}$ could take place, hydride transfer would be the RDS. Clearly, switching the metal center from Fe(II) to Co(III) changes the CO_2 hydrogenation RDS from H_2 splitting to hydride transfer.

In order to pinpoint the pivotal features that are responsible for the dramatic difference in the CO_2 hydrogenation reactivity between $\mathbf{R}_{Fe}$ and $\mathbf{R}_{Co}$, in light of the BEP principle, we first analyzed the reaction-driving forces. Our calculation showed that the H_2-splitting process is nearly thermoneutral for $\mathbf{R}_{Fe}$, whereas it is highly exergonic for $\mathbf{R}_{Co}$ (Figure 2.5). More importantly, we found that the

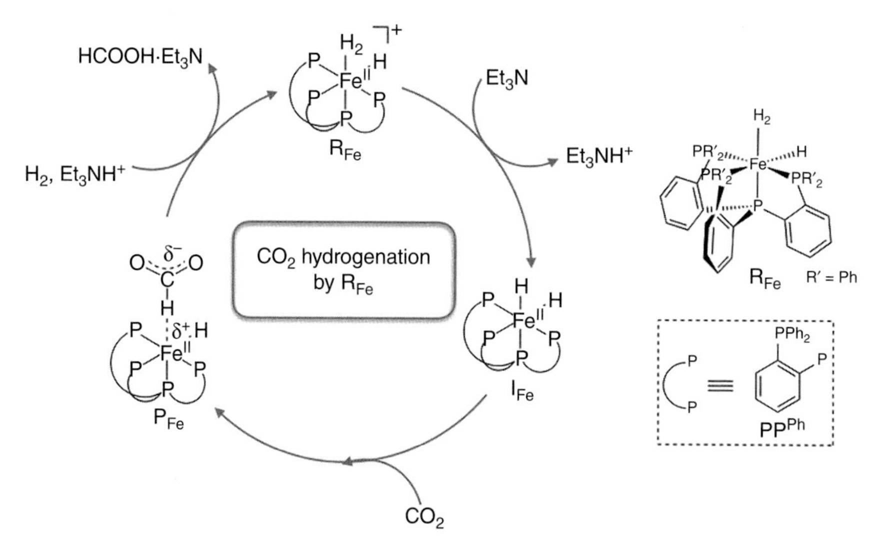

Figure 2.4 Proposed catalytic cycle exhibited by iron complex R_{Fe}.

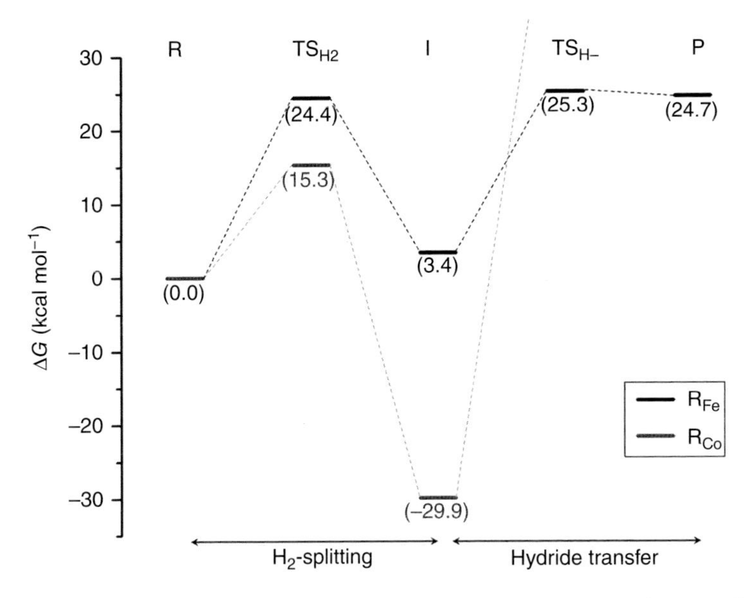

Figure 2.5 DLPNO-CCSD(T) free energy profile of key reaction steps for R_{Fe} and R_{Co} complexes.

H_2-splitting barriers for $\mathbf{R}_{Fe}$ and $\mathbf{R}_{Co}$ correlate with their distinct reaction-free energies (ΔG). Therefore, the BEP principle provides an explanation why the H_2-splitting reactivity of $\mathbf{R}_{Co}$ substantially surpasses that of $\mathbf{R}_{Fe}$. The H_2-splitting driving force for $\mathbf{R}_{Fe}$ and $\mathbf{R}_{Co}$ reflects the differential bonding strength between the metal hydride interaction (M—H^-) in the intermediate **I** and the metal–H_2 interaction (M—H_2) in the reactant **R**, as NEt_3H^+ is the common product. The computed H_2 binding enthalpies for the two metal centers differ by only 5 kcal mol^{-1} and therefore cannot explain the wide reaction energy range of over 30 kcal mol^{-1} (Figure 2.5). Thus, the M—H^- bonding strength in **I**, which can be quantified by its hydricity or hydride affinity dictates the H_2-splitting driving force. Hydricity, $\Delta G°_{H^-}$(MH), measures the ability of a metal hydride complex (e.g. **I**) to donate its hydride as $MH \rightarrow M^+ + H^-$ [30]. According to this definition, a less positive value of $\Delta G°_{H^-}$(MH) corresponds to a higher hydride donating ability or lower hydricity. Obviously, a metal center that can form stronger bonds with the hydride ligand should be a poor hydride donor and feature a higher hydricity. The calculated hydricities for $\mathbf{I}_{Fe}$ and $\mathbf{I}_{Co}$ are 58 and 100 kcal mol^{-1}, respectively. Clearly, the drastically different hydricity between dihydride complexes $\mathbf{I}_{Fe}$ and $\mathbf{I}_{Co}$ rationalizes their distinct H_2-splitting driving forces. Similarly, the hydricity also correlates with the hydride transfer reactivity of a metal hydride complex [25, 31]. In our case, the overwhelming hydricity of $\mathbf{I}_{Co}$ prohibits hydride transfer (**I** → **P**, Figure 2.5) and the moderate hydricity of $\mathbf{I}_{Fe}$ renders an accessible hydride transfer barrier. Although the metal centers in $\mathbf{I}_{Fe}$ and $\mathbf{I}_{Co}$ possess the same number of *d* electrons, the high hydricity of $\mathbf{I}_{Co}$ could be attributed to the higher oxidation state of the cobalt ion and hence greater overall charge on the complex. This indicates that hydride complexes containing a low-valent cobalt center would be a good hydrogenation catalyst as exemplified by [Co^I(dmpe)H] discussed above [25, 32]. On the other hand, charge compensation by using anionic ligands is also expected to increase the catalytic activity as shown by a recent example of a Co(III) complex, [Cp^*Co^{III}(6,6′-O-bpy)(H_2O)] [33]. Taken together, we can propose a general strategy to enhance the catalytic activity of the existing non-noble catalysts, $\mathbf{1}_{Fe}$ and $\mathbf{1}_{Co}$. For low-hydricity catalysts (e.g. complex $\mathbf{1}_{Fe}$), pulling electron density from the metal center to strengthen the M—H^- bond and thereby to increase its hydricity would lower the H_2-splitting RDS barrier. However for high-hydricity catalysts (e.g. complex $\mathbf{1}_{Co}$), a diminished hydride transfer RDS barrier may be achieved by pushing electron density to the metal center to weaken the M—H^- interaction and thus to lower its hydricity.

2.3 Newly Designed Catalysts and Their Reactivity

To verify the above-mentioned design strategies, we computationally investigated the reactivity of CO_2 hydrogenation of a series of new Co(III) and Fe(II) complexes (Figure 2.6) [34]. Note here that similar Si- and C-anchor ligands with a slightly different ligand substituent (R = *i*Pr) have already been reported in the literature [35]. Complexes [Co(H)(η^2-H_2)(CP_3^{Ph})] (CP_3^{Ph} = tris(2-(diphenylphosphino)-phenyl)methyl) ($\mathbf{R}_{Co/C}$) and [Co(H)(η^2-H_2)(SiP_3^{Ph})] (SiP_3^{Ph} = tris(2-(diphenylphosphino)phenyl)silyl) ($\mathbf{R}_{Co/Si}$) are obtained by replacing the phosphine anchor in their parent complex $\mathbf{R}_{Co}$ with anionic C^- and Si^-, intended for pushing electron density to the Co(III) center. To pull electron density from the Fe(II) center, electron-withdrawing —NO_2 groups were added to the *para*-positions of the phenyl rings in complex $\mathbf{R}_{Fe}$, leading to complex [Fe(H)(η^2-H_2)(PP_3^{PhNO2})]$^+$ (PP_3^{PhNO2} = tris(2-(diphenylphosphino)-4-nitrophenyl)phosphine) ($\mathbf{R}_{Fe/NO2}$). For comparison, we also studied complexes

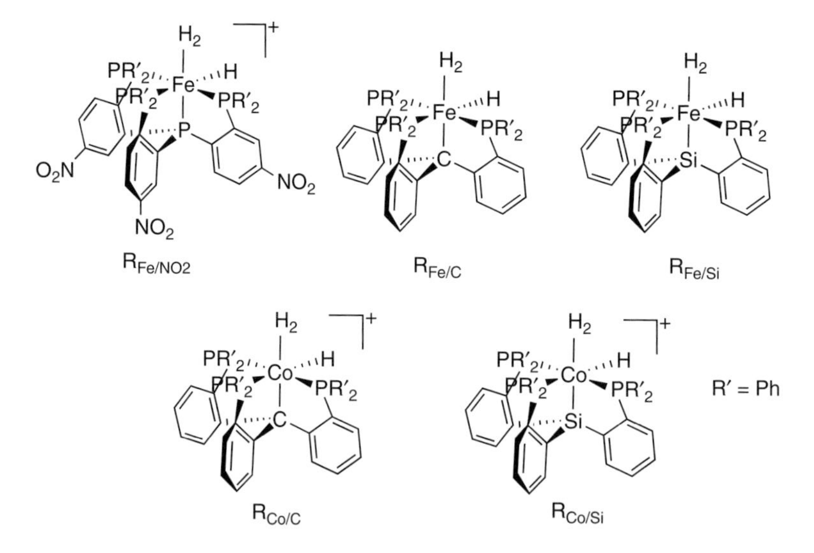

Figure 2.6 Newly designed Fe(II)- and Co(III)-based complexes.

Table 2.1 Intrinsic reaction barriers ($\Delta G^{\ddagger}$) for Co(III)- and Fe(II)-based complexes.

Complex	$\Delta G^{\ddagger}$ (DLPNO-CCSD(T)) (kcal mol^{-1})	
	H_2 splitting	Hydride transfer
$\mathbf{R}_{Co}$	4.4	>30
$\mathbf{R}_{Co/C}$	6.6	8.3
$\mathbf{R}_{Co/Si}$	11.4	11.8
$\mathbf{R}_{Fe}$	14.1	9.7
$\mathbf{R}_{Fe/C}$	16.2	0.3
$\mathbf{R}_{Fe/Si}$	19.3	1.7
$\mathbf{R}_{Fe/NO2}$	8.5	11.8

$\mathbf{R}_{Fe/C}$ and $\mathbf{R}_{Fe/Si}$ featuring anionic anchors C^- and Si^-, respectively. Our calculation showed that the use of an anionic anchor ligand in complexes $\mathbf{R}_{Co/C}$ and $\mathbf{R}_{Co/Si}$ radically lowers the hydride transfer RDS barrier compared to $\mathbf{R}_{Co}$, although the H_2-splitting barriers for both the complexes increase (Table 2.1). Specifically, the hydride transfer step can occur with accessible barriers for complexes $\mathbf{R}_{Co/C}$ and $\mathbf{R}_{Co/Si}$; in contrast, an activation-free energy of over 30 kcal mol^{-1} is found for this transformation mediated by complex $\mathbf{R}_{Co}$ (Figure 2.5). A similar situation was observed for complexes $\mathbf{R}_{Fe/C}$ and $\mathbf{R}_{Fe/Si}$ relative to $\mathbf{R}_{Fe}$, viz the H_2-splitting barriers increase and the hydride transfer barriers decrease. As a consequence, the catalytic activity of $\mathbf{R}_{Fe/C}$ and $\mathbf{R}_{Fe/Si}$ is expected to be much lower because H_2 splitting is the RDS. In fact, $[Fe(H)(\eta^2\text{-}H_2)(SiP_3{}^{iPr})]$ ($SiP_3{}^{iPr}$ = tris(2-(dipisopropylphosphino)phenyl)silyl), an analog to complex $\mathbf{R}_{Fe/Si}$, has been recently proposed to follow a different mechanism other than heterolytic H_2 splitting at the metal center because of the prohibitively high barrier [36]. In comparison to complexes $\mathbf{R}_{Fe/C}$ and $\mathbf{R}_{Fe/Si}$, the electron-withdrawing effect in complex $\mathbf{R}_{Fe/NO2}$ lowers the H_2-splitting barrier and enlarges the hydride transfer barrier (Table 2.1), and the RDS is switched from H_2 splitting to the hydride transfer. All in all, our theoretical results substantiated our proposed ligand design strategy. Clearly, a single modification of the ligand affects the barriers of both key steps; hence, one needs to strike an exquisite balance between them in order to develop more efficient catalysts.

2.4 Correlation Between Hydricity and Reactivity

We have already shown that tuning the hydricity of dihydride intermediates by ligand modifications has appreciable effects on the reactivity of both H_2-splitting and hydride transfer steps for all iron and cobalt complexes under investigation. Now it seems intriguing to rationalize why a single parameter, hydricity, modulates the barriers of both pivotal steps. First, we observe a good correlation between the H_2-splitting barrier ($\Delta G^{\ddagger}$) and the reaction-free energy (ΔG) (Figure 2.7a), nicely complying with the BEP principle. For instance, introducing the electron-donating moieties to $\mathbf{R}_{Co/C}$, $\mathbf{R}_{Co/Si}$, $\mathbf{R}_{Fe/C}$, and $\mathbf{R}_{Fe/Si}$ drops their

H_2-splitting driving forces. In contrast, the driving force of $\mathbf{R}_{Fe/NO2}$ gets enhanced. Second, the H_2-splitting driving force is predominantly dictated by the M—H^- interaction strength or the hydricity of the resulting dihydride complex (*vide supra*), implying a direct correlation between the hydricity ($\Delta G°_{H^-}$) and the H_2-splitting driving force (ΔG) as shown in Figure 2.7b. As such, $\Delta G°_{H^-}$ should correlate with $\Delta G^{\ddagger}$ ($R^2 = 0.81$, Figure 2.7c). Therefore, an H_2-splitting process to generate a low-hydricity species will encounter a higher barrier and vice versa. For hydride transfer, a correlation between the hydricity and the hydride transfer barrier can also be expected, as the hydricity directly measures the ease of breaking the M—H^- bond. Indeed, we observe a good linear relationship ($R^2 = 0.92$, Figure 2.8b) between them. In line with the BEP principle, the hydride transfer barrier may correlate with its driving force ($R^2 = 0.92$, Figure 2.8a). Furthermore, the hydride transfer reaction energy is also largely dependent on the M—H^- bonding strength, as $HCOO^-$ or $HCOOH{\cdot}NEt_3$ is the common product. Thus, one parameter, the hydricity of dihydride intermediate **I**, controls the barrier of the two crucial steps of CO_2 hydrogenation. The high hydricity of $\mathbf{I}_{Co}$ hampers the hydride transfer process, whereas the electron-rich metal centers in $\mathbf{I}_{Co/C}$ and $\mathbf{I}_{Co/Si}$ with substantially lowered hydricities can perform the reaction with moderate barrier. A similar effect has been observed for complexes $\mathbf{I}_{Fe/C}$ and $\mathbf{I}_{Fe/Si}$. The electron-withdrawing effect in $\mathbf{R}_{Fe/NO2}$ enhances

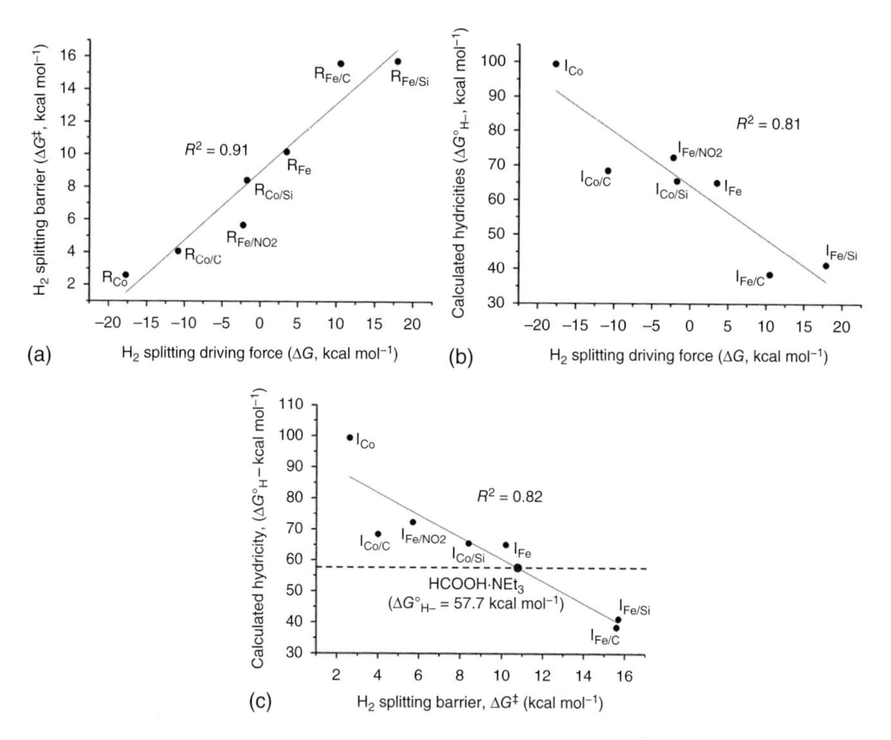

Figure 2.7 Correlation plots for the H_2-splitting step: (a) correlation between barrier and driving force; (b) correlation between driving force and hydricity; and (c) correlation between hydricity and barrier.

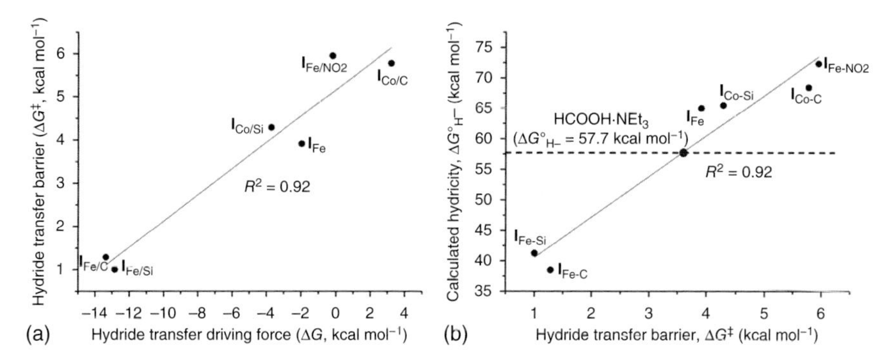

Figure 2.8 Correlation plots for the hydride transfer step: (a) correlation between barrier and driving force; (b) correlation between hydricity and barrier.

the hydricity and therefore increases the hydride transfer barrier. Altogether, newly designed cobalt-based complexes $\mathbf{R}_{Co/Si}$ and $\mathbf{R}_{Co/C}$ possess accessibly low barriers for both hydride transfer and H_2-splitting steps and appear to be promising catalysts for CO_2 hydrogenation.

As shown in Figures 2.7c and 2.8b, the free energy barriers for both the H_2-splitting and the hydride transfer steps elegantly correlate with the hydricity of the active intermediate **I**. Thus, if the dihydride intermediate **I** features a relatively high hydricity, the hydride transfer process may encounter a high barrier as observed for $\mathbf{I}_{Co}$. In contrast, a catalyst with a relatively low hydricity may undergo facile hydride transfer, whereas the corresponding H_2-splitting step proceeds slowly as shown for $\mathbf{I}_{Fe/C}$ and $\mathbf{I}_{Fe/Si}$. For an efficient catalyst, its hydricity cannot be too high or too low. Based on our calculations, its hydricity should be roughly close to that of the product HCOOH·NEt_3 (57.7 kcal mol^{-1}) [29]. As can be seen in Figures 2.7c and 2.8b, in which the hydricity of HCOOH·NEt_3 is shown with a horizontal line, species $\mathbf{I}_{Fe}$ and $\mathbf{I}_{Co/Si}$ are closest to that line and hence can achieve the aforementioned balance between the two key steps. The existence of an optimal hydricity value for efficient catalysts could be nicely illustrated in Figure 2.9, which depicts the correlations of both H_2-splitting and hydride transfer barriers ($\Delta G^{\ddagger}$) with the hydricity ($\Delta G^{\circ}{}_{H^-}$). Two linear-fitted lines for H_2-splitting and hydride transfer cross at a point that represents a hydricity value of 59.7 kcal mol^{-1}, which is very close to our estimated optimal value (57.7 kcal mol^{-1}). The optimal value further complements our earlier prediction of $\mathbf{I}_{Fe}$ and $\mathbf{I}_{Co/Si}$ to be the most efficient catalytic intermediates as they are closest to the crossing point in Figure 2.9.

2.5 Concluding Remarks

Using the CO_2 hydrogenation reaction as an example, we have demonstrated how insights into the RDS step of a catalytic process obtained from computational modeling can be translated into efficient non-noble metal-based catalyst design. Our systematic investigation on the key reaction steps of a series of Fe(II) and Co(III) complexes show a good correlation between the reaction barrier and the

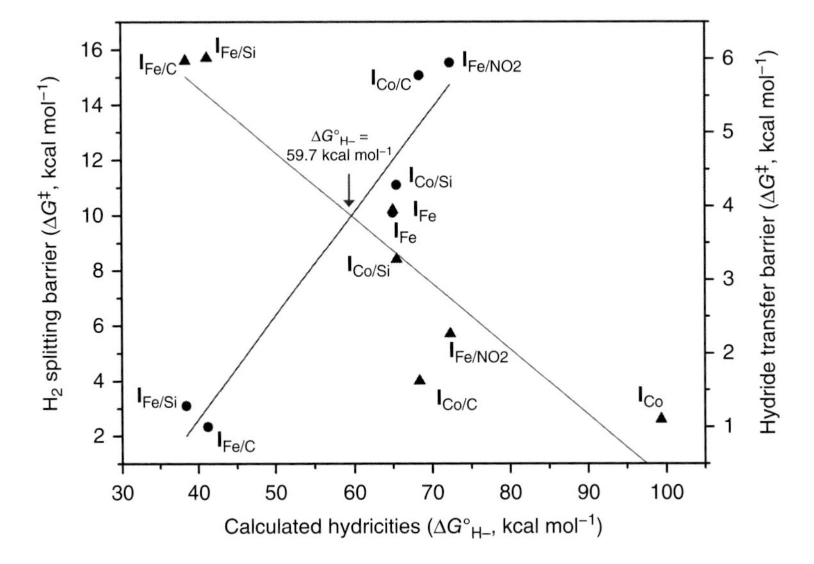

Figure 2.9 Plot correlating both the barriers (H_2-splitting and hydride transfer) and calculated hydricity.

hydricity of the catalytically active metal hydride species. Specifically, we have shown that enhancing the electron-donating power of the supporting ligands is likely to accelerate the hydride transfer step by weakening the metal hydride interaction (e.g. $\mathbf{R}_{Co/C}$ and $\mathbf{R}_{Co/Si}$). In contrast, lowering the hydricity decelerates the H_2-splitting process as found for $\mathbf{R}_{Fe/NO2}$. All observations reflect that the electronic requirements for the H_2-splitting and hydride transfer processes are just opposite. This poses the requirement of a delicate balance between the two processes for an efficient CO_2 hydrogenation catalyst.

The current example showcases the importance of insights obtained by a high-quality computational study into the complex catalytic mechanism directing rational catalyst design. Particularly, precise knowledge about the rate-limiting steps and their governing factors seems to be absolutely crucial. However, the presence of multiple key steps with comparable energetics imposes major challenges in obtaining the necessary details. Highly accurate estimation of the reaction energetics helps overcome those challenges, for which highly correlated *ab initio* calculations (e.g. DLPNO-CCSD(T)) are required. Such computations can produce energetics within the chemical accuracy (~2 kcal mol^{-1}); therefore, the RDSs in a complex catalytic process can be unambiguously determined. Thus, a deep understanding of the reaction mechanism obtained through high-level quantum-chemical calculations opens up new design and development possibilities for first-row transition metal catalysts.

Acknowledgments

We gratefully acknowledge the financial support from the Max-Planck Society.

References

1 Houk, K.N. and Cheong, P.H.-Y. (2008). *Nature* 455: 309–313.
2 Saeidi, S., Amin, N.A.S., and Rahimpour, M.R. (2014). *J. CO2 Util.* 5: 66–81.
3 Ziebart, C. and Beller, M. (2014). *Transformation and Utilization of Carbon Dioxide*, vol. 5 (ed. B.M. Bhanage and M. Arai), 73–102. Berlin Heidelberg: Springer-Verlag.
4 McQuarrie, D.A. and Simon, J.D. (1997). *Physical Chemistry: A Molecular Approach*, 1190. Sausalito, CA: University Science Books.
5 Laidler, K.J. and King, M.C. (1983). *J. Phys. Chem.* 87: 2657–2664.
6 Hohenberg, P. and Kohn, W. (1964). *Phys. Rev.* 136: B864.
7 Harvey, J.N. (2006). *Annu. Rep. Prog. Chem., Sect. C* 102: 203–226.
8 (a) Riplinger, C. and Neese, F. (2013). *J. Chem. Phys.* 138: 034106. (b) Riplinger, C., Sandhoefer, B., Hansen, A., and Neese, F. (2013). *J. Chem. Phys.* 139: 134101.
9 Evans, M.G. and Polanyi, M. (1938). *Trans. Faraday Soc.* 34: 11–23.
10 Valentine, J.S. (2007). *Biological Inorganic Chemistry: Structure and Reactivity*, 1e (ed. H.B. Gray, E.I. Stiefel, J.S. Valentine and I. Bertini), 319. University Science Books.
11 Appel, A.M., Bercaw, J.E., Bocarsly, A.B. et al. (2013). *Chem. Rev.* 113: 6621–6658.
12 (a) Tans, P. and Keeling, R. NOAA/ESRL (www.esrl.noaa.gov/gmd/ccgg/trends/) and Scripps Institution of Oceanography scrippsco2.ucsd.edu/ (accessed 22 August 2016). (b) Friedlingstein, P., Andrew, R.M., Rogelj, J. et al. (2014). *Nat. Geosci.* 7: 709–715.
13 (a) Li, H. and Liao, J.C. (2013). *Energy Environ. Sci.* 6: 2892. (b) Glueck, S.M., Gümüs, S., Fabian, W.M.F., and Faber, K. (2009). *Chem. Soc. Rev.* 39: 313.
14 (a) Langanke, J., Wolf, A., Hofmann, J. et al. (2014). *Green Chem.* 16: 1865. (b) Ola, O., Mercedes, Maroto-Valer, M., and Mackintosh, S. (2013). *Energy Procedia* 37: 6704–6709.
15 (a) Benson, E.E., Kubiak, C.P., Sathrum, A.J., and Smieja, J.M. (2009). *Chem. Soc. Rev.* 38: 89–99. (b) Mondal, B., Song, J., Neese, F., and Ye, S. (2015). *Curr. Op. Chem. Biol.* 25: 103–109.
16 (a) Savéant, J.-M. (2008). *Chem. Rev.* 108: 2348–2378. (b) Federsel, C., Jackstell, R., and Beller, M. (2010). *Angew. Chem. Int. Ed.* 49: 6254. (c) Kumar, B., Llorente, M., Froehlich, J. et al. (2012). *Annu. Rev. Phys. Chem.* 63: 541. (d) Costentin, C., Robert, M., and Savéant, J.-M. (2013). *Chem. Soc. Rev.* 42: 2423.
17 (a) Jessop, P.G., Ikariya, T., and Noyori, R. (1995). *Chem. Rev.* 95: 259–272. (b) Jessop, P.G., Joó, F., and Tai, C.-C. (2004). *Coord. Chem. Rev.* 248: 2425. (c) Leitner, W. (1995). *Angew. Chem. Int. Ed.* 34: 2207. (d) Wang, W., Wang, S., Ma, X., and Gong, J. (2011). *Chem. Soc. Rev.* 40: 3703.
18 Klankermayer, J., Wesselbaum, S., Beydoun, K., and Leitner, W. (2016). *Angew. Chem. Int. Ed.* 128: 7416–7467.
19 Inoue, Y., Izumida, H., Sasaki, Y., and Hashimoto, H. (1976). *Chem. Lett.* 5: 863–864.

20 Gassner, F. and Leitner, W. (1993). *J. Chem. Soc., Chem. Commun.* 1465.
21 Tanaka, R., Yamashita, M., and Nozaki, K. (2009). *J. Am. Chem. Soc.* 131: 14168–14169.
22 (a) Filonenko, G.A., van Putten, R., Schulpen, E.N. et al. (2014). *ChemCatChem* 6: 1526. (b) Filonenko, G.A., Conley, M.P., Copéret, C. et al. (2013). *ACS Catal.* 3: 2522–2526.
23 Tai, C.-C., Chang, T., Roller, B., and Jessop, P.G. (2003). *Inorg. Chem.* 42: 7340–7341.
24 Ziebart, C., Federsel, C., Anbarasan, P. et al. (2012). *J. Am. Chem. Soc.* 134: 20701.
25 Jeletic, M.S., Mock, M.T., Appel, A.M., and Linehan, J.C. (2013). *J. Am. Chem. Soc.* 135: 11533–11536.
26 (a) Ahlquist, M.R.S.G. (2010). *J. Mol. Catal. A: Chem.* 324: 3. (b) Yang, X. (2011). *ACS Catal.* 1: 849–854.
27 Hou, C., Jiang, J., Zhang, S. et al. (2014). *ACS Catal.* 4: 2990–2997.
28 Ogo, S., Kabe, R., Hayashi, H. et al. (2006). *Dalton Trans.* 4657–4663.
29 Mondal, B., Neese, F., and Ye, S. (2015). *Inorg. Chem.* 54: 7192–7198.
30 (a) DuBois, D.L. and Berning, D.E. (2000). *Appl. Organomet. Chem.* 14: 860. (b) Qi, X.-J., Fu, Y., Liu, L., and Guo, Q.-X. (2007). *Organometallics* 26: 4197–4203.
31 Muckerman, J.T., Achord, P., Creutz, C. et al. (2012). *Proc. Natl. Acad. Sci. U.S.A.* 109: 15657–15662.
32 Federsel, C., Ziebart, C., Jackstell, R. et al. (2012). *Chem. Eur. J.* 18: 72–75.
33 Badiei, Y.M., Wang, W.-H., Hull, J.F. et al. (2013). *Inorg. Chem.* 52: 12576–12586.
34 Mondal, B., Neese, F., and Ye, S. (2016). *Inorg. Chem.* 55: 5438.
35 (a) Whited, M.T., Mankad, N.P., Lee, Y. et al. (2009). *Inorg. Chem.* 48: 2507. (b) Creutz, S.E. and Peters, J.C. (2014). *J. Am. Chem. Soc.* 136: 1105. (c) Moret, M.-E. and Peters, J.C. (2011). *J. Am. Chem. Soc.* 133: 18118. (d) Moret, M.-E. and Peters, J.C. (2011). *Angew. Chem. Int. Ed.* 50: 2063. (e) Lee, Y., Mankad, N.P., and Peters, J.C. (2010). *Nat. Chem.* 2: 558. (f) Ye, S., Bill, E., and Neese, F. (2016). *Inorg. Chem.* 55: 3468.
36 Fong, H. and Peters, J.C. (2015). *Inorg. Chem.* 54: 5124.

3

Catalysis with Multinuclear Complexes

Neal P. Mankad

University of Illinois at Chicago, Department of Chemistry, 845 West Taylor Street, Chicago, IL 60607, USA

3.1 Introduction

In the fields of organometallic chemistry and homogeneous catalysis, single-site catalysis is the dominant paradigm. In other words, typical catalytic reactions are conceptualized and designed to occur at a single metal site, which is solely responsible for managing all the bond-breaking and bond-forming events and shifting all the redox equivalents necessary for a given transformation. Great synthetic efforts are spent constructing elaborate ligand scaffolds that are meant to prevent multinuclear reaction pathways, thereby favoring single-site mechanisms where individual reaction steps are well documented and catalytic concepts are well understood [1]. This single-site approach has proven itself to be versatile and prolific, but it is inherently limited because it explores an arbitrarily limited part of chemical space. Complementary approaches involving the cooperation of two or more metal sites are comparatively underexplored. Several motivations have energized a growing community of researchers to pursue such strategies.

First and foremost, because the catalytic chemistry of binuclear and multinuclear platforms is underdeveloped, it stands to reason that new modes of reactivity and/or selectivity will emerge upon exploring this void in chemical space. Often these discoveries will complement traditional single-site catalysts and, therefore, will add to the available synthetic toolbox. Rate enhancements and selectivity amplification are also possible outcomes.

Secondly, for many important catalytic applications, the single-site approach requires or favors the use of noble metals. Presumably, this trend stems from the ability of noble metals to facilitate multielectron redox chemistry under mild conditions, thereby allowing them to engage in redox-active reaction steps (e.g. oxidative addition, reductive elimination) while simultaneously being able to manage key redox-neutral steps (e.g. migratory insertion, β-hydride elimination) during catalysis. By contrast, base metals tend to favor single-electron chemistry that precludes their ability to participate in these classic single-site catalytic concepts [2]. A set of illustrative examples occurs in Group 10 of the periodic table. The 3d member, Ni, has the +2 and +3 forms as its commonly available oxidation states. The heavier congeners, Pd and Pt, have the +2 and +4 forms as

Non-Noble Metal Catalysis: Molecular Approaches and Reactions,
First Edition. Edited by Robertus J. M. Klein Gebbink and Marc-Etienne Moret.

their commonly available oxidation states. Clearly, then, earth-abundant Ni will favor one-electron chemistry in single-site scenarios, whereas the noble metal counterparts in the same group (Pd and Pt) will favor two-electron chemistry. The advancement of non-noble metal replacements for noble metal catalysts, thus, requires exploration of novel strategies that allow non-noble metals to participate in multielectron redox processes under mild conditions. Multimetallic redox cooperation represents one promising approach toward this end.

Thirdly, knowledge of biological systems and the "design" principles evolved therein is advanced by the study of multimetallic assemblies and their reactivity. Bimetallic and multimetallic cooperation is known to be crucial to certain biological redox transformations of simple small-molecule substrates, such as H_2O, O_2, CO_2, N_2, N_2O, and NO, that are activated at metallocofactor sites within the metalloproteins [3]. In some cases, the role of multimetallic cooperation is established [4], while in other cases, the intimate nature of the cooperation remains elusive [5, 6]. Understanding synthetic multimetallic systems and their chemical behavior stands to guide hypotheses regarding these biological systems, which are constrained to make use of only bioavailable non-noble metals.

This chapter focuses on bimetallic and multimetallic reaction pathways and their applications to catalysis. Although not intended to be comprehensive, the chapter pays focus to emerging examples and to bimetallic systems, where conceptualization is simpler and where mechanistic understanding often has outpaced that for multimetallic systems with >2 metals. Bimetallic and bifunctional chemistry involving ≥2 transition metal sites is considered, whereas examples involving main group elements are excluded except for selected examples. Similarly, examples featuring at least one noble metal center in the multimetallic assembly are excluded from consideration. The topics of cooperative behavior with multiple p-block sites and with noble metal-containing multimetallic assemblies have been reviewed elsewhere [7, 8].

3.2 Stoichiometric Reaction Pathways

The design of single-site catalytic systems relies on the deep understanding that is already available regarding stoichiometric reaction pathways of single metal sites. These fundamental reaction steps (e.g. oxidative addition, reductive elimination, migratory insertion, and β-hydride elimination) form the foundation for homogeneous catalysis and provide a conceptual framework for designing new single-site reactions. Similarly, to develop bimetallic and multimetallic transformations, it is imperative to understand the available reaction pathways for this concept. This section focuses on established and emerging stoichiometric reaction pathways involving either reactive metal–metal bonds or bifunctional chemistry of closely associated metal sites lacking direct interaction in the absence of substrate. Such reaction pathways provide the foundation for designing multimetallic catalytic transformations.

3.2.1 Bimetallic Binding and Activation of Substrates

A simple type of single-site reaction that is crucial to many catalytic processes is the binding of an unsaturated substrate, with accompanying activation of

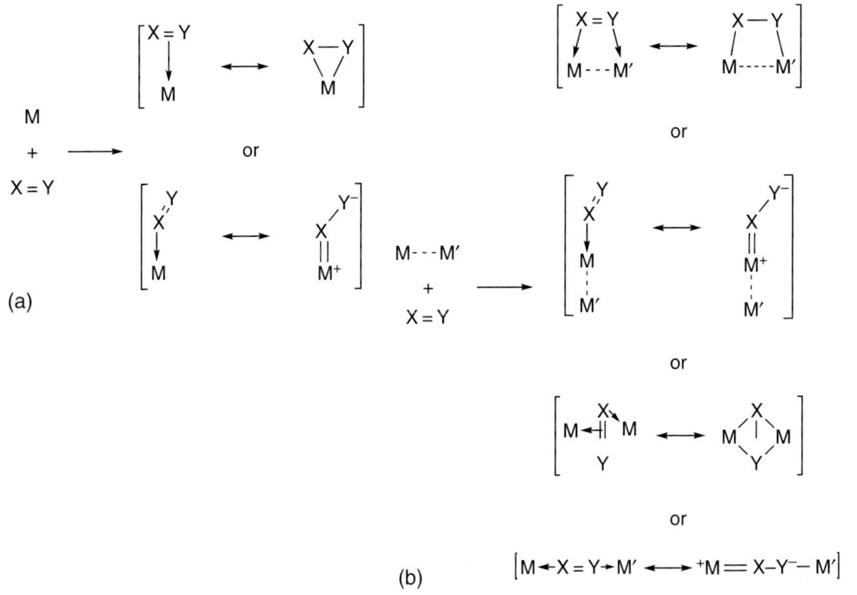

Figure 3.1 Binding and activation of unsaturated substrates (a) at a single metal site and (b) at bimetallic reaction centers.

the substrate through π-backbonding from the metal center. Depending on the nature of the metal and substrate combination, this binding can occur either side-on or end-on (Figure 3.1a). Similar binding and activation are available for bimetallic reaction centers. Depending on the nature of the system, this behavior can occur through bifunctional substrate activation by both metal sites or through substrate binding to a reactive metal site that interacts with a supporting metal site (Figure 3.1b).

3.2.1.1 Small-Molecule Activation

A well-known example of bimetallic substrate binding involves O_2 activation, either to bimetallic complexes with binucleating ligands or to separate single-site complexes that cooperate upon substrate binding. In both cases, O_2 activation yields various types of $[M_2O_2]$ complexes featuring a range of different transition metals, including biomimetic Cu and Fe examples [9]. Another example of bifunctional substrate activation is Floriani's landmark report of CO_2 binding by Co…M cooperation (M = Li, Na, and K) [10]. Anionic Co(I) salen complexes with inner-sphere alkali metal cations were found to bind and activate CO_2, providing bent CO_2 adducts **1** (Figure 3.2). This bifunctional CO_2 activation was proposed to involve a nucleophilic Co(I) center engaging the electrophilic carbon of CO_2 simultaneous with the electrophilic alkali metal center engaging a nucleophilic oxygen of CO_2. Evidence for bifunctional cooperation being crucial to CO_2 binding included the fact that CO_2 loss was observed upon addition of a crown ether. The kinetic stability of the CO_2 adducts and the extent of CO_2 activation were both modulated by the identity of the alkali metal site, with the Li derivative providing the most kinetically stable adduct (as judged by stability toward vacuum) with the most activated CO_2 unit (as determined by IR spectroscopy). More

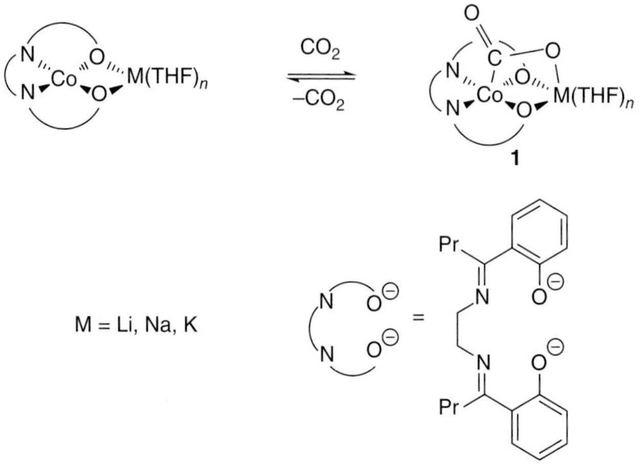

Figure 3.2 Bifunctional CO_2 binding and activation by Co…M cooperation, reported by Floriani. Source: Gambarotta et al. 1982 [10]. Reproduced with permission of ACS.

recently, a dicopper reaction center was found to induce coupling of two CO_2 units to generate oxalate [11]. Similar coupling behavior has been observed for ketone coordination to a Zr–Co reaction center [12], which will be discussed in more detail below.

Other recent examples of small-molecule binding at bimetallic reaction centers include N_2 activation at reduced Co centers bound to Lewis-acid-supporting metals. Thomas reported the [Zr–Co–N_2] species **2**, with the Co center serving as a reactive metal site whose properties were modulated by the Lewis acidic Zr center (Figure 3.3a) [13]. The bimetallic N_2 complex was accessible at milder reduction potentials than for analogs lacking the Zr site, and the N_2 fragment was found to be less strongly activated than in Co-only analogs because of the electron-withdrawing nature of the Co–Zr dative bonding interaction. Lu reported closely related [M–Co–N_2] analogs **3** in which the supporting metal could be varied systematically (Figure 3.3b) [14]. Reduction potential and N_2 activation were found to be correlated with one another and to be dependent on the supporting metal site, with N_2 activation decreasing (slightly) across the 3d period.

3.2.1.2 Alkyne Activation

Homobimetallic complexes have long been known to activate alkynes, with alkyne coordination to $Co_2(CO)_8$ during the Pauson–Khand reaction, being a leading example [15]. Complexes with metal–metal multiple bonds have recently emerged as promising candidates, as well. Both Theopold and Kempe have demonstrated alkyne binding to Cr–Cr quintuply bonded systems, generating [C_2Cr_2] products **4** in which the alkyne moiety has been activated significantly, consistent with a [2 + 2] metallocycloaddition formulation (Figure 3.4a) [16, 17]. The related Mo–Mo quintuply bonded complex **5** was reported by Tsai to engage in [2+2+2] metallocycloaddition with two alkyne equivalents, generating aromatic [C_4Mo_2] product **6** (Figure 3.4b) [18]. In all of these cases, the alkyne binding and activation are conceptually similar to

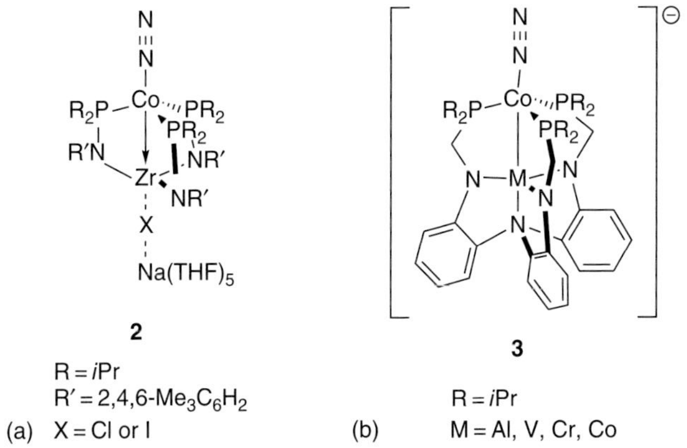

Figure 3.3 Bimetallic N_2 binding and activation by Co–M complexes, reported by (a) Thomas and (b) Lu. (a) Source: Greenwood et al. 2009 [13]. Reproduced with permission of ACS. (b) Clouston et al. 2015 [14]. Reproduced with permission of ACS.

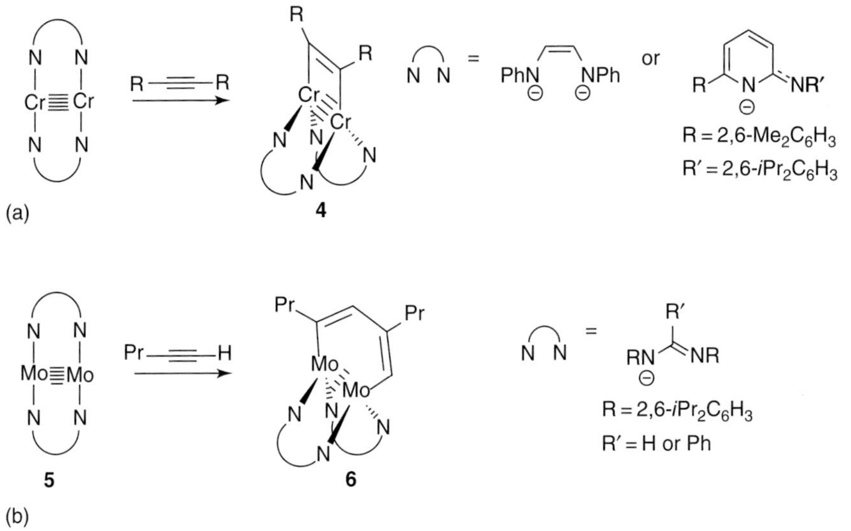

Figure 3.4 Bimetallic alkyne binding and activation by quintuply bonded homobimetallics: (a) [2+2] cycloadditions reported by Theopold and Kempe [16, 17] and (b) [2+2+2] cycloaddition reported by Tsai. (b) Source: Chen et al. 2012 [18]. Reproduced with permission of John Wiley & Sons.

C–C elongation and pyramidalization observed upon the formation of classical single-site alkyne adducts with metallocyclopropene character (Figure 3.1a).

3.2.2 Bimetallic Analogs of Oxidative Addition and Reductive Elimination

Beyond coordination and activation of substrates, many crucial catalytic transformations rely on reaction steps by which reactive metal sites mediate bond-making and bond-breaking events within the coordination sphere.

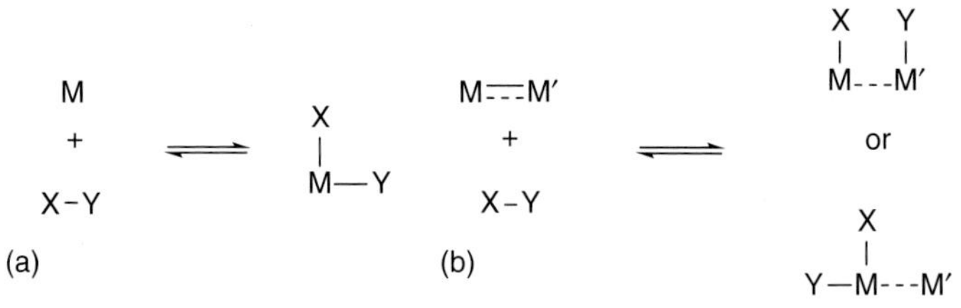

Figure 3.5 Oxidative addition and reductive elimination reactions (a) at a single metal site and (b) at bimetallic reaction centers.

Specifically, single-site oxidative addition is the canonical device by which catalysts break the bonds of substrates, and single-site reductive elimination is the canonical device by which catalysts couple two ligands to generate new bonds in products. These two reaction steps, which are related to one another by microscopic reversibility (Figure 3.5a), are two-electron redox process: oxidative bond cleavage increases metal oxidation state by two units and reductive bond elimination decreases metal oxidation state by two units. Bimetallic analogs of oxidative addition and reductive elimination, which can either involve direct cooperation of two metal sites or occur at one reactive metal site whose reactivity is modulated by a supporting metal (Figure 3.5b), is a promising strategy to circumvent the need for two-electron redox cycling by single-site noble metal systems.

3.2.2.1 E—H Addition and Elimination

The simplest substrate for such reactions is H_2. Hoffmann established by computational methods that dinuclear reductive elimination of H_2 from homobimetallic systems has a high barrier because of its symmetry-forbidden nature [19]. However, the barrier can be lowered through use of lower symmetry systems, such as heterobimetallics, or through nonconcerted addition/elimination pathways. Levina et al. reported the elimination of H_2 from the reaction of a hydridic [Ni–H] species and a protic [W–H] species (Figure 3.6a) [20]. An unusual [Ni...H–H...W] intermediate **7** was observed by NMR spectroscopy at low temperature before elimination of H_2 upon warming, which also generated the heterobimetallic isocarbonyl species **8**. Mankad showed that such polarized

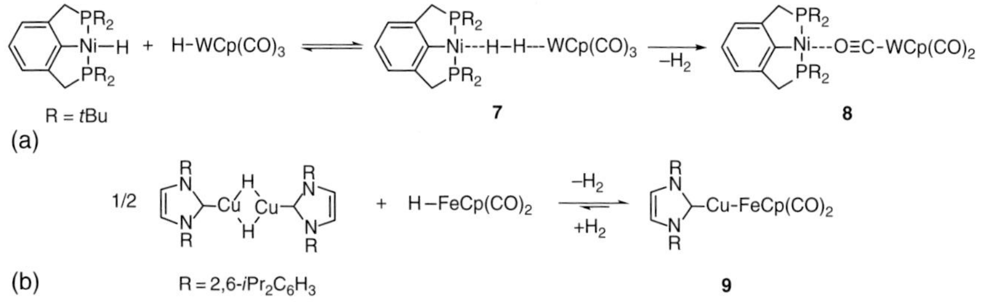

Figure 3.6 Bimetallic H_2 addition and elimination reactions with polarized heterobimetallic systems reported by (a) Peruzzini and (b) Mankad. (a) Source: Levina et al. 2010 [20]. Reproduced with permission of ACS. (b) Mazzacano and Mankad 2013 [21]. Reproduced with permission of ACS.

heterobimetallic systems can both evolve and consume H_2 (Figure 3.6b) [21]. Rapid elimination of H_2 was observed from the reaction of a hydridic [Cu–H] species and a protic [Fe–H] species, which produced the heterobimetallic [Cu–Fe] species **9**. Indirect evidence for the reverse reaction, i.e. bimetallic H_2 activation by **9**, was obtained from observed H/D scrambling behavior [21] and from limited activity in catalytic hydrogenation [22], which will be discussed further below. The importance of heterolytic H_2 addition/elimination in such systems is evident from the fact that the reactivity ceases upon attenuation of [Fe–H] acidity [23].

Several recent examples of E–H bimetallic addition/elimination reactions have also emerged. The same Mankad complex **9** that is capable of H_2 addition/elimination was also found to engage in reversible B–H addition/elimination (Figure 3.7a) [21]. Elimination of pinacolborane was observed readily during reaction of a hydridic [Cu–H] species with $CpFe(CO)_2Bpin$ (pin, pinacolate). Indirect evidence for the reverse reaction, B–H activation of pinacolborane, was obtained by trapping the transient [Cu–H] species with CO_2 [21, 26]. Binuclear reductive elimination of C—H bonds has been known for some time, including a landmark mechanistic study by Halpern [27]. The more challenging reaction, C–H cleavage, was reported to occur by discandium cooperation by Diaconescu and coworkers (Figure 3.7b) [24]. Reduction of a [Sc–I] complex in aromatic solvents resulted in the formation of 1 : 1 mixtures of [Sc–Ar] and [Sc–H] products **10** and **11**, indicating the cooperative activation of the arene C—H bond across two reduced Sc centers. Finally, reversible Si–H activation/elimination was observed with a dinickel(I) species by Uyeda and

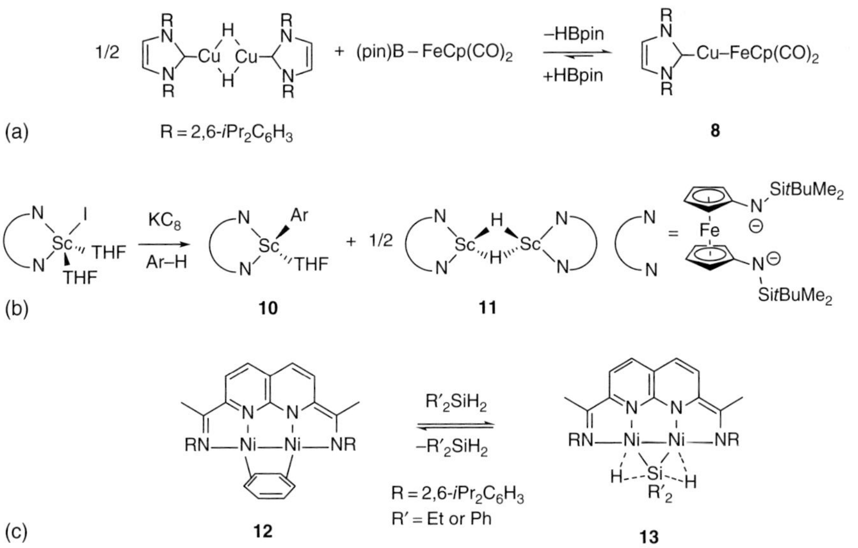

Figure 3.7 Bimetallic E-H addition and elimination reactions reported by (a) Mankad, (b) Diaconescu, and (c) Uyeda (pin, pinacolate). (a) Source: Mazzacano and Mankad 2013 [21]. Reproduced with permission of ACS. (b) Huang et al. 2014 [24]. Reproduced with permission of ACS. (c) Steiman and Uyeda 2015 [25]. Reproduced with permission of ACS.

coworker (Figure 3.7c) [25]. Addition of secondary silanes to the nickel–nickel bonded precursor **12** produced adduct **13**, which was formulated as having two resonance contributors: a silane σ-complex and a μ-silylene dihydride complex.

3.2.2.2 C—X Activation and C—C Coupling

C–X activation and C–C coupling are processes of importance to organic chemistry. Bimetallic C–X activation reactions have been known for some time with alkyl halides, particularly for "early-late" heterobimetallic complexes with polar metal–metal bonds [28]. A recent example is a series of studies from Mankad on complex **9**, which was shown to activate R–X substrates such as CH_3I and $PhCH_2Cl$ to generate (NHC)CuX (NHC, *N*-heterocyclic carbene) and $RFeCp(CO)_2$ products [29, 30]. The "radical clock" substrates cyclopropylmethyl bromide and iodide were used to show that the R–X activation processes are two-electron reactions for bromide and chloride electrophiles but have competing radical pathways for iodide electrophiles (Figure 3.8a). The ratio of ring-closed **14** and ring-opened **15** resulting from **9** differed from that arising from $Na[FeCp(CO)_2]$, indicating that R–X activation occurred from Cu—Fe-bonded **9** rather than from dissociation of $[FeCp(CO)_2]^-$ and subsequent S_N2 attack. A bimetallic transition state was proposed based on experimentally calibrated quantum chemical calculations. Bimetallic C–C coupling reactions are less well documented. A well-defined example was reported by Agapie, who demonstrated stoichiometric C–C reductive elimination of fluorenone upon carbonylation of dinickel species **16** and of fluorene upon reaction with dihaloalkanes (Figure 3.8b) [31]. Bimetallic C–X oxidative addition and C–C reductive elimination have also been implicated in Ni-catalyzed coupling reactions on the basis of kinetics measurements and ligand redistribution experiments [32, 33].

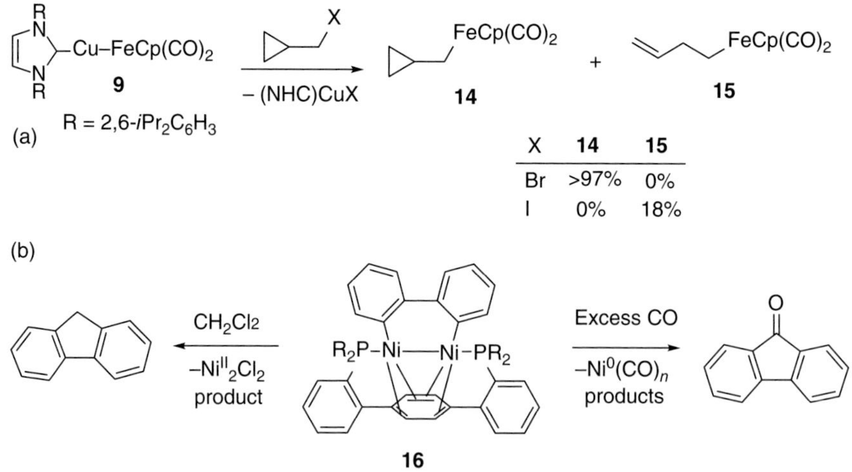

X	14	15
Br	>97%	0%
I	0%	18%

Figure 3.8 Bimetallic C–X reactions: (a) C–X activation by Mankad and (b) C–C coupling by Agapie (NHC, *N*-heterocyclic carbene). (a) Source: Karunananda et al. 2015 [30]. Reproduced with permission of ACS. (b) Velian et al. 2010 [31]. Reproduced with permission of ACS.

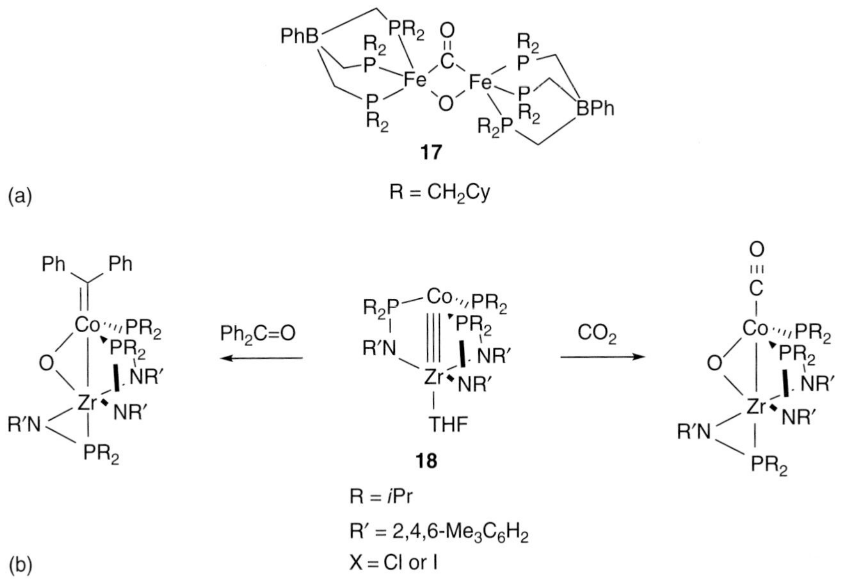

(a)

17

R = CH_2Cy

$Ph_2C{=}O$ ← 18 → CO_2

18

R = *i*Pr

R′ = 2,4,6-$Me_3C_6H_2$

X = Cl or I

(b)

Figure 3.9 Bimetallic C=O cleavage products reported by (a) Peters and (b) Thomas [12, 34]. (a) Source: Lu et al. 2007 [35]. Reproduced with permission of ACS.

3.2.2.3 C=O Cleavage

Cleavage of multiple bonds by oxidative addition is particularly challenging. Several examples of C=O cleavage using bimetallic cooperation have been reported recently. Peters reported the formation of [Fe_2(μ-O)(μ-CO)] complex **17** by reaction of a masked Fe(I) precursor with CO_2 (Figure 3.9a) [35]. Selectivity for this C=O cleavage product relative to other observed products including a (μ-oxalate) species was shown to be modulated by controlling mononuclear vs binuclear reaction pathways [36]. Thomas has shown that [Zr–Co] complex **18** is capable of C=O cleavage reactions of CO_2 and benzophenone, generating (μ-oxo)(carbonyl) and (μ-oxo)(carbene) bimetallic products, respectively (Figure 3.9b) [12, 34]. In the latter case, a tetrametallic radical coupling product involving C—C bond formation between two benzophenone moieties was identified at intermediate stages of the reaction. Bimetallic deoxygenation of CO_2 has also been reported recently using U…K cooperation and with Ti=Ti doubly bonded reaction centers [37, 38].

3.3 Application in Catalysis

Several of the stoichiometric reactions described above have been implicated in catalytic transformations involving bimetallic cooperation. Some recent examples of bimetallic catalysis are highlighted in this section, with a focus on cases that incorporate at least one of the reaction types discussed above. Catalytic transformations involving reactivity of metal–metal bonds are discussed separately from bimetallic systems that lack metal–metal bonds. This is because these two subclasses of bimetallic catalysis entail distinct design considerations, which is important to consider as new examples continue to emerge through rational design.

3.3.1 Catalysis with Reactive Metal–Metal Bonds

3.3.1.1 Bimetallic Alkyne Cycloadditions

As discussed above, metal–metal bonded complexes of various bond multiplicity are known to coordinate alkynes, often resulting in cycloaddition reactivity. Although the well-known $Co_2(CO)_8$-mediated Pauson–Khand reaction is stoichiometric in Co and requires noble metals for catalytic turnover [15], both Tsai and Uyeda have developed catalytic variants of alkyne trimerization reactions involving alkyne coordination and subsequent cycloaddition. Tsai demonstrated that the Mo–Mo quintuply bonded species **5** is a catalyst for the trimerization of 1-pentyne, producing 1,3,5-tri-*n*-propylbenzene in high yield as the sole regioisomer (Figure 3.10a) [18]. The [2+2+2] adduct **6** was also found to be a viable catalyst for this process, consistent with it serving as a catalytic intermediate before incorporation of a third alkyne equivalent. Regioselectivity for the 1,3,5-substituted product was thought to arise from the steric bulk of the amidinate supporting ligands directing the approach of the third alkyne substrate, although only one alkyne was tested and so generality was not established.

Uyeda and coworker reported that the Ni—Ni-bonded species **12** is also a catalyst for alkyne trimerization [39]. Several substrates were tested, all giving 1,2,4-trisubstituted arenes as the major products in high yields (Figure 3.10b). The 1,3,5-trisubstituted isomers were among the minor products. For certain alkynes, examples of both the initial alkyne adduct and the metallocyclopentadiene intermediate were isolated and characterized. Both involved coordination to one of the Ni centers, with the distal Ni engaging in a secondary coordination interaction with the incipient π-system. Regioselectivity was thought to be dictated by the distal Ni center biasing the electronic environment of the

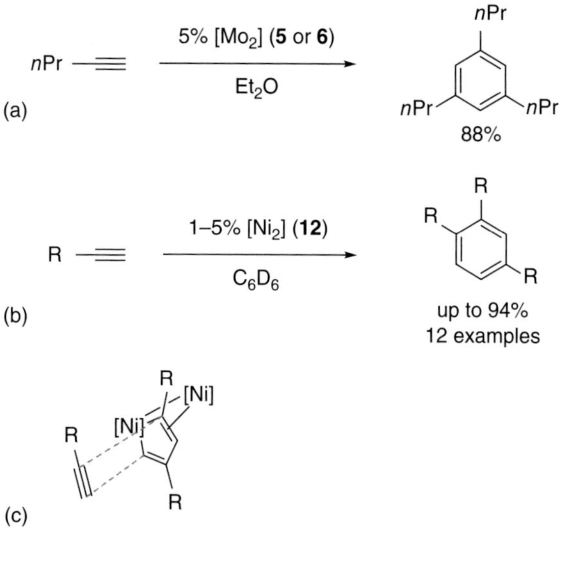

(a) *n*Pr—≡ → 5% [Mo_2] (**5** or **6**), Et_2O → 1,3,5-tri-*n*Pr-benzene, 88%

(b) R—≡ → 1–5% [Ni_2] (**12**), C_6D_6 → 1,2,4-R₃-benzene, up to 94%, 12 examples

(c)

Figure 3.10 Bimetallic catalysis for alkyne cyclotrimerization: (a) sterically driven dimolybdenum catalysis studied by Tsai. Source: Chen et al. 2012 [18]. Reproduced with permission of John Wiley & Sons. (b) Dinickel catalysis studied by Uyeda. Pal and Uyeda 2015. [39]. Reproduced with permission of ACS. (c) Secondary Ni…π interaction proposed to dictate regioselectivity during catalysis. See Figures 3.4 and 3.7 for catalyst structures.

metallocyclopentadiene by interacting with one of the two π-bonds, thus steering the third alkyne equivalent in a regiospecific manner (Figure 3.10c). Consistent with this proposal, mononickel catalysts with extremely similar ligand scaffolds gave poor regioselectivities under the same conditions (in addition to exhibiting low activities).

3.3.1.2 Bimetallic Oxidative Addition/Reductive Elimination Cycling

Many crucial processes catalyzed by noble metals involve site–site oxidative addition/reductive elimination cycling. Analogous bimetallic processes hold the promise of achieving similar transformations with non-noble metal catalysts. Two recent examples were demonstrated by Mankad and involve derivatives of catalyst **9**, which was discussed above for its reversible H–H and B–H addition/elimination reactivity. Leveraging a stoichiometric Fe-mediated borylation reaction reported previously by Hartwig and coworkers [40], Mankad and coworker developed a catalytic variant using [Cu–Fe] species **9** (Figure 3.11a) [21]. In the proposed mechanism, the metal–metal bonded catalyst undergoes bimetallic oxidative addition with pinacolborane, as described above. The resulting boryliron species is capable of photochemical arene borylation, a reaction typically conducted with single-site Ir catalysts [41]. Catalytic turnover is then achieved through the bimetallic reductive elimination of H_2. Here, the redox-active steps require bimetallic cooperation, whereas the redox-neutral arene borylation (known to involve σ-bond metathesis) occurs at Fe alone. Thus, a stoichiometric Fe-mediated transformation was rendered catalytic through use of bimetallic oxidative addition/reductive elimination cycling.

In addition to dehydrogenative borylation, species **9** and related heterobimetallic complexes also are capable of hydrogenative processes. Mankad and coworker subsequently reported heterobimetallic catalysis for semi-hydrogenation of diarylalkynes (Figure 3.11b) [22]. Such transformations are often in the realm of single-site Rh catalysts. High selectivities for *trans*-hydrogenation products were observed in all cases. Although [Cu–Fe] complexes such as **9** do show catalytic activity, as do [Cu–Ru] and [Ag–Fe] derivatives, optimal results were achieved with a [Ag–Ru] catalyst. The proposed mechanism involves bimetallic oxidative addition of H_2, as described above. The resulting [Cu–H] or [Ag–H] species undergoes 1,2-insertion of the alkyne substrate, which is a well-documented process for [Cu–H] complexes [42]. Catalytic turnover is achieved through bimetallic reductive elimination of a *cis*-alkene. (Under catalytic conditions, the ultimate *trans*-alkene product was then proposed to form through a single-site isomerization mechanism not shown here.) Once again, the redox-neutral 1,2-insertion process occurred at a single metal site, but catalytic turnover was only achieved through use of bimetallic oxidative addition/reductive elimination cycling.

3.3.2 Bifunctional and Tandem Catalysis without Metal–Metal Bonds

3.3.2.1 Cooperative Activation of Unsaturated Substrates

The seminal work of Floriani on cooperative CO_2 activation discussed above has inspired many subsequent studies. In addition to further studies on stoichiometric CO_2 activation highlighted above, some progress has been made recently on

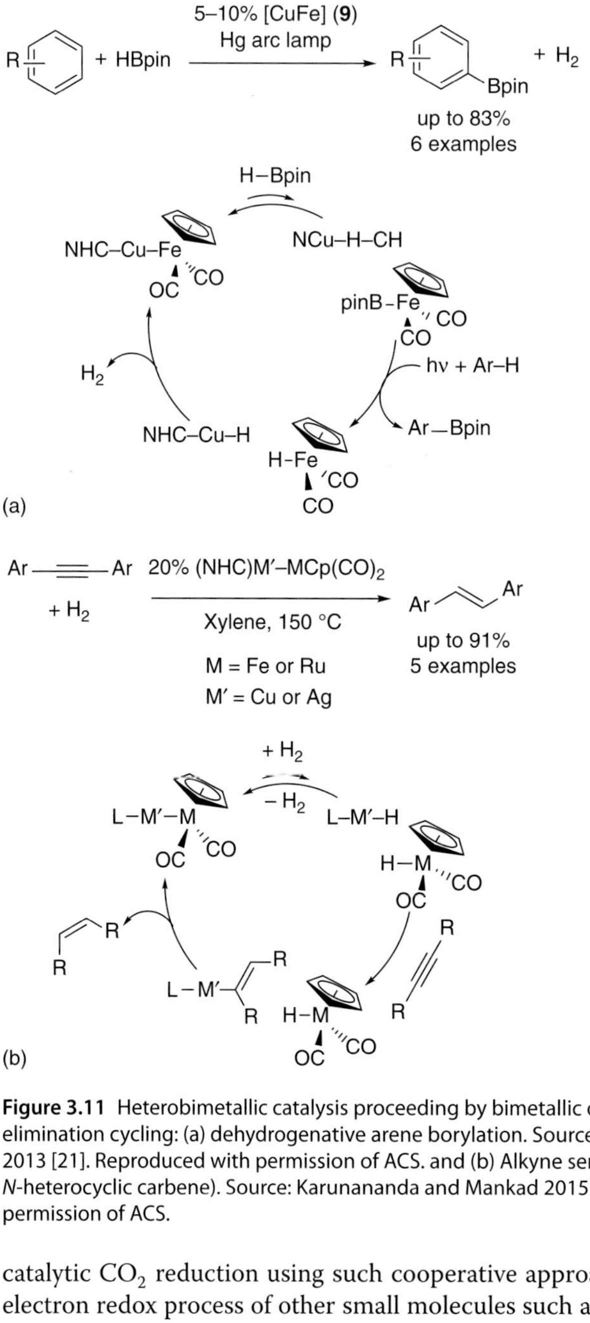

Figure 3.11 Heterobimetallic catalysis proceeding by bimetallic oxidative addition/reductive elimination cycling: (a) dehydrogenative arene borylation. Source: Mazzacano and Mankad 2013 [21]. Reproduced with permission of ACS. and (b) Alkyne semi-hydrogenation (NHC, *N*-heterocyclic carbene). Source: Karunananda and Mankad 2015 [22]. Reproduced with permission of ACS.

catalytic CO_2 reduction using such cooperative approaches, in addition multi-electron redox process of other small molecules such as formic acid and nitrite. Hazari and Schneider and coworkers reported the pincer-ligated Fe catalyst **19** for formic acid dehydrogenation (Figure 3.12a) [43]. Although the Fe catalyst by itself was able to achieve turnover numbers (TONs) of about 1000 for this

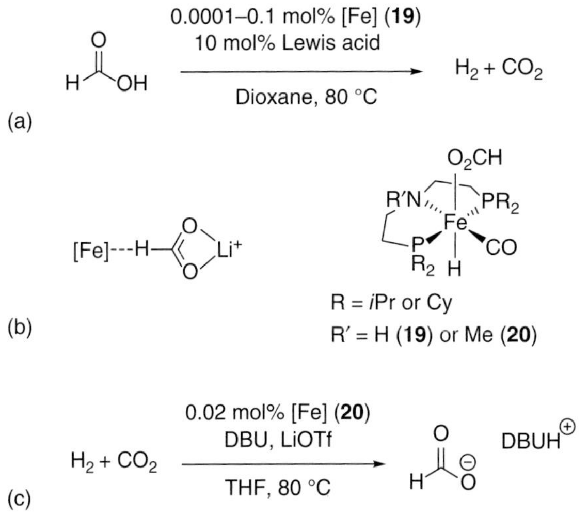

Figure 3.12 Lewis-acid-assisted catalysis for (a) formic acid dehydrogenation reported by Hazari and Schneider. Source: Bielinski et al. [43] Reproduced with permission of ACS. (b) Proposed Li^+-assisted decarboxylation and (c) CO_2 hydrogenation reported by Hazari and Bernskoetter [44].

(a) HCOOH → $H_2 + CO_2$ (0.0001–0.1 mol% [Fe] (**19**), 10 mol% Lewis acid; Dioxane, 80 °C)

(b) [Fe]---H–CO_2···Li^+; Fe complex with O_2CH, R′N, PR_2, CO, H. R = *i*Pr or Cy; R′ = H (**19**) or Me (**20**)

(c) $H_2 + CO_2$ → HCO_2^- $DBUH^+$ (0.02 mol% [Fe] (**20**), DBU, LiOTf; THF, 80 °C)

process, addition of Lewis acid additives such as $LiBF_4$ increased TONs to about 1 000 000. This is the highest known TON for formic acid dehydrogenation for a non-noble metal system. A Lewis-acid-assisted mechanism for decarboxylation of Fe-coordinated formate was proposed to play a key role in increasing TON (Figure 3.12b). Hazari and Bernskoetter subsequently reported that the same catalyst and its tertiary amine analog **20** were catalysts for the reverse reaction, CO_2 hydrogenation to formate (Figure 3.12c) [44]. Once again, Lewis acid additives increased TONs by an order of magnitude, with TONs up to 60 000 being achieved with assistance from LiOTf. This is the highest known TON for CO_2 hydrogenation by a non-noble metal system. In this case, no evidence for cooperative CO_2 activation was found, but instead Li^+ was proposed to assist with formate displacement from Fe during catalytic turnover. An apparent bimetallic effect on product selectivity during catalytic CO_2 reduction was identified by Mankad, but again, cooperative CO_2 activation was not proposed in the hypothetical auto-tandem mechanism [26].

Peters reported a heterobimetallic Co/Mg electrocatalyst for the selective three-proton/two-electron reduction of nitrite to N_2O (Figure 3.13a) [45]. In addition to catalytic activity, several catalytic intermediates were isolated and characterized in this system, allowing for the cooperative NO_2^- activation mode to be established definitively. The NO_2^- coordination in this system (Figure 3.13b) closely mimics the original CO_2 adducts of Floriani (see above). Lastly, Bouwman reported a dicopper electrocatalyst for CO_2 conversion to oxalate (Figure 3.13c) [46]. A key to the proposed mechanism is the dinuclear C–C coupling of two coordinated CO_2 ligands to generate a bridging oxalate unit (Figure 3.13d).

In addition to these examples, several interesting polymerization systems exist that utilize cooperative binding effects to control polymerization rates and stereoselectivities [47–50].

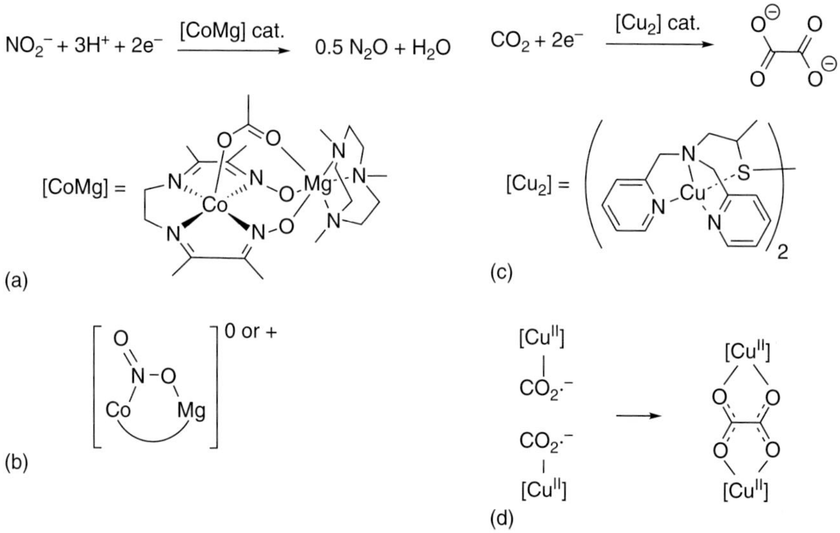

Figure 3.13 Cooperative small-molecule reduction: (a) electrocatalytic nitrite reduction and (b) proposed nitrite binding studied by Peters and (c) electrocatalytic CO_2 reduction and (d) proposed dicopper-mediated CO_2 coupling studied by Bouwman. (a) Source: Uyeda and Peters 2013 [45]. Reproduced with permission of ACS. (b) Angamuthu et al. 2010 [46]. Reproduced with permission of AAAS.

3.3.2.2 Cooperative Processes with Bimetallic Oxidative Addition and/or Reductive Elimination

Although the stoichiometric and catalytic processes discussed so far involving bimetallic oxidative addition and/or reductive elimination invoke the involvement of metal–metal bonds, this is not a strict requirement. Indeed, the famous frustrated Lewis pair systems of the p-block are capable of stoichiometric and catalytic processes involving the addition of H_2 and other substrates despite spatial separation between the two reaction centers [7]. Such a strategy was used with d-block elements by Coates to develop catalytic carbonylations of epoxides to generate β-lactones. The initial reports of this transformation utilized a Cr/Co heterobimetallic system [51, 52], although in subsequent studies, the electrophilic Cr(III) site could be replaced with a Al(III) Lewis acid [53]. The proposed mechanism for this process (Figure 3.14a) features a bimetallic epoxide-opening step that can be considered a bimetallic oxidative addition, as well as a bimetallic lactone-closing step that can be considered a bimetallic reductive elimination.

Sorensen reported the use of bimetallic cooperation to catalyze the acceptor less dehydrogenation of alkanes [54], a transformation typically conducted with single-site Ir catalysts. The Sorensen system was designed to utilize two cocatalysts that could each mediate a hydrogen atom abstraction process (Figure 3.14b), thereby resulting in the net dehydrogenation of the saturated substrate. A polyoxotungstate cocatalyst initiates alkane oxidation by abstracting a single hydrogen atom from its photoexcited state, generating an alkyl radical and a polytungsten species that is a strong hydrogen atom donor. A cobaloxime cocatalyst completed alkane oxidation by abstracting a second hydrogen atom from the transient

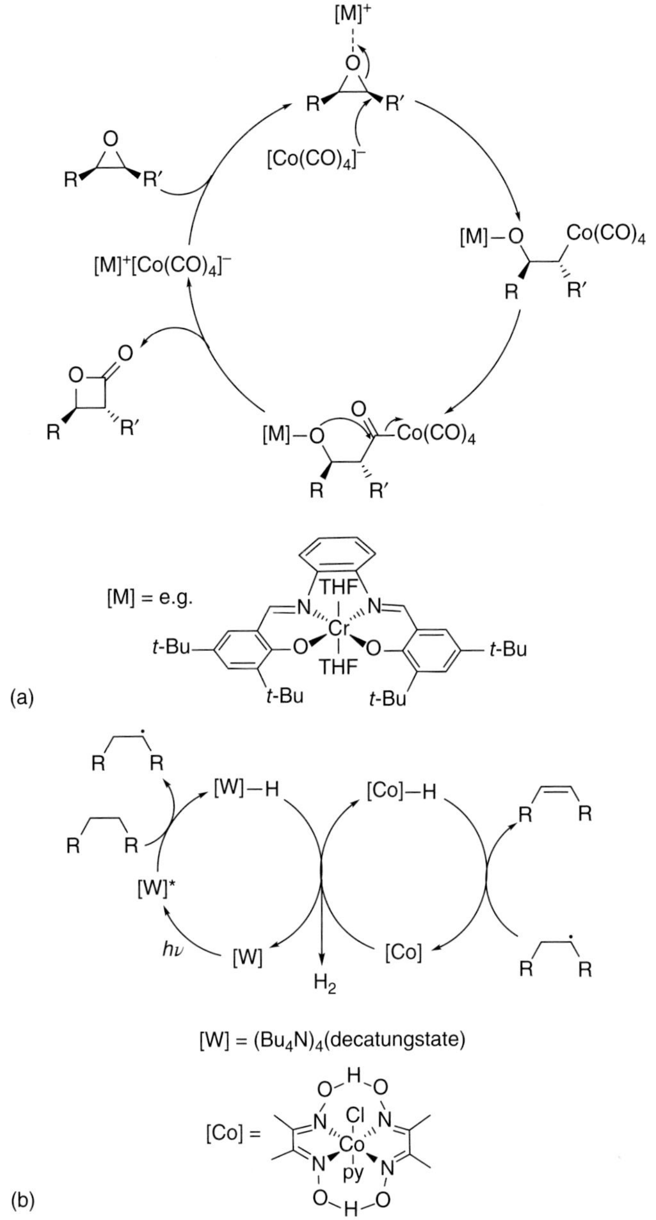

[M]+
O
R R′
[Co(CO)4]−
O
R R′
[M]+[Co(CO)4]−
[M]–O Co(CO)4
R R′
O
O
R R′
O
[M]–O Co(CO)4
R R′

[M] = e.g.
THF
N Cr N
t-Bu O O t-Bu
THF
t-Bu t-Bu

(a)

R R
[W]–H [Co]–H
R R
R R
[W]*
hν [W] [Co]
H2
R R

[W] = (Bu4N)4(decatungstate)

O H O
N Cl N
[Co] = Co
N py N
O H O

(b)

Figure 3.14 Cooperative mechanisms for (a) epoxide carbonylation studied by Coates. (a) Source: Schmidt et al. 2005 [51]. Reproduced with permission of ACS. (b) Alkane dehydrogenation studied by Sorensen [54]. The * indicates a photoexcited state.

alkyl radical, generating a cobalt hydride species that can act as a hydrogen atom acceptor. Catalytic turnover is enabled through these two metal hydride species reacting together to generate H_2, which can be considered a bimetallic reductive elimination event.

3.4 Polynuclear Complexes

Moving beyond binuclear catalysis to systems with ≥3 metal sites is a promising research direction. In addition to offering more tunable catalyst features, such systems should, in principle, be able to mediate multielectron redox transformations more readily [55]. Because of the complex nature of such systems, however, it is often challenging to study individual reaction steps and develop a solid conceptual underpinning. As a result, although many non-noble metal clusters are catalytically active, mechanistic understanding is often incomplete. Of the many examples of non-noble multimetallic clusters that catalyze important transformations, a few selected examples are shown in Table 3.1. For further reading, an excellent review is available [8].

Recent efforts by several groups have aimed at better understanding the intimate nature of multimetallic cooperation in such systems [6]. A particularly fascinating study, entitled "Testing the Polynuclear Hypothesis," was reported by Betley [56]. The triiron(II) complex **21** was shown to engage in both two- and four-electron redox processes (Figure 3.15). The two-electron reduction of 1,2-diphenylhydrazine resulted in the formation of the $[Fe_3(\mu_3\text{-}NPh)]$ complex **22**. The four-electron reduction of azobenzene yielded

Table 3.1 Selected examples of non-noble multimetallic clusters and their catalytic activity.[a)]

Entry	Cluster	Catalytic application
1	$[HMFe(CO)_9]^-$ (M = Cr, Mo, W)	Alkene isomerization
2	$HFeCo_3(CO)_{12}$	Hydrogenation
3	$(OC)_4Mn(\mu\text{-}AsMe_2)Fe(CO)_4$	Hydrogenation
4	$Mo_2[Co_3(CO)_9CCO_2]_4$	Hydrogenation
5	$FeCoCu(diethanolamine)(NCS)_2(MeOH)_2$	Alkane oxidation
6	$Fe_3Co(CO)_{13}$	Methanol homoligation
7	$FeCu(\mu\text{-}Ph_2Ppy)(CO)_3Cl$	Alcohol carbonylation
8	$FeCo_2(\mu\text{-}PPh)(CO)_9$	Hydroformylation
9	$[Mo_3CuS_4(dmpe)_3Cl_4]^+$	Cyclopropanation
10	$HFeCo_3(CO)_{11}(PPh_3)$	Pauson–Khand
11	$[ClCuZn(Ph_2PC_{10}H_5O)_2]_2$	Conjugate addition
12	$FeCo_2(\mu_3\text{-}PPh)(CO)_9$	Hydrosilylation
13	$Mo_2Fe_4Co_2S_8(SPh)_6(OMe)_3]^{3-}$	Hydrodesulfurization

a) These and other examples are available in Ref. [8].

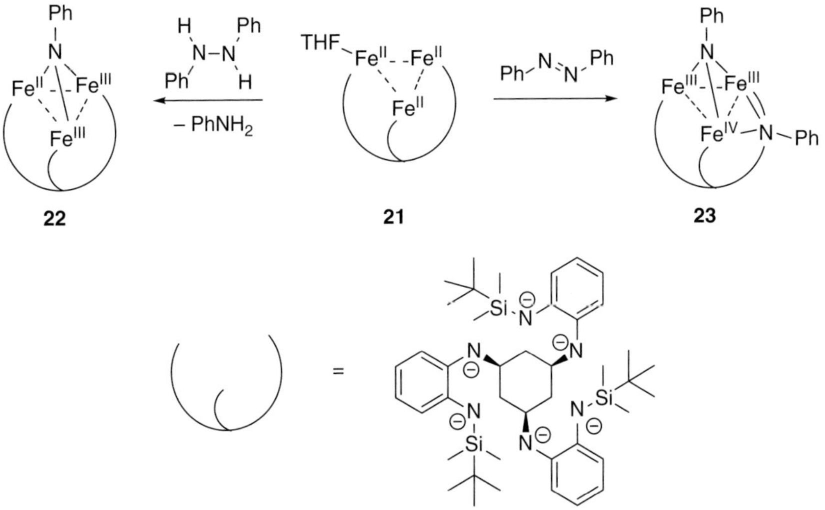

Figure 3.15 Two- and four-electron redox processes exhibited by a trinuclear iron cluster studied by Betley. Source: Powers and Betley 2013 [56]. Reproduced with permission of ACS.

the $[Fe_3(\mu_3\text{-}NPh)(\mu_2\text{-}NPh)]$ complex **23**. Physical measurements ruled out the participation of ligand-based redox chemistry in these reactions, implying that the triiron core itself was being oxidized from $[Fe^{II}Fe^{II}Fe^{II}]$ to $[Fe^{II}Fe^{III}Fe^{III}]$ and $[Fe^{III}Fe^{III}Fe^{IV}]$, respectively, in the two- and four-electron reactions. Collectively, these results highlight the rich redox chemistry available to multinuclear clusters in comparison to their single-site analogs.

3.5 Outlook

Recent work by several research groups highlighted above have provided deeper understanding of fundamental bimetallic reaction steps, such as bifunctional activation of unsaturated substrates, bimetallic oxidative addition, and bimetallic reductive elimination. Collectively, these bimetallic reaction steps provide a toolbox that is used to break and form both polar and nonpolar chemical bonds; this toolbox complements the mononuclear processes that are well established in organometallic systems. Recently, several successful examples of catalysis that incorporate these fundamental bimetallic reaction steps have been explored, and many more examples are likely to emerge from this active area of research. Binuclear and multinuclear catalytic processes present many challenges beyond those faced by mononuclear systems, including an increased number of side reactions and an entropic penalty of bringing two reactive sites together around a common substrate. Nonetheless, the studies highlighted here provide indications that overcoming these challenges stands to open parts of chemical space that have previously been underexplored.

Acknowledgments

Financial support to the author is provided by the National Science Foundation (CHE-1664632) and an Alfred P. Sloan Research Fellowship.

References

1 Gildner, P.G. and Colacot, T.J. (2015). *Organometallics* 34: 5497.
2 Chirik, P.J. and Wieghardt, K. (2010). *Science* 327: 794–795.
3 Appel, A.M., Bercaw, J.E., Bocarsly, A.B. et al. (2013). *Chem. Rev.* 113: 6621–6658.
4 Jeoung, J.H. and Dobbek, H. (2007). *Science* 318: 1461–1464.
5 Hoffman, B.M., Lukoyanov, D., Yang, Z.-Y. et al. (2014). *Chem. Rev.* 114: 4041–4062.
6 Kanady, J.S., Tsui, E.Y., Day, M.W., and Agapie, T. (2011). *Science* 333: 733–736.
7 Stephan, D.W. (2015). *J. Am. Chem. Soc.* 137: 10018–10032.
8 Buchwalter, P., Rosé, J., and Braunstein, P. (2015). *Chem. Rev.* 115: 28–126.
9 Que, L. and Tolman, W.B. (2002). *Angew. Chem. Int. Ed.* 41: 1114–1137.
10 Gambarotta, S., Arena, F., Floriani, C., and Zanazzi, P.F. (1982). *J. Am. Chem. Soc.* 104: 5082–5092.
11 Pokharel, U.R., Fronczek, F.R., and Maverick, A.W. (2014). *Nat. Commun.* 5: 5883.
12 Marquard, S.L., Bezpalko, M.W., Foxman, B.M., and Thomas, C.M. (2013). *J. Am. Chem. Soc.* 135: 6018–6021.
13 Greenwood, B.P., Forman, S.I., Rowe, G.T. et al. (2009). *Inorg. Chem.* 48: 6251–6260.
14 Clouston, L.J., Bernales, V., Carlson, R.K. et al. (2015). *Inorg. Chem.* 54: 9263–9270.
15 Blanco-Urgoiti, J., Añorbe, L., Pérez-Serrano, L. et al. (2004). *Chem. Soc. Rev.* 33: 32–42.
16 Shen, J., Yap, G.P.A., Werner, J.-P., and Theopold, K.H. (2011). *Chem. Commun.* 47: 12191–12193.
17 Noor, A., Sobgwi Tamne, E., Qayyum, S. et al. (2011). *Chem. Eur. J.* 17: 6900–6903.
18 Chen, H.-Z., Liu, S.-C., Yen, C.-H. et al. (2012). *Angew. Chem. Int. Ed.* 51: 10342; *Angew. Chem. Int. Ed.* 2012, 124: 10488.
19 Trinquier, G. and Hoffmann, R. (1984). *Organometallics* 3: 370–380.
20 Levina, V.A., Rossin, A., Belkova, N.V. et al. (2010). *Angew. Chem. Int. Ed.* 50: 1367; *Angew. Chem. Int. Ed.* 2010, 123: 1403.
21 Mazzacano, T.J. and Mankad, N.P. (2013). *J. Am. Chem. Soc.* 135: 17258–17261.
22 Karunananda, M.K. and Mankad, N.P. (2015). *J. Am. Chem. Soc.* 137: 14598–14601.
23 Mazzacano, T.J. and Mankad, N.P. (2015). *Chem. Commun.* 51: 5379–5382.

24 Huang, W., Dulong, F., Khan, S.I. et al. (2014). *J. Am. Chem. Soc.* 136: 17410–17413.
25 Steiman, T.J. and Uyeda, C. (2015). *J. Am. Chem. Soc.* 137: 6104–6110.
26 Bagherzadeh, S. and Mankad, N.P. (2015). *J. Am. Chem. Soc.* 137: 10898–10901.
27 Nappa, M.J., Santi, R., Diefenbach, S.P., and Halpern, J. (1982). *J. Am. Chem. Soc.* 104: 619.
28 Gade, L.H. (2000). *Angew. Chem. Int. Ed.* 39: 2658–2678.
29 Jayarathne, U., Mazzacano, T.J., Bagherzadeh, S., and Mankad, N.P. (2013). *Organometallics* 32: 3986–3992.
30 Karunananda, M.K., Parmelee, S.R., Waldhart, G.W., and Mankad, N.P. (2015). *Organometallics* 34: 3857–3864.
31 Velian, A., Lin, S., Miller, A.J.M. et al. (2010). *J. Am. Chem. Soc.* 132: 6296–6297.
32 Breitenfeld, J., Ruiz, J., Wodrich, M.D., and Hu, X. (2013). *J. Am. Chem. Soc.* 135: 12004–12012.
33 Xu, H., Diccianni, J.B., Katigbak, J. et al. (2016). *J. Am. Chem. Soc.* 138: 4779.
34 Krogman, J.P., Foxman, B.M., and Thomas, C.M. (2011). *J. Am. Chem. Soc.* 133: 14582–14585.
35 Lu, C.C., Saouma, C.T., Day, M.W., and Peters, J.C. (2007). *J. Am. Chem. Soc.* 129: 4–5.
36 Saouma, C.T., Lu, C.C., Day, M.W., and Peters, J.C. (2013). *Chem. Sci.* 4: 4042.
37 Cooper, O., Camp, C., Pécaut, J. et al. (2014). *J. Am. Chem. Soc.* 136: 6716–6723.
38 Kilpatrick, A.F.R., Green, J.C., and Cloke, F.G.N. (2015). *Organometallics* 34: 4816–4829.
39 Pal, S. and Uyeda, C. (2015). *J. Am. Chem. Soc.* 137: 8042–8045.
40 Waltz, K.M., He, X., Muhoro, C., and Hartwig, J.F. (1995). *J. Am. Chem. Soc.* 117: 11357–11358.
41 Mkhalid, I.A.I., Barnard, J.H., Marder, T.B. et al. (2010). *Chem. Rev.* 110: 890–931.
42 Mankad, N.P., Laitar, D.S., and Sadighi, J.P. (2004). *Organometallics* 23: 3369–3371.
43 Bielinski, E.A., Lagaditis, P.O., Zhang, Y. et al. (2014). *J. Am. Chem. Soc.* 136: 10234–10237.
44 Zhang, Y., MacIntosh, A.D., Wong, J.L., Bielinski, G.A. et al. (2015). *Chem. Sci.* 6: 4291.
45 Uyeda, C. and Peters, J.C. (2013). *J. Am. Chem. Soc.* 135: 12023–12031.
46 Angamuthu, R., Byers, P., Lutz, M. et al. (2010). *Science* 327: 313–315.
47 Ahmed, S.M., Poater, A., Childers, M.I. et al. (2013). *J. Am. Chem. Soc.* 135: 18901–18911.
48 Radlauer, M.R., Day, M.W., and Agapie, T. (2012). *J. Am. Chem. Soc.* 134: 1478–1481.
49 McInnis, J.P., Delferro, M., and Marks, T.J. (2014). *Acc. Chem. Res.* 47: 2545–2557.
50 Garden, J.A., Saini, P.K., and Williams, C.K. (2015). *J. Am. Chem. Soc.* 137: 15078–15081.

51 Schmidt, J.A.R., Lobkovsky, E.B., and Coates, G.W. (2005). *J. Am. Chem. Soc.* 127: 11426–11435.
52 Kramer, J.W., Lobkovsky, E.B., and Coates, G.W. (2006). *Org. Lett.* 8: 3709–3712.
53 Mulzer, M., Whiting, B.T., and Coates, G.W. (2013). *J. Am. Chem. Soc.* 135: 10930–10933.
54 West, J.G., Huang, D., and Sorensen, E.J. (2015). *Nat. Commun.* 6: 10093.
55 Muetterties, E.L. (1977). *Science* 196: 839–848.
56 Powers, T.M. and Betley, T.A. (2013). *J. Am. Chem. Soc.* 135: 12289–12296.

4

Copper-Catalyzed Hydrogenations and Aerobic N—N Bond Formations: Academic Developments and Industrial Relevance

Paul L. Alsters and Laurent Lefort

InnoSyn B.V., Urmonderbaan 22, 6167 RD Geleen, the Netherlands

4.1 Introduction

In the fine chemical domain, (homogeneous) copper catalysis holds a prominent place in the formation of carbon–carbon and carbon–heteroatom bonds, both of which are of crucial importance in the synthesis of pharmacological, agrochemical, and other industrially relevant compounds [1]. Heterogeneous copper catalysts have a long history in several industrially important hydrogenation processes (see Section 4.3). The remarkable versatility of copper catalysis is further demonstrated by its use in various industrial aerobic oxidation processes [2]. In this chapter, we focus on homogeneously Cu-catalyzed hydrogenations and (aerobic) N—N bond formations, two research areas that hold promise for industrial application. The use of copper as a catalyst bears several industrially attractive features. Despite increasing demand, security of supply is high because there are major copper-producing countries in every continent and copper recycling is gaining interest [3]. When used in catalytic amounts, the price of the metal does not significantly contribute to the cost of goods [4]. Especially in aerobic oxidations, the copper catalyst is usually added as an inexpensive salt or even as metallic copper. Frequently, oxidative transformations do not require addition of a ligand to the copper catalyst, and when a ligand is required, it is usually a simple, readily available *N*-donor (see Section 4.2). Copper has been classified by the European Medicines Agency (EMA) as a class 2 metal (low safety concern) [5]. Use of a ligand may be advantageous to reduce copper contamination in the final product to a very low level, with the ligand functioning as a metal scavenger. Decomplexation of copper from the final product may also be achieved with other scavengers, such as ethylenediamine and Na_2EDTA (EDTA, ethylenediaminetetraacetic acid) [4]. Copper can be removed from wastewater by commonly applied techniques such as hydroxide precipitation [6]. Elimination of copper from aqueous effluents is important because copper is one of the most toxic elements to aquatic species, despite its concomitant role as an essential trace metal for all living organisms [7]. Although the efficient removal of copper from the final product and waste streams may be hampered by the high catalyst load-

Non-Noble Metal Catalysis: Molecular Approaches and Reactions,
First Edition. Edited by Robertus J. M. Klein Gebbink and Marc-Etienne Moret.

ing typically employed, the overall benefits outweigh the downsides, especially for Cu-catalyzed N—N bond-forming reactions that eliminate the need for hazardous reagents (hydrazines and azides) and reduce both the number of steps and the amount of waste.

4.2 Cu-Promoted N—N Bond Formation

Paul L. Alsters

In the following, we provide an overview of Cu-promoted conversions that lead to the generation of a N—N bond. The latter may also include N=N double bonds, unless indicated otherwise. Here, "promoted" is used as a collective term that covers use of both catalytic and stoichiometric amounts of copper. The emphasis is on aerobic transformations, i.e. those that require dioxygen (O_2) as a terminal oxidant. In the vast majority of N—N bond-forming reactions, O_2 serves as a hydrogen acceptor in what can be classified formally as a cross-dehydrogenative coupling of two N–H-bearing fragments (Eqs. (4.1) and (4.2)):

$$R^1R^1NH + HNR^2R^2 + 0.5\,O_2 \rightarrow R^1R^1N\text{–}NR^2R^2 + H_2O \quad (4.1)$$

$$R^1NH_2 + H_2NR^2 + O_2 \rightarrow R^1N = NR^2 + 2\,H_2O \quad (4.2)$$

A few examples of nonoxidative Cu-promoted N—N bond formations are included. This choice is justified for various reasons: (i) despite the nonoxidative overall conversion, O_2 is still required; (ii) although the actual N—N bond-forming step proceeds nonoxidatively, the overall multistep conversion also includes an oxidative step; and (iii) the reported nonoxidative conversion may likely be combined with a subsequent in situ oxidative step (e.g. an aromatization) by using aerobic conditions, thus expanding the synthetic scope of the approach. In most examples, Cu plays the role of a catalyst. Some Cu-mediated, stoichiometric N—N bond-forming transformations are also included in the overview. Their inclusion is rationalized by the fact that they can often be tuned toward a catalytic protocol by switching to aerobic conditions. These allow reoxidation of the reduced copper species generated by the overall reaction, thus allowing the oxidized copper species to reenter the catalytic cycle. It should be noted that in reported catalytic cycles based on a Cu^{II} species as the active oxidant in the N—N bond-forming step, it is often not clear or unambiguously demonstrated whether this requires transfer of two electrons to one Cu^{II} center (thus generating one Cu^0), or to two Cu^{II} centers (thus generating two Cu^I).

Besides ambiguity about the precise nature of the copper species involved in the redox cycle, the proposed mechanisms for multistep N—N bond-generating transformations are sometimes (highly) speculative rather than based on sound experimental evidence. In addition, the actual roles of copper and even dioxygen are often not evident from the nature of the overall N—N bond-forming transformation. For example, although the overall reaction involves elimination rather than dehydrogenation, copper is undergoing redox changes in the proposed mechanism and aerobic conditions are used, despite stoichiometric amounts of Cu. To illustrate this, we refer to the Cu-mediated formation of pyrazolines from oxime esters and *N*-sulfonylimines. Overall, this reaction is formally a condensation with the elimination of a carboxylic acid. Although this suggests that copper merely acts as a Lewis acid, the reaction is reported to proceed with redox changes at copper under aerobic conditions via an oxidative cyclization (Scheme 4.1) [8].

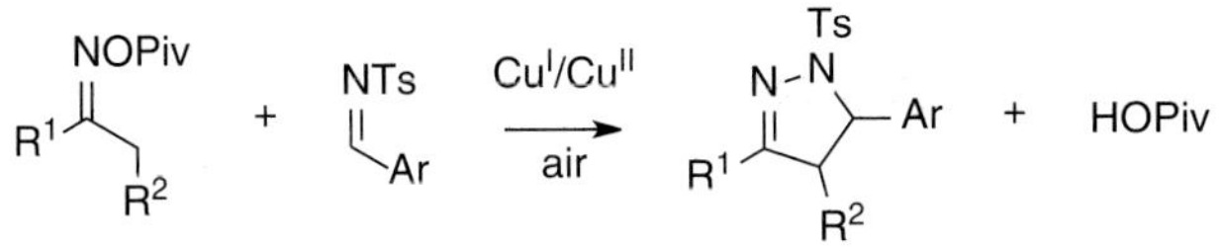

NOPiv R^1 R^2 + NTs Ar $\xrightarrow[\text{air}]{Cu^I/Cu^{II}}$ Ts N-N Ar R^1 R^2 + HOPiv

Scheme 4.1 Pyrazoline formation via oxidative coupling of oxime esters and *N*-sulfonylimines (Piv, pivalate; Ts, tosylate).

Because the mechanism of Cu-promoted N—N bond-forming reactions is often not yet clear, but the overall transformation can be unambiguously classified by inter alia distinguishing changes in oxidation state of the reactants and products, parts of the following are organized by describing the various N—N bond-forming steps in terms of the overall transformation involved. Although, as explained above, such a classification based on the overall transformation may not reflect the mechanism of the N—N bond formation, it offers the advantage of providing easy guidance to select coupling partners for N—N bond generation without the understanding of the actual mechanism.

4.2.1 Noncyclization N—N or N=N Bond Formations

In this section, we focus on N—N bond-generating reactions that are not necessarily accompanied by ring closure. Occasionally, (large) rings may be formed as a result of an intramolecular N—N bond formation, but this is not an inherent feature of the transformation.

4.2.1.1 N—N Single-Bond-Forming Reactions

Even though the (intermolecular) dehydrogenative coupling of two N—H fragments may be considered to constitute an archetypal N—N bond formation (Eq. (4.1)), it is of limited organic synthetic importance for the formation of azines and hydrazines from imines and amines, respectively. Industrially, the state-of-the-art chlorine-free production of hydrazine is based on the Pechiney–Ugine–Kuhlmann process with methyl ethyl ketone, ammonia, and hydrogen peroxide as raw materials. The actual N—N bond-forming step required for ultimate hydrazine formation proceeds dehydratively between NH_3 and 3-methyl-3-ethyl-1,2-oxaziridine. The latter is generated in situ by oxygenation of its imine precursor with H_2O_2 in the presence of an activator [9]. This noncatalyzed, nonaerobic industrial process is out of the scope of this chapter.

Stoichiometric CuCl in pyridine mediated the high-yield aerobic formation of tetraphenylhydrazine from diphenylamine [10, 11], but only a modest yield was obtained from *N*-methylaniline [10]. This system was also used for the cyclodehydrogenation of *N*-benzylidene-*o*-phenylenediamines to the corresponding 1,1′-bibenzimidazoles (Scheme 4.2) [12].

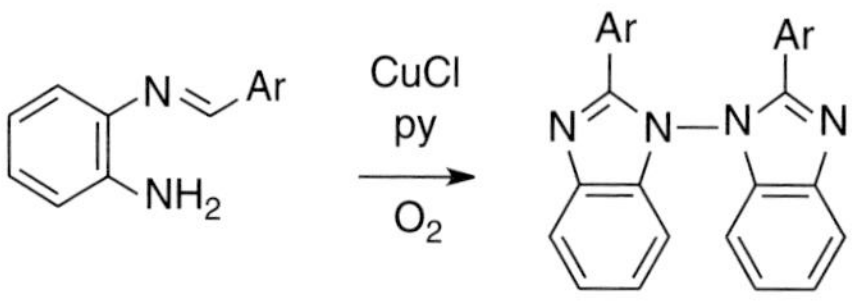

Scheme 4.2 1,1′-Bibenzimidazole formation via aerobic cyclodehydrogenation of *N*-benzylidene-*o*-phenylenediamines.

The latter reaction was supposed to proceed via arylimine radicals ($ArNH^{\bullet}$) generated upon hydrogen abstraction from $ArNH_2$ by Cu^{II}. Interestingly, this N—N bond-forming transformation does not involve homocoupling of benzimidazole intermediates [12]. The scope of the aerobic homocoupling of *N*-alkylanilines was greatly expanded by switching to a mixed CuBr/CuO catalyst system with TMEDA (tetramethylethylenediamine) as a ligand [13]. It is observed that there is still an unmet need for catalytic dehydrogenative N—H coupling methods that allow the preparation of unsymmetrical hydrazines, partly substituted hydrazines, and hydrazines devoid of aryl substituents.

Aerobic protocols for the preparation of azines from benzonitrile have been described. These involve oxidative dimerization under the CuX catalysis (X = Cl; I) of iminometal halides obtained by the addition of MeLi or Grignard reagents to the nitrile (Scheme 4.3; nitrile approach) [14, 15]. Instead of dioxygen, *tert*-butyl peroxybenzoate was also particularly efficient as an oxidant [14]. Several catalytic protocols for the preparation of benzophenone azine or substituted derivatives thereof by Cu-catalyzed aerobic dehydrogenative coupling of the corresponding imine precursors are available (Scheme 4.3; ketone approach) [16–18]. A patent describes the CuCl-catalyzed aerobic homocoupling of pinacolone imine to pinacolone azine as part of a process for the manufacture of hydrazodicarbonamide [19]. A zeolite was added to trap the water generated during the reaction. The patent emphasizes the use of O_2 instead of Cl_2 and the hydrazine-free feedstock as advantageous aspects. Despite these advantages, and even though the foregoing examples demonstrate that azines as hydrazine precursors can be generated efficiently via Cu-catalyzed aerobic imine couplings, the Pechiney–Ugine–Kuhlmann process based on more expensive H_2O_2 instead of O_2 apparently outcompetes aerobic methods [20]. The fact that this process does not require a catalyst is most likely a significant differentiator.

Aerobic *N*-nitrosation of aromatic and aliphatic secondary amines was achieved with substoichiometric DBU (1,8-diazabicyclo[5.4.0]undec-7-ene) and catalytic $Cu(OTf)_2$, using nitromethane as the nitroso source [21]. The mildly basic conditions may favor the use of this method instead of conventional *N*-nitrosation that requires acidic conditions.

4.2.1.2 N=N Double Bond-Forming Reactions

The Cu-catalyzed aerobic dehydrogenative coupling of anilines to azobenzenes (or 1,2-diaryldiazenes) is widely applied as a laboratory-scale method.

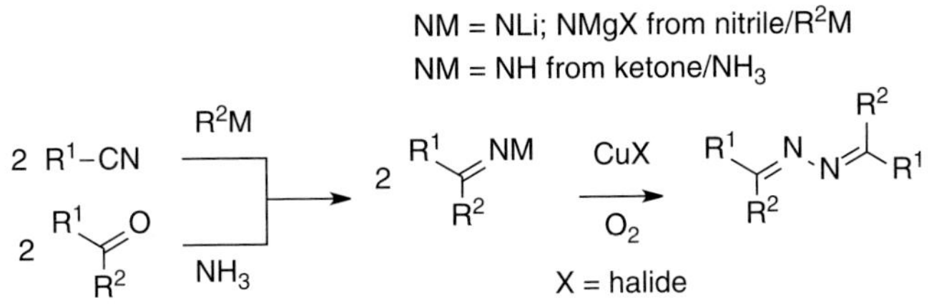

NM = NLi; NMgX from nitrile/R^2M
NM = NH from ketone/NH_3

2 R^1–CN R^2M
2 $R^1C(O)R^2$ NH_3
→ 2 $R^1C(=NM)R^2$ —CuX, O_2→ $R^1(R^2)C=N–N=C(R^2)R^1$
X = halide

Scheme 4.3 Approaches to obtain azines from nitriles or ketones.

Industrially, azobenzenes are a very important class of dyes. Their manufacture is, however, not based on aerobic dehydrogenative aniline coupling, but usually on diazotization of an aniline followed by coupling of the diazonium species to an arene [22]. The latter so-called "coupling reaction" only proceeds efficiently with electron-rich arenes (phenols or anilines). In contrast, aerobic dehydrogenative aniline coupling allows preparation of a wide variety of both symmetrical and unsymmetrical azobenzenes, thanks to the development of various catalytic protocols. Combined with the ready availability of the required Cu catalyst components, this explains the prominent place that dehydrogenative coupling has gained as a synthetic tool for azobenzene preparation.

Synthetic and mechanistic aspects of azobenzene formation via Cu-catalyzed aerobic dehydrogenative aniline coupling have been dealt within the very extensive review by Kozlowski and coworkers on aerobic copper catalysis [23]. Most commonly, the reaction is thought to involve arylimine radicals that are generated via one-electron oxidation by Cu^{II}, which is regenerated aerobically from the thus generated Cu^{I} species. The resulting hydrazobenzene (1,2-diarylhydrazine) intermediate undergoes facile subsequent dehydrogenation to yield the final azobenzene product. This ease of the latter dehydrogenation explains the lack of protocols for selective dehydrogenative N–H coupling toward partly substituted hydrazines, as noted above under Section 4.2.1.1.

Seminal Russian and Japanese work in the 1950s was based on the use of CuCl as a catalyst in pyridine [24–28]. This system has still been used recently for the preparation of various (symmetric) substituted azobenzenes [29–31], including a ferrocene-containing macrocyclic azobenzene [32]. Electron-withdrawing and ortho-substituents in the aniline are not well tolerated and strongly reduce the yield. The system also allowed the aerobic coupling of *N*-hydroxyanilines to yield azoxybenzenes. An arene with two *N*-hydroxyaniline moieties could be polymerized to the corresponding poly(azoxyarylene) [33]. Of practical interest is the much more recent finding that pyridine can be replaced by acetonitrile [34]. Occasionally, CuBr is substituted for CuCl under otherwise similar conditions [35]. The most widely applied protocol for aerobic azobenzene formation from anilines appears to be the one developed by Jiao and coworker [36]. It is based on CuBr as the catalyst with ~3 equiv. of pyridine as the ligand and toluene as the solvent at slightly elevated temperature (Scheme 4.4).

Besides being suitable for the dehydrogenative homocoupling of anilines devoid of sterically or electronically deactivating groups [36–39], it has also allowed the preparation of more challenging classes of azobenzenes, such

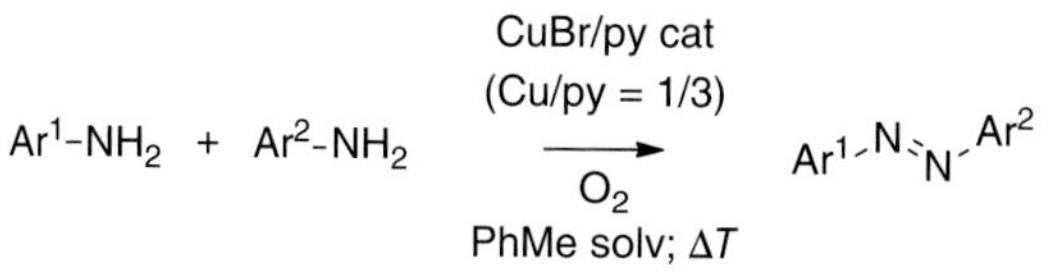

Scheme 4.4 Jiao protocol for copper-catalyzed aerobic dehydrogenative coupling of anilines to aromatic azo compounds.

as those with electron-withdrawing substituents [40–47], ortho-substituents [40, 42, 43, 45–48], or even unsymmetric azobenzenes [40, 42, 45, 49–51]. To obtain sufficiently high selectivity for the latter class, a large excess of the less reactive aniline coupling partner has to be used.[1] Another broadly applicable system is based on $CuCl_2$ as the catalyst and stoichiometric *n*-BuMgBr as a strong base capable of deprotonating aniline. Except for 3-nitroaniline, excellent yields of symmetric azobenzenes were obtained in THF (tetrahydrofuran) as a solvent at room temperature with a range of anilines [52]. Still another aerobic system with a broad scope was developed recently. It employs red copper as the catalyst in the presence of substoichiometric pyridine and stoichiometric NH_4Br, using toluene as the solvent at elevated temperature. Apart from the broad range of symmetric and unsymmetric azobenzenes with a large variety of substituents, it also allowed the formation of a bridged *Z*-azobenzene that changed color from yellow to red upon UV irradiation that triggered photoisomerization to the usual *E*-configuration [53]. An unusual Cu-catalyzed aerobic coupling to a diazo product has been observed for 3-methyl-4-oxa-5-azahomoadamantane. For this substrate, the initial dehydrogenative coupling was followed by a spontaneous isomerization to yield **1** (Scheme 4.5) [54].

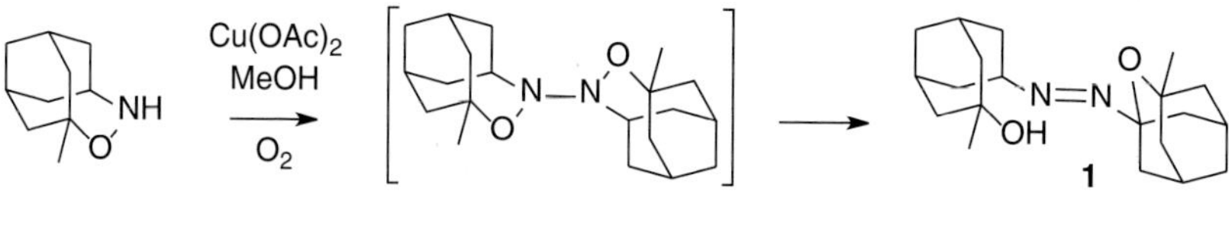

Scheme 4.5 Unusual diazo compound formation via aerobic dehydrogenative coupling of 3-methyl-4-oxa-5-azahomoadamantane followed by isomerization.

4.2.2 Cyclization N—N Bond Formations

The following deals with cyclizations that are accompanied by the formation of a N—N bond as part of the ring. With few exceptions, the latter is a five-membered aromatic *N*-heterocycle. As already addressed above, given the uncertainty about the actual mechanism of many of these reactions, we will organize various cyclizations according to a description of the overall transformation from reactant(s) to product under a given set of reaction conditions. In a one-pot protocol with staged reaction conditions, we describe the overall transformation

1 In one case, an unymmetric azobenzene was prepared without reported yield using an excess of aniline as most reactive coupling partner [40].

starting from the reactant(s) present under the reaction conditions required to achieve the N—N bond-forming cyclization. Three classes to describe the overall cyclization process will be distinguished:

- *Dehydrogenative cyclizations (DCs)*: formation of the cyclic product requires formal loss of x H_2 from the reactant(s), with O_2 serving as a hydrogen acceptor to yield water;
- *Eliminative cyclizations (ECs)*: formation of the cyclic product from the reactant(s) is accompanied by elimination of y HL molecular entities different from H_2 (L = leaving group);
- *Eliminative dehydrogenative cyclization (EDC)*: the cyclic product formation requires formal extrusion of both x H_2 and y HL from the reactant(s).

Note that these overall transformations accompanied by ring formation may be the result of both intra- and intermolecular reactions. For practical reasons, we will also classify cyclizations as "dehydrogenative" when these involve (usually N-) metalated species that aerobically cyclize under the formation of a metal (hydr)oxide instead of water. Despite their overall nonoxidative nature that does not inherently call for aerobic conditions, ECs have been included for reasons outlined earlier.

To illustrate that these three classes are mere descriptions of the overall transformation without necessarily bearing a relationship with the actual mechanism of the ring-closing step, we refer to two papers that deal with the formation of 1,2,4-triazoles from amidines. The amidine prepared as its sodium salt by the addition of $NaNH_2$ to benzonitrile has been oxidatively coupled to form 3,5-diphenyl-1,2,4-triazole on treatment with stoichiometric CuCl and O_2. The coupling was suggested to proceed via initial bimolecular Cu-mediated dehydrogenative N—N bond formation to yield an intermediate diazabutadiene-l, 4-diamine that subsequently cyclizes to the 1,2,4-triazole under elimination of NH_3 (Scheme 4.6, path A) [15]. In a more recent paper, the same 1,2,4-triazole was obtained from its amidine precursor by another Cu-catalyzed aerobic coupling protocol. In this case, the cyclization was thought to involve prior bimolecular condensation under the elimination of NH_3 to generate *N*-[amino(phenyl)methylene]benzamidine as an intermediate. The consecutive

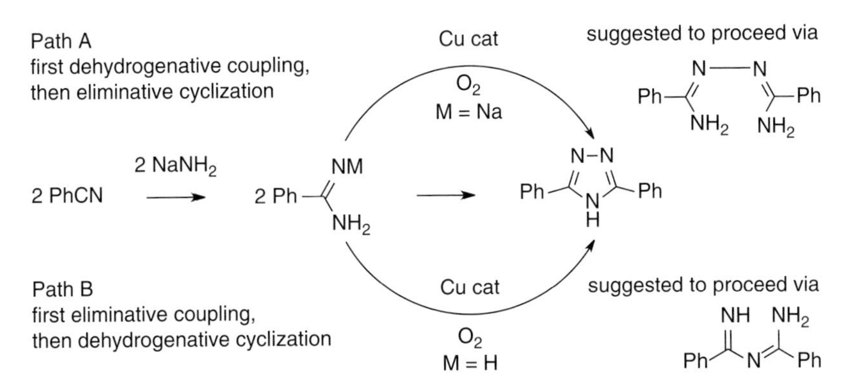

Scheme 4.6 Two mechanisms proposed for 1,2,4-triazole formation from an amidine.

N—N bond formation proceeds on the latter species via a Cu-catalyzed dehydrogenation, yielding 3,5-diphenyl-1,2,4-triazole (Scheme 4.6, path B) [55]. Thus, although the proposed order of the elimination and dehydrogenation steps is reversed in these two papers while dealing with a similar overall transformation, we will label both approaches as EDC.

Despite frequent ambiguity about the actual mechanism, true cyclization steps under concomitant formation of a N—N bond as part of the *N*-heterocycle have been well demonstrated for various substrates. For the most widely reported type of cyclization that yields a five-membered aromatic *N*-heterocycle via N—N single-bond generation, Scheme 4.7 collects various tautomeric precursors of N—N bond-containing heterocycles that are generated in a DC or a EC step. As indicated in this scheme, the iminoenamine tautomer is often assumed to form a Cu^{II} chelate that generates the heterocycle by reductive elimination of Cu^0 or 2 Cu^I (not shown).

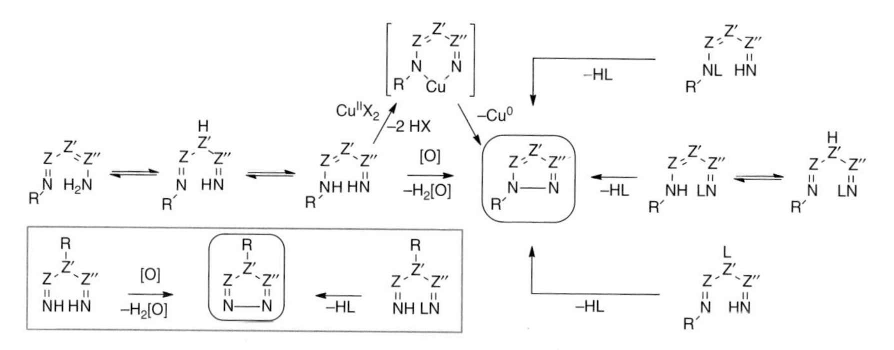

Scheme 4.7 Tautomers that yield N—N bond containing five-membered aromatic heterocycles (within rounded rectangles) in a DC (left to right arrows) or EC (right to left arrows) step. Square box: 4-substituted product. L denotes a leaving group; Z, Z′, and Z″ denote any atom or set of atoms that is in line with the indicated resonance structures.

N—N bond-generating cyclizations have proven to be a powerful synthetic approach to access a number of aromatic *N*-heterocycles with significant industrial relevance, notably because of their frequent occurrence as structural entities in agrochemicals and especially pharmaceuticals. Pyrazoles, 1,2,3-triazoles, and 1,2,4-triazoles have all been successfully synthesized via this strategy. Compared to traditional approaches to construct these heterocycles, Cu-catalyzed aerobic N—N bond-generating cyclizations frequently offer several advantages that make them particularly attractive for industrial use. Besides synthetic advantages outlined below, this alternative approach may benefit from the replacement of hazardous reagents such as carcinogenic hydrazines or explosive azides by safer alternatives. Also, the required copper catalysts and ligands are often readily available on scale. The importance of triazoles as a class of heterocyclic compounds with a broad range of biological activities urged authors to stress the need for more sustainable preparations of triazoles [56], and it is felt that current developments in Cu-catalyzed aerobic N—N bond formations are highly relevant in this context. Still on the down side, catalyst loading is usually fairly large (mol% range), which may evoke considerable workup and downstream processing efforts to meet regulatory requirements. In addition, many reported aerobic laboratory protocols are carried out with a dioxygen concentration well above the OLC (oxygen limit concentration), thus calling for significant additional development efforts to secure safe scale-up [57, 58]. A suitable reactor technology, such as continuous flow reactors, may help to that end [59–61].

4.2.2.1 Dehydrogenative Cyclizations

Chart 4.1 collects the product classes to be dealt with the below. They are indicated with lowercase, bold, italic letters between square brackets.

Five-membered C_3N_2 heterocycles: A broad range of tetra-substituted 1*H*-pyrazole-4-carboxylate and 4-benzoyl derivatives [**a**] were prepared via dehydrogenative coupling of enamines with nitriles. This modular, highly atom-efficient and regioselective approach provides a valuable alternative to conventional synthetic methods based on toxic and carcinogenic hydrazines. Although the original protocol employed stoichiometric copper [62–64], a catalytic amount could be used in the presence of cocatalytic 2-picolinic acid combined with 1 bar O_2 as atmosphere [65, 66]. Under stoichiometric conditions, this approach has also been used for the preparation of a 3-(trifluoromethyl)-1*H*-pyrazole-4-carboxylate from gaseous CF_3CN [65]. Pharmaceutical researchers from Janssen R&D have developed a facile synthesis of 3-alkyl-, alkenyl-, or aryl-substituted 1*H*-indazoles [**b**] via addition of various organometallic reagents to 2-aminobenzonitriles, followed by $Cu(OAc)_2$-catalyzed aerobic DC of the resulting ketimines [67]. Both unsubstituted and alkyl or aryl-substituted amino groups were well tolerated. Remarkably, an *N*-methylindazole was even generated from 2-(dimethylamino)benzonitrile via demethylation during N—N bond formation.

Five-membered C_2N_3 heterocycles: An almost century-old example of a dehydrogenative aerobic N—N bond-forming cyclization is the synthesis of 2-aryl-2*H*-benzo[1,2,3]triazoles [**c**] from 2-aminoazobenzenes under the influence of Cu^{II}. It was already then recognized that the reaction could be made catalytic in copper by blowing air through the solution [68]. From the corresponding azo precursors, the stoichiometric method has also been used for the preparation of 2-aryl-5-amino-7-hydroxy-2*H*-[1,2,3]triazolo[4,5-d]pyrimidines [**d**] [69] and quinoxalines containing 2-substituted 2*H*-benzo[1,2,3]triazole moieties [**e1**, **e2**] [70]. Despite the stoichiometric usage of Cu^{II},[2] air was continuously bubbled through the reaction mixture in the latter case. In the presence of pyridine, triazoles [**c**] were obtained from 2-aminoazobenzenes with catalytic loading of CuCl [71]. Closely related is the ammoniacal $CuSO_4$-mediated formation of 5-amino-2-phenyl-2*H*-1,2,3-triazole-4-carboxylate derivatives [**f**, Ar = Ph; E = C(O)NH_2 or C(NH)NH_2; R = H] by

2 A ~1/1 M Cu^{II}/substrate ratio was employed in Ref. [70]. This corresponds to the stoichiometric ratio when Cu^{II} acts as a 2 electron oxidant, generating Cu^0. The equimolar ratio is obviously substoichiometric in case of Cu^{II} acting as a 1 electron oxidant that yields Cu^I, which may be aerobically reoxidized to Cu^{II}.

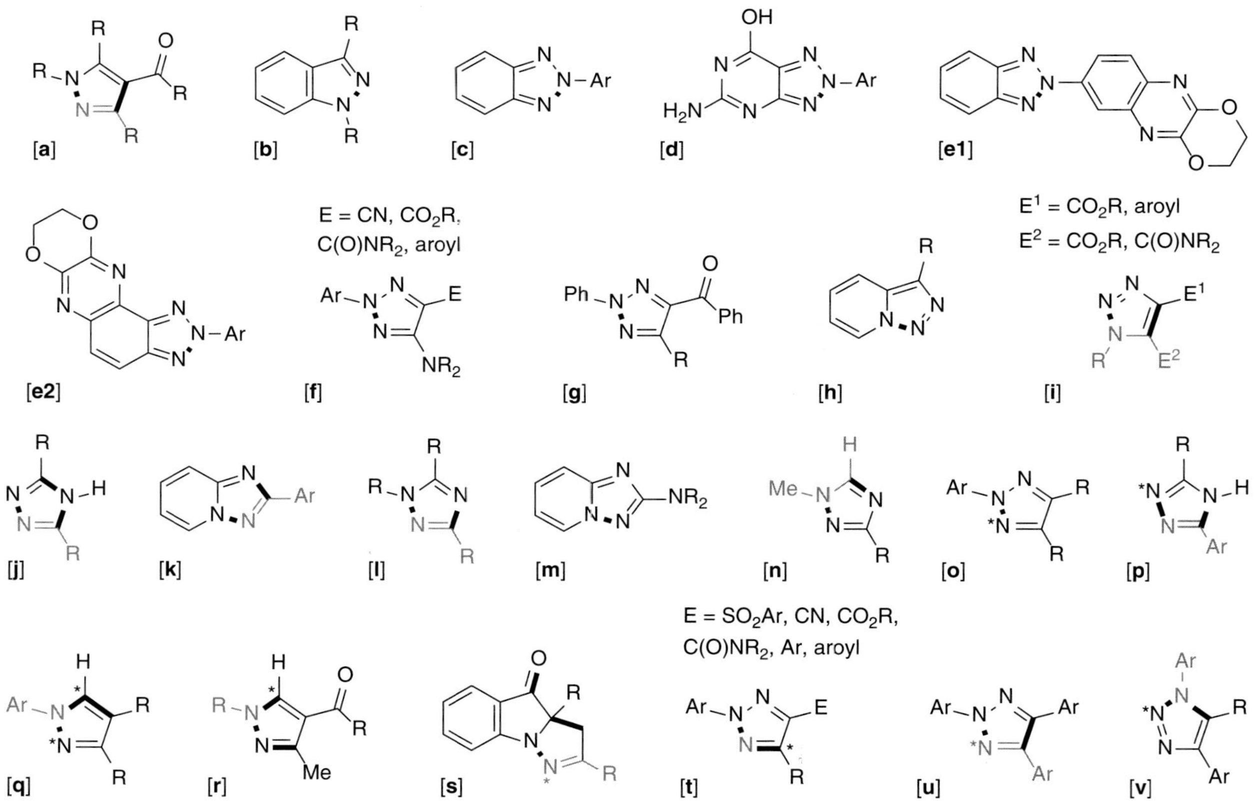

[a] [b] [c] [d] [e1]

$E^1 = CO_2R$, aroyl
$E^2 = CO_2R$, $C(O)NR_2$

E = CN, CO_2R, $C(O)NR_2$, aroyl

[e2] [f] [g] [h] [i]

[j] [k] [l] [m] [n] [o] [p]

E = SO_2Ar, CN, CO_2R, $C(O)NR_2$, Ar, aroyl

[q] [r] [s] [t] [u] [v]

Chart 4.1 A structural overview of aromatic *N*-heterocyclic product classes accessible via DC, EC, and EDC steps. Bold bonds are generated in these steps, with nitrogen atoms involved in N—N bond formation also being highlighted in bold. For intermolecular cyclizations, structural segments originating from different reactants are distinguishable by gray vs black drawings, or the use of italic vs nonitalic atom labels. Atoms labeled with an asterisk carried an atom or group other than hydrogen that is eliminated from the reactant during EC or EDC.

DC of the phenylazomalonate amidine precursors [72]. Ammoniacal Cu^{II} was also successfully used for the preparation of various 5-substituted 4-benzoyl-2-phenyl-2*H*-1,2,3-triazoles [**g**] via DC of α-imino phenylhydrazones (containing a nonsubstituted imine nitrogen atom) [73]. A similar synthetic strategy was used for the preparation of a broad range of 5-amino-4-aroyl-2-aryl-2*H*-1,2,3-triazole and 5-amino-2-aryl-2*H*-1,2,3-triazole-4-carboxylic acid derivatives [**f**] (as the corresponding nitriles, esters, or amides). The required α-imino arylhydrazone cyclization precursors were obtained by the addition of amines to the nitrile group of α-arylhydrazonoyl cyanides. Although initial protocols were employing stoichiometric Cu^{II} in pyridine [74, 75], substoichiometric Cu^{II} was also found to work well under aerobic conditions in THF or MeCN as solvent in a one-pot protocol [76, 77]. A range of various [1,2,3]triazolo[1,5-*a*]pyridines [**h**] were readily obtained from 2-acylpyridines via their hydrazones that were dehydrogenatively cyclized under aerobic conditions using simple copper salts as catalysts and water or an organic solvent as the medium. Basic conditions were required, either by using $Cu(OAc)_2$ as a catalyst in an organic solvent (with acetate serving as proton acceptor) or by adding NaOH to an aqueous solution containing $Cu(NO_3)_2$ as the catalyst [78, 79]. Basicity is thought to facilitate the generation of the (not yet ring-closed) diazo isomer of the [1,2,3]triazolo[1,5-*a*]pyridine via double-proton abstraction from the hydrazone NH_2 group combined with double one-electron reduction of cupric ions [78]. In what is overall a double DC, *N*-substituted glycine esters and amides cyclized with α-diazo compounds to generate 1,4,5-trisubstituted 1*H*-1,2,3-triazoles [**i**] [80]. $CuBr_2$ was used as the catalyst in the presence of excess DBU under an O_2 atmosphere. Experimental support was obtained for initial dehydrogenation to yield the *N*-substituted glyoxylate imine from the glycine precursor, followed by a second dehydrogenative aromatization of the [3+2] cycloaddition product of the imine and the diazo compound.

The synthesis of 1,2,4-triazoles via new strategies based on Cu-catalyzed N—N bond formation has received significant attention in recent years, mostly driven by the pharmaceutical importance of this heterocyclic unit [81]. An early stoichiometric dehydrogenative example within the domain of inorganic synthesis entails the synthesis of Cu^{I} 3,5-disubstituted 1,2,4-triazolates [**j**] by one-pot treatment of nitriles, ammonia, and Cu^{II} salts that serve as an oxidant and Lewis acidic nitrile activators to facilitate nucleophilic attack of *N*-nucleophiles (such as NH_3 and intermediate amidines). Evidence was obtained that the actual N—N bond formation proceeds via two-electron oxidation of Cu^{II} bound 1,3,5-triazapentadienes, of which an example was isolated [82]. Related mechanistic principles have been applied to generate catalytic dehydrogenative, aerobic protocols of organic synthetic interest. Via these approaches, a large variety of 1,2,4-triazoles were generated by coupling of nitriles to amidines or 2-aminopyridines, using readily accessible copper catalysts. Coupling of 2-aminopyridines (yielding 2-substituted [1,2,4]triazolo[1,5-*a*]pyridines [**k**]) or *N*-arylbenzamidines (yielding 1,3,5-trisubstituted 1,2,4-1*H*-triazoles [**l**]) required the presence of 1,10-phen (phenanthroline) as the ligand of the CuX (X = Br; I) catalyst and ZnI_2 as the cocatalyst [83, 84], whereas amidines devoid of *N*-substituents required the presence of excess Cs_2CO_3 [83]. Building on this approach, a one-pot synthesis was developed to construct triazolopyridines [**k**] from benzonitriles prepared in situ via ammoniacal aerobic CuI/bpy/TEMPO (bpy = 2,2′-bipyridine; TEMPO, (2,2,6,6-tetramethylpiperidin-1-yl)oxyl)-catalyzed oxidation of the benzyl alcohol [85]. Using a large excess of benzonitrile as the reaction medium, CuX/1,10-phen (X = Br; Cl) without ZnI_2 sufficed as the catalyst for the preparation of triazolo[1,5-*a*]pyridines [**k**, Ar = Ph], as patented by Hoffmann-La Roche [86, 87]. From a process safety and scalability point of view, it is important to note that good yields were obtained when air was replaced by (elevated or atmospheric pressure) O_2/N_2 (5 : 95), i.e. well below the OLC [87].

Coupling of nitriles to amidines or 2-aminopyridines was also catalyzed by heterogeneous copper–zinc supported on Al_2O_3–TiO_2, which was active under ligand-, base-, and additive-free conditions [88]. Besides, amidines, including *N*-arylamidines, were successfully coupled to nitriles under the formation of the corresponding di- or tri-substituted 1,2,4-triazoles [**l**, R–N = H–N or Ar–N] with an aerobic system based on excess Na_2CO_3 and $Cu(OAc)_2$ as the catalyst, the latter requiring 1,10-phen as a ligand in the case of *N*-arylamidines. The use of toluene instead of DMSO or 1,2-dichlorobenzene as a solvent is another advantageous aspect of this system [89]. In a related one-pot method, 2,4-disubstituted 1,3,5-triazapentadienes were first prepared by Cs_2CO_3-mediated coupling of two amidine units under the elimination of NH_3, followed by their copper-powder-catalyzed N—N bond-generating aerobic cyclization. Both symmetric and unsymmetric 3,5-disubstituted 1,2,4-triazoles [**l**, R–N = H–N] were obtained, the latter by coupling two different amidines [90]. Researchers from Hoffmann-La Roche attempted to extend the DC of 2-aminopyridines with nitriles to cyanamides, driven by the relevance of the intended 2-amino-[1,2,4]triazolo[1,5-*a*]pyridine scaffold in medicinal chemistry. Although direct intermolecular coupling of dimethylcyanamide to 2-aminopyridine according to the published protocol proved to be low yielding [83], a high yield was obtained in a two-step approach involving intramolecular CuBr/1,10-phen-catalyzed aerobic DC of the (2-pyridinyl)guanidine [91]. The latter was prepared via sodium *tert*-butoxide-mediated addition of 2-aminopyridine to dimethylcyanamide. Besides various 2-amino-[1,2,4]triazolo [1,5-*a*]pyridines [**m**], pyrimido-, pyridazido-, and pyrazidotriazoles were also prepared via this approach. The researchers highlighted the economic and environmental benefits of this approach compared to prior multistep sequences. Finally, 3-substituted 1-methyl-1*H*-1,2,4-triazoles [**n**] were obtained in an unusual double-dehydrogenative intermolecular cyclization of amidines with *N*,*N*-dimethylformamide (DMF), using $CuCl_2$ as the catalyst in the presence of K_3PO_4 as the base to facilitate formyl hydrolysis [92].

4.2.2.2 Eliminative Cyclizations

Five-membered C_3N_2 heterocycles: The above-mentioned formation of pyrazolines from *O*-pivaloyl oxime esters and *N*-sulfonylimines (Scheme 4.1) constitutes a recent example of a Cu-mediated cyclization with the (formal) elimination of pivalic acid [8].

Five-membered C_2N_3 heterocycles: An early example of a Cu-mediated N—N bond-generating cyclization accompanied by elimination of a smaller molecular fragment concerns the formation of 1,2,3-triazoles from osazones, i.e. 1,2-dihydrazones (usually derived from carbohydrates) [93]. The resulting triazoles are called "osotriazoles." Cyclization occurs with osazones derived from arylhydrazines and is accompanied by the elimination of the corresponding arylamine [94]. A large range of arylosotriazoles [**o**] has been prepared, including parent 2-phenyl-2*H*-1,2,3-triazole obtained from the osazone of glyoxal [95], and several reviews dealing with their preparation, mechanism of formation, and applications have appeared [96–99]. Although the elimination of an arylamine during osotriazole formation suggests that Cu^{II} merely acts as a Lewis acid, the actual mechanism is complex and involves reduction of Cu^{II}, as evidenced by the precipitation of metallic copper during the reaction [100]. The first step may involve an overall two-electron oxidation of the 1,2-dihydrazone by Cu^{II} to yield 1,2-bis(arylazo)ethylene via dehydrogenation [96]. The latter species are known to yield 1,2,3-triazoles in a pericyclic reaction [101]. A one-electron radical mechanism involving the elimination of a phenylimine radical has also been proposed [100]. Besides 1,2,3-triazoles, 1,2,4-triazoles [**p**] have also been accessed following an EC approach. The latter involved the addition of the amino group of $HONH_2$ to the cyano group of a nitrile, thereby generating an amidoxime, followed by $Cu(OAc)_2$-catalyzed consecutive addition of the amidoxime to another (aromatic) nitrile and dehydrative cyclization to yield the triazole. The proposed mechanism does not invoke any redox changes of Cu^{II}, which serves merely as a Lewis acid [102].

4.2.2.3 Eliminative Dehydrogenative Cyclizations

Five-membered C_3N_2 heterocycles: A broad range of 1,3- or 1,3,4-substituted pyrazoles [**q**] were obtained by aerobic oxidative condensation of oxime acetates containing an α-CH_2 group with arylamines and paraformaldehyde [103]. The transformation was catalyzed by Cs_2CO_3 and CuBr, with the Cu catalyst thought to induce N–O cleavage of the oxime acetate to generate a Cu^{II} enamide species. In addition, it is involved in the final dehydrogenation of the penultimate pyrazoline obtained by Cu-catalyzed N—N bond formation. 1,3,4-Substituted pyrazoles [**r**] could also be generated from α-(1,3-dithian-2-yl) enamine ketones by cyclization with primary aryl- or alkylamines via EDC. $CuBr_2$ was used as a catalyst in the presence of excess NEt_3 under air. Besides N—N bond-forming cyclization, the elimination of propane-1,3-dithiol is also supposed to involve Cu^{II} [104]. In a complex sequence that combines an overall EDC with oxygenation, pyrazolo[1,5-*a*]indol-4-one derivatives [**s**] were obtained aerobically from 2-substituted indoles and α-CH_2 group-containing oxime acetates [105]. The $CuCl_2$ catalyst is assumed to participate in transforming the oxime acetate into an azirine, which via its vinyl nitrene isomer undergoes a [3+2] cycloaddition to the C=N bond of a hydroperoxide intermediate obtained via aerobic oxygenation of the indole.

Five-membered C_2N_3 heterocycles: Stoichiometric protocols for the DC of isolated α-imino phenylhydrazones and related tautomeric structures already referred above have been translated into aerobic catalytic protocols based on an EDC strategy [72, 73]. To that end, α-arylhydrazonoketones were condensed with NH_4OAc, followed by in situ $CuBr_2$-catalyzed oxidative cyclization of the resulting α-imino arylhydrazones, thus generating the corresponding 2-aryl-2*H*-1,2,3-triazoles [**t**] [106]. Via a complex cascade involving C–H functionalization and C—C/N—N bond formation, bisarylhydrazones derived from (het)arylcarbaldehydes and arylhydrazines were aerobically cyclized to 2,4,5-triaryl-2*H*-1,2,3-triazoles [**u**] under $Cu(OAc)_2$ catalysis. It was shown that the cascade generates the azobenzene derived from the arylhydrazine as a by-product, probably by the elimination of arylimine radicals, followed by their dimerization to the hydrazobenzene and subsequent dehydrogenation of the latter [107]. Alternatively, 2,4,5-triaryl-2*H*-1,2,3-triazoles [**u**] were obtained in a one-pot protocol from bisarylhydrazones generated by the alkylation of arylhydrazines with benzylbromides followed by in situ CuBr-catalyzed aerobic dehydrogenation [108]. 1,4-Disubstituted or 1,4,5-trisubstituted 1,4-diaryl-1*H*-1,2,3-triazoles [**v**] were formed from anilines and *N*-tosylhydrazones derived from 1-(het)arylalkan-1-ones in a $Cu(OAc)_2$-mediated EDC under air. Despite stoichiometric usage of copper, the presence of air was important to obtain good yields. Remarkably, both working under N_2 or pure O_2 instead of air lowered the yield [109]. It was shown that the actual N—N bond-forming cyclization occurs on the α-anilino derivative of the *N*-tosylhydrazone. This derivative was supposed to be generated by an aza-Michael addition of the aniline to a 1-tosyl-2-vinyldiazene, the latter resulting from Cu^{II}-mediated dehydrogenation of the *N*-tosylhydrazone via a Cu enamido intermediate.

With respect to 1,2,4-triazole synthesis based on EDC, mentioned under Section 4.2.2, aerobic, Cu-mediated, homocoupling of the amidine prepared (as its sodium salt) from $NaNH_2$ and benzonitrile to yield 3,5-diphenyl-1,2,4-triazole constitutes an early example [15]. The scope of aerobic benzamidine homocoupling has been widened by the use of $Cu(OTf)_2$/1,10-phen as the catalyst and $K_3[Fe(CN)_6]$ as the cocatalyst. By adding $(NH_4)_2CO_3$, benzimidates could also be homocoupled with this system [55].

4.3 Cu-Catalyzed Homogeneous Hydrogenation

Laurent Lefort

Copper is a metal with a long history in catalytic hydrogenation. As early as in the 1920s, Cu-based heterogeneous catalysts were discovered for the production of methanol via hydrogenation of CO [110, 111]. The short lifetime of these catalysts due to a fast poisoning by impurities in the feed and/or sintering prevented their commercial use for roughly 40 years. But in 1960s, Imperial Chemical Industries Ltd. (ICI) developed a new low-pressure/low-temperature process based on a novel preparation of the Cu catalysts (Cu–ZnO) [112, 113]. Such catalysts are still used today for the production of methanol on million tons scale. Another example are the copper–chromite ($CuO–Cr_2O_3$) catalysts for the hydrogenation of esters [114, 115]. Discovered in the 1930s by Adkins and Connor [116], they remain the catalysts of choice for the reduction of fatty acid esters to the corresponding alcohols at an industrial scale [117].

It is worth to mention that the first-ever metal-catalyzed homogeneous hydrogenation was obtained with a Cu catalyst. Indeed, in 1938, Calvin described the reduction of para-quinone with $Cu(OAc)_2$ in quinoline [118]. Despite this flying start, catalytic hydrogenation by homogeneous Cu complexes, which is the focus of this chapter, has lagged behind. Recent progresses have been achieved in this field as part of the current trend to develop cheap catalysts based on non-noble metals. In this section, we will focus solely on catalytic systems using H_2 as the reducing agent. The section is divided by the class of substrates to be hydrogenated. For the interested reader, several recent reviews cover copper-based reductions either with stoichiometric amount of Cu and/or with other reducing agents [119–123].

In some studies [124, 125], a distinction is made between "catalytic hydrogenation" and "catalytic hydride reduction." In "catalytic hydrogenation," a molecule of H_2 delivers both hydrogen atoms to a molecule of the substrate via the assistance of the homogeneous catalyst. In "catalytic hydride reduction," the homogeneous catalyst in the form of a metal hydride delivers only one hydride to the substrate. A proton source from the reaction media is therefore needed to complete the catalytic cycle. In the case of Cu complexes, the activation of H_2 proceeds exclusively in a heterolytic manner, leading to the formation of a copper hydride and the release of a proton in the reaction media in the form of an alcohol (Scheme 4.8). The copper hydride is subsequently added to an unsaturated substrate. This step, referred to as a hydrocupration, generates an organocopper species that undergoes protonolysis and liberates the product. In theory, a Cu-catalyzed reduction with H_2 should be called "catalytic hydride reduction." However, both H atoms are ultimately originating from H_2 with

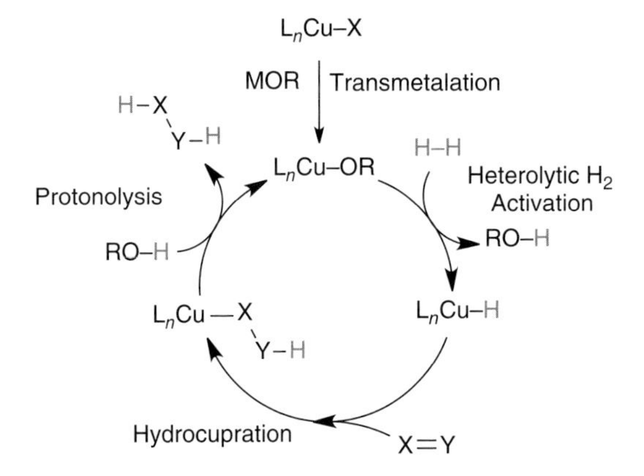

Scheme 4.8 General catalytic cycle for Cu-based hydrogenation.

alcohol acting as a temporary storage of the proton. Consequently, in this chapter, we will use the term "hydrogenation" also in view of the fact that we cover only cases where H_2 is the reducing agent.

4.3.1 Hydrogenation of CO_2 to Formate and Derivatives

In this section, we focus on the hydrogenation of CO_2 to formate. This transformation is indeed the first one where a homogeneous Cu complex acted as a hydrogenation catalyst with a significant activity [126]. In 1970, researchers from the Shell research center in Emeryville tested a range of metal complexes for the conversion of CO_2 in the presence of hydrogen and dimethylamine into DMF. $[(PPh_3)_3CuCl]$ was shown to be active under relatively mild conditions with a turnover number (TON) higher than 900 (Table 4.1, entry 1). D_2 experiments suggested that DMF is formed by the nucleophilic attack of dimethylamine onto a Cu formate.

Although only stoichiometric, the contributions of Sneeden and coworker are worth mentioning because they deal with the well-known hexameric copper hydride $[HCu(PPh_3)]_6$ [127]. These authors showed that this Cu complex reacts with CO_2 in benzene to form the formato copper complex, $[(PPh_3)Cu(OCOH)]$. The reactivity of the latter was studied, and it was concluded to be a "stable end product" rather than an intermediate in the catalytic hydrogenation of CO_2. In a subsequent study [132], the same authors reported the formation of substoichiometric amounts of ethyl formate during the hydrogenation of CO_2 in EtOH/Et_3N by dpm-Cu complexes, $[(CuCl)_2dpm]_2$ and $[Cu_3Cl_2dpm_3]Cl$ (dpm = bis[diphenylphosphino]methane) (Table 4.1, entry 2).

Ikariya and coworkers showed that simple copper (I or II) salts such as acetate, nitrate, and methoxide catalyze the hydrogenation of CO_2 to formate with a TON up to 160 [128]. DBU was initially used as a base and appeared to be significantly better than any other bases tested, including those with similar pK_a. A Cu-DBU complex, $[Cu(DBU)_2I]$, was isolated and demonstrated to have the same activity than CuI. However, no intermediates were identified nor were the exact reasons for the superiority of DBU clearly explained (Table 4.1, entry 3).

Appel and coworkers showed that a cationic Cu-triphos complex (triphos, 1,1,1-tris-(diphenylphosphinomethyl)ethane), $[Cu(triphos)(CH_3CN)]^+$, is active in the hydrogenation of CO_2 with TON up to 500 [129]. Here also, DBU appeared to be the base of choice. A screening of alternative bases demonstrated that there is no correlation between the pK_a of the base and the catalytic activity. Surprisingly, bases with the strongest ability to coordinate with the Cu center led to the more active catalyst. In a follow-up study, the same authors were able to identify a cationic dicopper hydride $[(triphos\text{-}Cu)_2H]^+$ formed upon reaction of $[Cu(triphos)(CH_3CN)]^+$ with H_2 in the presence of a base [133]. Without a base, the reaction of this dicopper hydride with CO_2 was determined to be too slow to be catalytically relevant. Therefore, the authors propose that a coordinating base such as DBU dissociates the dicopper hydride into $[Cu(triphos)(DBU)]^+$ and an unobserved mononuclear hydride, [Cu(triphos)H]. The latter reacts quickly with CO_2 to form Cu(triphos) formate, a species that is insoluble in CH_3CN, the reaction media. Here again, the base plays a critical role and exchanges rapidly

Table 4.1 Cu catalysts for the hydrogenation of CO_2.

Entry (References)	Cu catalyst	Reaction	Reaction conditions	Results
1 [126]	$(PPh_3)_3CuCl$	$H_2 + CO_2 + Me_2NH \rightarrow$ DMF	125 °C, 27 bar CO_2, 27 bar H_2, 17 h, benzene	TON > 900
2 [127, 128]	$[(CuCl)_2dpm]_2$ $[Cu_3Cl_2dpm_3]Cl$	$H_2 + CO_2 + EtOH/Et_3N \rightarrow HCOOEt$	120 °C, 15 bar CO_2, 15 bar H_2, 17 h, $EtOH/Et_3N$ (5 : 1 vol:vol)	Substoichiometric
3 [129]	$Cu(OAc)_2H_2O$	$H_2 + CO_2 + DBU \rightarrow HCOO^-; HDBU^+$	100 °C, 20 bar CO_2, 20 bar H_2, 21 h, 1,4-dioxane	TON = 160
4 [130, 131]	$Cu(triphos)(CH_3CN)^+; PF_6^-$	$H_2 + CO_2 + DBU \rightarrow HCOO^-; HDBU^+$	100–140 °C, 40 bar CO_2, 40 bar H_2, 21 h, CH_3CN	TON = 500

with the formate to form $[Cu(triphos)(DBU)]^+$ as the resting state of the catalytic cycle (Table 4.1, entry 4).

4.3.2 Hydrogenation of Carbonyl Compounds

In the late 1980s, Stryker and coworkers published a series of articles dedicated to the reduction of α,β-unsaturated carbonyl compounds using first stoichiometric amounts and later catalytic amounts of the phosphine-stabilized copper hydride hexamer, $[(Ph_3P)CuH]_6$, that is later known as the Stryker reagent [124, 130, 131, 134]. The authors initially demonstrated that the addition of a stoichiometric amount of the Cu hydride complex to α,β-unsaturated ketones and esters led to the selective reduction of the C—C double bond under mild conditions [124]. Considering that $[(Ph_3P)CuH]_6$ could be obtained in high yield by the hydrogenolysis of $[Cu(OtBu)]_4$ in the presence of PPh_3, they reasoned that the copper(I) enolate formed during the 1,4-conjugate reduction of α,β-unsaturated carbonyl compounds would also react in a similar manner, generating a new hydride and possibly allowing the transformation to be catalytic. Slow conversion of cyclohexenone to cyclohexanone was achieved under 5 bar of H_2 by catalytic amounts of $[(Ph_3P)CuH]_6$ or its precursors, $[Cu(OtBu)]_4/PPh_3$. Remarkably, at 13 bar H_2, the completely reduced product, i.e. cyclohexanol, was obtained. In a follow-up study, the same authors showed that no $[(Ph_3P)CuH]_6$ could be recovered at the end of the reaction and that most of the copper was in the form of a black precipitate [135]. However, addition of an excess PPh_3 maintained the copper in solution, and up to 80% of it could be recovered as $[(Ph_3P)CuH]_6$ at the end of the reaction. Additionally, α,β-unsaturated ketones with substituents on either carbon of the double bond could now be reduced into the corresponding saturated ketone with high yields. Forcing the reaction conditions (higher H_2 pressure, extended reaction time) led to the formation of the fully reduced product, i.e. the saturated alcohol. In the absence of added PPh_3, the Cu species active in the reduction of α,β-unsaturated ketones would reduce to a certain extent nonfunctionalized olefins such as 1-hexene added to the reaction mixture. This was completely suppressed upon addition of PPh_3 (Table 4.2, entry 1).

Ten years later, Stryker and coworkers published two additional papers where they replaced PPh_3 by different phosphine ligands including multidentate phosphines [135, 142]. The new Cu(I) complexes were prepared either by ligand exchange with $[(Ph_3P)CuH]_6$ or by direct synthesis from CuCl and *t*BuONa under H_2 atmosphere. Small variations in the ligand structure appeared to have a drastic effect on the catalyst activity and selectivity. Remarkably, dimethylphenylphosphine (PMe_2Ph)-based catalysts were able to catalytically reduce nonconjugated ketones but also α,β-unsaturated ketones and aldehydes to the corresponding allylic alcohols – i.e. basically, the complementary chemoselectivity than the previously observed. This catalyst was also very chemoselective, reducing ketones while leaving alkenes, alkynes, and dienes untouched (Table 4.2, entry 2).

In 2007, Shimizu et al. reported the copper-catalyzed enantioselective hydrogenation of aryl ketones [143]. A screening of the Cu precursor and the chiral

Table 4.2 Cu catalysts for the hydrogenation of carbonyl compounds.

Entry (References)	Cu catalyst	Reaction (representative example)	Reaction conditions	Results
1 [124, 135]	$[(Ph_3P)CuH]_6$ with or w/o PPh_3, $(CuOtBu)_4/PPh_3$	α,β-Unsaturated ketones to saturated ketones and/or alcohols 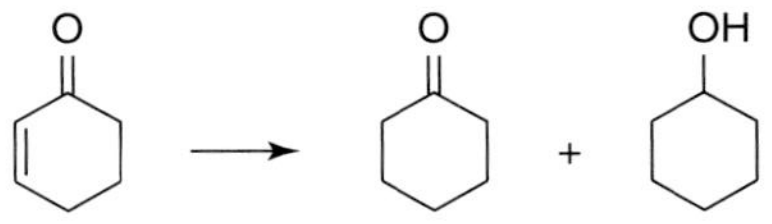	25 °C, 10–70 bar H_2, 1–75 h, benzene or THF	TON = 40–50
2 [136, 137]	$[(Ph_3P)CuH]_6$ with 6 equiv. of PMe_2Ph CuCl, *t*BuONa, PMe_2Ph (1 : 1 : 6)	Saturated ketones to alcohols α,β-Unsaturated ketones or aldehyde to allylic alcohols 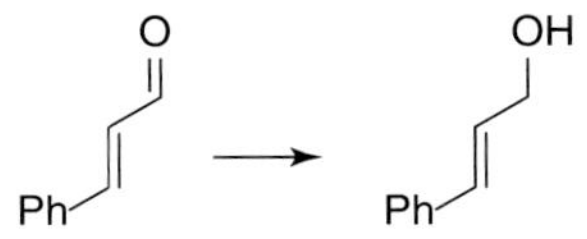	25 °C, 1–70 bar H_2, 10–60 h, benzene or THF *t*BuOH as additive	TON = 40–50
3 [138]	$[Cu(NO_3)(P(3,5\text{-xylyl})_3)_2]$ 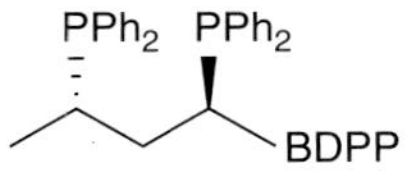 BDPP	Aromatic and heteroaromatic ketones to alcohols, α,β-unsaturated aldehyde to allylic alcohols 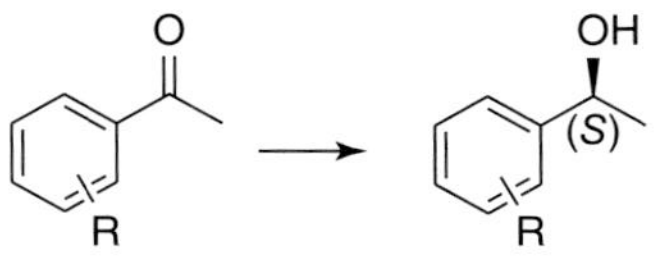	30 °C, 50 bar H_2, 16 h, *i*-PrOH $P(3,5\text{-xylyl})_3$ and *t*-BuONa as additives	TON up to 3000 Ee up to 91%

(Continued)

Table 4.2 (Continued)

Entry (References)	Cu catalyst	Reaction (representative example)	Reaction conditions	Results
4 [139]	$[Cu(NO_3)(PPh_3)_2]$ 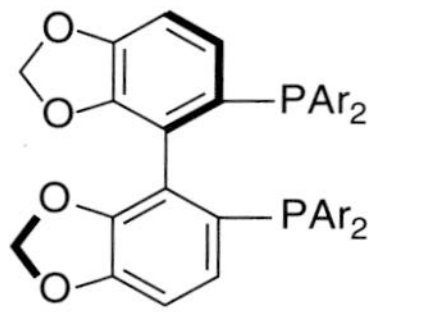 Ar = C_6H_2-3,5-*t*-Bu_2-4-OMe DTBM-SegPhos	α,β-Unsaturated ketones to saturated ketones 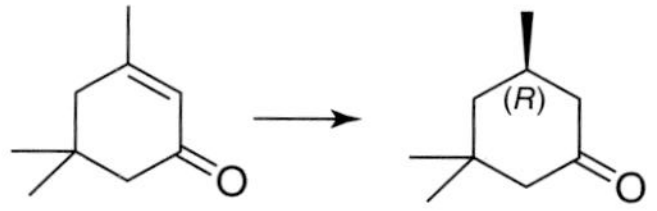	30 °C, 50 bar H_2, 16 h, *i*-PrOH *t*-BuONa as additive	TON up to 90 Selectivity 90% Ee: 96%
5 [140]	$Cu(OAc)_2$ 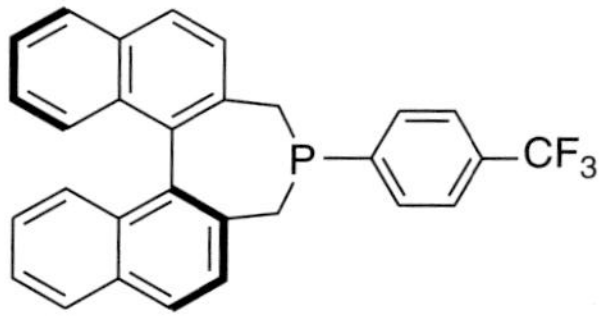	Ketones to alcohols 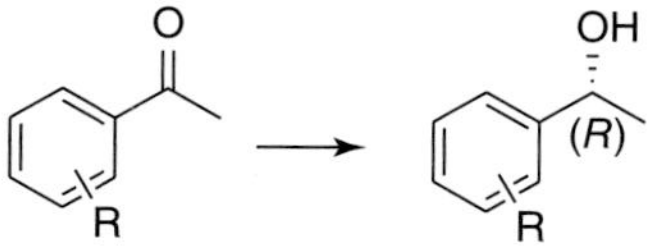	10 °C, 50 bar H_2, 16 h, *i*-PrOH *t*-BuOK as additive	TON up to 300 Ee up to 89%
6 [141]	$Cu(OAc)_2$ 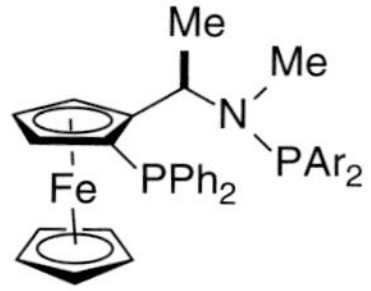 Ar = 3,5-diMe-C_6H_3 Me-Bophoz	Ketones to alcohols 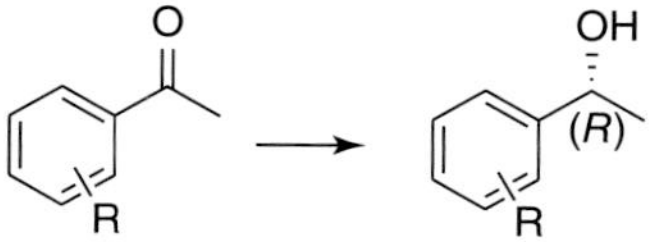	15 °C, 20 bar H_2, 16–45 h, *i*-PrOH P(3,5-xylyl)$_3$ and *t*-BuONa as additives	TON up to 60 Ee up to 96%

ligand led to the identification of $[Cu(NO_3)(P(3,5\text{-xylyl})_3)_2]$/(*S*,*S*)-BDPP (BDPP, 2,4-bis(diphenylphosphino)pentane) as the most active and enantioselective catalyst (Table 4.2, entry 3). Extra $P(3,5\text{-xylyl})_3$ and a strong alkoxide base were used as additives to ensure a good catalytic activity. Modest to high *ee*'s were obtained for a range of substituted acetophenones at a substrate-to-catalyst ratio (S/C) of 500. Two years later, the same group extended the scope of this catalyst to the hydrogenation of aldehydes, heteroaromatic ketones, and α,β-unsaturated ketones and aldehydes [136, 137]. In the case of α,β-unsaturated aldehydes, they obtained excellent chemoselectivities in favor of the allylic alcohols. For α,β-unsaturated ketones, switching to (*R*)-(−)-5,5′-bis[di(3,5-di-*tert*-butyl-4-methoxyphenyl)phosphino]-4,4′-bi-1,3-benzodioxole ((*R*)-DTBM-SEGPHOS) as chiral ligand allowed the enantioselective 1,4-reduction toward the saturated ketone with ee's up to 99% (Table 4.2, entry 4).

In 2011, Beller and coworkers showed that $Cu(OAc)_2$ in combination with monodentate binaphthophosphepine ligands were also efficient catalysts for the asymmetric hydrogenation of ketones (Table 4.2, entry 5) [138]. Here again, a strong base such as an alkoxide or a hydroxide was needed for catalytic activity. The chiral ligand was also shown to exert an important influence on both catalyst activity and enantioselectivity. Although the reactions are run in *i*-PrOH as a solvent, no reaction takes place in the absence of H_2, ruling out the possibility of transfer hydrogenation. A range of aromatic ketones were converted with high yields at a substrate-to-catalyst ratio of 300 with *ee*'s ranging from 59% to 89%.

Two years later, Johnson and coworkers used high-throughput screening to identify new chiral ligands for Cu-catalyzed asymmetric hydrogenation of aryl and heteroaryl ketones [144]. Out of roughly 60 chiral phosphines tested, ligands from the BIPHEP (2,2′-bis[di(aryl)phosphino]-6,6′-dimethoxy-1,1′-biphenyl) and SEGPHOS (5,5′-bis[di(aryl)phosphino]-4,4′-bi-1,3-benzodioxole) families already known in Cu-catalyzed hydrosilylation were the best together with a bidentate phosphine, Me-Bophoz (1-[2-diarylphosphinoferrocenyl](*N*-methyl)(*N*-diphenylphosphino)ethylamine) (Table 4.2, entry 6). After an unsuccessful attempt to increase the enantioselectivity obtained with the parent ligands by condition and additive screening, the authors explored structural modifications of the Bophoz ligands. Up to 14 Bophoz ligands were prepared, resulting in the identification of a new ligand affording a better ee. One equivalent of an ancillary achiral triarylphosphine is required for catalyst activity as well as 15 equiv. of an alkoxide base. A range of aromatic and heteroaromatic ketones were hydrogenated with the optimized catalyst with ee up to 96% – albeit at a rather low substrate-to-catalyst ratio of 66.

4.3.3 Hydrogenation of Olefins and Alkynes

It has been known since the mid-1970s that stoichiometric amounts of copper hydrides are able to mediate the stereoselective reduction of alkynes to cis-alkenes [139–141, 145, 146]. The first copper hydrides were ill-defined species generated *in situ* by the reaction of a Cu(I) halide with a borohydride [145], a Grignard reagent [146], MgH_2 [147], or an aluminum hydride [148]. In 1990, Stryker and coworkers was the first one to report the use of a well-defined

Table 4.3 Cu catalysts for the semihydrogenation of alkynes.

Entry (References)	Cu catalyst	Reaction (representative example)	Reaction conditions	Results
1 [125]	$[(Ph_3P)CuCl]_4$		100 °C, 5 bar H_2, 3 h, Toluene *i*-PrOH and *t*-BuOLi as additives	TON up to 50 Cis-selectivity up to 99%
2 [150]	[MesCu] Ar–N N⁺ HO PF_6^- Ar = 2,4,6-triMe-C_6H_2	R_1—≡—R_2 ⟶ R_1 R_2 R_1, R_2 = alkyl, aryl	100 °C, 100 bar H_2, 18 h, THF *n*-BuLi as additives	TON up to 20 Cis-selectivity up to 99.9%
3 [151]	[(SIMes)CuCl] Ar–N N–Ar Ar = 2,4,6-triMe-C_6H_2 SIMes		100 °C, 1 bar H_2, 12 h, octane/1,4-dioxane (4 : 1) *t*-BuONa as additives	TON up to 20 Cis-selectivity up to 99%

copper hydride $[(Ph_3P)CuH]_6$ for this transformation [149]. In all cases, the reduction was assumed to proceed via hydrocupration generating an alkenyl copper intermediate that had to be hydrolyzed to liberate the desired cis-olefin. Considering these early reports, it is surprising that the first Cu-catalyzed homogeneous semihydrogenation of alkynes was only reported in 2015 by Semba et al. [147] These authors used $[(PPh_3)CuCl]_4$ as a copper precursor in the presence of LiO *t*-Bu as a base and *i*-PrOH as a proton source (Table 4.3, entry 1). The catalytic system was first optimized with 1-phenyl-1-hexyne, showing that a hydrogen pressure as low as 5 bar could be used and that the reaction did not occur in the absence of H_2, therefore ruling out *i*-PrOH as the reducing agent. Internal aliphatic and aromatic alkynes were converted in high yields into their corresponding cis-olefins at a rather low substrate-to-catalyst ratio of 50. No reaction was observed with a terminal alkyne. The same year, Teichert and coworkers showed that the Cu-catalyzed semihydrogenation of a range of internal aryl- and diaryl-substituted alkynes could be accomplished in the absence of protic additives (Table 4.3, entry 2) [148]. Using a *N*-heterocyclic carbene (NHC) ligand with a pending hydroxyl group, they were able to avoid the use of an alcohol as a proton source. Indeed, the release of the desired olefin occurs with intermolecular protonation of the putative vinyl copper intermediate. The reaction carried out with D_2 confirmed that both H atoms are delivered to the alkyne from the hydrogen gas. Although this catalytic system affords high yields of cis-olefin, it requires a high pressure of hydrogen (100 bar), a low substrate-to-catalyst ratio (S/C = 10–20), and a long reaction time (18 hours). Finally, Sawamura and coworkers recently reported a related NHC-base Cu catalyst for the semireduction of internal alkynes [149]. The main advantage of this catalytic system is that it is based on a commercially available NHC, namely, SIMes (Table 4.3, entry 3), and it operates at 1 bar H_2. Remarkably, the unligated Cu precursor, CuCl, is almost as active as the Cu-NHC, whereas the unsaturated NHC, IMes-based catalyst was not active. A strong *t*-butoxide base is allegedly required to generate a Cu hydride via heterolytic cleavage of Cu(*t*-BuO). The nature of the base and the solvent (the best one being a 4 : 1 mixture of octane/1,4-dioxane) is crucial for catalyst activity. High yields of cis-olefins are obtained for a range of internal alkyne at a rather low substrate to a catalyst ratio of 10.

4.4 Conclusions

Few metals other than copper combine broad, synthetically relevant catalytic versatility with industrially relevant features such as security of supply, low cost, acceptable toxicity, scalable catalyst, and waste handling. This chapter deals with two very distinct types of copper catalysis that have undergone significant recent progress, i.e. (aerobic) N—N bond-forming transformations and homogeneously catalyzed hydrogenations. Although the latter is still fairly embryonic, the former has already reached a level of applicability that raised industrial interest, especially in medicinal chemistry for the synthesis of five-membered aromatic

N-heterocycles (pyrazoles and triazoles). The field continues to evolve, and new Cu-catalyzed transformations are being discovered, as illustrated by the first Cu-catalyzed aerobic synthesis of 1,2,4-oxadiazoles by N—O bond formation [125] or the recent series of publications on semihydrogenation of alkynes [150, 151].

References

1 Evano, G. and Blanchard, N. (eds.) (2013). *Copper-Mediated Cross-Coupling Reactions*. Hoboken, NJ: Wiley.
2 Hruszkewycz, D., McCann, S., and Stahl, S.S. (2016). *Liquid Phase Aerobic Oxidation Catalysis: Industrial Applications and Academic Perspectives* (ed. S.S. Stahl and P.L. Alsters), 67–83. Weinheim: Wiley-VCH.
3 OECD (2015). *Material Resources, Productivity and the Environment*, OECD Green Growth Studies, 119–129. Paris: OECD Publishing.
4 Kunz, K. and Lui, N. (2013). *Copper-Mediated Cross-Coupling Reactions* (ed. G. Evano and N. Blanchard), 725–743. Hoboken, NJ: Wiley.
5 European Medicines Agency, Committee for Medicinal Products for Human Use (2008). Guideline on the Specification Limits for Residues of Metal Catalysts or Metal Reagents, Doc. Ref. EMEA/CHMP/SWP/4446/2000, London, 21 February 2008.
6 Cheremisinoff, P.N. (1995). *Handbook of Water and Wastewater Treatment Technology*, 413–448. New York, NY: Marcel Dekker.
7 Woody, C.A. and O'Neal, S.L. (2012). *Effects of Copper on Fish and Aquatic Resources*. The Nature Conservancy.
8 Wu, Q., Zhang, Y., and Cui, S. (2014). *Org. Lett.* 16: 1350.
9 Schirmann, J.-P. and Bourdauducq, P. (2002). Hydrazine. In: *Ullmann's Encyclopedia of Industrial Chemistry*. Weinheim: Wiley-VCH.
10 Kajimoto, T., Takahashi, H., and Tsuji, J. (1982). *Bull. Chem. Soc. Jpn.* 55: 3673.
11 Zhang, Y., Tang, Q., and Luo, M. (2011). *Org. Biomol. Chem.* 9: 4977.
12 Speier, G. and Párkányi, L. (1986). *J. Org. Chem.* 51: 218.
13 Yan, X.-M., Chen, Z.-M., Yang, F., and Huang, Z.-Z. (2011). *Synlett* 4: 569.
14 Love, B.E. and Tsai, L. (1992). *Synth. Commun.* 22: 165.
15 Kauffmann, T., Albrecht, J., Berger, D., and Legler, J. (1967). *Angew. Chem. Int. Ed.* 6: 633; (1967). *Angew. Chem.* 79: 620.
16 Meyer, R. and Pillon, D. (1959). Preparation of Benzophenone-azine. Rhone-Poulenc. US Patent 2,870,206.
17 Nawata, T., Sakaguchi, S., Kohzaki, T. et al. (1988). Process for Producing Benzophenone-azine. Mitsubishi Gas Chemical Company. US Patent 4,751,326.
18 Laouiti, A., Rammah, M.M., Rammah, M.B. et al. (2012). *Org. Lett.* 14: 6.
19 Jautelat, M. and Leidinger, W. (2002). Verfahren zur Herstellung von Hydrazodicarbonamid (HDC) über Ketimine. Bayer AG. EP 1174420.
20 Hayashi, H. (1990). *Catal. Rev.* 32: 229.
21 Sakai, N., Sasaki, M., and Ogiwara, Y. (2015). *Chem. Commun.* 51: 11638.

22 Hunger, K., Mischke, P., and Rieper, W. (2002). Azo dyes, 1. General. In: *Ullmann's Encyclopedia of Industrial Chemistry*. Weinheim: Wiley-VCH.
23 Allen, S.A., Walvoord, R.R., Padilla-Salinas, R., and Kozlowski, M.C. (2013). *Chem. Rev.* 113: 6234.
24 Terent'ev, A.P. and Mogilianskii, Y.D. (1955). *Dokl. Akad. Nauk SSSR* 103: 91.
25 Terent'ev, A.P. and Mogilianskii, Y.D. (1958). *Zh. Obshch. Khim.* 28: 1959.
26 Terent'ev, A.P. and Mogilianskii, Y.D. (1961). *Zh. Obshch. Khim.* 31: 326.
27 Kinoshita, K. (1959). *Bull. Chem. Soc. Jpn.* 32: 777.
28 Kinoshita, K. (1959). *Bull. Chem. Soc. Jpn.* 32: 780.
29 Strueben, J., Gates, P.J., and Staubitz, A. (2014). *J. Org. Chem.* 79: 1719.
30 Köhl, I. and Lüning, U. (2014). *Synthesis* 46: 2376.
31 Hammerich, M. and Herges, R. (2015). *J. Org. Chem.* 80: 11233.
32 Muraoke, T., Kinbara, K., and Aida, T. (2006). *Nature* 440: 512.
33 Ding, Y., Padias, A.B., and Hall, H.K. (1999). *Polym. Bull.* 42: 689.
34 Lu, W. and Xi, C. (2008). *Tetrahedron Lett.* 49: 4011.
35 Badiei, Y.M., Krishnaswamy, A., Melzer, M.M., and Warren, T.H. (2006). *J. Am. Chem. Soc.* 128: 15056.
36 Zhang, C. and Jiao, N. (2010). *Angew. Chem. Int. Ed.* 49: 6174; (2010). *Angew. Chem.* 122: 6310.
37 Chu, C.-C., Chang, Y.-C., Tsai, B.-K. et al. (2014). *Chem. Asian J.* 9: 3390.
38 Lohse, M., Nowosinski, K., Traulsen, N.L. et al. (2015). *Chem. Commun.* 51: 9777.
39 Shoji, T., Kamata, N., Maruyama, A. et al. (2015). *Bull. Chem. Soc. Jpn.* 88: 1338.
40 Xiong, F., Qian, C., Lin, D. et al. (2013). *Org. Lett.* 15: 5444.
41 Karanam, M. and Choudhury, A.R. (2013). *Cryst. Growth Des.* 13: 4803.
42 Qian, C., Lin, D., Deng, Y. et al. (2014). *Org. Biomol. Chem.* 12: 5866.
43 Ryu, T., Min, J., Choi, W. et al. (2014). *Org. Lett.* 16: 2810.
44 Seth, K., Nautiyal, M., Purohit, P. et al. (2015). *Chem. Commun.* 51: 191.
45 Liang, Y.-F., Li, X., Wang, X. et al. (2015). *ACS Catal.* 5: 1956.
46 Geng, X. and Wang, C. (2015). *Org. Biomol. Chem.* 13: 7619.
47 Geng, X. and Wang, C. (2015). *Org. Lett.* 17: 2434.
48 Dong, M., Babalhavaeji, A., Hansen, M. et al. (2015). *Chem. Commun.* 51: 12981.
49 Son, J.-Y., Kim, S., Jeon, W.H., and Lee, P.H. (2015). *Org. Lett.* 17: 2518.
50 Sakano, T., Ohashi, T., Yamanaka, M., and Kobayashi, K. (2015). *Org. Biomol. Chem.* 13: 8359.
51 Xia, C., Wei, Z., Shen, C. et al. (2015). *RSC Adv.* 5: 52588.
52 Zhang, M., Zhang, R., Zhang, A.-Q. et al. (2009). *Synth. Commun.* 39: 3428.
53 Wang, J., He, J., Zhi, C. et al. (2014). *RSC Adv.* 4: 16607.
54 Sasano, Y., Murakami, K., Nishiyama, T. et al. (2013). *Angew. Chem. Int. Ed.* 52: 12624; (2013). *Angew. Chem.* 125: 12856.
55 Sudheendran, K., Schmidt, D., Frey, W. et al. (2014). *Tetrahedron* 70: 1635.
56 Ferreira, V.F., da Rocha, D.R., da Silva, F.C. et al. (2013). *Expert Opin. Ther. Pat.* 23: 319.
57 Schmid, A., Kollmer, A., Sonnleitner, B., and Witholt, B. (1999). *Bioproc. Eng.* 20: 91.

58 Osterberg, P.M., Niemeier, J.K., Welch, C.J. et al. (2015). *Org. Process Res. Dev.* 19: 1537.
59 Gemoets, H.P.L., Hessel, V., and Noël, T. (2016). *Liquid Phase Aerobic Oxidation Catalysis: Industrial Applications and Academic Perspectives* (ed. S.S. Stahl and P.L. Alsters), 399–419. Weinheim: Wiley-VCH.
60 Gavriilidis, A., Constantinou, A., Hellgardt, K. et al. (2016). *React. Chem. Eng.* 1: 595.
61 Hone, C.A., Roberge, D.M., and Kappe, C.O. (2017). *ChemSusChem* 10: 32.
62 Neumann, J.J., Suri, M., and Glorius, F. (2010). *Angew. Chem. Int. Ed.* 49: 7790; (2010). *Angew. Chem.* 122: 7957.
63 Neumann, J., Suri, M., and Glorius, F. (2011). *Pyrazole Synthesis by Coupling of Carboxylic Acid Derivatives and Enamines.* Westfälische Wilhelms-Universität Münster. WO 2011120861.
64 Hwang, J.Y., Kim, H.-Y., Park, D.-S. et al. (2013). *Bioorg. Med. Chem. Lett.* 23: 6467.
65 Suri, M., Jousseaume, T., Neumann, J., and Glorius, F. (2012). *Green Chem.* 14: 2193.
66 Glorius, F. and Suri, M. (2013). *Pyrazole Synthesis by Coupling of Carboxylic Acid Derivatives and Enamines.* Westfälische Wilhelms-Universität Münster. DE 102012100961.
67 Chen, C.-Y., Tang, G., He, F. et al. (2016). *Org. Lett.* 18: 1690.
68 Schmidt, M.P. and Hagenbočker, A. (1921). *Ber. Dtsch. Chem. Ges.* 54B: 2191.
69 Benson, F.R., Hartzel, L.W., and Savell, W.L. (1950). *J. Am. Chem. Soc.* 72: 1816.
70 Sabnis, R.W. and Rangnekar, D.W. (1992). *J. Heterocycl. Chem.* 29: 65.
71 Terpugova, M.P., Amosov, Y.I., and Kotlyarevskii, I.L. (1982). *Izv. Akad. Nauk SSSR. Ser. Khim.* 5: 1166.
72 Richter, E. and Taylor, E.C. (1956). *J. Am. Chem. Soc.* 78: 5828.
73 Hirsch, B. and Ciupe, J. (1963). *Chimia* 17: 159.
74 Schäfer, H., Gewald, K., Bellmann, P., and Gruner, M. (1991). *Monatsh. Chem.* 122: 195.
75 Bel'skaya, N.P., Demina, M.A., Sapognikova, S.G. et al. (2008). *ARKIVOC* 16: 9.
76 Gavlik, K.D., Lesogorova, S.G., Sukhorukova, E.S. et al. (2016). *Eur. J. Org. Chem.* 2700.
77 Gavlik, K.D., Sukhorukova, E.S., Shafran, Y.M. et al. (2017). *Dyes Pigm.* 136: 229.
78 Hirayama, T., Ueda, S., Okada, T. et al. (2014). *Chem. Eur. J.* 20: 4156.
79 Mori, H., Sakamoto, K., Mashito, S., and Matsuoka, Y. (1993). *Chem. Pharm. Bull.* 41: 1944.
80 Li, Y.-J., Li, X., Zhang, S.-X. et al. (2015). *Chem. Commun.* 51: 11564.
81 Hagen, T.J. and Zhang, Z. (2013). *Heterocyclic Chemistry in Drug Discovery* (ed. J.J. Li), 373–395. Hoboken, NJ: Wiley.
82 Zhang, J.-P., Lin, Y.-Y., Huang, X.-C., and Chen, X.-M. (2005). *J. Am. Chem. Soc.* 127: 5495.
83 Ueda, S. and Nagasawa, H. (2009). *J. Am. Chem. Soc.* 131: 15080.

84 Zhang, L., Tang, D., Gao, J. et al. (2016). *Synthesis* 48: 3924.
85 Yin, W., Wang, C., and Huang, Y. (2013). *Org. Lett.* 15: 1850.
86 Flohr, A., Groebke Zbinden, K., and Koerner, M. (2013). *Triazolopyridine Compounds as PDE10A Inhibitors*. F. Hoffmann-La Roche AG. WO 2013041472.
87 Bartels, B., Fantasia, S.M., Flohr, A. et al. (2013). *Process for the Preparation of 2-Phenyl-[1,2,4]triazolo[1,5-A]pyridine Derivatives*. F. Hoffmann-La Roche AG. WO 2013117610.
88 Meng, X., Yu, C., and Zhao, P. (2014). *RSC Adv.* 4: 8612.
89 Wang, F., You, Q., Wu, C. et al. (2015). *RSC Adv.* 5: 78422.
90 Xu, H., Jiang, Y., and Fu, H. (2013). *Synlett* 24: 125.
91 Bartels, B., Bolas, C.G., Cueni, P. et al. (2015). *J. Org. Chem.* 80: 1249.
92 Huang, H., Guo, W., Wu, W. et al. (2015). *Org. Lett.* 17: 2894.
93 Hann, R.M. and Hudson, C.S. (1944). *J. Am. Chem. Soc.* 66: 735.
94 Henseke, G. and Müller, G. (1962). *J. Prakt. Chem.* 4: 47.
95 Riebsomer, J.L. (1948). *J. Org. Chem.* 13: 815.
96 El Khadem, H. (1963). *Adv. Carbohydr. Chem.* 18: 99.
97 Simon, H. and Kraus, A. (1970). *Fortschr. Chem. Forsch.* 14: 430.
98 El Khadem, H.S. and Fatiadi, A.J. (2000). *Adv. Carbohydr. Chem. Biochem.* 55: 175.
99 El Ashry, E.S.H. and Rashed, N. (2000). *Curr. Org. Chem.* 4: 609.
100 El Khadem, H.S. (1998). *Carbohydr. Res.* 313: 255.
101 George, M.V., Mitra, A., and Sukumaran, K.B. (1980). *Angew. Chem. Int. Ed.* 19: 973; (1980). *Angew. Chem.* 92: 1005.
102 Xu, H., Ma, S., Xu, Y. et al. (2015). *J. Org. Chem.* 80: 1789.
103 Tang, X., Huang, L., Yang, J. et al. (2014). *Chem. Commun.* 50: 14793.
104 Wang, S., Li, Y., Bi, X., and Liu, Q. (2015). *Synlett* 26: 1895.
105 Huang, H., Cai, J., Ji, X. et al. (2016). *Angew. Chem. Int. Ed.* 55: 307; (2016). *Angew. Chem.* 128: 315.
106 Wu, L., Guo, S., Wang, X. et al. (2015). *Tetrahedron Lett.* 56: 2145.
107 Guru, M.M. and Punniyamurthy, T. (2012). *J. Org. Chem.* 77: 5063.
108 Hu, J.-R., Zhang, W.-J., and Zheng, D.-G. (2013). *Tetrahedron* 69: 9865.
109 Chen, Z., Yan, Q., Liu, Z. et al. (2013). *Angew. Chem. Int. Ed.* 52: 13324; (2013). *Angew. Chem.* 125: 13566.
110 Herman, R.G. (1991). *Studies in Surface Science and Catalysis*, vol. 64 (ed. L. Guczi), 265–349. Elsevier.
111 Palo, D.R., Dagle, R.A., and Holladay, J.D. (2007). *Chem. Rev.* 107: 3992.
112 Waugh Catalysis, K.C. (1992). *Today* 15: 51.
113 Waugh, K.C. (2012). *Catal. Lett.* 142: 1153.
114 Pritchard, J., Filonenko, G.A., van Putten, R. et al. (2015). *Chem. Soc. Rev.* 44: 3808.
115 Prasad, R. and Singh, P. (2011). *Bull. Chem. React. Eng. Catal.* 6 (2): 63.
116 Adkins, H. and Connor, R. (1931). *J. Am. Chem. Soc.* 53: 1091.
117 Noweck, K. and Grafahrend, W. (2006). Fatty alcohols. In: *Ullmann's Encyclopedia of Industrial Chemistry*. Weinheim: Wiley-VCH.
118 Calvin, M. (1938). *Trans. Faraday Soc.* 34: 1181.
119 Jordan, A.J., Lalic, G., and Sadighi, J.P. (2016). *Chem. Rev.* 116: 8318.

120 Dhayal, R.S., van Zyl, W.E., and Liu, C.W. (2016). *Acc. Chem. Res.* 49: 86.
121 Deutsch, C., Krause, N., and Lipshutz, B.H. (2008). *Chem. Rev.* 108: 2916.
122 Díez-González, S. and Nolan, S.P. (2008). *Acc. Chem. Res.* 41 (2): 349.
123 Rendler, S. and Oestreich, M. (2007). *Angew. Chem. Int. Ed.* 46: 498; (2007). *Angew. Chem.* 119: 504.
124 Koenig, T.M., Daeuble, J.F., Brestensky, D.M., and Stryker, J.M. (1990). *Tetrahedron Lett.* 31: 3237.
125 Kuram, M.R., Kim, W.G., Myung, K., and Hong, S.Y. (2016). *Eur. J. Org. Chem.* 438.
126 Haynes, P., Slaugh, L.H., and Kohnle, J.F. (1970). *Tetrahedron Lett.* 11: 365.
127 Denise, B. and Sneeden, R.P.A. (1981). *J. Organomet. Chem.* 221: 111.
128 Watari, R., Kayaki, Y., Hirano, S.-I. et al. (2015). *Adv. Synth. Catal.* 357: 1369.
129 Zall, C.M., Linehan, J.C., and Appel, A.M. (2015). *ACS Catal.* 5 (9): 5301.
130 Mahoney, W.S., Brestensky, D.M., and Stryker, J.M. (1988). *J. Am. Chem. Soc.* 110: 291.
131 Brestensky, D.M. and Stryker, J.M. (1989). *Tetrahedron Lett.* 30: 5677.
132 Beguin, B., Denise, B., and Sneeden, R.P.A. (1981). *J. Organomet. Chem.* 208: C18.
133 Zall, C.M., Linehan, J.C., and Appel, A.M. (2016). *J. Am. Chem. Soc.* 138 (31): 9968.
134 Mahoney, W.S. and Stryker, J.M. (1989). *J. Am. Chem. Soc.* 111: 8818.
135 Chen, J.-X., Daeuble, J.F., Brestensky, D.M., and Stryker, J.M. (2000). *Tetrahedron* 56: 2153.
136 Shimizu, H., Sayo, N., and Saito, T. (2009). *Synlett* 1295.
137 Shimizu, H., Nagano, T., Sayo, N. et al. (2009). *Synlett* 3143.
138 Junge, K., Wendt, B., Addis, D. et al. (2011). *Chem. Eur. J.* 17: 101.
139 Yoshida, T. and Negishi, E. (1974). *J. Chem. Soc., Chem. Commun.* 762.
140 Crandall, J.K. and Collonges, F. (1976). *J. Org. Chem.* 41: 4089.
141 Ashby, E.C., Lin, J.J., and Goel, A.B. (1978). *J. Org. Chem.* 43: 757.
142 Chen, J.-X., Daeuble, J.F., and Stryker, J.M. (2000). *Tetrahedron* 56: 2789.
143 Shimizu, H., Igarashi, D., Kuriyama, W. et al. (2007). *Org. Lett.* 9: 1655.
144 Krabbe, S.W., Hatcher, M.A., Bowman, R.K. et al. (2013). *Org. Lett.* 15: 4560.
145 Masure, D., Coutrot, P., and Normant, J.F. (1982). *J. Organomet. Chem.* 226: C55.
146 Daeuble, J.F., McGettigan, C., and Stryker, J.M. (1990). *Tetrahedron Lett.* 31: 2397.
147 Semba, K., Kameyama, R., and Nakao, Y. (2015). *Synlett* 26: 318.
148 Pape, F., Thiel, N.O., and Teichert, J.F. (2015). *Chem. Eur. J.* 21: 15934.
149 Wakamatsu, T., Nagao, K., Ohmiya, H., and Sawamura, M. (2016). *Organometallics* 35: 1354.
150 Salnikov, O.G., Liu, H.-J., Fedorov, A. et al. (2017). *Chem. Sci.* 8: 2426.
151 Kaeffer, N., Liu, H.-J., H-K Lo, A., and Fedorov, C.C. (2018). *Chem. Sci.* 9: 5366.

5

C=C Hydrogenations with Iron Group Metal Catalysts

Tim N. Gieshoff and Axel Jacobi von Wangelin

University of Regensburg, Institute of Organic Chemistry, Department of Chemistry, Universitätsstraße 31, 93040 Regensburg, Germany
University of Hamburg, Department of Chemistry, Martin Luther King Pl 6, 20146 Hamburg, Germany

5.1 Introduction

Metal-catalyzed hydrogenations of C=C bonds are key operations in many organic synthesis endeavors and technical manufacture of chemicals. Heterogeneous catalysts dominate industrial hydrogenation processes with numerous examples in all areas of applications such as the petrochemical valorization of alkene- and arene-cracking products or the large-scale hydrogenation of vegetable oils [1]. Molecular catalysts in a homogeneous phase are often employed where the desired reaction requires especially high selectivity, e.g. enantioselectivity, which can be induced by a rational ligand design and rationalized through a deeper mechanistic understanding. Important examples of technical C=C hydrogenations are the synthesis of the anti-Parkinson drug levodopa, the anti-inflammatory drug naproxen, or the flavor citronellol before its conversion to (−)-menthol (Figure 5.1) [2].

Various molecular sources of hydrogen atoms, as well as metal-centered hydrogen activation and delivery mechanisms, are known in the literature [3]. Gaseous dihydrogen, H_2, is the most widely available, cleanest, and most versatile source of hydrogen, especially on larger scales [1a]. This chapter focuses on C=C hydrogenations with gaseous dihydrogen in the presence of molecular iron group metal catalysts.

Most developments of active hydrogenation catalysts in the homogeneous phase involve the increasingly rare noble metals such as rhodium, iridium, ruthenium, palladium, and platinum [4]. Applications of these catalysts to various syntheses of organic molecules have documented their high activities, high selectivities, wide substrate scopes, and high functional group compatibilities [4]. Furthermore, the mode of action of such processes is rather well understood because of the advent of modern spectroscopic and theoretical tools. However, modern economic and environmental constraints have recently prompted the search for alternative metal catalysts. The high abundance and accessibility, low costs, and low toxicities make iron group metals (iron, cobalt, and nickel)

Non-Noble Metal Catalysis: Molecular Approaches and Reactions,
First Edition. Edited by Robertus J. M. Klein Gebbink and Marc-Etienne Moret.

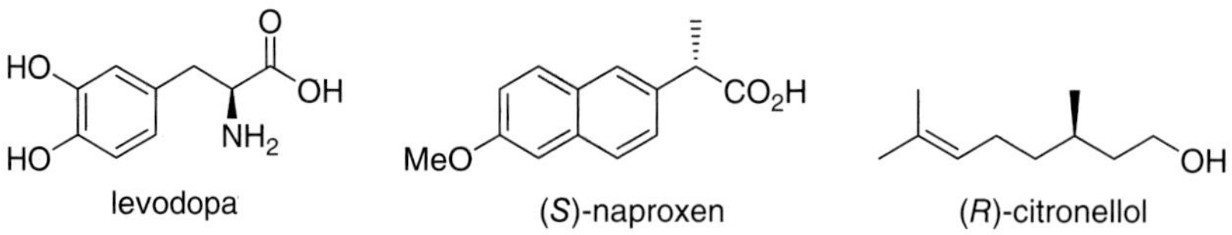

levodopa (S)-naproxen (R)-citronellol

Figure 5.1 Technical products from homogeneous C=C hydrogenations.

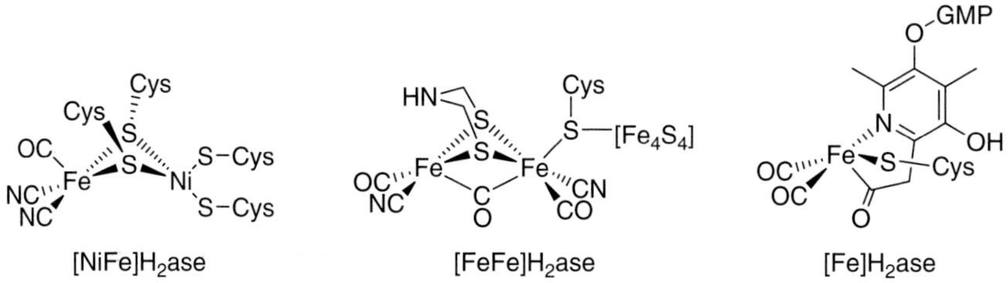

[NiFe]H_2ase [FeFe]H_2ase [Fe]H_2ase

Figure 5.2 Active sites of various hydrogenases (Cys, cysteine; GMP, guanosyl-5′-monophosphate).

an especially attractive class of hydrogenation catalysts, which have received only very little attention in the past decades [5]. Further stimulus to study such catalysts comes from the facts that many of the largest technical hydrogenations (Haber–Bosch process, Fischer–Tropsch process, or gas-to-liquid [GTL] process; plant oil hydrogenation to margarine; and adiponitrile reduction) and many biological hydrogenations are catalyzed by iron group metals. Figure 5.2 illustrates the active sites of natural hydrogenase enzymes that reversibly oxidize dihydrogen [6].

The field of hydrogenations catalyzed by base metals has rapidly developed in the past decade with many new molecular catalysts reported in the literature (Figure 5.3) [7]. Especially, C=C hydrogenation methods with tridentate pincer

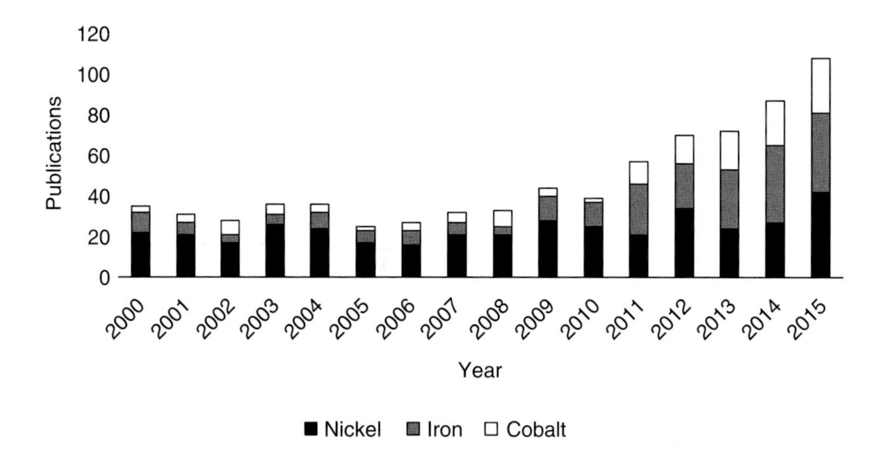

Figure 5.3 Publications per year for the search terms "nickel," "cobalt," "iron," and "hydrogenation" [7]. Source: Courtesy of www.webofknowledge.com.

complexes have been highly successful. The following sections provide an overview of the most important developments in the field.

5.2 Iron

5.2.1 Introduction

Iron-catalyzed hydrogenations of alkenes and alkynes have been known for many decades. Heterogeneous iron catalysts such as Raney®-iron or Urushibara-iron were reported to partially hydrogenate alkynes at high temperatures [8]. Ziegler-type hydrogenation catalysts were developed immediately following the observation of the famous "nickel effect" in Ziegler–Natta polymerizations in the 1960s. The Ziegler-type iron catalysts based on the reaction of an iron(III) salt with a triorganoaluminum compound (AlR_3) were successfully applied to hydrogenations of largely unfunctionalized alkenes at ambient hydrogen pressures and temperatures [9]. Various theories were postulated to describe the true nature of this type of bimetallic catalyst, but generalization is difficult because of the vast number of different catalyst compositions, conditions of preparations, and observed catalytic activities [10].

In the past decade, an increasing amount of conceptually similar bimetallic catalysts formed from simple iron salt precursors and simple main group metal reductants have been reported to be active in hydrogenations of various alkenes and alkynes. These protocols mainly aimed at the *in situ* preparation of active catalyst mixtures that operate in the absence of a complex or expensive ligand. Common reducing agents include Grignard reagents and group 13 hydrides (Scheme 5.1) [11]. In many cases, spectroscopic and kinetic studies were indicative of the formation of iron nanoparticles that act as the active catalysts. Good activities were mostly observed for the hydrogenation of alkenes and alkynes at ambient conditions. However, the presence of a fairly basic or nucleophilic reductant limits the application of such *in situ*-prepared catalysts to substrates that are void of acidic and highly electrophilic substituents.

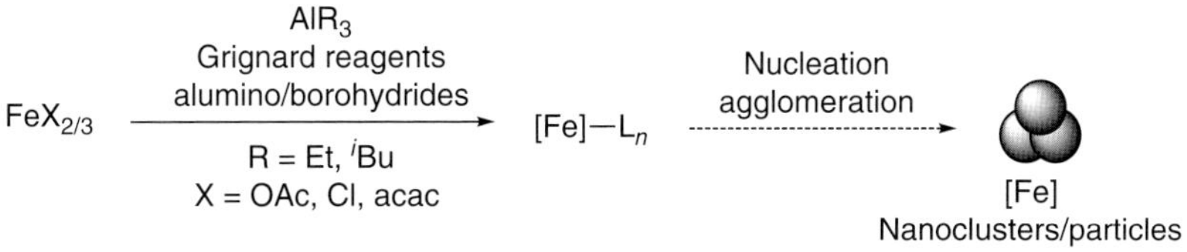

$FeX_{2/3}$ —[AlR_3; Grignard reagents; alumino/borohydrides; R = Et, ^{i}Bu; X = OAc, Cl, acac]→ $[Fe]{-}L_n$ ---[Nucleation agglomeration]--→ [Fe] Nanoclusters/particles

Scheme 5.1 Generation of Ziegler-type hydrogenation catalysts and similar bimetallic catalysts.

Metal nanoparticle catalysis is a hybrid concept that combines the best of two worlds: the higher stability and facile downstream separation of heterogeneous catalysts and the high dispersion, high modularity, high activity, and easier mechanistic investigations of homogeneous catalysts [12]. Only very few applications of well-defined iron nanoparticles to the hydrogenation of alkenes and alkynes were reported. Monodisperse iron nanoparticles of 1.5 ± 0.2 nm were synthesized by decomposition of $\{Fe(N[Si(CH_3)_3]_2)_2\}_2$ at 3 bar H_2 and 150 °C and fully characterized. At an elevated pressure of H_2 (10 bar), mono- and disubstituted alkenes and alkynes were hydrogenated in excellent yields [13]. Highly selective amine-linked, polystyrene-supported iron nanoparticles were synthesized by thermal decomposition of $Fe(CO)_5$ to give an active catalyst for hydrogenations operated in flow reactors [14]. Iron nanoparticles supported on hydroxyl-functionalized graphene were synthesized and applied to alkene hydrogenations [15].

The early use of molecular iron catalysts in hydrogenations is mainly associated with iron carbonyl derivatives that were intensively studied in the 1960s. Iron pentacarbonyl was reported to convert methyl linoleate to methyl stearate under high temperature conditions [16]. Later, UV irradiation was shown to enhance the catalyst activity, most likely through the more facile dissociation of CO ligands to give the active low-valent species $Fe(CO)_3$ [17]. It is important to note that under thermal and UV treatment, iron carbonyls can form not only several homogeneous species but also iron nanoparticles; thus, an unambiguous distinction between homogeneous and heterogeneous mechanisms is difficult [14, 15]. Several hydrogenation protocols employing well-defined molecular iron catalysts were developed in the late decades of the twentieth century, but satisfyingly high catalyst stabilities and reactivities were only observed with the introduction of *P*- and *N*-based pincer ligands.

5.2.2 Pincer Complexes

Peters and coworker reported the synthesis of a series of cationic alkyliron(II) *P,P,P*-pincer complexes synthesized from the reaction of an iron(II)chloride pincer complex with the corresponding alkyl lithium or Grignard reagents (Scheme 5.2) [18]. These first-generation catalysts show moderate activities in the hydrogenation of largely nonfunctionalized substrates such as styrene, 1-hexene, ethylene, cyclooctene, and 2-pentyne. Catalyst **2** is slightly more active, yet the turnover frequencies are between 1.6 and 24 h^{-1}. Competitive oligomerizations and polymerizations are observed in the hydrogenation of terminal alkynes. Although catalyst activity and substrate scope are rather limited, the underlying reaction mechanism was thoroughly studied. Several plausible intermediates of a catalytic hydrogenation cycle could be isolated upon trapping with phosphine ligands. A trihydridoiron(IV) complex **4** is formed upon oxidative addition of dihydrogen. The reversibility of H_2 activation was proven by conversion of **4** to the monohydridoiron(II) complex **5** in the absence of H_2 (Scheme 5.3). Based on these results, the authors proposed the key role of an iron(II/IV) redox process in which an olefin enters in the iron hydride bond of a monohydridoiron(II) species. Addition of hydrogen via formal oxidative addition or σ-bond metathesis is followed by reductive elimination with the release of alkane and reformation of the monohydridoiron(II) intermediate. The determination of reaction orders in the substrate, iron catalyst, and H_2 suggests that the oxidative addition of H_2 at the alkyliron species is rate determining.

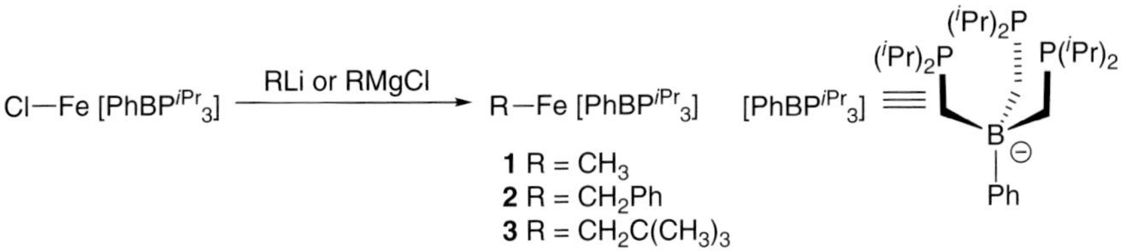

Scheme 5.2 Synthesis of iron(II)-alkyl *P,P,P* pincer complexes.

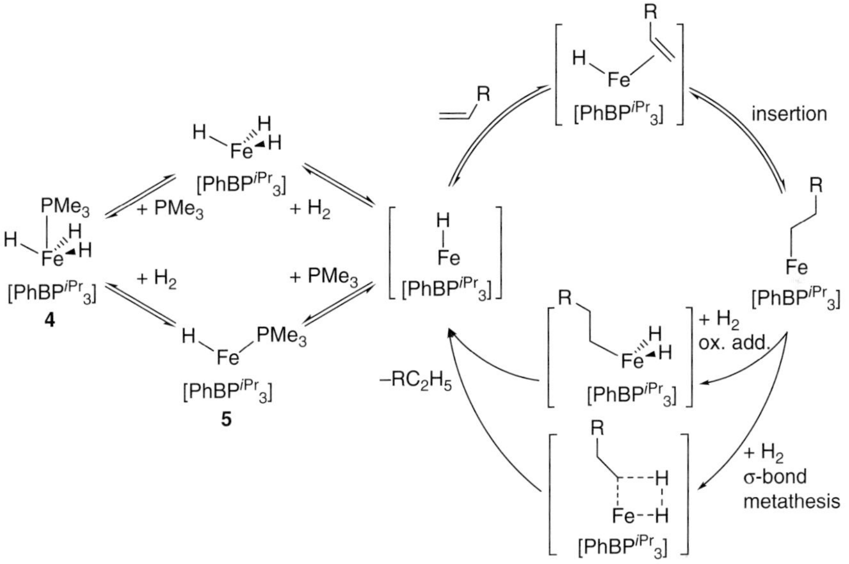

Scheme 5.3 Mechanism of hydrogenation proposed by Peters et al. [18].

Budzelaar and coworkers introduced bis(imino)pyridine iron(II) complexes to the field of alkene hydrogenations. These complexes were activated according to a Ziegler protocol with tri(isobutyl)aluminum. Excellent activities in the hydrogenation of 1-octene were observed [19]. Shortly after, Chirik and coworkers prepared the bis(imino)pyridine *N,N,N*-pincer iron complex **7** by the reduction of corresponding iron(II) halide complexes **6** with sodium amalgam or sodium triethylborohydride, which affects ligand reduction rather than iron center reduction (Scheme 5.4) [20]. Catalyst 7 contains a dianionic biradical form of the ligand, which coordinates the iron(II) center as supported by Mössbauer spectroscopy and computational studies (Scheme 5.4) [21]. The active catalyst is generated upon dissociation of the labile dinitrogen ligands, which gives a tricoordinated iron complex similar to $Fe(CO)_3$. The coordination of an olefin is followed by oxidative addition of dihydrogen. Reductive elimination gives the desired hydrogenation product and regenerates the active catalyst (Scheme 5.5).

Catalyst 7 exhibits excellent activities in the hydrogenation of a diverse set of alkenes and exceeds the productivity (turnover frequency, TOF) of some common precious metal catalysts (Table 5.1) [20]. Largely nonfunctionalized

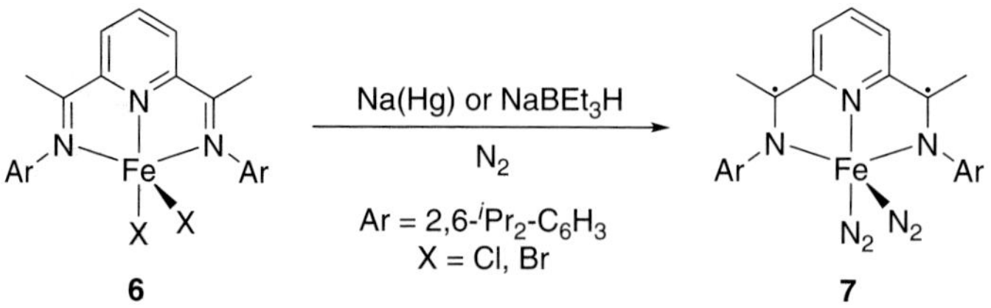

Scheme 5.4 Synthesis of bis(imino)pyridine iron complex **7**.

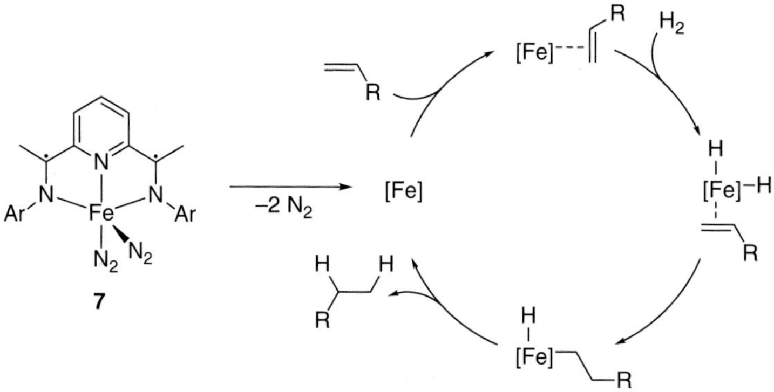

Scheme 5.5 Mechanistic proposal of alkene hydrogenation with **7** according to Chirik et al. Source: Bart et al. 2004 [20]. Reproduced with permission of ACS.

Table 5.1 Comparison of **7** with various precious metal catalysts in the hydrogenation of 1-hexene.

Entry	Catalyst	Time (min)	TOF (h^{-1})
1	7	12	1814
2	10% Pd/C	12	366
3	$(PPh_3)_3RhCl$	12	10
4	$[(cod)Ir(PCy_3)py]PF_6$	12	75

mono- and disubstituted alkenes, styrenes, and oxygen- and nitrogen-containing alkenes are cleanly hydrogenated under mild conditions. Nonproductive carbonyl and primary amine coordination to the catalyst compete with the olefin coordination; therefore, hydrogenation rates decrease in the presence of such functional groups [22].

The high activity of the complex prompted the synthesis of a small library of similar complexes by the same group (Figure 5.4). By decreasing steric bulk of the *N*-aryl substituents (**8**) and the introduction of an electron-donating group in the para-position of the pyridine (**9**), the catalytic activities in the hydrogenation of ethyl-3,3-dimethylacrylate are greatly improved [23]. Largely nonfunctionalized and sterically hindered tri- and tetrasubstituted alkenes can be cleanly converted. Substitution of the imine moieties by strongly σ-donating *N*-heterocyclic

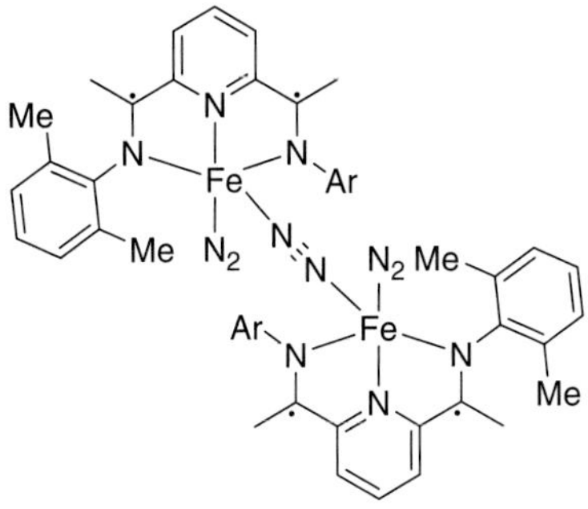

Figure 5.4 Modified bis(imino)pyridine and bis(NHC)pyridine iron complexes by Chirik et al. Source: Bart et al. 2004 [20]. Reproduced with permission of ACS.

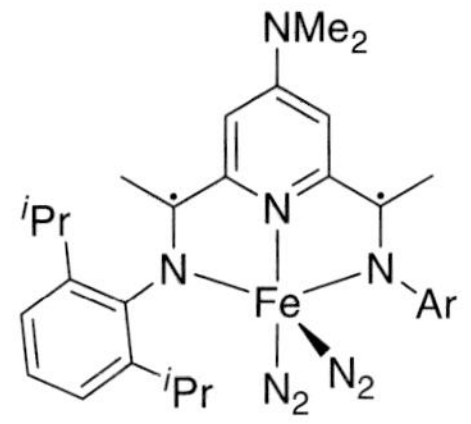

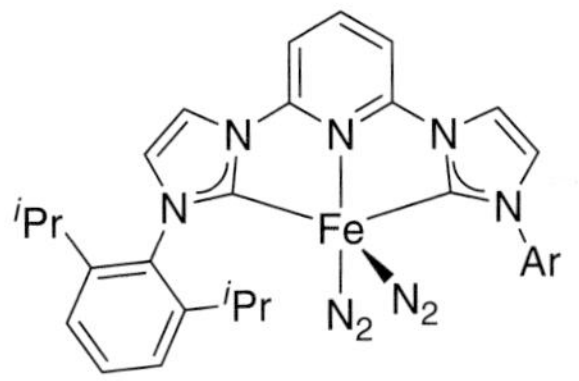

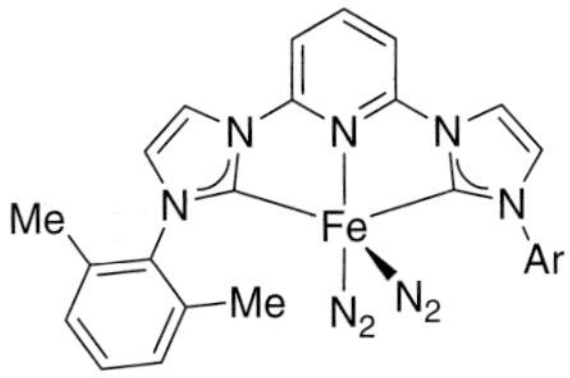

carbenes (NHCs) further increases the electron density on the metal [24]. The resultant *C,N,C*-pincer ligands show only little redox activity so that the corresponding bis(dinitrogen) iron complexes **10** and **11** contain iron(0) centers, which is supported by Mössbauer spectroscopy, X-ray absorption spectroscopy, and density functional theory (DFT) calculations [25]. The latter complex (**11**) is a competent catalyst for the hydrogenation of 2,3-dimethylindene (Table 5.2) and thus represents one of the most active molecular iron catalysts reported at that time [23b, 26].

Table 5.2 Comparison of complexes **7–11** in the hydrogenation of sterically hindered substrates.

		% Conversion (reaction time) with catalyst				
Entry	Substrate	7	8	9	10	11
1	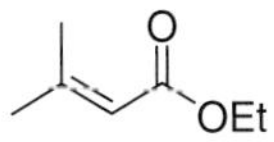	65 (24 h)	>95 (7 h)	>95 (1 h)	>95 (1 h)	35 (1 h)
2	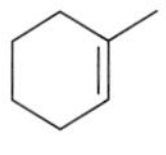	0 (24 h)	2 (24 h)	3 (24 h)	20 (24 h)	>95 (12 h)
3	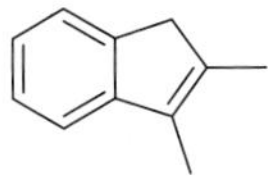	3 (48 h)	<1 (48 h)	3 (48 h)	4 (48 h)	68 (48 h)

In 2013, Milstein and coworkers developed the new acridine-based *P,N,P* pincer iron complex **12**, which was accomplished by reaction of iron(II) bromide with the bis(phosphine) ligand and subsequent reduction with sodium borohydride in acetonitrile (Scheme 5.6) [27]. The resultant complex can be viewed as the nitrogen analog of a Xantphos iron complex. Complex **12** was employed in the semihydrogenations of alkynes to alkenes, which generally bear the challenge of chemoselectivity and stereoselectivity. (*E*)-Alkenes were selectively formed at an elevated H_2 pressure and elevated temperature (Table 5.3). The reaction conditions tolerate nitriles, ketones, and esters. Over-reduction to alkanes was observed only in few examples.

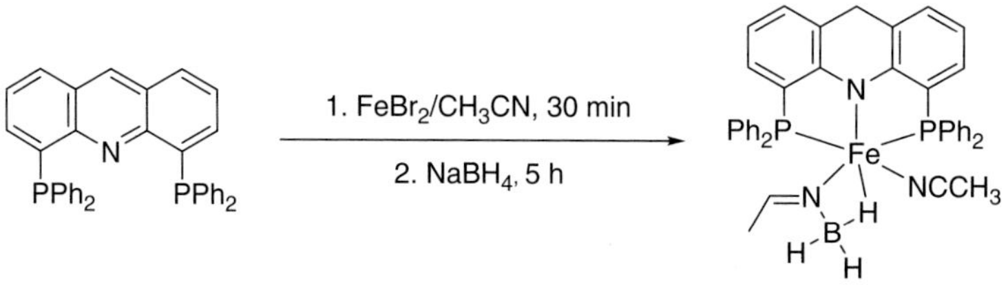

1. $FeBr_2/CH_3CN$, 30 min
2. $NaBH_4$, 5 h

Scheme 5.6 Synthesis of *P,N,P* pincer iron complex **12** by Milstein et al. Source: Srimani et al. 2013 [27]. Reproduced with permission of John Wiley & Sons.

The authors rationalize the observed high (*E*) selectivity with an isomerization of the initially formed (*Z*)-alkene under reaction conditions. Experiments in the absence of dihydrogen proved that **12** quantitatively converts (*Z*)-stilbene to (*E*)-stilbene. The initial formation of (*Z*)-stilbene under hydrogenation conditions was observed at lower temperatures.

Table 5.3 Selected examples of the (*E*)-selective alkyne hydrogenation with **12**.

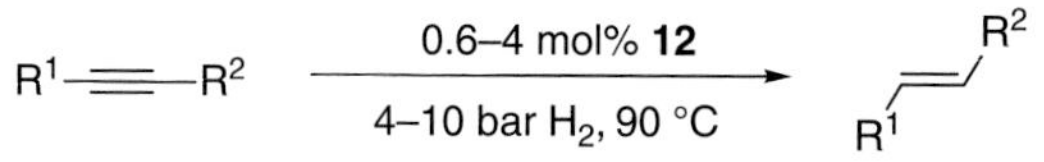

R^1–≡–R^2 → (0.6–4 mol% **12**; 4–10 bar H_2, 90 °C) → (*E*)-R^1CH=CHR^2

Entry	Substrate	% Yield (alkane)	*E:Z*
1	Ph–≡–Ph	99	100 : 0
2	Ph–≡	99	—
3	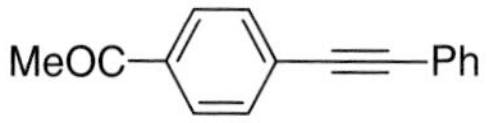 MeOC–C_6H_4–≡–Ph	99	64 : 36
4	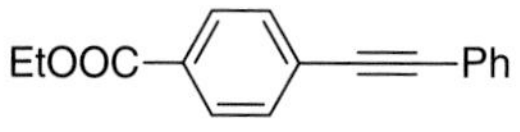 EtOOC–C_6H_4–≡–Ph	89 (11)	99 : 1
5	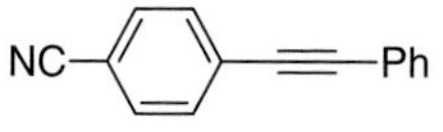 NC–C_6H_4–≡–Ph	94	99 : 1
6	Me_3Si–≡–Ph	76 (24)	99 : 1

Table 5.4 Hydrogenation of various *N*-heterocycles with **14**.

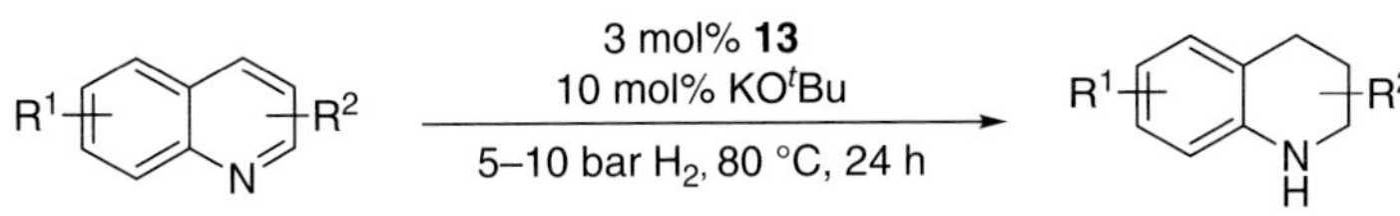

Reaction: R^1/R^2-substituted N-heterocycle → (3 mol% **13**, 10 mol% KO^{t}Bu; 5–10 bar H_2, 80 °C, 24 h) → R^1/R^2-substituted tetrahydro N-heterocycle (N–H)

Entry	Substrate	% Yield
1	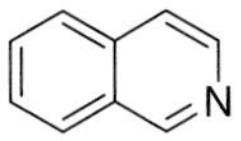 isoquinoline	66
2	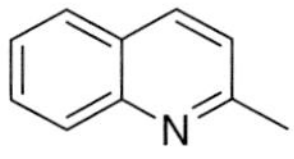 2-methylquinoline	92
3	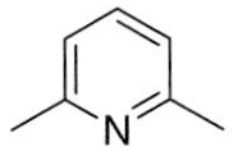 2,6-dimethylpyridine	60

The *P,N,P*-pincer iron complex **13** with a saturated backbone was shown to be active in the dehydrogenation of methanol and the hydrogenation of unactivated esters. The mechanism operates under basic conditions via reversible addition of H_2 to the bifunctional Fe–N moiety [28]. The deprotonated complex **14** was also applied to hydrogenations and dehydrogenations of *N*-heterocycles (Table 5.4) and hydrogenations of styrenes at mild conditions (Table 5.5) by Jones and coworkers [29]. Isolation of dihydridoiron(II) complex **16** was achieved upon borane trapping, which indicates the key role of **15** as a catalytic intermediate of this hydrogenation mechanism (Scheme 5.7). Consistently, N-methylation of the ligand results in complete inhibition of hydrogenation activity. DFT calculations support the notion of a step-wise mechanism with initial Fe-centered hydride transfer to the alkene followed by proton transfer from the amine. Such a

Table 5.5 Hydrogenation of various styrenes with **14**.

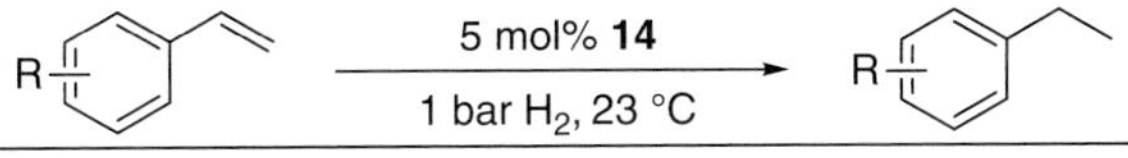

Reaction: R-substituted styrene → (5 mol% **14**; 1 bar H_2, 23 °C) → R-substituted ethylbenzene

Entry	Substrate	Time (h)	% Yield
1	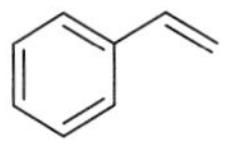 styrene	24	100
2	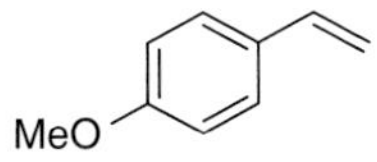 MeO-substituted styrene	168	100
3	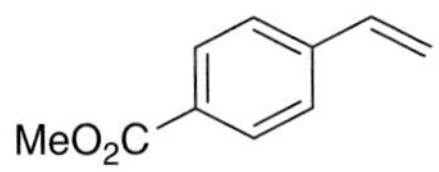 MeO_2C-substituted styrene	0.7	100

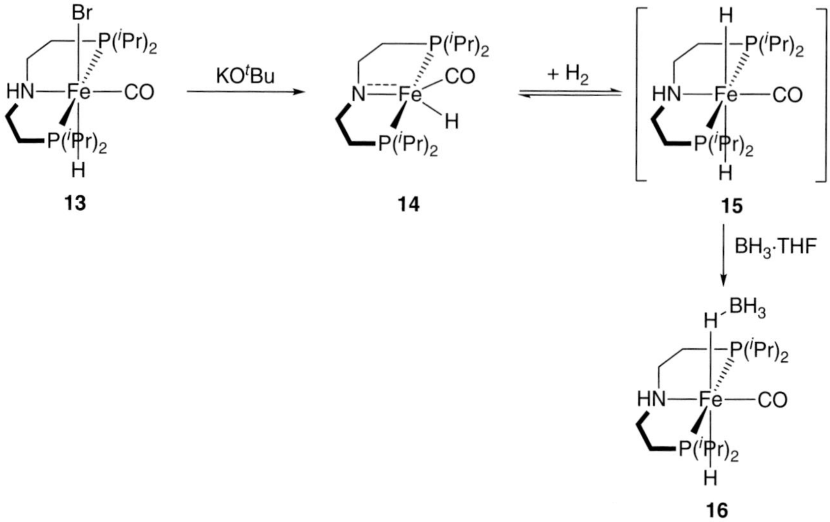

Scheme 5.7 Aliphatic *P,N,P*-complex **13** and possible key intermediates in hydrogenation.

mechanistic scenario should facilitate hydrogenations of polar double-bond systems. Indeed, experiments with catalyst **14** demonstrate its high activity in the hydrogenation of functionalized electron-deficient styrenes and inertness toward 1-hexene.

5.2.3 Others

Arene ferrates were studied in catalytic alkene hydrogenations by von Wangelin and coworkers. Monoanionic bis(anthracene) ferrate (**17**) shows good activity for styrenes. However, the cobaltate derivative is much more active and tolerates various functional groups [30]. Such homoleptic arene complexes were first prepared by Ellis and coworkers by the reduction of metal(II) bromides with potassium/anthracene and can be viewed as homogeneous sources of metal species in the oxidation state –I (Scheme 5.8) [31]. Mechanistic studies were mostly performed with cobaltate, which undergoes rapid π-ligand exchange with various alkenes as an initiating step under hydrogenation conditions.

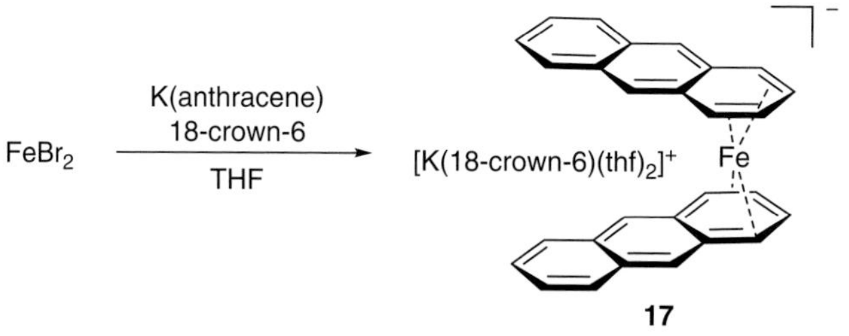

Scheme 5.8 Synthesis of potassium bis(anthracene)ferrate **17**.

Table 5.6 Comparative hydrogenation with iron complexes **17**, **18**, and **19**.

R⁄⁄ → R (5 mol% [Fe]; 2 bar H_2, 20 °C, 24 h)

		% Yield with [Fe]		
Entry	**Substrate**	**17**	**18**	**19**
1		72	90	6
2	8	15	84	72

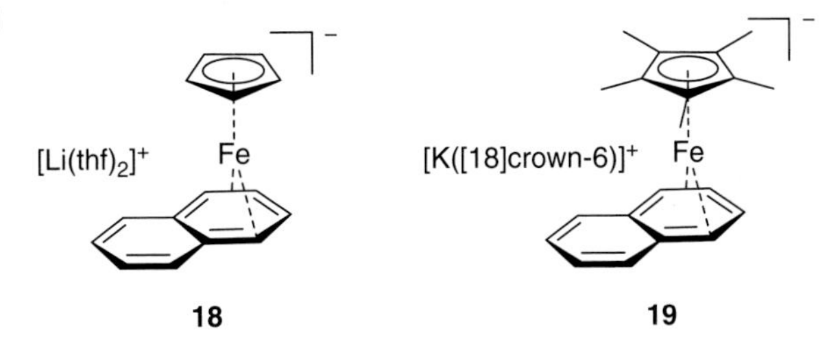

Figure 5.5 Cyclopentadienyl iron complexes **18** and **19**.

In addition, the same group prepared a library of homoleptic and heteroleptic arene/alkene metalates and compared their activity in the hydrogenation of alkenes (Table 5.6) [32]. In this series, iron complexes, initially reported by Jonas (**18**) and Wolf (**19**), were studied in alkene hydrogenation (Figure 5.5), presenting a moderate-to-good activity in the hydrogenation of styrene and 1-dodecene (Table 5.6) [33].

5.3 Cobalt

5.3.1 Introduction

Cobalt-based hydrogenation catalysts were developed mostly parallel to their iron counterparts when probing the activities of the less expensive first-row transition metals in comparison with the established noble metal systems. Heterogeneous Raney-cobalt was reported to hydrogenate styrene already in 1958, but its activity is far inferior to Raney-nickel. However, superior selectivity was observed in the hydrogenation of nitriles and oximes [34]. With the advent of Ziegler–Natta polymerization catalysis, various combinations of simple cobalt salt precursors with triorganoaluminum additives were applied to catalytic hydrogenation of simple alkenes such as 1-hexene and cyclohexene [9]. The differences in catalyst precursors, additives, conditions, and preparation methods have so far prevented a unified proposal of the operating catalyst species [10]. Recently, Finke and coworkers demonstrated that active cobalt clusters form from a cobalt(II)neodecanoate/Et_3Al catalyst system by a careful choice of poisoning tests as well as analysis by mass spectrometry, X-ray absorption fine structure, and transmission electron microscopy (Scheme 5.9) [35].

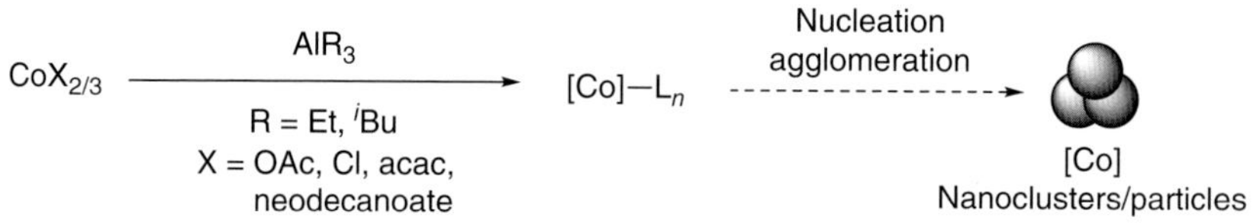

$$CoX_{2/3} \xrightarrow[\substack{R = Et,\ {}^iBu \\ X = OAc,\ Cl,\ acac, \\ neodecanoate}]{AlR_3} [Co]{-}L_n \xrightarrow{\text{Nucleation agglomeration}} \underset{\text{Nanoclusters/particles}}{[Co]}$$

Scheme 5.9 Generation of Ziegler-type Co/Al hydrogenation catalysts.

Although defined cobalt nanoparticles were prepared by various methods and their surface, size, and composition were carefully analyzed, extended catalytic studies into hydrogenation reactions remained scarce until the very recent past. Beller and coworkers prepared Co_3O_4/Co core/shell nanoparticles featuring nitrogen-doped graphene layers on alumina by pyrolysis of cobalt(II) acetate/phenanthroline and applied them in the hydrogenation of *N*-heteroarenes [36]. Cobalt nanoparticles on charcoal were synthesized by thermal treatment of dicobalt octacarbonyl and were also shown to be active in the hydrogenation of alkenes [37].

Several cobalt-catalyzed C=C hydrogenation protocols were developed with homogeneous cobalt carbonyl complexes following the seminal discovery of the hydroformylation of olefins with syngas (CO/H_2) [38]. Improvements in selectivity and activity were achieved by introducing monodentate phosphine ligands in the late 1970s, which enabled clean hydrogenations of alkynes and alkenes [39]. Later, the use of bidentate ligands and their applications to stereoselective hydrogenations of functionalized olefins were reported [40]. The majority of recent reports emphasized tridentate amine/pyridine-based pincer ligand motifs.

5.3.2 Pincer Complexes

Concurrent with the development of bis(imino)pyridine iron catalysts, the related cobalt complexes were initially studied in olefin polymerization reactions and shortly after also in olefin hydrogenations, both under Ziegler conditions in the presence of triisobutylaluminum as an activator [19, 41]. The structurally defined alkylcobalt complex **22** was synthesized by sequential alkylation of bis(imino)pyridinecobalt(II) dichloride (**20**) with methyl lithium/MAO (MAO, methylaluminoxane) and trimethylsilylmethyl lithium (Scheme 5.10) [42]. Strong redox participation of the ligand accounts for the formulation of **22** as a low-spin Co(II) center with a ligand-centered radical anion [43]. Under reaction conditions, complex **22** affects dihydrogen activation upon release of tetramethylsilane. The resultant monohydridocobalt intermediate was identified by ^1H and ^{13}C nuclear magnetic resonance (NMR) and is believed to be the catalytically active species in the hydrogenation of terminal and internal disubstituted alkenes (Scheme 5.11).

According to DFT calculations, dihydrogen splitting does not operate by an oxidative addition event but through σ-bond metathesis, which is the

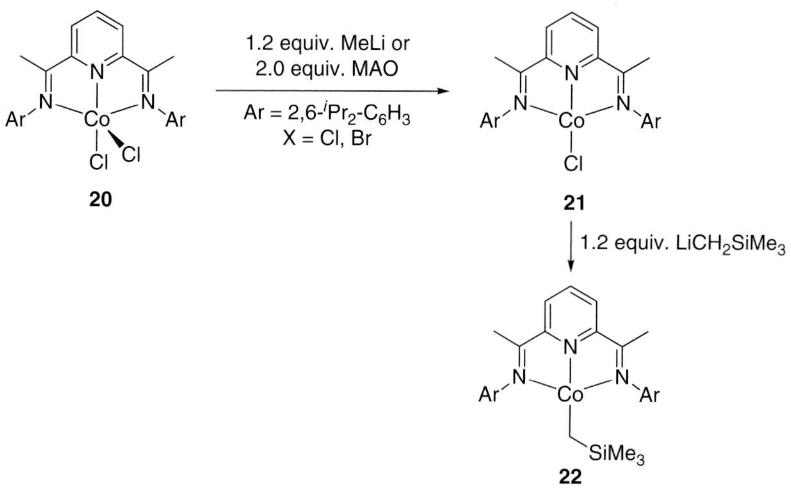

Scheme 5.10 Synthesis of alkyl bis(imino)pyridine cobalt complex **22**.

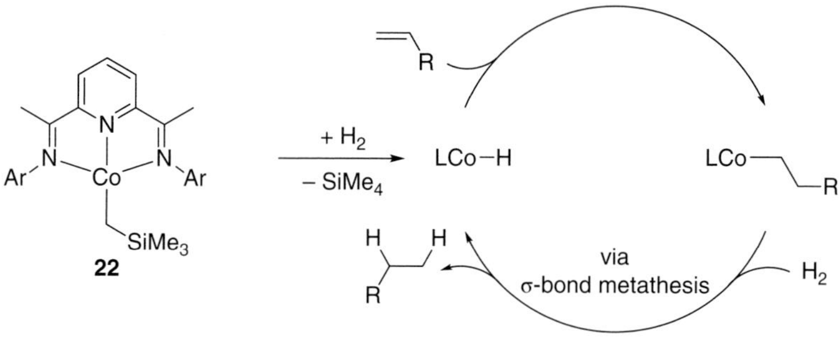

Scheme 5.11 Mechanistic proposal of catalytic hydrogenation with **22**.

rate-limiting step. A chiral version of complex **22** was prepared by replacing one aryl group with a chiral *sec*-alkylamine moiety. Initially developed for oligomerization reactions by Bianchini et al., the chiral bis(imino)pyridine (methyl)cobalt(I) complex **(*S*)-25** was prepared from **(*S*)-23** (Scheme 5.12) [44]. Complex **(*S*)-25** was successfully applied to the hydrogenation of prochiral alkenes; notably, no further directing groups are required to achieve high enantiomeric excess. In general, the stereoselective hydrogenation of such minimally functionalized alkenes is a challenging task because of the lack of chelating coordination modes, the absence of directing groups, and the lack of a strong stereochemical bias through bulky substituents. Higher enantiomeric excess values were observed for electron-rich styrenes, which were explained by their lower reactivity (Table 5.7).

The origin of enantioselectivity in the hydrogenation of benzannulated exocyclic and endocyclic cycloalkenes was rationalized in a comprehensive report. The authors documented that the 1,2-alkene addition is the enantiodetermining and rate-limiting step. Isomerization of the starting material is competitive if the alkylcobalt complex can adopt a conformation that can undergo syn-β-hydride

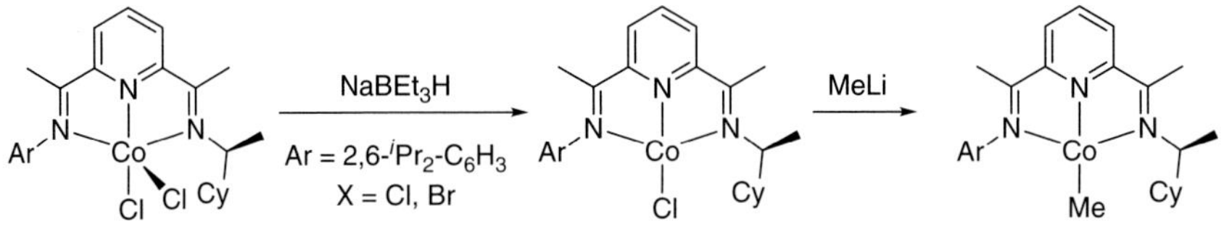

(S)-23 —[$NaBEt_3H$; Ar = 2,6-$^i Pr_2$-C_6H_3; X = Cl, Br]→ (S)-24 —[MeLi]→ (S)-25

Scheme 5.12 Synthesis of chiral bis(imino)pyridine methyl cobalt (*S*)-25.

Table 5.7 Hydrogenation of prochiral styrenes with (*S*)-25.

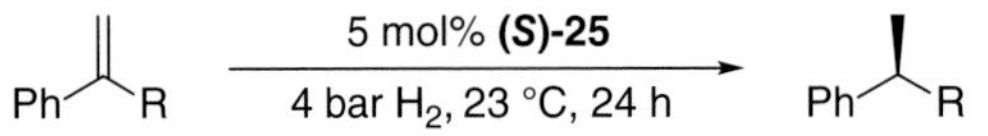

Reaction conditions: 5 mol% **(S)-25**, 4 bar H_2, 23 °C, 24 h

Entry	Substrate	% Yield	% ee
1	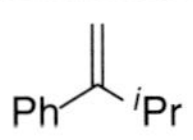 Ph, ^{i}Pr	87	90
2	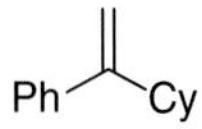 Ph, Cy	70	80
3	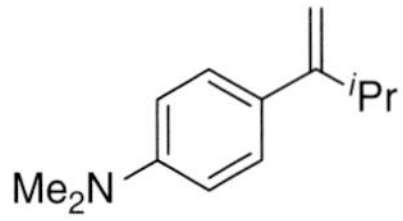 Me_2N, ^{i}Pr	>98	96
4	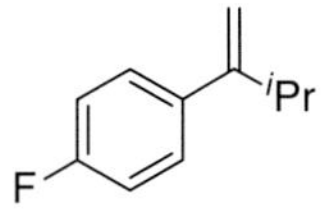 F, ^{i}Pr	>98	78

elimination as exemplified in the hydrogenation of 1-methylene-indane. Rapid isomerization and hydrogenation results in the exclusive formation of the non-expected enantiomer (Table 5.8) [45]. Careful choice of the chiral moiety is necessary to suppress dehydrogenative C–H insertion of the ligand side chains to give a cobaltacycle. Cyclometalation of **(*S*)-25** is reversible under hydrogenation conditions, whereas the *tert*-butyl homologue gives almost exclusively the inactive form **(*R*)-26**, which undergoes very slow conversion to the active hydridocobalt complex (Scheme 5.13).

In 2016, a related ligand design was embedded within the chiral oxazoline iminopyridine cobalt complex **27** by Lu and coworkers (Figure 5.6) [46]. In contrast to the earlier works by Budzelaar and Chirik, precatalyst activation was achieved *in situ* by the addition of sodium triethylborohydride. Application in the stereoselective hydrogenation of 1,1-diphenylethenes revealed a higher hydrogenation activity with higher enantiomeric excess under mild conditions compared to **(*S*)-25** in some examples, making this ligand an interesting modulation of the typical bis(imino)pyridines for further investigations (Table 5.9).

Table 5.8 Enantioselective hydrogenation of exo- and endocyclic alkenes with **(*S*)-25**.

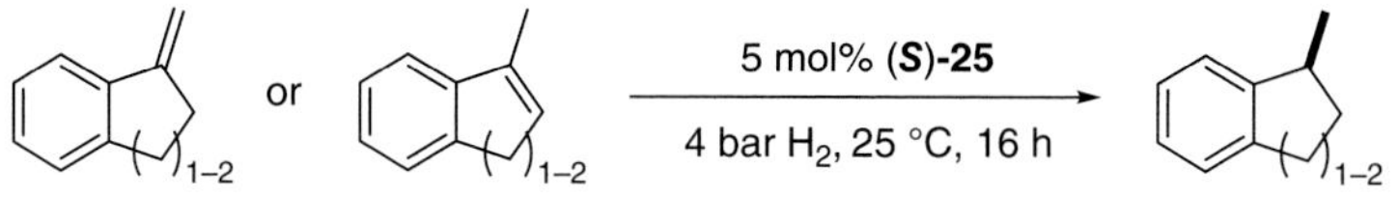

5 mol% **(*S*)-25**; 4 bar H_2, 25 °C, 16 h

Entry	Substrate	Product	% Yield (ee)
1	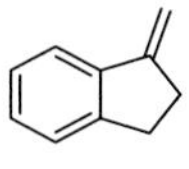	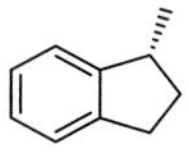	84 (74)
2	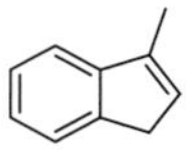		88 (89)
3	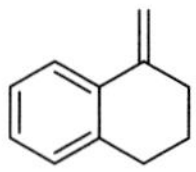	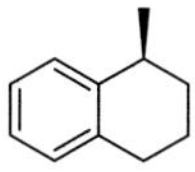	87 (53)
4	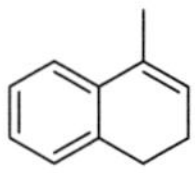	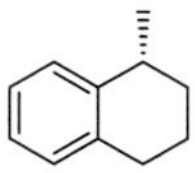	96 (93)

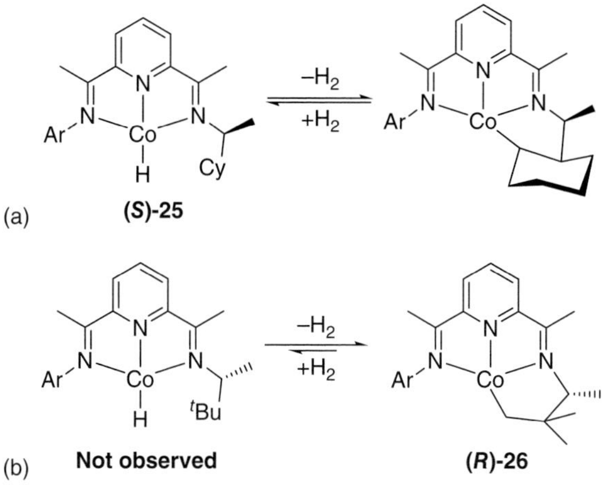

Scheme 5.13 Competing cyclometalation of active cobalt hydride intermediates with catalyst **(*S*)-25** (a) and **(*R*)-26** (b).

Figure 5.6 Oxazoline iminopyridine complex **27**.

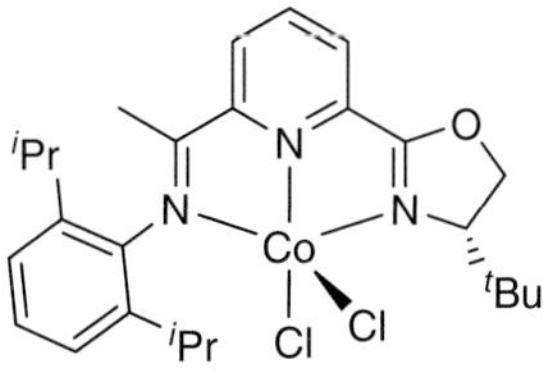

Table 5.9 Hydrogenation of 1,1-diarylethenes with **27**.

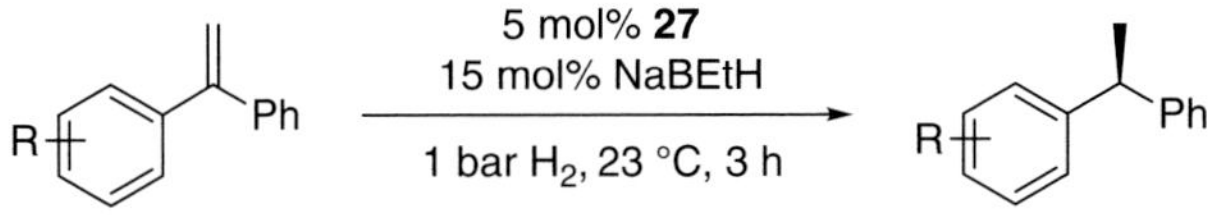

5 mol% **27**, 15 mol% NaBEtH, 1 bar H_2, 23 °C, 3 h

Entry	Substrate	% Yield	% ee
1	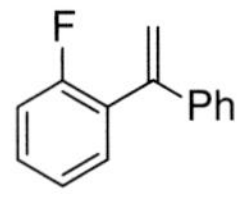 F, Ph	>99	60
2	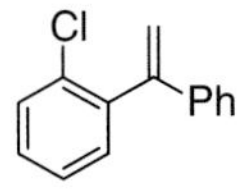 Cl, Ph	>99	90
3	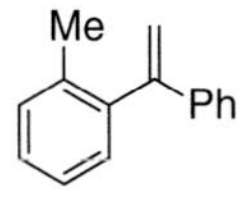 Me, Ph	>99	80
4	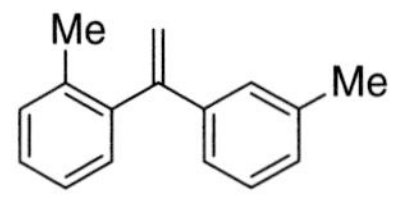 Me, Me	>99	86

In analogy to their observations in iron-catalyzed hydrogenations, the Chirik group has modified the bis(imino)pyridine ligand by replacing the imines with strongly σ-donating NHCs to enhance the electron density at the metal center (Figure 5.7) [47]. Again, this ligand modification results in increased reactivity compared to its bis(imino)pyridine analog. Therefore, **28** constitutes one of the most active base metal hydrogenation precatalysts for sterically hindered alkenes (Table 5.10).

Upon H_2 addition, **28** readily forms the hydride complex **29**, which is most likely the active catalyst under hydrogenation conditions. Interestingly, **29** undergoes hydrogen migration from cobalt to the electrophilic 4-pyridyl position, which has not been observed with the analogous iron complex **11** (Scheme 5.14). A combined computational and spectroscopic study favors the presence of a pyridine-centered radical ligand, which is responsible for the observed H atom migration and the redox-noninnocence of the *C,N,C*-pincer ligand.

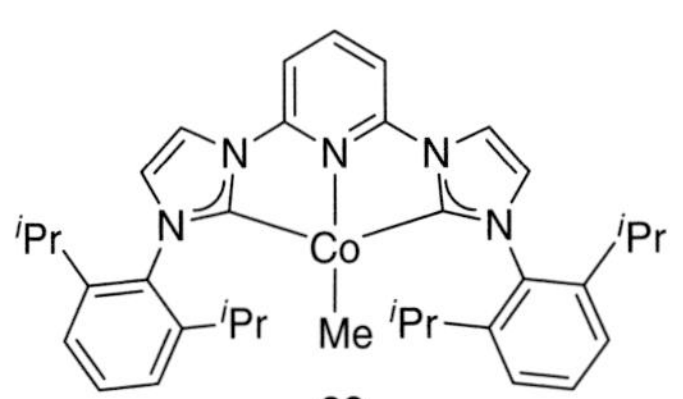

Figure 5.7 Bis(arylimidazol-2-ylidene)pyridine cobalt methyl **28**.

Table 5.10 Hydrogenation of sterically hindered alkenes with **28**.

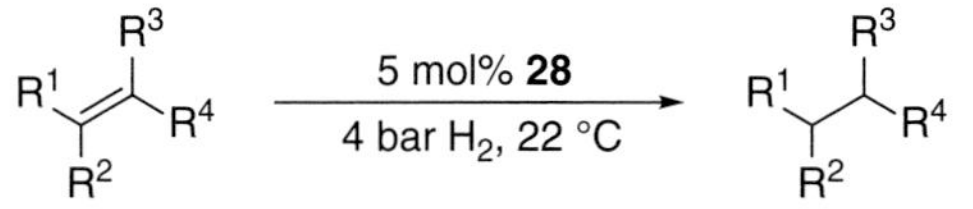

5 mol% **28**, 4 bar H_2, 22 °C

Entry	Substrate	% Yield
1	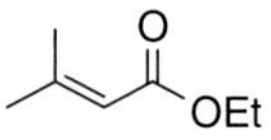 O, OEt	>95 (1 h)
2	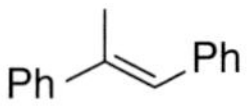 Ph, Ph	>95 (5 h)
3	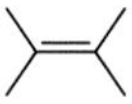	15 (24 h)

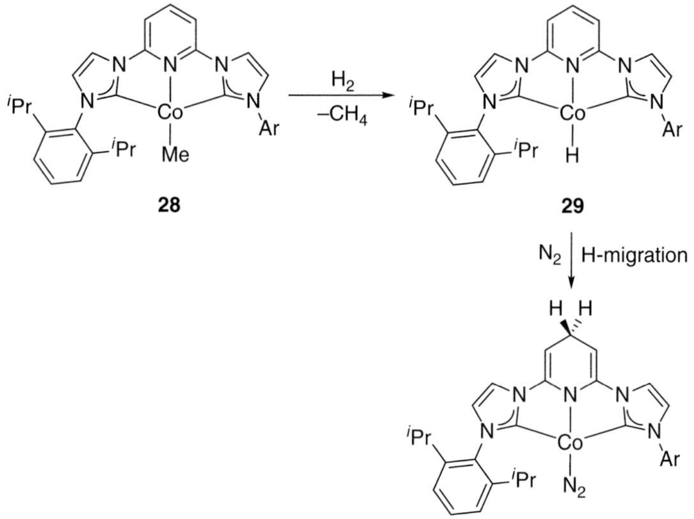

Scheme 5.14 Reactivity of **28** with dihydrogen and sequential H-atom migration.

In 2012, Hanson and coworkers applied cobalt precatalysts containing aliphatic *P,N,P*-pincer ligands to olefin and carbonyl hydrogenation reactions [48]. Upon activation with the strong Brookhart acid $H[BAr^F_4]\cdot(Et_2O)_2$ (Ar^F = 3,5-bis(trifluoromethyl)phenyl), the inactive precursor **30** is converted to the cationic hydrogenation catalyst **31** (Scheme 5.15). Similar to the alkylcobalt complex **22** in Scheme 5.11, **31** forms an active hydridocobalt species under hydrogenation conditions upon release of tetramethylsilane as evidenced by crossover and trapping experiments.

Hydrogenation of a wide scope of alkenes, imines, and ketones was performed with *in situ*-generated **31**. The base-free operation enables the tolerance of various functional groups (e.g. carboxylic acids and ketones). The presence of alcohol

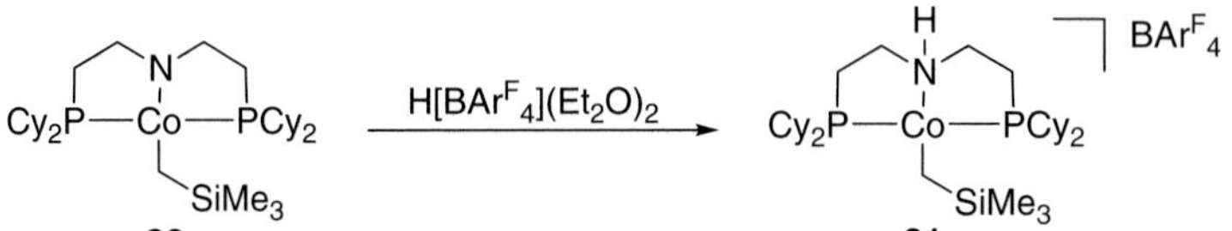

Scheme 5.15 Formation of cationic pincer catalyst **31** by protonation of **30**.

functions or water does not affect the catalytic activity, which attests to the high stability of **31**. In contrast to catalysis by the analogous iron complex **14**, a bifunctional mechanism with amine participation was excluded because of the similar activity of the *N*-methylated complex. In accordance with this assumption, **31** is also a competent catalyst in the hydrogenation of nonpolar C=C bonds, whereas **14** fails to hydrogenate 1-hexene. **31** shows high chemoselectivity in the hydrogenation of less hindered C=C bonds and in the presence of carbonyl groups (Table 5.11). Elevated temperatures also enable the clean hydrogenation of carbonyl compounds.

In 2014, Peters and coworker applied new *P,B,P*-pincer cobalt complexes to hydrogenation reactions [49]. The bis(phosphino)boranecobalt(I) complex **32** was synthesized by complexation of cobalt(II) bromide and sequential reduction with Na/Hg (Scheme 5.16). Importantly, **32** is capable of activating two equivalents of dihydrogen in a reversible manner to form the dihydridoboratocobalt dihydride **33**. Under mild hydrogenation conditions, simple olefins such as 1-octene and styrene are hydrogenated with a turnover frequency of 1000 h^{-1}, but the complex fails to convert internal olefins (i.e. cyclooctene and norbornene).

Table 5.11 Alkene hydrogenation with precatalyst **30**.

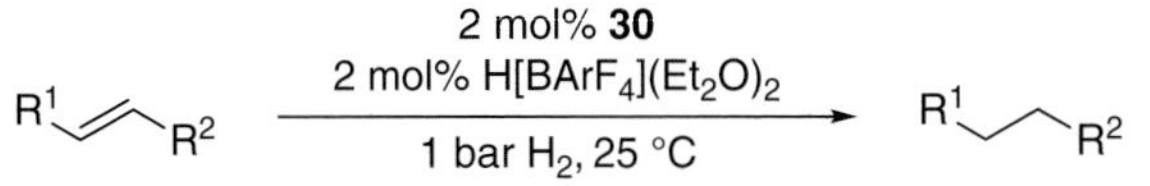

2 mol% **30**, 2 mol% $H[BArF_4](Et_2O)_2$, 1 bar H_2, 25 °C

Entry	Substrate	Product	Time (h)	% Yield
1	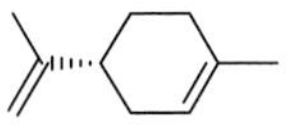	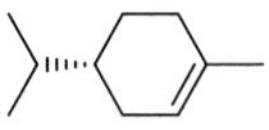	40	80
2	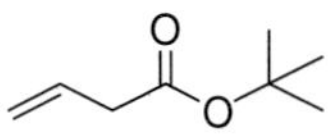	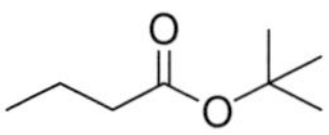	24	99
3	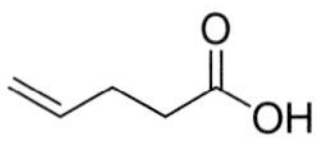	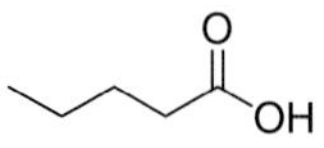	24 (60 °C)	99
4	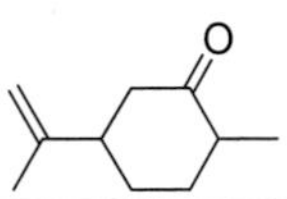	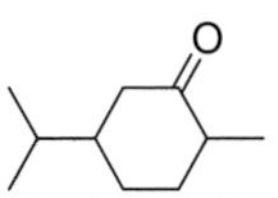	42	99

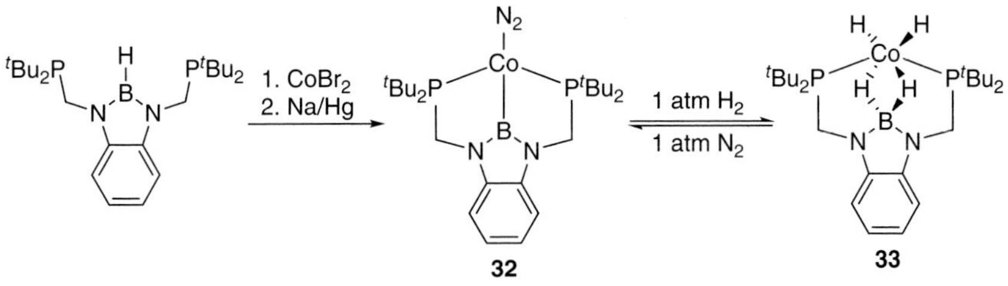

Scheme 5.16 Synthesis and reversible hydrogen addition of *P,B,P* pincer complex **32**.

5.3.3 Others

A different ligand design of the precatalyst was used by Wolf and von Wangelin in their application of the heteroatom-free bis(anthracene)cobaltate complexes (**34**) to hydrogenation reactions of alkenes, alkynes, and carbonyls (Scheme 5.17) [30]. The complex was first reported by Ellis and Brennessel in 2002 and constitutes a convenient metal (–I) source that contains labile hydrocarbon ligands [50]. According to the authors, the catalyst is stabilized by the presence of various π-coordinating compounds that, under hydrogenation conditions, are the corresponding substrates (e.g. olefins). NMR studies documented the fast ligand exchange of anthracene by styrene, cod, and other simple alkenes (Scheme 5.18). Longer reaction times and elevated pressure also lead to the hydrogenation of the anthracene ligand. The absence of π-acidic ligands results in particle formation and catalyst deactivation, although the resultant nanoparticles are still moderately effective catalysts for the hydrogenation of simple alkenes and styrenes. Catalyst **34** was applied to a wide scope of alkenes (1–4 bar H_2, r.t.), ketones, and imines (10 bar H_2, 60 °C) and showed comparable activity to the cobalt catalyst **31**.

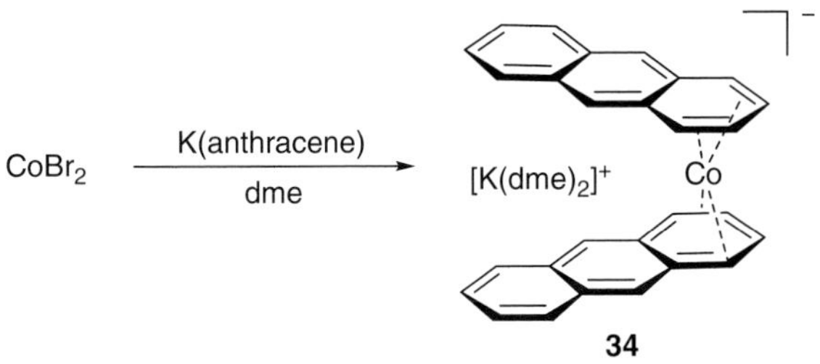

Scheme 5.17 Synthesis of potassium bis(anthracene) cobaltate **34**.

In an extended study, the same groups synthesized a library of heteroleptic bis(arene) and bis(alkene)cobaltate complexes (Figure 5.8) [32]. Despite only small stereoelectronic differences between these complexes, the nature of the π-hydrocarbon ligand drastically influences the catalytic reactivity (Table 5.12). The authors observed a decreasing reactivity for complexes with more strongly coordinating alkene substrates (**37**, **38**).

A structurally unique complex class was reported by Stryker and coworkers in 2013 [51]. By reaction of cobalt(II) chloride with a sterically demanding

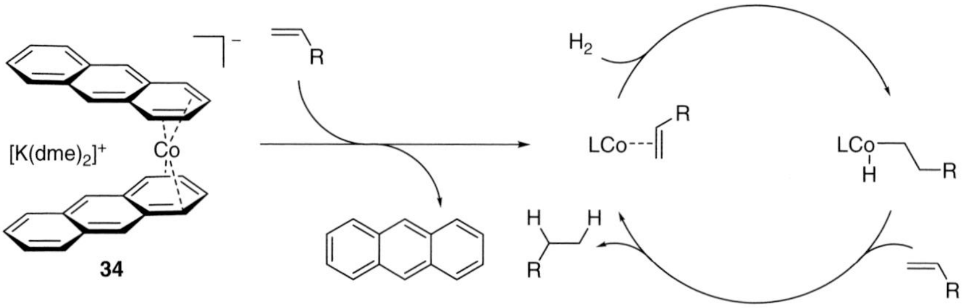

Scheme 5.18 Proposed mechanism for catalytic hydrogenation with **34**.

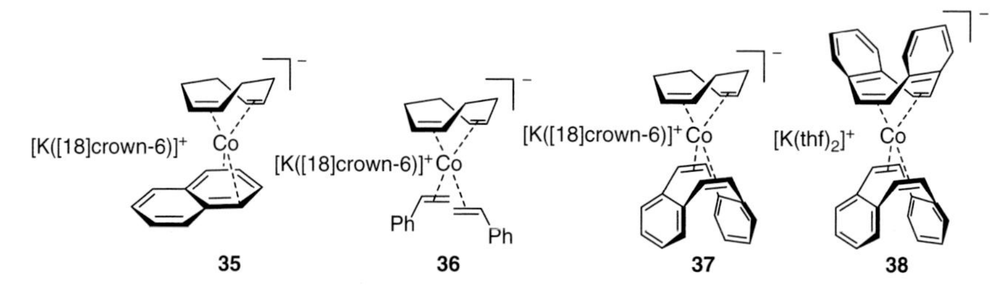

Figure 5.8 Heteroleptic arene/alkene cobaltate complexes **35–38**.

Table 5.12 Comparative hydrogenation of alkenes with **34–38**.

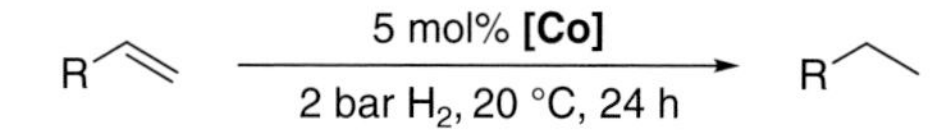

5 mol% **[Co]**, 2 bar H_2, 20 °C, 24 h

Entry	Substrate	% Yield with [Co]				
		34	**35**	**36**	**37**	**38**
1	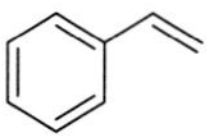	94	99	72	36	0
2	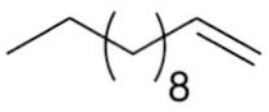	58	93	85	71	0

lithium phosphoranimide and sequential reduction with sodium amalgam, the tetrameric cobalt(I) cluster **39** is formed (Scheme 5.19). The square-planar complex can be viewed as a simplest ligand-supported mimetic of metallic surface arrays. The authors reported the good activity of **39** in the hydrogenation of allylbenzene and diphenylacetylene using only 0.5 mol% of catalyst.

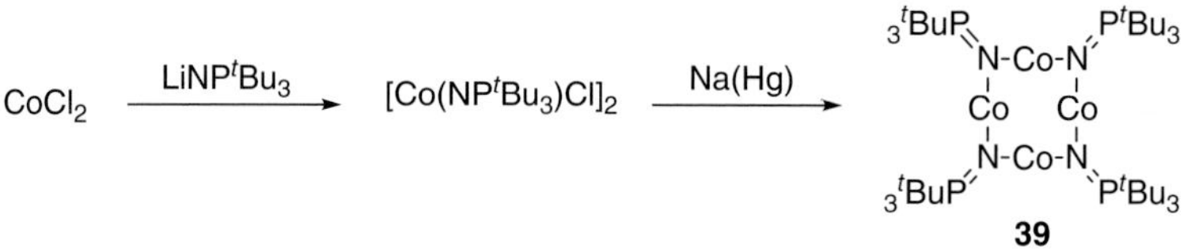

Scheme 5.19 Synthesis of tetrameric cobalt complex **39**.

In a landmark publication, Chirik and coworkers reported the use of chiral bidentate bis(phosphine)cobalt catalysts in highly stereoselective hydrogenations of largely unfunctionalized alkenes and dehydroamino acids [52]. Remarkably, the authors were able to identify very potent catalysts by high-throughput screening, which allowed the fast comparison of a vast number of chiral bidentate phosphine ligands in cobalt-catalyzed enantioselective alkene hydrogenations (Table 5.13).

After identification of ^{iPr}DuPhos as a suitable ligand, complex **40** was isolated and applied in the hydrogenation of enamides with excellent yield and moderate-to-good enantioselectivity (Scheme 5.20). Soon after, the same group prepared related nonchiral bis(phosphine)cobalt(II) dialkyl complexes (Figure 5.9), which prove very active in alkene hydrogenations [53].

Complex **42** effectively catalyzes the hydrogenation of terminal and disubstituted C=C bonds. Notably, the authors reported catalyst activation in the presence of hydroxyl-containing substrates, which enables the hydrogenation of trisubstituted alkenes under mild conditions (Table 5.14). The hydroxyl functionality is proposed to act as a directing group to facilitate olefin

Table 5.13 Selected bis(phosphine) ligands in enantioselective cobalt-catalyzed hydrogenation.

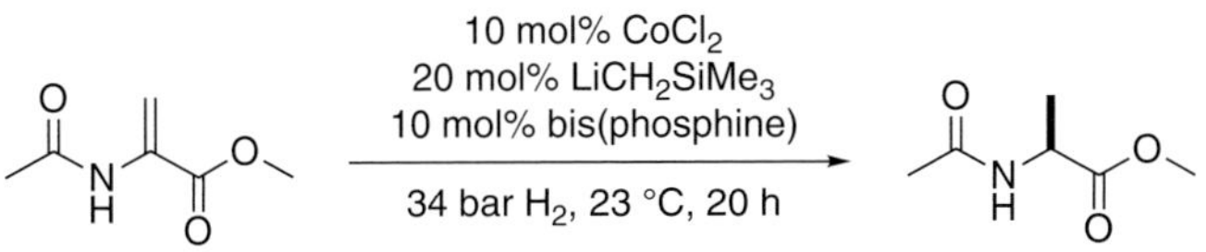

10 mol% $CoCl_2$, 20 mol% $LiCH_2SiMe_3$, 10 mol% bis(phosphine); 34 bar H_2, 23 °C, 20 h

Entry	Bis(phosphine)	% Yield	% ee (major isomer)
1	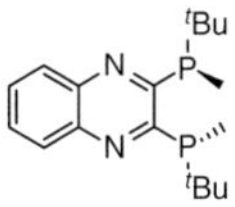 (*R*,*R*)-QuinoxP	93.2	96.4 (*R*)
2	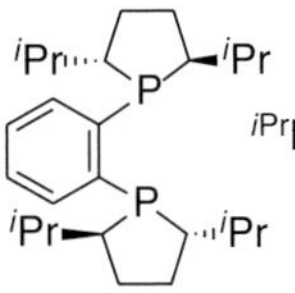 ^{iPr}DuPhos	92.3	94.2 (*S*)
3	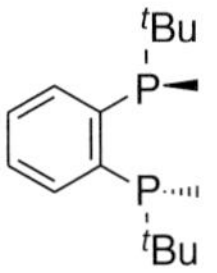 (*R*,*R*)-BenzP	>99	93.4 (*R*)
4	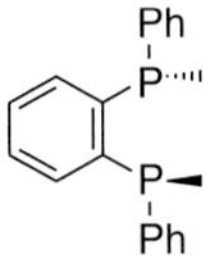 (*S*,*S*)-1,2-$(MePhP)_2C_6H_4$	>99	77.0 (*S*)

Scheme 5.20 Enantioselective alkene hydrogenation with **(*R*,*R*)-40** (a) and **(*S*,*S*)-40** (b).

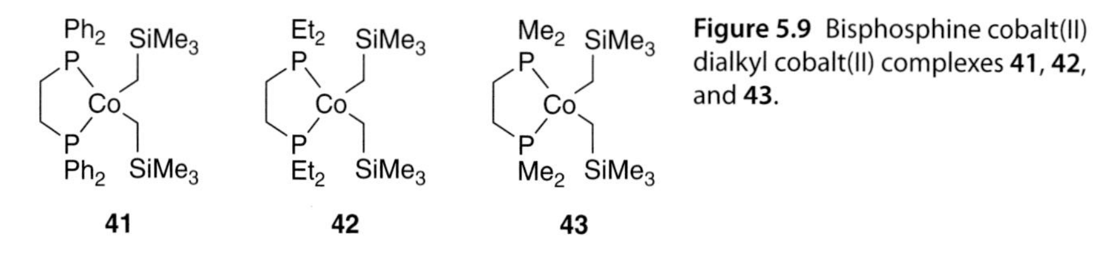

Figure 5.9 Bisphosphine cobalt(II) dialkyl cobalt(II) complexes **41, 42,** and **43.**

Table 5.14 Hydrogenation of oxygen-containing alkenes with **42.**

5 mol% **42**, 4 bar H_2, 25 °C

Entry	Substrate	Product	Time (h)	% Yield
1	OMe	OMe	14	<5
2	OH	OH	14	97
3	OMe	OMe	14	85
4	OH	OH	4	>99

coordination. This effect is intramolecular in nature; an intermolecular activation by addition of alcohol is unsuccessful. Under hydrogenation conditions, the authors proposed hydrogenolysis of both alkyl moieties and formation of a dihydridocobalt(II) complex. Insertion of olefin gives a monohydridocobalt(II) alkyl complex, which can reductively eliminate the resulting alkane upon generation of a cobalt(0) complex (Scheme 5.21). The latter proposal was supported by trapping a cyclooctadienecobalt(0) complex upon reaction of **42** with cyclooctadiene under dihydrogen atmosphere.

5.4 Nickel

5.4.1 Introduction

Heterogeneous nickel catalysts in various forms are very well established for C=C hydrogenation reactions. Most prominent are applications of Raney-nickel

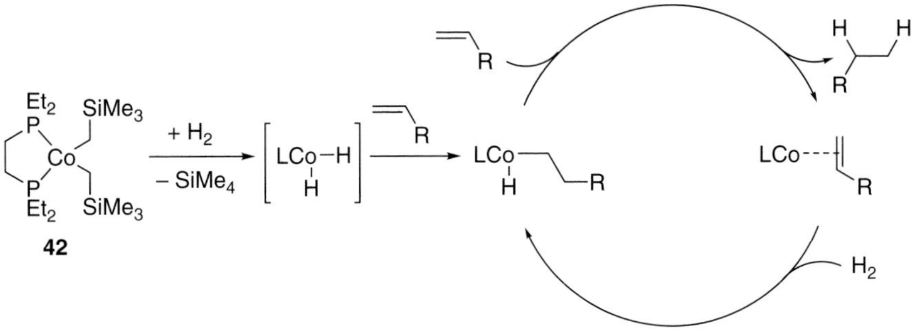

Scheme 5.21 Proposed olefin hydrogenation mechanism for precatalyst **42**.

catalysts, which were developed already in 1926 [54]. The high catalyst activities at room temperature led to numerous implementations in industrial processes and organic synthesis programmes. The broad range of catalyzed reactions includes hydrogenation of C=C bonds, nitriles, nitro compounds, and other unsaturated functional groups. Similar activities were often achieved with the nonpyrophoric Urushibara-nickel catalysts (mostly Fe/Zn) [55]. Reports of olefin hydrogenations with Ziegler-type Ni/Al catalysts followed in the 1960s [10]. Today, Raney-nickel catalysts display the widest scope in hydrogenation reactions. However, the heterogeneous nature, rather undefined composition and surface properties, and the high reactivity with many functional groups have stimulated significant efforts toward the design of homogeneous catalysts that allow facile control over activity, selectivity, and physical properties through rational ligand design. Still only very few powerful homogeneous nickel-catalyzed C=C hydrogenations have been reported, despite the decades experience with the related Reppe and Wilke chemistry [56]. Early examples include the hydrogenation of methyl linoleate with bis(triphenylphosphine)nickel(II) halides in 1967 [57]. Bidentate bis(phosphine) nickel(II) complexes were studied in the hydrogenation of 1-octene in 1998 [58]. The more recent applications of homogeneous nickel catalysts to hydrogenation reactions are summarized below.

5.4.2 Pincer Complexes

One of the rare examples in homogeneous nickel hydrogenation chemistry was reported by Sánchez and coworkers in 2004. A set of aminosalen-type *O,N,N*-pincer nickel(II) complexes was evaluated under hydrogenation conditions (Figure 5.10) [59]. The chiral, air-stable complexes are active in the

Figure 5.10 Aminosalen-type nickel(II) complexes **43**, **44**, and **45**.

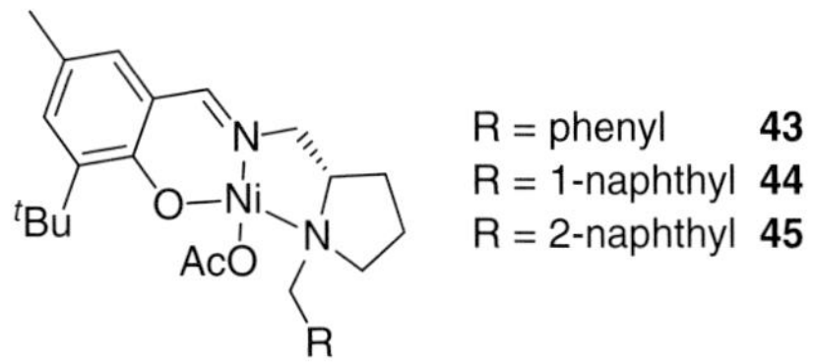

R = phenyl **43**
R = 1-naphthyl **44**
R = 2-naphthyl **45**

Table 5.15 Hydrogenation of alkenes with **43**.

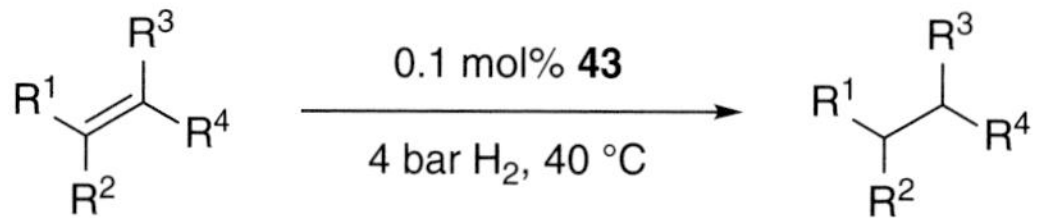

Reaction: $R^1R^2C=CR^3R^4$ → $R^1R^2CH–CHR^3R^4$; 0.1 mol% **43**, 4 bar H_2, 40 °C

Entry	Substrate	TOF (h^{-1})
1	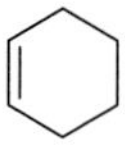 cyclohexene	4020
2	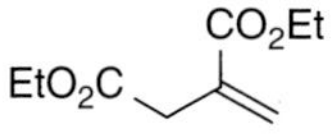 EtO_2C–CH_2–C(CO_2Et)=CH_2	2400
3	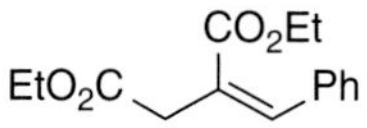 EtO_2C–CH_2–C(CO_2Et)=CH–Ph	220

hydrogenation of alkenes and imines (Table 5.15). No stereoselectivity is induced in the reactions of prochiral alkenes. In a comparative study, the authors showed similar activities of **43–45** with the analogous palladium complexes in terms of turnover frequencies.

Parallel to their work on cobalt catalysts, Hanson and coworkers prepared the *P,N,P*-nickel(II) hydride complex **46** by reduction of the corresponding nickel(II) bromide complex and sequential protonation (Scheme 5.22) [60]. The catalytic activity in the hydrogenation of terminal alkenes under 4 bar H_2 at 80 °C is only moderate. A bifunctional mechanism that would involve alkane generation by an intramolecular protonation of the alkylnickel intermediates by the NH function (as observed with iron complex **14**) was excluded based on the observation that no methane is released from the thermal treatment of the catalytically equally active methylnickel(II) complex **47** (Scheme 5.23). The authors proposed a purely metal-centered mechanism via reversible alkene 1,2-insertion into the Ni—H bond, dihydrogen addition, and reductive elimination.

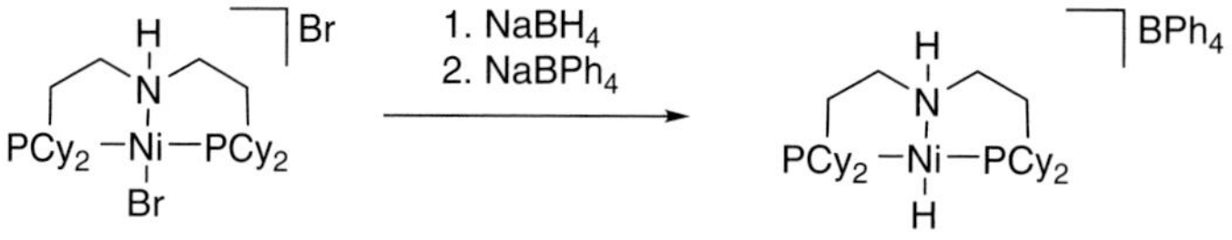

H–N, PCy_2–Ni–PCy_2, Br; Br → (1. $NaBH_4$; 2. $NaBPh_4$) → H–N, PCy_2–Ni–PCy_2, H; BPh_4; **46**

Scheme 5.22 Synthesis of *P,N,P*-nickel(II) hydride **46**.

In 2012, Peters and coworker reported the synthesis and comprehensive study of *P,B,P*-pincer nickel(II) complexes [61]. The boryl bis(phosphine)nickel **48** reversibly adds dihydrogen to give the square-planar borohydridonickel(II) hydride **49** (Scheme 5.24). The heterolysis of H_2 occurs at the nickel—boron bond where nickel acts as a Lewis base, which formally accepts a proton. The

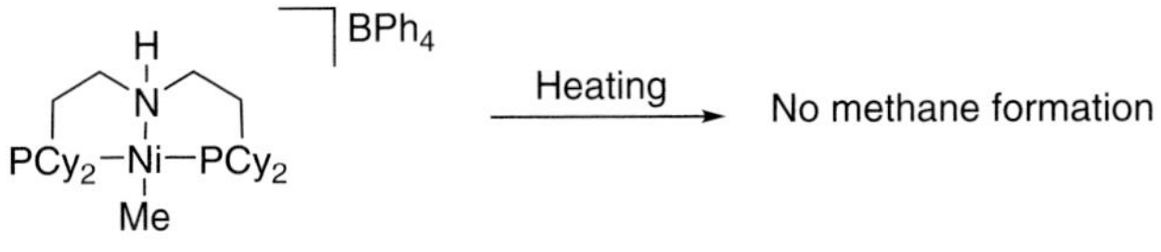

No methane formation

Scheme 5.23 Stability of *P,N,P* nickel(II) methyl complex **47**.

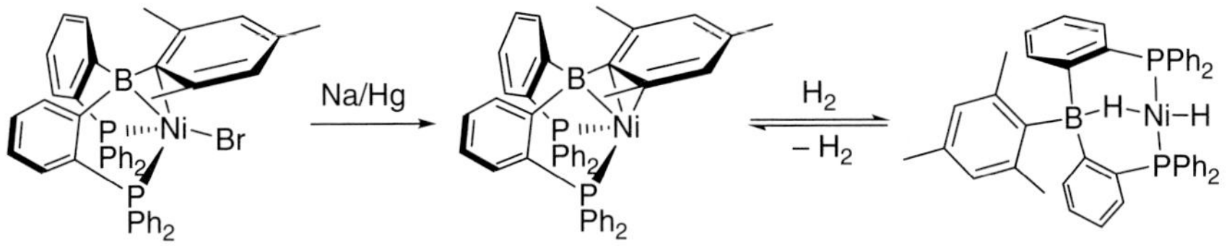

Scheme 5.24 Synthesis of borylnickel complex **48** and reversible addition of dihydrogen.

Lewis acidic boryl ligand stabilizes the formal dihydridonickel complex **49**, allowing reversible hydrogen activation at room temperature. Complex **48** was successfully applied to hydrogenations of styrene under mild conditions with a TOF of about 20 h^{-1}.

Two years later, the same group reported the *P,B,P*-nickel(I) hydride complex **50** with similar hydrogenation activity (Scheme 5.25) [62]. The proposed mechanism involves reversible olefin insertion into the Ni—H bond with consecutive hydrogenolysis. The authors demonstrated the beneficial effect of the boryl ligand in **50** in comparison with isoelectronic and isostructural phenyl and amino functions (Table 5.16).

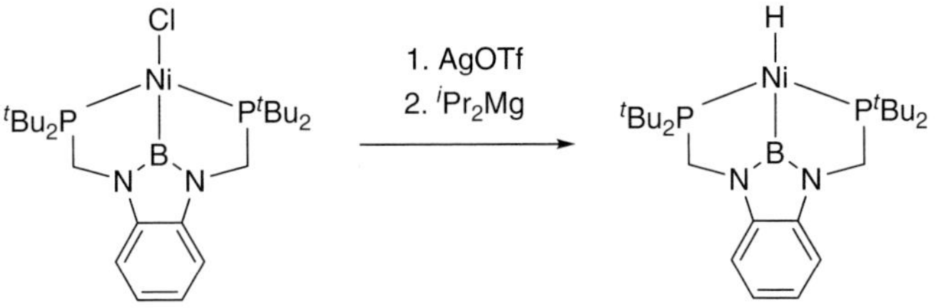

Scheme 5.25 Synthesis of *P,B,P* nickel(I) hydride **50**.

5.4.3 Others

Bidentate bisphosphine nickel(II) complexes were shown to be active in hydrogenation earlier [58, 63]. In an effort to expand the general mechanistic understanding of diphosphinenickel catalysis, a library of 24 diarylphosphine nickel(II) complexes was tested in the hydrogenation of 1-octene [64]. In general, catalytic activity could be enhanced by introducing electron-donating groups in the aryl moiety and increasing the steric bulk. The most active complex **51** achieves a high TOF of 4500 h^{-1} at 50 bar H_2 pressure and 50 °C (Figure 5.11).

Table 5.16 Hydrogenation of terminal alkenes with the nickel complexes **46** and **50**.

		TOF (h^{-1}) (% Yield)	
Entry	**Substrate**	**46**	**50**
1	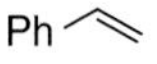 Ph	0.4 (100)	25 (100)
2	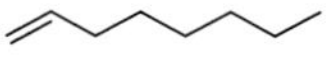	0.4 (70)	25 (64)
3	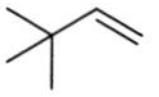	0.2 (97)	5 (100)

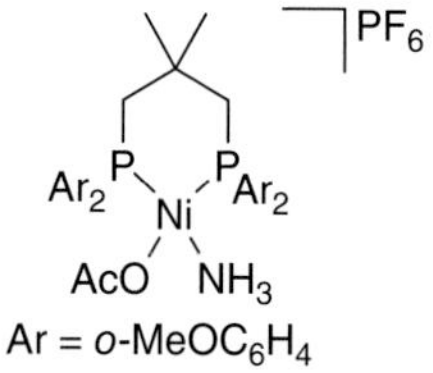

PF_6
Ar_2P Ni PAr_2
AcO NH_3
Ar = *o*-$MeOC_6H_4$
51

Figure 5.11 The bidentate diphosphine nickel(II) complex **51**.

Stryker and coworkers reported the preparation of structurally unique square tetrametallic cobalt and nickel complexes in 2013 (Scheme 5.26) [51]. This class of planar metal(I) complexes can be depicted as simple models of a ligand-supported metal surface. The tetrameric nickel complex **52** shows equal hydrogenation activity as the cobalt complex in reactions of allylbenzene and diphenylacetylene under mild conditions.

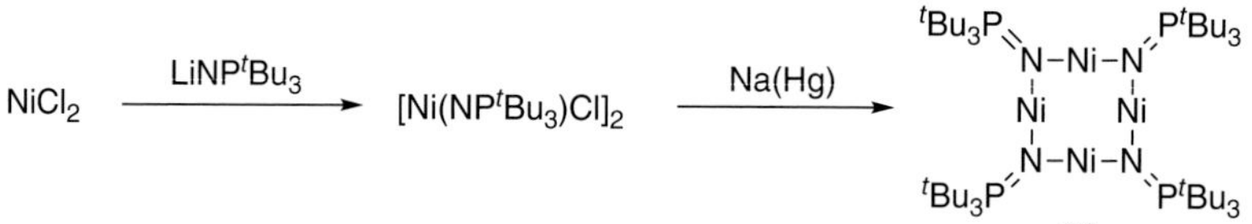

$NiCl_2$ → ($LiNP^tBu_3$) → $[Ni(NP^tBu_3)Cl]_2$ → ($Na(Hg)$) → **52**

$^tBu_3P$ P^tBu_3
N–Ni–N
Ni Ni
N–Ni–N
$^tBu_3P$ P^tBu_3
52

Scheme 5.26 Synthesis of the tetrameric nickel(I) complex **52**.

5.5 Conclusion

In the past decade, multifaceted studies of iron group metal catalysts have demonstrated their great potential in olefin hydrogenations (and hydrofunctionalizations). Pincer-type complexes have emerged as the most powerful class of catalysts because of their modular properties and the effective cooperation of redox events at the metal and the ligand backbone. Catalytic activities rivaling those of the established noble metal catalysts have been achieved in several cases. Cobalt catalysts have been especially productive in terms of substrate scope and mechanistic understanding. Enantioselective hydrogenations of simple alkenes and dehydroamino acid derivatives have been realized with chiral cobalt catalysts bearing pyridyldiimine and diphosphine ligands, respectively. These transformations constitute key strategic steps in various manufacturing routes of fine chemicals and pharmaceuticals and could trigger a paradigm shift toward the replacement of current noble metal technologies.

Generally, the fluxional coordination chemistry of 3d transition metals, the participation of single-electron transfer steps, the presence of free radical intermediates, and the high sensitivity of low-valent 3d metals toward air and moisture are major challenges in the usage of iron group metal catalysts. These aspects often require special handling procedures, the application of sophisticated spectroscopic tools to study the reaction mechanisms, and might even exclude the employment of certain functional molecules. However, the recent tremendous progress in the field has significantly expanded the window of opportunities. The key to success is a detailed understanding of the underlying mechanisms and the knowledge-based design of proper ligand architectures. The iron group metal age is only beginning to dawn.

Acknowledgments

We gratefully acknowledge the financial support from the Deutsche Forschungsgemeinschaft through the Emmy Noether and Heisenberg programs and a research grant. T.N.G. was a doctoral fellow of the Evonik Foundation.

References

1 (a) Keim, W., Behr, A., and Schmitt, G. (1986). *Grundlagen der industriellen Chemie. Technische Produkte und Prozesse*. Berlin: Salle, Frankfurt am Main. (b) Veldsink, J.W., Bouma, M.J., Schön, N.H., and Beenackers, A.A.C.M. (1997). *Catal. Rev.* 39: 253.
2 (a) Knowles, W.S. (1986). *J. Chem. Educ.* 63: 222. (b) Ohta, T., Takaya, H., Kitamura, M. et al. (1987). *J. Org. Chem.* 52: 3174. (c) Heydrich, G., Gralla, G., Ebel, K. et al. (2009). WO2009033870 A1.
3 Wang, D. and Astruc, D. (2015). *Chem. Rev.* 115: 6621.
4 (a) de Vries, J.G. and Elsevier, C.J. (eds.) (2006). *The Handbook of Homogeneous Hydrogenation*. Weinheim: Wiley-VCH. (b) Johnson, N.B., Lennon, I.C., Moran, P.H., and Ramsden, J.A. (2007). *Acc. Chem. Res.* 40: 1291.
5 White, M.C. (2016). *Adv. Synth. Catal.* 358: 2364.
6 (a) Shima, S., Chen, D., Xu, T. et al. (2015). *Nat. Chem.* 7: 995. (b) Arpe, H.-J. (2010). *Industrial Organic Chemistry*. Weinheim: Wiley-VCH.
7 Determined by title search on www.webofknowledge.com for "iron", "cobalt", "nickel" "and" "hydrogenation".
8 (a) Thompson, A.F. and Wyatt, S.B. (1940). *J. Am. Chem. Soc.* 62: 2555. (b) Taira, S.-I. (1962). *Bull. Chem. Soc. Jpn.* 35: 840.

9 (a) Sloan, M.F., Matlack, A.S., and Breslow, D.S. (1963). *J. Am. Chem. Soc.* 85: 4014. (b) Breslow, D.S. and Matlack, A.S. (1963). US3113986 A.

10 Alley, W.M., Hamdemir, I.K., Johnson, K.A., and Finke, R.G. (2010). *J. Mol. Catal. A: Chem.* 315: 1.

11 (a) Guo, N., Hu, M.-Y., Feng, Y., and Zhu, S.-F. (2015). *Org. Chem. Front.* 2: 692. (b) Gieshoff, T.N., Villa, M., Welther, A. et al. (2015). *Green Chem.* 17: 1408. (c) Gieshoff, T.N., Welther, A., Kessler, M.T. et al. (2014). *Chem. Commun.* 50: 2261. (d) Frank, D.J., Guiet, L., Kaslin, A. et al. (2013). *RSC Adv.* 3: 25698. (e) Welther, A., Bauer, M., Mayer, M., and von Wangelin, A.J. (2012). *ChemCatChem* 4: 1088. (f) Rangheard, C., de Julián Fernández, C., Phua, P.-H. et al. (2010). *Dalton Trans.* 39: 8464. (g) Phua, P.-H., Lefort, L., Boogers, J.A.F. et al. (2009). *Chem. Commun.* 3747.

12 Astruc, D., Lu, F., and Aranzaes, J.R. (2005). *Angew. Chem. Int. Ed.* 44: 7852.

13 (a) Kelsen, V., Wendt, B., Werkmeister, S. et al. (2013). *Chem. Commun.* 49: 3416. (b) Lacroix, L.-M., Lachaize, S., Falqui, A. et al. (2008). *J. Appl. Phys.* 103, 07D521.

14 Hudson, R., Hamasaka, G., Osako, T. et al. (2013). *Green Chem.* 15: 2141.

15 Stein, M., Wieland, J., Steurer, P. et al. (2011). *Adv. Synth. Catal.* 353: 523.

16 Frankel, E.N., Emken, E.A., Peters, H.M. et al. (1964). *J. Org. Chem.* 29: 3292.

17 Schroeder, M.A. and Wrighton, M.S. (1976). *J. Am. Chem. Soc.* 98: 551.

18 Daida, E.J. and Peters, J.C. (2004). *Inorg. Chem.* 43: 7474.

19 Knijnenburg, Q., Horton, A.D., van der Heijden, D. et al. (2003). WO2003042131 A1.

20 Bart, S.C., Lobkovsky, E., and Chirik, P.J. (2004). *J. Am. Chem. Soc.* 126: 13794.

21 Bart, S.C., Chlopek, K., Bill, E. et al. (2006). *J. Am. Chem. Soc.* 128: 13901.

22 Trovitch, R.J., Lobkovsky, E., Bill, E., and Chirik, P.J. (2008). *Organometallics* 27: 1470.

23 (a) Russell, S.K., Darmon, J.M., Lobkovsky, E., and Chirik, P.J. (2010). *Inorg. Chem.* 49: 2782. (b) Yu, R.P., Darmon, J.M., Hoyt, J.M. et al. (2012). *ACS Catal.* 2: 1760.

24 Danopoulos, A.A., Wright, J.A., and Motherwell, W.B. (2005). *Chem. Commun.* 784.

25 Darmon, J.M., Yu, R.P., Semproni, S.P. et al. (2014). *Organometallics* 33: 5423.

26 Chirik, P.J. (2015). *Acc. Chem. Res.* 48: 1687.

27 Srimani, D., Diskin-Posner, Y., Ben-David, Y., and Milstein, D. (2013). *Angew. Chem. Int. Ed.* 52: 14131.

28 Alberico, E., Sponholz, P., Cordes, C. et al. (2013). *Angew. Chem. Int. Ed.* 52: 14162.

29 (a) Chakraborty, S., Brennessel, W.W., and Jones, W.D. (2014). *J. Am. Chem. Soc.* 136: 8564. (b) Xu, R., Chakraborty, S., Bellows, S.M. et al. (2016). *ACS Catal.* 6: 2127.

30 Gärtner, D., Welther, A., Rad, B.R. et al. (2014). *Angew. Chem. Int. Ed.* 53: 3722.

31 Brennessel, W.W., Jilek, R.E., and Ellis, J.E. (2007). *Angew. Chem. Int. Ed.* 46: 6132.

32 Büschelberger, P., Gärtner, D., Reyes-Rodriguez, E. et al. (2017). *Chem. Eur. J.* 23: 3139.
33 (a) Frings, A.J. (1988). Synthese und Reaktionen neuer Organoeisenverbindungen Ph.D. dissertation. Germany: University of Bochum; (b) Jonas, K. (1990). *Pure Appl. Chem.* 62: 1169. (c) Schnöckelborg, E.-M., Khusniyarov, M.M., de Bruin, B. et al. (2012). *Inorg. Chem.* 51: 6719.
34 (a) Aller, B.V. (1958). *J. Appl. Chem.* 8: 492. (b) Banwell, M.G., Jones, M.T., Reekie, T.A. et al. (2014). *Org. Biomol. Chem.* 12: 7433.
35 Alley, W.M., Hamdemir, I.K., Wang, Q. et al. (2011). *Langmuir* 27: 6279.
36 Chen, F., Surkus, A.-E., He, L. et al. (2015). *J. Am. Chem. Soc.* 137: 11718.
37 Son, S.U., Park, K.H., and Chung, Y.K. (2002). *Org. Lett.* 4: 3983.
38 (a) Wender, I., Levine, R., and Orchin, M. (1950). *J. Am. Chem. Soc.* 72: 4375. (b) Friedman, S., Metlin, S., Svedi, A., and Wender, I. (1959). *J. Org. Chem.* 24: 1287.
39 (a) Hidai, M., Kuse, T., Hikita, T. et al. (1970). *Tetrahedron Lett.* 11: 1715. (b) Pregaglia, G.F., Andreetta, A., Ferrari, G.F., and Ugo, R. (1971). *J. Organomet. Chem.* 30: 387. (c) Ferrari, G.F., Andreetta, A., Pregaglia, G.F., and Ugo, R. (1972). *J. Organomet. Chem.* 43: 209. (d) Feder, H.M. and Halpern, J. (1975). *J. Am. Chem. Soc.* 97: 7186.
40 Ohgo, Y., Takeuchi, S., Natori, Y., and Yoshimura, J. (1981). *Bull. Chem. Soc. Jpn.* 54: 2124.
41 Kooistra, T.M., Knijnenburg, Q., Smits, J.M.M. et al. (2001). *Angew. Chem. Int. Ed.* 40: 4719.
42 Knijnenburg, Q., Horton, A.D., van der Heijden, H. et al. (2005). *J. Mol. Catal. A: Chem.* 232: 151.
43 Knijnenburg, Q., Hetterscheid, D., Kooistra, T.M., and Budzelaar, P.H.M. (2004). *Eur. J. Inorg. Chem.* 1204.
44 (a) Monfette, S., Turner, Z.R., Semproni, S.P., and Chirik, P.J. (2012). *J. Am. Chem. Soc.* 134: 4561. (b) Bianchini, C., Mantovani, G., Meli, A. et al. (2003). *Eur. J. Inorg. Chem.* 1620.
45 (a) Hopmann, K.H. (2013). *Organometallics* 32: 6388. (b) Friedfeld, M.R., Shevlin, M., Margulieux, G.W. et al. (2016). *J. Am. Chem. Soc.* 138: 3314.
46 Chen, J., Chen, C., Ji, C., and Lu, Z. (2016). *Org. Lett.* 18: 1594.
47 Yu, R.P., Darmon, J.M., Milsmann, C. et al. (2013). *J. Am. Chem. Soc.* 135: 13168.
48 Zhang, G., Scott, B.L., and Hanson, S.K. (2012). *Angew. Chem. Int. Ed.* 51: 12102.
49 Lin, T.-P. and Peters, J.C. (2013). *J. Am. Chem. Soc.* 135: 15310.
50 Brennessel, W.W., Young, V.G., and Ellis, J.E. (2002). *Angew. Chem. Int. Ed.* 41: 1211.
51 Camacho-Bunquin, J., Ferguson, M.J., and Stryker, J.M. (2013). *J. Am. Chem. Soc.* 135: 5537.
52 Friedfeld, M.R., Shevlin, M., Hoyt, J.M. et al. (2013). *Science* 342: 1076.
53 Friedfeld, M.R., Margulieux, G.W., Schaefer, B.A., and Chirik, P.J. (2014). *J. Am. Chem. Soc.* 136: 13178.
54 Raney, M. (1927). US1628190 A.

55 Urushibara, Y., Nishimura, S., and Uehara, H. (1955). *Bull. Chem. Soc. Jpn.* 28: 446.

56 Keim, W. (1990). *Angew. Chem. Int. Ed. Engl.* 29: 235.

57 Itatani, H. and Bailar, J.C. (1967). *J. Am. Chem. Soc.* 89: 1600.

58 Angulo, I.M., Kluwer, A.M., and Bouwman, E. (1998). *Chem. Commun.* 2689.

59 González-Arellano, C., Gutiérrez-Puebla, E., Iglesias, M., and Sánchez, F. (2004). *Eur. J. Inorg. Chem.* 1955.

60 Vasudevan, K.V., Scott, B.L., and Hanson, S.K. (2012). *Eur. J. Inorg. Chem.* 4898.

61 Harman, W.H. and Peters, J.C. (2012). *J. Am. Chem. Soc.* 134: 5080.

62 Lin, T.-P. and Peters, J.C. (2014). *J. Am. Chem. Soc.* 136: 13672.

63 Angulo, I.M., Bouwman, E., van Gorkum, R. et al. (2003). *J. Mol. Catal. A: Chem.* 202: 97.

64 Mooibroek, T.J., Wenker, E.C.M., Smit, W. et al. (2013). *Inorg. Chem.* 52: 8190.

6

Base Metal-Catalyzed Addition Reactions Across C—C Multiple Bonds

Rodrigo Ramírez-Contreras and Bill Morandi

Max-Planck-Institut für Kohlenforschung, Kaiser-Wilhelm-Platz 1, 45470 Mülheim an der Ruhr, Germany

6.1 Introduction

Catalytic derivatization of alkenes and alkynes via addition of functional groups across the C—C multiple bonds is of great importance for the field of organic synthesis because organic building blocks such as amines, alcohols, alkyl halides, or carbonyls can be prepared conveniently in fewer synthetic steps compared to more traditional approaches [1]. Furthermore, unsaturated carbon scaffolds, and alkenes in particular, are not only easily accessible on bulk scale from petroleum but can also be readily prepared through a plethora of synthetic methods.

The term "base metal" refers to earth-abundant, relatively inexpensive metals that include late first-row transition metals, but it also includes main group metals such as Pb, Al, and Zn [2]. This chapter describes landmark research published from 2006 to date, a span of 10 years, in the area of late first-row (Mn to Cu) transition metal-catalyzed additions to nonpolar C—C multiple bonds exclusively. First-row metals not only provide a more economical alternative to noble transition metals, so profusely used in catalysis in general, but also provide mechanistically complementary approaches to unsolved problems in organic synthesis.

Given the vast amount of work done in this area, and the limited space that these pages provide, cycloaddition and intramolecular reactions are not discussed in this chapter. In Sections 6.2 and 6.3, several recent examples of reactions that proceed through one-electron processes are presented, as these types of mechanisms are a common theme among late first-row transition metals. Section 6.4 highlights reactions that operate through two-electron mechanisms. Finally, hydrosilylation reactions, which proceed through a variety of mechanisms, will be examined in Section 6.5.

Non-Noble Metal Catalysis: Molecular Approaches and Reactions,
First Edition. Edited by Robertus J. M. Klein Gebbink and Marc-Etienne Moret.

6.2 Catalytic Addition to Alkenes Initiated Through Radical Mechanisms

6.2.1 Hydrogen Atom Transfer as a General Approach to Hydrofunctionalization of Unsaturated Bonds

Hydrogen atom transfer (HAT) is an elementary mechanistic step that can be regarded as a type of proton-coupled electron transfer (PCET) reaction in which H^+ and e^- move together in the form of a hydrogen atom [3]. Several olefin hydrogenation reactions catalyzed by first-row transition metal hydrides operate through HAT mechanisms [4]. In such instances, $H^{\cdot}$ is transferred from M—H to an unsaturated C—C bond to form a metalloradical and a carbon-centered radical that can trap a second $H^{\cdot}$ to generate the formally hydrogenated product. Isayama and Mukaiyama reported in 1989 their work on a hydration reaction of olefins catalyzed by a Co^{II} diketonate complex in the presence of molecular oxygen as the oxidant and silane as the reductant (Scheme 6.1) [5]. In this process, the carbon-centered radical generated by HAT from [Co]–H is intercepted by molecular oxygen. Further reduction of the reaction intermediates results in the formation of the unprotected alcohol. This seminal report sets an important precedent for olefin hydrofunctionalization reactions that has inspired a plethora of research laboratories to develop very powerful reactions initiated by HAT mechanisms (Scheme 6.2). We have selected some recent examples that demonstrate the power of this approach to realize that transformations are not easily achievable using classical ionic pathways.

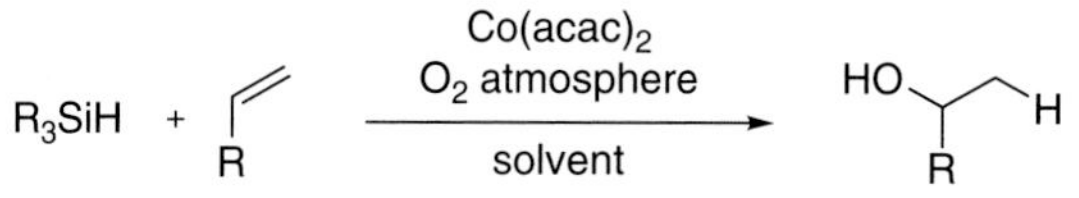

$$R_3SiH + \text{(alkene, R)} \xrightarrow[\text{solvent}]{Co(acac)_2,\ O_2 \text{ atmosphere}} \text{HO–CH(R)–CH}_2\text{–H}$$

Scheme 6.1 Mukaiyama hydration of olefins.

6.2.2 Hydrazines and Azides via Hydrohydrazination and Hydroazidation of Olefins

6.2.2.1 Co- and Mn-Catalyzed Hydrohydrazination

Carreira and coworkers found that a mixture of Co^{III} complexes **1a** and **1b** could efficiently catalyze the hydrohydrazination of terminal olefins in the presence of $PhSiH_3$ in ethanol using di-*t*-butylazidocarboxylate (**2**) as the hydrazine source (Scheme 6.3) [7, 8].

It was found that the steric bulk around the N=N unit is crucial in order to avoid the competing reduction of **2** to its corresponding hydrazine. Monosubstituted olefins afforded the respective products in yields above 70%, with high Markovnikov regioselectivity. Styrenes can be used as substrates for this reaction, although slow addition of the substrate and low catalyst loadings are necessary in this particular case in order to minimize undesired polymerization side reactions. Hydrohydrazination products from 2-vinyl-substituted Boc- and Ts-protected pyrroles, and indole, thiophene, and furan were obtained in yields within the range of 68–80%. 1,1-Disubstituted, α,β-unsaturated, endocyclic

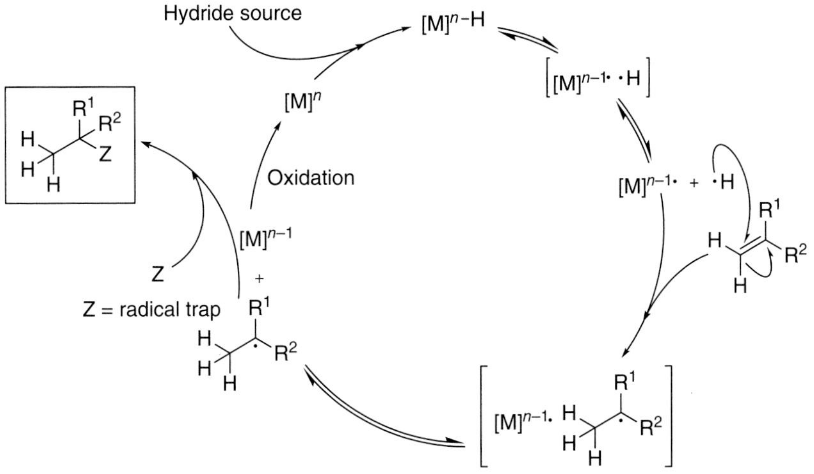

Scheme 6.2 Hydrogen atom transfer (HAT) mechanism for the Markovnikov hydrofunctionalization of olefins [6].

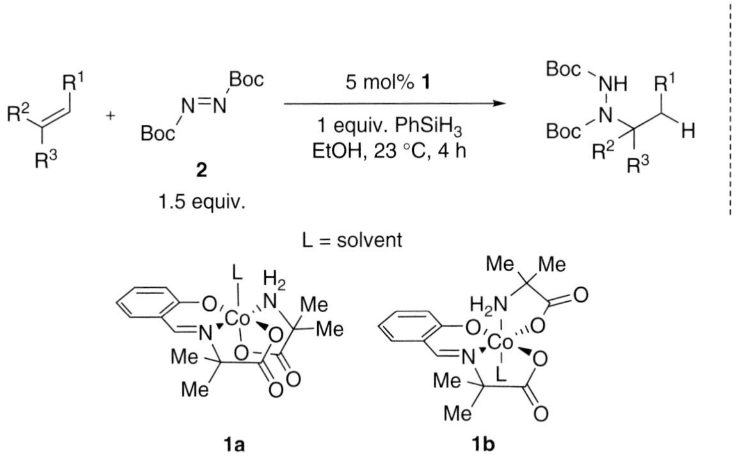

Scheme 6.3 Co-catalyzed hydrohydrazination reaction.

olefins and styrene derivatives provided the respective products in the range within 60–80%. In particular, exclusive Markovnikov regioselectivity is observed in the case where styrenes are used as substrates. Propargylic hydrazines can be accessed from enynes with a protecting group installed on the alkyne portion of the molecule (Scheme 6.4) [9]. With this type of substrates, care should be taken to quench the reaction mixture once it reaches completion in order to avoid over-reduction of the alkyne moiety. Conjugated dienes can be used as well, although the regioselectivity in this set of substrates is more dependent on the particular structure of each substrate. In all the cases presented, hydrohydrazination is favored over Diels–Alder cycloaddition.

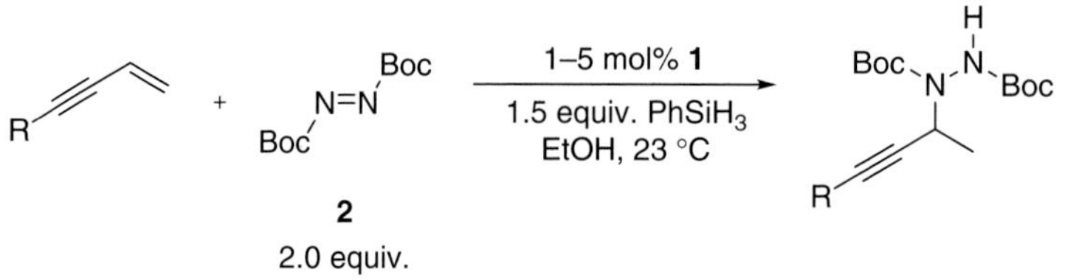

Boc N=N Boc + R (enyne) → Boc N N(H) Boc

2
2.0 equiv.

1–5 mol% **1**
1.5 equiv. $PhSiH_3$
EtOH, 23 °C

Scheme 6.4 Preparation of propargylic hydrazines from enynes by catalytic hydrohydrazination.

The instance of α,β-disubstituted olefins proved to be more challenging for the Co catalyst system. In this case, the rate of the product-forming reaction pathway becomes comparable to that of the parasitic reduction of **2**, hence lowering the yield of the target product below 50%. In an attempt to circumvent this problem, and based on the literature precedent of Mukaiyama's Mn-catalyzed hydration system [7, 10], $Mn(dpm)_3$ (**3**) was tested as a potential catalyst candidate (Scheme 6.5). Under optimized conditions, an overall increase in reactivity was observed when **3** was used relative to the Co(salen) system **4**, particularly in the case of endocyclic and tetrasubstituted olefins. This increased reactivity, however, comes at the expense of a decrease in the Markovnikov/anti-Markovnikov selectivity of the reaction – 5.5 : 1 with 80% yield under optimum conditions when 4-phenyl-1-butene is used as the substrate. It was also found that this system is compatible with the use of poly(methoxyhydroxysilane), a more economical hydride source, although at the expense of an increased reaction time relative to $PhSiH_3$.

6.2.2.2 Cobalt- and Manganese-Catalyzed Hydroazidation of Olefins

A method for the direct hydroazidation of unactivated alkenes, a discernible extension of the work on hydrohydrazination described in Section 6.2.2.1, was first reported by the Carreira group (Scheme 6.6a) [7, 11]. Initial studies on the catalytic hydroazidation revealed that the mixture of **1a/b** and **3** – which were shown to be the competent catalysts for hydrohydrazination of unactivated olefins – was not adequate. Modest yields of the hydroazidation product obtained from test reactions using the mixture **1a/b** inspired further catalyst design. It was eventually discovered that a complex formed by the treatment of $Co(NO_3)_2{\cdot}6H_2O$ with ligand **5** followed by oxidation with H_2O_2 afforded an efficient hydroazidation catalyst [11]. More conveniently, such a complex can also be prepared *in situ* from $Co(BF_4)_2{\cdot}6H_2O$, a Co^{II} source that bears more weakly coordinating counter anions, and using *t*-BuOOH as an initiator. It can be speculated that a cationic Co^{III} species is formed in which the remaining coordination sites are taken up by solvent molecules, although no explicit mention was made in the original contribution. Such a formulation would be consistent with the reported observation of poor solubility of the complex in solvents of lower dielectric constant [7]. In terms of the catalytic reaction, both $PhSiH_3$ and tetramethyldisiloxane were shown to be appropriate hydride sources, with the latter being a more economical option which, in some instances, performs slightly below par with the more standard $PhSiH_3$. The reaction is conducted

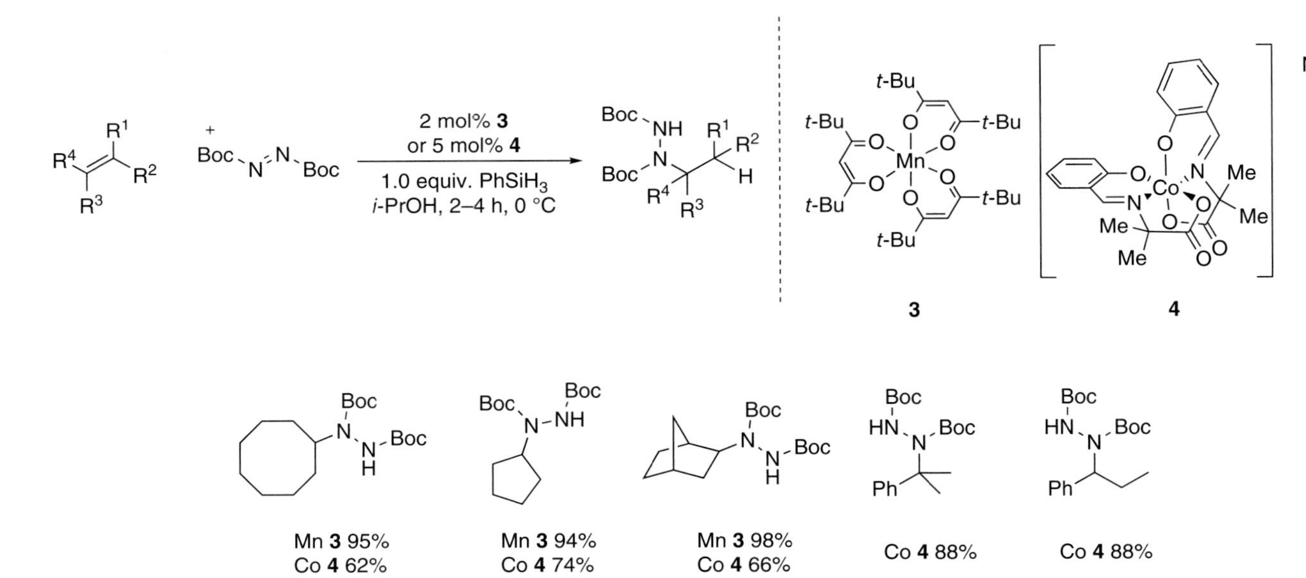

Scheme 6.5 Mn- and Co-catalyzed hydrohydrazination of olefins. Selected examples presented.

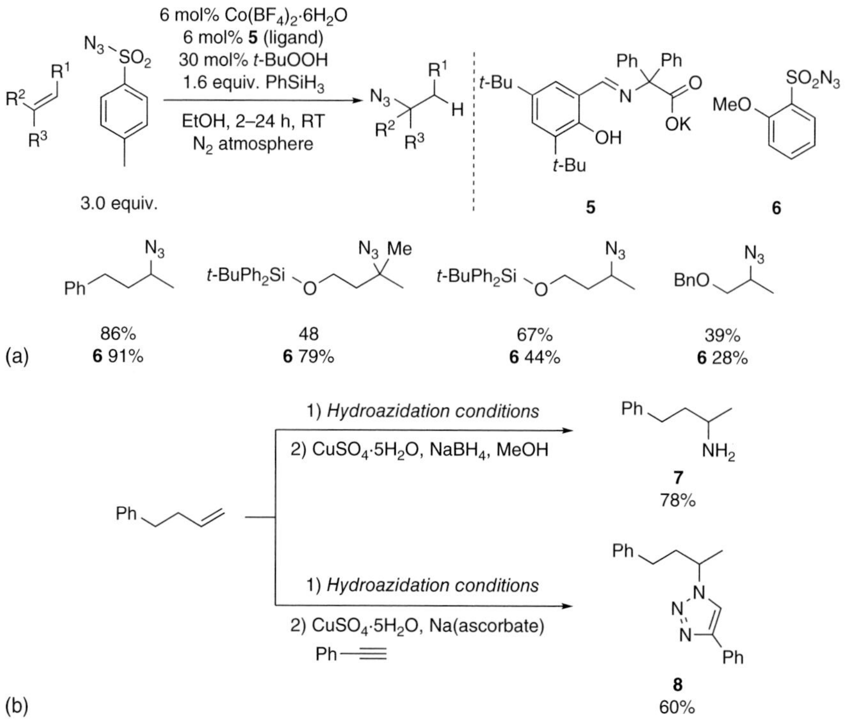

Scheme 6.6 (a) Co-catalyzed hydroazidation of unactivated olefins. (b) One-pot synthesis of amines and triazoles via catalytic hydroazidation.

under an inert gas atmosphere in order to preclude Mukaiyama-type side reactions with adventitious oxygen. The method afforded products with very high Markovnikov regioselectivity, and in terms of scope, substrates with free hydroxyl groups were not tolerated, but performs well once a silyl protecting group is installed. Esters and ketones are tolerated, as well as 1,1-disubstituted olefins. It was not possible to achieve full conversion when trisubstituted olefins were used, although useful yields (48%) of the respective products could be obtained when $PhSiH_3$ was used as the reductant. Substrates that contained an ester, phenyl, or alkyne group conjugated with the double bond did not afford any product. Attempts to improve the yields of more challenging substrates led to the design of the more reactive azide **6**. In test reactions using 4-phenyl-1-butene, it was found that by the substitution of TsN_3 for **6** in an otherwise identical experimental setup, the amount of azide source could be reduced by half to obtain yields of 90%. It was also found that in the case of the more challenging α-methyl disubstituted allylic ethers, yields were improved from c. 40% to more than 76%. Nevertheless, substrates that were unreactive when TsN_3 was used as the azide source were still found to be unreactive when **6** was employed.

One-pot syntheses of amines or triazoles – by *in situ* reduction or click reaction with an alkyne and the azide product, respectively – are possible given the mild reaction conditions and reagents used in the hydroazidation reaction (Scheme 6.6b, products **7** and **8**).

6.2.3 Co-Catalyzed Hydrocyanation of Olefins with Tosyl Cyanide

The cobalt-catalyzed hydrocyanation of olefins [12] represents a complementary approach to the direct addition of hydrogen cyanide [13] and avoids the potential hazards associated with the direct use of HCN gas (Scheme 6.7). The catalyst system consists of a Co^{II} salen complex (**9**) that is treated with $PhSiH_3$ to form the catalytically active Co^{III}—H *in situ* in degassed ethanolic solvent and *p*-toluenesulfonyl cyanide as the CN source. The reaction affords the product of net hydrocyanation with exclusive Markovnikov regioselectivity in good to excellent yields. Styrene derivatives and indene provided the products of cyanation at the benzylic position in moderate-to-good yields, whereas cyclic 1,2-disubstituted and α,β-unsaturated olefins did not provide the desired products.

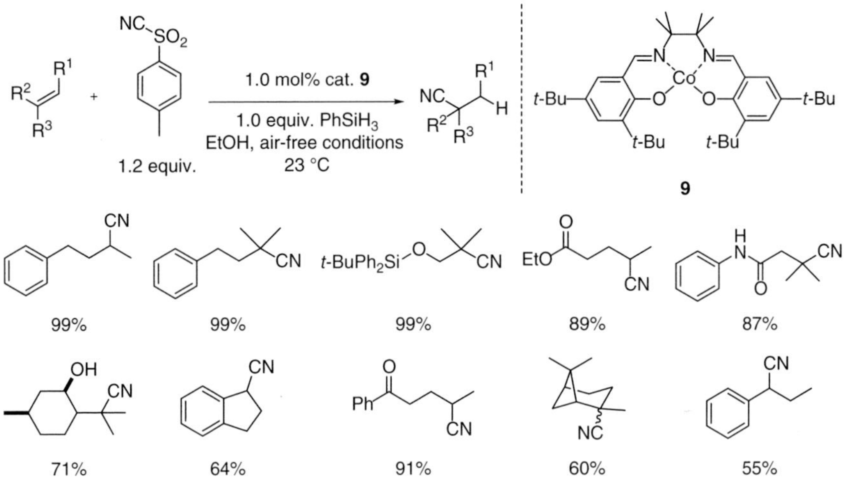

1.0 mol% cat. **9**, 1.0 equiv. $PhSiH_3$, EtOH, air-free conditions, 23 °C; $NC-SO_2$-tolyl 1.2 equiv.

Yields: 99%, 99%, 99%, 89%, 87%, 71%, 64%, 91%, 60%, 55%

Scheme 6.7 Cobalt-catalyzed hydrocyanation of olefins.

6.2.4 Co-Catalyzed Hydrochlorination of Olefins with Tosyl Chloride

Alkene hydrochlorination is a reaction that takes place at useful rates only with olefins that lead to stable carbocationic intermediates, or with strained alkenes. This lack of generality is a problem that was first addressed and solved by the Carreira group [14]. In this system, unactivated terminal and monosubstituted olefins can be efficiently hydrochlorinated using a cobalt hydride catalyst generated *in situ* from $Co(BF_4)_2$ and $PhSiH_3$ in the presence of **5** as a ligand and *t*-BuOOH as an initiator and *p*-toluenesulfonyl chloride (Scheme 6.8). Terminal alkenes with aromatic rings in the allylic or homoallylic position are good substrates for this reaction, whereas styrene derivatives did not yield product. The proposed catalytic cycle invokes a concerted olefin hydrocobaltation step with Markovnikov regioselectivity, followed by homolytic Co—C bond cleavage to form a carbon-based radical that is trapped by *p*-toluenesulfonyl chloride to form the hydrochlorinated product. Alternatively, a mechanism can be envisaged where HAT takes place, consistent with the observed Markovnikov

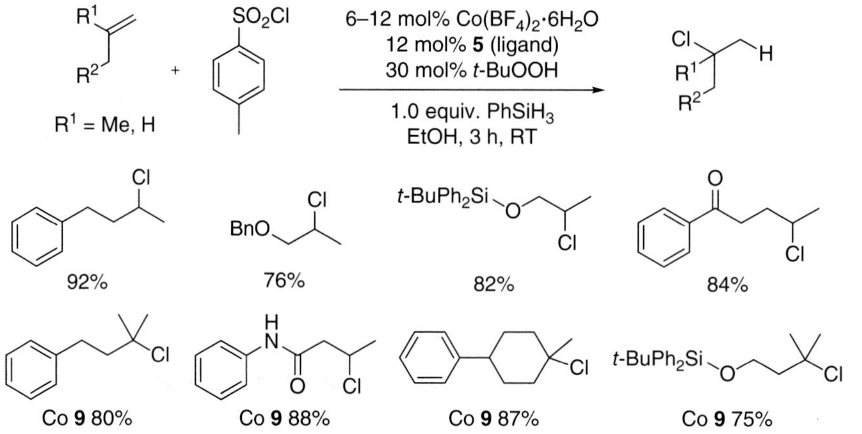

Scheme 6.8 Cobalt-catalyzed hydrochlorination of olefins.

regioselectivity, to form $Co^{II\cdot}$ and a carbon-based radical that is subsequently trapped with *p*-toluenesulfonyl chloride [6, 15].

6.2.5 $Fe^{III}/NaBH_4$-Mediated Additions of Unactivated Alkenes

Boger and coworkers have reported a series of studies on the hydrofunctionalization of unactivated double bonds mediated by stoichiometric amounts of Fe^{III} oxalate, using $NaBH_4$ as the hydride source (Scheme 6.9) [16]. This system was used in combination with several different types of radical traps, several of which are conveniently available as alkali metal salts of anions commonly found in the laboratory, namely, SCN^-, OCN^-, N_3^-, and NO_2^- to afford the respective addition products in good yields. The radical trap regioselectivity is for sulfur in the case of KSCN and for nitrogen in the case of KOCN. Unactivated terminal alkenes and styrenes were shown to be competent substrates for this transformation. This system displays Mukaiyama-type reactivity when O_2 is used in combination with catalytic amounts of Fe^{II} phthalocyanine.

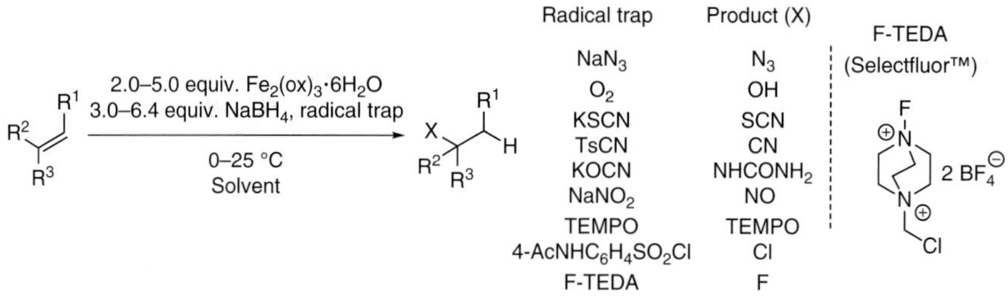

Radical trap	Product (X)
NaN_3	N_3
O_2	OH
KSCN	SCN
TsCN	CN
KOCN	$NHCONH_2$
$NaNO_2$	NO
TEMPO	TEMPO
$4\text{-}AcNHC_6H_4SO_2Cl$	Cl
F-TEDA	F

Scheme 6.9 Stoichiometric hydrofunctionalization of alkenes mediated by $Fe_2(ox)_3\cdot 6H_2O$.

Boger's $Fe^{III}/NaBH_4$ system also supports a radical hydrofluorination reaction that uses F-TEDA (also known by its trade name Selectfluor™) as the fluorine atom source (Scheme 6.9), albeit stoichiometric in Fe [17]. This precedent demonstrated the feasibility of radical fluorine addition across double bonds, a concept that was further developed shortly thereafter by Shigehisa et al., who devised a Co-catalyzed radical hydrofluorination reaction of unactivated alkenes (Scheme 6.10) [18]. In this case, the catalytically active hydride species is generated *in situ* from compound **9** and $(Me_2SiH)_2O$. *N*-Fluoro-2,4,6-trimethylpyridinium tetrafluoroborate functions as both the optimum fluorine source and the stoichiometric oxidant. The reaction affords products with Markovnikov regioselectivity exclusively, in yields up to 80%, and tolerates esters, amides, silyl ethers, alcohols, acetals, nitro, and tosylate functionalities. Nevertheless, this method does not tolerate amine, carboxylic acid, phenol, or alkyne functionalities – which were interestingly compatible with Boger's stoichiometric system – and is low yielding when the substitution/steric hindrance around the double bond increases. Mechanistic experiments revealed the intermediacy of radical species, and once again, although no explicit mention is made in the original work, a case can be made for a HAT-initiated catalytic cycle. It is worth to note that the reaction needs to proceed under air-free conditions to avoid a Mukaiyama-type parasitic reaction with O_2 that could lead to the formation of hydration products.

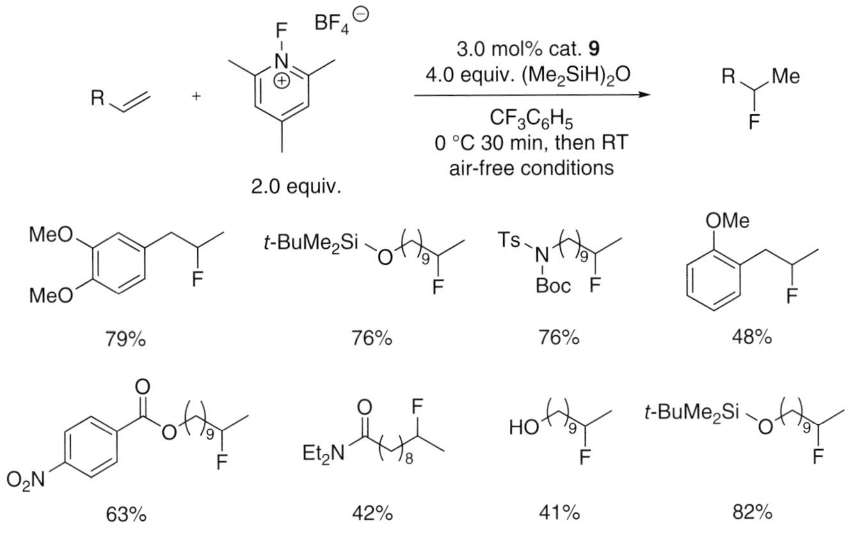

3.0 mol% cat. **9**
4.0 equiv. $(Me_2SiH)_2O$
$CF_3C_6H_5$
0 °C 30 min, then RT
air-free conditions
2.0 equiv.
79% 76% 76% 48%
63% 42% 41% 82%

Scheme 6.10 Co-catalyzed hydrofluorination of unactivated olefins.

6.2.6 Co-Catalyzed Markovnikov Hydroalkoxylation of Unactivated Olefins

Although several examples of inter- and intramolecular hydroalkoxylation reactions exist in the literature, limited functional group tolerance is a shortcoming that needed to be addressed. The cobalt-catalyzed hydroalkoxylation reaction reported by Shigehisa, Hiroya, and coworkers represents an improvement upon the limitations found on previous methods in terms of functional group tolerance (Scheme 6.11) [19, 20]. In this method, the reaction is conducted in alcoholic solvent, which is also the alkoxide source, in the presence of a fluoropyridinium reagent (**10** or **11**), used as the stoichiometric oxidant, $PhSiH_3$ or $(Me_2SiH)O$, used as the hydride sources, and the salen Co^{II} catalyst **9**. It was found that the optimum reaction conditions for a given substrate were not necessarily the same for another substrate, and therefore, the conditions had to be fine-tuned for each particular compound. In general, lower yields of the desired product were observed for all cases of *t*-butanol addition as a result of olefin isomerization. Silyl ether, acetal, nitro, amide, tosylate, and halogen functional groups were tolerated, and rather modest yields of a pyridine-substituted substrate were obtained, presumably as a result of its strong coordinating ability. Finally, a stoichiometric amount of the alcohol could be used if the reaction is carried out in benzotrifluoride.

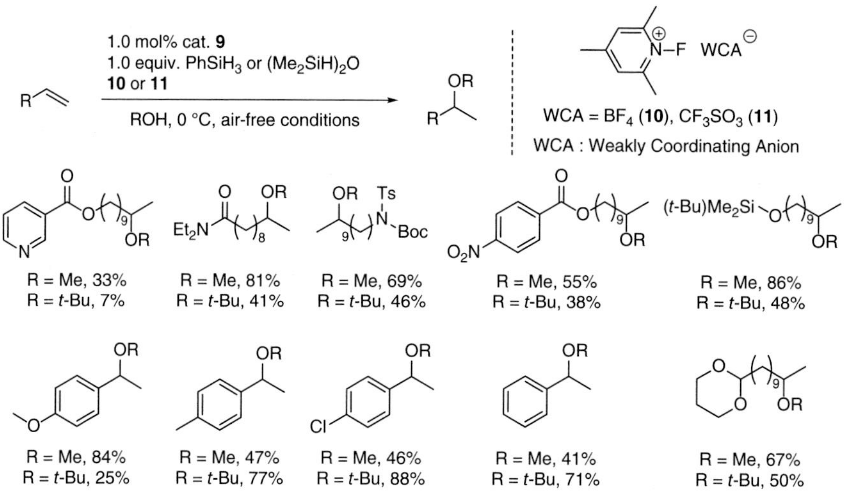

1.0 mol% cat. **9**
1.0 equiv. $PhSiH_3$ or $(Me_2SiH)_2O$
10 or **11**
ROH, 0 °C, air-free conditions

WCA = BF_4 (**10**), CF_3SO_3 (**11**)
WCA : Weakly Coordinating Anion

R = Me, 33%; R = *t*-Bu, 7%
R = Me, 81%; R = *t*-Bu, 41%
R = Me, 69%; R = *t*-Bu, 46%
R = Me, 55%; R = *t*-Bu, 38%
R = Me, 86%; R = *t*-Bu, 48%
R = Me, 84%; R = *t*-Bu, 25%
R = Me, 47%; R = *t*-Bu, 77%
R = Me, 46%; R = *t*-Bu, 88%
R = Me, 41%; R = *t*-Bu, 71%
R = Me, 67%; R = *t*-Bu, 50%

Scheme 6.11 Co-catalyzed Markovnikov hydroalkoxylation of olefins.

In terms of the mechanism, it is proposed that the Co^{II} pre-catalyst is oxidized to the active Co^{III} species by **10/11**, concomitant with the formation of R_3SiF and a Co^{III}—H species that adds across the olefin double bond with Markovnikov regioselectivity. The resulting organometallic species undergoes a homolytic Co—C bond cleavage to form a carbon-based radical that is then oxidized to a carbocationic intermediate by a Co^{III} species. This carbocation is then trapped by a molecule of alcohol to form the protonated product. Based on recent evidence for HAT reactions from Co^{III}—H species [6, 15], a HAT transfer step to the olefin to form a carbon-based radical is also plausible. It can be speculated that the low yields reported could be attributed in part to over-reduction of the substrate or to parasitic hydrofluorination/hydrodefluorination side reactions because carbocationic species were invoked in the catalytic cycle, although no further elaboration was provided.

6.2.7 Fe-Catalyzed Hydromethylation of Unactivated Olefins

Another example of a HAT-initiated reaction is the chemoselective hydromethylation of unactivated olefins mediated by stoichiometric $Fe(acac)_3$ (acac = acetylacetonate), using $PhSiH_3$ as the hydride source in the presence of MeOH [21]. The HAT step generates a carbon-based radical that is trapped by $H_2C{=}NNHSO_2$(*n*-Oct), generated *in situ*, to affect the C—C connection. The reaction can be made catalytic if $B(OMe)_3$ is added to the reaction mixture, in which case the amount of $Fe(acac)_3$ can be dropped down to 0.3 or 0.5 equiv. The resulting hydrazine is reductively cleaved in methanol to form the desired hydromethylated product. The reaction has to be conducted under oxygen-free conditions to avoid the formation of Mukaiyama-type oxidation products [5]. This method can be applicable to late-stage functionalization of complex organic architectures, for instance, hydromethylation of Gibberellic acid (Scheme 6.12), which is a transformation that would be otherwise challenging to achieve. This also happens to illustrate one of the limitations of the method, namely, the low yields obtained when sterically congested substrates are used. Additionally, styrenes mainly afford homodimerization products.

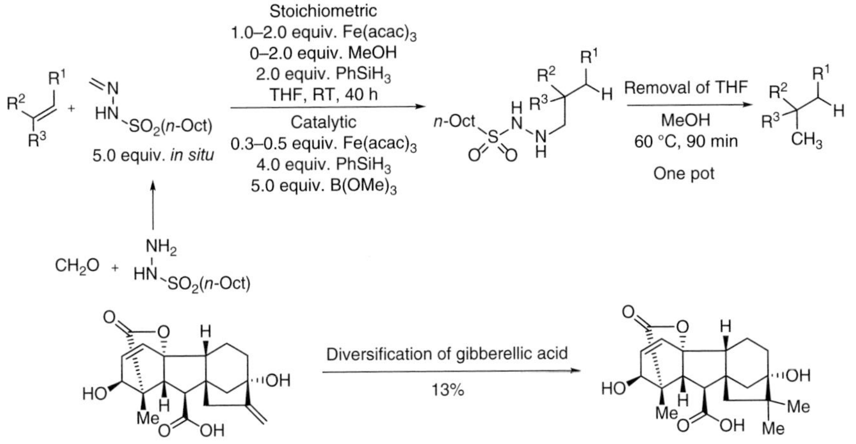

Stoichiometric
1.0–2.0 equiv. $Fe(acac)_3$
0–2.0 equiv. MeOH
2.0 equiv. $PhSiH_3$
THF, RT, 40 h

Catalytic
0.3–0.5 equiv. $Fe(acac)_3$
4.0 equiv. $PhSiH_3$
5.0 equiv. $B(OMe)_3$

R^1 R^2 R^3 + HN SO_2(*n*-Oct) 5.0 equiv. *in situ*

CH_2O + NH_2 HN SO_2(*n*-Oct)

n-Oct S O O N H N H R^1 R^2 R^3 H

Removal of THF
MeOH
60 °C, 90 min
One pot

R^1 R^2 R^3 H CH_3

Diversification of gibberellic acid
13%

HO O O H OH Me H O OH

Me Me

Scheme 6.12 Fe-catalyzed hydromethylation of olefins.

6.2.8 Hydroamination of Olefins Using Nitroarenes to Obtain Anilines

Baran and coworkers reported a method for a radical-mediated C—N connection using nitro(hetero)arenes and olefins, resulting in net hydroamination across the alkene double bond [22]. This method employs $Fe(acac)_3$ as the Fe^{III} source in loadings of 30 mol% and $PhSiH_3$ as the hydride source. The reaction is likely to proceed through HAT from the putative metal hydride to the olefin to form a carbon-centered radical that is trapped by the nitro(hetero)arene. A final reduction step is performed using Zn and aqueous HCl to afford the target amine in yields of c. 50%. The method displays high Markovnikov regioselectivity, which allows for the facile construction of sterically hindered amines that would be otherwise challenging to access. The method was shown to be general in terms of the reaction conditions required and was shown to simplify the construction of hindered amine drug intermediates (Scheme 6.13).

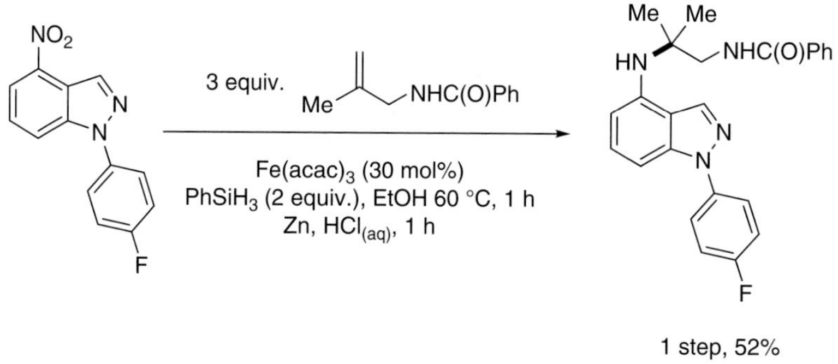

3 equiv. Me NHC(O)Ph

$Fe(acac)_3$ (30 mol%)
$PhSiH_3$ (2 equiv.), EtOH 60 °C, 1 h
Zn, $HCl_{(aq)}$, 1 h

1 step, 52%

Scheme 6.13 One-step synthesis of an intermediate of a glucocorticoid receptor modulator.

This method displays very good tolerance to diverse heteroaromatic rings; halides and pseudohalides; and amino, hydroxyl, and carbonyl groups. In particular, the tolerance of the latter group allows for intramolecular cascade cyclizations. Nevertheless, 2-nitropyridines, free (thio)phenols, and nitroimidazoles were not particularly well tolerated, and nitroalkanes gave much lower yields than nitroarenes. Another limiting aspect of the method is the need to use more than 1 equiv. of the olefin, in which case the method is best suited when the nitro compound is more valuable.

In a recent study on HAT-initiated hydrogenations of olefins, it was found that Ph(*i*-PrO)SiH_2 is one of the several solvolysis products of $PhSiH_3$ and that it is consumed at a much faster rate than the rest of the silanols in the reduction reaction [23]. This observation led to the implementation of Ph(*i*-PrO)SiH as a more reactive hydride source in HAT-mediated reactions. For instance, use of Ph(*i*-PrO)SiH_2 as the hydride source allowed for the Baran hydroamination reaction to proceed at room temperature with a much lower pre-catalyst loading of 1 mol% of $Fe(acac)_3$ in a mixture of isopropanol and ethyl acetate to obtain good yields at the expense of an increased reaction time, or at a higher yield than the seminal report if the reaction is conducted at 60 °C [23]. A further development of the Fe-catalyzed hydroamination reaction is the replacement of the iron source for the amino-bis(phenolate) Fe^{III} complex **12** in a 2 mol% loading, in conditions otherwise identical to Baran's work (Figure 6.1) [24].

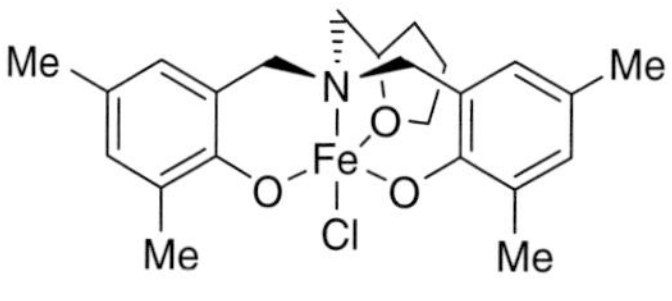

12

Figure 6.1 Amino-bis(phenolate) Fe^{III} catalyst precursor for the hydroamination reaction of olefins.

6.2.9 Dual-Catalytic Markovnikov Hydroarylation of Alkenes

The Markovnikov hydroarylation of olefins is a method that was developed by a combination of two catalytic processes, one in which carbon-centered radicals are generated by a HAT step from a Co^{III}—H to an olefin and a second step in which such radicals are trapped by a Ni^{II}(X)(Ar) complex to finally generate the target hydroarylation product [25]. It can be speculated that the C—C bond formation step proceeds through a Ni^{III}(X)(R)(Ar) species [26]. The catalytically active Co^{III}—H is formed by oxidation of the salen Co^{II} complex **13** with the fluoropyridinium salt **10**, followed by hydride transfer from Ph(*i*-PrO)SiH_2. This particular silane was also a competent reductant for Ni^{II} complex **14**. The scope of the reaction is limited to terminal olefins for the reason that internal olefins displayed a tendency to undergo isomerization under the optimized reaction conditions. In terms of the scope of electrophiles, electron-poor arenes seemed to perform better in general, presumably as a result of increased rates of Ar–X oxidative addition to Ni [25]. The performance of the system improved as the distance between the substituent and the double bond increased and organoboronates, halides, hydroxyl groups, and heteroaromatic groups were tolerated (Scheme 6.14). Preliminary mechanistic experiments involving radical clocks suggested the intermediacy of radical species, although further mechanistic work would be necessary to clearly establish the reaction pathway.

6.3 Other Catalytic Additions to Unsaturated Bonds Proceeding Through Initial $R^{\cdot}$ ($R \neq H$) Attack

In this section, we will present relevant examples of late first-row transition metal-catalyzed reactions proceeding through the initial addition of a main group-element radical R · ($R \neq H$) to a carbon–carbon multiple bond. In particular, transformations that introduce functional groups relevant to medicinal chemistry, such as the CF_3 and NR_3, will be described.

6.3.1 Cu-Catalyzed Trifluoromethylation of Unactivated Alkenes

Trifluoromethylation and trifluoromethylazidation of olefins can be obtained by direct, Cu-catalyzed incorporation of the trifluoromethyl and azido group into unactivated terminal olefins [27, 28]. These methodologies are described in detail in Chapter 13.

6.3.2 Mn-Catalyzed Aerobic Oxidative Hydroxyazidation of Alkenes

1,2-Azido alcohols are versatile building blocks that can be used in the synthesis of a wide variety of different molecules via manipulation of the azide or hydroxyl groups. Such molecules include amino alcohols, aziridines, lactones, triazoles, among others. Traditionally, 1,2-azido alcohols have been accessed through azide ring opening of epoxides [29], transfer hydrogenation of α-azido ketones, or reduction and substitution of α-bromo ketones [30]. Jiao and coworkers has

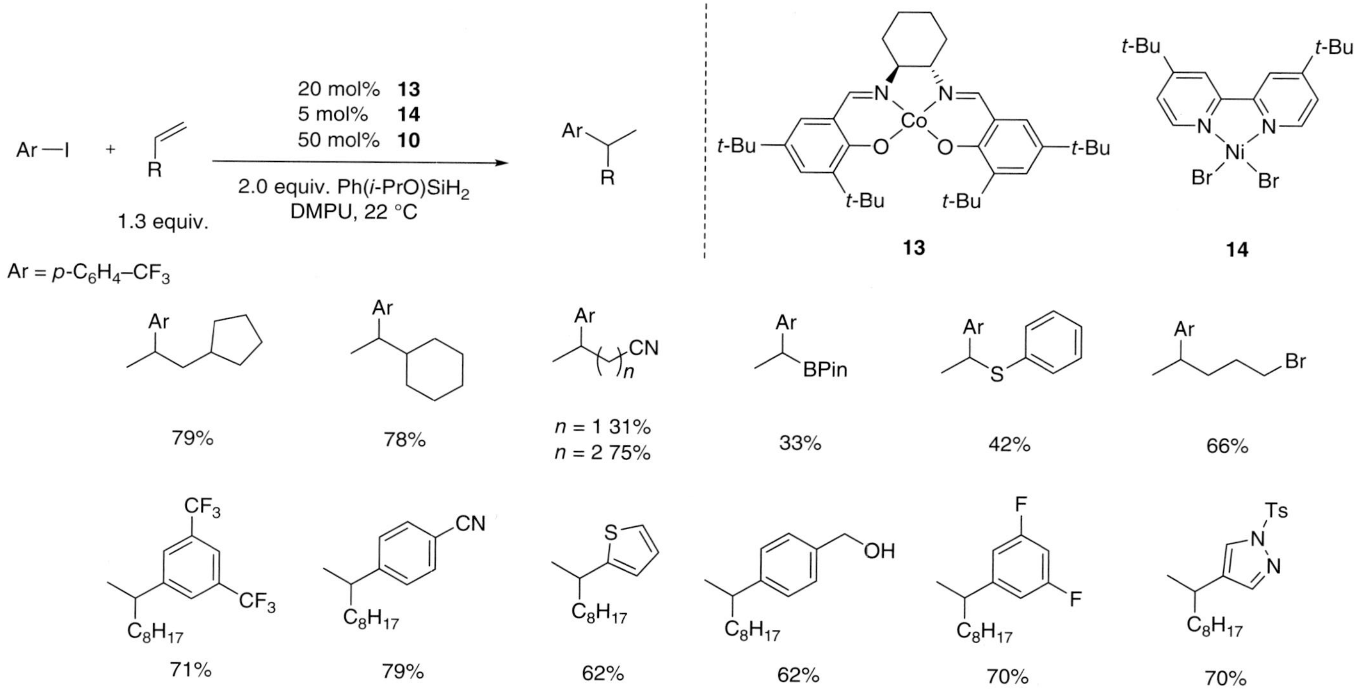

Ar—I + (olefin, R) 1.3 equiv.

20 mol% **13**, 5 mol% **14**, 50 mol% **10**; 2.0 equiv. Ph(*i*-PrO)SiH$_2$, DMPU, 22 °C → Ar(CH$_3$)CH–R

Ar = *p*-C$_6$H$_4$–CF$_3$

79%; 78%; *n* = 1 31%, *n* = 2 75%; 33%; 42%; 66%

71%; 79%; 62%; 62%; 70%; 70%

Scheme 6.14 Dual-catalytic hydroarylation of olefins. Isolated yields of representative examples of products are presented.

reported a direct and highly regioselective synthesis of 1,2-azidoalcohols from olefins and $TMSN_3$ catalyzed by Mn^{II}, which also employs atmospheric oxygen as the terminal oxidant, reminiscent of Mukaiyama-type hydrations, and uses PPh_3 as the reductant [31].

Styrenes substituted with both electron-donating and electron-withdrawing groups afforded the target products in yields above 70% (Scheme 6.15). Methyl substitution on the ring caused the yields to drop slightly, from 92% for *p*-methyl to 78% for *o*-methyl. When conjugated dienes and enynes are employed, the method displays high regioselectivity for the terminal ene position and affords good yields of the products. A variety of linear and cyclic aliphatic alkenes performed well. Heteroarenes and functional groups susceptible to oxidation were tolerated. Finally, 2-vinylbenzoic acids afforded, interestingly, cyclic peroxyalcohols.

Based on density functional theory (DFT) calculations, it is proposed that the reaction proceeds through azide radicals – generated by oxidation mediated by a Mn^{III} species – which are then trapped by the olefin to form a carbon-based radical, which in turn reacts with oxygen to form an organic superoxo radical. The superoxo species is sequentially reduced to the peroxo species by Mn^{II}, followed by a final reduction step to the alcohol by PPh_3 (Scheme 6.16).

6.3.3 Fe-Catalyzed Aminohydroxylation of Alkenes

Inspired by recent literature in the area of Fe-catalyzed aminohydroxylation, most notably the work of Yoon, Xu, and coworkers [32], the Morandi group has designed a method that permits direct access to N-unprotected 1-amino-2-alcohols, which are an important subclass of scaffolds relevant in medicinal chemistry, from alkenes [33]. The reaction is conducted in either neat alcohol or a mixture of water/acetonitrile and is catalyzed by the Fe^{II} phthalocyanine complex **15** at room temperature. $PivONH_3OTf$ is used as the electrophilic amino group source. This particular choice of amino group source circumvents the common problem of catalyst poisoning encountered when other N-unprotected aminating agents are used. The reaction afforded 1-amino-2-alcohols as the only regioisomer in good yields in the case where styrenes were used as substrates, although aliphatic alkenes could also be used albeit at the expense of decreased yields. Styrenes substituted with electron-donating and mildly electron-withdrawing groups were tolerated, and amino ethers can be conveniently accessed when alcoholic solvent is used instead of water (Scheme 6.17). Aminohydroxylation of 1,1-disubstituted olefins afforded products with a tertiary alcohol center, which is susceptible to elimination reactions in the acidic reaction medium to form allylic amines. This side reaction can, in some cases, substantially erode the yield of the target product (Scheme 6.17). The crude products can be derivatized directly without need for further purification, and if isolation of pure products is desired, direct purification of the unprotected amines is also possible.

A drastic decrease in the yield of product to c. 20% when the reaction is conducted in the presence of (2,2,6,6-tetramethylpiperidin-1-yl)oxyl (TEMPO) suggests the involvement of radical species in the reaction pathway. Nevertheless,

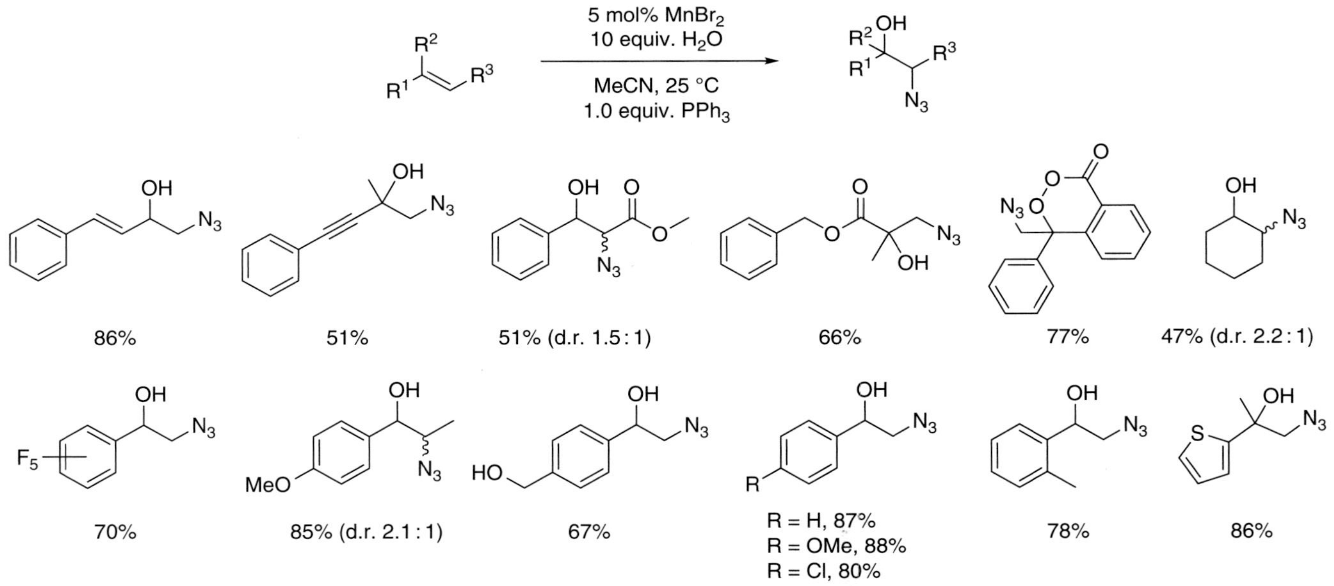

5 mol% $MnBr_2$
10 equiv. H_2O
MeCN, 25 °C
1.0 equiv. PPh_3

86%

51%

51% (d.r. 1.5 : 1)

66%

77%

47% (d.r. 2.2 : 1)

70%

85% (d.r. 2.1 : 1)

67%

R = H, 87%
R = OMe, 88%
R = Cl, 80%

78%

86%

Scheme 6.15 Mn-catalyzed, aerobic oxidative hydroxyazidation of olefins. Representative examples of products are presented.

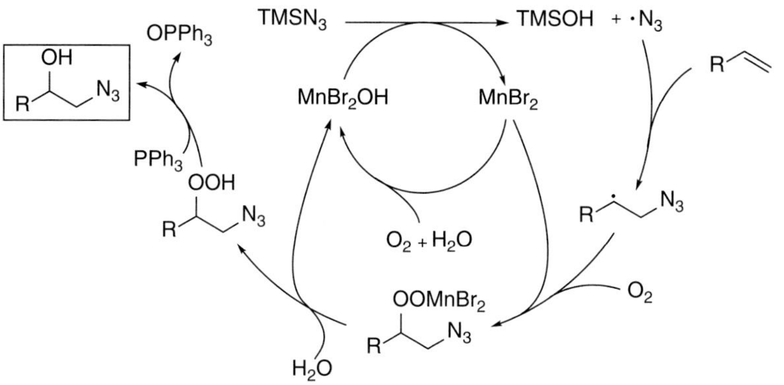

Scheme 6.16 Proposed catalytic cycle for the Mn-catalyzed, aerobic oxidative hydroxyazidation of olefins.

when cyclopropane-substituted olefins were subjected to the reaction conditions, no ring-opening products were detected, which would be consistent with either the intermediacy of short-lived radical species, or a polar mechanism being operative. It was postulated that an alternative pathway could involve the nucleophilic ring opening of a protonated aziridine intermediate (Scheme 6.18).

6.4 Catalytic Addition to Alkenes Initiated Through Polar Mechanisms

A plethora of metal hydrides of main group and transition elements undergo addition reactions of the M—H bond across C—C multiple bonds to form organometallic species, which can be further transformed, given the reactivity of the newly formed C—M bond. This type of addition reaction, known as hydrometallation, is usually an intermediate mechanistic step in the context of a larger catalytic cycle. Transformations involving hydrometallation reactions have found a myriad of applications in catalytic functionalization [34]. In this section, we present recent examples of first-row transition metal-catalyzed addition reactions across unsaturated bonds proceeding through an initial hydrometallation step – or, in one example, a carbometallation step – including hydroamination, hydrocyanation, and hydrosilylation reactions.

6.4.1 Cu-Catalyzed Hydroamination of Alkenes and Alkynes

Efficient methods for the regio- and enantioselective hydroamination of styrenes and unactivated alkenes have been developed independently by the groups of Buchwald, and Hirano and Miura [35]. Both methods rely on ligated copper hydride catalysts and hydroxylamino-*N*-OBz as the electrophilic source of nitrogen (Scheme 6.19). In the case of the Buchwald system, the catalytically active copper hydride catalyst is formed *in situ* from diethoxysilane, 2 mol% of $Cu(OAc)_2$, and a chiral ligand. Enantioselectivities of up to 99% ee were attained

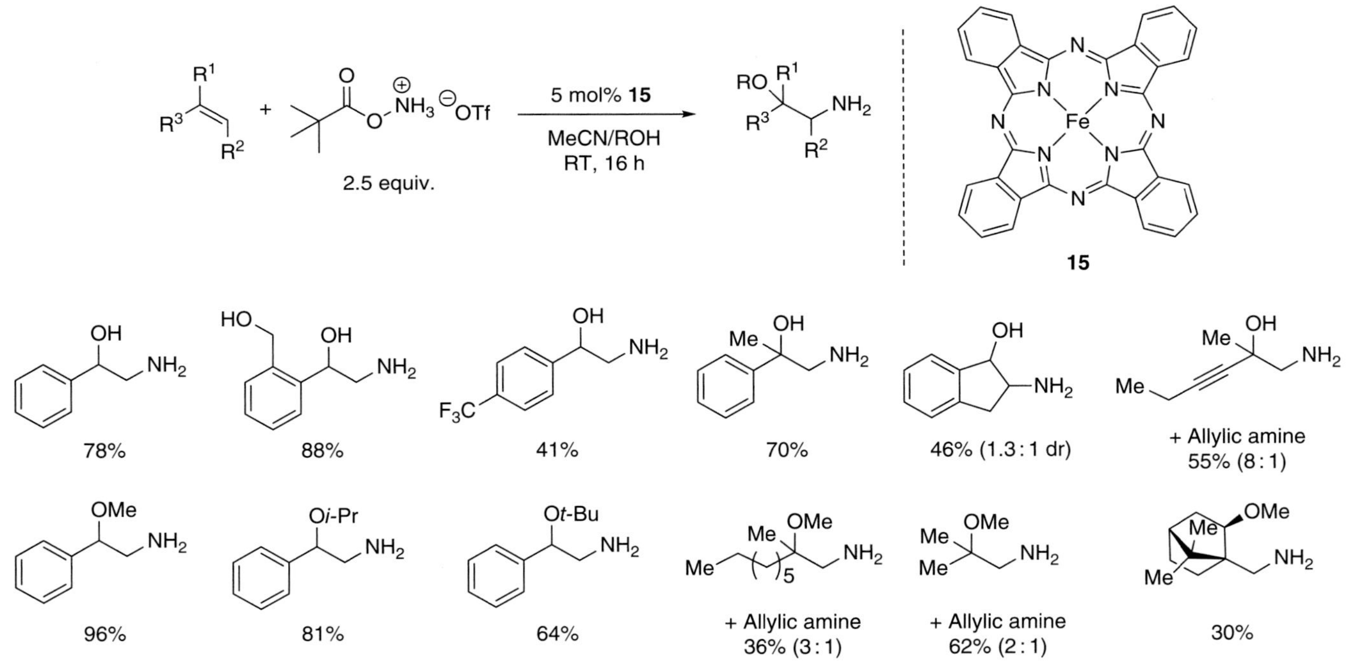

5 mol% **15**

MeCN/ROH
RT, 16 h

2.5 equiv.

15

78%　88%　41%　70%　46% (1.3 : 1 dr)　+ Allylic amine 55% (8 : 1)

96%　81%　64%　+ Allylic amine 36% (3 : 1)　+ Allylic amine 62% (2 : 1)　30%

Scheme 6.17 Aminohydroxylation of olefins catalyzed by Fe^{II} phthalocyanine complex **15**. Ratios between target product and allylic amine presented in parenthesis.

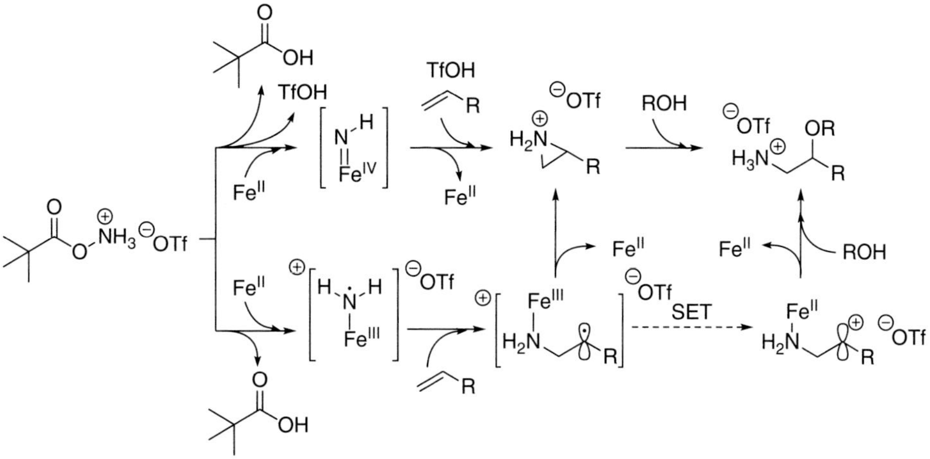

Scheme 6.18 Proposed reaction mechanisms for the aminohydroxylation reaction catalyzed by **15**.

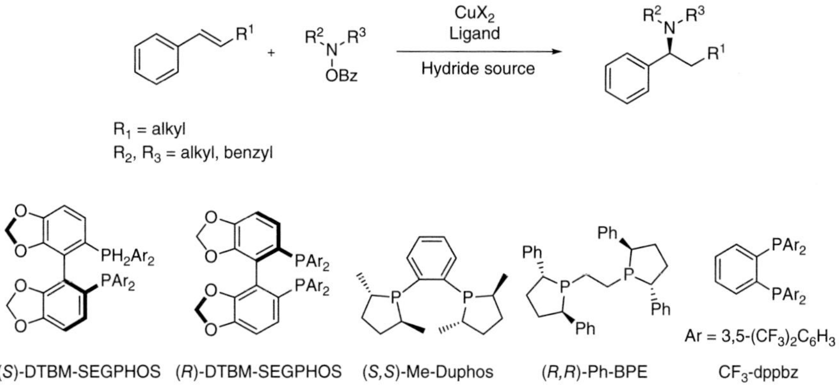

Scheme 6.19 The Cu-catalyzed hydroamination of styrenes.

when the (*R*)-(−)-5,5′-bis[di(3,5-di-*tert*-butyl-4-methoxyphenyl)phosphino]-4,4′-bi-1,3-benzodioxole ((*R*)-DTBM-SEGPHOS) ligand was used. This method performs extremely well when β-substituted and β,β-disubstituted styrenes are used as substrates and displays excellent Markovnikov regioselectivity. In contrast to this, anti-Markovnikov selectivity is observed when aliphatic alkenes are used. This selectivity can be traced back to the stabilization of the resulting radical by delocalization over the π-system of styrene, whereas in the case of aliphatic alkenes, sterics play a more important role [36]. In the latter case, β-chiral amines can be accessed when 1,1-disubstituted alkenes are used (Scheme 6.20) [35d]. α-Aminosilanes can be obtained by an analogous procedure using vinyl silanes [35e].

A further development of the hydroamination reaction is the copper-catalyzed asymmetric hydroamination of unactivated internal olefins developed by the Buchwald group (Scheme 6.21) [37].

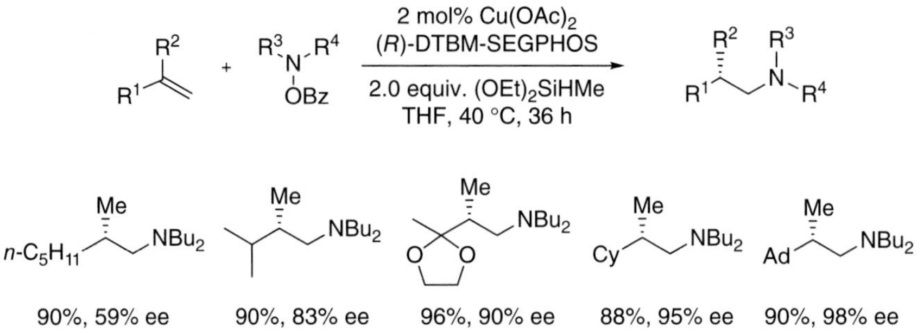

2 mol% $Cu(OAc)_2$
(*R*)-DTBM-SEGPHOS
2.0 equiv. $(OEt)_2SiHMe$
THF, 40 °C, 36 h

90%, 59% ee   90%, 83% ee   96%, 90% ee   88%, 95% ee   90%, 98% ee

Scheme 6.20 Hydroamination of unactivated, 1,1-disubstituted alkenes. Selected examples of products are presented.

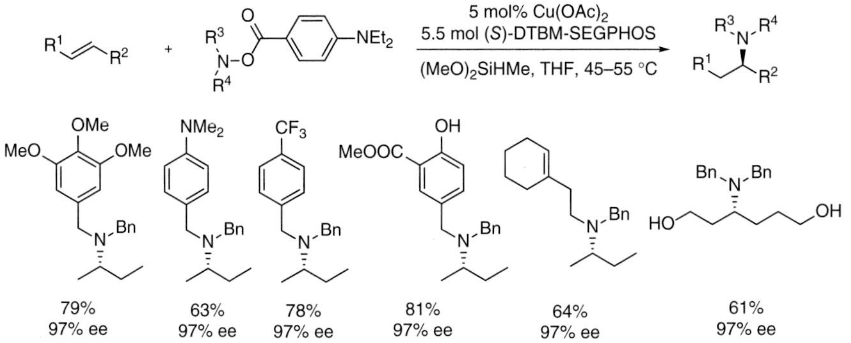

5 mol% $Cu(OAc)_2$
5.5 mol (*S*)-DTBM-SEGPHOS
$(MeO)_2SiHMe$, THF, 45–55 °C

79% 97% ee   63% 97% ee   78% 97% ee   81% 97% ee   64% 97% ee   61% 97% ee

Scheme 6.21 Copper-catalyzed asymmetric hydroamination of unactivated internal olefins.

The Hirano and Miura system uses polymethylhydrosiloxane as the source of hydride, LiO*t*-Bu as an additive, and CF_3-dppbz ligand to obtain racemic products. Good-to-moderate enantioselectivities were obtained when (*S,S*)-Me-Duphos and (*R,R*)-Ph-BPE (BPE = *bis*(2,5-diphenylphospholano) ethane) ligands were employed [35c]. This method also accommodated well for β-substituted and β,β-disubstituted styrenes.

An additional variant of the hydroamination reaction of alkenes is the direct hydroamination of alkynes to form enamines and the reductive hydroamination of alkynes to obtain linear or branched alkylamines (Scheme 6.22) [38]. In both instances, the alkyne first undergoes a syn-hydrocupration reaction to form a vinyl copper intermediate that can be intercepted directly with the hydroxylamino-*N*-OBz reagent to form the enamine. Alternatively, the vinyl copper species can be subjected to alcoholysis to form the corresponding alkene, which can then undergo a second hydrocupration step followed by reaction with the *N*-benzoate to form the desired amine. Both terminal and internal alkynes can be used as substrates for the reductive hydroamination reaction, whereas only internal alkynes could be used in the direct hydroamination reaction. The factors controlling the regioselectivity of the carbocupration step for both the direct and the reductive hydroamination reactions are the same as for the case of alkenes, Markovnikov regioselectivity is observed for aryl

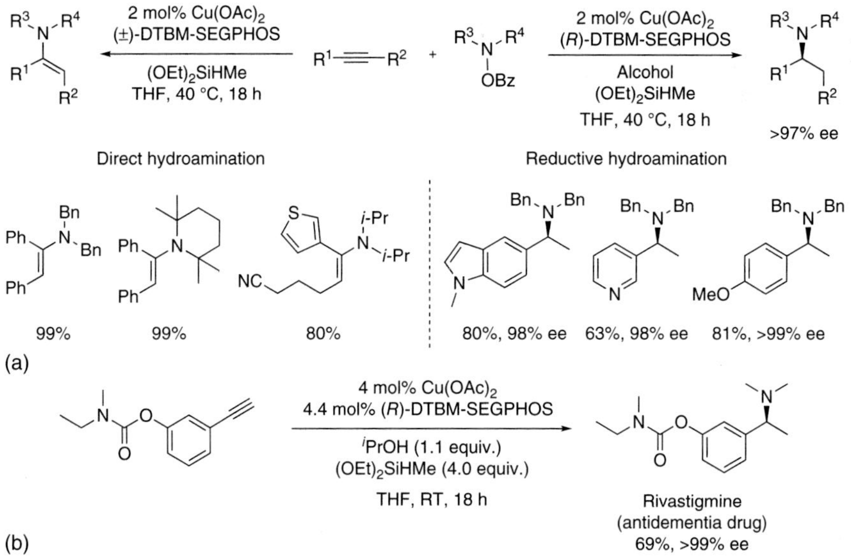

Scheme 6.22 (a) Direct and reductive hydroamination reactions of alkynes. (b) Application to the synthesis of the pharmaceutical target molecule rivastigmine.

alkynes and anti-Markovnikov selectivity for aliphatic alkynes. The reductive hydroamination reaction tolerates heterocycles, hydroxyl groups, aryl hylides, acetals, and ketals and affords products in good yields and enantioselectivities upward of 97% ee.

6.4.2 Ni-Catalyzed, Lewis-acid-Assisted Carbocyanation of Alkynes

Nakao, Hiyama, and coworkers reported the first examples of carbocyanation reactions of alkynes using aryl nitriles, catalyzed by a $Ni(cod)_2/PR_3$ catalyst system. Although reactivity was observed under catalytic conditions, it was found that the rate of arylcyanation of alkynes using electron-rich nitriles was normally rather slow, taking from one to several days to attain useful conversions [39a]. The reason for this phenomenon is thought to be the buildup of negative charge on the aryl *ipso*-carbon during the oxidative addition step of the C_{ar}—CN bond to the Ni^0 fragment, which would also be the rate-determining step [40]. Based on the literature precedent for the utility of Lewis acids in improving the rates of hydrocyanation reactions of olefins using Ni catalyst systems [41], it was postulated that the addition of a Lewis acid cocatalyst could accelerate the rate of carbocyanation reactions. It was indeed found that addition of Al- and B-based Lewis acid cocatalysts to the reaction mixtures significantly increased the reaction rates in most cases, an effect that could be traced back not only to increased rates of C—CN oxidative addition but also to increased rates of product-forming C—CN reductive elimination (Scheme 6.23) [39]. Syn-carbocyanation is favored, and when unsymmetrical alkynes are used as substrates, the formation of the isomer with the cyano group *gem* to the bulkier R group is preferred. This regioselectivity decreases as the difference in size between the alkyne R groups diminishes [39e].

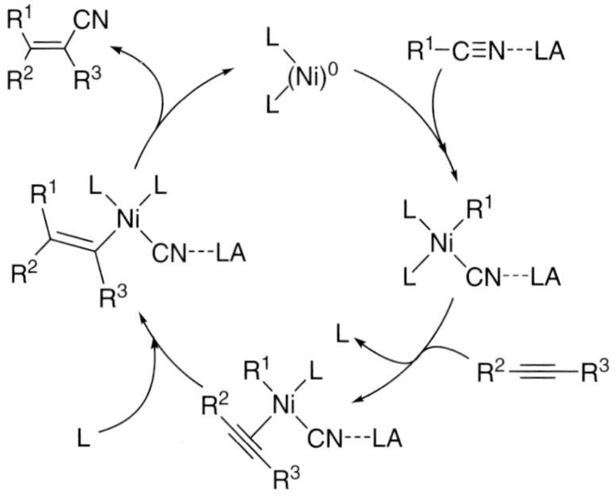

Scheme 6.23 Proposed mechanism for the Ni-catalyzed, Lewis-acid-assisted carbocyanation of alkynes.

The choice of optimum reaction parameters, including Lewis acid, phosphine ligands, temperature, and solvent, are substrate dependent. Given the diversity of R group sizes that can be present in any given set of substrates, the choice of phosphine with the appropriate balance of stereoelectronic properties is critically important, but it was noted that in general, the use of electron-rich phosphines is necessary in order to attain maximum efficiency. One notable exception to this would be the use of $P(4\text{-}CF_3\text{–}C_6H_4)_3$ in the carbocyanation reaction using allyl cyanide (Table 6.1). A summary of reaction conditions for a selected group of substrates is presented in Table 6.1. The system was extended to vinyl, allyl, and alkyl nitriles to afford the respective carbocyanation products in good yields.

6.4.3 Ni-Catalyzed Transfer Hydrocyanation

Inspired by the work of Nakao and Hiyama, a nickel-catalyzed, HCN-free hydrocyanation/retrocyanation reaction has been developed by the Morandi research group on the basis of the concept of shuttle catalysis [42, 43]. In this reaction, HCN is formally transferred from a simple nitrile to an olefin, circumventing the potential hazards associated with the direct addition of hydrogen cyanide gas to alkenes. The catalyst system consists of a DPEphos-ligated Ni and operates through a Ni^0–Ni^{II} cycle in the presence of cocatalytic $AlMe_2Cl$, used to assist both the oxidative addition and reductive elimination of C—CN onto and from the metal complex. This is an equilibrium reaction that is driven to completion by extrusion of gaseous products, isobutylene, for example, when the target product is the nitrile – the hydrocyanation reaction – or by the release of ring strain from norbornene or norbornadiene as sacrificial acceptors of HCN when the target product is the olefin – the retrocyanation reaction – and allows for controlled interconversion of the cyano and the alkene functional groups (Scheme 6.24). Anti-Markovnikov selectivity was observed for transfer hydrocyanation to styrenes, vinylic heterocycles, and sterically congested aliphatic olefins, although minor amounts of the Markovnikov product could be detected in the reaction

Table 6.1 Ni-catalyzed, Lewis-acid-assisted carbocyanation of alkynes.

$$R^1{-}CN + R^2{-}C{\equiv}C{-}R^3 \xrightarrow[\text{solvent, }\Delta]{\text{1 mol\% Ni(cod), 2 mol\% Phosphine ligand, 4 mol\% Lewis acid}} (R^1)(R^2)C{=}C(CN)(R^3)$$

R^1	R^2	R^3	Temperature (°C)	Time (h)	Solvent	Ligand	Lewis acid	Product (%)
MeO–C_6H_4–	*n*-Pr	*n*-Pr	50	16	Toluene	$PPhMe_2$	$AlMe_2Cl$	96
MeOOC–C_6H_4–	*n*-Pr	*n*-Pr	80	25	Toluene	$PPhMe_2$	$AlMe_2Cl$	93
Me_2N–C_6H_4–	*n*-Pr	*n*-Pr	80	21	Toluene	$PPhMe_2$	$AlMe_2Cl$	87
2,6-dimethylphenyl	*n*-Pr	*n*-Pr	100	134	Toluene	$PPhMe_2$	$AlMe_2Cl$	78
Cl–C_6H_4–	*p*-Anis	$SiMe_3$	60	37	Toluene	$PPh_2(i\text{-}Pr)$	$AlMe_2Cl$	$R^1(R^2)C{=}C(CN)R^3$ (73%), $R^1(R^3)C{=}C(CN)R^2$ (5%)
Ph–CH=CH–	*n*-Pr	*n*-Pr	80	20	Toluene	PMe_3 (4 mol%)	BPh_3 (8 mol%)	94
Me_3Si–CH=CH–	*n*-Pr	*n*-Pr	80	13	Toluene	P(cyclopentyl)$_3$ (10 mol%) 5 mol% $Ni(cod)_2$	BPh_3 (20 mol%)	89
allyl	Me	Ph	80	8	CH_3CN	$P(4\text{-}CF_3\text{–}C_6H_4)_3$ (20 mol%) 10 mol% $Ni(cod)_2$	NA	$R^1(R^2)C{=}C(CN)R^3$ / $R^1(R^3)C{=}C(CN)R^2$ 43 total 94 : 6

Selected examples are presented.

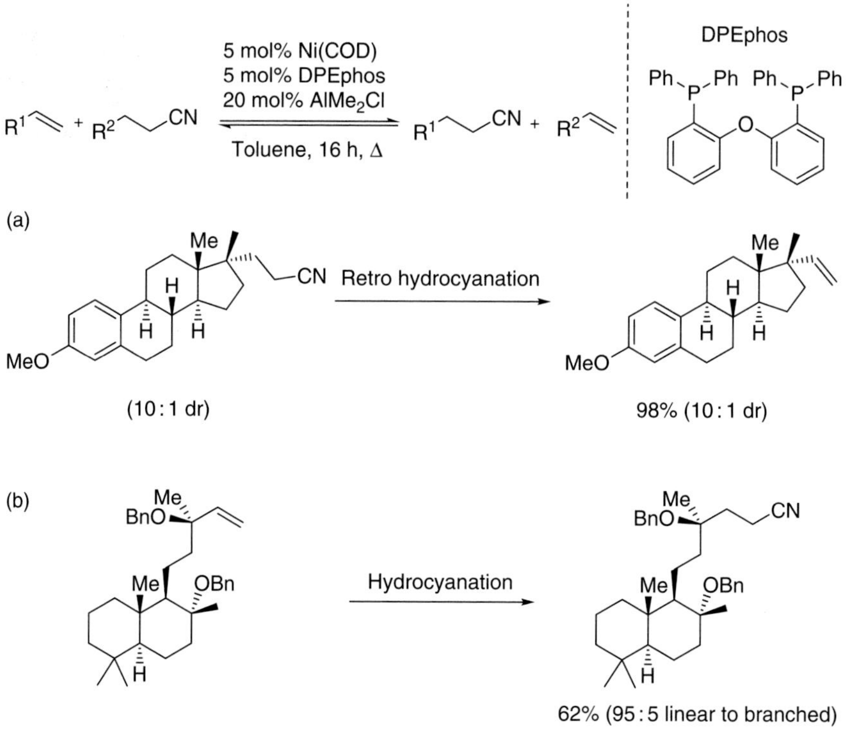

Scheme 6.24 Catalytic transfer hydrocyanation reaction. (a) Late-stage functionalizations of an estrone derivative by retro-hydrocyanation. (b) Late-stage functionalizations of a sclareol derivative by hydrocyanation.

mixtures, whereas significant isomerization is observed when aliphatic alkenes are used as substrates. The reactivity was also extended to internal alkynes to yield the corresponding vinyl nitriles in good yields with moderate-to-good regioselectivities. The reaction tolerates F, Cl, ester, ketone, ether, OBn, NTs, and OTBS functional groups (TBS = *t*-butyldimethylsilyl), which allows for the late-stage functionalization of complex molecular architectures; nevertheless, unprotected alcohols and dienes are not compatible.

In terms of the retrocyanation reaction, the method afforded the target alkenes from simple aliphatic nitriles with little or no isomerization, and a range of styrenes can be prepared from primary, secondary, and tertiary nitriles (Scheme 6.24). The method can also be applied to the late-stage synthesis of structurally complex alkenes.

6.5 Hydrosilylation Reactions

6.5.1 Fe-Catalyzed, Anti-Markovnikov Hydrosilylation of Alkenes with Tertiary Silanes and Hydrosiloxanes

The Chirik group has developed a highly efficient anti-Markovnikov hydrosilylation reaction of unactivated terminal olefins, catalyzed by well-defined Fe^0

complexes of a bis(imino)pyridine pincer architecture (PDI) (**16a,b**), that uses readily available tertiary alkyl and alkoxy silanes – Et_3SiH, $(Me_3SiO)_2MeSiH$, and $(EtO)_3SiH$ (Scheme 6.25) [44]. The ability to use tertiary silanes is desirable given the fact that such silanes are more economical relative to their primary and secondary congeners and also because the hydrosilylation products contain no reactive Si—H bonds that could hinder downstream applications [45].

The reaction is well behaved; the products of olefin hydrogenation, isomerization, or polymerization were not detected; and it is compatible with styrene, allyl polyether, and unsubstituted terminal olefins. The individual catalyst loadings and reaction temperatures and times depend on the identity of the substrate and could fall in the range of 200–7000 ppm of catalyst, temperatures within the range of 23–60 °C, and reaction times of a few minutes to an hour. Although a Fe^0—Fe^{II} catalytic cycle was proposed at the time this work was published, more recent spectroscopic and structural studies of Fe bis(imino)pyridine complexes of the type **16** have shown that a more accurate description of the complex is that of a bis(imino)-pyridine dianion coordinating an Fe^{II} ion (Scheme 6.25) [46]. For this reason, it is likely that a different catalytic pathway may be operational.

A relevant industrial application of the hydrosilylation reaction is the cross-linking of silicon polymers to obtain new, heavier molecular weight compounds. In such applications, catalyst recovery can pose a difficult problem depending on the characteristic of the resulting material, and it is therefore important to find alternatives to the more traditional catalyst systems based on Pt. In a test reaction, 500 ppm of catalyst **16** was used to cross-link a neat mixture of the silicon fluids SL6020 and SL6100 (Momentive Performance Materials) (Scheme 6.25) to obtain a solid cross-linked material within two hours at 23 °C.

6.5.2 Highly Chemoselective Co-Catalyzed Hydrosilylation of Functionalized Alkenes Using Tertiary Silanes and Hydrosiloxanes

A recent contribution in this area details a highly chemo- and regioselective hydrosilylation reaction of functionalized olefins using a bis(carbene) pincer-ligated Co^I complex **17** [47]. In this method, 5 mol% of **17**, 1.0 equiv. of a tertiary silane – Et_3SiH, $(Me_3SiO)_2MeSiH$, $(MeO)_3SiH$, or Me_2PhSiH – and 1.0 equiv. of the functionalized olefin in benzene solution exclusively afford the anti-Markovnikov products in good yields. The reaction tolerates a range of sensitive functional groups including nitrile, hydroxyl, ketone, aldehyde, ester, and internal olefin (Scheme 6.26).

Steric constraint is a critical factor that modulates the regioselectivity in cases where more than one olefinic double bond was present, and the chemoselectivity in the case of more reactive groups such as an aldehyde. In the absence of dimethyl substitution α to the aldehyde carbonyl group, a mixture of c. 1 : 1 silanol and alkylsilane was obtained. The reaction is proposed to proceed via a Co^I–Co^{III} catalytic cycle through a Chalk–Harrod mechanism (Scheme 6.27) [48].

6.5.3 Alkene Hydrosilylation Using Tertiary Silanes with α-Diimine Ni Catalysts

A Ni-based catalyst system capable of utilizing tertiary alkoxysilanes of the type $HSiMe_n(OR)_{3-n}$ (R = OEt, $OSiMe_3$) in an anti-Markovnikov hydrosilylation of

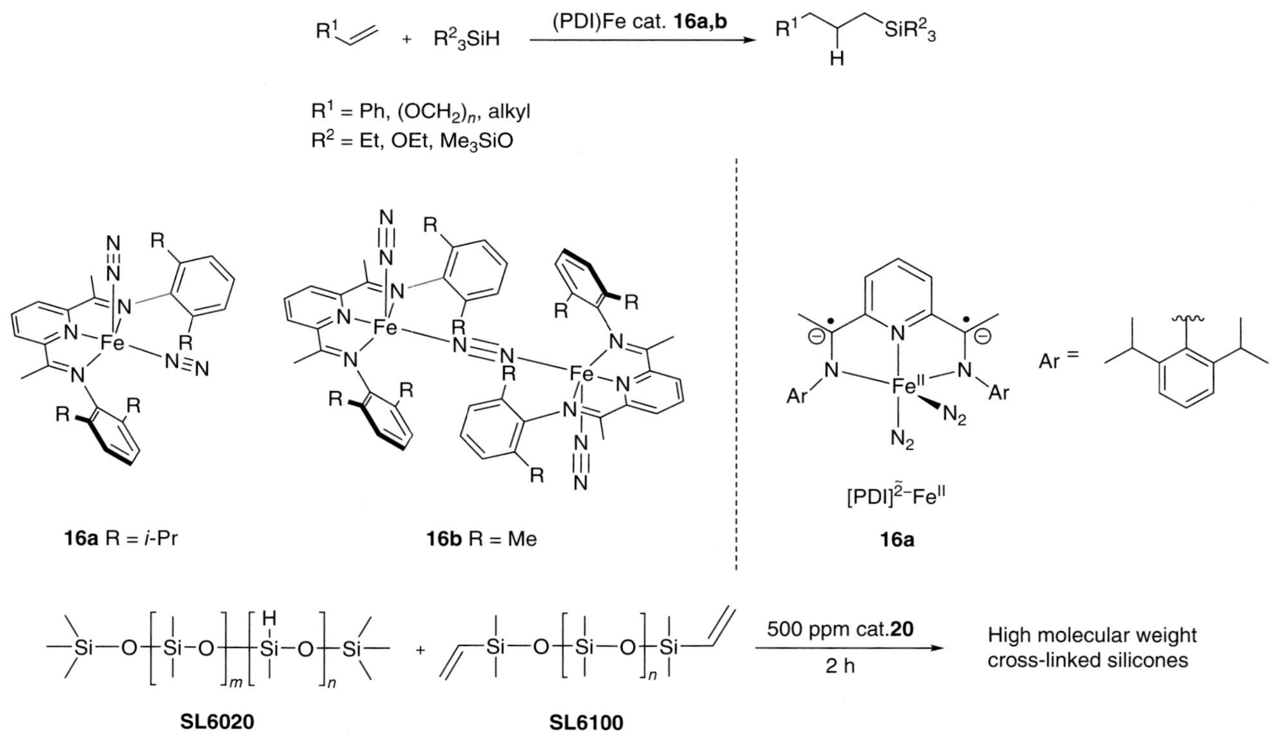

R^1-CH=CH$_2$ + R^2_3SiH —(PDI)Fe cat. **16a,b**→ $R^1CH_2CH_2SiR^2_3$

R^1 = Ph, $(OCH_2)_n$, alkyl
R^2 = Et, OEt, Me_3SiO

16a R = *i*-Pr

16b R = Me

$[PDI]^{\tilde{2}-}$–Fe^{II}

16a

Ar =

SL6020 + SL6100 —500 ppm cat.**20**, 2 h→ High molecular weight cross-linked silicones

SL6020

SL6100

Scheme 6.25 Anti-Markovnikov hydrosilylation reaction catalyzed by the Fe^0 bis(imino)pyridine pincer complex **16**.

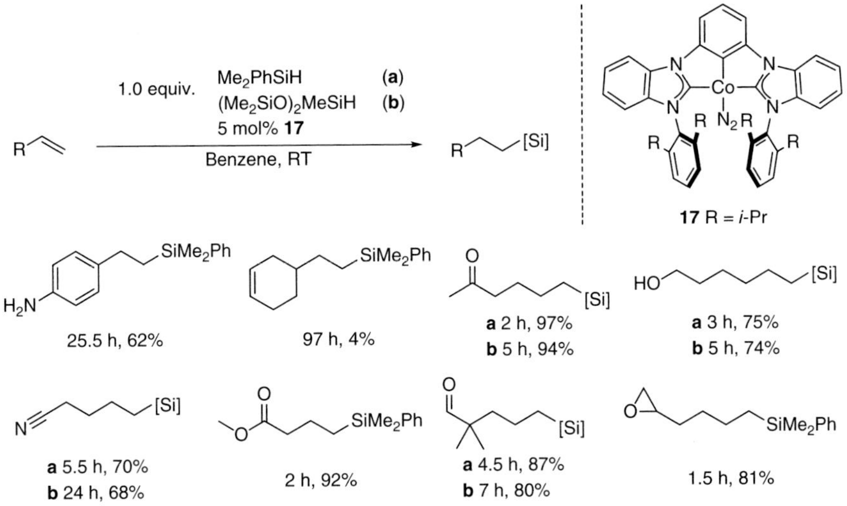

Scheme 6.26 Anti-Markovnikov hydrosilylation reaction of functionalized olefins catalyzed by the bis(carbene) pincer-ligated Co^{I} catalyst **17**. Representative examples of substrates presented.

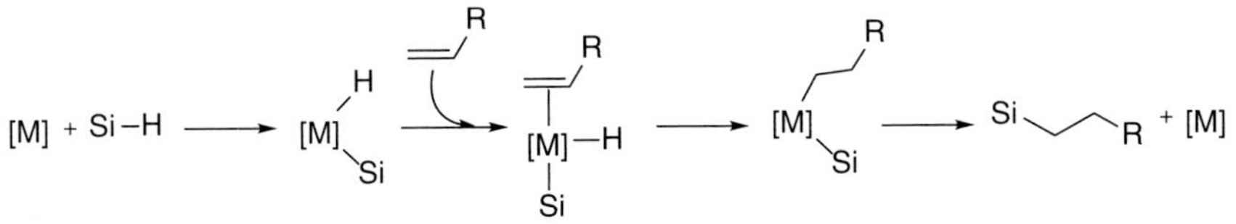

Scheme 6.27 Chalk–Harrod mechanism for the hydrosilylation reaction.

alkenes has also been reported by the Chirik group [49]. The catalyst precursor was found to be a dimeric $[(\alpha\text{-DI})Ni—H]_2$ species **18**, which is formed *in situ* in a preactivation step from an equimolar mixture of Ni(2-ethylhexanoate)$_2$ and an α-diimine (α-DI) ligand and 6 equiv. of $(EtO)_3SiH$. The hydrosilylation of 1-octene proceeds quantitatively with high regioselectivity at 23 °C for 23 hours in neat silane and olefin, with loadings of 1 mol% Ni catalyst, demonstrating less activity with respect to compounds **16** and **17**. The reaction time can be reduced to four hours if the temperature is increased to 40 °C. The system was also capable of cross-coupling oligomeric silicone fluids to form solid polymers in two hours at 80 °C using a catalyst load of 100 ppm.

DFT calculations suggest that compound **18** contains Ni^{II} centers with one-electron-reduced α-DI ligands (see Chapter 1 for a detailed discussion of redox-active ligands). The mechanism is thought to proceed through dissociation of compound **18** into its monomers. Olefin insertion into the Ni—H bond takes place to form a Ni–alkyl complex that can isomerize and "chain walk" as demonstrated by labeling experiments, followed by σ-bond metathesis with $R_3Si—H$ to form the product and regenerate the Ni—H species. Alternatively, $R_3Si—H$ can undergo oxidative addition to form a Ni^{III} complex from which product-forming reductive elimination takes place (Scheme 6.28).

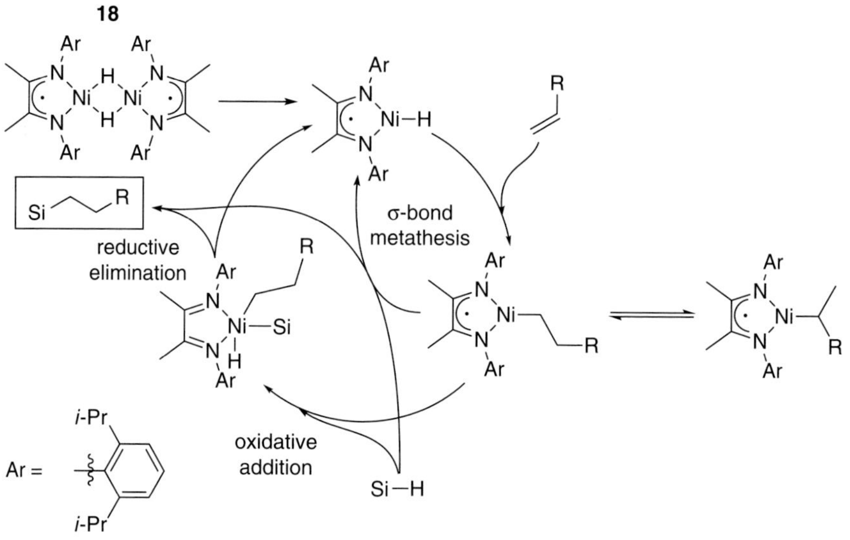

Scheme 6.28 Proposed mechanism for the anti-Markovnikov hydrosilylation reaction of alkenes catalyzed by (α-DI)Ni—H.

6.5.4 Chemoselective Alkene Hydrosilylation Catalyzed by Ni Pincer Complexes

Another hydrosilylation base metal catalyst that displays good chemo- and regioselectivity is the bis(amino)amido pincer-ligated Ni^{II} complex **19** (Scheme 6.29) [50]. This is a highly active catalyst for olefin hydrosilylation that can achieve full conversion to the anti-Markovnikov hydrosilylated product with catalyst loadings as low as 0.01 mol% – which translates into turnover number (TON) of around 10 000 – when the reaction is run in neat silane. It was noted, however, that the reaction run in neat silane is quite exothermic, which leads to some formation of dialkylsilanes. This problem is avoided by diluting the reaction mixture with an appropriate solvent. It was found that secondary silanes, Ph_2H_2Si in particular, worked best, whereas primary and secondary silanes gave consistently lower yields (c. 30% yield of the desired product).

The reaction demonstrated good tolerance toward epoxide, ester, unprotected primary amino, sulfonylamido, ketone, and aldehyde functional groups. Although both terminal and internal double bonds were found to be reactive, hydrosilylation proceeds faster in the case of the former, which allowed for selective hydrosilylation of a terminal double bond on a substrate-containing both types. The reaction does not tolerate carboxylic acids, alcohols, or allyl halides. In the case of the ketone- and aldehyde-containing substrates, the use of *N,N*-dimethylformamide as the solvent, and increasing the catalyst loading to 2 mol%, further improved the chemoselectivity (Scheme 6.29). The system is also capable of accomplishing an isomerization/hydrosilylation reaction. It was found that by treating 2-, 3-, or 4-octene with 2.0 equiv. of Ph_2H_2Si and 10 mol% of **23**, *n*-octyl-$SiHPh_2$ can be obtained (Scheme 6.29).

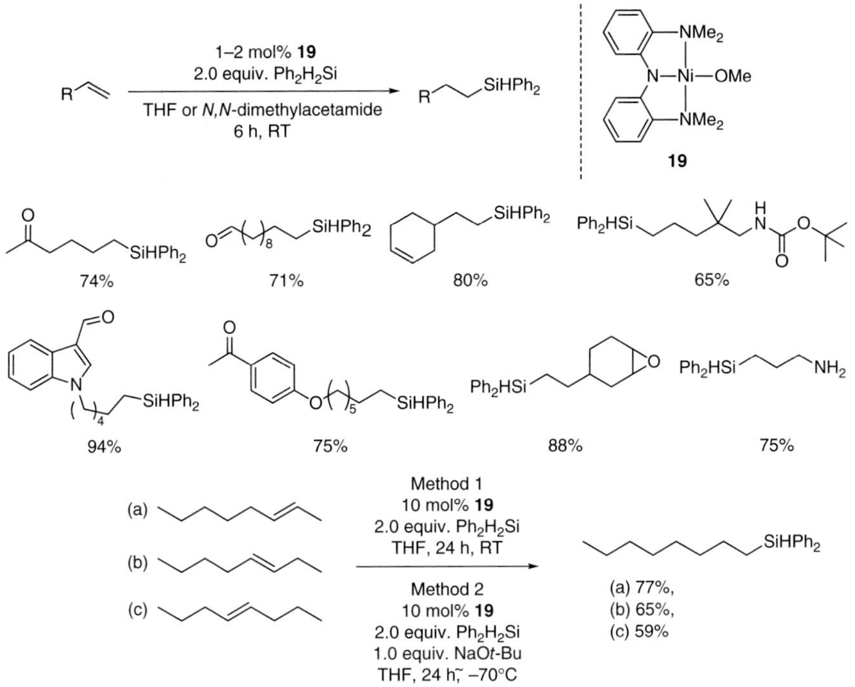

Scheme 6.29 Anti-Markovnikov hydrosilylation and isomerization/hydrosilylation reactions of olefins catalyzed by the bis(amino)amido pincer-ligated Ni^{II} catalyst **19**. Representative examples of substrates presented.

6.5.5 Fe- and Co-Catalyzed Regiodivergent Hydrosilylation of Alkenes

A catalyst system for the regiodivergent hydrosilylation of olefins, based on a phosphino-iminopyridine pincer platform, has been recently reported (Scheme 6.30, compounds **20** and **21**) [51]. The anti-Markovnikov hydrosilylation reaction is catalyzed by the Fe^{II} pincer complex **20**, uses $PhSiH_3$, and proceeds with high regioselectivity to form the target products in good yields. The reaction can be run either in neat olefin or in tetrahydrofuran (THF) solution and requires the use of $NaHBEt_3$ as an activator. The scope of substrates for this reaction demonstrated a limited functional group tolerance and included a protected alcohol, ether, acetal, and chloride functional groups (Scheme 6.30).

The complementary Markovnikov hydrosilylation reaction is catalyzed by the Co^{II} complex **21**, which likewise proceeds in good yields and excellent regioselectivity, does not necessitate the use of $NaHBEt_3$, and was similar in scope with the reaction catalyzed by **20** (Scheme 6.30). Styrenes did not react in a regioselective manner and afforded 1 : 1 mixtures of the hydrosilylated regioisomers.

6.5.6 Co-Catalyzed Markovnikov Hydrosilylation of Terminal Alkynes and Hydroborylation of α-Vinylsilanes

The pyridinebis(oxaline) Co^{II} pincer complex **22** is an efficient catalyst for the highly regioselective Markovnikov hydrosilylation of terminal alkynes devoid of directing groups with Ph_2SiH_2 to obtain α-vinylsilanes (Scheme 6.31a) [52].

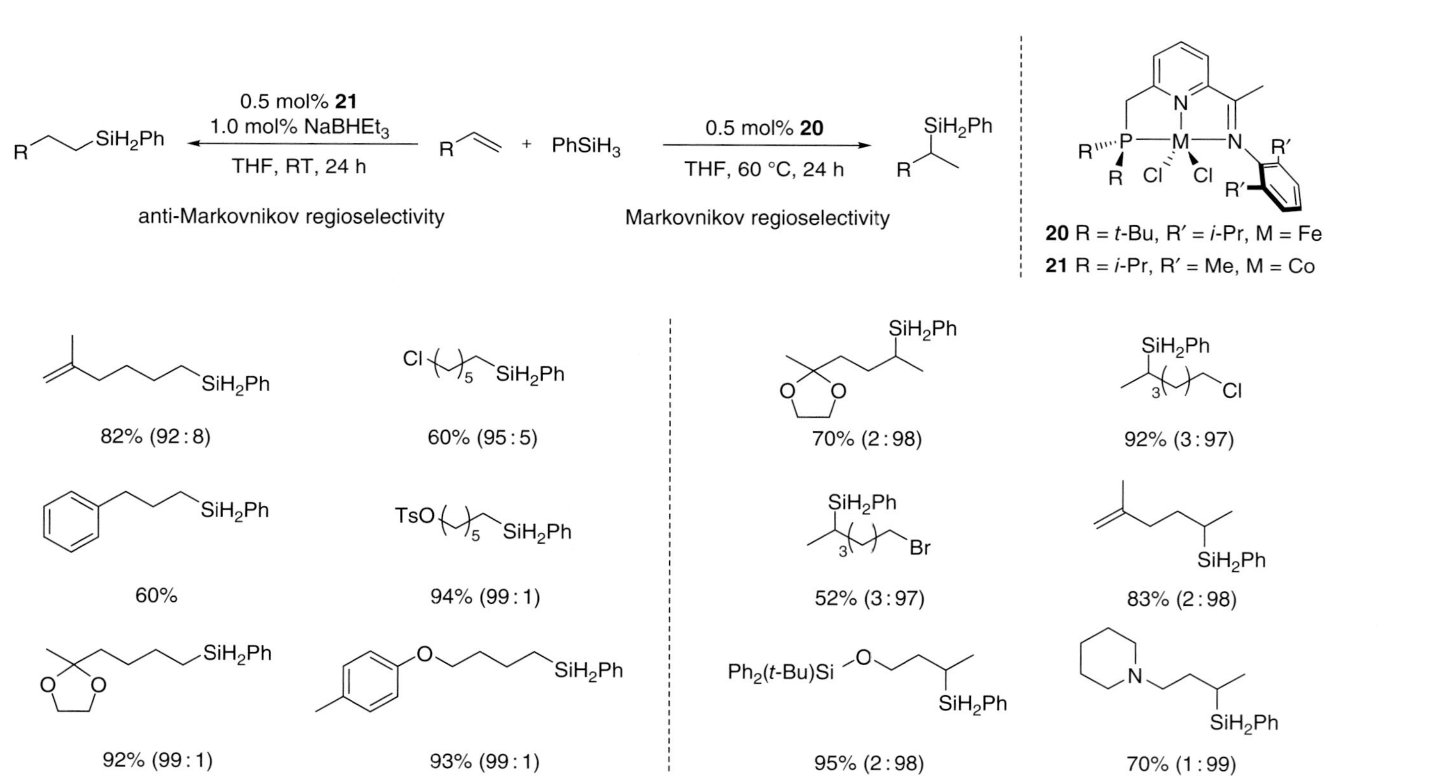

Scheme 6.30 Selected examples of products of regiodivergent hydrosilylation of alkenes catalyzed by Fe^{II} and Co^{II} phosphino-iminopyridine pincer complexes **20** and **21**. Ratios of anti-Markovnikov to Markovnikov products are presented in parenthesis.

This reaction affords the hydrosilylated products in yields upward of 65% with excellent Markovnikov regioselectivity in most cases. This hydrosilylation reaction tolerates ester, tertiary amino, pyridine, thiophene, ketone, F, Cl, and Br functional groups (Scheme 6.31). Substrates with ortho substituents were slightly less reactive than the meta and para congeners and required increased catalyst loading and reaction time – 2.0 mol% and two hours vs the standard 0.5 mol% and one hour. A decrease in regioselectivity was observed for alkyl acetylenes relative to phenylacetylenes. A complementary approach to the synthesis of α-vinylsilanes from alkyl acetylenes, with much higher Markovnikov regioselectivity, is a Cu^{I}-JohnPhos system that uses $PhMe_2SiB(pin)$ as the silylating agent (Scheme 6.31b) [53]. This system tolerates hydroxyl, chloride, cyano, tertiary amino, and ester groups.

The Cu-catalyzed reaction is proposed to operate via a $[Cu^{I}]$–silyl complex that adds across the triple bond to form a vinyl cuprate, which is subsequently alcoholized to form the target product and CuOMe. The catalytically active $[Cu^{I}]$–silyl complex is regenerated by reaction of CuOMe with $R_3SiB(pin)$, which also forms B(pin)OMe as a by-product [53]. The Co-catalyzed hydrosilylation reaction is proposed to involve a similar reaction sequence.

gem-Borosilanes can be prepared from α-vinylsilanes via a Markovnikov hydroboration reaction with HB(pin) and catalyzed by the bis(imino)pyridine Co^{II} pincer complex **23** (Scheme 6.32) [52, 54].

6.5.7 Fe and Co Pivalate Isocyanide-Ligated Catalyst Systems for Hydrosilylation of Alkenes with Hydrosiloxanes

Although in the preceding sections hydrosilylation reactions catalyzed by well-defined catalyst systems have been discussed, examples of active Fe and Co anti-Markovnikov hydrosilylation catalysts formed *in situ* from their corresponding M^{II} pivalates have been recently reported (Scheme 6.33) [55]. This reaction proceeds in the presence of an adamantyl isocyanide ligand and 1.3 equiv. of $Me_3SiOSiHMe_2$, although $PhSiH_3$ and Ph_2SiH_2 could also be used, but at the expense of increased reaction times. It was found that the hydrosilylation reaction of styrene was better behaved when the Fe catalyst was used, whereas the Co catalyst provided better results with α-methylstyrenes and aliphatic alkenes. The scope of substrates reported for this reaction was rather limited but included an ester, epoxide, chloride, and fluoride functional groups, which could be susceptible to reduction. Both metal systems were successfully used in cross-linking reactions of model polysiloxane oligomers to form heavier molecular weight silicones (Scheme 6.33). $(MPDE)_2Fe$ (MPDE = η^5-3-methylpentadienyl) and $(COT)_2Fe$ (COT = *cyclo*-octatetraene) have also been demonstrated to be very competent catalyst precursors for the anti-Markovnikov hydrosilylation of styrene with either hydride-terminated polydimethylsiloxane (PDMS) or $PhMe_2SiH$, in the presence of an isocyanide ligand [55b].

Although $(EtO)_2SiH$, $(MeO)_3SiH$, $(EtO)_2MeSiH$, and $(MeO)_2SiH$ were also tested, it was found that they did not perform as efficiently as PMDS; however, it was discovered that they could act as activating agents. It was observed that before the treatment of the metal pivalate with a hydroalkoxysilane, in the presence of the isocyanide ligand, the hydrosilylation reaction was completed

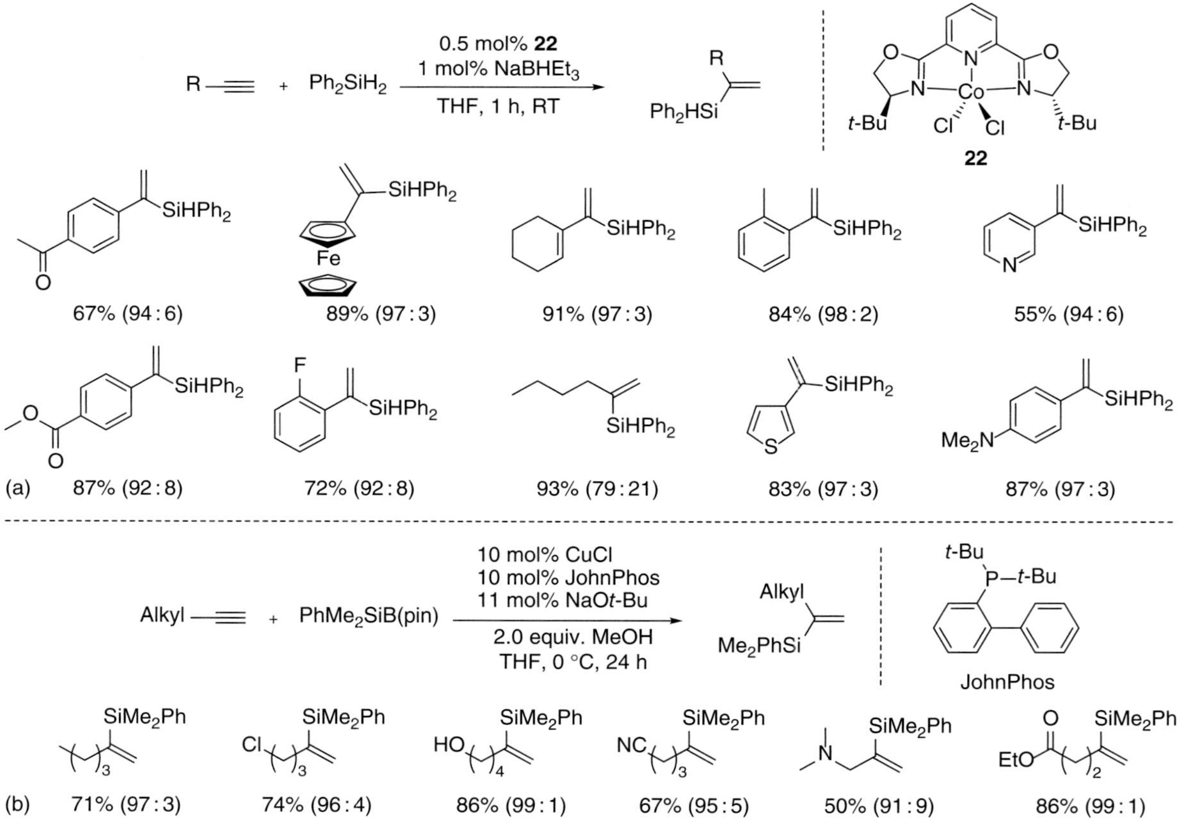

(a) R—≡ + Ph_2SiH_2 → (0.5 mol% **22**, 1 mol% $NaBHEt_3$; THF, 1 h, RT) → Ph_2HSi(R)C=CH_2

Catalyst **22**

Product	Yield (ratio)
4-acetylphenyl–SiHPh2	67% (94 : 6)
ferrocenyl–SiHPh2	89% (97 : 3)
cyclohexenyl–SiHPh2	91% (97 : 3)
2-methylphenyl–SiHPh2	84% (98 : 2)
3-pyridyl–SiHPh2	55% (94 : 6)
4-(methoxycarbonyl)phenyl–SiHPh2	87% (92 : 8)
2-fluorophenyl–SiHPh2	72% (92 : 8)
n-butyl–SiHPh2	93% (79 : 21)
3-thienyl–SiHPh2	83% (97 : 3)
4-(dimethylamino)phenyl (Me_2N)–SiHPh2	87% (97 : 3)

(b) Alkyl—≡ + $PhMe_2SiB(pin)$ → (10 mol% CuCl, 10 mol% JohnPhos, 11 mol% NaO*t*-Bu; 2.0 equiv. MeOH, THF, 0 °C, 24 h) → Me_2PhSi(Alkyl)C=CH_2

Ligand: JohnPhos

Product	Yield (ratio)
$SiMe_2Ph$, alkyl chain $(CH_2)_3$	71% (97 : 3)
$SiMe_2Ph$, Cl$(CH_2)_3$	74% (96 : 4)
$SiMe_2Ph$, HO$(CH_2)_4$	86% (99 : 1)
$SiMe_2Ph$, NC$(CH_2)_3$	67% (95 : 5)
$SiMe_2Ph$, Me_2NCH_2	50% (91 : 9)
$SiMe_2Ph$, EtO(C=O)$(CH_2)_2$	86% (99 : 1)

Scheme 6.31 (a) Markovnikov hydrosilylation reaction of terminal alkynes catalyzed by the bisoxazoline Co^{II} pincer complex **22**. (b) Markovnikov hydrosilylation reaction of alkyl acetylenes catalyzed by Cu^{I}. Ratios of α/(*E*)-β isomers are presented in parenthesis.

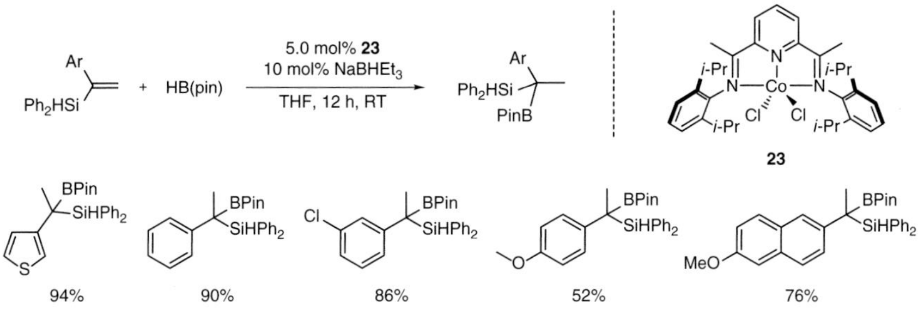

Scheme 6.32 Markovnikov hydroboration reaction of α-vinylsilanes.

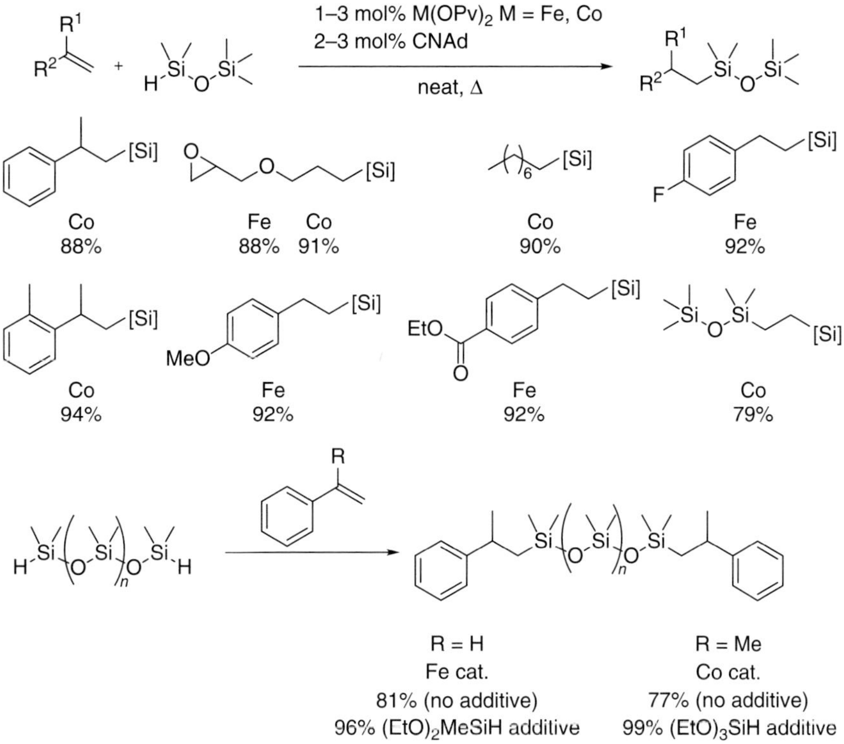

Scheme 6.33 Anti-Markovnikov hydrosilylation of olefins catalyzed by Fe and Co pivalates ligated by adamantyl isocyanide. Selected examples are presented.

in a reduced amount of time and at lower temperature relative to the reactions without pretreatment. This preactivation step allowed for a slight improvement in the yields of polysiloxane cross-coupling products.

6.6 Conclusion

This chapter has presented recent developments in the area of alkene and alkyne functionalization reactions via addition to C—C multiple bonds catalyzed by late first-row transition metals. These metals can operate, in low oxidation states, through the same mechanistic pathways as their noble metal congeners, namely, two-electron processes such as migratory insertion, oxidative addition, and reductive elimination. Nevertheless, the ability of first-row transition metals to access a wider range of oxidation states more easily facilitates one-electron processes, such as HAT, which is a key step in the hydrofunctionalization of alkenes. This aspect in particular illustrates the importance of base metal chemistry, which should be underscored not only insofar as noble metal replacement is concerned, but also because it is complementary to that of noble metals. The further harnessing and development of their one-electron chemistry represents a rich and exciting field of research that will play a major role in the future of alkene and alkyne functionalization reactions.

References

1 Beller, M. and Bolm, C. (eds.) (2008). *Transition Metals for Organic Synthesis: Building Blocks and Fine Chemicals.* Weinheim, Germany: Wiley-VCH Verlag GmbH.
2 Daintith, J. (ed.) (2000). *Dictionary of Chemistry*, 61. Oxford, UK: Oxford University Press.
3 (a) Mayer, J.M. (2011). *Acc. Chem. Res.* 44: 36. (b) Siewert, I. (2015). *Chem. Eur. J.* 21: 15078.
4 Choi, J., Tang, L., and Norton, J.R. (2007). *J. Am. Chem. Soc.* 129: 234.
5 Isayama, S. and Mukaiyama, T. (1989). *Chem. Lett.* 18: 1071.
6 Bullock, R.M. and Samsel, E.G. (1990). *J. Am. Chem. Soc.* 112: 6886.
7 Waser, J., Gaspar, B., Nambu, H., and Carreira, E.M. (2006). *J. Am. Chem. Soc.* 128: 11693.
8 Waser, J. and Carreira, E.M. (2004). *J. Am. Chem. Soc.* 126: 5676.
9 Waser, J., González-Gómez, J.C., Nambu, H. et al. (2005). *Org. Lett.* 7: 4249.
10 Waser, J. and Carreira, E.M. (2004). *Angew. Chem. Int. Ed.* 43: 4099; *Angew. Chem. Int. Ed.* **2004**, *116*, 4191.
11 Waser, J., Nambu, H., and Carreira, E.M. (2005). *J. Am. Chem. Soc.* 127: 8294.
12 Gaspar, B. and Carreira, E.M. (2007). *Angew. Chem. Int. Ed.* 46: 4519; *Angew. Chem. Int. Ed.* **2007**, *119*, 4603.
13 (a) Brown, E.S. and Rick, E.A. (1969). *J. Chem. Soc. D: Chem. Commun.* 112b. (b) Taylor, B.W. and Swift, H.E. (1972). *J. Catal.* 26: 254. (c) Backvall, J.-E. and Andell, O.S. (1984). *J. Chem. Soc., Chem. Commun.* 260.
14 Gaspar, B. and Carreira, E.M. (2008). *Angew. Chem. Int. Ed.* 47: 5758; *Angew. Chem. Int. Ed.* **2008**, *120*, 5842.
15 Crossley, S.W.M., Obradors, C., Martinez, R.M., and Shenvi, R.A. (2016). *Chem. Rev.* 116: 8912.
16 (a) Leggans, E.K., Barker, T.J., Duncan, K.K., and Boger, D.L. (2012). *Org. Lett.* 14: 1428. (b) Ishikawa, H., Colby, D.A., and Boger, D.L. (2008). *J. Am. Chem. Soc.* 130: 420. (c) Ishikawa, H., Colby, D.A., Seto, S. et al. (2009). *J. Am. Chem. Soc.* 131: 4904.
17 Barker, T.J. and Boger, D.L. (2012). *J. Am. Chem. Soc.* 134: 13588.

18 Shigehisa, H., Nishi, E., Fujisawa, M., and Hiroya, K. (2013). *Org. Lett.* 15: 5158.

19 Shigehisa, H., Aoki, T., Yamaguchi, S. et al. (2013). *J. Am. Chem. Soc.* 135: 10306.

20 Shigehisa, H., Kikuchi, H., and Hiroya, K. (2016). *Chem. Pharm. Bull.* 64: 371.

21 Dao, H.T., Li, C., Michaudel, Q. et al. (2015). *J. Am. Chem. Soc.* 137: 8046.

22 Gui, J., Pan, C.-M., Jin, Y. et al. (2015). *Science* 348: 886.

23 Obradors, C., Martinez, R.M., and Shenvi, R.A. (2016). *J. Am. Chem. Soc.* 138: 4962.

24 Zhu, K., Shaver, M.P., and Thomas, S.P. (2016). *Chem. Asian J.* 11: 977.

25 Green, S.A., Matos, J.L.M., Yagi, A., and Shenvi, R.A. (2016). *J. Am. Chem. Soc.* 138: 12779.

26 Tasker, S.Z., Standley, E.A., and Jamison, T.F. (2014). *Nature* 509: 299.

27 Parsons, A.T. and Buchwald, S.L. *Angew. Chem. Int. Ed.* 2011, 50: 9120; *Angew. Chem. Int. Ed.* **2011**, *123*, 9286.

28 Wang, F., Qi, X., Liang, Z. et al. (2014). *Angew. Chem. Int. Ed.* 53: 1881; *Angew. Chem. Int. Ed.* **2014**, *126*, 1912.

29 Larrow, J.F., Schaus, S.E., and Jacobsen, E.N. (1996). *J. Am. Chem. Soc.* 118: 7420.

30 (a) Watanabe, M., Murata, K., and Ikariya, T. (2002). *J. Org. Chem.* 67: 1712. (b) Patonay, T., Konya, K., and Juhasz-Toth, E. (2011). *Chem. Soc. Rev.* 40: 2797.

31 Sun, X., Li, X., Song, S. et al. (2015). *J. Am. Chem. Soc.* 137: 6059.

32 (a) Lu, D.-F., Zhu, C.-L., Jia, Z.-X., and Xu, H. (2014). *J. Am. Chem. Soc.* 136: 13186. (b) Michaelis, D.J., Shaffer, C.J., and Yoon, T.P. (2007). *J. Am. Chem. Soc.* 129: 1866. c Williamson, K.S. and Yoon, T.P. (2012). *J. Am. Chem. Soc.* 134: 12370.

33 Legnani, L. and Morandi, B. (2016). *Angew. Chem. Int. Ed.* 55: 2248. *Angew. Chem. Int. Ed.* **2016**, *128*, 2288.

34 (a) Zaidlewicz, M., Wolan, A., and Budny, M. (2014). Hydrometallation of C=C and C≡C bonds. Group 3. In: *Comprehensive Organic Synthesis II* (ed. P. Knochel and G.A. Molander), 877. Amsterdam, UK: Elsevier. (b) Makabe, H. and Negishi, E.-I. (2003). *Handbook of Organopalladium Chemistry for Organic Synthesis*, 2789. Wiley. (c) Trost, B.M. and Ball, Z.T. (2005). *Synthesis* 853. (d) Negishi, E.-I. (2007). *Bull. Chem. Soc. Jpn.* 80: 233. (e) Greenhalgh, M.D., Jones, A.S., and Thomas, S.P. (2015). *Chem. Cat. Chem.* 7: 190. (f) Montgomery, J. (2013). Organonickel chemistry. In: *Organometallics in Synthesis: Fourth Manual* (ed. B.H. Lipshutz), 319. Hoboken, NJ: Wiley.

35 (a) Pirnot, M.T., Wang, Y.-M., and Buchwald, S.L. (2016). *Angew. Chem. Int. Ed.* 55: 48; *Angew. Chem. Int. Ed.* **2016**, *128*, 48. (b) Zhu, S., Niljianskul, N., and Buchwald, S.L. (2013). *J. Am. Chem. Soc.* 135: 15746. (c) Miki, Y., Hirano, K., Satoh, T., and Miura, M. (2013). *Angew. Chem. Int. Ed.* 52: 10830; *Angew. Chem. Int. Ed.* **2013**, *125*, 11030. (d) Zhu, S. and Buchwald, S.L. (2014). *J. Am. Chem. Soc.* 136: 15913. (e) Niljianskul, N., Zhu, S., and Buchwald, S.L. (2015). *Angew. Chem. Int. Ed.* 54: 1638; *Angew. Chem. Int. Ed.* **2015**, *127*, 1658.

36 Dang, L., Zhao, H., Lin, Z., and Marder, T.B. (2007). *Organometallics* 26: 2824.

37 Yang, Y., Shi, S.-L., Niu, D. et al. (2015). *Science* 349: 62.
38 Shi, S.-L. and Buchwald, S.L. (2015). *Nat. Chem.* 7: 38.
39 (a) Nakao, Y., Oda, S., and Hiyama, T. (2004). *J. Am. Chem. Soc.* 126: 13904. (b) Nakao, Y., Yada, A., Ebata, S., and Hiyama, T. (2007). *J. Am. Chem. Soc.* 129: 2428. (c) Hirata, Y., Yukawa, T., Kashihara, N. et al. (2009). *J. Am. Chem. Soc.* 131: 10964. (d) Akira, Y., Tomoya, Y., Hiroaki, I. et al. (2010). *Bull. Chem. Soc. Jpn.* 83: 619. (e) Nakao, Y., Oda, S., Yada, A., and Hiyama, T. (2006). *Tetrahedron* 62: 7567.
40 García, J.J., Brunkan, N.M., and Jones, W.D. (2002). *J. Am. Chem. Soc.* 124: 9547.
41 Tolman, C.A., Seidel, W.C., Druliner, J.D., and Domaille, P.J. (1984). *Organometallics* 3: 33.
42 Fang, X., Yu, P., and Morandi, B. (2016). *Science* 351: 832.
43 Bhawal, B.N. and Morandi, B. (2016). *ACS Catal.* 6: 7528.
44 (a) Tondreau, A.M., Atienza, C.C.H., Weller, K.J. et al. (2012). *Science* 335: 567. (b) Hojilla Atienza, C.C., Tondreau, A.M., Weller, K.J. et al. (2012). *ACS Catal.* 2: 2169.
45 Yilgör, E. and Yilgör, I. (1165). *Prog. Polym. Sci.* 2014 (39).
46 (a) Bart, S.C., Chłopek, K., Bill, E. et al. (2006). *J. Am. Chem. Soc.* 128: 13901. (b) Chirik, P.J. and Wieghardt, K. (2010). *Science* 327: 794.
47 Ibrahim, A.D., Entsminger, S.W., Zhu, L., and Fout, A.R. (2016). *ACS Catal.* 6: 3589.
48 (a) Chalk, A.J. and Harrod, J.F. (1965). *J. Am. Chem. Soc.* 87: 16. (b) Meister, T.K., Riener, K., Gigler, P. et al. (2016). *ACS Catal.* 6: 1274.
49 Pappas, I., Treacy, S., and Chirik, P.J. (2016). *ACS Catal.* 6: 4105.
50 Buslov, I., Becouse, J., Mazza, S. et al. *Angew. Chem. Int. Ed.* 2015, 54: 14523; *Angew. Chem. Int. Ed.* **2015**, *127*, 14731.
51 Du, X., Zhang, Y., Peng, D., and Huang, Z. (2016). *Angew. Chem. Int. Ed.* 55: 6671; *Angew. Chem. Int. Ed.* **2016**, *128*, 6783.
52 (a) Zuo, Z., Yang, J., and Huang, Z. (2016). *Angew. Chem. Int. Ed.* 55: 10839; *Angew. Chem.* **2016**, *128*, 10997. (b) Oestreich, M., Hartmann, E., and Mewald, M. (2013). *Chem. Rev.* 113: 402.
53 Wang, P., Yeo, X.-L., and Loh, T.-P. (1254). *J. Am. Chem. Soc.* 2011: 133.
54 Yoshida, H. (2016). *ACS Catal.* 6: 1799.
55 (a) Noda, D., Tahara, A., Sunada, Y., and Nagashima, H. (2016). *J. Am. Chem. Soc.* 138: 2480. (b) Sunada, Y., Noda, D., Soejima, H. et al. (2015). *Organometallics* 34: 2896.

7

Iron-Catalyzed Cyclopropanation of Alkenes by Carbene Transfer Reactions

Daniela Intrieri, Daniela M. Carminati, and Emma Gallo

Chemistry Department of Milano University, Via Golgi 19, Milano 20136, Italy

7.1 Introduction

The development of sustainable chemical processes respecting the 12 basic principles of "green chemistry" formulated by Anastas and Warner in 1998 [1] is an urgent scientific task. Accordingly, the production of valuable chemicals by using eco-friendly catalytic procedures is of paramount importance, considering the impact that catalysis has in numerous fields of chemical science. In this context, the use of first-row transition metal complexes to promote catalytic reactions is constantly increasing in view of their recognized advantages such as their high earth abundance, economic convenience, low toxicity, and versatile catalytic behavior [2].

Among the first-row transition metals, iron, being the most abundant metal on earth after aluminum, is currently receiving a great deal of attention in the design of catalytic methodologies respectful toward the environment [3]. Iron derivatives are cost-effective, environmentally benign, can be efficiently recycled (the global end-of-life recycling rate of iron is above 50%) [4], and exhibit very good catalytic efficiency in several synthetic transformations [5]. Numerous reviews summarize the catalytic applications of iron complexes in organic synthesis, testifying to the increasing interest of the scientific community in replacing classic noble transition metal catalysts with iron-based catalytic systems whose potential is still underexplored [6–11]. Figure 7.1 shows the approximate number of publications on the catalytic use of iron complexes in the 2000–2017 period (according to the SciFinder database).

Among the large range of synthetic transformations, iron complexes catalyze the one-pot reaction of an alkene with a diazo compound ($R^1R^2C{=}N_2$) to synthesize cyclopropane-containing compounds, which show an intrinsic reactivity because of the high strain of the three-membered ring. Thus, subsequent ring-opening reactions are responsible for the synthesis of other high added-value compounds [12–15]. which often show biological and pharmaceutical characteristics [16, 17]. Such reactions can also be the mechanism of action through which drug substances containing strained three-membered rings produce their pharmacological effect [18].

Non-Noble Metal Catalysis: Molecular Approaches and Reactions,
First Edition. Edited by Robertus J. M. Klein Gebbink and Marc-Etienne Moret.

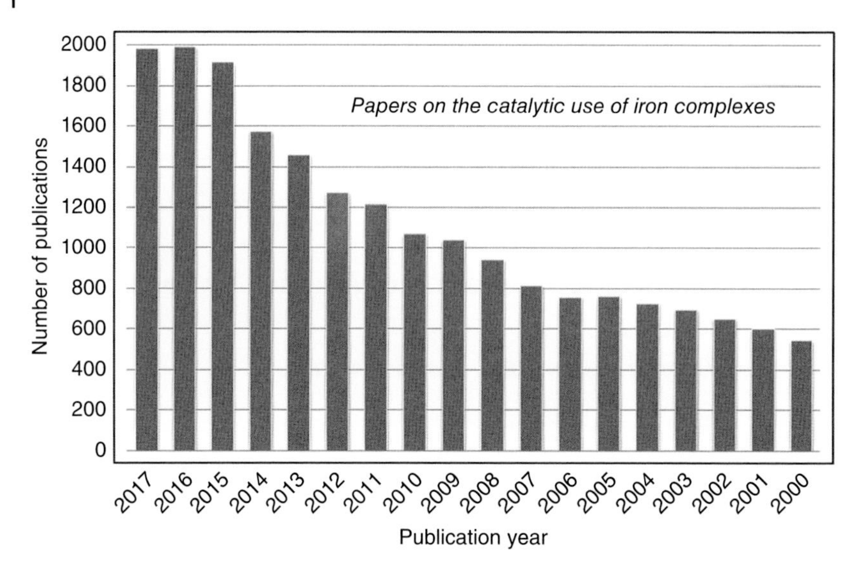

Figure 7.1 Approximate number of publications dealing with the catalytic use of iron complexes in the 2000–2017 period. Source: Courtesy of SciFinder.

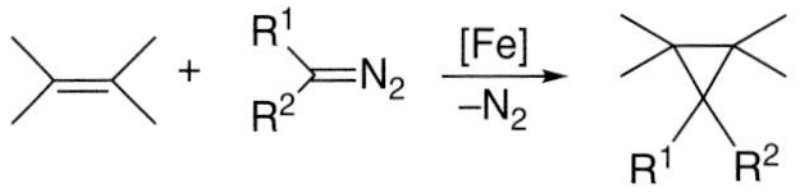

Scheme 7.1 General synthesis of iron-catalyzed cyclopropanation of alkenes by diazo compounds.

In order to introduce a cyclopropane unit into an organic skeleton using environmentally benign technologies, it is mandatory to make a careful selection of starting materials [19–21]. In this context, diazo compounds represent a class of sustainable, atom-efficient carbene sources, which produce eco-friendly N_2 as the only stoichiometric by-product of the carbene transfer reaction (Scheme 7.1) [22, 23]. In addition, diazo derivatives can now be handled on a large scale by managing them under continuous flow technologies [24–27], thus furthering the increase of procedure eco-tolerability and opening new doors to industrial applications.

Even if the use of continuous flow conditions enlarges the practical employment of diazo-based processes, when the diazo derivative cannot be safely handled in a laboratory, it is replaced by other carbene sources that perform cyclopropanation by using simpler experimental methodologies.

This chapter aims to give an overview of results referring to the iron-catalyzed cyclopropane formation by carbene transfer reactions. The catalytic activity of iron biocatalysts is not discussed herein. We sincerely apologize in advance if some important contributions have been unintentionally omitted.

7.2 Achiral Iron Porphyrin Catalysts

The large interest of the scientific community in synthesizing iron porphyrin-based cyclopropanation catalysts [28] is due to several reasons: (i) native iron–heme-containing enzymes catalyze biological cyclopropanations [29], (ii) the sustainability of inexpensive, durable iron metal is coupled with the low toxicity of porphyrin ligands, (iii) the good reactivity/chemical stability relationship of iron porphyrins is responsible for excellent catalytic performance with different substrates and experimental conditions, and (iv) porphyrin ligands can be "decorated" by well-designed substituents to fine-tune the catalytic properties of iron derivatives.

Porphyrins coordinate a metal center in their dianionic form by using the four nitrogen atoms of the tetrapyrrolic core and can be differently functionalized by introducing substituents on meso and β-pyrrolic positions to obtain a large class of porphyrin ligands including chiral ones (Figure 7.2).

Inspired by the catalytic activity of biological systems that display an iron(II) center, synthetic iron(II) porphyrin derivatives were employed to promote the cyclopropanation reaction of alkenes by diazo reagents. Complex Fe^{II}(TTP) (**1**) (TTP, dianion of *meso*-tetrakis-(4-tolyl) porphyrin) was used to promote the cyclopropanation of various styrenes [30, 31]. The catalytic performance was good in terms of cyclopropane yields and TON (turnover number) values; however, chemo- and diastereoselectivities were modest to good (Scheme 7.2).

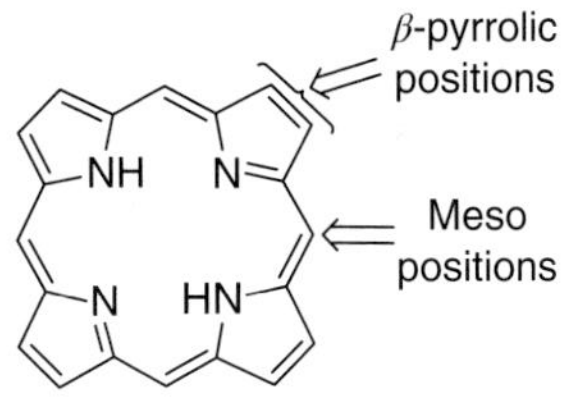

Figure 7.2 The core structure of a porphyrin molecule.

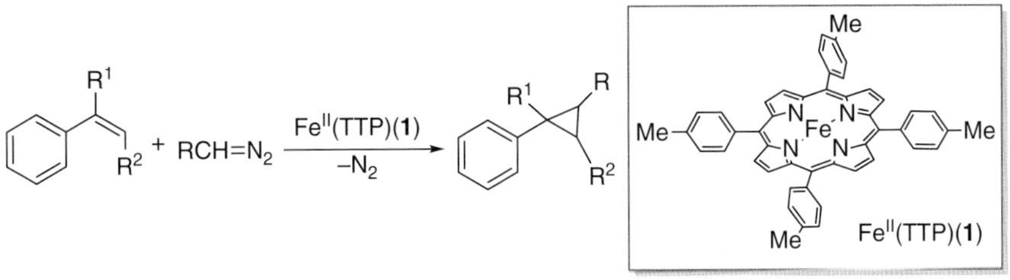

Scheme 7.2 Fe^{II}(TTP) (**1**)-catalyzed alkene cyclopropanation.

It is important to underline that the reaction diastereo- and chemoselectivities are two aspects of fundamental importance to determine the efficiency of a catalytic system because alkene cyclopropanation can occur with different *cis*/*trans*-cyclopropane ratios [32] and the concomitant formation of side products derived from diazo dimerization (Scheme 7.3). To avoid the formation of dimerization derivatives, the diazo reagent is usually added slowly to the catalytic mixture by using, for example, a syringe pump. This experimental strategy maintains the local concentration of the diazo compound very low with the consequent limitation of carbene-coupling reactions.

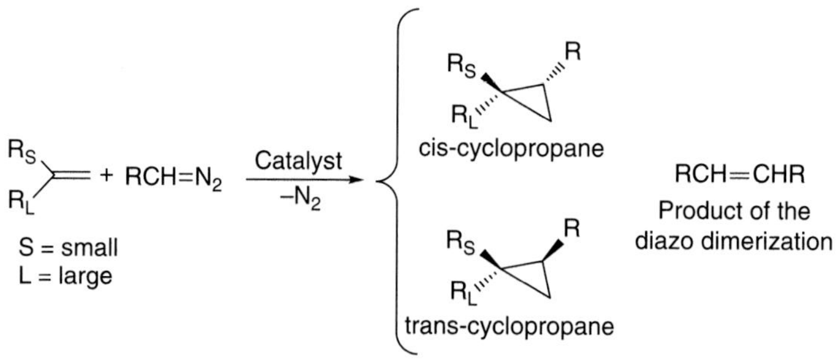

Scheme 7.3 Cyclopropane diastereoisomers and products of diazo dimerization.

Cyclopropanations mediated by **1** generally occur with a trans-diastereoselectivity, whereby a putative iron carbene intermediate is involved as the key species of the operating mechanism (Scheme 7.4).

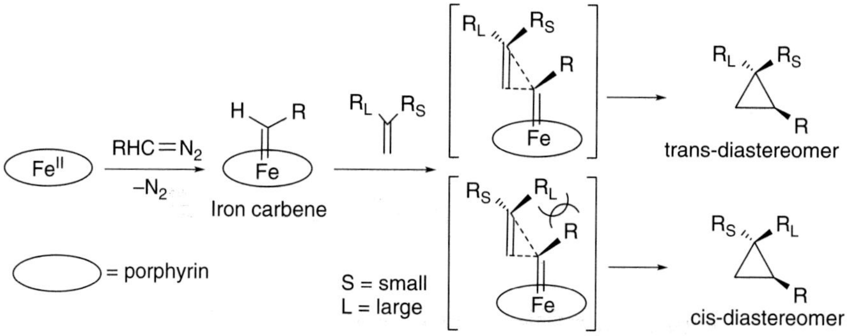

Scheme 7.4 Suggested alkene cyclopropanation mechanism.

It is generally assumed that the alkene approaches the supposed carbene intermediate in an *end-on* manner, and then, the reaction of the unsaturated substrate with the iron carbene involves two different transition states whose relative stability depends on the steric interaction between the large R_L group on the alkene and the R substituent on the carbene moiety. This mechanistic suggestion explains the usually observed trans-diastereoselectivity and also the low impact that the steric hindrance of substituents on meso-positions of the porphyrin ligand has on the reaction stereocontrol.

To better support the mechanism shown in Scheme 7.4, many efforts have been made to synthesize iron carbene complexes by reacting iron(II) porphyrins with diazo compounds, but the extreme instability of iron carbene complexes prevented the isolation of a large class of derivatives. The chemical reactivity of iron carbene complexes is due to the facility of dimerization decomposition pathways, which can be limited by tuning the electronic and steric features of substituents on the carbene carbon atom. The chemical structure of *mono*-substituted Fe(porp)(CHR) (porp, generic porphyrin) complexes were proposed on the basis of spectroscopic data [31], but they have not been isolated in a pure form until now. Conversely, many reports are present in the literature on the synthesis and characterization of halocarbene (Fe(porp)(CX_2), Fe(porp)(C(X)X′) and Fe(porp)(C(R)X)) [33, 34] and disubstituted alkyl or alkoxycarbene complexes (Fe(porp)(CR_2), Fe(porp)(C(R)R′), and Fe(porp)(C(R)CO_2R′)) [35]. The X-ray crystal structure of Fe(F_{20}TPP)(CPh_2) (**2**) (F_{20}TPP, dianion of *meso*-tetrakis-(pentafluorophenyl)porphyrin), as an example of stable iron carbene porphyrin complexes, is shown in Scheme 7.5 [35].

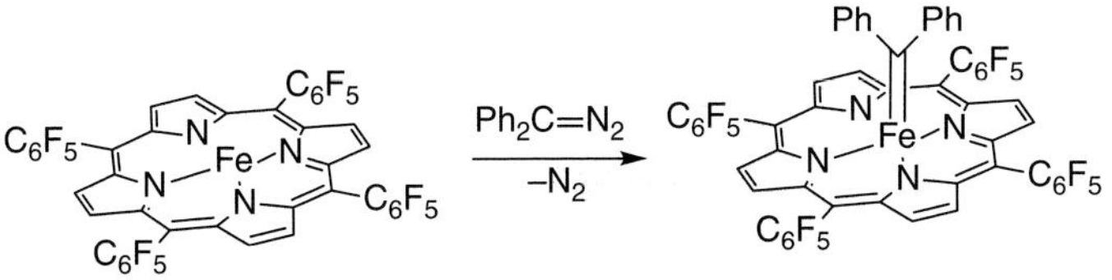

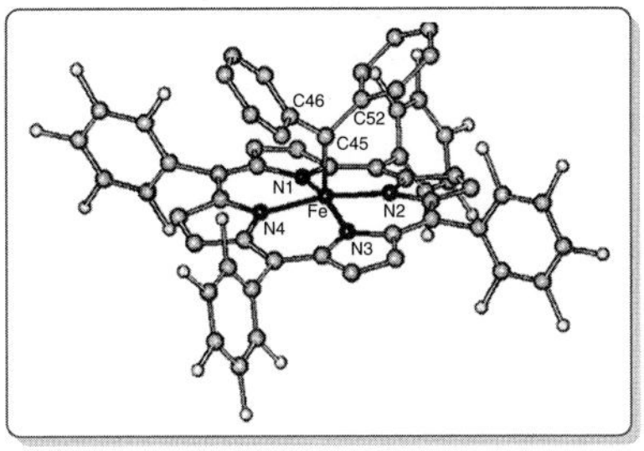

Scheme 7.5 Synthesis and molecular structure of Fe(F_{20}TPP)(CPh_2) (**2**). Reprinted with permission from [35]. Copyright 2016 American Chemical Society.

Complex **2** was synthesized by the reaction of Fe(F_{20}TPP) (**3**) with $Ph_2C{=}N_2$ and displayed a short iron–carbene carbon atom distance of 1.767(3) Å. Complex **2** transferred the carbene moiety to styrene only in the presence of the electron donor 2,6-dichloropyridine as the axial ligand, while efficiently catalyzing the cyclopropanation of styrene by ethyl diazoacetate (EDA). The structure of Fe(F_{20}TPP)(CPh_2) (**2**) was also theoretically studied by using DFT calculations (DFT, density functional theory), and the obtained data indicated that its chemical stability is due to the high steric hindrance of the phenyl carbene substituents and the electronic deficiency of the porphyrin ligand [36].

A large application of Fe(II)-based systems is hampered by the high instability of iron(II) porphyrin complexes; thus to bypass the related practical problems, active iron(II) species are synthesized *in situ* by reacting an iron(III) porphyrin complex with a reducing agent. The nature of the required reductant is strictly dependent on the electronic characteristics of the porphyrin ligand; strong reducing species, such as cobaltocene ($CoCp_2$), are usually required except when very electron-poor porphyrin ligands constitute the iron catalyst. In the last case, the reaction proceeds without the addition of a reductive cocatalyst, thanks to the mildly reducing properties of the diazo compound [37], which plays a double role as a carbene source and a reducing agent. It is important to underline that until now, this last hypothesis has not been confirmed by the isolation of an iron(II) complex from the stoichiometric reaction of an iron(III) porphyrin with a diazo compound.

The Fe^{III}(porp)Cl (**8–14**)/$CoCp_2$ combination was used for the synthesis of cyclopropanes **4–7** by using EDA as the carbene source [30, 38], which was also employed in the absence of the reducing cocatalyst when either Fe(F_{20}TPP)Cl (**9**) [30] or Fe(TDCPP)Cl (**11**) [38] (TDCPP, dianion of *meso*-tetrakis-(2,6-dichlorophenyl) porphyrin) was the reaction catalyst (Scheme 7.6). The syntheses of cyclopropanes **4–7** occurred in good yields with a trans/cis-diastereomeric ratio up to 78 and the contemporary formation of diethyl maleate as the side product of the carbene-coupling reaction.

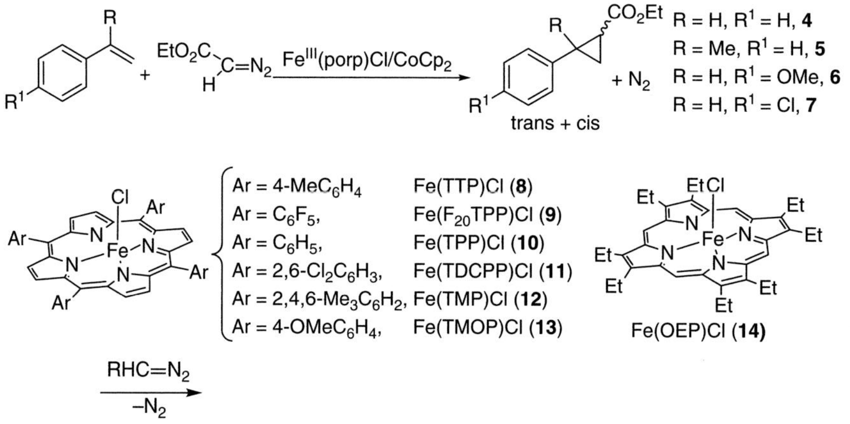

R = H, R^1 = H, **4**
R = Me, R^1 = H, **5**
R = H, R^1 = OMe, **6**
R = H, R^1 = Cl, **7**

Ar = 4-MeC_6H_4 Fe(TTP)Cl (**8**)
Ar = C_6F_5, Fe(F_{20}TPP)Cl (**9**)
Ar = C_6H_5, Fe(TPP)Cl (**10**)
Ar = 2,6-$Cl_2C_6H_3$, Fe(TDCPP)Cl (**11**)
Ar = 2,4,6-$Me_3C_6H_2$, Fe(TMP)Cl (**12**)
Ar = 4-$OMeC_6H_4$, Fe(TMOP)Cl (**13**)
Fe(OEP)Cl (**14**)

Scheme 7.6 Cyclopropanation of styrenes catalyzed by the Fe^{III}(porp)Cl/$CoCp_2$ catalytic system.

The direct cyclopropanation of alkenes is not feasible with every diazo derivative because some of them are too reactive to be safely handled in a laboratory. For example, using diazomethane (CH_2N_2) or trifluoromethyl diazomethane (CF_3CHN_2) as the carbene source is not trouble-free; thus, alternative synthetic procedures have been developed to add a "H_2C:" or a "CF_3HC:" fragment to a double bond to form the cyclopropane functionality.

Morandi and Carreira reported cyclopropanation of various styrenes under biphasic conditions using the nitrosoamine Diazald as a methylene carbene "CH_2:" source. The water-soluble Diazald decomposes when treated with a base forming CH_2N_2, which migrates from the aqueous into the organic phase where it reacts with both electron-rich and electron-poor alkenes yielding the desired cyclopropane product molecule. The reaction in the organic phase was efficiently catalyzed by 0.1 mol% of Fe(TPP)Cl (**10**) (TPP, dianion of *meso*-tetrakis-(phenyl)porphyrin), which tolerates the biphasic reaction medium, strong alkaline conditions (6 M KOH), and the presence of air well (Scheme 7.7) [39, 40].

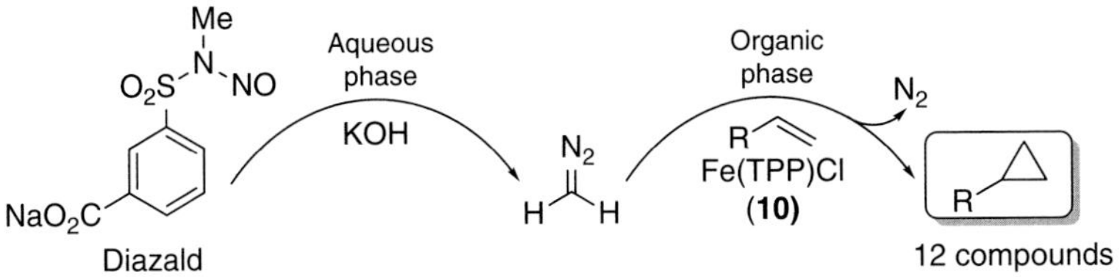

Scheme 7.7 Cyclopropanation of alkenes in a biphasic medium.

It is important to underline that an efficient phase separation is required to avoid the presence in the organic phase of water traces, which can provoke the decomposition of the supposed active iron carbene intermediate "Fe(TPP)(CH_2)". It was suggested that this last species can be formed by the reaction of the *in situ*-formed Fe^{II}(TPP) complex with diazomethane. Also in this case, diazomethane should play both the role of the carbene precursor and the reductant of the starting iron(III) precatalyst **10**.

The experimental procedure reported by Morandi and Carreira was also employed in the presence of a catalyst formed by a Fe(porp)Cl core installed into a dendrimer, which can modulate the catalytic performance of the active site by the creation of a tailored microenvironment [41]. The cyclopropanation of 4-chloro-α-methyl styrene proceeds at a similar rate to that observed in the presence of **10** to pave the way for a more extensive use of dendrimer-containing catalysts exhibiting several practical advantages such as an easy recovery of the catalytic species by precipitation, nano-, or ultrafiltration.

As mentioned above, the insertion of a "CF_3HC:" fragment into an organic skeleton can be technically challenging by using the corresponding diazo reagent; therefore, Carreira and coworkers explored the reactivity of trifluoroethylamine hydrochloride ($F_3CCH_2NH_2{\cdot}HCl$) as an inexpensive and safe carbene precursor. The reaction was effective for the cyclopropanation of differently substituted styrenes [42], dienes, and enynes [43], as illustrated in Scheme 7.8.

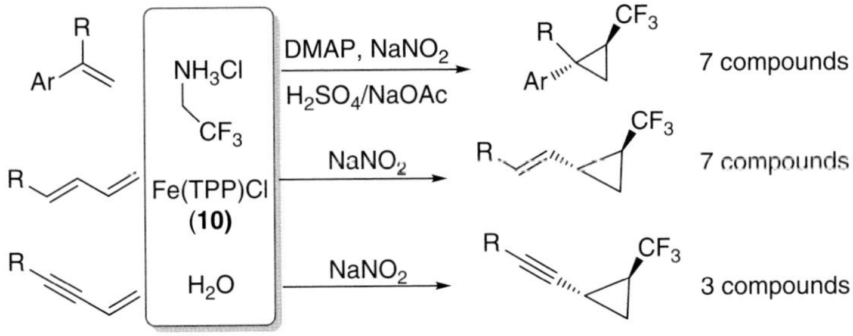

Scheme 7.8 Cyclopropanation of styrenes, dienes, and enynes by using trifluoroethylamine hydrochloride as the carbene source.

All the reactions were run in water by employing Fe^{III}(TPP)Cl (**10**) as the catalyst and the unsaturated substrate as the limiting reagent; it should be noted that the cyclopropanation of styrenes also required the presence of a H_2SO_4/NaOAc buffer and 4-(dimethylamino)pyridine (DMAP) to proceed. Reactions shown in Scheme 7.8 occurred with very good yields (up to 95%) and in the majority of cases with a complete trans-diastereoselectivity.

The sulfur ylide reagent (2,2,2-trifluoroethyl)diphenyl-sulfonium triflate ($Ph_2S^+CH_2CF_3\cdot^-OTf$) (Tf, triflyl) is also an effective precursor of a "CF_3HC:" fragment, which can be transferred to several alkenes in the presence of **10** and the base CsF in dimethylacetamide (DMA) (Scheme 7.9) [44].

Scheme 7.9 Cyclopropanation of alkenes by using $Ph_2S^+CH_2CF_3\cdot^-OTf$ as the carbene source.

Electron-rich, electron-neutral, and electron-poor alkenes have been efficiently converted into the corresponding trans-trifluoromethylated cyclopropanes in yields up to 98%; conversely, the catalytic protocol has not been successful when internal alkenes were the chosen reagents. For this reason, the authors propose the catalytic mechanism illustrated in Scheme 7.10 on the basis of a mechanistic investigation and previously reported studies.

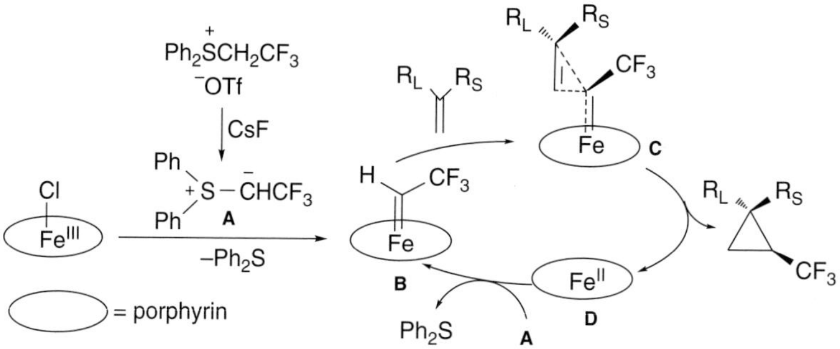

Scheme 7.10 Proposed mechanism for the cyclopropanation of alkenes by using $Ph_2S^+CH_2CF_3\cdot^-OTf$ as the carbene source.

It was suggested that $Ph_2S^+CH_2CF_3\cdot^-OTf$ decomposes in a basic medium forming the ylide **A**, which reacts with the starting iron(III) porphyrin complex to form intermediate carbene **B**. The interaction of **B** with the alkene is then responsible for the formation of both the cyclopropane and the iron(II) species **D**, which can react with ylide **A** to restart the catalytic cycle. Like other similar mechanisms where internal alkenes are inert, an *end-on* approach of the alkene to the carbene species was proposed in accord with the observed trans-selectivity determined by the interaction of the sterically encumbered R_L with the CF_3 substituent (Scheme 7.10).

In view of the positive results achieved by using the ylide reagent $Ph_2S^+CH_2CF_3\cdot^-OTf$ as the carbene precursor, Lin, Xiao, and coworkers [45] also investigated the activity of Fe^{III}(TPP)Cl (**10**) in promoting the reaction between the difluoroethylsulfonium salt $Ph_2S^+CH_2CF_2H\cdot{}^-OTf$ with terminal alkenes. The procedure, shown in Scheme 7.11, was efficient in forming differently substituted difluoromethyl cyclopropanes with excellent diastereoselectivities and yields.

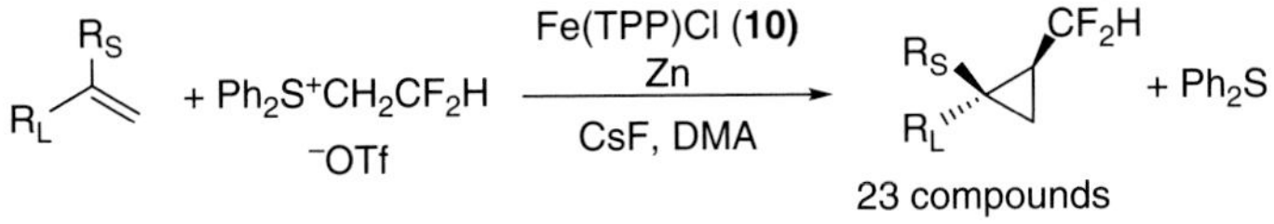

R_S, R_L alkene + $Ph_2S^+CH_2CF_2H$ ^-OTf → [Fe(TPP)Cl (**10**), Zn / CsF, DMA] → R_S, R_L, CF_2H cyclopropane + Ph_2S

23 compounds

Scheme 7.11 Fe(TPP)Cl-catalyzed cyclopropanation of alkenes by using $Ph_2S^+CH_2CF_2H\cdot{}^-OTf$ as the carbene source.

The reaction proceeded only in the presence of zinc, which is probably necessary to reduce the precatalyst Fe^{III}(TPP)Cl into an iron(II) species, which is able to react with $Ph_2S^+CH_2CF_2H\cdot{}^-OTf$ in generating the catalytically active carbene intermediate. The authors suggested a catalytic mechanism analogous to that illustrated in Scheme 7.10.

Considering the importance of using water as the reaction solvent and the usual instability of diazo regents in an aqueous medium, alternative synthetic routes employing water-stable carbene precursors have been developed. Carreira and coworkers reported the use of glycine ethyl ester hydrochloride, in place of EDA, to transfer a "(EtO_2C)CH" functionality to a double bond. The reaction worked well in water in open air irrespective of the electronic features of the starting alkene, and the corresponding *trans*-cyclopropyl ester was formed in good yields and selectivities (Scheme 7.12) [46].

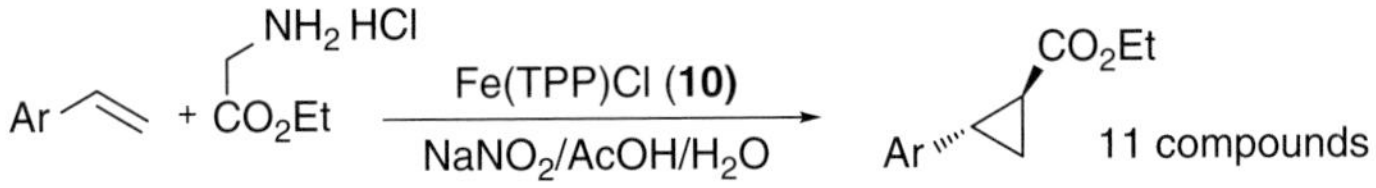

Ar alkene + NH_2 HCl / CO_2Et → [Fe(TPP)Cl (**10**) / $NaNO_2/AcOH/H_2O$] → Ar, CO_2Et cyclopropane   11 compounds

Scheme 7.12 Cyclopropanation of alkenes by using glycine ethyl ester hydrochloride as the carbene source.

A similar strategy was adopted by Koenigs and coworkers [47], who reported the generation of diazo acetonitrile by treating aminoacetonitrile hydrochloride with sodium nitrite ($NaNO_2$) in a 1 : 1 H_2O/CH_2Cl_2 mixture. The slow addition of $NaNO_2$ to a $(NC)CH_2NH_2\cdot HCl$ solution was responsible for the continuous formation of the "(NC)HC:" carbene moiety which, in the presence of Fe(TPP)Cl (**10**), reacts with alkenes forming the desired cyclopropanes. The methodology was efficient for the cyclopropanation of 29 different styrenes and α-substituted styrenes (Scheme 7.13).

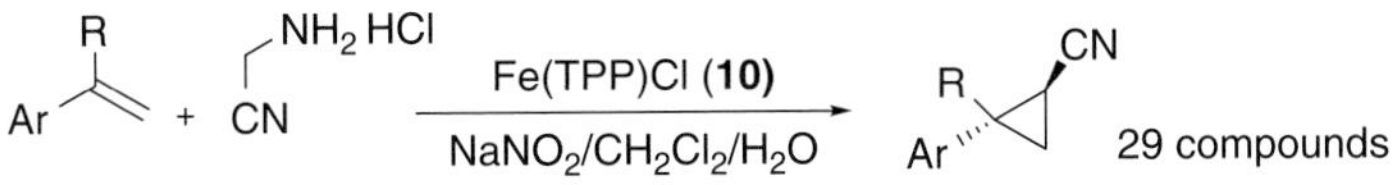

R, Ar alkene + NH_2 HCl / CN → [Fe(TPP)Cl (**10**) / $NaNO_2/CH_2Cl_2/H_2O$] → R, Ar, CN cyclopropane   29 compounds

Scheme 7.13 Cyclopropanation of alkenes by using aminoacetonitrile hydrochloride as the carbene source.

The reported methodology is of particular importance in view of the very hazardous nature of diazo acetonitrile, (NC)HC=N_2, which hampers its safe handling in carbene transfer reactions.

7.3 Chiral Iron Porphyrin Catalysts

Considering the importance to obtain cyclopropane-containing compounds in a pure chiral form, great efforts have been made to synthesize chiral iron porphyrin catalysts, which efficiently promote enantioselective carbene transfer reactions [48]. In 1999, Gross et al. reported on the catalytic activity of D_2-symmetrical porphyrin iron(III) catalyst **15** in the asymmetric cyclopropanation of styrene by EDA (Scheme 7.14) [49]. The catalytic efficiency of **15** was modest and the cyclopropyl ester was formed with the trans/cis-diastereomeric ratio of 6.6, and trans (*1R,2R*) and cis (*1R,2S*) isomers were obtained with 15% and 23% ee, respectively (ee, enantiomeric excess). Better results were achieved by using D_4-symmetrical chiral porphyrins **16–18** (Scheme 7.14).

The chiral iron(III) porphyrin (Halt)FeCl (**16**) (Halt, Halterman porphyrin), reported by Che and coworkers in 2006, is active in synthesizing the cyclopropanes shown in Figure 7.3 [50]. The desired compounds were obtained with good trans-diastereo- and enantioselectivities.

Considering that these reactions do not require the addition of a reductive cocatalyst to proceed, the authors proposed that complex **16** is electron-poor enough to be directly reduced by EDA to an iron(II) complex, which can be transformed into an active carbene intermediate. This suggestion explains the drastic catalytic efficiency decrease, which was observed for the reaction run in open air where oxidative degradation of the very reactive iron(II) "(Halt)Fe" complex can take place. The authors did not report direct experimental evidence of the formation of an iron carbene species, except when an axial ligand was added to the reaction mixture. Upon treatment of complex **16** with EDA in the presence of pyridine (py) or 1-methylimidazole (MeIm), the formation of type **B** complexes (Halt)Fe($CHCO_2Et$)(py) and (Halt)Fe($CHCO_2Et$)(MeIm) was detected by electrospray mass spectrometry (ESMS) (Scheme 7.15).

The addition of an axial ligand to the catalytic mixture is also responsible for an improvement in the reaction trans-diastereoselectivity, indicating a possible formation of a six-coordinate monocarbene species **B** during the catalytic cycle. The proposed reaction mechanism is shown in Scheme 7.15, and, in accord with the high reactivity of terminal alkenes, it also suggests an *end-on* approach of the double bond to active intermediate **B**.

Complex **16** is able to promote the cyclopropanation of styrenes by using diazoacetophenone (PhCO)CHN_2 [51] and trifluoromethyl diazomethane (CF_3CHN_2) [52], forming several cyclopropane derivatives as shown in Scheme 7.16.

Cyclopropanes from diazoacetophenone were formed in yields of up to 67%, trans/cis-diastereomeric ratios of up to 96 : 4, and ee_{trans}'s of up to 80%. Also in this case, the observed trans-diastereoselectivity was ascribed to the formation

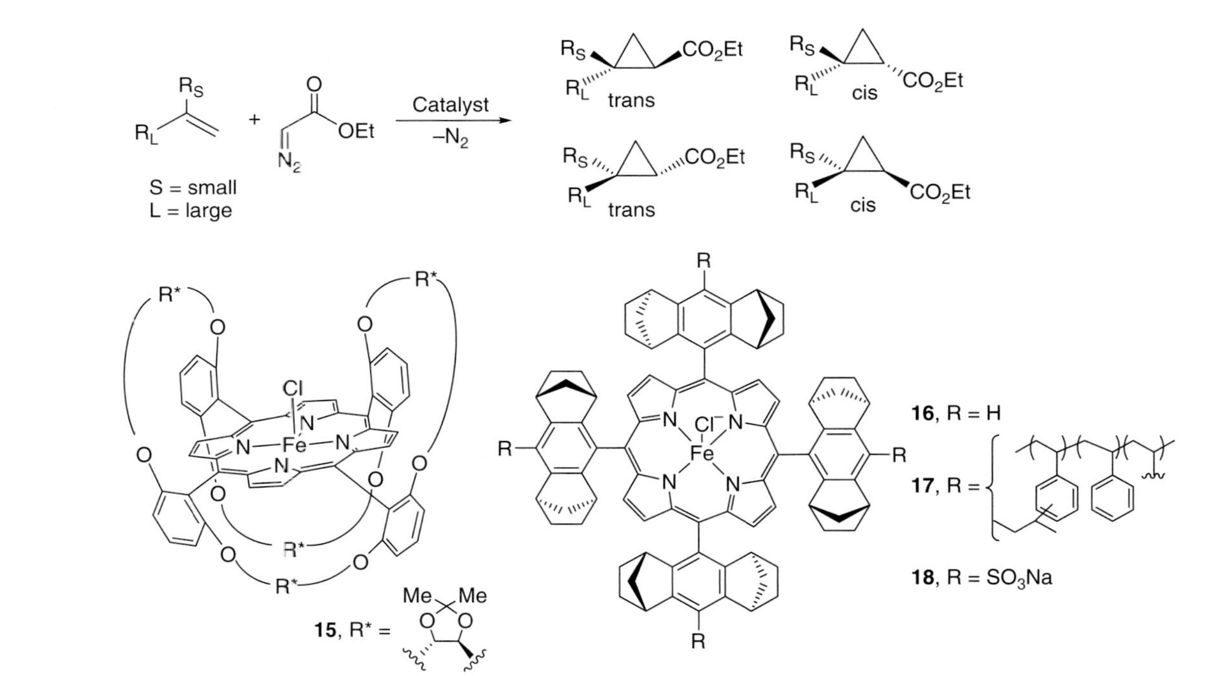

Scheme 7.14 D_2-symmetrical porphyrin iron(III) catalyst **15** and iron porphyrins **16–18** used in asymmetric alkene cyclopropanations.

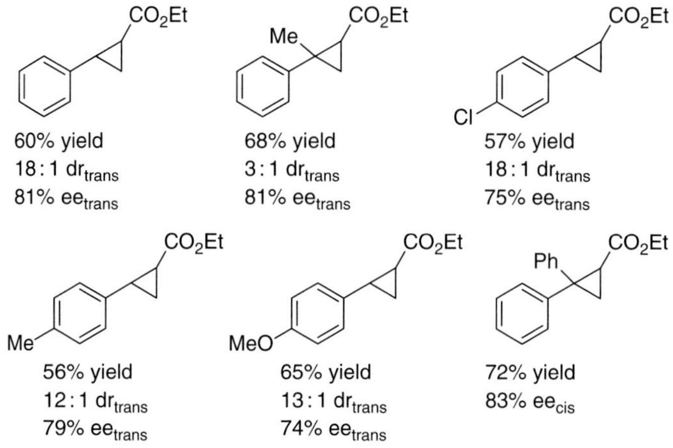

60% yield
18 : 1 dr_{trans}
81% ee_{trans}

68% yield
3 : 1 dr_{trans}
81% ee_{trans}

57% yield
18 : 1 dr_{trans}
75% ee_{trans}

56% yield
12 : 1 dr_{trans}
79% ee_{trans}

65% yield
13 : 1 dr_{trans}
74% ee_{trans}

72% yield
83% ee_{cis}

Figure 7.3 Cyclopropanes synthesized in the presence of catalyst **16**.

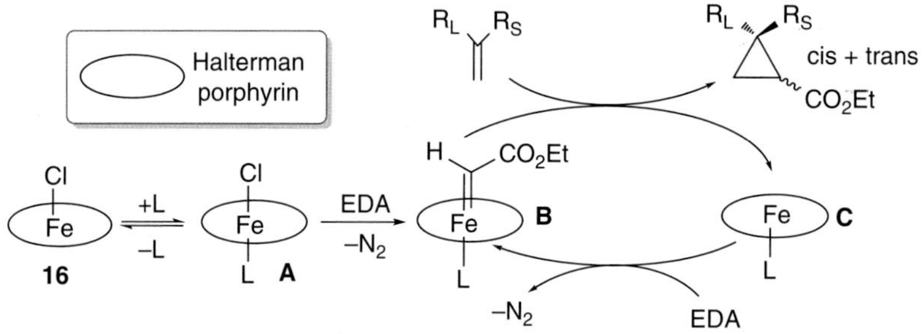

Scheme 7.15 Proposed mechanism of complex **16**-catalyzed alkene cyclopropanation by EDA.

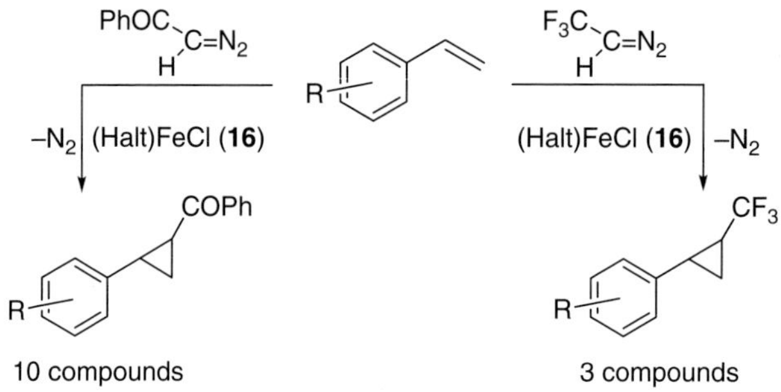

10 compounds

3 compounds

Scheme 7.16 Cyclopropanation of styrenes by (PhCO)CHN_2 and CF_3CHN_2 catalyzed by complex **16**.

of a late transition state. An almost quantitative trans-diastereoselectivity (trans/cis = 99 : 1) was always observed by using CF_3CHN_2 as the carbene source; however, enantioselectivities lower than those observed using diazoacetophenone were achieved (up to 69% ee_{trans}). Complex (Halt)FeCl (**16**) is also able to promote the decomposition of a biologically active diazo derivative, such

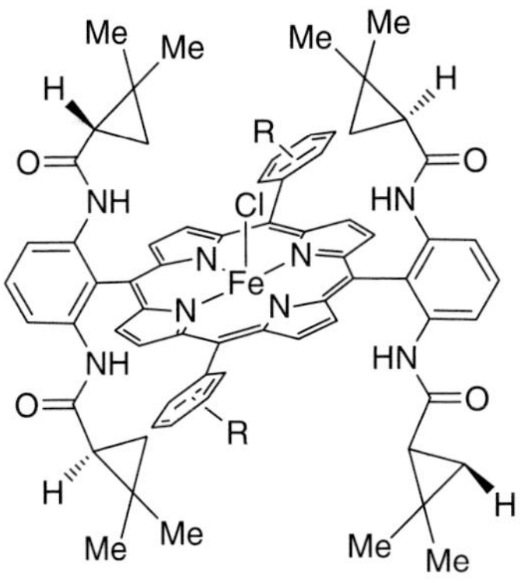

19, R = 3,5-(^{t}Bu)$_2$

Figure 7.4 D_2-symmetrical iron(III) chiral porphyrin **19**.

as N- and O-protected 6-diazo-5-oxo-L-norleucine (DON), whose reaction with styrene formed the corresponding cyclopropane in 95% yield, 95 : 5 dr_{trans} (dr, diastereomeric ratio), and 80% ee_{trans} [53].

The anchoring of complex **16** to a polymeric structure yielded heterogeneous catalyst **17** (Scheme 7.14), which is as efficient as **16** in terms of reaction yields and diastereoselectivities, but resulted in a marked decrease of the reaction enantioselectivity (dr_{trans} up to 97 : 3, ee_{trans} up to 56%) [52].

Finally, the introduction of SO_3Na substituents on the skeleton of the Halterman porphyrin provided a water-soluble ligand, and the corresponding catalyst **18** (Scheme 7.14) was used to catalyze the reaction between styrene and EDA in an aqueous medium, where the corresponding cyclopropane was obtained in 85% yield, 92 : 8 dr_{trans}, and 83% ee_{trans} [54]. The reaction was performed in the presence of $CoCp_2$ to transform complex **18** into a claimed active iron(II) intermediate.

D_2-symmetrical porphyrin ligands were extensively used by Zhang and coworker to form chiral cobalt(II) complexes active in cyclopropanations [55], in which they are more active than their iron counterparts. Iron(III) porphyrin **19** shown in Figure 7.4 was employed to promote the reactions between styrene derivatives and EDA, but even if high trans-diastereoselectivities were observed (up to trans/cis = 93 : 7), the reaction enantiocontrol was only modest (ee_{trans} up to 28%) [56].

Better results were obtained by Gallo, Boitrel, and coworkers using an iron(III) methoxy derivative of a C_2-symmetrical porphyrin ligand (Scheme 7.15) [57, 58]. Iron catalyst **20** was employed in the cyclopropanation of several alkenes by using an equimolar alkene/diazo compound ratio or a slight excess of the diazo reagent. These experimental conditions are feasible because carbene dimerization is not well promoted by **20** and they are relevant when expensive alkenes are chosen as the starting material. The procedure works well using terminal alkenes and not excessively encumbered diazo compounds, pointing out a dependence of the catalytic efficiency on the degree of the steric bulk close to the porphyrin plane.

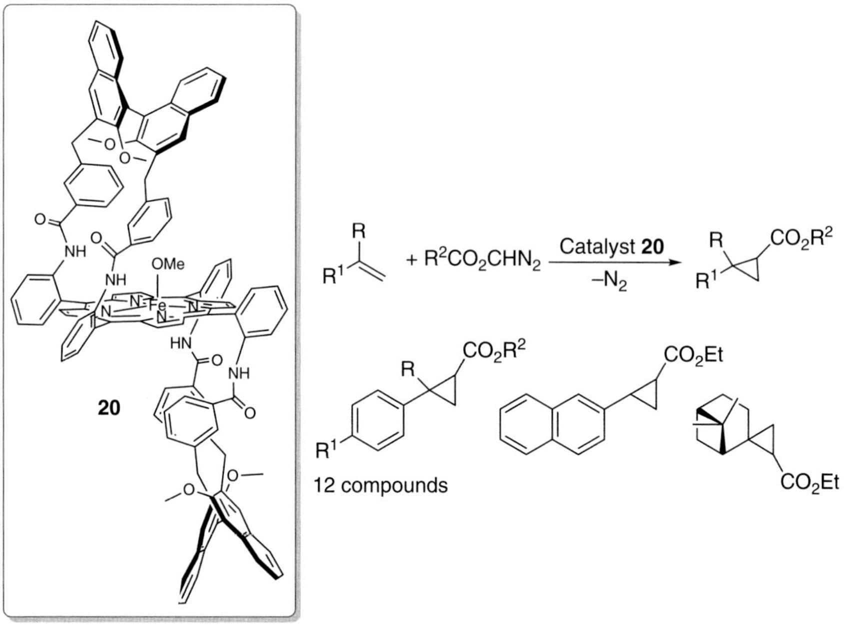

Scheme 7.17 C_2-symmetrical iron(III) chiral porphyrin **20**.

Cyclopropanes shown in Scheme 7.17 were obtained with excellent trans-diastereoselectivities (trans/cis-diastereomeric ratio up to 99 : 1) and good enantioselectivities (up to 87% ee_{trans}). It should be noted that some cyclopropanations were promoted by a very low catalyst loading of 0.01 mol% (TON = 10 000), where the diazo conversion occurred in short reaction times, providing very interesting TOF (turnover frequency) values up to 120 000 h^{-1}.

Experimental and DFT studies indicated an important catalytic role of the methoxy axial ligand, which fine-tunes the reactivity on the trans-position of the iron metal center. The theoretical analysis indicated that the excellent trans-diastereoselectivity is due to the interaction of the incoming alkene with the carbene moiety embedded in the tridimensional environment defined by the two arms surrounding the porphyrin plane.

It should be noted that when the chiral binaphthyl moieties of **20** were replaced by amino acid residues [59], the resulting complexes **21** and **22** (Figure 7.5) were unable to efficiently control the reaction enantioselectivity. DFT studies suggested that, even if the chiral pockets of complexes **21** and **22** are close to the active catalytic iron metal, they are too mobile to discriminate an enantiomeric pathway.

7.4 Iron Phthalocyanines and Corroles

Along with iron porphyrin catalysts, other porphyrinoid complexes, such as iron phthalocyanines and corroles, have been tested as cyclopropanation

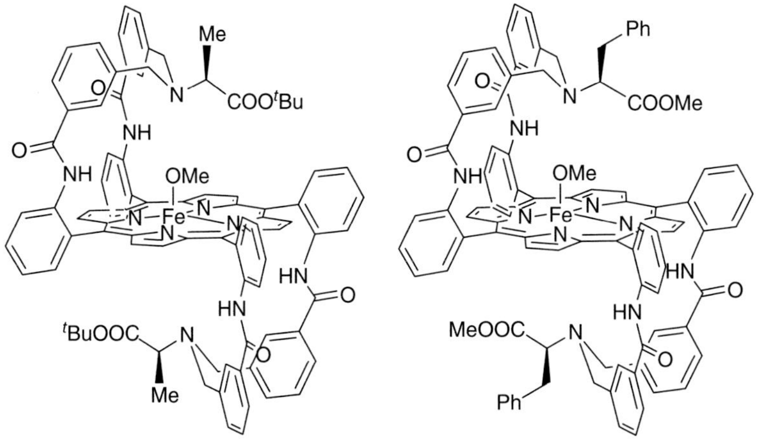

Figure 7.5 C_2-symmetrical iron(III) chiral porphyrins **21** and **22**.

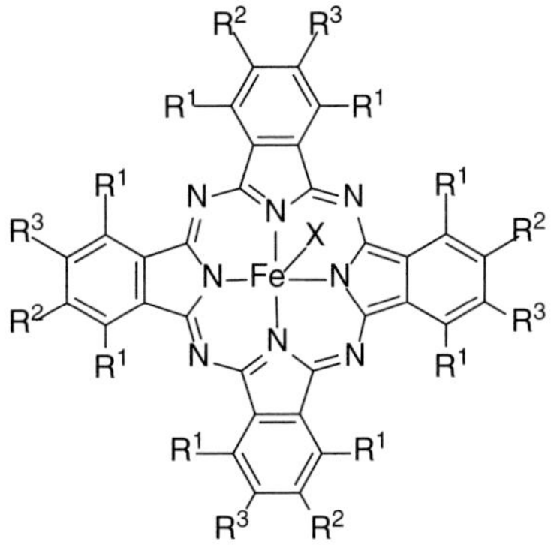

23, $R^1 = R^2 = R^3 = H$, X = none
24, $R^1 = R^2 = R^3 = H$, X = Cl
25, $R^1 = R^2 = R^3 = Cl$, X = none
26, $R^1 = R^2 = R^3 = F$, X = Cl
27, $R^1 = H$, $R^2 = H$ or ^{t}Bu, $R^3 = H$ or ^{t}Bu, X = none
28, $R^1 = H$, $R^2 = H$ or *m*-OC_6H_4-CF_3, $R^3 =$ *m*-OC_6H_4-CF_3, or H, X = Cl

Figure 7.6 Iron phthalocyanine complexes **23–28** used as cyclopropanation catalysts.

catalysts. Iron phthalocyanines display a prominent chemical stability, and their insolubility in several organic solvents permits an easy recovery and recycle. Sain and coworkers reported in 2004 that Fe(II) phthalocyanine complex **23** (Figure 7.6) catalyzes, in a heterogeneous medium, the cyclopropanation of 4-methylstyrene by EDA in 55% reaction yield [60].

Catalyst **23** is also active in promoting the reaction between 4-methylstyrene and trimethylsilyl diazomethane (Me_3SiCHN_2), giving the corresponding *C*-silyl cyclopropanes with trans-diastereoselectivity (Scheme 7.18) [61]. Complex **23**

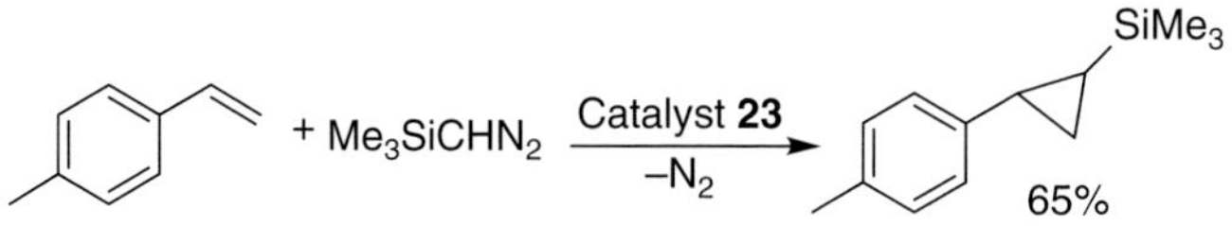

+ Me_3SiCHN_2 — Catalyst **23**, $-N_2$ → $SiMe_3$ cyclopropane, 65%

Scheme 7.18 Synthesis of *C*-silyl cyclopropanes catalyzed by complex **23**.

was recovered at the end of the reaction by a simple filtration and reused several times without losing its catalytic efficiency.

The effect of the ligand electronic characteristics and metal oxidation state on the catalytic efficiency of the iron phthalocyanine complex was studied by reacting styrene with EDA in the presence of catalysts **24–28** (Figure 7.6, Scheme 7.19) [62].

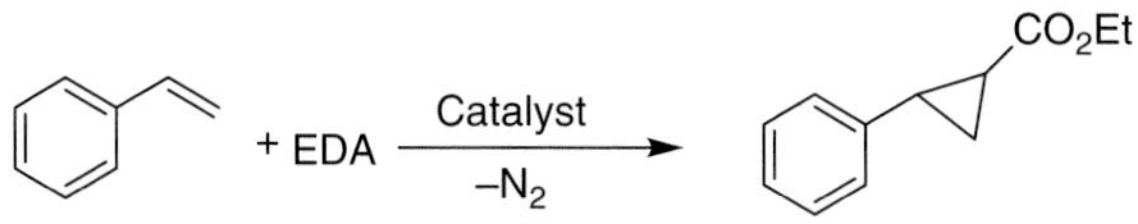

+ EDA — Catalyst, $-N_2$ → CO_2Et cyclopropane

iron(III) catalyst **24**, 62% yield, 51% conversion, trans/cis = 2.4
iron(II) catalyst **25**, 78% yield, 75% conversion, trans/cis = 2.1
iron(III) catalyst **26**, 89% yield, 76% conversion, trans/cis = 5
iron(II) catalyst **27**, 80% yield, 73% conversion, trans/cis = 2.8
iron(III) catalyst **28**, 90% yield, 99% conversion, trans/cis = 2

Scheme 7.19 Cyclopropanation of styrene by EDA using catalysts **24–28**.

The best performance was observed in the presence of iron(III) catalyst **28** bearing electron-withdrawing substituents on the phthalocyanine skeleton. However, the diastereoselectivity of the reaction catalyzed by **28** was only modest (trans/cis = 2). No great differences were observed by using iron(III) or iron(II) catalysts, suggesting that the same intermediates are formed during the reaction independent of the oxidation state of the starting precatalyst.

As discussed until now, the generally assumed mechanistic proposal for the alkene cyclopropanation catalyzed by iron(III) porphyrinoids is based on the formation of an active iron carbene species formed by the reaction of a carbene source with an *in situ*-formed iron(II) complex. Thus, the study of the catalytic activity of iron precatalysts showing metal oxidation states different than (II) and (III) was of great importance. In this regard, the interest in using corroles as porphyrinoid ligands was because of their electronic features, which permit the stabilization of central metal atoms in high oxidation states. Iron(IV) corrole Fe^{IV}(tpfc) (**29**) (tpfc, dianion of *tris*(pentafluorophenyl) corrole) was synthesized by treating the trianionic corrole ligand with $FeCl_2$ in air and completely characterized including the determination of its molecular structure by X-ray diffraction (Scheme 7.20) [63, 64].

Complex **29** was also obtained by starting from $Fe^{III}(tpfc)(OEt_2)_2$ (**30**) (OEt_2, diethyl ether), which was used as a precursor of two other iron corroles, $Fe^{III}(tpfc)(py)_2$ (**31**) and $(Fe^{IV}(tpfc))_2O$ (**32**), as shown in Scheme 7.21 [65].

Complexes **29–32** were employed to promote the styrene cyclopropanation by using the two different diazo reagents shown in Scheme 7.22 [65].

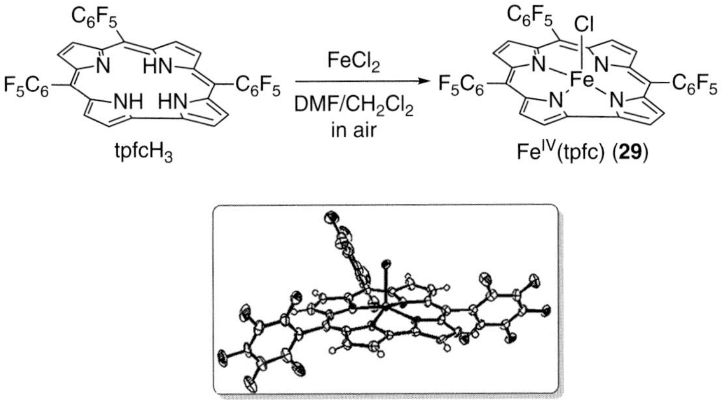

Scheme 7.20 Synthesis and molecular structure of Fe^{IV}(tpfc) (**29**). Simkhovich et al. 2000 [63]. Copyright 2000. Reprinted with permission from American Chemical Society.

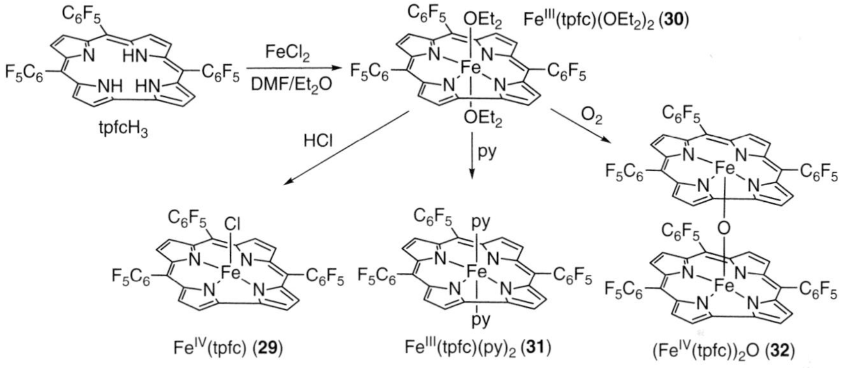

Scheme 7.21 Synthesis of Fe^{IV}(tpfc) (**29**), $Fe^{III}(tpfc)(OEt_2)_2$ (**30**), $Fe^{III}(tpfc)(py)_2$ (**31**), and $(Fe^{IV}(tpfc))_2O$ (**32**).

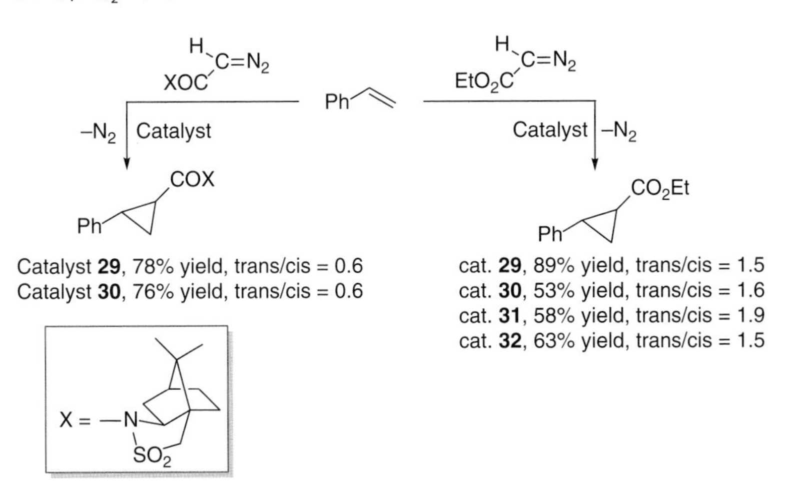

Scheme 7.22 Cyclopropanation of styrene catalyzed by iron corroles **29–32**.

The analysis of these reactions indicated that iron(III) and (IV) corroles show a similar catalytic efficiency in accord with the formation of the same active intermediate. The authors proposed that the diazo reagent reacts with an iron(III) species, which is either the catalyst employed (reactions catalyzed by **30** or **31**) or formed by the reduction of the starting iron(IV) complex by the diazo compound (reactions catalyzed by **29** or **32**). The lack of the influence of the iron oxidation state on the catalytic performance was evident when sterically hindered $XCOCH{=}N_2$ was used as the starting material (Scheme 7.22). Almost identical yields and trans/cis-diastereomeric ratios were obtained when either Fe^{IV}(tpfc) (**29**) or $Fe^{III}(tpfc)(OEt_2)_2$ (**30**) was used as the cyclopropanation catalyst.

7.5 Iron Catalysts with N or N,O Ligands

Tetraaza macrocyclic ligands show many structural similarities with porphyrins; therefore, they were used to synthesize iron(II) complexes in order to compare their catalytic activity with those of iron porphyrin catalysts. Figure 7.7 displays the structures of achiral and chiral iron(II) tetraaza macrocyclic complexes **33** and **34–36**, respectively [31, 66].

Styrene was reacted with EDA in the presence of complex **33**, but only 20% of EDA was converted; conversely, aryldiazoacetates such as mesityldiazomethane and *p*-tolyldiazomethane were transformed into the corresponding cyclopropanes **I** and **II** in 43% and 16% yield, respectively (Figure 7.8) [31]. Even if cyclopropane **I** was formed in a modest yield, a cis-diastereoselectivity was achieved conversely to what was observed in the cyclopropanation of the other alkene substrates.

Better results were observed by catalyzing styrene cyclopropanation with complexes **34–36** [66]. In particular, good yields were observed in the synthesis of cyclopropanes **III** and **V** catalyzed by **34**, whereas complex **36** was more efficient in promoting the formation of **V** than **III** (Figure 7.8). The nonexceptional diastereoselectivity observed in all tested cases can be explained with the formation of a well-accessible carbene intermediate because of a low steric hindrance around the active metal center. It was proposed that the poorly encumbered N_4-core of tetraaza iron complex does not discriminate between the cis and trans approaches of the incoming alkene substrate with the

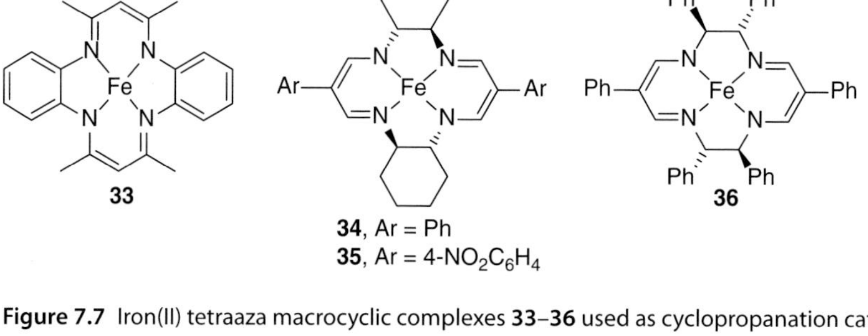

Figure 7.7 Iron(II) tetraaza macrocyclic complexes **33–36** used as cyclopropanation catalysts.

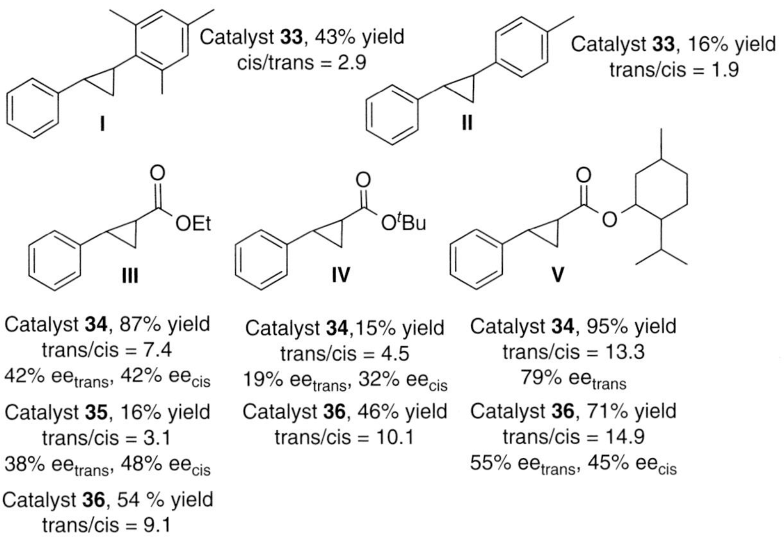

Catalyst **33**, 43% yield
cis/trans = 2.9

I

Catalyst **33**, 16% yield
trans/cis = 1.9

II

OEt

III

Catalyst **34**, 87% yield
trans/cis = 7.4
42% ee_{trans}, 42% ee_{cis}

Catalyst **35**, 16% yield
trans/cis = 3.1
38% ee_{trans}, 48% ee_{cis}

Catalyst **36**, 54 % yield
trans/cis = 9.1

O^tBu

IV

Catalyst **34**,15% yield
trans/cis = 4.5
19% ee_{trans}, 32% ee_{cis}

Catalyst **36**, 46% yield
trans/cis = 10.1

V

Catalyst **34**, 95% yield
trans/cis = 13.3
79% ee_{trans}

Catalyst **36**, 71% yield
trans/cis = 14.9
55% ee_{trans}, 45% ee_{cis}

Figure 7.8 Cyclopropanes synthesized in the presence of catalysts **33–36**.

consequence of a low degree of reaction stereocontrol. The best result in terms of reaction productivity and stereocontrol was registered for the styrene cyclopropanation with (−)menthyl diazoacetate catalyzed by **34** leading to compound **V** (Figure 7.8). The latter was formed in 95% yield, high trans-diastereoselectivity (trans/cis = 13.3), and good enantioselectivity (ee_{trans} = 79%).

Chiral C_1-, C_2-symmetric terpyridine ligands **37–42** were also employed to synthesize iron derivatives to test them as cyclopropanation promoters. These ligands were employed to synthesize iron(II) derivatives by reacting them with $FeCl_2$, whereas only less sterically encumbered ligands **41** and **42** were reacted with $FeCl_3$ to form the corresponding iron(III) derivatives. All iron complexes were completely characterized, including X-ray crystal structure analysis of Fe^{III}(**41**)Cl_3 (Figure 7.9) [67].

All obtained complexes were employed to catalyze the cyclopropanation of styrene by EDA and were active only in the presence of a stoichiometric excess of AgOTf (Ag/Fe ratio of 3 : 1), which converted the starting iron complex into the active catalyst (Scheme 7.23).

Best results in terms of diastereoselectivity (trans/cis = 76 : 24) and yield (78%) were recorded by using Fe^{II}(**37**)Cl_2, whereas the most sterically encumbered Fe^{II}(**39**)Cl_2 complex promotes the cyclopropane formation with the best enantiocontrol (ee_{trans} = 65%, ee_{cis} = 67%). Very poor results were achieved in the presence of iron(III) complexes Fe^{III}(**41**)Cl_3 and Fe^{III}(**42**)Cl_3, both in terms of diastereo- and enantioselectivities (Scheme 7.23).

Chiral spiro-*bis*oxazoline ligands shown in Figure 7.10 were used in combination with $Fe(ClO_4)_2{\cdot}4H_2O$ and $NaBAr_F$ ($NaBAr_F$, sodium tetrakis-[3,5-*bis* (trifluoromethyl)phenyl]borate) to induce asymmetric intramolecular cyclopropanation of diazoesters giving chiral [3.1.0]bicycloalkane lactones [68].

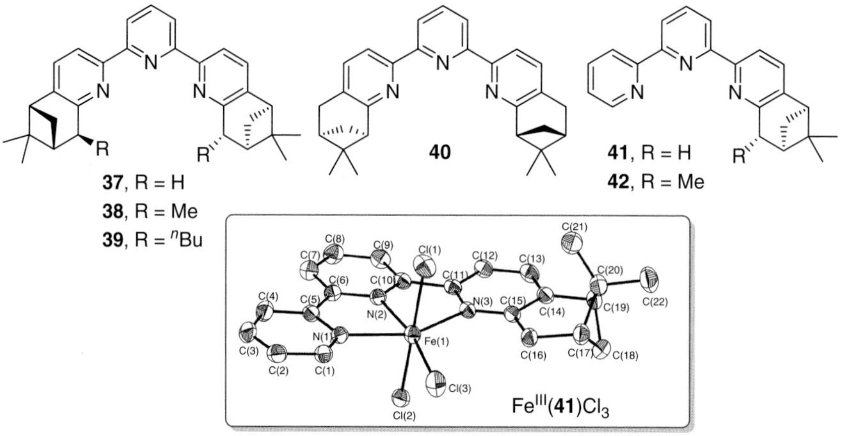

37, R = H
38, R = Me
39, R = ^{n}Bu
40
41, R = H
42, R = Me
Fe^{III}(**41**)Cl_3

Figure 7.9 Ligands **37–42** used in Fe-mediated cyclopropanation catalysis. Reprinted from Ref. [67]. Copyright 2009 with permission from Elsevier.

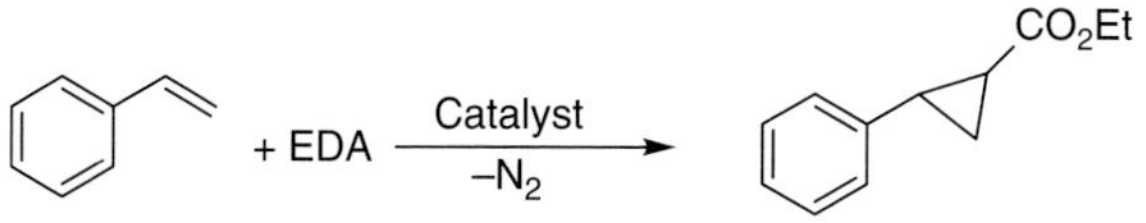

+ EDA — Catalyst, $-N_2$ → CO_2Et

Fe^{II}(**37**)Cl_2, 78% yield, trans/cis = 76 : 24, 36% ee_{trans}, 33% ee_{cis}
Fe^{II}(**38**)Cl_2, 71% yield, trans/cis = 68 : 32, 54% ee_{trans}, 54% ee_{cis}
Fe^{II}(**39**)Cl_2, 65% yield, trans/cis = 65 : 35, 65% ee_{trans}, 67% ee_{cis}
Fe^{II}(**40**)Cl_2, 41% yield, trans/cis = 72 : 28, 4% ee_{trans}, 5% ee_{cis}
Fe^{II}(**41**)Cl_2, 48% yield, trans/cis = 58 : 42, 4% ee_{trans}, 7% ee_{cis}
Fe^{II}(**42**)Cl_2, 60% yield, trans/cis = 66 : 34, 2% ee_{trans}, 5% ee_{cis}
Fe^{III}(**41**)Cl_3, 39% yield, trans/cis = 59 : 41, 5% ee_{trans}, 5% ee_{cis}
FeIII(**42**)Cl_3, 55% yield, trans/cis = 65 : 35, 4% ee_{trans}, 6% ee_{cis}

Scheme 7.23 Cyclopropanation of styrene by EDA catalyzed by iron complexes derived from ligands **37–42**.

These ligands were employed for the cyclopropanation of 2-methylallyl 2-diazo-2-phenylacetate (Scheme 7.24, Ar = Ph), and best results were obtained by using ligand **43**. Then, the study of the reaction scope in the presence of the **43**/$Fe(ClO_4)_2 \cdot 4H_2O$/$NaBAr_F$ catalytic system permitted the isolation of 14 compounds derived from the intramolecular cyclopropanation of differently substituted α-diazoesters (Scheme 7.24a). All the tested substrates were transformed into the desired compounds in good yields (86–96%); the reaction enantioselectivity was high (up to 96% ee) when electron-withdrawing substituents were present on the aryl group, whereas the electron-rich aryl fragments were responsible for a lowering of the reaction enantiocontrol. Thus, to expand the reaction scope, the effect of the electronic and steric nature of the allylic portion of the starting material was also investigated. These studies indicated that terminal alkenes (R^2 = R^3 = H) react more efficiently than the internal ones, and the presence of the R^1 substituent is necessary to induce the

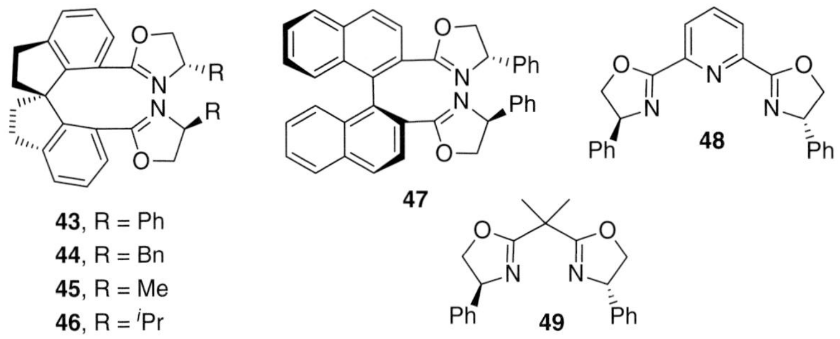

O N R N R O

43, R = Ph
44, R = Bn
45, R = Me
46, R = ^{i}Pr

O N Ph N Ph O
47

O N N O Ph 48 Ph

O N N O Ph 49 Ph

Figure 7.10 Chiral spiro-*bis*oxazoline ligands **43–49** used in combination with $Fe(ClO_4)_2{\cdot}4H_2O$ and $NaBAr_F$ for asymmetric intramolecular cyclopropanations.

cyclopropanation with good enantioselectivity, in fact when $R^1 = H$ only 6% ee was registered (Scheme 7.24b).

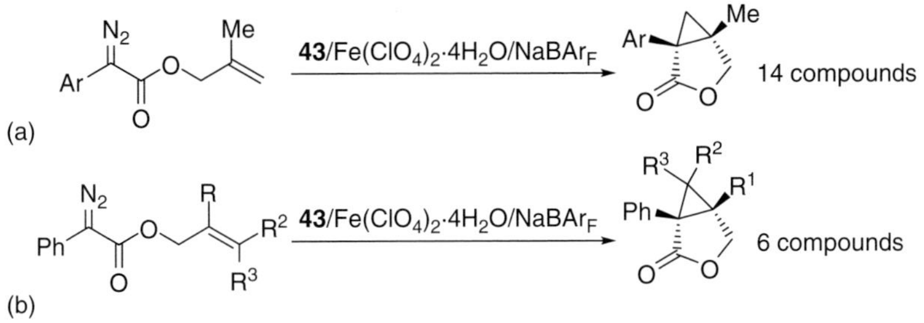

(a) N_2 Me Ar O O — **43**/$Fe(ClO_4)_2{\cdot}4H_2O$/$NaBAr_F$ → Ar Me O O — 14 compounds

(b) N_2 R Ph O O R^2 R^3 — **43**/$Fe(ClO_4)_2{\cdot}4H_2O$/$NaBAr_F$ → R^3 R^2 R^1 Ph O O — 6 compounds

Scheme 7.24 Asymmetric synthesis of chiral [3.1.0]bicycloalkane lactones catalyzed by **43**/$Fe(ClO_4)_2{\cdot}4H_2O$/$NaBAr_F$.

The **43**/$Fe(ClO_4)_2{\cdot}4H_2O$/$NaBAr_F$ catalytic system was also active in promoting the intramolecular cyclopropanation of indoles [69], and the corresponding bicyclic ring systems were achieved in yields up to 94% and excellent stereocontrol (up to >99.9% ee) (Scheme 7.25).

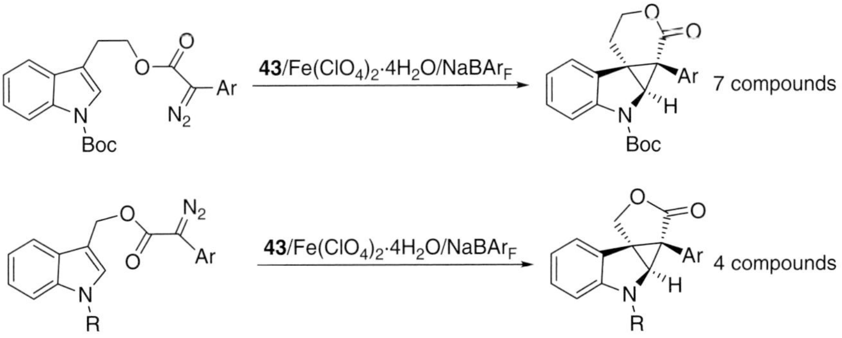

N Boc O O Ar N_2 — **43**/$Fe(ClO_4)_2{\cdot}4H_2O$/$NaBAr_F$ → O O Ar H N Boc — 7 compounds

N R O N_2 O Ar — **43**/$Fe(ClO_4)_2{\cdot}4H_2O$/$NaBAr_F$ → O O Ar H N R — 4 compounds

Scheme 7.25 Asymmetric intramolecular cyclopropanation of indoles catalyzed by **43**/$Fe(ClO_4)_2{\cdot}4H_2O$/$NaBAr_F$.

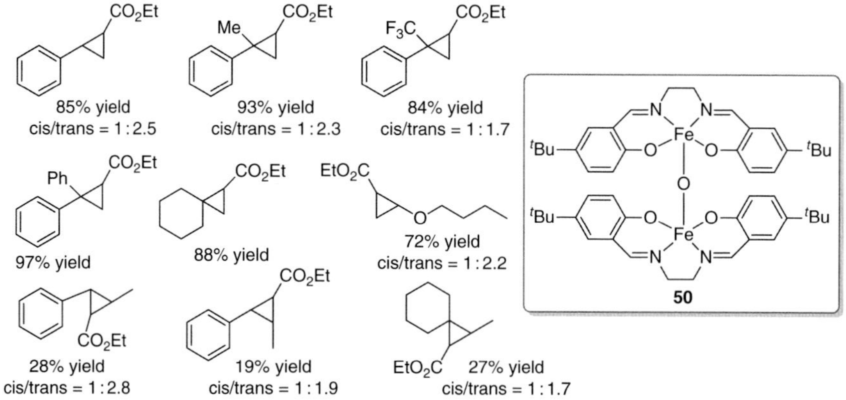

Figure 7.11 Complex **50**-catalyzed synthesis of cyclopropanes.

Finally, also Schiff bases (SB) were used as ligands to synthesize iron(III) complexes active in cyclopropanation reactions. μ-Oxo-*bis*[(SB)iron(III)] complex **50** was used to synthesize a series of cyclopropanes by using EDA as the carbene source [70]. Although the reaction was effective for the cyclopropanation of both terminal and internal alkenes, better yields were obtained by using terminal alkenes as starting materials. In fact, the cyclopropanes of internal alkenes, such as *trans*- and *cis*-β-methylstyrene, and ethylidenecyclohexane were obtained only in low yields (Figure 7.11).

It should be noted that with catalyst **50**, cyclopropanations occur without the contemporary formation of dimerization products derived from the coupling reactions of the carbene moiety, as it is usually observed in the presence of other iron catalysts. In addition, the reaction can be performed in open air to indicate that air-stable active intermediates are formed during the catalytic reactions.

7.6 The $[Cp(CO)_2Fe^{II}(THF)]BF_4$ Catalyst

Lewis acid complex $[Cp(CO)_2Fe^{II}(THF)]BF_4$ (**51**) (Cp, cyclopentadienyl; THF, tetrahydrofuran) was reported by Hossain and coworkers in 1992 to catalyze the cyclopropanation of styrene and α-methylstyrene with a cis-diastereoselectivity which, as stated above, is not common with other iron complexes [71]. The reaction scope was investigated by the same authors, and catalyst **51** was found active in the cyclopropanation of other alkenes, producing the cis-isomer as the major reaction product, except in one case (Figure 7.12) [72].

Concerning the reaction mechanism, Hossain and coworkers proposed that the first step of the cycle is the elimination of the THF ligand from **51**, leading to unsaturated complex **A**, which could possibly coordinate the diazo reagent to form intermediate **B**. Then, molecular nitrogen is lost with the consequent formation of an active carbene intermediate **C**, which is responsible for the transfer of the carbene moiety to the double bond of the alkene to yield the desired cyclopropane (Scheme 7.26) [72]. The observed cis-selectivity was attributed to the

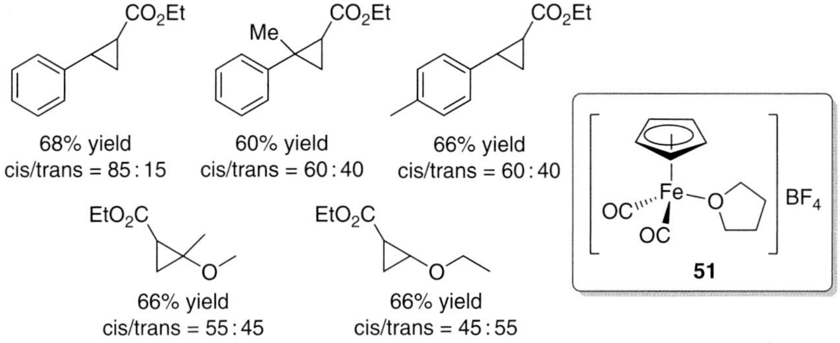

Figure 7.12 Complex **51**-catalyzed synthesis of cyclopropanes.

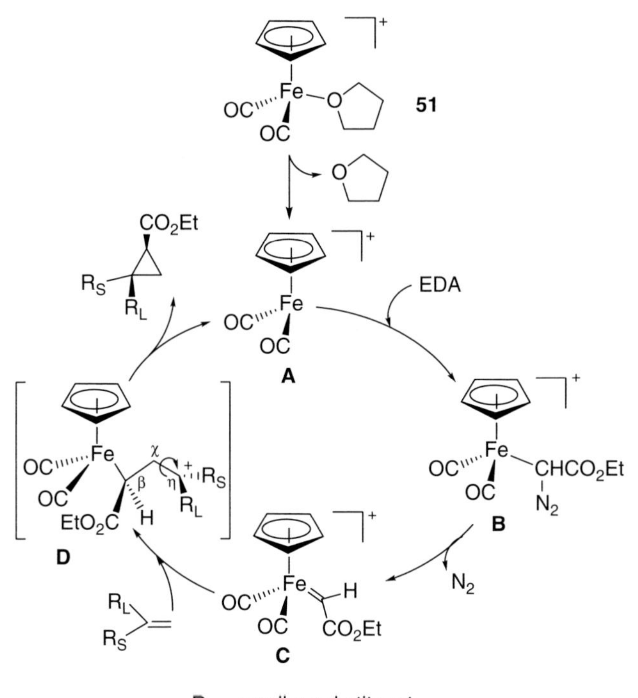

Scheme 7.26 Mechanistic proposal for the **51**-catalyzed cyclopropanation.

formation of intermediate **D**, which presents the short-lived C_γ-carbocation that collapses before the occurrence of a C_β—C_γ bond rotation.

Catalyst **51** was anchored on silica, thanks to the high affinity of the catalyst for the solid support, and the so-obtained heterogeneous catalyst displayed a good catalytic activity in the reaction of styrene with EDA, leading to the corresponding cyclopropane in good yields [73]. The catalyst was active for six consecutive runs, but an inversion of the diastereoselectivity was observed with respect to that observed in homogeneous conditions. In fact, although in the presence of homogeneous catalyst **51** a cis/trans ratio of 84 : 16 was registered, the heterogeneous catalyst was responsible for the formation of a slight excess of the trans-isomer (trans/cis = 1.2 : 1).

In order to understand the behavior of the supported form of catalyst **51**, a detailed study of the catalytic reaction was performed. This study supports the hypothesis that the observed inversion of the reaction diastereoselectivity is because of the steric and electronic effects of the silica support. The intermediate carbene species **C** is stabilized by the interaction with silica and, consequently, intermediate **D** lives long enough to partially scramble the stereochemistry before the ring closure reaction, leading to the cyclopropane molecule [74]. The interaction of carbene with the alkene substrate occurs through the formation of a late transition state resulting in a higher trans/cis-diastereomeric ratio. Unfortunately, intermediate **C** is too reactive to be isolated; thus, the geometrical and electronic characteristics of the carbene iron bond were studied by using DFT, which indicated a strong double-bonded character of the iron–carbene carbon bond [75].

7.7 Conclusions

The aim of this chapter is to give the reader an overview of the application of iron complexes in promoting carbene transfer reactions leading to cyclopropane-containing molecules. The literature analysis indicates that the field is essentially dominated by the employment of iron porphyrinoid complexes that display a remarkable superiority over other iron complexes, both in terms of reaction productivity and stereocontrol. The study of the ligand influence on the catalytic efficiency was performed by synthesizing a large class of porphyrinoids, and their electronic and steric characteristics were fundamental to fine-tune the transfer of the carbene moiety with high chemo- and stereoselectivities. Conversely, other ligands largely used in other classes of catalytic transformations (i.e. aza-macrocyclics or Schiff bases) did not exhibit much activity in iron-based catalytic cyclopropanations.

The formation of an active iron carbene intermediate is always proposed independent of the nature of the ligand coordinated to the iron metal center. However, scarce information on the structure and electronic nature of key species formed during the catalytic cycle has been reported until now. Only some iron carbene species have been isolated and characterized, and available data on their nature and reactivity do not clearly indicate a general mechanism for iron-catalyzed cyclopropanations.

To date, catalytic iron systems have been underexplored, and a lot of work can be done to generally apply them in cyclopropanation reactions. It is important to stress that to attain this goal, it will be mandatory to clarify the involved catalytic mechanisms to furnish fundamental information on synthetic strategies devoted to design efficient catalytic complexes. This dual approach could be beneficial in developing new catalytic procedures that are able to promote cis-diastereoselectivity, which is seldom attained using the iron catalysts available today.

In conclusion, the importance in expanding the application of sustainable catalytic systems to the synthesis of fine chemicals justifies the increase in interest from the scientific community to study iron-based procedures; thus, a huge advancement in this research area can be envisaged in the near future.

References

1 Anastas, P. and Warner, J. (1998). *Green Chemistry: Theory, Practice*. Oxford University Press.
2 Su, B., Cao, Z.-C., and Shi, Z.-J. (2015). *Acc. Chem. Res.* 48: 886.
3 Egorova, K.S. and Ananikov, V.P. (2016). *Angew. Chem. Int. Ed.* 55: 12150–12162; (2016). *Angew. Chem.* 128: 12334.
4 Graedel, T.E., Allwood, J., Birat, J.-P. et al. (2011). *J. Ind. Ecol.* 15: 355.
5 Enthaler, S., Junge, K., and Beller, M. (2008). *Angew. Chem. Int. Ed.* 47: 3317; (2008). *Angew. Chem.* 120: 3363.
6 Ollevier, T. and Keipour, H. (2015). *Top. Organomet. Chem.* 50: 259.
7 Yang, X.-H., Song, R.-J., Xie, Y.-X., and Li, J.-H. (2016). *ChemCatChem* 8: 2429.
8 Bauer, I. and Knölker, H.-J. (2015). *Chem. Rev.* 115: 3170.
9 Wang, C. and Wan, B. (2012). *Chin. Sci. Bull.* 57: 2338.
10 Bolm, C., Legros, J., Le Paih, J., and Zani, L. (2004). *Chem. Rev.* 104: 6217.
11 Gopalaiah, K. (2013). *Chem. Rev.* 113: 3248–3296.
12 Ganesh, V. and Chandrasekaran, S. (2016). *Synthesis* 48: 4347.
13 Pons, A., Poisson, T., Pannecoucke, X. et al. (2016). *Synthesis* 48: 4060.
14 Rassadin, V.A. and Six, Y. (2016). *Tetrahedron* 72: 4701.
15 Vshyvenko, S., Reed, J.W., Hudlicky, T., and Piers, E. (2014). *Comprehensive Organic Synthesis II*, 2e, 999–1076. Amsterdam: Elsevier.
16 Donaldson, W.A. (2001). *Tetrahedron* 57: 8589.
17 Chen, D.Y.K., Pouwer, R.H., and Richard, J.-A. (2012). *Chem. Soc. Rev.* 41: 4631.
18 Talele, T.T. (2016). *J. Med. Chem.* 59: 8712.
19 Díaz-Requejo, M.M. and Pérez, P.J. (2008). *Chem. Rev.* 108: 3379.
20 Doyle, M.P., Duffy, R., Ratnikov, M., and Zhou, L. (2010). *Chem. Rev.* 110: 704.
21 Chanthamath, S. and Iwasa, S. (2016). *Acc. Chem. Res.* 49: 2080.
22 Mertens, L. and Koenigs, R.M. (2016). *Org. Biomol. Chem.* 14: 10547.
23 Ford, A., Miel, H., Ring, A. et al. (2015). *Chem. Rev.* 115: 9981.
24 Roda, N.M., Tran, D.N., Battilocchio, C. et al. (2015). *Org. Biomol. Chem.* 13: 2550.
25 Müller, S.T.R. and Wirth, T. (2015). *ChemSusChem* 8: 245.
26 Deadman, B.J., Collins, S.G., and Maguire, A.R. (2015). *Chem. Eur. J.* 21: 2298.
27 Castano, B., Gallo, E., Cole-Hamilton, D.J. et al. (2014). *Green Chem.* 16: 3202.
28 Intrieri, D., Carminati, D.M., and Gallo, E. (2016). Recent advances in metal porphyrinoid-catalyzed nitrene and carbene transfer reactions. In: *Handbook*

of Porphyrin Science, vol. 38 (ed. K.M. Kadish, R. Guilard and K. Smith), 1–99. Hackensack, NJ: World Scientific Publishing Company.
29 Wessjohann, L.A., Brandt, W., and Thiemann, T. (2003). *Chem. Rev.* 103: 1625.
30 Wolf, J.R., Hamaker, C.G., Djukic, J.-P. et al. (1995). *J. Am. Chem. Soc.* 117: 9194.
31 Hamaker, C.G., Mirafzal, G.A., and Woo, L.K. (2001). *Organometallics* 20: 5171.
32 Caballero, A., Prieto, A., Díaz-Requejo, M.M., and Pérez, P.J. (2009). *Eur. J. Inorg. Chem.* 1137.
33 Mansuy, D., Lange, M., Chottard, J.C. et al. (1978). *Angew. Chem. Int. Ed.* 17: 781; (1978). *Angew. Chem.* 90: 828.
34 Guerin, P., Battioni, J.-P., Chottard, J.-C., and Mansuy, D. (1981). *J. Organomet. Chem.* 218: 201.
35 Li, Y., Huang, J.-S., Zhou, Z.-Y. et al. (2002). *J. Am. Chem. Soc.* 124: 13185–13193.
36 Khade, R.L. and Zhang, Y. (2015). *J. Am. Chem. Soc.* 137: 7560.
37 Salomon, R.G. and Kochi, J.K. (1973). *J. Am. Chem. Soc.* 95: 3300.
38 Tagliatesta, P. and Pastorini, A. (2003). *J. Mol. Catal. A: Chem.* 198: 57.
39 Morandi, B. and Carreira, E.M. (2012). *Science* 335: 1471.
40 Kaschel, J., Schneider, T.F., and Werz, D.B. (2012). *Angew. Chem. Int. Ed.* 51: 7085; (2012). *Angew. Chem.* 124: 7193.
41 Vinš, P., de Cózar, A., Rivilla, I. et al. (2016). *Tetrahedron* 72: 1120.
42 Morandi, B. and Carreira, E.M. (2010). *Angew. Chem. Int. Ed.* 49: 938; (2010). *Angew. Chem.* 122: 950.
43 Morandi, B., Cheang, J., and Carreira, E.M. (2011). *Org. Lett.* 13: 3080.
44 Duan, Y., Lin, J.-H., Xiao, J.-C., and Gu, Y.-C. (2016). *Org. Lett.* 18: 2471.
45 Duan, Y., Lin, J.-H., Xiao, J.-C., and Gu, Y.-C. (2017). *Chem. Commun.* 53: 3870.
46 Morandi, B., Dolva, A., and Carreira, E.M. (2012). *Org. Lett.* 14: 2162.
47 Hock, K.J., Spitzner, R., and Koenigs, R.M. (2017). *Green Chem.* 19: 2118.
48 Intrieri, D., Carminati, D.M., and Gallo, E. (2016). *Dalton Trans.* 45: 15746.
49 Gross, Z., Galili, N., and Simkhovich, L. (1999). *Tetrahedron Lett.* 40: 1571.
50 Lai, T.-S., Chan, F.-Y., So, P.-K. et al. (2006). *Dalton Trans.* 4845.
51 Nicolas, I., Roisnel, T., Maux, P.L., and Simonneaux, G. (2009). *Tetrahedron Lett.* 50: 5149.
52 Le Maux, P., Juillard, S., and Simonneaux, G. (2006). *Synthesis* 1701.
53 Le Maux, P., Nicolas, I., Chevance, S., and Simonneaux, G. (2010). *Tetrahedron* 66: 4462.
54 Nicolas, I., Maux, P.L., and Simonneaux, G. (2008). *Tetrahedron Lett.* 49: 5793.
55 Cui, X. and Zhang, X.P. (2014). *Wiley Ser. React. Intermed. Chem. Biol.* 7: 491.
56 Chen, Y. and Zhang, X.P. (2007). *J. Org. Chem.* 72: 5931.
57 Carminati, D.M., Intrieri, D., Caselli, A. et al. (2016). *Chem. Eur. J.* 22: 13599.
58 Intrieri, D., Le Gac, S., Caselli, A. et al. (2014). *Chem. Commun.* 50: 1811.
59 Carminati, D.M., Intrieri, D., Le Gac, S. et al. (2017). *New J. Chem.* 41: 5950.
60 Sharma, V., Jain, S., and Sain, B. (2004). *Catal. Lett.* 94: 57.
61 Sharma, V.B., Jain, S.L., and Sain, B. (2006). *Catal. Commun.* 7: 454.

62 Liu, H.-H., Wang, Y., Shu, Y.-J. et al. (2006). *J. Mol. Catal. A: Chem.* 246: 49.
63 Simkhovich, L., Galili, N., Saltsman, I. et al. (2000). *Inorg. Chem.* 39: 2704–2705.
64 Gross, Z., Simkhovich, L., and Galili, N. (1999). *Chem. Commun.* 599.
65 Simkhovich, L., Mahammed, A., Goldberg, I., and Gross, Z. (2001). *Chem. Eur. J.* 7: 1041.
66 Du, G., Andrioletti, B., Rose, E., and Woo, L.K. (2002). *Organometallics* 21: 4490.
67 Yeung, C.-T., Sham, K.-C., Lee, W.-S. et al. (2009). *Inorg. Chim. Acta* 362: 3267–3273.
68 Shen, J.-J., Zhu, S.-F., Cai, Y. et al. (2014). *Angew. Chem. Int. Ed.* 53: 13188–13191; (2014). *Angew. Chem.* 126: 13404.
69 Xu, H., Li, Y.-P., Cai, Y. et al. (2017). *J. Am. Chem. Soc.* 139: 7697.
70 Edulji, S.K. and Nguyen, S.T. (2003). *Organometallics* 22: 3374.
71 Seitz, W.J., Saha, A.K., Casper, D., and Hossain, M.M. (1992). *Tetrahedron Lett.* 33: 7755.
72 Seitz, W.J., Saha, A.K., and Hossain, M.M. (1993). *Organometallics* 12: 2604.
73 Redlich, M., Mahmood, S.J., Mayer, M.F., and Hossain, M.M. (2000). *Synth. Commun.* 30: 1401.
74 Casper, D.J., Sklyarov, A.V., Hardcastle, S. et al. (2006). *Inorg. Chim. Acta* 359: 3129.
75 Meng, Q., Wang, F., Qu, X. et al. (2007). *THEOCHEM* 815: 157.

8

Novel Substrates and Nucleophiles in Asymmetric Copper-Catalyzed Conjugate Addition Reactions

Ravindra P. Jumde, Syuzanna R. Harutyunyan, and Adriaan J. Minnaard

University of Groningen, Faculty of Mathematics and Natural Sciences, Nijenborgh 4, 9747 AG Groningen, The Netherlands

8.1 Introduction

Carbon–carbon bond formation is at the center of chemical synthesis that leads to a nearly unlimited structural diversity of carbon compounds and as such is the key to new pharmaceuticals, polymers, and a wide range of organic materials and catalysts [1]. At the same time, the pivotal role of molecular chirality in nature places the availability of chiral products in high enantiomeric purity as a core objective of modern synthetic organic chemistry [2]. Conjugate additions (CA) of a carbon nucleophile to an acceptor-substituted alkene (Michael acceptor) are vital transformations in this endeavor [3]. Since the initial discovery of the catalytic power of Cu(I) salts for CA in 1941, Cu(I)-based catalysts have become a mainstay in CA of hard organometallics (Scheme 8.1) [4]. The following 80 years have seen the realization of many remarkable asymmetric catalytic systems with a wide range of organometallics [3, 5]. The scope of nucleophiles was expanded to a uniquely wide range of organometallic reagents, in an approximate order of reactivity: $R_2Zn < R_3Al < RMgX$. Typical Michael acceptors used in copper-catalyzed asymmetric CA reactions are cyclic and acyclic ketones, lactones, and esters. There have been many reviews, book chapters, and monographs devoted to this topic in recent years [3, 5].

O, X, R, +, ⊖ Nu, Chiral catalyst, O, Nu, X, *, R

Scheme 8.1 Copper(I)-catalyzed addition of nucleophiles to α,β-unsaturated carbonyl compounds.

In this chapter, we will discuss very recent developments in the field of copper-catalyzed asymmetric CA reactions, involving novel electrophilic substrates, namely α-substituted α,β-unsaturated carbonyl compounds and alkenyl-heteroarenes.

Non-Noble Metal Catalysis: Molecular Approaches and Reactions,
First Edition. Edited by Robertus J. M. Klein Gebbink and Marc-Etienne Moret.

8.2 Catalytic Asymmetric Conjugate Additions to α-Substituted α,β-Unsaturated Carbonyl Compounds

In the large majority of studies on (copper)-catalyzed asymmetric conjugate addition (ACA) reactions, the Michael acceptor lacks α-substitution (2-substitution), a situation similar for all organometallic reagents. Nevertheless, the development of methodology that allows the use of 2-substituted enones is of critical importance for natural product synthesis as will be illustrated in this paragraph. As such, the products of this reaction, 2,3-disubstituted ketones, can be obtained equally well by conjugate addition to unsubstituted enones followed by alkylation of the resulting enolate. The use of in particular 2-alkyl cyclopentenone and 2-alkyl cyclohexenone, however, leads after conjugate addition and subsequent alkylation to building blocks with two adjacent chiral centers, one of them being a quaternary center. The stereochemistry of this quaternary center is steered by the initially formed center at the 3-position (though not completely, *vide infra*). Although this chiral induction is a matter of substrate control, in other words leads to a fixed stereochemical relation between the centers, this is not necessarily a limitation as interchanging the initially present 2-substituent and the electrophile used leads at least in principle to the other stereoisomer. As for most ligands hold that both enantiomers are available, this provides accessibility to all stereoisomers.

Several years ago, a first report describing the copper-catalyzed conjugate addition to 2-methyl cyclohexenone (**1**) originated from Vuagnoux-d'Augustin and Alexakis and comprised the enantioselective addition of Me_3Al and Et_3Al (Scheme 8.2) [6]. This knowledge was subsequently used by Helmchen and coworkers in a synthesis of pumiliotoxin C (**3**) [7].

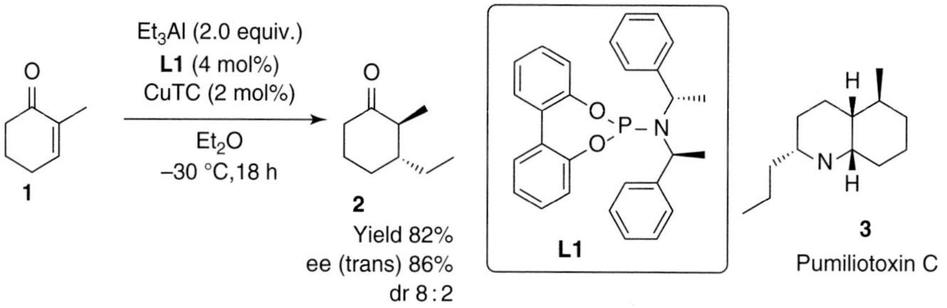

Scheme 8.2 Copper-catalyzed asymmetric conjugate addition of alkylaluminum reagents. (TC, thiophene-2-carboxylate).

In 2014, two reports appeared on the use of Grignard reagents in this type of reaction. Mauduit, Alexakis, and coworkers reported the successful application of Cu(I)-*N*-heterocyclic carbene (NHC) complexes in the asymmetric addition of Grignard reagents to 2-methyl cyclopentenone (**4**) and the corresponding hexenone (Scheme 8.3) [8]. The resulting magnesium enolates were subsequently alkylated to provide a quaternary stereocenter vicinal to the initially formed stereocenter.

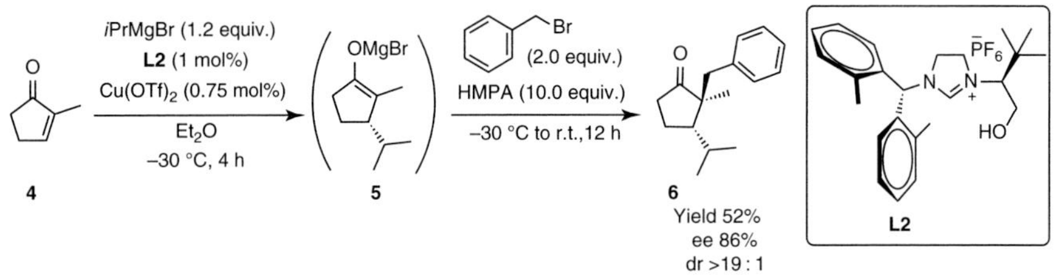

iPrMgBr (1.2 equiv.)
L2 (1 mol%)
$Cu(OTf)_2$ (0.75 mol%)
Et_2O
−30 °C, 4 h
OMgBr
Br
(2.0 equiv.)
HMPA (10.0 equiv.)
−30 °C to r.t.,12 h
4
5
6
Yield 52%
ee 86%
dr >19 : 1
$\bar{P}F_6$
HO
L2

Scheme 8.3 Copper-NHC-catalyzed asymmetric conjugate addition of Grignard reagents to cyclic enones.

From the results, it became clear that the reaction performed best with α-branched Grignard reagents such as isopropylmagnesium bromide, leading to high enantioselectivities. The intermediate magnesium enolate (**5**) is not very nucleophilic but a combination of hexamethylphosphoramide (HMPA) as a decomplexing agent and a reactive electrophile such as benzyl bromide smoothly led to the expected alkylation product (**6**). The electrophile was added trans with respect to the 3-substituent, as expected, and the diastereoselectivity was excellent, probably because of the steric bulk of the isopropyl group.

We subsequently reported the asymmetric conjugate addition of Grignard reagents to 2-methyl cyclopentenone with a copper catalyst based on Rev-JosiPhos (**L3**) and enolate alkylation with a variety of electrophiles [9]. This result is remarkable in two aspects as **L3** stood out from a lengthy list of related ferrocene-diphosphine ligands and in addition performed much better with 2-substituted cyclopentenone (**4**) than with unsubstituted cyclopentenone or (substituted) cyclohexenone (Scheme 8.4). In addition, the catalyst system gave just a moderate enantioselectivity with methylmagnesium bromide and

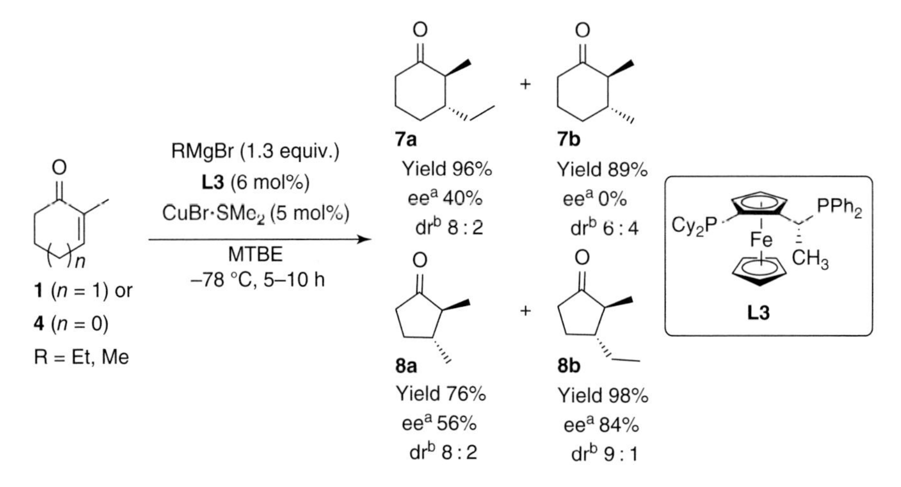

Scheme 8.4 Cu(I)-catalyzed asymmetric conjugate addition of Grignard reagents.[a]ee of the major trans diastereomer, [b]dr, trans:cis ratio. The absolute configuration of the products was established by comparison of the optical rotation with literature values (MTBE, methyl-*tert*-butylether).

α-branched Grignard reagents, but a high enantioselectivity with linear Grignard reagents. This means that the method is complementary to that of Mauduit and Alexakis.

Protonation of the formed enolate gave already reasonable diastereomeric ratios in the range of 6 : 4 to 8 : 2 in favor of the trans compound. When HMPA was added with benzyl bromide as the electrophile, the reaction reached almost full conversion with high diastereoselectivity. This approach could, however, not be consistently extended to other electrophiles as the isolated yields varied strongly. N, N′-Dimethylpropyleneurea (DMPU) (1,3-dimethyltetrahydropyrimidine-2(1*H*)-one) was used as a versatile alternative to HMPA in somewhat larger amounts. We were pleased to see that this procedure consistently gave excellent conversions and good isolated yields and diastereomeric ratios for a variety of electrophiles (Scheme 8.5).

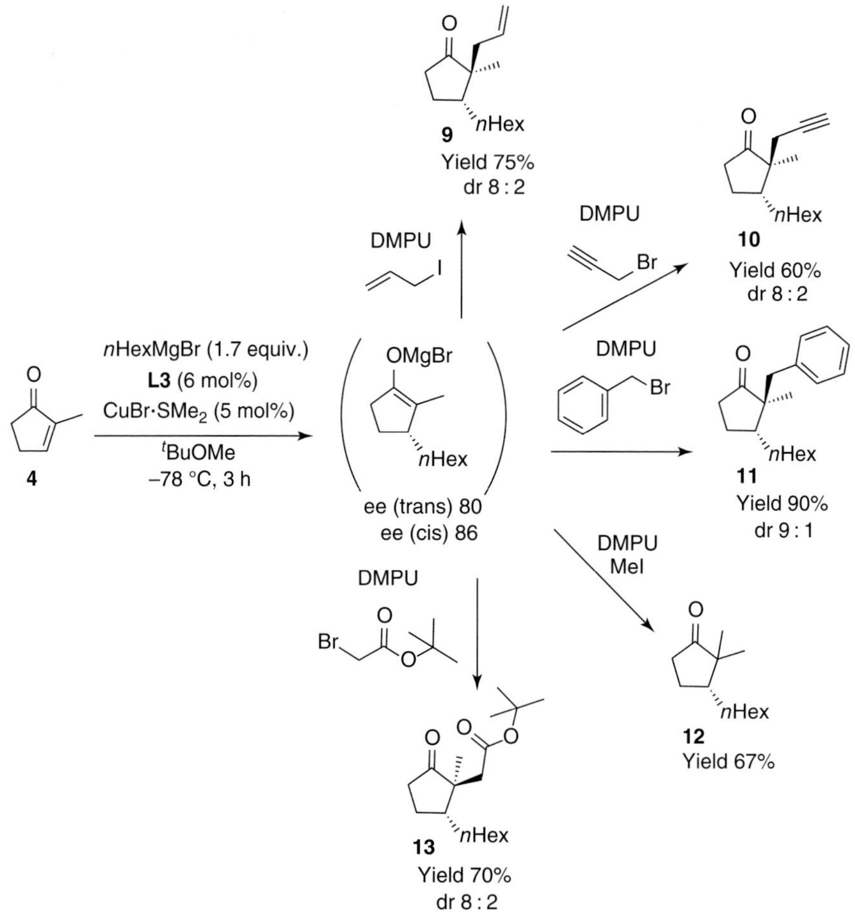

4
*n*HexMgBr (1.7 equiv.)
L3 (6 mol%)
$CuBr{\cdot}SMe_2$ (5 mol%)
^{t}BuOMe
−78 °C, 3 h
OMgBr
*n*Hex
ee (trans) 80
ee (cis) 86
DMPU
I
9
*n*Hex
Yield 75%
dr 8 : 2
DMPU
Br
10
*n*Hex
Yield 60%
dr 8 : 2
DMPU
Br
11
*n*Hex
Yield 90%
dr 9 : 1
DMPU
MeI
12
*n*Hex
Yield 67%
DMPU
Br
13
*n*Hex
Yield 70%
dr 8 : 2

Scheme 8.5 Conjugate addition followed by α-alkylation in MTBE/DMPU (3.5 equiv. of the electrophile and 10 equiv. of DMPU were used).

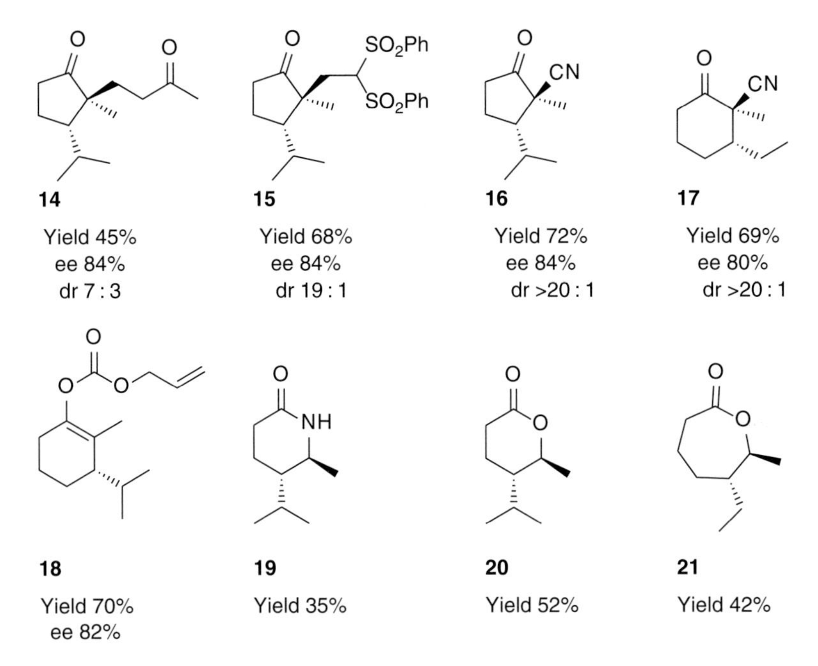

Figure 8.1 A selection of the products obtained via asymmetric conjugate addition–enolate trapping according to Germain and Alexakis.

This work was followed recently by a comprehensive study on the copper-NHC catalyzed conjugate addition of Grignard reagents to 2-substituted cyclopentenone and cyclohexenone by Germain and Alexakis, thereby defining the current state of the art [10]. The scope of Grignard reagents and Michael acceptors was studied and expanded (Figure 8.1). Although α-branched Grignard reagents gave higher enantioselectivities with 2-substituted cyclopentenones, both linear and β- or γ-branched Grignard reagents were most suited for the ACA to 2-methylcyclohexenone. For the sequential ACA–enolate trapping reaction, several unexplored electrophiles were used, thereby giving rise to highly functionalized cyclic ketones with contiguous α-quaternary and β-tertiary centers (**14–17**, Figure 8.1). The usefulness of this strategy was further illustrated by conversion of the products to lactams and lactones (**18–21**) via Beckmann and Bayer–Villiger reactions, respectively.

As already preluded upon by the authors of the aforementioned papers, the products of the sequential ACA–enolate alkylation are tailor-made for application in natural product synthesis, in particular for terpenoids and steroids, because the methyl-bearing quaternary stereocenters can be integrated in bicyclic and polycyclic ring systems. Although this has not been effectuated thus far, several groups have already explicitly referred to this opportunity in their reports on the total synthesis of natural products that use a nonasymmetric conjugate addition to 2-methyl cyclopentenone. In the synthesis of hyperforin by Ting and Maimone [11], as well as in the synthesis of the paxilline indole diterpenes by Pronin and coworkers [12], this approach can readily be used (Scheme 8.6).

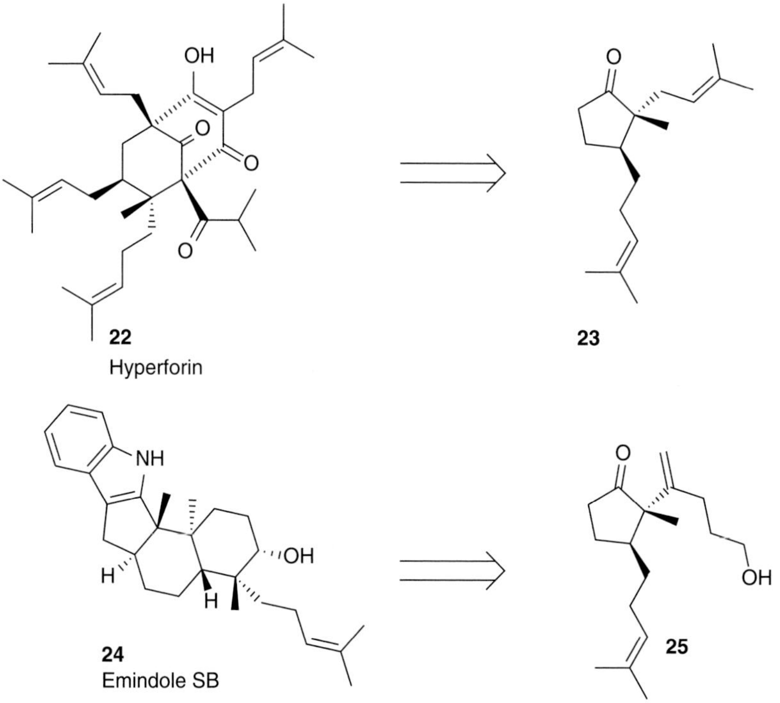

22
Hyperforin

23

24
Emindole SB

25

Scheme 8.6 Natural product syntheses in which sequential asymmetric conjugation–enolate trapping can be readily applied.

8.3 Catalytic Asymmetric Conjugate Additions to Alkenyl-heteroarenes

Nitrogen-containing heteroarenes (Figure 8.2a) represent a large fraction of all known active pharmaceutical ingredients (APIs) and approximately half of these molecules are chiral [13]. In N-containing heteroarenes, the embedded C=N bond exhibits electron-withdrawing properties comparable to that exhibited by carbonyls (Figure 8.2b) [14]. As a result, alkenyl-heteroarenes can, for instance, be exploited as Michael acceptors in CA of carbon nucleophiles (Figure 8.2c). Addition of carbon nucleophiles to conjugated vinyl-substituted heteroaromatic compounds, leading mainly to achiral molecules, is well known [14, 15]. However, although catalytic asymmetric C—C bond formation by CA of, for instance, organometallics is a routine procedure for additions to common Michael acceptors (e.g. enones, enals, or enoates), examples of catalytic asymmetric additions to alkenyl-heteroarenes have started to appear only recently [16]. The lack of methodologies for the latter is often related to the intrinsically low reactivity of these molecules toward nucleophilic addition compared to common Michael acceptors. In addition, because of the Lewis basic nature of the nitrogen in the heterocycle, it is more likely to coordinate with the metal catalyst, which could be detrimental for the outcome of reactions where a chiral ligand is involved. Nevertheless, recent reports suggest that this coordination can be excluded by

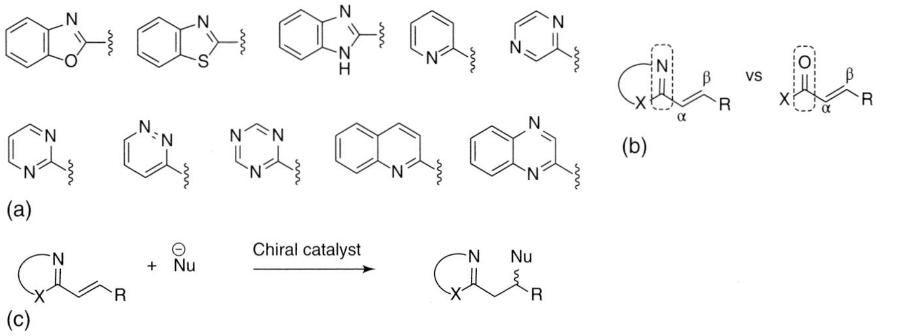

Figure 8.2 (a) Different heteroarenes, (b) alkenyl-heteroarene vs α,β-unsaturated carbonyls, and (c) conjugate addition of a nucleophile to alkenyl-heteroarenes.

using a proper catalytic system, and the low reactivity issue of these molecules can be tackled either by the use of strong nucleophiles and/or by external activation. Here, we summarize the seminal achievements in the field since 2008, focusing on asymmetric copper-catalyzed CA to alkenyl-heteroarenes, while occasionally referring to older examples and other transition metal-catalyzed protocols for better comparison and understanding.

8.3.1 A Brief Overview of Asymmetric Nucleophilic Conjugate Additions to Alkenyl-heteroarenes

Although there are many examples of non-enantioselective nucleophilic additions to vinyl-heteroarenes [14, 15], examples of enantioselective transition metal-catalyzed nucleophilic additions to β-substituted heteroarenes started emerging very recently [16]. The first breakthrough was in 2009 when Lam and coworkers reported highly enantioselective Cu-catalyzed addition of small hydride nucleophiles to β,β-disubstituted alkenyl-heteroarenes (Scheme 8.7a) [16a]. Following this report, other transition metal-catalyzed reactions started to appear. In 2010, the same group reported highly enantioselective Rh-catalyzed

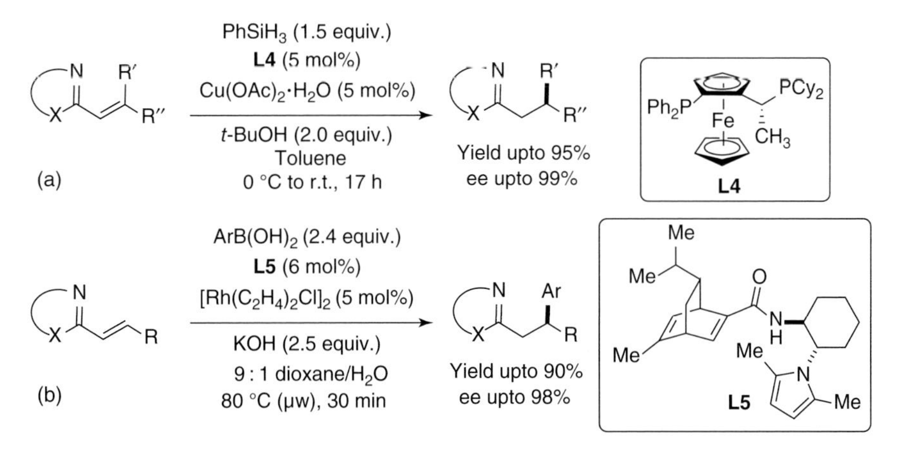

Scheme 8.7 Different strategies for conjugate nucleophilic addition to alkenyl-heteroarenes.

addition of carbon nucleophiles (aryl boronic acids) to β-monosubstituted alkenyl-heteroarenes using microwave conditions (Scheme 8.7b) [16b,c].

From the above examples, it is clear that the reactivity issue of alkenyl-heteroarene can be tackled either by using small nucleophiles like hydrides in the presence of a chiral copper complex [16a] or using activated substrates at higher temperatures in the presence of chiral rhodium catalysts [16b,c].

8.3.2 Copper-Catalyzed Asymmetric Nucleophilic Conjugate Additions to Alkenyl-heteroarenes

Copper-catalyzed methodologies for ACA to β-substituted alkenyl-heteroarenes are scarce because of their markedly lower reactivity when compared to common Michael acceptors. To circumvent the lack of reactivity toward carbon nucleophiles, in 2009, Lam and coworkers performed the conjugate reduction of β,β-disubstituted alkenyl-heteroarenes using a copper–biphosphine catalytic system and $PhSiH_3$ as a hydride source [16a]. Different biphosphine ligands were tested, and the Josiphos family was found to be optimal. In the presence of $Cu(OAc)_2{\cdot}H_2O$ (5 mol%), Josiphos ligand **L4** (5 mol%), $PhSiH_3$ (1.5 equiv.), and *t*-BuOH (2.0 equiv.), a range of β,β-disubstituted alkenyl-heteroarenes were reduced enantioselectively (benzoxazoles **26a–d**, oxazole **26e**, benzothiazole **26f**, pyridines **26g–i**, quinoline **26j**, and pyrazine **26k**) to the corresponding products (**27a–k**) (Scheme 8.8). The process can also tolerate different functionalities

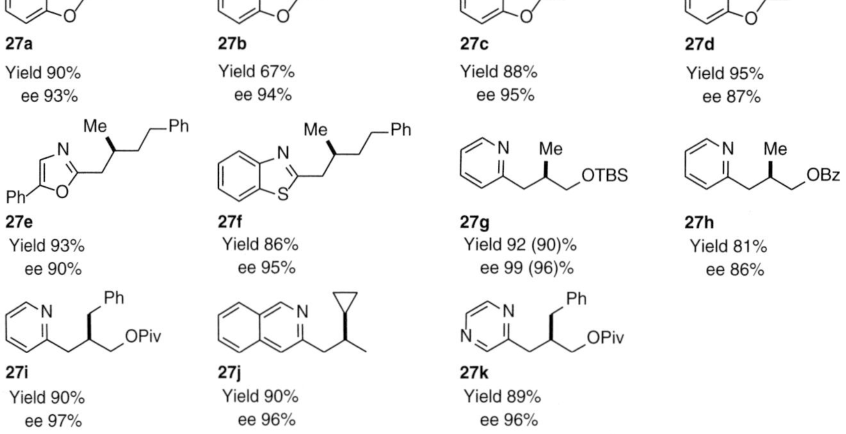

Scheme 8.8 Enantioselective copper-catalyzed reduction of β,β-disubstituted alkenyl-heteroarenes.

at the β-position, including simple aliphatic, phenyl, benzyl, oxygenated alkyl, and also hindered α-branched cyclopropane groups. Moreover, the reaction can be run with a low loading of the catalyst (2 mol%), providing product **27g** with comparable yield and enantioselectivity (values in parentheses) as that of 5 mol% catalyst loading.

The experiments to explore the origin of the reactivity suggest that alkene reduction by copper hydride can occur without assistance of the directing effect from the nitrogen atom (Scheme 8.9). In this experiment, product **27l** was isolated with high enantioselectivity and moderate yield after four days. On the other hand, these experiments are also suggestive of the importance of the conjugation of the alkene to the C=N moiety, as no reduction was observed in case of pyridine substrate **26m**.

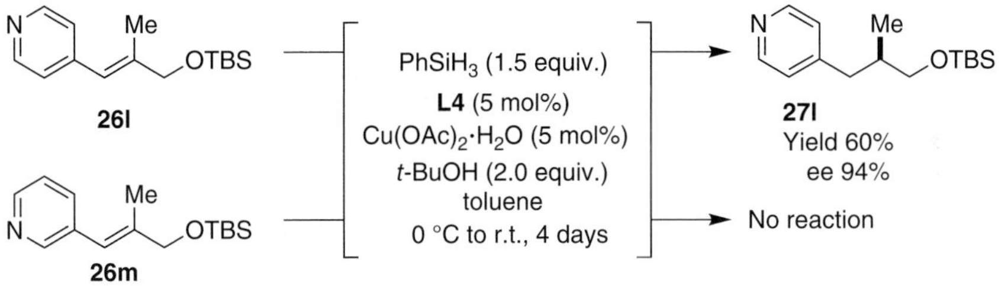

26l; 26m; $PhSiH_3$ (1.5 equiv.), **L4** (5 mol%), $Cu(OAc)_2{\cdot}H_2O$ (5 mol%), *t*-BuOH (2.0 equiv.), toluene, 0 °C to r.t., 4 days; 27l Yield 60%, ee 94%; No reaction

Scheme 8.9 Reactivity comparison between 4- and 3-alkenylpyridine.

Following Lam's procedure, Yun and coworkers reported in 2012 another example of copper-catalyzed asymmetric conjugate reduction of 2-alkenylbenzoxazole **26n**, with a pinacol boronic ester at the β-position (Scheme 8.10). The corresponding reduced product **27n** was isolated in good yield and enantioselectivity. The presence of the boronic ester group could potentially be used for further modification toward the preparation of more complex heteroarenes [17].

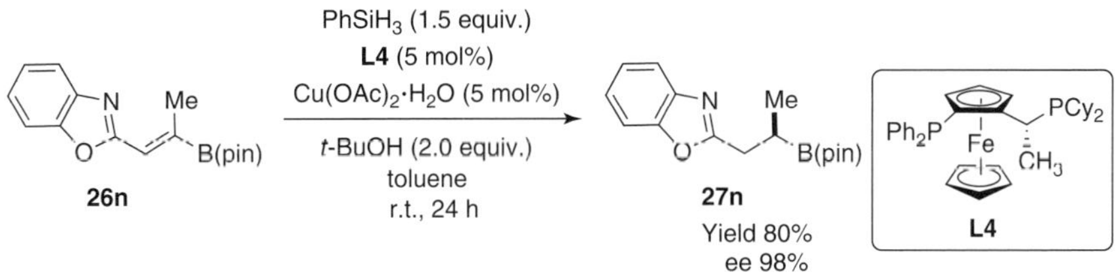

26n; $PhSiH_3$ (1.5 equiv.), **L4** (5 mol%), $Cu(OAc)_2{\cdot}H_2O$ (5 mol%), *t*-BuOH (2.0 equiv.), toluene, r.t., 24 h; 27n Yield 80%, ee 98%; L4

Scheme 8.10 Copper-catalyzed asymmetric conjugate reduction of a 2-alkenylbenzoxazole.

Following the success of catalytic enantioselective conjugate reduction of alkenyl-heteroarenes, in 2012, Lam and coworkers reported highly enantioselective copper-catalyzed reductive coupling reactions of vinyl-heteroarenes with ketones [16c]. In this transformation, the first step is the copper-catalyzed CA of a hydride, followed by the coupling of the resulting intermediate with the corresponding ketones. The protocol includes a copper–bisphosphine complex as the catalyst, $PhSiH_3$ as the stoichiometric reductant, and various ketone electrophiles (Scheme 8.11). Among the different bisphosphine ligands screened, ligands **L6**, **L7**, and **L8** provided the highest yield, diastereoselectivity, and enantioselectivity. A variety of vinyl-heteroarenes (quinoline **28a**, **b**, pyridine **28c**, **h**, isoquinoline **28d**, pyrimidine **28e**, **k**, quinazoline **28f**, thiazole **28g**, oxazole **28i**, and triazine **28j**) were reductively coupled with different acyclic ketones such as alkyl, aryl, or heteroaryl (products **30a–g**), cyclic ketones such as indanones (product **30e**), tetralone (product **30f**), 4-chromanone (product **30g**), and 4-thiochromanone (product **30h**). Interestingly, this process is not only applicable to vinyl-heteroarenes because β-substituted alkenyl-heteroarenes

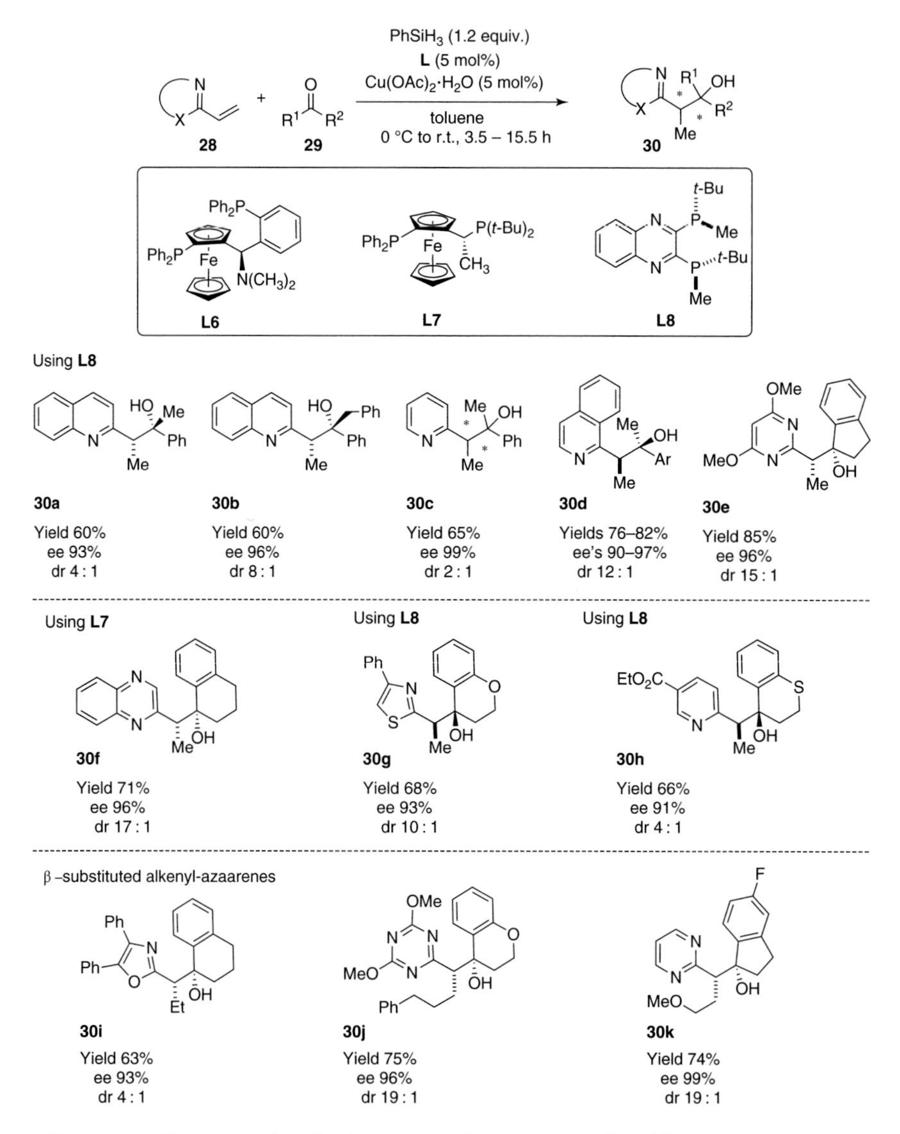

Scheme 8.11 Copper-catalyzed reductive coupling reactions of vinyl-heteroarenes with ketones.

(**28i–k**) can also undergo the reductive coupling with different ketones (product **30i–k**).

Subsequent to the procedure of copper-catalyzed enantioselective reductive coupling of vinyl-heteroarenes with ketones, Lam and coworkers also demonstrated the effective use of *N*-Boc-protected aldimines as electrophiles in an analogous process in 2015 (Scheme 8.12) [18]. The diastereo- and enantioselective reductive coupling of different vinyl-heteroarenes (**28**) with various *N*-Boc aldimines (**31**) was catalyzed by chiral copper–bisphosphine complexes in the presence of 1,1,3,3-tetramethyldisiloxane (TMDS) as a stoichiometric hydride source. A variety of reductively coupled chiral heterocyclic products (quinoline **32a**, quinoxaline **32b**, pyridine **32c**, pyridazine **32d**, benoxazole **32e**, benzothiazole **32f**, **h**, and thiazole **32g**) were prepared with moderate-to-good yield, diastereoselectivity, and enantioselectivity (Scheme 8.11). The Boc protecting group of product **32a** (Ar = 2-naphthyl) was removed under acidic conditions (TMSCl in MeOH) to furnish free chiral α-amine product **32a**.

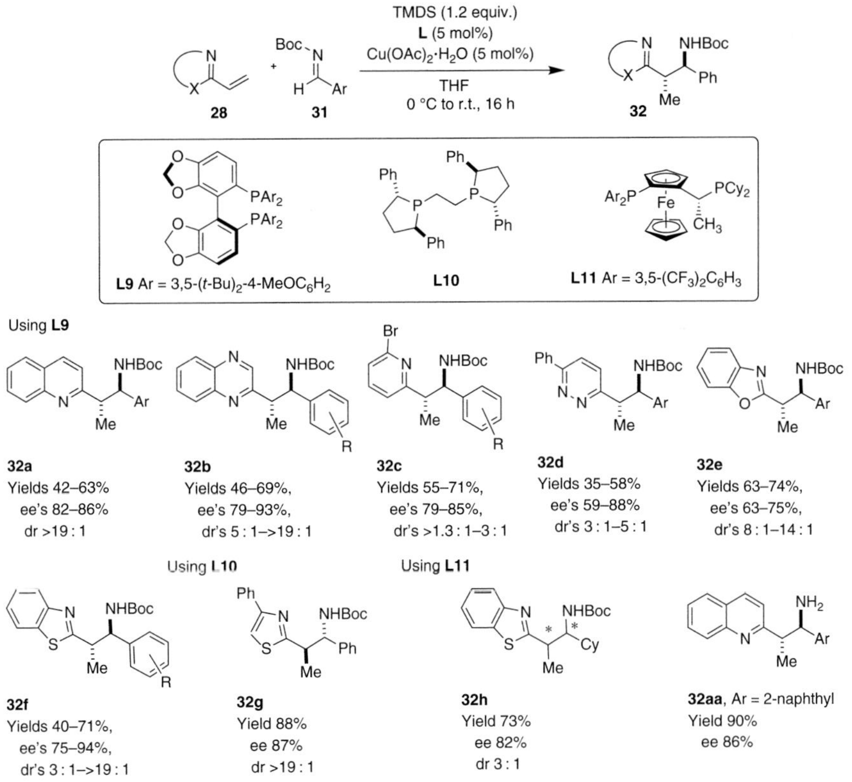

TMDS (1.2 equiv.), **L** (5 mol%), $Cu(OAc)_2 \cdot H_2O$ (5 mol%), THF, 0 °C to r.t., 16 h: **28** + **31** → **32**

L9 Ar = 3,5-(*t*-Bu)$_2$-4-MeOC$_6$H$_2$ | **L10** | **L11** Ar = 3,5-(CF$_3$)$_2$C$_6$H$_3$

Using **L9**

32a Yields 42–63%, ee's 82–86%, dr >19 : 1

32b Yields 46–69%, ee's 79–93%, dr's 5 : 1–>19 : 1

32c Yields 55–71%, ee's 79–85%, dr's >1.3 : 1–3 : 1

32d Yields 35–58%, ee's 59–88%, dr's 3 : 1–5 : 1

32e Yields 63–74%, ee's 63–75%, dr's 8 : 1–14 : 1

32f Yields 40–71%, ee's 75–94%, dr's 3 : 1–>19 : 1

Using **L10**

32g Yield 88%, ee 87%, dr >19 : 1

Using **L11**

32h Yield 73%, ee 82%, dr 3 : 1

32aa, Ar = 2-naphthyl, Yield 90%, ee 86%

Scheme 8.12 Copper-catalyzed reductive coupling reactions of vinyl-heteroarenes with *N*-Boc aldimines.

Lam and coworkers reported another copper-catalyzed transformation to prepare highly functionalized heterocycles by borylative coupling of vinyl-heteroarenes with *N*-Boc imines [19]. Different alkylboronates were prepared by three component coupling of various vinyl-heteroarenes (**28**) and a range of aldimines (**31**) with bis(pinacolato)diboron in the presence of $CuF(PPh)_3 \cdot 2MeOH$ and the 1,1′-ferrocenediyl-bis(diphenylphosphine) ligand **L12**, followed by the *in situ* transformation to primary alcohol products by $NaBO_3 \cdot 4H_2O$ in moderate-to-good yields and diastereoselectivities (pyridine (**33a**), quinoline (**33b, i**), quinoxaline (**33c**), pyrimidine (**33d, e, l, m–p**), thiazole (**33f, q–s**), benzoxazole (**33g**), and benzothiazole (**33h**); see Scheme 8.13). It was demonstrated that the Boc group of product **33b** can be removed under acidic conditions (TMSCl in MeOH at 40 °C) to provide bishydrochloride salt **33ba** in nearly quantitative yield.

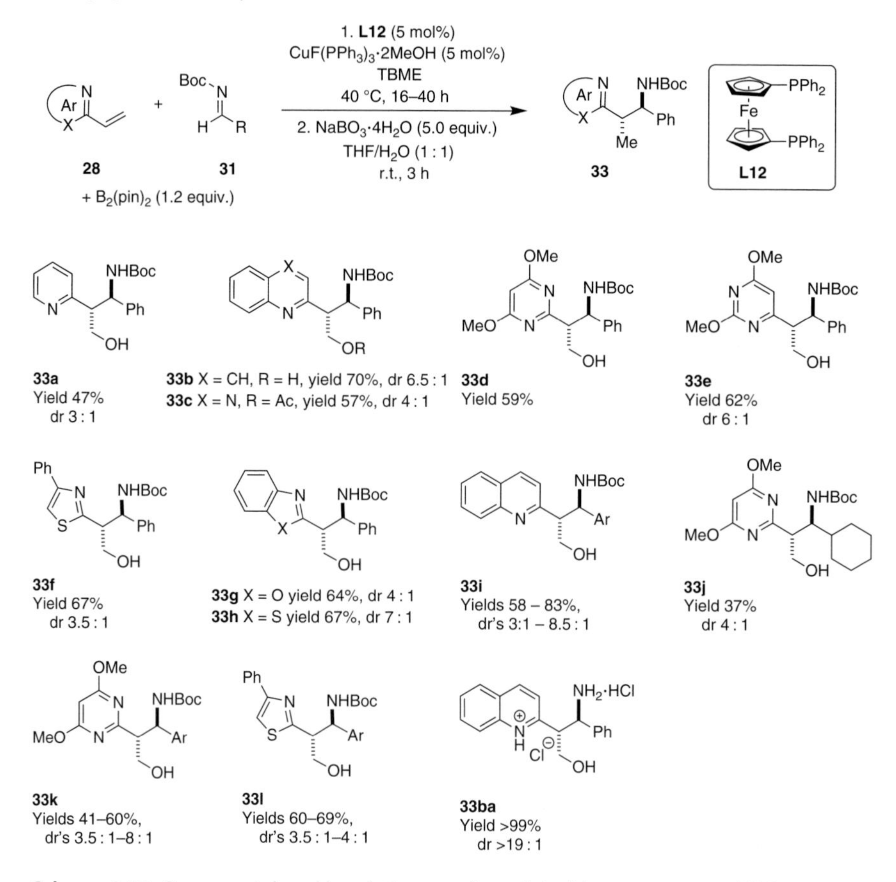

Scheme 8.13 Copper-catalyzed borylative coupling of vinyl-heteroarenes and *N*-Boc aldimines.

All the discussed copper-catalyzed methodologies are limited to reductions [16a, 17] and reductive/borylative couplings [18, 19]. Addition of carbon nucleophiles (mainly arylation) was only reported using precious transition metal catalysis (Rh and Pd) [16b,c,d]. The lack of methodologies for nucleophilic addition to β-substituted alkenyl-heteroarenes is rooted in their lower reactivity because of the relatively weak activation from the heteroaromatic moiety as well as steric hindrance from the β-substituent.

One of the solutions to tackle the reactivity issue is to use stronger organometallic nucleophiles, such as Grignard reagents, in combination with a chiral copper catalyst. Grignard reagents are among the most commonly used, inexpensive organometallics in synthetic chemistry [20], especially in copper-catalyzed conjugate in addition to conventional Michael acceptors.

In 2015, our group decided to address the reactivity issue of β-substituted alkenyl-heteroarenes by using highly reactive Grignard reagents as nucleophiles in the presence of a chiral copper catalyst [21]. Until then, the only attempt at CA of a Grignard reagent to alkenyl-heteroarenes was reported in 1998, using a nickel catalyst, phenyl magnesium chloride, and 4-(1-alkenyl)pyridines (Scheme 8.14) [22]. Even though the reaction did not appear to be ligand accelerated and the product was isolated with poor enantioselectivity (0–15% ee), this early report suggested that catalytic enantioselective addition of Grignard reagents is possible.

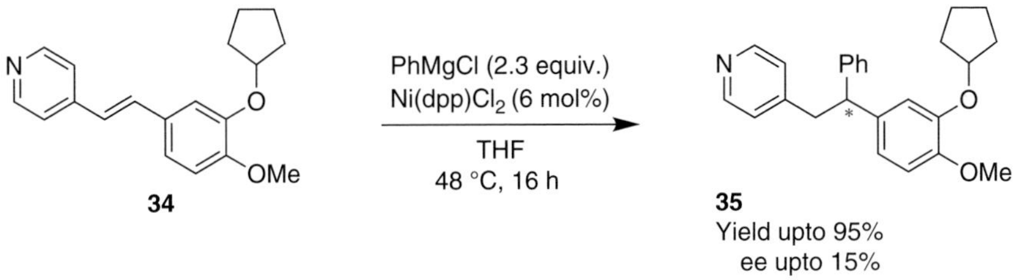

34

PhMgCl (2.3 equiv.)
$Ni(dpp)Cl_2$ (6 mol%)
THF
48 °C, 16 h

35
Yield upto 95%
ee upto 15%

Scheme 8.14 Ni-catalyzed Grignard addition.

Our early attempts toward the addition of EtMgBr to 2-styrylbenzoxazole (**36a**) in the presence of $CuBr{\cdot}SMe_2$ at −25 °C resulted in a complex mixture after 24 hours, without the expected addition product (Table 8.1, entry 1). Also, when using chiral ferrocenyl ligand **L3**, no trace of the product was observed (Table 8.1, entry 2). These initial results show that stronger nucleophiles, in this case Grignard reagents, alone cannot tackle the low reactivity of β-substituted alkenyl-heteroarenes toward nucleophilic addition but that external activation is also necessary.

Activation of electrophilic substrates toward nucleophilic addition is commonly carried out with Lewis acids (LA) [23]. As our previous work [24] had already established that the compatibility of LA with Grignard reagents is not an issue, we anticipated that the LA activation of alkenyl-heteroarenes could be a viable approach to address the reactivity issue and decided to apply it for the activation of **36a**.

The first attempt to use $BF_3{\cdot}OEt_2$ in the presence of $CuBr{\cdot}SMe_2$ resulted in no conversion toward the product at −78 °C, but addition of chiral ligand **L3** provided the desired product **37a** in 59% yield and 87% enantioselectivity (Table 8.1, entries 3 and 4). From ligand- and solvent-screening studies (entries 6–12), ligand **L3** and diethyl ether turned out to be optimal, providing product **37a** with 94% yield and 96% enantioselectivity. Among the different LAs tested ($BF_3{\cdot}OEt_2$, $TiCl_4$, TMSCl, and $MgBr_2$), $BF_3{\cdot}OEt_2$ provided the best results. The optimization studies revealed that several bisphosphine ligands (**L3, L13, 14**), in combination

Table 8.1 Cu(I)-catalyzed enantioselective addition of Grignard reagent to 2-styryl-1,3-benzoxazole (**36a**).

EtMgBr (1.5 equiv.)
L (6–12 mol%)
$CuBr{\cdot}SMe_2$ (5–10 mol%)
$BF_3{\cdot}OEt_2$ (1.5 equiv.)
Solvent
−78 °C to −25 °C, 24 h

36a → **37a**

L3 R^1 = Cy, R^2 = Ph
L4 R^1 = Ph, R^2 = Cy
L13 Ar = Ph
L14 Ar = 4-Me-C_6H_4

Entry	L	Solvent	$BF_3{\cdot}OEt_2$ (equiv.)	Temperature (°C)	% Yield	% ee
1	—	*t*BuOMe	0	−25	Complex mixture	—
2	**L3**	*t*BuOMe	0	−25	Complex mixture	—
3	—	Toluene	1.5	−78	0	—
4	**L3**	Toluene	1.5	−78	59	87
6	**L4**	Et_2O	1.5	−78	35	53
7	**L13**	Toluene	1.5	−78	36	91
8	**L14**	Toluene	1.5	−78	45	92
9	**L3**	*t*BuOMe	1.5	−78	55	93
10	**L3**	CH_2Cl_2	1.5	−78	67	94
11	**L3**	THF	1.5	−78	57	50
12	**L3**	Et_2O	1.5	−78	94	96

with a Cu(I) salt, are capable of effectively promoting the reaction, whereas several different solvents studied (except Tetrahydrofuran, THF) were well tolerated by the alkylation protocol (entry 4 and entries 9–12).

Further evaluation of the substrate scope was carried out following the optimized reaction conditions: $CuBr{\cdot}SMe_2$ (5 mol%), **L3** (6 mol%), Grignard reagent (1.5 equiv.), and $BF_3{\cdot}OEt_2$ (1.5 equiv.) at −78 °C in Et_2O solvent for 18 hours (Scheme 8.15). To study the influence of different types of substrate modifications, a variety of substrates were subjected to the alkylation protocol using EtMgBr.

The stereoelectronic effect of the β-substituent was evaluated using different benzoxazole-derived substrates **36a,b**. All substrates bearing electron-rich and electron-poor substituents provided their corresponding products **37a,b** with consistently high enantioselectivity. At the same time, the reactivity of the substrates proved to be strongly dependent on the nature of these substituents at the β-position, providing products **37a,b** with moderate-to-excellent isolated yields. Remarkably, heteroaromatic substrates other than benzoxazoles, such as benzothiazoles (**36c,d**), oxazoles (**36e**), pyrimidines (**36f**), triazine (**36g**), and quinoline (**36h**), all furnished the corresponding products (**37c–h**) with high yields and enantiopurity when subjected to the alkylation protocol.

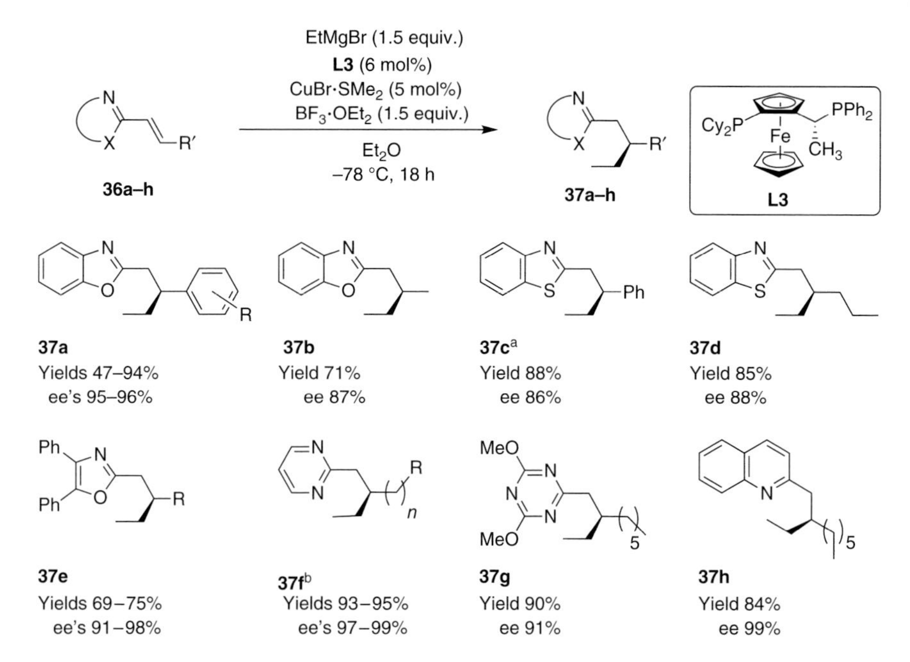

Scheme 8.15 Enantioselective addition of Grignard reagents to alkenyl-heteroarenes (heterocycle scope). [a] 3.0 equiv. of EtMgBr and 2.2 equiv. of $BF_3 \cdot OEt_2$ were used in this case. [b] Using *t*BuOMe or toluene instead of diethyl ether provides the addition product **37f** (R = —$(CH_2)_4CH_3$) with 92% and 80% isolated yields, respectively, and 99% ee.

An extensive nucleophile scope was carried out on two structurally different substrates: **36a** (benzoxazole) and **36f** (pyrimidine). A range of Grignard reagents was added successfully, providing the corresponding products (**38a–i**) with good-to-excellent enantioselectivities (Scheme 8.16). The Grignard reagents include alkyl-Grignards with different chain lengths (Et, *n*-Bu, and *n*-hexyl), sterically demanding α-, β-, γ-branched Grignards, functionalized Grignards having a terminal olefinic or trimethylsilyl moiety, and also an aryl-Grignard (PhMgBr). Interestingly, as other methodologies of nucleophilic addition to conjugated alkenyl-heteroarenes were restricted to reduction or arylation, this catalytic system represents the first example that enables the addition of a wide variety of alkyl Grignard reagents as well as phenyl Grignard. Moreover, the reported reaction can be scaled up to 1.0 mmol, the catalyst loading can be reduced from 5 to 1 mol%, and the catalyst can be recovered and reused after the reaction, in all cases providing the product with the same levels of yield and enantioselectivity as originally.

8.4 Conclusion

In conclusion, this chapter describes the progress in asymmetric copper(I)-catalyzed nucleophilic conjugate addition to novel electrophilic substrates, including (i) α-substituted α,β-unsaturated carbonyl compounds and

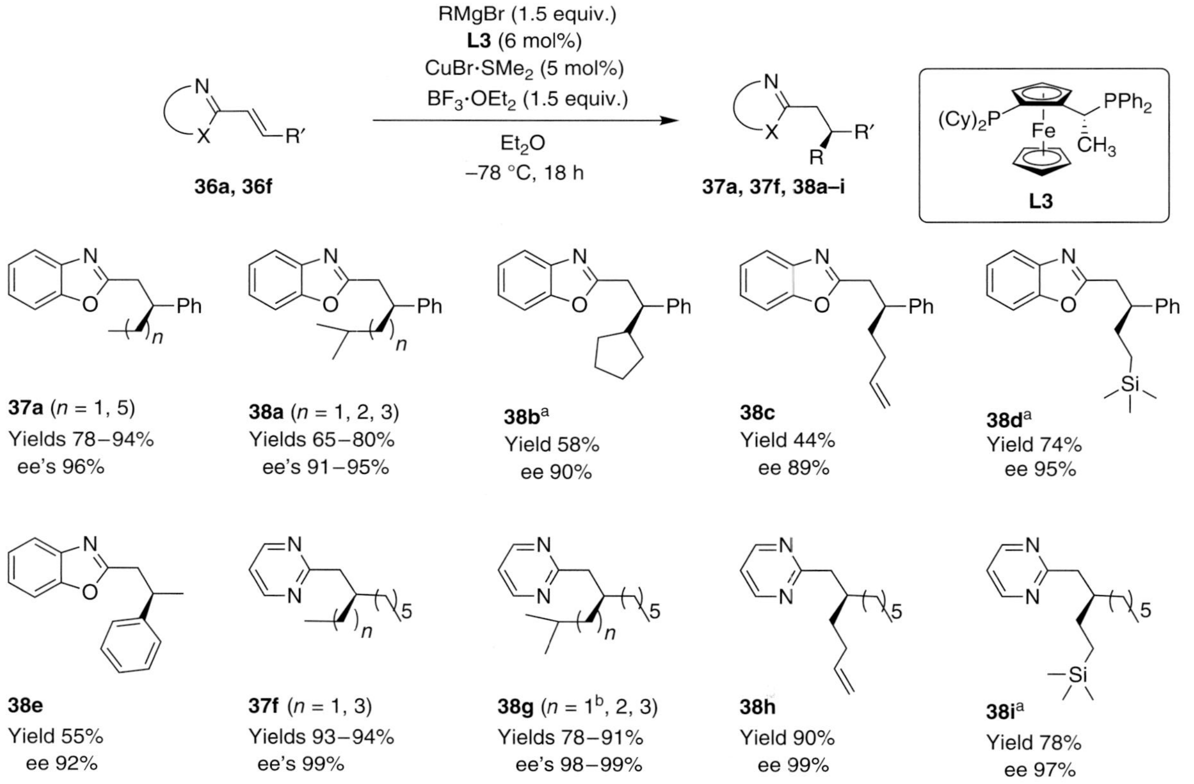

RMgBr (1.5 equiv.)
L3 (6 mol%)
$CuBr{\cdot}SMe_2$ (5 mol%)
$BF_3{\cdot}OEt_2$ (1.5 equiv.)
Et_2O
−78 °C, 18 h

36a, 36f → **37a, 37f, 38a–i**

$(Cy)_2P$ Fe PPh_2 CH_3
L3

37a (n = 1, 5)
Yields 78–94%
ee's 96%

38a (n = 1, 2, 3)
Yields 65–80%
ee's 91–95%

38b[a]
Yield 58%
ee 90%

38c
Yield 44%
ee 89%

38d[a]
Yield 74%
ee 95%

38e
Yield 55%
ee 92%

37f (n = 1, 3)
Yields 93–94%
ee's 99%

38g (n = 1[b], 2, 3)
Yields 78–91%
ee's 98–99%

38h
Yield 90%
ee 99%

38i[a]
Yield 78%
ee 97%

Scheme 8.16 Grignard scope on benzoxazole- and pyrimidine-derived substrates. [a] 3.0 equiv. of EtMgBr and 2.0 equiv. of $BF_3{\cdot}OEt_2$ were used in this case. [b] Solvent mixture Et_2O/CH_2Cl_2 (2 : 1) was used in this case.

(ii) alkenyl-heteroarenes. In the former cases, the possibility of the formation of challenging contiguous quaternary- and tertiary-stereocenters was shown by sequential asymmetric addition/Mg enolate trapping, whereas in the latter cases, copper-catalyzed conjugate reduction and conjugate addition of carbon-based nucleophiles to rarely considered Michael acceptors such as alkenyl-heteroarenes has been accomplished. Both strategies are important tools for preparing multifunctional chiral enones and for introducing a wide range of functionality adjacent to the heterocyclic moiety. Imminent challenges in the field are (i) the development of a general methodology for methylations, (ii) the synthesis of very challenging quaternary stereocenters, and (iii) the understanding of the exact mechanism of these Cu(I)-catalyzed protocols.

References

1 Marek, I. (ed.) (2014). *Comprehensive Organic Synthesis*, Carbon–Carbon Bond Formation, 2e, vol. 3. Elsevier B.V.

2 (a) Jacobsen, E.N., Pfaltz, A., and Yamamoto, H. (eds.) (1999). *Comprehensive Asymmetric Catalysis*, vol. 1–3. Springer. (b) Ojima, I. (ed.) (2010). *Catalytic Asymmetric Synthesis*, 3e. Wiley. (c) Gruttadauria, M. and Giacalone, F. (eds.) (2011). *Catalytic Methods in Asymmetric Synthesis: Advanced Materials, Techniques, and Applications*. Wiley.

3 (a) Alexakis, A., Krause, N., and Woodward, S. (2014). *Copper-catalyzed Asymmetric Conjugate Addition*. Wiley-VCH. (b) Ortiz, P., Lanza, F., and Harutyunyan, S.R. (2016). *Progress in Enantioselective Cu(I)-catalyzed Formation of Stereogenic Centers* (ed. S.R. Harutyunyan), 99–134. Springer.

4 Kharasch, M.S. and Tawney, P.O. (1941). *J. Am. Chem. Soc.* 63: 2308.

5 (a) Tomioka, K. and Nagaoka, Y. (1999). *Comprehensive Asymmetric Catalysis* (ed. E.N. Jacobsen, A. Pfaltz and H. Yamamoto), 1105–1120. Springer. (b) Hayashi, T. and Yamasaki, K. (2003). *Chem. Rev.* 103: 2829. (c) Tomioka, K. (2004). *Comprehensive Asymmetric Catalysis*, Supplement 2 (ed. E.N. Jacobsen, A. Pfaltz and H. Yamamoto), 109–124. Springer. (d) Harutyunyan, S.R., den Hartog, T., Geurts, K. et al. (2008). *Chem. Rev.* 108: 2824. (e) Alexakis, A., Backvall, J.E., Krause, N. et al. (2008). *Chem. Rev.* 108: 2796. (f) Jerphagnon, T., Pizzuti, M.G., Minnaard, A.J., and Feringa, B.L. (2009). *Chem. Soc. Rev.* 38: 1039. (g) Hayashi, T. and Yamasaki, K. (2003). *Chem. Rev.* 103: 2829. (h) Howell, G.P. (2012). *Org. Process Res. Dev.* 16: 1258.

6 Vuagnoux-d'Augustin, M. and Alexakis, A. (2007). *Chem. Eur. J.* 13: 9647.

7 Gärtner, M., Qu, J., and Helmchen, G. (2012). *J. Org. Chem.* 77: 1186.

8 Germain, N., Guénée, L., Mauduit, M., and Alexakis, A. (2014). *Org. Lett.* 16: 118.

9 Calvo, B.C., Madduri, A.V.R., Harutyunyan, S.R., and Minnaard, A.J. (2014). *Adv. Synth. Catal.* 356: 2061.

10 Germain, N. and Alexakis, A. (2015). *Chem. Eur. J.* 21: 8597.

11 Ting, C.P. and Maimone, T.J. (2015). *J. Am. Chem. Soc.* 137: 10516.

12 George, D.T., Kuenstner, E.J., and Pronin, S.V. (2015). *J. Am. Chem. Soc.* 137: 15410.

13 (a) Blacker, J.A. and Williams, M.T. (eds.) (2011). *Pharmaceutical Process Development: Current Chemical and Engineering Challenges*. Royal Society of Chemistry. (b) Vitaku, E., Smith, D.T., and Njardarson, J.T. (2014). *J. Med. Chem.* 57: 10257. (c) Wu, Y.-J. (2012). *Prog. Heterocycl. Chem.* 24: 1.
14 Best, D. and Lam, H.W. (2014). *J. Org. Chem.* 79: 831.
15 Klumpp, D.A. (2012). *Synlett* 23: 1590.
16 (a) Rupnicki, L., Saxena, A., and Lam, H.W. (2009). *J. Am. Chem. Soc.* 131: 10386. (b) Pattison, G., Piraux, G., and Lam, H.W. (2010). *J. Am. Chem. Soc.* 132: 14373. (c) Saxena, A., Choi, B., and Lam, H.W. (2012). *J. Am. Chem. Soc.* 134: 8428.
17 Jung, H.-Y., Feng, X., Kim, H., and Yun, J. (2012). *Tetrahedron* 68: 3444.
18 Choi, B., Saxena, A., Smith, J.J. et al. (2015). *Synlett* 26: 350.
19 Smith, J.J., Best, D., and Lam, H.W. (2016). *Chem. Commun.* 52: 3770.
20 Richey, H.G. (1999). *Grignard Reagents: New Developments*. Wiley.
21 Jumde, R.P., Lanza, F., Veenstra, M., and Harutyunyan, S.R. (2016). *Science* 352: 433.
22 Houpis, I.N., Lee, J., Dorziotis, I. et al. (1998). *Tetrahedron* 54: 1185.
23 (a) Yamamoto, Y., Yamamoto, S., Yatagai, H. et al. (1982). *J. Org. Chem.* 47: 119. (b) Yamamoto, H. (ed.) (2000). *Lewis Acids in Organic Synthesis*, vol. 1–2. Wiley-VCH. (c) Marcantoni, E. and Petrini, M. (2014). *Comprehensive Organic Synthesis*, 2e, vol. 1 (ed. P. Knochel and G.A. Molander), 344–364. Elsevier B.V.
24 Rong, J., Oost, R., Desmarchelier, A. et al. (2015). *Angew. Chem. Int. Ed.* 54: 3038; (2015). *Angew. Chem.* 127: 3081.

9

Asymmetric Reduction of Polar Double Bonds

Raphael Bigler, Lorena De Luca, Raffael Huber, and Antonio Mezzetti

ETH Zürich, Department of Chemistry and Applied Biosciences, Vladimir-Prelog-Weg 1-5/10, CH-8093 Zürich, Switzerland

9.1 Introduction

This chapter discusses the base-metal-catalyzed asymmetric hydrogenation of polar C=X double bonds of prostereogenic substrates (X = O or N), and hence primarily ketones and imines, with a literature coverage until June 2018. Relevant non-enantioselective reactions (such as hydrogenation of aldehydes, esters, and nitriles) are briefly mentioned, but not reactions involving CO_2. The catalytic methods discussed are asymmetric direct (H_2) hydrogenation (AH), asymmetric transfer hydrogenation (ATH), and asymmetric hydrosilylation (AHS). The products of these reactions, secondary alcohols and amines [1], are of paramount importance for the pharmaceutical and agrochemical industry [2]. However, only a handful of production processes relying on asymmetric hydrogenation of polar double bonds are currently in operation, also because only catalysts based on precious metals, mostly ruthenium, rhodium, and iridium, fulfill the requirements. In fact, for industrial application, the ideal catalyst should be commercially available, highly enantioselective, extremely active, and robust with high turnover numbers [3]. Cheap and nontoxic base metal catalysts would soften these requisites. However, complexes of 3d metals are intrinsically less stable than their heavier analogues, which challenges ligand and catalyst design. The coordination chemistry aspects connected with the issue of stability are highlighted throughout this chapter. Also, parallel to the challenges, this chapter explicitly addresses the potential chances of 3d metal catalysis.

9.1.1 Catalytic Approaches for Polar Double Bond Reduction

The main approaches for the enantioselective reduction of polar C=X bonds are direct hydrogenation with H_2 (AH), transfer hydrogenation (ATH), and hydrosilylation (AHS). This chapter is organized accordingly, but some catalysts are active in more than one reaction type. AH is completely atom economic, but the handling of H_2 requires more precautions than ATH, which is thermoneutral and operationally simple and safe on a small scale. Therefore, the latter has been broadly studied in academia [4]. However, ATH is reversible, and the

Non-Noble Metal Catalysis: Molecular Approaches and Reactions,
First Edition. Edited by Robertus J. M. Klein Gebbink and Marc-Etienne Moret.

enantiomeric excess of the alcohol product erodes as equilibrium is approached, which requires strictly controlled conditions. In view of its complete atom efficiency and the manageable safety measures, AH is the method of choice for industry [3], as it does not require a hydrogen donor solvent and, being irreversible, can be run with high concentration of substrate or even without solvent. However, ATH in ^{i}PrOH is feasible on a large scale if the by-product acetone can be efficiently removed by distillation [5].

In ATH, either ^{i}PrOH or formic acid/formate salts can be used as reducing agents. As ^{i}PrOH has a similar oxidation potential as the reaction product, ATH with ^{i}PrOH is under equilibrium conditions [6]. This restricts its application to substrates for which the equilibrium favors the optically active alcohol, such as electron-poor or sterically demanding ketones, or requires high dilution conditions, which, in turn, hampers large-scale application. Finally, most systems are only active in the presence of base, which is problematic for base-sensitive substrates. Alternatively, formic acid can be used as the reducing agent. Removing the resulting CO_2 under gas flow renders the reaction irreversible. As drawbacks, the reaction is not thermoneutral, formic acid is corrosive and inhibits many catalysts, and its decomposition produces flammable H_2 and poisonous carbon monoxide upon decarboxylation and dehydration, respectively [6].

AHS is a viable alternative to AH and ATH in view of the mild reaction conditions used [1]. The silane is formally added across a carbonyl double bond to afford the corresponding silyl ether, which is cleaved in a second step under basic or acidic conditions or in the presence of fluoride. Silanes are attractive reducing agents as they are stable, diverse, easy to handle, and only mildly hydridic [7]. As the reaction is irreversible, high conversions to the silyl ether can be achieved, the only common side product being the silyl enol ether. However, the silyl ethers have to be hydrolyzed to the alcohols, which generates significant amounts of silicon-based waste. Hence, the AHS of ketones or imines is rarely used in industry [3]. A general feature of hydrosilylation is its mechanistic diversity as discussed below.

9.1.2 The Role of Hydride Complexes

All the three reactions (AH, ATH, and AHS) can, at least in principle, involve hydride transfer to the carbonyl carbon atom, and hydride complexes are commonly invoked as intermediates. Such complexes are intrinsically less stable for 3d than for 4d and 5d transition metals, which can be attributed to weaker metal hydride bonds [8]. Common strategies to stabilize such complexes involve the use of bulky, soft ligands such as phosphines or cyclopentadienyl, and the contents of this chapter confirm that the expanding use of such ligands is primarily responsible for the recent surge of base metal hydrogenation catalysts. However, as compared to noble metals, the mechanistic understanding of the reduction of carbonyl functionalities by first-row transition catalyst is still limited [9]. This particularly affects AHS, for which several different mechanisms are possible [7]. Thus, benzaldehyde insertion into a Ni—H bond has been observed, whereas a Fe(II) hydride complex has been found to hydrosilylate PhCHO without involvement of the hydride ligand, possibly via a Lewis-acid mechanism [10].

Additionally, silanes are known to react with ketones even in the absence of metals when activated by an oxygenated base [11]. Considering that many of the AHS catalysts discussed below use acetate complexes, a mechanistic possibility is that hypervalent, five-coordinate hydrosilicates might be competent for hydride transfer [7]. The observation that hydride complexes are not necessarily involved in hydrosilylation may explain why many AHS catalysts contain hard nitrogen ligands rather than phosphines or Cp.

9.1.3 Ligand Choice and Catalyst Stability

Stabilizing complexes of 3d metals, and in particular hydrides, is a formidable task that requires the whole toolbox of coordination chemistry. Thus, many of the catalysts discussed below contain polydentate ligands, which stabilize the complexes by means of the chelate (or macrocyclic) effect. Tetra-, tri-, and bidentate ligands, as well as cyclopentadienyl ligands, are ubiquitous among base metal catalysts. Further, soft, strong-field ligands favor the formation of hydride complexes for all transition metals. In the specific case of base metals, they also promote the low-spin configuration. It should be appreciated that this is not just about the easy characterization of diamagnetic complexes (at least for the d^6 electron configuration). In fact, low-spin complexes generally feature stronger metal–ligands bond, as nicely illustrated by octahedral complexes. In low-spin octahedral (O_h) complexes, the e_g orbitals, which have σ* character, are empty up to the d^6 electron configuration but are partially occupied in high-spin ones, which weakens the metal–ligand bond. Also, the t_{2g} orbitals are metal–ligand bonding if π-acceptor ligands are present, which further strengthens the M—L bonds and hence stabilizes the complexes.

A further advantage of P-containing ligands is the wealth of information offered by ^{31}P NMR (nuclear magnetic resonance) spectroscopy, which facilitates the synthesis and characterization of new (diamagnetic) complexes. The next sections show that the availability of appropriate coordination compounds is pivotal for the success of a 3d metal in the reduction of polar bonds. However, other strong-field ligands have been successfully used in asymmetric catalysis (either as chiral or ancillary ligands, see below), in particular carbon monoxide and isonitriles.

9.2 Manganese

Until 2017, in which the first Mn(I) catalysts for the ATH and AH hydrogenation of ketones appeared, manganese qualified as the "missing element" in the asymmetric hydrogenation reactions with base metals [12]. In early 2017, Zirakzadeh et al. reported a Mn(I) hydride complex containing a chiral PNP (2,6-bis(dialkylphosphinomethyl)-pyridine) pincer (Scheme 9.1a), which catalyzes the ATH of acetophenones in basic 2-propanol (S : C : B = 100 : 1 : 4, B is ^{t}BuOK, up to 96% conversion after two hours) with up to 86% ee [13]. Clarke and coworkers reported the first Mn(I) catalyst for the AH of ketones just weeks later [14].

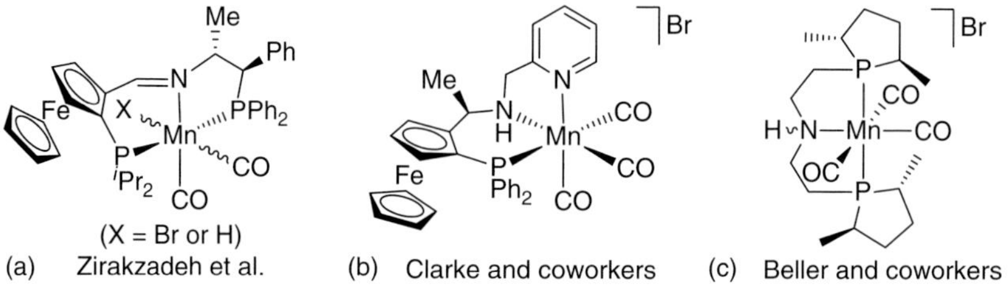

(X = Br or H)
(a) Zirakzadeh et al. (b) Clarke and coworkers (c) Beller and coworkers

Scheme 9.1 Manganese(I) chiral ATH (a) and AH (b, c) precatalysts.

The PNN pincer complex in Scheme 9.1b, in which the central amine donor enforces the facial configuration, hydrogenates ketones under relatively mild conditions (50 bar H_2, 50 °C) in basic ethanol (S : C : B = 100 : 1 : 10, B is ^{t}BuOK). High steric hindrance is the key to high enantioselectivity, as acetophenone gives only 20% ee, whereas 2-methoxyphenyl *tert*-butyl ketone reaches 91% ee. Esters are also hydrogenated at 75 °C *ceteris paribus*. The use of tridentate, phosphine-containing pincer ligands is an evident analogy with pincer Fe(II) catalysts, with which Mn(I) shares the d^6 electron configuration (see below). At variance with the above complexes, which share the rigid ferrocenyl framework motif, Beller and coworkers has used a more flexible PN(H)P backbone combined with phospholanes as rigid stereogenic units (Scheme 9.1c) [15]. Remarkably, the tricarbonyl Mn(I) derivative $[Mn(CO)_3(PN(H)P]Br$ catalyzes the AH of dialkyl ketones with up to 84% ee (92 : 8 *e.r.*) under 30 bar H_2.

Achiral Mn(I) complexes are among the most active hydrosilylation catalysts known [16]. This advantage has been exploited by Huang and coworkers, who reported the first manganese-catalyzed asymmetric hydrosilylation in 2017 [17]. The catalyst is a Mn(II) analogue of the iron(II) species depicted in Scheme 9.19 and, under the same conditions, hydrosilylated aryl alkyl ketones quantitatively with up to 93% ee (96.5 : 3.5 *e.r.*). The first manganese-based hydroboration of ketones was recently described by Gade and coworkers with the tridentate monoanionic boxmi ligands (see Scheme 9.18) [18]. Catalyst $[Mn(CH_2SiMe_3)(boxmi)]$ (5 mol%) hydroborated acetophenones with pinacolborane in toluene quantitatively and with enantioselectivity above 99% ee.

9.3 Iron

The significance of iron catalysis to organic synthesis has grown dramatically in the last years [19]. The interest in such catalysts can be explained with the abundance, low price, and low toxicity of iron, which make it particularly appealing for the pharmaceutical industry, in particular for reduction and hydroelementation reactions [20], as well as for enantioselective transformations in general [21, 22]. Chiral nitrogen ligands are ubiquitous with iron [21, 23], but their combination with phosphorus donors has been pivotal to success in asymmetric reductions [24]. Thus, either tridentate PNP pincer ligands or tetradentate ligands with an N_2P_2 donor set play a major role in the iron(II)-catalyzed reduction of polar double bonds. Chiral phosphines have been introduced at a very early stage, too. Additionally, half-sandwich cyclopentadienyl complexes have also been exploited, as substantiated below. Although iron hydride complexes are often invoked to explain these Fe-catalyzed reductions, it should be noted that such complexes tend to be unstable and are difficult to detect and characterize, which often makes their involvement in catalysis rather speculative [25]. Iron-catalyzed transfer hydrogenation is discussed below before direct (H_2) hydrogenation because of the higher success of the first method. Some catalysts are active with both methods and are accordingly discussed in both sections.

9.3.1 Iron Catalysts in Asymmetric Transfer Hydrogenation (ATH)

In 2004, Gao and coworkers published a seminal paper on the ATH of ketones based on a system formed *in situ* from equimolar amounts of the trinuclear iron(0) cluster $[HFe_3(CO)_{11}]^-$ and a potentially tetradentate diamino P(NH)(NH)P ligand (Scheme 9.2) [26]. The catalyst, which is a chiral variant of Vancheesan's transfer hydrogenation system [27], reduces ketones with moderate-to-high enantioselectivity (16–93% ee) and tolerates bulky substituents, α,α-diphenylacetone giving the highest selectivity (98% ee).

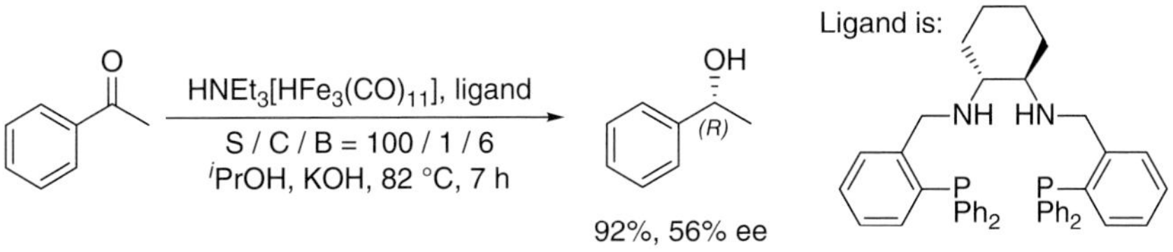

$HNEt_3[HFe_3(CO)_{11}]$, ligand
S / C / B = 100 / 1 / 6
^{i}PrOH, KOH, 82 °C, 7 h
OH (*R*)
92%, 56% ee
Ligand is:
NH HN
P Ph$_2$ P Ph$_2$

Scheme 9.2 Gao's pioneering iron(II)-based catalytic system.

However, the activity was modest (TOF = 0.5–33 h^{-1}), and the nature of the active catalyst (i.e. binding mode of the ligand) was not elucidated. The analogous diimino PNNP ligand was used by Beller and coworkers under identical conditions for the ATH of phosphinoyl imines with high enantioselectivity (89–98% ee) [28].

The field of stable, well-defined iron complexes as isolated precatalysts was pioneered by Morris and coworkers, who prepared the first iron precatalyst for the ATH of ketones [29] by reaction of iron(II) salts with a PNNP ligand derived from (*R*,*R*)-cyclohexane-1,2-diamine previously used with ruthenium (Scheme 9.3) [30].

Independently of the nature of the iron(II) salt (i.e. $FeCl_2$ or $[Fe(OH_2)_6](BF_4)_2$), the reaction in MeCN afforded the trans-complex $[Fe(MeCN)_2(PNNP)]^{2+}$ [29].

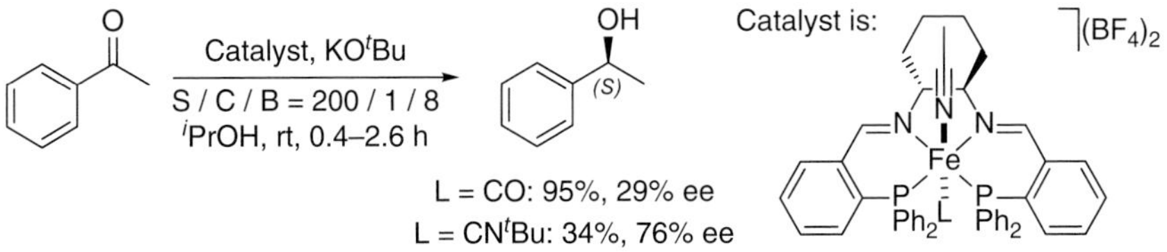

Catalyst, KO^{t}Bu
S / C / B = 200 / 1 / 8
^{i}PrOH, rt, 0.4–2.6 h
OH (*S*)
L = CO: 95%, 29% ee
L = CN^{t}Bu: 34%, 76% ee
Catalyst is:
$(BF_4)_2$
N N N Fe P P Ph$_2$ L Ph$_2$

Scheme 9.3 Morris' first-generation iron/PNNP ATH catalyst.

This versatile precursor readily reacts with CO or *tert*-butyl isocyanide to afford mixed acetonitrile–carbonyl or acetonitrile–isonitrile complexes that are highly active in the ATH of ketones (TOF up to 995 h^{-1}), but only moderately enantioselective (18–76% ee). The corresponding complexes containing the diamino P(NH)(NH)P ligand were not reported. Successive studies failed to disclose the catalytically active species under ATH conditions [31] and suggested that iron(0) nanoparticles with a ligand-functionalized shell are operating [32].

As the steric repulsion between the substituent on the mutually *cis* phosphines in *trans*-$[Fe(L)(MeCN)(PNNP)]^{2+}$ may account for the instability of the complexes, methylene instead of 1,2-phenylene was used to bridge the phosphine and imine donors. In this second generation of catalysts, the 5–5–5 chelate gives a larger P–Fe–P angle than the original 6–5–6 PNNP ligand (109.81(8) vs 100.24(8)° in the bis(acetonitrile) complexes) [33]. Its complexes were prepared in a template synthesis using Matt's phosphonium dimer [34] as the source of the formylphosphine component and (*R*,*R*)-1,2-diphenylethylene-1,2-diamine as the stereogenic element (Scheme 9.4, left). The mixed acetonitrile–carbonyl complex (Scheme 9.4, right) catalyzes the ATH of 11 prochiral aryl alkyl (and of 2 dialkyl and 1 alkenyl alkyl) ketones with turnover frequency (TOF) up to 4900 h^{-1} in ^{i}PrOH with potassium *tert*-butoxide as the base (8 equiv. vs catalyst) [35].

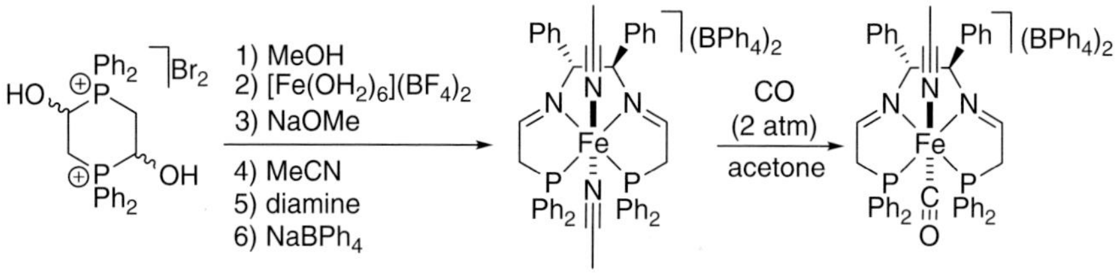

Scheme 9.4 Morris' second-generation iron/PNNP ATH catalysts.

Most optically active alcohols were obtained with good enantioselectivity (80–94% ee), the highest (99% ee) being achieved with *tert*-butyl phenyl ketone, a challenging ATH substrate [36]. In analogy to Beller's seminal work [28], these second-generation complexes are also active in the ATH of phosphinoyl imines and give chiral amines with excellent enantioselectivity (95–99% ee) [37]. The template approach allowed screening different diamines, substituents on phosphorus, and ancillary ligands. The bromocarbonyl complexes perform similarly in ATH as their acetonitrile–carbonyl analogues [38] but are easier to prepare and more stable because of the push–pull interaction between the π-donating bromide and the π-accepting carbonyl ligand. As diamine linker, (L)-1,2-diphenylethylene-1,2-diamine was superior to (L)-cyclohexane-1,2-diamine in the ATH of acetophenone both in terms of activity (20 000 and 4900 h^{-1}, respectively) and enantioselectivity (81% and 60% ee, respectively) [38]. Substitution of the phenyl substituents on phosphorus by alkyl groups was detrimental [39], but *meta*-xylyl groups improved the enantioselectivity to 90% ee without loss of activity [40].

Mechanistic studies indicate that, in the presence of base, deprotonation occurs at both α-positions of the imines [41], followed by slow reduction to a catalytically active amido–enamido complex (Scheme 9.5) [42]. Poisoning experiments with mercury and substoichiometric trimethylphosphine [42] as well as density functional theory (DFT) calculations with a simplified model [43] further support such a homogeneous mechanism.

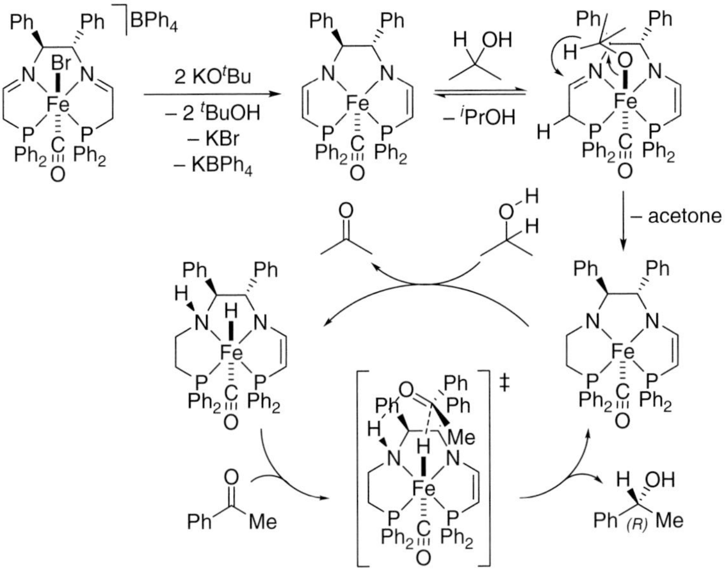

Scheme 9.5 Activation mechanism of Morris' second-generation iron/PNNP ATH catalysts.

The transfer of the NH proton and hydride to the *si*-enantioface of acetophenone, reminiscent of the bifunctional *concerto* mechanism operating with [RuCl(Ts-dpen)(arene)] complexes (Ts-dpen = *N*-tosylated 1,2-diphenylethylene-1,2-diamine) [44], affords (*R*)-1-phenylethanol as the major enantiomer [45]. A π–π interaction between the aryl ring on the ketone and the enamido moiety of the complex is pivotal for enantiodiscrimination. An excess of base is required to suppress complete reduction of the P–N–N–P backbone, which gives a catalytically inactive diamido complex [45].

Taking advantage of these findings, Morris and coworkers developed the third-generation catalyst, which features amino/imino ligands and is highly active (TOF up to $200\,s^{-1}$) (Scheme 9.6). Acetophenone is reduced with up to 98% ee, but the enantioselectivity is modest with other ketones (24–99% ee), whereas activated imines are reduced with high enantioselectivity [46]. A dicyclohexylphosphino group increases the enantioselectivity in part, but at the cost of lower activity (R = Cy, R′ = Ph) [47].

Attempts to use these third-generation catalysts under direct hydrogenation conditions have shown that H_2 splitting is feasible, but less effective than ^{i}PrOH activation. The AH of acetophenone is unselective (35% ee) and slow even under forcing conditions (50 °C, 20 atm H_2, 10 hours) [48]. Thus, ATH remains the method of choice with these catalysts.

Alternative approaches to prepare Fe(II) precatalyst that withstand the relatively harsh ATH conditions exploit strong-field ligands other than phosphines

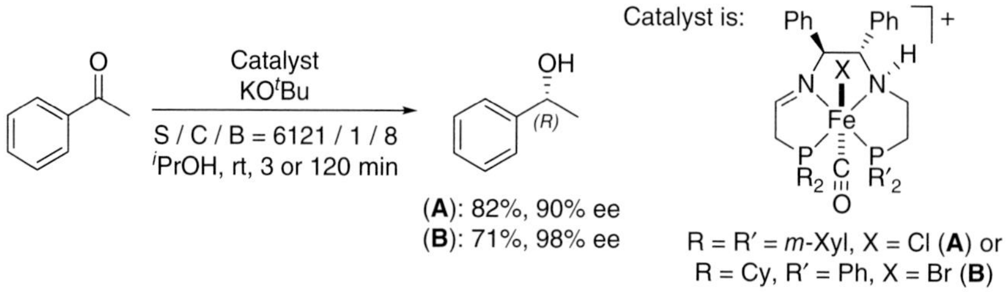

Scheme 9.6 Morris' third-generation iron/PNNP ATH catalysts.

and CO, and/or the macrocyclic effect. Thus, Reiser and coworkers has used chiral bidentate isonitrile ligands (BINC) to prepare the low-spin Fe(II) complexes *trans*-[$FeCl_2(BINC)_2$]. These catalyze the ATH of aryl alkyl ketones with moderate activity (turnover number [TON] up to 20, TOF up to 20 h^{-1}) and enantioselectivity (30–91% ee) (Scheme 9.7) [49]. The proposed mechanism involves β-hydrogen transfer from coordinated *iso*-propoxide to isonitrile to form a Fe—CH=NR intermediate, which then transfers the hydride to the substrate.

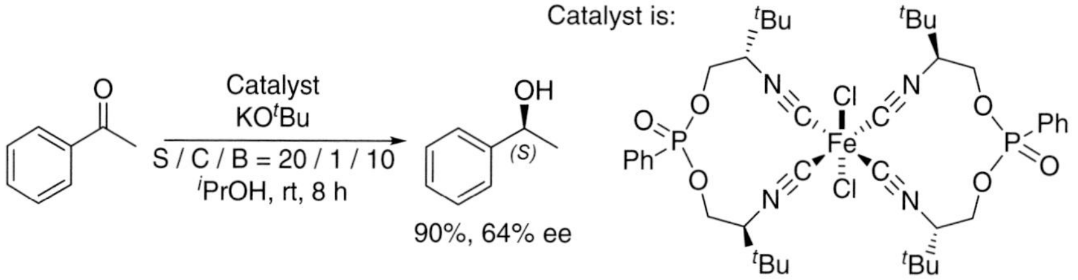

Scheme 9.7 Reiser's iron(II) ATH catalyst based on chiral bidentate isonitrile ligands.

Cyclopentadienyl was also investigated as strong-field ligand for ATH catalysts. Wills and coworkers prepared chiral "cyclone" complexes that reduce acetophenone with up to 25% ee (Scheme 9.8) [50]. Beyond the modest performance in ATH of this specific catalyst, half-sandwich complexes are interesting for the direct hydrogenation (AH) of ketones and imines (see Section 9.3.2).

In a first attempt of exploiting the macrocyclic effect for catalyst stabilization, Gao and coworkers prepared a potentially hexadentate $(NH)_4P_2$ macrocycle and used it in combination with the trinuclear iron cluster [$Fe_3(CO)_{12}$] (and NH_4Cl as promoter) to reduce aryl alkyl ketones to optically active alcohols with outstanding enantioselectivity (90–99% ee) at low catalyst loadings (0.5–0.02 mol%) (Scheme 9.9) [51]. With *para*-chloroacetophenone, the TOF reached up to 1940 h^{-1}. The same catalyst also catalyzes the direct hydrogenation with H_2 (50 atm) in MeOH as a solvent instead of ^{i}PrOH, but with reduced activity (see Section 9.3.2).

To combine the advantages of the macrocycle with that of the N_2P_2 donor set, Mezzetti and coworkers developed chiral, C_2-symmetric macrocyclic ligands based on an enantiomerically pure bis(formyl)-substituted diphosphine

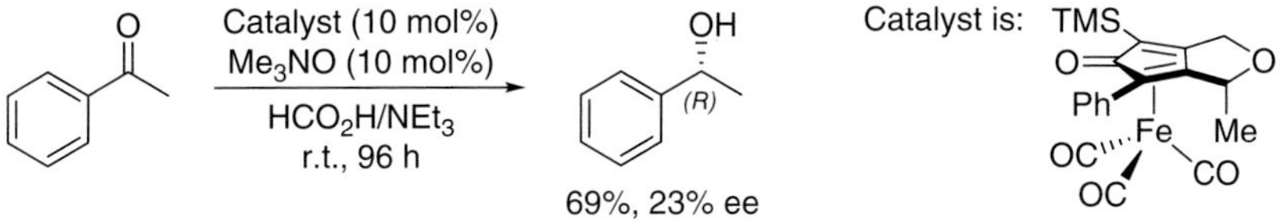

Scheme 9.8 Wills' "cyclone"-type catalyst in the ATH of acetophenone.

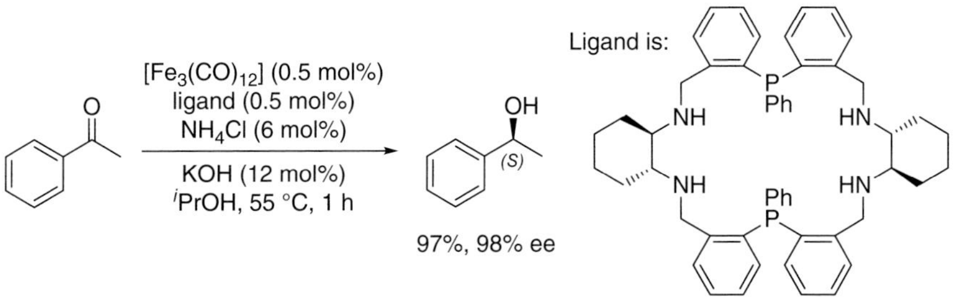

Scheme 9.9 Gao's chiral $(NH)_4P_2$ macrocycle/Fe(0) in situ catalyst for the ATH of ketones.

prepared either from Jugé's [52] oxazaphospholidine borane [53] or, more conveniently, from Han's [54] menthyl *H*-phosphinate [55]. Only one enantiomer of cyclohexane-1,2-diamine gives the macrocyclization product, apparently because of the conformation of the monoimine intermediate. The corresponding bis(acetonitrile) N_2P_2 complexes $[Fe(MeCN)_2(N_2P_2)]^{2+}$ (Scheme 9.10) are diamagnetic but decompose under ATH conditions [56], as observed with Morris' first-generation catalysts [32].

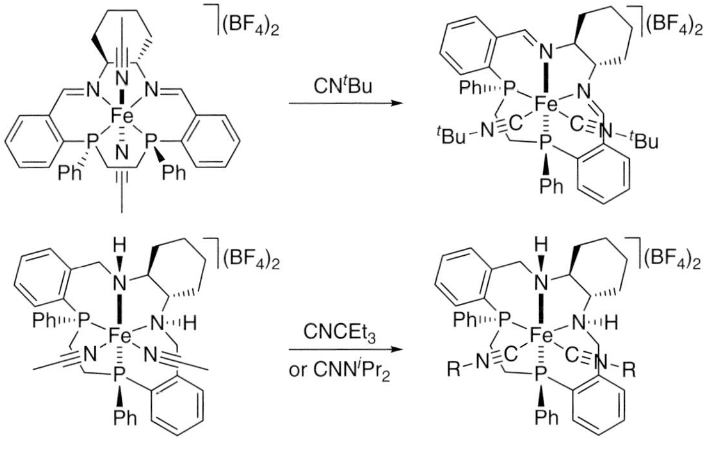

Scheme 9.10 Mezzetti's chiral N_2P_2 macrocyclic ATH precatalysts.

Exchanging the MeCN ligands by strong-field ligands such as isonitriles gave robust catalysts with reproducible but moderate enantioselectivity. Thus, the CN^tBu derivative reduced acetophenone with 84% ee [56]. As the imine groups undergo reduction to amine under ATH conditions, the $(NH)_2P_2$ macrocycles were prepared by reduction of the imine moieties with $LiAlH_4$. Complexation with $[Fe(OH_2)_6]BF_4$ in $MeCN/CH_2Cl_2$ in the presence of a catalytic amount of

DBU to epimerize the *N*-stereocenters, followed by ancillary ligand substitution with isonitriles, gave diastereomerically pure *cis*-β bis(isonitrile) complexes. With bulky isonitriles such as $CNCEt_3$ and CNN^iPr_2, a broad scope of substrates was reduced with excellent enantioselectivity and significantly increased activity (TOF up to 6650 h^{-1}) with respect to the unsaturated macrocycle [57]. Aryl alkyl ketones X-$C_6H_4C(O)$(alkyl) with different aromatic substitution patterns (X = Me, Cl, OMe, and CF_3) are reduced with 94–99% ee, as well as acylpyridines and acylthiophenes. Also, a phosphinoyl imine was reduced with 98% ee using the $CNCEt_3$ catalyst. Scaling up the system to 100 mmol of substrate allows to reduce the catalyst loading (S/C = 10 000/1) [55]. Although they do not challenge the superb activity of Morris' catalysts, these iron(II)/N_2P_2 macrocyclic catalysts are presently the most enantioselective systems for a broad scope of aryl alkyl ketones. Also, they were used to prepare the hydride complex $[FeH(CNCEt_3)(N_2P_2)]BF_4$, which catalyzes the base-free ATH of benzil to the corresponding benzoins with yields up to 83% and high enantioselectivity (up to 95% ee). This is the first example of highly enantioselective benzil hydrogenation [58]. As in Reiser's BINC ligands, the isonitrile ligands appear pivotal to stabilize the catalysts as a tunable alternative to the CO ligand, which plays an analogous role in Morris' third-generation complexes (Scheme 9.6).

9.3.2 Iron Catalysts in Asymmetric Direct (H_2) Hydrogenation (AH)

In 2007, following the concept of ligand–metal bifunctional catalysis [44], Casey and Guan reported the use of Knölker's complex [59], an iron analogue of Shvo's catalyst [60], as the first well-defined iron catalyst for the direct hydrogenation of ketones at low pressure (3 atm H_2) [61], but the development of chiral variants for asymmetric direct hydrogenation has met with little success so far. In 2011, Berkessel et al. prepared the first chiral analog of a Knölker-type catalyst by substituting one carbonyl ligand with an enantiopure phosphoramidite ligand (Scheme 9.11) [62]. Upon photolytically induced CO dissociation, the complex reduced acetophenone under H_2 pressure (10 atm), but with low enantioselectivity (31% ee), possibly because of the formation of a 1.00 : 0.69 mixture of diastereomeric hydride complexes.

Piarulli and coworkers reported a Knölker-type Fe(0) complex based on chiral cyclopentadienones decorated with an axially chiral 1,1′-binaphthalene motif, which catalyzed the AH of aryl alkyl ketones with moderate activity (TON up to 50, TOF up to 2.8 h^{-1}) and enantioselectivity (46–77% ee) (Scheme 9.12) [63].

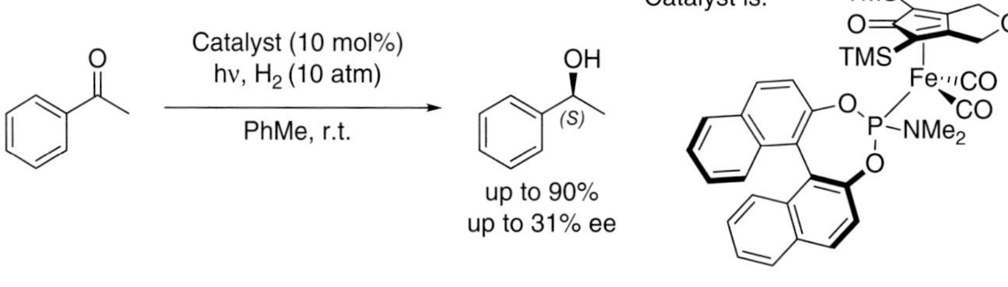

Catalyst (10 mol%)
hv, H_2 (10 atm)
PhMe, r.t.
OH
(S)
up to 90%
up to 31% ee
Catalyst is:
TMS
O
TMS
Fe
CO
CO
O
P
NMe2
O

Scheme 9.11 Berkessel's photoactivated Shvo-type, half-sandwich Fe(0) AH catalyst.

Related half-sandwich complexes (Scheme 9.8) were less efficient in the ATH of acetophenone [50].

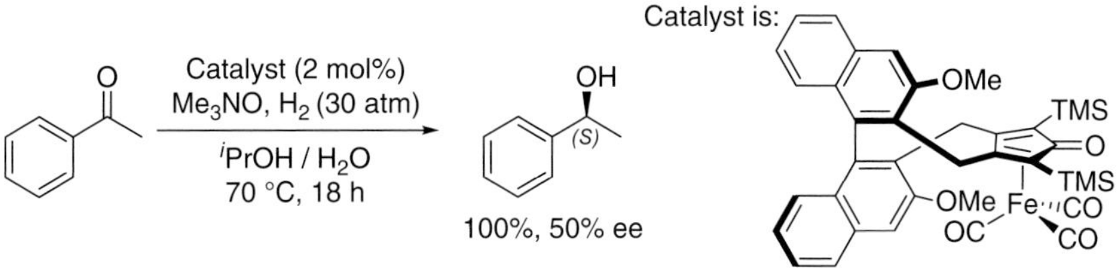

Scheme 9.12 Piarulli's and Gennari's half-sandwich Fe(0) AH catalyst.

Beller and coworkers circumvented the lack of available chiral half-sandwich complexes by combining Knölker's complex with axially chiral hydrogen phosphates as enantiopure Brønsted acids (Scheme 9.13). The resulting system catalyzes the AH of nonactivated imines, quinoxalines, and 2*H*-1,4-benzoxazines to the corresponding amines with good-to-excellent enantioselectivity (58–98% ee) [64–67].

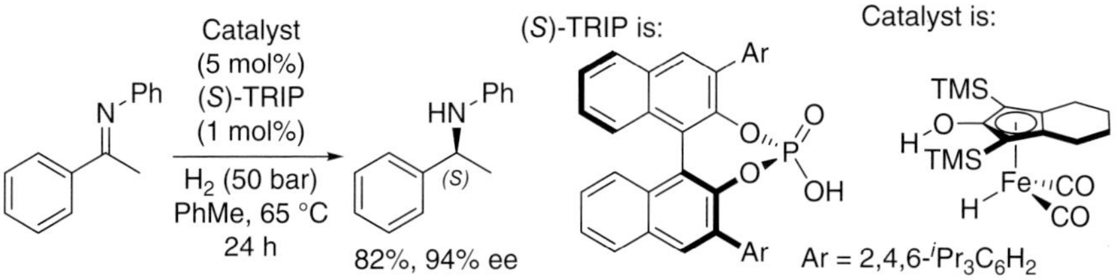

Scheme 9.13 Beller's AH catalyst based on an achiral catalyst/chiral anion approach.

Mechanistic studies suggested that TRIP activates the imine toward the attack of the hydride, whereas its conjugate base assists heterolytic H_2 splitting (Scheme 9.14). DFT calculations showed that (*S*)-TRIP is hydrogen bonded to the CpOH motive in Knölker's complex during the reaction, hence acting as a base in the heterolytic splitting of hydrogen, as a Brønsted acid in the activation of the imine toward hydride transfer, and as a ligand for iron(II) in the resting state [68].

Also, hybrid P/N ligands have found application in the AH of ketones. In 2011, Gao and coworkers used the hexadentate macrocycle in Scheme 9.9 for direct hydrogenation (50 atm H_2) with MeOH as a solvent instead of ^{i}PrOH [69]. The activity was lower than in ATH (TOF up to 40 h^{-1}), but a broad scope of aryl alkyl ketones and β-ketoesters was reduced with high enantioselectivity (>95% ee for most substrates). Substoichiometric catalyst poisons such as triphenylphosphine or 1,10-phenanthroline inhibited the reaction, which suggests that the active catalyst is most probably heterogeneous in nature.

Morris and coworkers showed that the Fe(II)/PNNP catalysts of the third generation (Scheme 9.6) are also active under AH conditions [49] but are less effective than in ATH (see Section 9.3.1) [46]. The AH of acetophenone gave 35% ee and was slow even under forcing conditions (50 °C, 20 atm H_2, 10 hours).

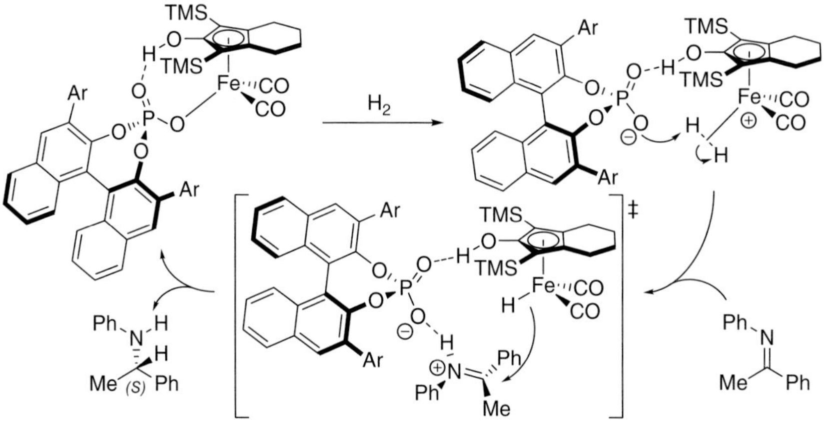

Scheme 9.14 Anion-assisted H_2 activation with Beller's half-sandwich catalyst.

Additionally, Morris and coworkers reported the first chiral Fe(II) AH catalyst based on an enantiopure tridentate pincer PNP ligand (**A**, Scheme 9.15), which, after activation with lithium aluminum hydride and *tert*-amyl alcohol, reduces ketones with excellent activity (TON up to 1980 h^{-1}) and moderate-to-good enantioselectivity (up to 85% ee for 1-acetonaphthone and 2-acetylthiophene, 80% ee for acetophenone) under 5 atm H_2 [70]. Similarly to ATH catalysts, the amino analogue **B** is significantly more enantioselective and gives 1-phenyethanol with 95% ee [71], whereas the ferrocenyl-based pincer ligand in Scheme 9.1a performs similarly to **A** [72].

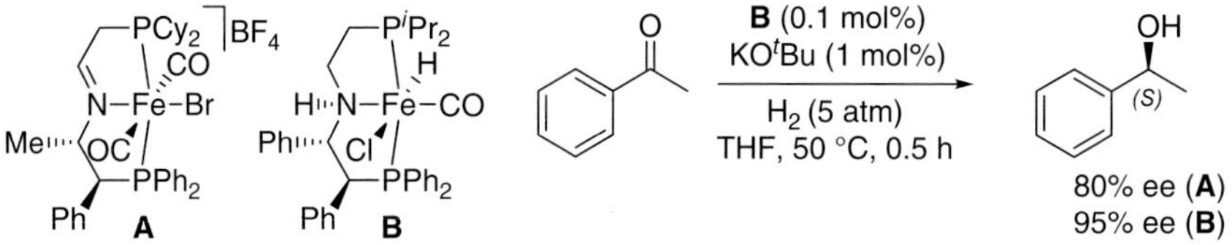

Scheme 9.15 PNP and PN(H)P pincer AH catalysts.

The structure of these catalysts is reminiscent of iron complexes with achiral PNP [73] and P(NH)P pincer ligands that catalyze hydrogenation and dehydrogenation reactions [10, 74, 75]. Overall, AH catalysts feature tridentate ligands, either as PNP or cyclopentadienyl ligand, in contrast to ATH catalysts, which are most efficient with tetradentate ligands (see Section 9.3.1). Despite recent success, however, the enantioselectivity of these AH catalysts is still susceptible of improvement.

9.3.3 Iron Catalysts in Asymmetric Hydrosilylation (AHS)

Many carbonyl derivatives are hydrosilylated in the presence of catalytic amounts of simple iron salts such as $FeCl_3$, or even without metal catalyst, which basically restricts the interest for catalytic hydrosilylation to its asymmetric variant

[19]. In 2007, Nishiyama and Furuta reported the first iron-catalyzed asymmetric hydrosilylation of ketones. The catalyst, formed *in situ* from ferrous acetate and the tridentate ligand (*S*,*S*)-^{t}Bu-bopa, reduced para-substituted aryl ketones with $HSi(OEt)_2Me$. After workup, the corresponding alcohols were formed with up to 79% ee (R = Ph) (Scheme 9.16) [76] and reached 88% ee with other phenyl ketones [77].

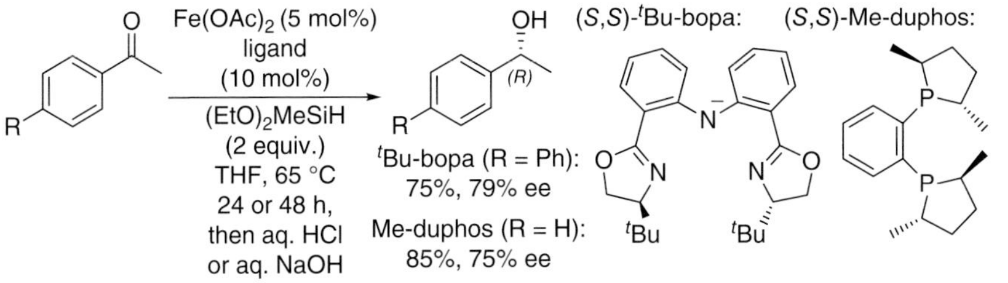

Fe(OAc)$_2$ (5 mol%), ligand (10 mol%), (EtO)$_2$MeSiH (2 equiv.), THF, 65 °C, 24 or 48 h, then aq. HCl or aq. NaOH

^{t}Bu-bopa (R = Ph): 75%, 79% ee
Me-duphos (R = H): 85%, 75% ee

(*S*,*S*)-^{t}Bu-bopa: (*S*,*S*)-Me-duphos:

Scheme 9.16 Nishiyama's catalysts for the asymmetric hydrosilylation of ketones.

The preformed high-spin Fe(III) complex [$FeCl_2$(Bn-bopa)] is catalytically active after reduction with zinc powder to an unidentified Fe(II) species [78]. The low-spin Fe(II) complex [$FeBr_2$(CO)(phebox)] quantitatively reduced 4-phenylacetophenone with $HSi(OEt)_2Me$ with Na(acac) as additive in hexane at 50 °C with 66% ee [79]. In the above reactions, the mechanisms remain a matter of speculation. In many cases, the use of oxygen-containing anions such as acetate or acetylacetonate is pivotal to give active catalysts (see below). The use of chiral phosphines in AHS has developed in parallel to that of nitrogen ligands. In 2008, Beller and coworkers reported asymmetric hydrosilylation catalysts formed *in situ* from iron(II) acetate (5 mol%) and a chiral diphosphine. Under conditions similar to Nishiyama's, (*S*,*S*)-Me-duphos gave the best enantioselectivity with hindered aryl alkyl ketones (48–99% ee) (Scheme 9.16) [80]. Polymethylhydrosiloxane (PMHS) was used instead of $HSi(OEt)_2Me$ without loss of enantioselectivity [81]. Gade's chiral, anionic NNN pincer derived from 1,3-bis(2-pyridylimino)isoindole (BPI) performed similarly [82].

Chirik and coworkers reported iron AHS catalysts with nitrogen-based ligands, in which the use of anionic ligands other than acetate increased the activity significantly (Scheme 9.17). The high-spin bis(neosyl) iron(II) complexes bearing either bidentate bis(oxazoline) or tridentate pybox ligands gave TONs of up to 330 with phenylsilane in Et_2O at 23 °C, but the enantioselectivity was low (up to 54% ee for acetophenone) [83].

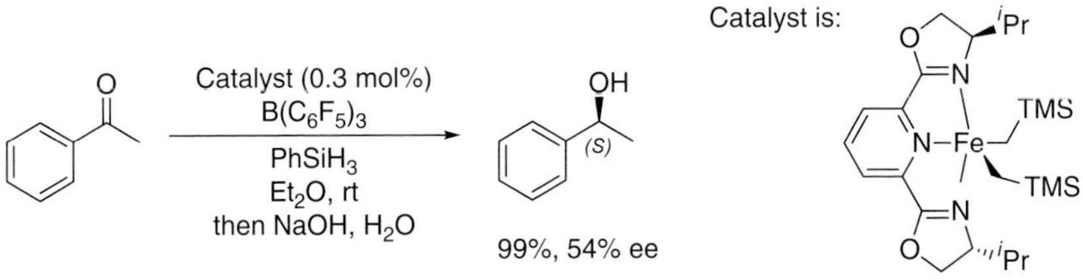

Catalyst (0.3 mol%), B(C$_6$F$_5$)$_3$, PhSiH$_3$, Et$_2$O, rt, then NaOH, H$_2$O

99%, 54% ee

Catalyst is:

Scheme 9.17 Chirik's iron(II) AHS catalyst based on chiral tridentate NNN ligands.

In a further effort to use well-defined Fe(II) complexes in catalysis, Gade and coworkers recently reported the neosyl and alkoxido complexes [Fe(Y)((*R*)-^{H}boxmi-Ph)] (Y = CH_2TMS or OR) that catalyze the asymmetric hydrosilylation of ketones with diethoxy(methyl)silane at low temperature with excellent enantioselectivity (≥95% ee for most substrates) (Scheme 9.18) [84]. Also, the catalyst is remarkably active and can be used even at low temperatures.

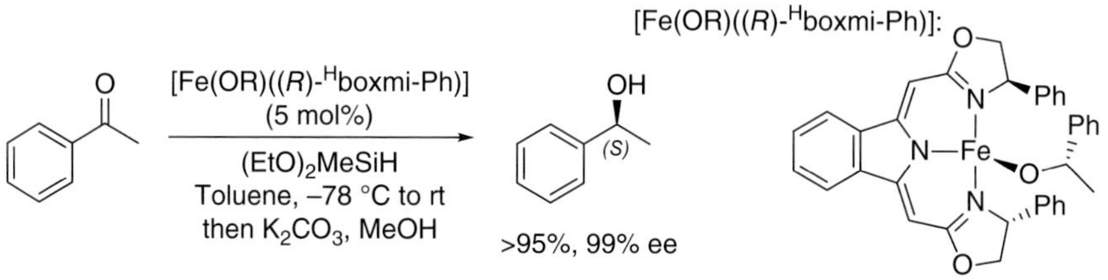

[Fe(OR)((*R*)-^{H}boxmi-Ph)] (5 mol%), $(EtO)_2MeSiH$, Toluene, −78 °C to rt, then K_2CO_3, MeOH; (*S*), >95%, 99% ee; [Fe(OR)((*R*)-^{H}boxmi-Ph)]

Scheme 9.18 Gade's iron(II) AHS catalyst with anionic NNN pincer ligands.

The same catalyst also efficiently reduces diaryl ketones (81% ee with (perfluorophenyl)phenyl ketone). The coordinatively unsaturated boxmi alkoxo complex is thought to undergo easy σ-bond metathesis with silane to give the silyl ether and a highly reactive transient, yet unobserved iron(II) hydride complex, which readily transfers hydride to the prochiral substrate. A kinetic study of the acetate complex [Fe(OAc)(boxmi)] in the AHS of acetophenone suggests a mechanism in which the catalyst is activated by slow reduction of the acetate ligand to alkoxo [85]. The alkoxo complex undergoes rate-determining σ-bond metathesis with the silane to give a highly reactive high-spin hydride complex. Coordination of the ketone and insertion of the carbonyl into the Fe—H bond regenerate the alkoxo complex and close the catalytic cycle. The high-spin hydride Fe(II) complexes remain elusive, though.

Following the trend toward well-defined catalysts, Huang and coworkers prepared [$FeBr_2$(NNN)] complexes (NNN is a chiral iminopyridine-oxazoline), which, upon activation with $NaBHEt_3$, catalyze the AHS of ketones with Ph_2SiH_2 (1 equiv.) at 25 °C (Scheme 9.19) [86]. Acetophenone gave 1-phenylethanol quantitatively with 83% ee, and the enantioselectivity reached 93% ee with bulkier ketones. The involvement of a hydride intermediate was postulated, but not demonstrated.

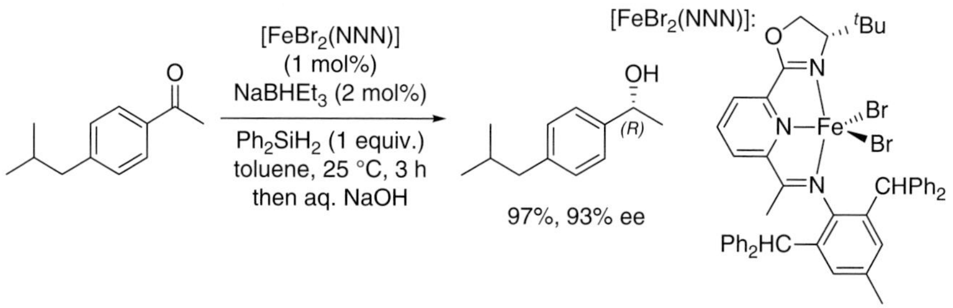

[$FeBr_2$(NNN)] (1 mol%), $NaBHEt_3$ (2 mol%), Ph_2SiH_2 (1 equiv.), toluene, 25 °C, 3 h, then aq. NaOH; (*R*), 97%, 93% ee; [$FeBr_2$(NNN)]

Scheme 9.19 Huang's NNN pincer AHS catalyst.

In general, conclusive evidence of the involvement of hydride complexes in iron-catalyzed AHS reactions is missing. In some cases, silane activation by acetate or analogous anionic species to give a more nucleophilic hypervalent hydrosilicate may play a role in these reactions, as suggested for the copper-catalyzed AHS of ketones (see Section 9.6.3).

9.4 Cobalt

Cobalt catalysts are among the first enantioselective catalysts reported, and salen-like ligands with a dianionic N_2O_2 donor set play a major role in these transformations [87]. Thanks to the involvement of three different oxidation states (I, II, and III), the chemistry of cobalt is mechanistically rich and unique, as illustrated by the first report of an AH cobalt(II) catalyst, which dates back to 1971 (see below) [88]. However, cobalt reduction catalysts tend to be chemoselective for C=C double bonds, which may explain why they play a marginal role in the AH of polar double bonds. Some examples of ATH and AHS catalysts are known, and cobalt excels in asymmetric BH_4^- reductions, which have been introduced by Mukaiyama and systematically studied by Yamada (see Section 9.4.4). The final part of the latter section briefly covers hydroboration.

9.4.1 Cobalt Catalysts in the AH of Ketones

In a pioneering study, Ohgo et al. reported that the achiral complex bis(dimethylglyoximato)cobalt(II) reduces benzil under atmospheric H_2 pressure in the presence of a chiral aminoalcohol such as quinine or quinidine as chiral auxiliary [88]. (*S*)-Benzoin was obtained almost quantitatively with 61.5% ee (Scheme 9.20). Under optimized conditions, the enantioselectivity reached 78% ee [89].

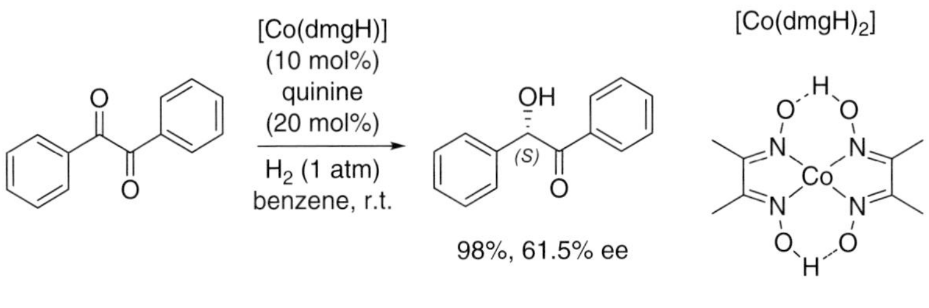

[Co(dmgH)] (10 mol%) quinine (20 mol%) → H_2 (1 atm) benzene, r.t.

98%, 61.5% ee

[Co(dmgH)$_2$]

Scheme 9.20 Ohgo's pioneering cobaloxime/quinine catalytic system for the AH of benzil.

The reaction does not involve a hydride attack onto the carbonyl group, as the Co(III) hydride formed by homolytic H_2 splitting undergoes H^+ abstraction in the presence of base. The resulting Co(I) species attacks benzil and forms a Co(III) alkyl complex (Scheme 9.21) [90]. The protonated quinine activates the substrate with enantioface recognition [89] in an *ante litteram* example of asymmetric catalysis by chiral Brønsted acids [91]. The cobalt–carbon bond is cleaved by backside attack of H^+ from protonated quinine to form the product. The resulting Co(III) complex may react with Co(I) to regenerate [$Co^{II}(dmgH)_2$].

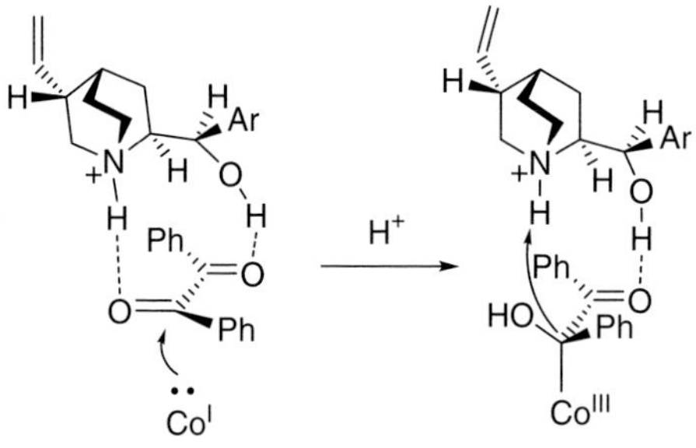

Scheme 9.21 Stereochemical course of benzil asymmetric hydrogenation by Ohgo's catalyst.

Hydride complexes of cobalt(I), which are formed under H_2 pressure from the carbonyl-containing species such as $[Co_2(CO)_6(P(neomenthyl)Ph_2)_2]$, catalyze the hydrogenation of the C=C double bond in enones (2 mol%, 30 atm H_2, 100 °C) with complete chemoselectivity, and are hence not applicable in the AH of polar double bonds [92]. Indeed, cobalt-catalyzed alkene hydrogenation is a developing field. Chirik and coworkers reported a chiral Co(I) bis(imino)pyridine catalyst for asymmetric alkene hydrogenation and isolated a cobalt hydride complex that is the resting state of the catalytic reaction [93]. However, hybrid phosphinoamino P(NH)P pincers enable metal–ligand bifunctional catalysis and shift the selectivity toward C=O reduction [94]. Taking advantage of this feature, Gao and coworkers has recently developed the first cobalt-catalyzed AH of ketones (Scheme 9.22) [95], but forcing conditions are required.

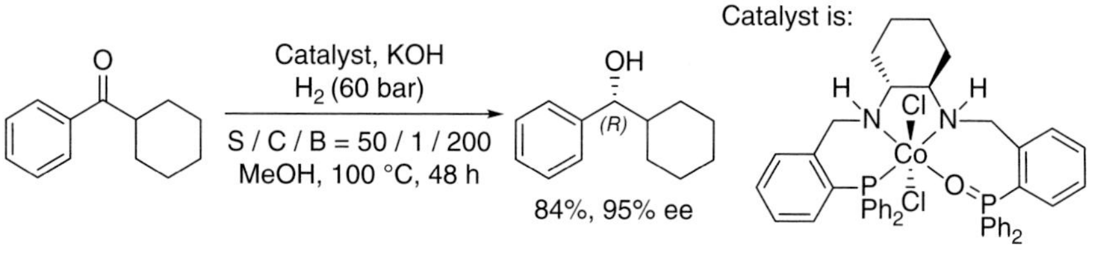

Scheme 9.22 Gao's cobalt(I) catalyst for the AH of ketones.

Notably, the catalyst bearing the semioxidized PNNP(O) ligand is more active than its PNNP analogue, and a large amount of KOH is required to activate the catalyst. The highest enantioselectivity is attained with bulky alkyl substituents.

9.4.2 Cobalt Catalysts in the ATH of Ketones

Cobalt-catalyzed ATH also plays a minor role. Overall, cobalt catalysts with chiral diamines are poorer ATH catalysts for β-ketoesters and ketones than their rhodium and iridium analogues [96]. Similarly, the PNNP ligand shown in Scheme 9.2 is less efficient in combination with cobalt than with iron, the best performance with propiophenone being 75% conversion after 83 hours and 61% ee [97].

9.4.3 Cobalt Catalysts in Asymmetric Hydrosilylation

In 1991, Brunner and Amberger reported the first AHS of acetophenone catalyzed by a system formed *in situ* from $[Co(py)_6](BPh_4)$ and pyridinyloxazoline ligands. Under solvent-free conditions, the enantioselectivity reached 56% ee, with silyl enol ether as by-product [98]. Nearly 20 years later, Nishiyama and coworkers achieved a breakthrough by combining $Co(OAc)_2$, bis(oxazolinylphenyl)amine ligand (Ph-bopa, see Scheme 9.16), and diethoxymethylsilane as superstoichiometric reducing agent [77]. A variety of aryl *n*-alkyl ketones were hydrosilylated with high enantiomeric excess (up to 98% ee) and in nearly quantitative yield. Interestingly, the cobalt-based catalyst is more enantioselective than the iron one (see Section 9.3.3). The dichloro complexes $[MCl_2(^{i}Pr\text{-bopa})]$ (M = Fe or Co) were isolated and structurally characterized, but turned out to be catalytically inactive. Thus, at present, the nature of the catalytically active species remains elusive.

Also, chiral diphosphines such as (*S*)-Xyl-P-Phos have been used as an efficient source of chirality in combination with $Co(OAc)_2$ hydrate and molecular sieves (Scheme 9.23a) [99]. With phenylsilane, 4-nitroacetophenone was reduced in 99% yield and with 95% ee, and the reaction was run in air. The substrate scope is limited to meta- or para-substituted electron-poor aryl alkyl ketones.

In 2012, Gade and coworkers showed that the high-spin d^7 complex $[Co^{II}(CH_2TMS)(BPI)]$ (BPI is a 1,3-bis(2-pyridylimino)isoindolate, Scheme 9.23b) catalyzes the AHS of aryl methyl ketones in the presence of $HSi(OEt)_2Me$ (2 equiv.). After basic hydrolysis, the alcohols were obtained in up to quantitative yield and 91% ee [100]. Small substituents (e.g. methyl) at the wingtips of the

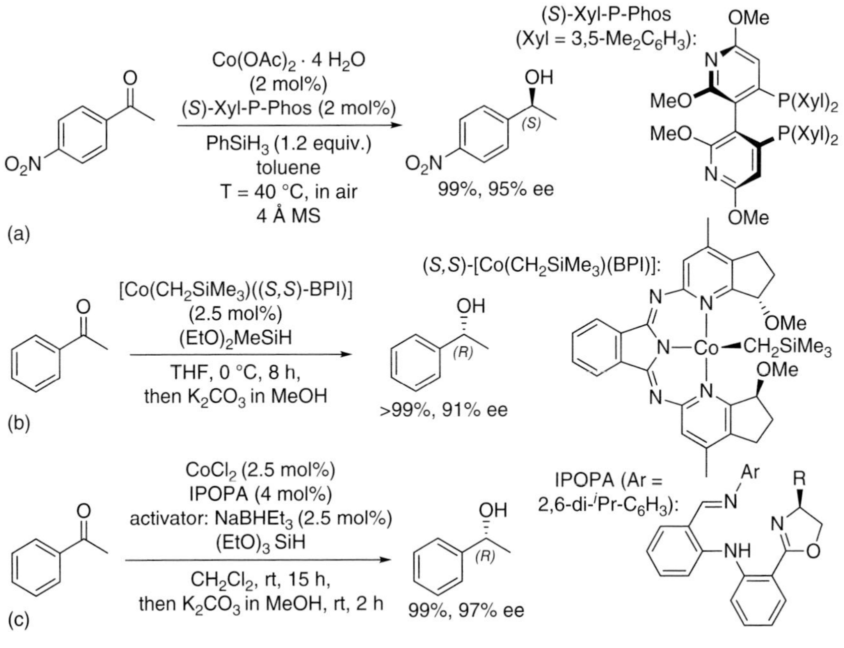

Scheme 9.23 Cobalt-based catalysts for the AHS of ketones.

ligand gave the highest enantioselectivity and increased the stability of the catalyst, which is extremely sensitive to air and moisture and thermally unstable. More recently, Lu and Lu has shown that an optimized NNN pincer ligand can be used in situ with $CoCl_2$ after activation with $NaBHEt_3$ (Scheme 9.23c) [101]. In the presence of $HSi(EtO)_3$, the resulting species, possibly a Co(II) amidohydride complex, gives 1-phenylethanol with 97% ee after workup, and the enantioselectivity exceeds 99% ee for its *o*-Me-substituted analogue.

9.4.4 Asymmetric Borohydride Reduction and Hydroboration

The asymmetric cobalt(II)-catalyzed reduction with BH_4^- has been discovered by Pfaltz and coworkers for the chemoselective reduction of the C=C bond of α,β-unsaturated carboxylic esters [102]. Shortly thereafter, Mukaiyama and coworkers extended the latter approach to the asymmetric reduction of the C=O bond in chromanones and related substrates with $NaBH_4$ (or KBH_4) and (β-oxoaldiminato)cobalt(II) complexes as catalysts [103]. The reactions were run for 120 hours at −20 °C, and the enantioselectivity attained 94% ee and high yield for 2,2-dimethyl-3,4-dihydronaphthalen-1(2*H*)-one (Scheme 9.24). The optical purity of the product increased from 90% to 94% ee upon changing from $NaBH_4$ to KBH_4 (see below). A small amount of alcohol was indispensable to achieve high enantioselectivity, and stirring $NaBH_4$ for 15 minutes with ethanol and tetrahydrofurfuryl alcohol (THFA) (1 equiv. each) in $CHCl_3$ gave a homogenous solution of a functionalized borohydride that reduced *n*-butyrophenone with increased enantioselectivity (97% ee) and shorter reaction times [104].

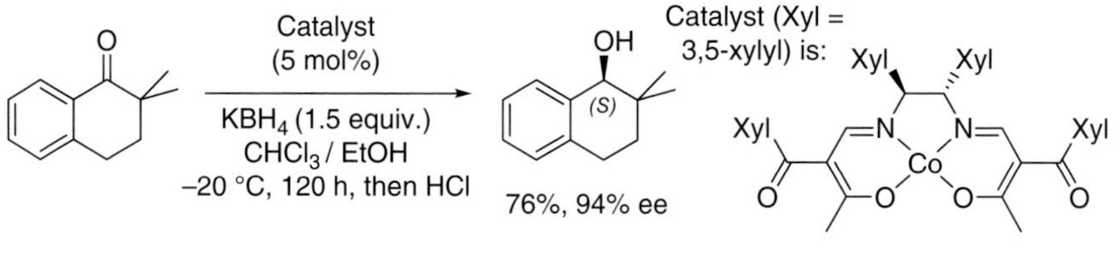

Scheme 9.24 Mukaiyama's β-ketoiminato cobalt(II) catalyst for asymmetric BH_4^- reduction.

With this system, *N*-diphenylphosphinoyl imines are reduced with up to 97% ee, whereas α,β-unsaturated esters and carboxamides undergo highly enantioselective 1,4-reduction [105]. In a mechanistic study, Yamada and coworkers proposed that, in the presence of MBH_4, the precatalyst [Co(ONNO)] reacts with chloroform to give the dichloromethyl hydrido cobalt(III) complex $M[CoH(CHCl_2)(ONNO)]$ [106]. This reaction is well documented and involves the H^+ abstraction from an intermediate Co(III) hydride as discussed above for Ohgo's Co-based AH system [88], followed by nucleophilic attack of the Co(I) complex onto $CHCl_3$ [107]. Further reaction with BH_4^- forms the active species, which was identified as an adduct of the anionic Co(III) hydride $[CoH(CHCl_2)(ONNO)]^-$ with the M^+ alkali counterion of MBH_4 [106]. During catalysis, the M^+ cation binds to the carbonyl oxygen atom of the ketone (Scheme 9.25), which is hence activated and directed analogously in a pattern that reminds the bifunctional catalysis discussed in Sections 9.3.1 and 9.3.2.

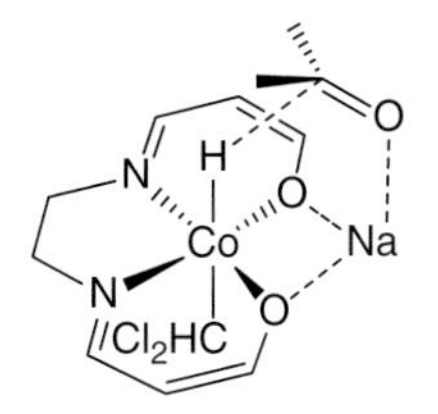

Scheme 9.25 Intermediate of the BH_4^- ketone reduction by β-ketoiminato cobalt(II) catalysts.

Accordingly, no enantioselectivity was observed when $Et_4N(BH_4)$ was used instead of MBH_4 (M = Li, Na, or K). Exchanging $CHCl_3$ with CH_3CCl_3 generates the corresponding 1-chlorovinyl Co(III) catalyst, which is more enantioselective than the Co–$CHCl_2$ analogue with challenging dialkylketones (where one alkyl group is bulky). Thus, 3-methyl-3-phenylbutan-2-one was reduced with 90% ee [108].

With chlorine-free solvents, methyl diazoacetate can be used instead of a chloroalkyl ligand to generate the corresponding carbene complex [109]. Related, bulkier complexes catalyze the BH_4^- reduction of 1,3-dicarbonyl compounds such as 1,3-diaryl-1,3-diketones to the 1,3-diols, which are useful stereogenic motifs for the synthesis of chiral ligands (Scheme 9.26) [110].

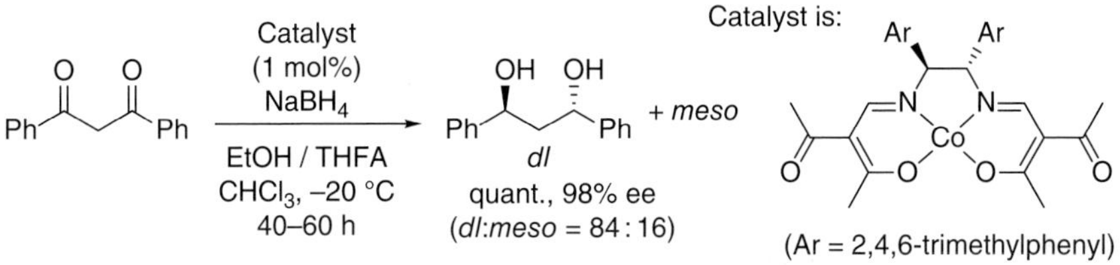

Catalyst (1 mol%) $NaBH_4$
EtOH / THFA
$CHCl_3$, −20 °C
40–60 h
dl
quant., 98% ee
(*dl*:*meso* = 84 : 16)
+ *meso*
Catalyst is:
(Ar = 2,4,6-trimethylphenyl)

Scheme 9.26 Borohydride reduction of 1,3-diketones by β-ketoiminato cobalt(II) catalysts.

The same complex catalyzes the enantioselective desymmetrization of 2-substituted 1,3-diaryl-1,3-diketones by reduction to the corresponding 3-hydroxypropanones in the presence of 1 equiv. of $NaBH_4$ [111], as well as the diastereospecific reduction of unsymmetrical 2-alkyl-1,3-diketones with high diastereo- and enantioselectivity [112].

Dynamic kinetic resolution was exploited in the BH_4^- reduction of 2-alkyl-substituted β-keto esters to the optically active anti-aldol derivatives [113], and of biaryl lactones to axially chiral biaryl compounds [114]. Impressively, the substrate scope also encompasses phosphinoyl imines (Scheme 9.27) [115]. These and further applications have been thoroughly reviewed [87].

Finally, Lu and coworkers has achieved the asymmetric hydroboration of aryl alkyl ketones with pinacolborane (HBpin) catalyzed by a $[CoCl_2(NNN)]$ complex (NNN is a chiral iminopyridine-oxazoline ligand of the class shown in Scheme 9.19) [116]. Yields were mostly close to or above 90%, and the enantioselectivity reached 99% ee (87% ee on a gram scale). To activate the precatalyst, $NaBHEt_3$ was used, but the nature of the active species was not investigated.

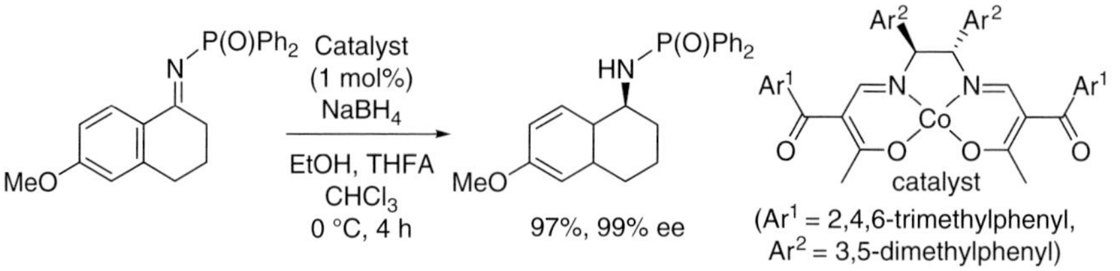

Scheme 9.27 Cobalt(II) β-ketoiminato catalysts for the BH_4^- reduction of phosphinoyl imines.

9.5 Nickel

Raney nickel [117, 118] or Ni nanoparticles [119] modified with chiral auxiliaries play a major role in the AH and ATH both of alkenes and of polar double bonds. In contrast, soluble nickel-based hydrogenation catalysts are very rare, which is surprising if one considers that, albeit still relatively weak, Ni—H bonds are stronger than Fe—H and Co—H ones [8], and hence, Ni(II) should give more stable hydrides than Fe(II) and Co.

9.5.1 Nickel Catalysts in Asymmetric H_2 Hydrogenation

In 2008, Hamada et al. reported homogeneous asymmetric nickel AH catalysts formed *in situ* from $Ni(OAc)_2$ and a chiral diphosphine (1 equiv.) that hydrogenated chirally labile α-amino-β-ketoester hydrochlorides under dynamic kinetic resolution conditions [120]. Josiphos-type ligands were most active and afforded yields of up to 98% and excellent diastereo- (anti/syn >99/1) and enantioselectivity (up to 95% ee) (Scheme 9.28). The same catalyst system hydrogenates α-aminoketone hydrochlorides to β-aminoalcohols with similar diastereo- and enantioselectivity [121].

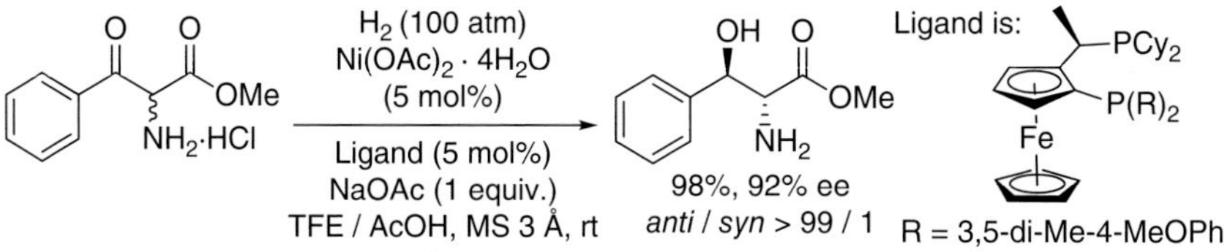

Scheme 9.28 Hamada's nickel(II) AH catalyst for chirally labile α-amino-β-keto esters.

These nickel catalysts are superior to their iridium analogues in terms of activity and enantioselectivity with sterically hindered and halogen-containing substrates, but a high catalyst loading (5–10 mol%) is required.

9.5.2 Nickel ATH Catalysts

A Ni complex generated *in situ* with a potentially pentadentate ligand with a N_2O_2P donor set catalyzes the ATH of ketones to the corresponding alcohols in high yield (up to 99%) and with enantioselectivity of up to 84% ee (Scheme 9.29) [97].

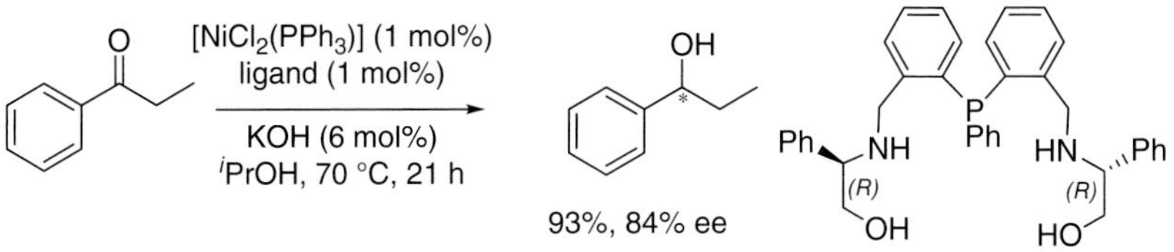

Scheme 9.29 Gao's nickel(II)/N_2O_2P catalyst for the ATH of ketones.

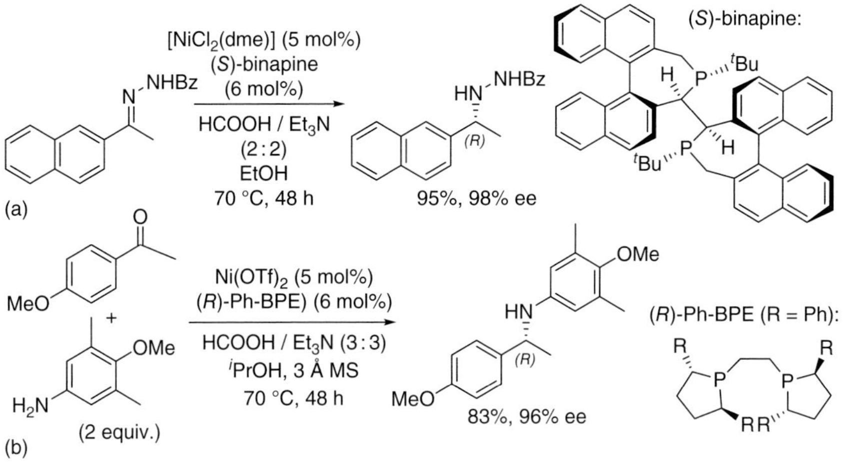

Scheme 9.30 Nickel(II)-catalyzed ATH of hydrazones and reductive amination of ketones.

As for C=N double bonds, Zhou and coworkers described a catalyst formed *in situ* from [$NiCl_2$(dme)] and electron-rich, bulky diphosphines for the ATH of hydrazones with formic acid/NEt_3 as hydrogen donor that gave TONs of 200, quantitative yield, and up to 98% ee (Scheme 9.30a) [122]. A 1,2-bis(phospholano)ethane-based catalyst hydrogenated benzosultams with more than 90% ee. Deuteration experiments indicated that the formyl H atom *H*COOH is donated as hydride, presumably after decarboxylation at nickel, to the imine carbon. Based on the above protocol, an *in situ* procedure has been developed, in which aryl alkyl ketones and arylamines or benzhydrizide are directly converted into the corresponding amine derivatives (Scheme 9.30b) [123].

9.5.3 Nickel AHS Catalysts

In 2012, Wu et al. reported a Ni(II)/Xyl-P-Phos catalytic system for the enantioselective hydrosilylation of electron-poor aryl alkyl ketones with $PhSiH_3$, which achieved good yields (up to 97%) and moderate-to-good enantioselectivity (up to 90% ee) (Scheme 9.31) [124]. Like its cobalt and copper analogues (see Sections 9.4.3 and 9.6.3), the catalyst was stable in air, and the addition of 4 Å molecular sieves dramatically enhanced both conversion (99%) and enantioselectivity (90% ee for *p*-nitroacetophenone). The use of NiF_2 instead of $Ni(OAc)_2 \cdot 4\ H_2O$ gave slightly lower enantioselectivity (see Section 9.6.3).

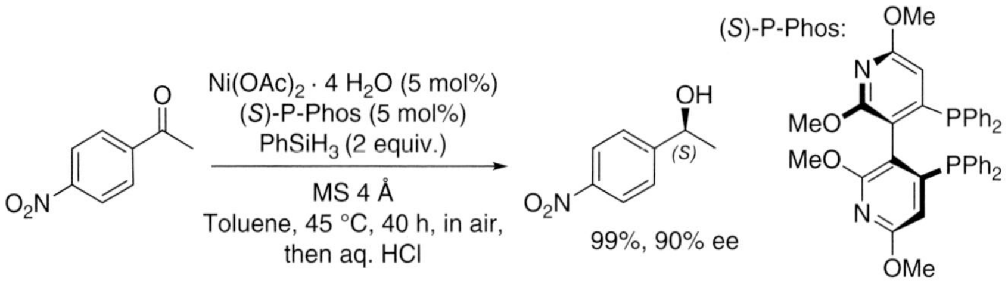

Scheme 9.31 Nickel(II)-catalyzed AHS of electron-poor aryl alkyl ketones.

9.5.4 Nickel-Catalyzed Asymmetric Borohydride Reduction

In 2014, Feng and coworkers reported a nickel-based Lewis acid catalyst for the asymmetric reduction of α-amino ketones with KBH_4 prepared *in situ* from $Ni(OTf)_2$ and a chiral *N,N'*-dioxide [125]. Electron-donating groups on the *N*-aryl substituent increased the enantioselectivity (Scheme 9.32). The β-amino alcohols, which are essential structural motifs of many natural products and drugs (such as β-blockers), were formed with up to 97% ee.

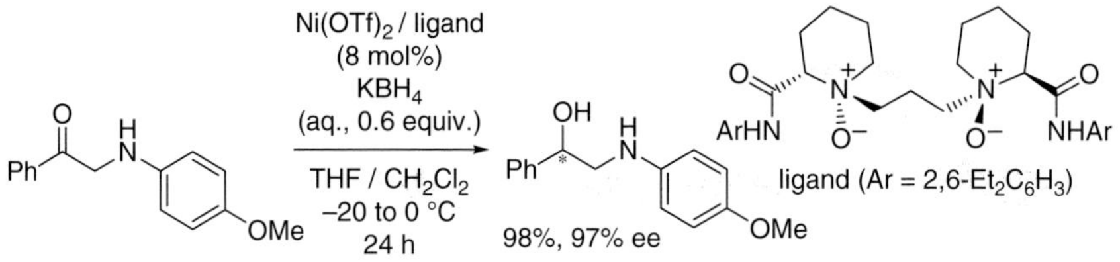

Scheme 9.32 Nickel(II)-catalyzed borohydride reduction of α-amino ketones.

9.5.5 Ni-Catalyzed Asymmetric Hydroboration of α,β-Unsaturated Ketones

As a straightforward approach to α-stereogenic allylic alcohols and convenient alternative to Corey–Bakshi–Shibata reduction or precious metal-catalyzed hydrogenation, Zhu and coworkers described an asymmetric 1,2-reduction of α,β-unsaturated ketones with pinacolborane catalyzed by $[Ni(COD)_2]$ modified with chiral *N,N*-bidentate ligands (Scheme 9.33) [126].

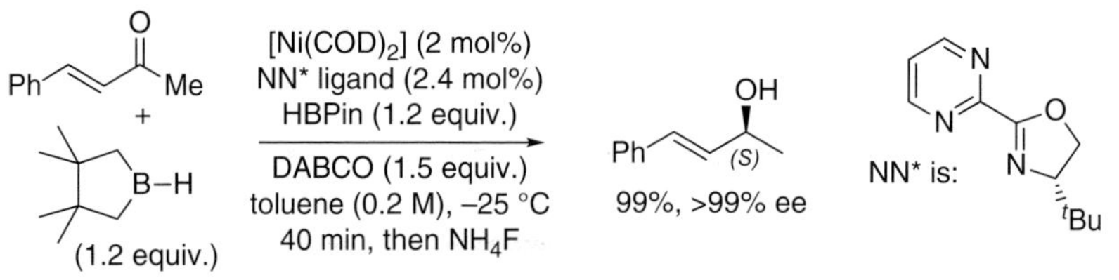

Scheme 9.33 Nickel-catalyzed asymmetric borane reduction of α,β-unsaturated ketones.

The addition of DABCO inhibited the background reaction (DABCO = 1,4-diazabicyclo[2.2.2]octane). The involvement of a Ni(II) hydride was suggested, but not proven, by analogy with other NiH catalysts [10].

9.6 Copper

In agreement with the increasing stability of the M—H bond toward the end of the 3d series [8], well-defined copper hydride complexes have been known since the early 1970s. Thus, Stryker's reagent $[CuH(PPh_3)]_6$ acts as a stoichiometric reducing agent in the 1,4-conjugate hydride addition onto α,β-unsaturated carbonyl compounds to give saturated ketones [127], and the reaction can be made catalytic with H_2 as hydrogen source [128]. With less bulky and more basic phosphines, and in particular PMe_2Ph, 1,2-reduction occurs [128, 129], showing the feasibility of the AH of ketones (see below). Still, AHS is by far the most developed copper-catalyzed reduction method (see Section 9.6.3) [130, 131].

9.6.1 Copper-Catalyzed AH

In 2007, Shimizu et al. reported the first example of copper-catalyzed AH of acetophenones with moderate-to-high enantioselectivity (up to 91% ee) and with high catalytic activity (S/C as high as 3000, but typically 500) [132]. A copper hydride is possibly generated from Cu(I) by heterolytic H_2 activation in the presence of phosphines and of NaO^tBu as the base [133]. The chiral diphosphine BDPP (2,4-bis(diphenylphosphino)pentane) gave the most efficient catalyst for ketones, and an excess of P(3,5-xylyl)$_3$ was pivotal to achieve high enantioselectivity (Scheme 9.34a). The substrate scope encompassed aryl and heteroaryl ketones as well as enones. Interestingly, enones gave either 1,2- or 1,4-reduction to the allylic alcohol or saturated ketone, respectively, depending on the chiral diphosphine used [134].

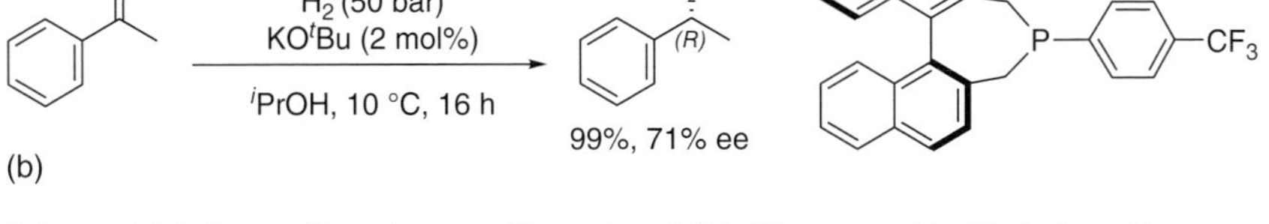

(a) [Cu(NO$_3$)(P(3,5-xylyl)$_3$)$_2$] (0.2 mol%), (*S*,*S*)-BDPP (0.2 mol%), P(3,5-xylyl)$_3$ (1.2 mol%); H_2 (50 bar), NaO^tBu (2.0 mol%), ^{i}PrOH, 30 °C, 16 h → 92%, 91% ee. (*S*,*S*)-BDPP

(b) [Cu(OAc)$_2$(L)$_2$] (0.3 mol%), H_2 (50 bar), KO^tBu (2 mol%); ^{i}PrOH, 10 °C, 16 h → 99%, 71% ee. L =

Scheme 9.34 Copper(I)- and copper(II)-catalyzed AH of ketones with chiral phosphines.

In 2011, Beller and coworkers reported a more general catalyst system based on monodentate phosphepine-type ligands (Scheme 9.34b). The isolated Cu(II) acetato complex $[Cu(OAc)_2(L)_2]$ reduced acetophenone with 71% ee in the presence of KO^tBu [135]. The highest enantioselectivity (89% ee) was achieved with *iso*-propyl phenyl ketone, and the catalyst requires hydrogen pressures in the range of Shimizu's system (50 bar). Starting from Shimizu's results, Hatcher and coworkers employed high-throughput screening of chiral diphosphines with the goal of disclosing a Cu-based catalyst that operates at lower H_2 pressures [136], which significantly increases the attractiveness of AH as compared to AHS (see Section 9.6.3). With electron-rich and electron-deficient acetophenones, as well as heteroaryl-substituted ketones, the combination of a chiral BoPhoz-type diphosphine with tris(3,5-xylyl)phosphine gave benzylic alcohols in good yields (65 to >95%) and enantioselectivity (up to 96% ee) under 20 bar H_2 (Scheme 9.35).

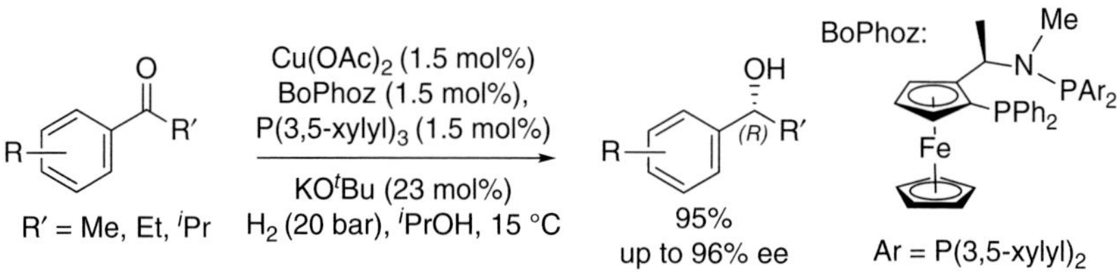

R′ = Me, Et, ^{i}Pr; $Cu(OAc)_2$ (1.5 mol%), BoPhoz (1.5 mol%), $P(3,5\text{-xylyl})_3$ (1.5 mol%); KO^tBu (23 mol%), H_2 (20 bar), ^{i}PrOH, 15 °C; (R); 95%, up to 96% ee; BoPhoz: Me, N, PAr_2, PPh_2, Fe; Ar = $P(3,5\text{-xylyl})_2$

Scheme 9.35 Johnson's copper-catalyzed AH of ketones.

Copper(II)/chiral diphosphine systems were used in the highly enantioselective hydrogenation of α-substituted ketones operating under dynamic kinetic resolution [137].

9.6.2 Copper-Catalyzed ATH of α-Ketoesters

The copper-catalyzed ATH of ketones with 2-propanol as hydrogen donor is not documented. List has shown that Hantzsch esters, which are powerful hydride donors, reduce α-ketoesters to the α-hydroxyesters in combination with chiral bisoxazoline Cu(II) complexes as Lewis acid catalysts (Scheme 9.36) [138].

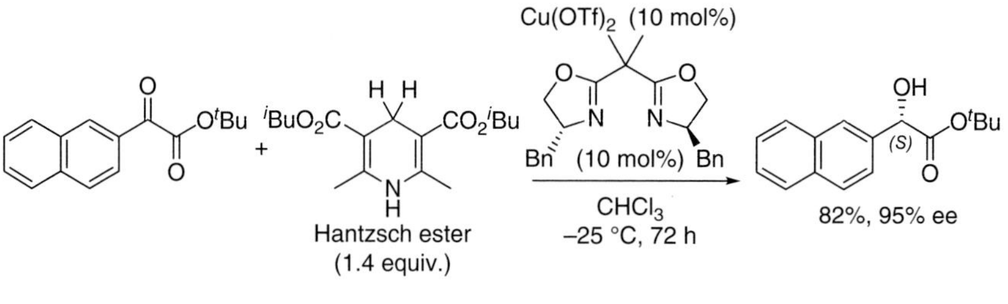

O^tBu + $^iBuO_2C$, $CO_2{}^iBu$, Hantzsch ester (1.4 equiv.); $Cu(OTf)_2$ (10 mol%), Bn (10 mol%) Bn; $CHCl_3$, −25 °C, 72 h; (S); 82%, 95% ee

Scheme 9.36 List's copper-catalyzed reduction of α-ketoesters with Hantzsch ester.

9.6.3 Copper-Catalyzed AHS of Ketones and Imines

Asymmetric hydrosilylation in general is by far the most developed copper-catalyzed reduction reaction [130, 131, 139] and is covered in this section only in its general lines. The AHS of ketones was pioneered by Brunner and Miehling, who reported in 1984 that H_2SiPh_2 hydrosilylated acetophenone in the presence of Cu(I) alkoxo complexes modified with chiral diphosphines, among which Norphos gave the highest enantioselectivity (39% ee; Norphos = trans-bis(diphenylphosphino)bicyclo[2.2.1]) [140]. A systematic study was triggered 15 years later by the discovery that the combination of Stryker's reagent with hydrosilanes as hydride source allows the 1,4-reduction of enones under mild conditions [130, 131, 139]. In 2001, Lipshutz adapted Buchwald's catalyst system for the 1,4-reduction of α,β-unsaturated esters [141] to the AHS of aryl ketones [142]. With PMHS as hydride source and a BIPHEP (2,2'-bis[di(aryl)phosphino]-6,6'-dimethoxy-1,1'-biphenyl) ligand, excellent yields and enantioselectivity were observed (Scheme 9.37a). This approach was extended to heteroaromatic ketones with a catalytic system based on CuCl/NaO^{t}Bu/(*R*)-DTBM-SEGPHOS (5,5′-bis[di(aryl)phosphino]-4,4′-bi-1,3-benzodioxole) [143]. Phosphinoyl benzimines undergo AHS with tetramethyldisiloxane (TMDS) as hydride donor under analogous conditions to give phosphinoyl amines (or the corresponding amines after deprotection with basic methanol) in quantitative yields and with excellent enantioselectivity (up to 99% ee) (Scheme 9.37b) [144].

Starting from this approach, a heterogenized, recyclable catalyst for the reduction of ketones, α,β-unsaturated ketones, esters, and lactones was developed by impregnation of a Cu(II) salt in a charcoal matrix, followed by reaction with the chiral ligand and NaOPh in the presence of PMHS [145].

Lipshutz et al. has observed that the structure of the silane strongly affects the reaction outcome of the AHS of ketones and suggested that the silane is an integral part of catalyst makeup [146]. Dagorne and coworkers have studied a similar system with $PhMeSiH_2$ as hydride donor [147]. As ketone reduction did not occur in the presence of a stoichiometric amount of [CuH(BINAP)], they proposed that a Cu(I)/silane adduct (either involving oxidative addition or not) may be competent for the insertion of the carbonyl group into the Cu—H bond.

A series of studies revealed the role that bases play in the activation of the silane. As strong bases are incompatible with certain substrates, this is a crucial point. Instead of a strong base, Riant and coworkers used BINAP-modified CuF_2 to catalyze the AHS of ketones with $PhSiH_3$ under aerobic and mild conditions.

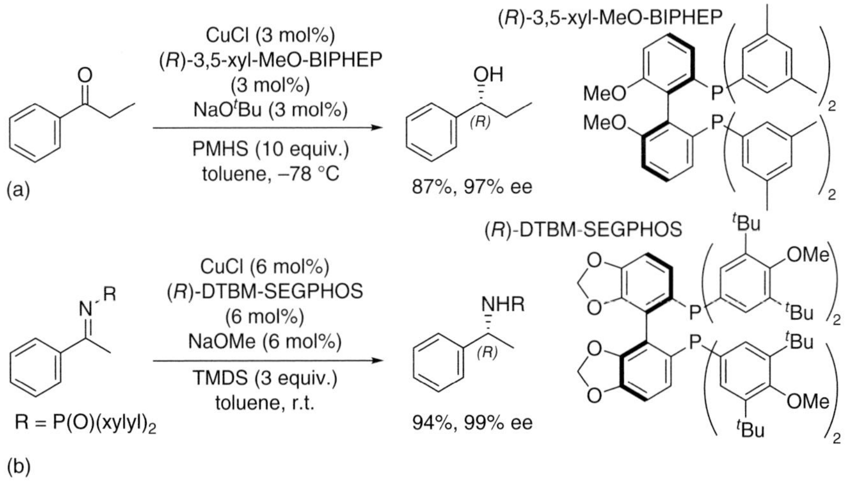

(a)
CuCl (3 mol%)
(*R*)-3,5-xyl-MeO-BIPHEP (3 mol%)
NaO^{t}Bu (3 mol%)
PMHS (10 equiv.)
toluene, –78 °C
87%, 97% ee
(*R*)-3,5-xyl-MeO-BIPHEP

(b)
R = P(O)(xylyl)$_2$
CuCl (6 mol%)
(*R*)-DTBM-SEGPHOS (6 mol%)
NaOMe (6 mol%)
TMDS (3 equiv.)
toluene, r.t.
94%, 99% ee
(*R*)-DTBM-SEGPHOS

Scheme 9.37 Lipshutz's copper-catalyzed AHS of ketones and phosphinoyl benzimines.

The catalyst loading was as low as 0.5 mol%, and acidic hydrolytic workup gave the corresponding secondary alcohols in good-to-quantitative yield and with up to 92% ee (Scheme 9.38) [148]. A similar system reduced unsymmetrical diaryl ketones with up to 98% ee [149].

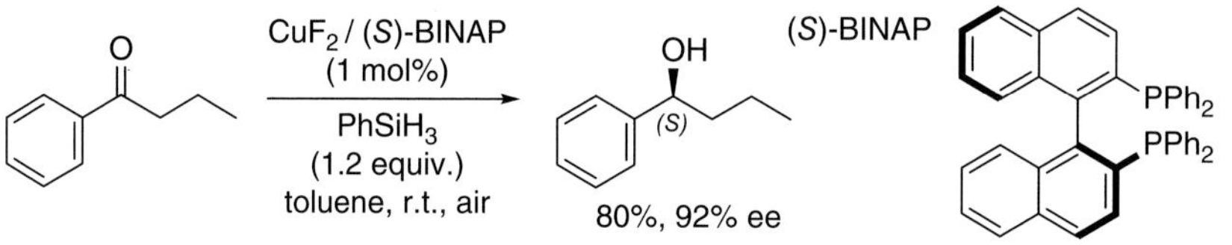

Scheme 9.38 Riant's CuF_2-catalyzed AHS of ketones under aerobic and mild conditions.

As a further advance, Beller and coworkers reported a strong base- and fluoride-free catalyst based on $Cu(OAc)_2$ and chiral monodentate phosphepine-derived ligands (which were later on used in AH, see Scheme 9.34) in combination with $PhSiH_3$ as hydride donor (Scheme 9.39) [150].

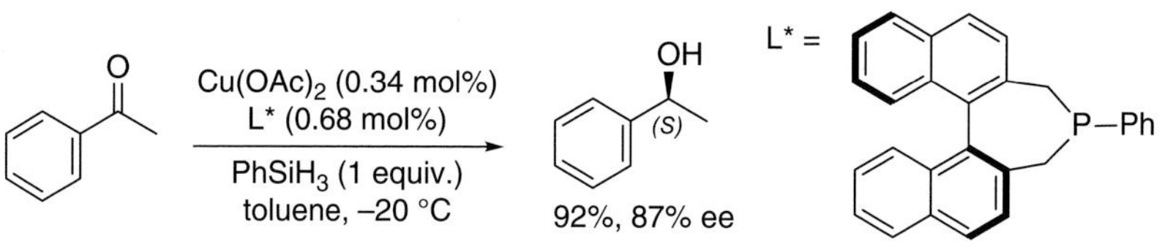

Scheme 9.39 Beller's Cu(OAc)/monodentate phosphine catalyst for the AHS of ketones.

The acetate is probably involved in silane activation, as suggested for the $Co(OAc)_2$-based AHS catalysts discussed above. Alternatively, $Cu(acac)_2$ can be used as a metal precursor in combination with PMHS [151]. Eventually, base-free $Cu(OAc)_2$/chiral diphosphine AHS catalysts were developed that reduce ketones containing haloalkyl [152], cycloalkyl [153], and heteroaryl groups [154], in the presence of $PhSiH_3$ with excellent enantioselectivity (up to 99% ee).

Finally, Cu(I) complexes of the type [CuCl(NHC)], where *N*-heterocyclic carbene (NHC) is a chiral monodentate Arduengo-type carbene ligand (Scheme 9.40), have been used in the copper-catalyzed AHS of linear aliphatic ketones such as 2-butanone at room temperature [155]. The system is very active and enantioselective but requires the addition of a base. The highest

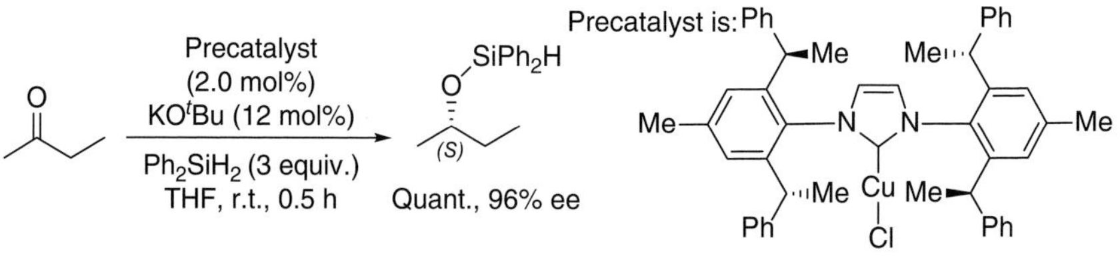

Scheme 9.40 Gawley's copper(I)/NCH catalyst for the AHS of ketones.

enantioselectivity (96% ee for 2-butanone) was achieved with disubstituted silanes R_2SiH_2 (R = Et or Ph).

It is generally accepted that a copper hydride complex (either used as such or formed *in situ* from the silane and the base or fluoride) promotes the insertion of the carbonyl group into the Cu—H bond, and the resulting copper alkoxide activates the Si—H bond by σ-bond metathesis. Kinetic studies have shown that, in the absence of a base (either oxygen-containing or fluoride), the carbonyl insertion and the successive σ-bond metathesis step feature similar kinetic constants [156]. Instead, in the presence of RO^- or fluoride, the silane is possibly converted into a more reactive five-coordinate hydrosilicate [7], with the effect that the first step, the insertion of the carbonyl group into the Cu—H bond, becomes rate determining [157]. Overall, these results suggest that the mechanism strongly depends on the structure of the silane and on the presence and nature of a base and may be the key to unravel the mechanistic complexity shown by AHS reactions in general.

9.7 Conclusion

The above discussion shows that, among the methods for the reduction of polar double bonds, asymmetric hydrosilylation has reached maturity, in particular with cobalt and copper catalysts. However, AHS is less atom-economic and mechanistically complex because of the combined effects of the structure of the silane and of the presence and nature of the base that is often necessary for its activation. In contrast, a better mechanistic understanding has been achieved for asymmetric transfer hydrogenation, in particular with Fe(II) catalysts. However, this technology is still cumbersome for industrial application because of the limited volume yield and the need for a solvent as hydrogen donor. Direct hydrogenation (AH), which is the reaction of choice for industry, remains the biggest challenge both because of the intrinsic low stability of 3d hydride complexes and because H_2 activation by base metals is still less efficient than with precious ones. However, results with iron, cobalt, and most recently manganese suggest that the judicious choice of the ligand may lead to applicable AH catalysts. In particular, phosphine ligands have been recognized as pivotal to stabilize hydride complexes. Also, chiral NHC ligands, which play an insignificant role with 3d metals presently, might be the source of future surprises.

References

1 Jacobsen, E.N., Pfaltz, A., and Imamoto, H. (eds.) (1999). *Comprehensive Asymmetric Catalysis*, vol. I, Chapter 6. Berlin: Springer-Verlag.
2 Blaser, H.U., Pugin, B., and Spindler, F. (2012). *Top. Organomet. Chem.* 42: 65.
3 Blaser, H.U., Spindler, F., and Studer, M. (2001). *Appl. Catal., A* 221: 119.
4 Wang, D. and Astruc, D. (2015). *Chem. Rev.* 115: 6621.

5 Blacker, A.J. and Thompson, P. (2010). *Asymmetric Catalysis on Industrial Scale: Challenges, Approaches and Solutions*, 2e (ed. H.U. Blaser and H.J. Federsel), 265. Weinheim: Wiley-VCH.
6 Gladiali, S. and Taras, R. (2008). Reduction of carbonyl compounds by hydrogen transfer. In: *Modern Reduction Methods* (ed. P.G. Andersson and I.J. Munslow), 135. Weinheim: Wiley-VCH.
7 Rendler, S. and Oestreich, M. (2008). *Modern Reduction Methods* (ed. P.G. Andersson and I.J. Munslow), 183. Weinheim: Wiley-VCH.
8 Uddin, J., Morales, C.M., Maynard, J.H., and Landis, C.R. (2006). *Organometallics* 25: 5566.
9 Chakraborty, S. and Guan, H.R. (2010). *Dalton Trans.* 39: 7427.
10 Chakraborty, S., Bhattacharya, P., Dai, H.G., and Guan, H.R. (2015). *Acc. Chem. Res.* 48: 1995.
11 Chuit, C., Corriu, R.J.P., Reye, C., and Young, J.C. (1993). *Chem. Rev.* 93: 1371.
12 Valyaev, D.A., Lavigne, G., and Lugan, N. (2016). *Coord. Chem. Rev.* 308: 191.
13 Zirakzadeh, A., de Aguira, S.R.M.M., Stöger, B. et al. (2017). *ChemCatChem* 9: 1744.
14 Widegren, M.B., Harkness, G.J., Slawin, A.M.Z. et al. (2017). *Angew. Chem. Int. Ed.* 56: 5825. (2017). *Angew. Chem.* 129: 5919.
15 Garbe, M., Junge, K., Walker, S. et al. (2017). *Angew. Chem. Int. Ed.* 56: 11237. (2017). *Angew. Chem.* 129: 11389.
16 Trovitch, R.J. (2017). *Acc. Chem. Res* 50: 2842.
17 Ma, X.C., Zuo, Z.Q., Liu, G.X., and Huang, Z. (2017). *ACS Omega* 2: 4688.
18 Vasilenko, V., Blasius, C.K., Wadepohl, H., and Gade, L.H. (2017). *Angew. Chem. Int. Ed.* 56: 8393. (2017). *Angew. Chem.* 129: 8513.
19 Bauer, I. and Knölker, H.J. (2015). *Chem. Rev.* 115: 3170.
20 Darcel, C. and Sortais, J.B. (2015). *Top. Organomet. Chem.* 50: 173.
21 Ollevier, T. and Keipour, H. (2015). *Top. Organomet. Chem.* 50: 259.
22 Darwish, M. and Wills, M. (2012). *Catal. Sci. Technol.* 2: 243.
23 Ollevier, T. (2016). *Catal. Sci. Technol.* 6: 41.
24 Castro, L.C.M., Li, H.Q., Sortais, J.B., and Darcel, C. (2015). *Green Chem.* 17: 2283.
25 Nakazawa, H. and Itazaki, M. (2011). *Top. Organomet. Chem.* 33: 27.
26 Chen, J.S., Chen, L.L., Xing, Y. et al. (2004). *Acta Chim. Sin.* 62: 1745.
27 Jothimony, K., Vancheesan, S., and Kuriacose, J.C. (1985). *J. Mol. Catal.* 32: 11.
28 Zhou, S.L., Fleischer, S., Junge, K. et al. (2010). *Angew. Chem. Int. Ed.* 49: 8121. (2010). *Angew. Chem.* 122: 8298.
29 Sui-Seng, C., Freutel, F., Lough, A.J., and Morris, R.H. (2008). *Angew. Chem. Int. Ed.* 47: 940. (2008). *Angew. Chem.* 120: 954.
30 Gao, J.X., Ikariya, T., and Noyori, R. (1996). *Organometallics* 15: 1087.
31 Prokopchuk, D.E., Sonnenberg, J.F., Meyer, N. et al. (2012). *Organometallics* 31: 3056.
32 Sonnenberg, J.F., Coombs, N., Dube, P.A., and Morris, R.H. (2012). *J. Am. Chem. Soc.* 134: 5893.

33 Mikhailine, A.A., Kim, E., Dingels, C. et al. (2008). *Inorg. Chem.* 47: 6587.
34 Matt, D., Ziessel, R., De Cian, A., and Fischer, J. (1996). *New J. Chem.* 20: 1257.
35 Mikhailine, A.A., Lough, A.J., and Morris, R.H. (2009). *J. Am. Chem. Soc.* 131: 1394.
36 Noyori, R. and Ohkuma, T. (2001). *Angew. Chem. Int. Ed.* 40: 40. (2001). *Angew. Chem.* 113: 41.
37 Mikhailine, A.A., Maishan, M.I., and Morris, R.H. (2012). *Org. Lett.* 14: 4638.
38 Mikhailine, A.A. and Morris, R.H. (2010). *Inorg. Chem.* 49: 11039.
39 Lagaditis, P.O., Lough, A.J., and Morris, R.H. (2010). *Inorg. Chem.* 49: 10057.
40 Sues, P.E., Lough, A.J., and Morris, R.H. (2011). *Organometallics* 30: 4418.
41 Lagaditis, P.O., Lough, A.J., and Morris, R.H. (2011). *J. Am. Chem. Soc.* 133: 9662.
42 Mikhailine, A.A., Maishan, M.I., Lough, A.J., and Morris, R.H. (2012). *J. Am. Chem. Soc.* 134: 12266.
43 Prokopchuk, D.E. and Morris, R.H. (2012). *Organometallics* 31: 7375.
44 Ikariya, T. (2011). *Bull. Chem. Soc. Jpn.* 84: 1.
45 Zuo, W.W., Prokopchuk, D.E., Lough, A.J., and Morris, R.H. (2016). *ACS Catal.* 6: 301.
46 Zuo, W.W., Lough, A.J., Li, Y.F., and Morris, R.H. (2013). *Science* 342: 1080.
47 Smith, S.A.M. and Morris, R.H. (2015). *Synthesis* 47: 1775.
48 Zuo, W.W., Tauer, S., Prokopchuk, D.E., and Morris, R.H. (2014). *Organometallics* 33: 5791.
49 Naik, A., Maji, T., and Reiser, O. (2010). *Chem. Commun.* 46: 4475.
50 Hopewell, J.P., Martins, J.E.D., Johnson, T.C. et al. (2012). *Org. Biomol. Chem.* 10: 134.
51 Yu, S.L., Shen, W.Y., Li, Y.Y. et al. (2012). *Adv. Synth. Catal.* 354: 818.
52 Jugé, S., Stephan, M., Merdès, R. et al. (1993). *J. Chem. Soc., Chem. Commun.* 531.
53 Bigler, R., Otth, E., and Mezzetti, A. (2014). *Organometallics* 33: 4086.
54 Chen, T.Q. and Han, L.B. (2015). *Synlett* 26: 1153.
55 Bigler, R. and Mezzetti, A. (2016). *Org. Process Res. Dev.* 20: 253.
56 Bigler, R. and Mezzetti, A. (2014). *Org. Lett.* 16: 6460.
57 Bigler, R., Huber, R., and Mezzetti, A. (2015). *Angew. Chem. Int. Ed.* 54: 5171. (2015). *Angew. Chem.* 127: 5260.
58 De Luca, L. and Mezzetti, A. (2017). *Angew. Chem. Int. Ed.* 56: 11949. (2017). *Angew. Chem.* 129: 12111.
59 Knölker, H.J., Baum, E., Goesmann, H., and Klauss, R. (1999). *Angew. Chem. Int. Ed.* 38: 2064. (1999). *Angew. Chem.* 111: 2196.
60 Shvo, Y., Czarkie, D., Rahamim, Y., and Chodosh, D.F. (1986). *J. Am. Chem. Soc.* 108: 7400.
61 Casey, C.P. and Guan, H.R. (2007). *J. Am. Chem. Soc.* 129: 5816.
62 Berkessel, A., Reichau, S., von der Höh, A. et al. (2011). *Organometallics* 30: 3880.
63 Gajewski, P., Renom-Carrasco, M., Vailati Facchini, S. et al. (2015). *Eur. J. Org. Chem.* 5526.

64 Zhou, S.L., Fleischer, S., Junge, K., and Beller, M. (2011). *Angew. Chem. Int. Ed.* 50: 5120. (2011). *Angew. Chem.* 123: 5226.
65 Fleischer, S., Zhou, S.L., Werkmeister, S. et al. (2013). *Chem. Eur. J.* 19: 4997.
66 Zhou, S.L., Fleischer, S., Jiao, H.J. et al. (2014). *Adv. Synth. Catal.* 356: 3451.
67 Fleischer, S., Werkmeister, S., Zhou, S.L. et al. (2012). *Chem. Eur. J.* 18: 9005.
68 Hopmann, K.H. (2015). *Chem. Eur. J.* 21: 10020.
69 Li, Y.Y., Yu, S.L., Wu, X.F. et al. (2014). *J. Am. Chem. Soc.* 136: 4031.
70 Lagaditis, P.O., Sues, P.E., Sonnenberg, J.F. et al. (2014). *J. Am. Chem. Soc.* 136: 1367.
71 Smith, S.A.M., Lagaditis, P.O., Lüpke, A. et al. (2017). *Chem. Eur. J.* 23: 7212.
72 Zirakzadeh, A., Kirchner, K., Roller, A. et al. (2016). *Organometallics* 35: 3781.
73 Zell, T. and Milstein, D. (2015). *Acc. Chem. Res.* 48: 1979.
74 Bornschein, C., Werkmeister, S., Wendt, B. et al. (2014). *Nat. Commun.* 5: 4111.
75 Bonitatibus, P.J., Chakraborty, S., Doherty, M.D. et al. (2015). *Proc. Natl. Acad. Sci. U.S.A.* 112: 1687.
76 Nishiyama, H. and Furuta, A. (2007). *Chem. Commun.* 760.
77 Inagaki, T., Phong, L.T., Furuta, A. et al. (2010). *Chem. Eur. J.* 16: 3090.
78 Inagaki, T., Ito, A., Ito, J., and Nishiyama, H. (2010). *Angew. Chem. Int. Ed.* 49: 9384. (2010). *Angew. Chem.* 122: 9574.
79 Hosokawa, S., Ito, J., and Nishiyama, H. (2010). *Organometallics* 29: 5773.
80 Shaikh, N.S., Enthaler, S., Junge, K., and Beller, M. (2008). *Angew. Chem. Int. Ed.* 47: 2497. (2008). *Angew. Chem.* 120: 2531.
81 Addis, D., Shaikh, N., Zhou, S.L. et al. (2010). *Chem. Asian J.* 5: 1687.
82 Langlotz, B.K., Wadepohl, H., and Gade, L.H. (2008). *Angew. Chem. Int. Ed.* 47: 4670. (2008). *Angew. Chem.* 120: 4748.
83 Tondreau, A.M., Darmon, J.M., Wile, B.M. et al. (2009). *Organometallics* 28: 3928.
84 Bleith, T., Wadepohl, H., and Gade, L.H. (2015). *J. Am. Chem. Soc.* 137: 2456.
85 Bleith, T. and Gade, L.H. (2016). *J. Am. Chem. Soc.* 138: 4972.
86 Zuo, Z.Q., Zhang, L., Leng, X.B., and Huang, Z. (2015). *Chem. Commun.* 51: 5073.
87 Pellissier, H. and Clavier, H. (2014). *Chem. Rev.* 114: 2775.
88 Ohgo, Y., Takeuchi, S., and Yoshimura, J. (1971). *Bull. Chem. Soc. Jpn.* 44: 583.
89 Ohgo, Y., Natori, Y., Takeuchi, S., and Yoshimura, J. (1974). *Chem. Lett.* 1327.
90 Waldron, R.W. and Weber, J.H. (1977). *Inorg. Chem.* 16: 1220.
91 Doyle, A.G. and Jacobsen, E.N. (2007). *Chem. Rev.* 107: 5713.
92 Le Maux, P., Massonneau, V., and Simonneaux, G. (1985). *J. Organomet. Chem.* 284: 101.
93 Friedfeld, M.R., Shevlin, M., Margulieux, G.W. et al. (2016). *J. Am. Chem. Soc.* 138: 3314.
94 Rösler, S., Obenauf, J., and Kempe, R. (2015). *J. Am. Chem. Soc.* 137: 7998. and Refs [5, 7] therein.
95 Zhang, D., Zhu, E.Z., Lin, Z.W. et al. (2016). *Asian J. Org. Chem.* 5: 1323.

96 ter Halle, R., Bréhéret, A., Schulz, E. et al. (1997). *Tetrahedron: Asymmetry* 8: 2101.
97 Li, Y.Y., Yu, S.L., Shen, W.Y., and Gao, J.X. (2015). *Acc. Chem. Res.* 48: 2587.
98 Brunner, H. and Amberger, K. (1991). *J. Organomet. Chem.* 417: C63.
99 Yu, F., Zhang, X.C., Wu, F.F. et al. (2011). *Org. Biomol. Chem.* 9: 5652.
100 Sauer, D.C., Wadepohl, H., and Gade, L.H. (2012). *Inorg. Chem.* 51: 12948.
101 Lu, X. and Lu, Z. (2016). *Org. Lett.* 18: 4658.
102 Leutenegger, U., Madin, A., and Pfaltz, A. (1989). *Angew. Chem. Int. Ed. Engl.* 28: 60. (1989). *Angew. Chem.* 101: 61.
103 Nagata, T., Yorozu, K., Yamada, T., and Mukaiyama, T. (1995). *Angew. Chem. Int. Ed. Engl.* 34: 2145. (1995). *Angew. Chem.* 107: 2309.
104 Sugi, K.D., Nagata, T., Yamada, T., and Mukaiyama, T. (1996). *Chem. Lett.* 737.
105 Yamada, T., Nagata, T., Ikeno, T. et al. (1999). *Inorg. Chim. Acta* 296: 86.
106 Iwakura, I., Hatanaka, M., Kokura, A. et al. (2006). *Chem. Asian J.* 1: 656.
107 Levitin, I.Y., Dvolaitzky, M., and Vol'pin, M.E. (1971). *J. Organomet. Chem.* 31: C37.
108 Tsubo, T., Chen, H.H., Yokomori, M. et al. (2012). *Chem. Lett.* 41: 780.
109 Ikeno, T., Iwakura, I., Shibahara, A. et al. (2007). *Chem. Lett.* 36: 738.
110 Ohtsuka, Y., Kubota, T., Ikeno, T. et al. (2000). *Synlett* 4: 535.
111 Ohtsuka, Y., Koyasu, K., Ikeno, T., and Yamada, T. (2001). *Org. Lett.* 3: 2543.
112 Ohtsuka, Y., Koyasu, K., Miyazaki, D. et al. (2001). *Org. Lett.* 3: 3421.
113 Ohtsuka, Y., Miyazaki, D., Ikeno, T., and Yamada, T. (2002). *Chem. Lett.* 31: 24.
114 Ashizawa, T., Tanaka, S., and Yamada, T. (2008). *Org. Lett.* 10: 2521.
115 Yamada, T., Nagata, T., Sugi, K.D. et al. (2003). *Chem. Eur. J.* 9: 4485.
116 Guo, J., Chen, J.H., and Lu, Z. (2015). *Chem. Commun.* 51: 5725.
117 Tai, A. and Sugimura, T. (2015). *Chiral Catalyst Immobilization and Recycling* (ed. D.E. De Vos, I.F.J. Vankelecom and P.A. Jacobs), 173. Weinheim: Wiley-VCH.
118 Osawa, T., Harada, T., and Takayasu, O. (2000). *Top. Catal.* 13: 155.
119 Alonso, F., Riente, P., and Yus, M. (2011). *Acc. Chem. Res.* 44: 379.
120 Hamada, Y., Koseki, Y., Fujii, T. et al. (2008). *Chem. Commun.* 6206.
121 Hibino, T., Makino, K., Sugiyama, T., and Hamada, Y. (2009). *ChemCatChem* 1: 237.
122 Xu, H.Y., Yang, P., Chuanprasit, P. et al. (2015). *Angew. Chem. Int. Ed.* 54: 5112. (2015). *Angew. Chem.* 127: 5201.
123 Yang, P., Lim, L.H., Chuanprasit, P. et al. (2016). *Angew. Chem. Int. Ed.* 55: 12083. (2016). *Angew. Chem.* 128: 12262.
124 Wu, F.F., Zhou, J.N., Fang, Q. et al. (2012). *Chem. Asian J.* 7: 2527.
125 He, P., Zheng, H.F., Liu, X.H. et al. (2014). *Chem. Eur. J.* 20: 13482.
126 Chen, F.L., Zhang, Y., Yu, L., and Zhu, S.L. (2017). *Angew. Chem. Int. Ed.* 56: 2022. (2017). *Angew. Chem.* 129: 2054. For alternative methods, see Refs [3–5] therein.
127 Mahoney, W.S., Brestensky, D.M., and Stryker, J.M. (1988). *J. Am. Chem. Soc.* 110: 291.
128 Mahoney, W.S. and Stryker, J.M. (1989). *J. Am. Chem. Soc.* 111: 8818.

129 Chen, J.X., Daeuble, J.F., and Stryker, J.M. (2000). *Tetrahedron* 56: 2789.
130 Deutsch, C., Krause, N., and Lipshutz, B.H. (2008). *Chem. Rev.* 108: 2916.
131 Lipshutz, B.H. (2014). *Copper Catalyzed Asymmetric Synthesis* (ed. A. Alexakis, N. Krause and S. Woodward), 179. Weinheim: Wiley-VCH.
132 Shimizu, H., Igarashi, D., Kuriyama, W. et al. (2007). *Org. Lett.* 9: 1655.
133 Goeden, G.V. and Caulton, K.G. (1981). *J. Am. Chem. Soc.* 103: 7354.
134 Shimizu, H., Nagano, T., Sayo, N. et al. (2009). *Synlett* 3143.
135 Junge, K., Wendt, B., Addis, D. et al. (2011). *Chem. Eur. J.* 17: 101.
136 Krabbe, S.W., Hatcher, M.A., Bowman, R.K. et al. (2013). *Org. Lett.* 15: 4560.
137 Zatolochnaya, O.V., Rodriguez, S., Zhang, Y.D. et al. (2018). *Chem. Sci.* 9: 4505.
138 Yang, J.W. and List, B. (2006). *Org. Lett.* 8: 5653.
139 Riant, O., Mostefaï, N., and Courmarcel, J. (2004). *Synthesis* 2943.
140 Brunner, H. and Miehling, W. (1984). *J. Organomet. Chem.* 275: C17.
141 Appella, D.H., Moritani, Y., Shintani, R. et al. (1999). *J. Am. Chem. Soc.* 121: 9473.
142 Lipshutz, B.H., Noson, K., and Chrisman, W. (2001). *J. Am. Chem. Soc.* 123: 12917.
143 Lipshutz, B.H., Lower, A., and Noson, K. (2002). *Org. Lett.* 4: 4045.
144 Lipshutz, B.H. and Shimizu, H. (2004). *Angew. Chem. Int. Ed.* 43: 2228. (2004). *Angew. Chem.* 116: 2277.
145 Lipshutz, B.H., Frieman, B.A., and Tomaso, A.E. (2006). *Angew. Chem. Int. Ed.* 45: 1259. (2006). *Angew. Chem.* 118: 1281.
146 Lipshutz, B.H., Noson, K., Chrisman, W., and Lower, A. (2003). *J. Am. Chem. Soc.* 125: 8779.
147 Issenhuth, J.T., Dagorne, S., and Bellemin-Laponnaz, S. (2006). *Adv. Synth. Catal.* 348: 1991.
148 Sirol, S., Courmarcel, J., Mostefai, N., and Riant, O. (2001). *Org. Lett.* 3: 4111.
149 Wu, J., Ji, J.X., and Chan, A.S.C. (2005). *Proc. Natl. Acad. Sci. U.S.A.* 102: 3570.
150 Junge, K., Wendt, B., Addis, A. et al. (2010). *Chem. Eur. J.* 16: 68.
151 Zhang, X.C., Wu, F.F., Li, S. et al. (2011). *Adv. Synth. Catal.* 353: 1457.
152 Yu, F., Zhou, J.N., Zhang, X.C. et al. (2011). *Chem. Eur. J.* 17: 14234.
153 Qi, S.B., Li, M., Li, S. et al. (2013). *Org. Biomol. Chem.* 11: 929.
154 Zhou, J.N., Fang, Q., Hu, Y.H. et al. (2014). *Org. Biomol. Chem.* 12: 1009.
155 Albright, A. and Gawley, R.E. (2011). *J. Am. Chem. Soc.* 133: 19680.
156 Vergote, T., Gharbi, S., Billard, F. et al. (2013). *J. Organomet. Chem.* 745–746: 133.
157 Vergote, T., Nahra, F., Merschaert, A. et al. (2014). *Organometallics* 33: 1953.

10

Iron-, Cobalt-, and Manganese-Catalyzed Hydrosilylation of Carbonyl Compounds and Carbon Dioxide

Christophe Darcel, Jean-Baptiste Sortais, Duo Wei, and Antoine Bruneau-Voisine

UMR 6226 CNRS-Université de Rennes, Centre for Catalysis and Green Chemistry, Institut des Sciences Chimiques de Rennes, Organometallics: Materials and Catalysis, Campus de Beaulieu, 263 Avenue du Général Leclerc, 35042 Rennes Cedex, France

10.1 Introduction

Since the beginning of the twenty-first century, there is a huge increase in the use of earth-abundant transition metals such as iron, manganese, or cobalt as powerful alternative catalysts to classical precious ones such as rhodium, palladium, or platinum in transformations for applied chemistry [1]. Indeed, considering the current important concerns about climate changes and the associated green chemistry principles, the substitution of these noble transition metals by more benign ones, such as the first-row transition metals, is highly desirable and is without any doubt one of the important challenges of the twenty-first century. On the other hand, the predicted twilight of the era of oil prompted the scientific community to find alternative methodologies to the usual ones, mainly by transforming highly oxygenated biomass into less functionalized chemical synthons similar to those produced from oil. Thus, new, efficient, and chemoselective reduction processes have to be found for both fine and bulk chemical transformations. Even if iron, manganese, and cobalt are valuable alternatives in such transformations, it is surprising that until recently, there were only scarce examples of large-scale applications [1, 2].

In this chapter, the fast and impressive improvement of numerous iron, manganese, and cobalt-catalyzed selective reductions of carbonyl and carboxylic derivatives via hydrosilylation will be reviewed, including recent transformations of inexpensive CO_2 as a C_1 building block.

10.2 Hydrosilylation of Aldehydes and Ketones

Chemo- and regioselective reduction reactions of carbonyl derivatives are some of the most relevant transformations in both bulk and fine chemistry, and transition metals such as ruthenium, iridium, and rhodium have dominated this area of research for decades [3]. For chemoselectivity issues,

Non-Noble Metal Catalysis: Molecular Approaches and Reactions,
First Edition. Edited by Robertus J. M. Klein Gebbink and Marc-Etienne Moret.

hydrosilylation is always a powerful alternative in reduction areas, particularly with inexpensive sources such as PMHS (polymethylhydrosiloxane) and TMDS (1,1,3,3-tetramethyldisiloxane).

10.2.1 Iron-Catalyzed Hydrosilylation

Iron-catalyzed hydrosilylations were intensively investigated during the past decade, and a selection of the most representative contributions will be reported herein. Indeed, the very first example of iron-catalyzed ketone hydrosilylation was described by Brunner and Fish in 1990, using the well-defined [Fe(Cp)(CO)(X)(L)] complexes **1** (X = Me, COMe; L = $Ph_2P(N(Me)CH(Me)Ph)$; 0.5–1 mol%), which catalyzes the reaction of acetophenone with 1 equiv. of diphenylsilane at 50–80 °C for 24 hours, quantitatively leading to the silylated ether Ph—CH(Me)($OSiHPh_2$) [4]. It must be underlined that no silylated enol ether is formed under such conditions. It was only two decades later that Beller and Nishiyama reported the first general iron catalytic hydrosilylation of carbonyl compounds. Thus, using an *in situ*-generated catalyst from $Fe(OAc)_2$ (5 mol%) and PCy_3 (10 mol%), functionalized aldehydes [5], and ketones [6] are efficiently reduced in the presence of 3 equiv. of PMHS in tetrahydrofuran (THF) at 65 °C for 16–20 hours (56 examples, 60–99% yields). Noticeably, ester, amino and cyano, and α,β-C=C unsaturated functional groups are tolerated. Interestingly, by the association of PCy_3 (1.1 mol%) with the air- and moisture-stable complex $[Bu_4N][Fe(CO)_3(NO)]$ (**2**) (1–2.5 mol%) [7], Plietker described a more active catalytic system for the hydrosilylation of various functionalized aldehydes and ketones using PMHS (aldehydes, 65–99%; ketones, 92–99%) at 30–50 °C for 14 hours [8]. *N,N,N′,N′*-tetramethylethylene-diamine (TMEDA, 10 mol%) [9] and sodium thiophenecarboxylate (**3**) (10 mol%; Figure 10.1) can also be used in association with $Fe(OAc)_2$ (5 mol%) to perform the hydrosilylation of ketones in the presence of 2 equiv. of $(EtO)_2MeSiH$ at 65 °C for 20–24 hours [10]. Furthermore, N-1-alkylated 2-(pyrazol-3-yl)pyridines such as **4** in combination with iron octanoate $[Fe(O_2C_8H_{15})_2]$ are also suitable catalyst precursors for the hydrosilylation of aldehydes and ketones [11]. Asymmetric reduction of ketones can also be performed using various chiral ligands. The chiral diphosphine (*S,S*)-Me-Duphos (**5**; Figure 10.1) in association with $Fe(OAc)_2$, in the presence of stoichiometric amounts of $(EtO)_2MeSiH$ or PMHS, catalyzes the enantioselective reduction of aromatic ketones at room temperature (rt.) (or 65 °C), and the corresponding alcohols can be obtained with yields

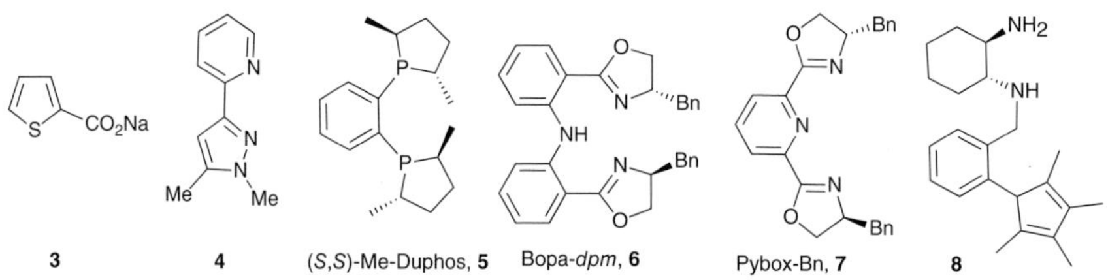

3 **4** (*S,S*)-Me-Duphos, **5** Bopa-*dpm*, **6** Pybox-Bn, **7** **8**

Figure 10.1 Ligands for iron-catalyzed hydrosilylation of ketones.

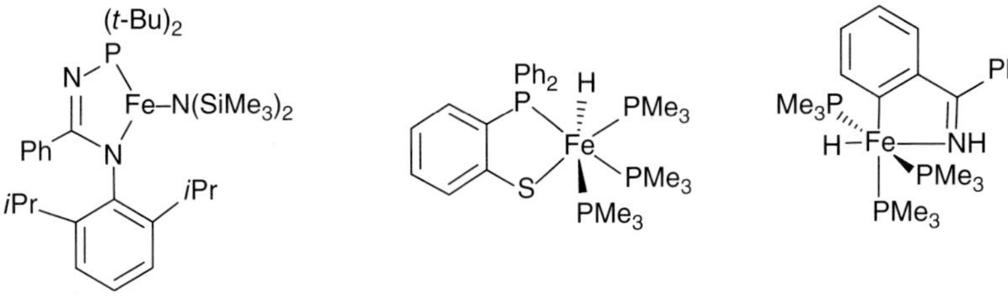

Figure 10.2 Well-defined iron complexes for catalyzed hydrosilylation of ketones and aldehydes.

and ee's up to 99% [12]. Chiral nitrogen ligands can also be used efficiently: *N,N,N*-bis(oxazolinylphenyl)-(BOPA) ligands (e.g. Bopa-dpm, **6**) (3 mol%) in combination with $Fe(OAc)_2$ (2 mol%) in the presence of $(EtO)_2MeSiH$ provides 88–99% yields and 50–88% ee's [13]. Notably, other chiral ligands such as Pybox-Bn (**7**) [14] or a cyclopentadienyl-bearing chiral diamine such as **8** [15] can also be used, resulting in ee's up to 79% and 37%, respectively (Figure 10.1).

Interestingly, well-defined iron complexes can also be efficient for the hydrosilylation reactions. Tilley reported the hydrosilylation of carbonyl derivatives using the highly air-sensitive, low-valent iron silylamide catalyst $[Fe(N(SiMe_3)_2)_2]$ (**9**) (0.01–2.7 mol%), in the presence of 1.6 equiv. of diphenylsilane at 23 °C for 0.3–20 hours with turnover frequencies (TOFs) up to 2400 h^{-1} for the reduction of 3-pentanone [16]. The related Fe(II) bis-(trimethylsilyl)amido complexes **10** with a coordinated *N*-phosphinoamidate ligand (0.015–1 mol%; Figure 10.2) can catalyze the hydrosilylation of a large range of aldehydes and ketones at rt. in the presence of 1 equiv. of phenylsilane with enhanced TOFs up to 23 600 h^{-1} for the reduction of acetophenone (vs 1266 h^{-1} with **9**). In comparison, the hydrido iron complex **11** bearing a P,S chelating ligand exhibited lower activities for the hydrosilylation of both aldehydes and ketones at 50 °C for two hours in the presence of 1.2 equiv. of $(EtO)_3SiH$ in THF (2 mol%, 5–95% isolated yields, TOFs up to 25 h^{-1}) [17]. Furthermore, cyclometallated imino iron hydrido complex **12** (0.3–0.6 mol%) was applied in hydrosilylation in the presence of 1.2 equiv. of $(EtO)_3SiH$ at 55 °C for 1–12 hours for aldehydes and ketones (65–92% yields) [18].

Pincer-type iron complexes were also used for the hydrosilylation of carbonyl compounds (Figure 10.3). In 2008, Chirik reported that bis(imino)pyridine (PDI) iron complexes such as $[Fe(iPrPDI)(N_2)_2]$ (**13**), which was already used successfully in the hydrosilylation of alkenes, is also efficient in the hydrosilylation of *p*-tolualdehyde and acetophenone (Ph_2SiH_2, 23 °C, one hour) [19]. Notably, under similar conditions, the iron dialkyl complex **14** (0.1 mol%) permits the reduction of a large variety of ketones including cyclohexenones, which chemoselectively lead to the corresponding cyclohexenols. Guan and coworkers reported the use of diphosphinite pincer iron hydride complexes such as **15** as catalysts (1 mol%) for the hydrosilylation of aromatic and aliphatic aldehydes (80–92% yields at 50–65 °C for 1–36 hours) and aromatic ketones (up to 88% yields at 50–80 °C for 4.5–48 hours) in the presence of 1.1 equiv. of $(EtO)_3SiH$

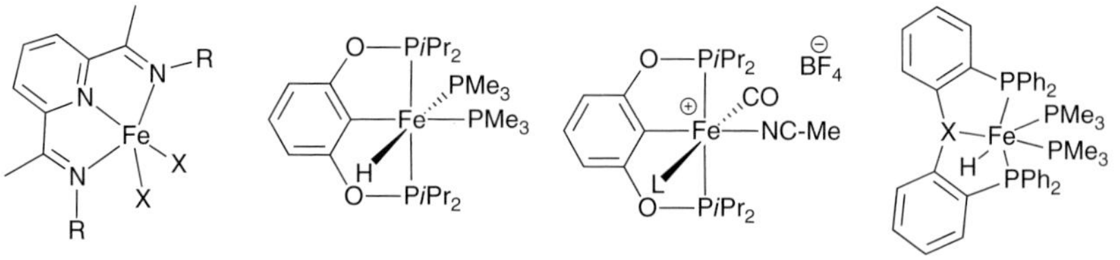

R = 2,6-*i*Pr$_2$C$_6$H$_3$, X = N$_2$ **13**
R = cyclohexyl, X = CH$_2$TMS **14**

15

L = PMe$_3$, **16**
L = CO, **17**

X = SiMe, **18**
X = CH, **19**

Figure 10.3 Well-defined iron pincer complexes for catalyzed hydrosilylation of ketones and aldehydes.

[20]. Similar activities are obtained with cationic pincer iron complexes **16**, **17** [21]. Similarly, tridentate PSiP [22] and PCP [23] ligand-based iron pincer complexes **18, 19** can be used under mild conditions in hydrosilylation. Thus, using 1 mol% of **18** and 1.5 equiv. of $(EtO)_3SiH$, the reduction of aldehydes and ketones can be performed at 60 °C in one and in six hours, respectively. Slightly higher activities are obtained with the complex **19** at 50 °C (aldehydes: 0.3–1 mol% **19**, 1–13 hours; ketones: 1 mol% **19**, 16 hours).

Interestingly, well-defined chiral iron complexes can be efficiently used for asymmetric hydrosilylation of ketones. In 2009, Chirik has shown that chiral tridentate (*S*,*S*)-(^{iPr}pybox)-Fe$(CH_2SiMe_3)_2$ complex **20** (0.3 mol%; Figure 10.4) can catalyze the asymmetric hydrosilylation of ketones in the presence of 2 equiv. of $PhSiH_3$ and 0.95 equiv. of $B(C_6F_5)_3$ at 23 °C in Et_2O for one hour, leading to the corresponding alcohols with ee's up to 54% [24]. Using (*S*,*S*)-phebox-*ip*-Fe$(CO)_2$Br complex **21** (2 mol%) in association with 2 mol% of Na(acac), Nishiyama and coworkers reported similar performances in the hydrosilylation of *p*-phenylacetophenone in the presence of 1.5 equiv. of $(EtO)_2MeSiH$ at 50 °C in hexane for 24 hours (66% ee) [25]. Interestingly, when using chiral iminopyridine–oxazoline iron complex **22** (1 mol%) in combination with 2 mol% of $NaBEt_3H$ as the catalyst, higher ee's are obtained in the hydrosilylation of arylketones (ee's up to 93%) in the presence of 1 equiv. of diphenylsilane at 25 °C; by contrast, alkyl ketones lead to low enantioselectivities (1–10% ee's) [26]. Using the *in situ*-generated catalyst from the ligand (*S*,*S*)-BOPA and $Fe(OAc)_2$ (**23**, 5 mol%) in the presence of 2 equiv. of $(EtO)_3SiH$, the reduction of aromatic ketones quantitatively leads to the corresponding alcohols with 32–95% ee after 48 hours at 65 °C, but low ee's are observed with alkyl ketones [27]. To date, Gade reported among the most efficient catalysts in terms of enantioselectivity and activity using isoindole-based iron catalysts: tetraphenyl-carbpi-Fe(OAc) complex **24** (5 mol%) can perform the hydrosilylation of ketones with moderate-to-good enantioselectivity (56–93%) by reaction with 2 equiv. of $(EtO)_2MeSiH$ at 40 °C for 40 hours [28]. The best results are obtained using the chiral iron alkoxide boxmi pincer complex **25** as the catalyst (5 mol%) in the presence of 2 equiv. of $(EtO)_2MeSiH$ in toluene for six hours in a temperature ranging from −78 °C to rt. It must be underlined that an unprecedented activity and stereoselectivity can be obtained (TOF = 240 h^{-1} at −40 °C, 73–99% ee for various alkyl aryl ketones) [29]. A detailed mechanism

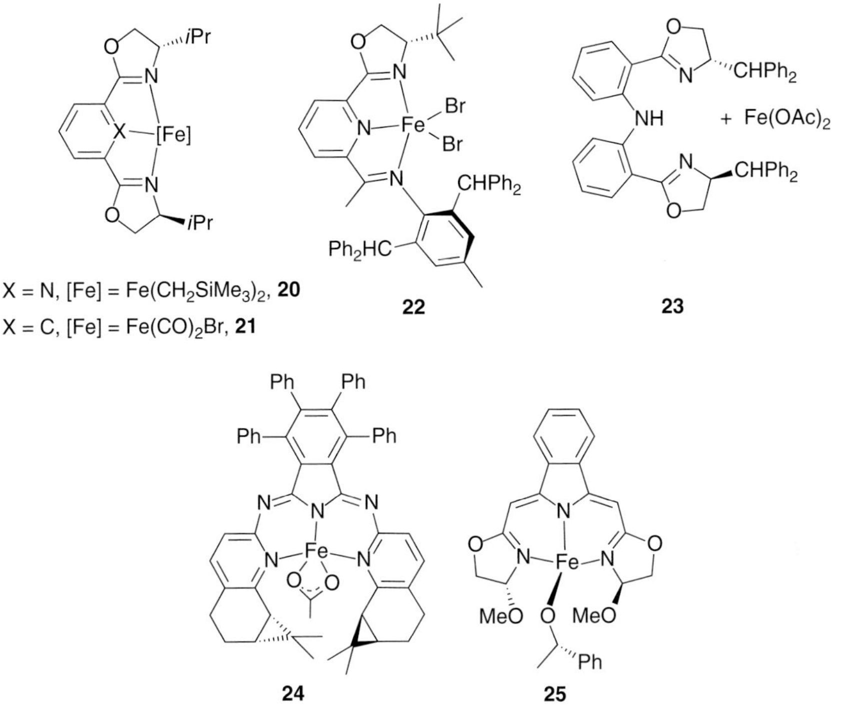

X = N, [Fe] = $Fe(CH_2SiMe_3)_2$, **20**
X = C, [Fe] = $Fe(CO)_2Br$, **21**

22

23

24

25

Figure 10.4 Chiral iron complexes for catalyzed asymmetric hydrosilylation.

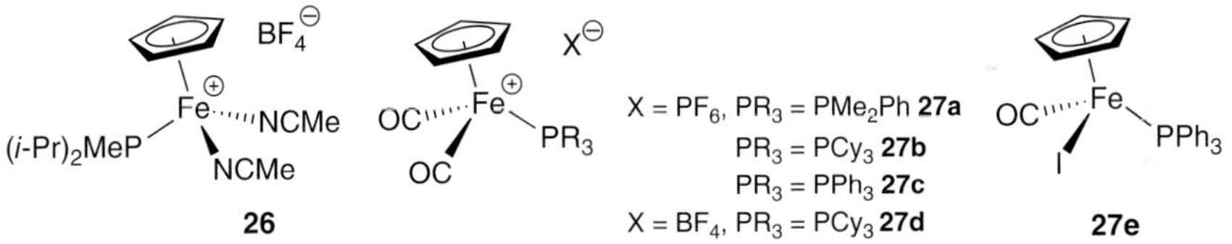

26

X = PF_6, PR_3 = PMe_2Ph **27a**
PR_3 = PCy_3 **27b**
PR_3 = PPh_3 **27c**
X = BF_4, PR_3 = PCy_3 **27d**

27e

Figure 10.5 Piano-stool cyclopentadienyl phosphine iron complexes.

study showed that the rate-determining step of the catalytic cycle is a σ-bond metathesis of the alkoxide complex with hydrosilane, which leads to an iron hydride species and generates the alkoxysilane compounds. The subsequent coordination and insertion of the ketone to the iron hydride complex then regenerates the catalytic alkoxy species [30].

Another important series of efficient iron complexes, cyclopentadienyl piano-stool iron(II) complexes, were extensively developed for the hydrosilylation of carbonyl derivatives. Following the pioneering contribution of Brunner [31], Nikonov and coworkers reported the activity of the cationic CpFe-phosphine complex **26** (5 mol%) for the hydrosilylation of benzaldehyde using $H_2SiMePh$ at 22 °C for three hours (Figure 10.5) [32].

A similar series of carbonyl complexes $[Fe(CO)_2(PR_3)][X]$ (**27**) were also described in hydrosilylation reactions (Figure 10.5) [33]. With **27a–c** (5 mol%) and 1.1 equiv. of $PhSiH_3$ at 30 °C in 16 hours under visible light irradiation

(24 W compact fluorescent lamps), aromatic aldehydes are reduced in very good conversions (92–98% in THF, 91–97% under neat conditions). Notably, PMHS (4 equiv.) can also promote the reduction under similar conditions in the presence of 5 mol% of complexes **27a, b** (conversions up to 95%). The neutral iron complex **27e** was found to be the most efficient of the series for the reduction of ketones such as acetophenone, when performed under neat conditions and visible light activation (1.2 equiv. $PhSiH_3$, 70 °C, 30 hours). It must also be underlined that the tetrafluoroborate complex **27d** exhibited similar or superior activities compared to the iodo- or hexafluorophosphate analogs (e.g. acetophenone: 98% conv., visible light activation, 70 °C, 16 hours with 1.2 equiv. of $PhSiH_3$, or 72 hours with 4 equiv. of PMHS). Noticeably, visible light activation has a crucial role for the removal of one CO ligand and the production of an unsaturated active species.

Piano-stool iron-NHC complexes (NHC, *N*-heterocyclic carbene) are also useful catalysts for hydrosilylation. In 2010, Royo and coworkers showed that tethered Cp-NHC iron complexes such as **28** (1 mol%) are suitable catalysts for the hydrosilylation of activated aldehydes ($(EtO)_2MeSiH$ (1.2 equiv.), 80 °C, 1–18 hours) [34]. Our group has shown that the neutral [Fe(I)(CO)(IMes)] (**29**) and the cationic $[Fe(CO)_2(IMes)][I]$ (**30**) complexes (under visible light irradiation) lead to efficient reductions of both aldehydes and ketones using 1 equiv. of $PhSiH_3$ (aldehydes, THF, 30 °C, 3 hours; ketones, toluene, 70 °C, 16 hours) [35]. It is important to notice that visible light activation is mandatory to generate the active catalyst from the cationic complex **30**, whereas neutral complex **29** works at 30 °C without any activation. Interestingly, when performing the reaction under solvent-free conditions and light irradiation, significant rate enhancements and better activities were observed even at lower temperatures (50 °C vs 70 °C) [36].

The structure of the NHC ligand also influences the activity of the corresponding $[Fe(Cp)(NHC)(CO)_2][I]$ precatalyst in the hydrosilylation of aldehydes and ketones (Figure 10.6). The complexes bearing 1,3-disubstituted imidazolidin-2-ylidene ligands $[Fe(Cp)(NHC)\ (CO)_2][I]$ such as **32** have exhibited moderate activity, as full conversions could be obtained only at 100 °C ($PhSiH_3$, 0.5–4 hours, neat conditions, without light activation) [37]. By contrast, the complex $[Fe(Cp)(NHC)(CO)_2]$ **33** (1 mol%) bearing an anionic six-membered ring NHC ligand incorporating a malonate backbone [38] and

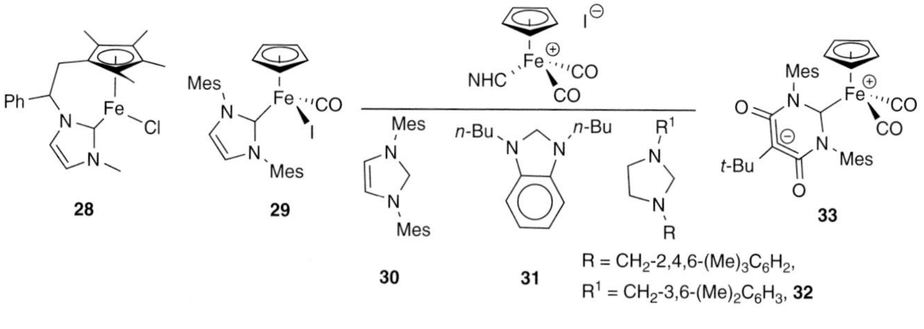

R = CH_2-2,4,6-$(Me)_3C_6H_2$,
R^1 = CH_2-3,6-$(Me)_2C_6H_3$, **32**

Figure 10.6 Diversity of the NHC ligands for Fe-NHC-catalyzed hydrosilylation.

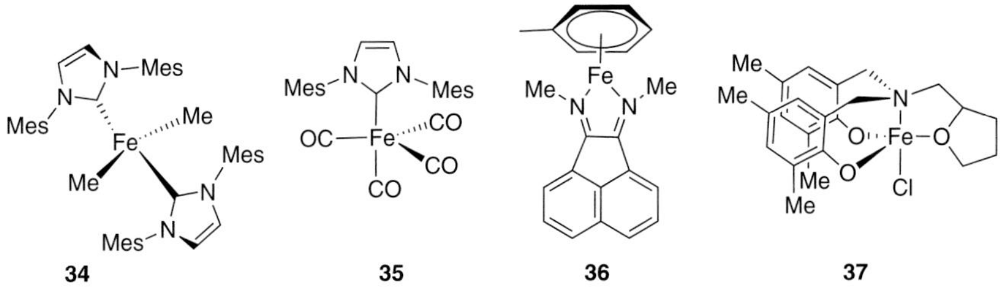

Figure 10.7 Selection of catalytically active iron complexes in hydrosilylation.

benzimidazole-based NHC iron cationic complexes such as **31** (2 mol%) efficiently promote the catalytic hydrosilylation of aromatic aldehydes ($PhSiH_3$ or Ph_2SiH_2, 30 °C, one to three hours) and of acetophenone derivatives ($PhSiH_3$, 70 °C, 16 hours), with similar activities to **30** [39].

In situ-generated catalysts from $Fe(OAc)_2$ and the imidazolium salt precursors (such as 1,3-bis(2,6-diisopropylphenyl)imidazolium chloride, IPr·HCl, or *N*-hydroxyethyl-imidazolium salts) in the presence of *n*-BuLi are also active catalysts for the reduction of ketones using PMHS [40]. Noticeably, the exact stoichiometry between the base and the imidazolium salts is crucial: indeed, simple alkoxide salts can also promote such catalytic hydrosilylations with trisubstituted silanes without the iron-based catalyst. Square planar NHC iron complex $[Fe(Me)_2(IMes)_2]$ **34** (0.1 mol%; Figure 10.7) exhibited high efficiency for the reduction of 2-acetonaphthone in the presence of $(EtO)_3SiH$ (1.1 equiv., 25 °C, five hours) or Ph_2SiH_2 (1.1 equiv., 40 °C, five hours) [41]. Hydrosilylation can also be promoted by iron(0): Royo has shown that the iron (0) NHC complex $[Fe(CO)_4(IMes)]$ **35** (1 mol%) can reduce aromatic aldehydes using phenylsilane (1.2 equiv.) at rt. for four hours with a broad functional group tolerance including reducible ketone, nitrile, and nitro moieties [42]. Another iron(0)-based catalyst **36** bearing a bis(arylamino)acenaphthene moiety permits the reduction of aldehydes and ketones with 1 equiv. of diphenylsilane at 70 °C for 0.5–18 hours [43].

Iron(III)-based complexes were recently reported as good candidates for the hydrosilylation. Indeed, 1 mol% of amine-bis(phenolate) iron complex **37** in the presence of 3 equiv. of triethoxysilane promotes the reduction of aldehydes and ketones at 80 °C for 3–24 hours [44]. Unexpectedly, an unusual chemoselectivity was observed using $FeCl_3 \cdot 6H_2O$ (10 mol%) in association with PMHS (2.7 equiv.) in 1,2-dichloroethane under microwave irradiation at 120 °C for one hour: aldehydes and ketones were selectively converted to the corresponding methylene compounds in yields up to 98% [45].

10.2.2 Cobalt-Catalyzed Hydrosilylation

Surprisingly, cobalt has been rarely described for use in the hydrosilylation of carbonyl compounds, which contrast with its use in hydrogenation [46] or C=C bond hydrosilylation [47]. In 1991, Brunner and Amberger reported the use of an *in situ*-generated chiral cobalt complex from $[Co(Py)_6][BPh_4]$ **38** (0.5 mol%)

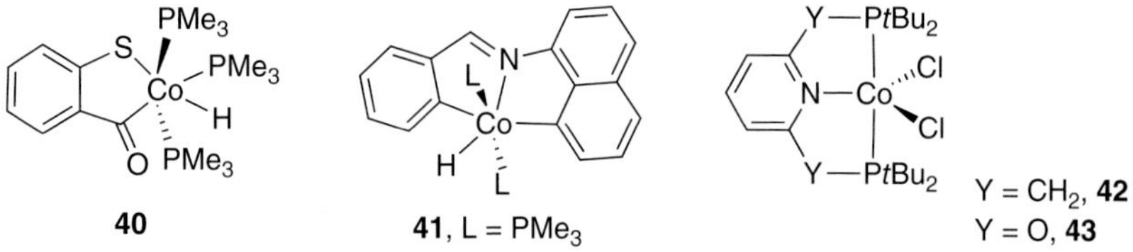

Figure 10.8 Cobalt complexes in hydrosilylation.

and the chiral pyridyloxazoline **39** (1.65 mol%; Figure 10.9) for the asymmetric hydrosilylation of acetophenone with ee's up to 53% by reaction with 1 equiv. of Ph_2SiH_2 at 0–20 °C for 18 hours [48].

Two decades later, cobalt hydride complexes were reported as active catalyst in the hydrosilylation of carbonyl derivatives. Li and coworkers described sulfur-coordinated acyl(hydrido)cobalt complex **40** (1 mol%; Figure 10.8) as a catalyst in the presence of 1.1 equiv. of $(EtO)_3SiH$ for the reduction of aldehydes at 40 °C for 2–8 hours (83–93% GC-yields) and ketones at 55 °C for 8–26 hours (67–88% yields) [49]. They also showed that the hydrido CNC pincer cobalt complex **41** (1–5 mol%) with 1.1 equiv. of $(EtO)_3SiH$ reduced both aldehydes and ketones at 60 °C for 3–24 hours (68–93% yields) [50]. The *in situ* catalysts prepared from 2,6-bis(di-*tert*-butyl-phosphinito)pyridine and 2,6-bis(di-*tert*-butylphosphinito)pyridine cobalt complexes **42** and **43**, respectively (1 mol%), in association with 2 mol% of $NaHBEt_3$, have higher activities in the hydrosilylation of carbonyl derivatives. **43** gives the best results as in the presence of 1 equiv. of $(EtO)_3SiH$, the reaction can take place at r. t. for 24 hours (aldehydes, 20–96% conv.; ketones, 25–92% conv.) [51].

Inspired by Brunner's pioneering results, Nishiyama reported the efficient hydrosilylation of arylalkylketones using *in situ*-generated chiral catalysts from BOPA ligands (e.g. Bopa-Ph **44**; Figure 10.9) (6 mol%) in combination with $Co(OAc)_2$ (5 mol%) in the presence of $(EtO)_2MeSiH$ at 65 °C for 24 hours (88–99% yields, 38–98% ee) [52]. Notably, better enantioselectivities are observed than with iron (ee's up to 88%, vide supra). Chan and coworkers reported high ee's using an *in situ*-generated catalyst (10 mol%) from $Co(OAc)_2$ $4H_2O$ and (*S*)-Xyl-P-Phos **45** in the presence of 1.2 equiv. of $PhSiH_3$ and 4 Å molecular sieves in toluene under air at 40–55 °C for 12–60 hours (75–96% ee) [53]. Using chiral 1,3-bis(2-pyridylimino)isoindolate cobalt alkyl complexes such as **46** as a catalyst (2.5 mol%), Gade reported the asymmetric reduction of arylmethylketones with ee's up to 90% at 15 °C for eight hours when using 2 equiv. of $(EtO)_2MeSiH$ [54].

10.2.3 Manganese-Catalyzed Hydrosilylation

The first example of Mn-catalyzed ketone hydrosilylation was reported by Yates in 1982 using $Mn_2(CO)_{10}$ **47** as the catalyst (2.4 mol%) in the presence of 1 equiv. of Et_3SiH in neat conditions under UV activation at 29 °C for 20 hours: the corresponding isopropyl triethylsilyl ether was obtained in 5% yield [55]. In this field, a breakthrough was made by Curtler: the reaction between the acyl manganese

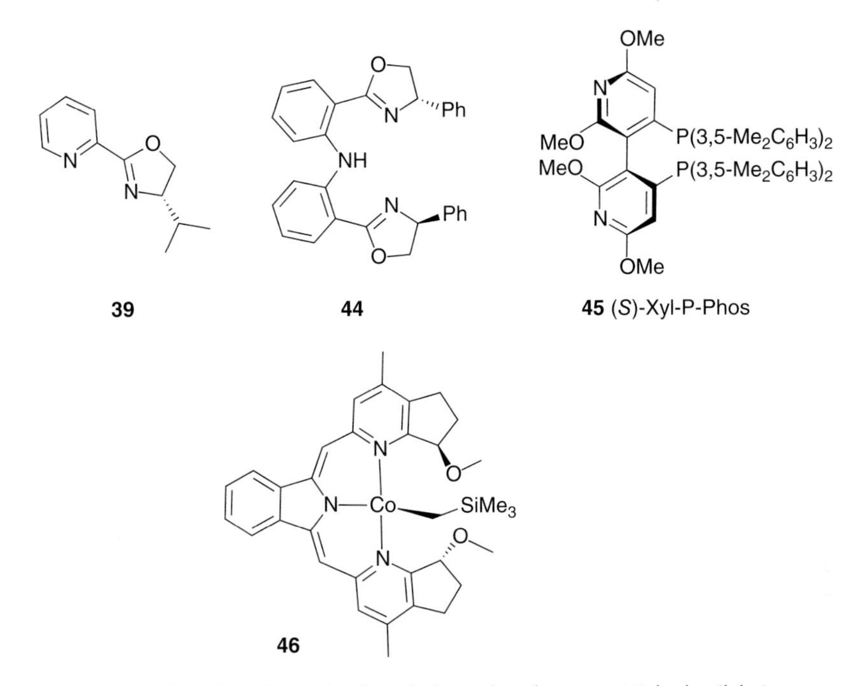

Figure 10.9 Ligands and complex for cobalt-catalyzed asymmetric hydrosilylation.

complex $Mn(CO)_5(COMe)$ **48** [56] and 1–2 equiv. of $HSiMePh_2$ at rt. leads to siloxyethyl complexes $(CO)_5MnCH(OSiMePh_2)CH_3$ **49** (81%) with a siloxyvinyl by-product $(CO)_5Mn—C(OSiMePh_2) = CH_2$ **50** (2%). Noteworthy, this reaction occurs without adding a catalyst, in an autocatalytic manner. Similarly, **48** can catalyze the hydrosilylation of $(\eta^5\text{-}C_5H_5)Fe(CO)_2(COR)$ complexes, leading to the corresponding $(\eta^5\text{-}C_5H_5)Fe(CO)_2(COSiHPh_2)(R)$ complex in up to 97% yield [57].

Half-sandwich 1-hydronaphthene-type manganese complex **51** and cationic naphthalene manganese complex **52** (Figure 10.10) also catalyze the hydrosilylation of alkyl and aryl ketones at rt. for 0.5–3 hours in the presence of diphenylsilane, **51** being the most active with a TOF of $100\,h^{-1}$ (hydrosilylation of acetophenone) [58]. Half-sandwich manganese cyclopentadienyl NHC complexes can also be suitable catalysts for hydrosilylation of aldehydes and ketones. Although cymanthrene, $CpMn(CO)_3$, showed very low activity (4% conv.), the corresponding complexes bearing NHC ligands exhibit higher activities

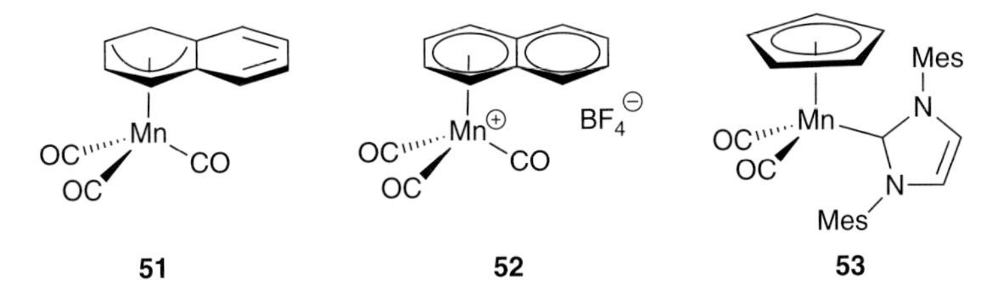

Figure 10.10 Efficient manganese complexes for catalytic hydrosilylation of ketones.

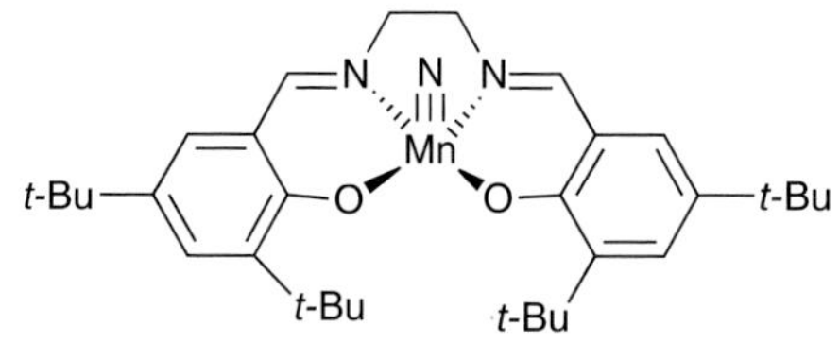

Figure 10.11 Multidentate ligand Mn complexes for catalytic hydrosilylation of ketones.

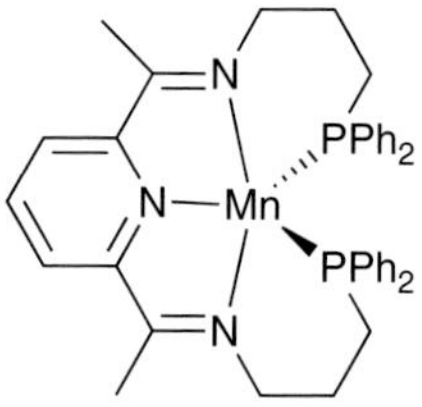

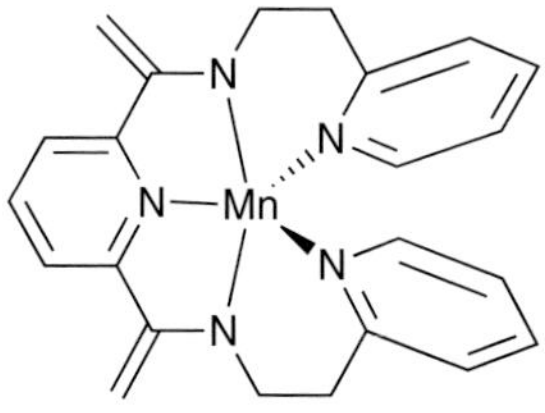

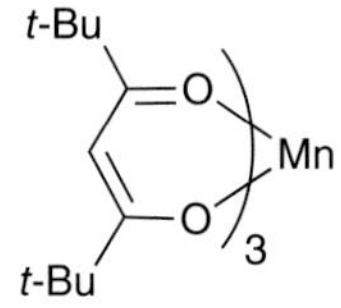

under UV irradiation (350 nm) in the presence of 1.5 equiv. of Ph_2SiH_2 at rt. for two hours, **53** being the most efficient catalyst (>99% conv. for the hydrosilylation of acetophenone). Notably, both the UV irradiation and the presence of the coordinated NHC ligand are crucial to obtain a significant catalytic activity. Interestingly, **53** offers a huge functional group tolerance (heterocyclic moieties, halides, conjugated and nonconjugated alkenyl, alkynyl nitrile, ester, etc.) [59].

Manganese complexes bearing multidentate ligands can also be the suitable catalysts in hydrosilylation. Indeed, salen–Mn(nitride) complex **54** (0.5 mol%; Figure 10.11) was shown to be efficient in the hydrosilylation of carbonyl derivatives in the presence of 0.5 equiv. of $PhSiH_3$ at 80 °C for one minute—three hours with yields up to 95% (TOFs up to 8800 h^{-1} for the reduction of p-NO_2-C_6H_4CHO); the reduction tolerates functional groups such as nitro or cyclopentadienyl, but not conjugated C=C bonds [60]. Trovitch developed hydrosilylation of ketones using redox noninnocent bis(imino)pyridine manganese complexes such as **55** and **56** [61]. Using 0.01–1 mol% of **55** in the presence of 1 equiv. of phenylsilane, aryl, and alkyl ketones can be hydrosilylated at 25 °C for 4 minutes–24 hours, leading to a mixture of tertiary and quaternary silanes Ph–SiH$(OCHR_2)_2$ and Ph–Si$(OCHR_2)_3$, respectively, in 80–99% conversions. It must be pointed out that this catalyst can reach an exceptional activity (TOF 76 800 h^{-1}) for the hydrosilylation of cyclohexanone. The paramagnetic bis(enamide)tris pyridine manganese complex **56** is also an efficient catalyst even if less efficient than **55**, with turnover numbers (TONs) and TOFs up to 14 170 and 2475 h^{-1}, respectively (PhCHO, 25 °C, one hour, 1 equiv. $PhSiH_3$).

In an original approach, Magnus has shown that aldehydes and cyclic five- and six-membered ketones can be reduced in the presence of phenylsilane at rt. in isopropanol under an atmosphere of oxygen using tris(dipivaloylmethanato) manganese(III) $(Mn(dpm)_3)$ (**57**) as the precatalyst [62]. This catalyst is less efficient for acyclic and macrocyclic ketones.

10.3 Reduction of Imines and Reductive Amination of Carbonyl Compounds

In this area, by contrast with the reduction via hydrogenation or hydrogen transfer, only scarce examples of iron complexes were reported as the catalyst under hydrosilylation conditions. Indeed, a general iron-catalyzed hydrosilylation of aldimines and ketimines is described using 2 mol% of [Fe(Cp)(IMes)$(CO)_2$][I] (**30**) in the presence of 2 equiv. of $PhSiH_3$; various aldimines lead to the corresponding amines under visible light irradiation and neat conditions at 30 °C for 30 hours (Scheme 10.1) [63]. Importantly, functional groups such as halides, ketones, esters, and alkenes are tolerated. Slightly harsher conditions are necessary to perform the hydrosilylation of ketimines: 5 mol% **29**, 24 hours, 100 °C (57–95% yields). Very recently, Mandal and coworkers reported that a highly active abnormal NHC iron complex **58** is able to reduce aldimines at rt. in dimethyl sulfoxide (DMSO) using a very low loading of catalyst (0.05 mol%) with TONs up to 17 000 [64].They also showed a broad functional group tolerance and extended this reduction to imines bearing N-alkylated O-protected sugars. To be able to reduce ketimines, 2 mol% of **58** is used.

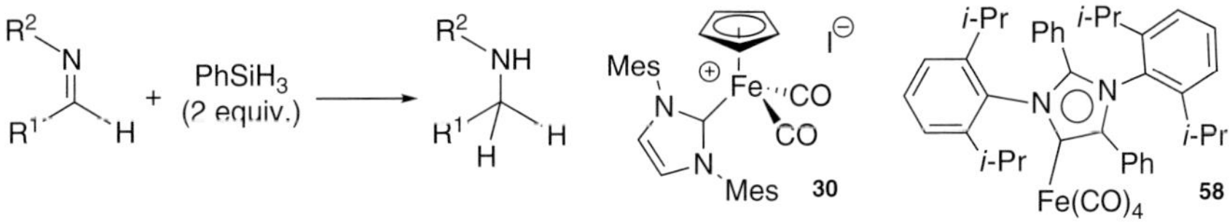

30 (2 mol%), neat conditions, 30 °C, 30 h, visible light irradation, 21 examples, 26–95% yields
58 (0.05 mol%), DMSO, r.t., 12 h, 25 examples, 59–96% yields

Scheme 10.1 Iron-catalyzed hydrosilylation of imines.

For large-scale preparation of amine derivatives, one of the most valuable ways is certainly the direct reductive amination (DRA) of aldehydes and ketones, using stoichiometric alkali reducing agents such as $LiAlH_4$ or $NaBH_4$. Alternatively, transition-metal-catalyzed DRA, including with earth-abundant ones, has been studied intensively during the past decade, mainly under hydrogenation conditions. Iron-catalyzed DRA can also be performed under hydrosilylation conditions. The first DRA of aldehydes with anilines was described in 2010 by Enthaler using $FeCl_3$ (5 mol%) in the presence of an excess of PMHS in THF at 60 °C for 24 hours (19–97% yields) [65]. Nevertheless, with alkylamines such as benzylamine, no reaction occurs. Using PMHS and phosphanyl-pyridine iron complex **59** (5 mol%), DRA of aromatic aldehydes and secondary amines in dimethylcarbonate (DMC) at 40 °C for 24 hours under visible light irradiation leads to the corresponding tertiary amines in 53–93% isolated yields (Scheme 10.2) [66]. Interestingly, ester, nitrile, ketone, and halide groups are tolerated. The phosphanyl-pyridine ligand has a crucial effect on the activity; indeed, with monophosphine complexes [CpFe$(CO)_2$$(PR_3)$][$BF_4$] **60** as catalysts, moderate conversions (35–58%) are observed.

Iron-catalyzed DRA is also possible starting from allylic or homoallylic alcohols and secondary and primary anilines [67]. Using 5 mol% of Fe(cod)$(CO)_3$ **61**

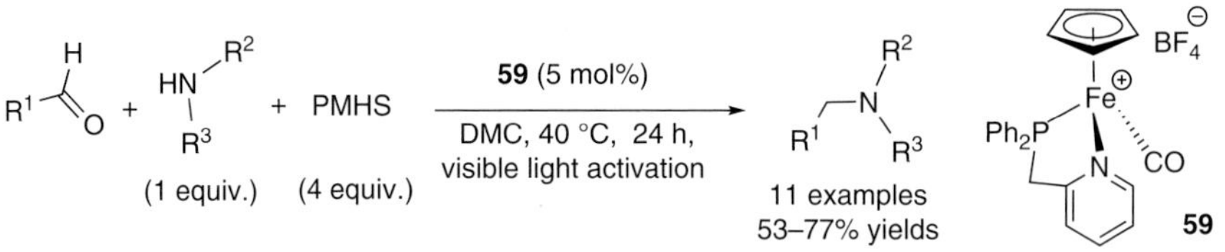

Scheme 10.2 Piano-stool phosphanyl-pyridine iron-catalyzed DRA.

(cod, cycloocta-1,5-diene) and 3 equiv. of PMHS in ethanol at 50–70 °C under visible light irradiation, the corresponding amines are produced in 31–95% yield, resulting from a formal DRA of (homo)allylic alcohols via a tandem isomerization/condensation/hydrosilylation reaction (Scheme 10.3).

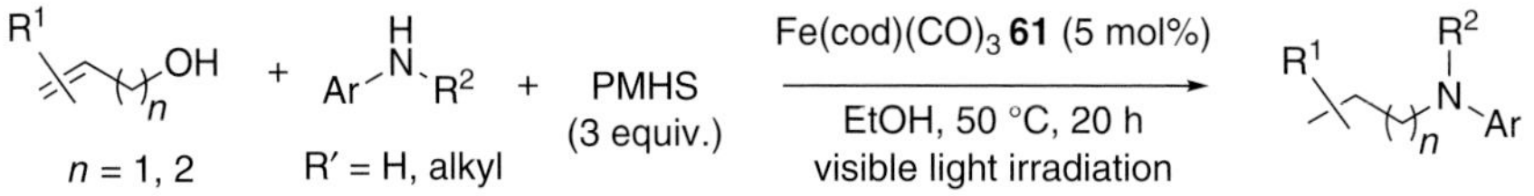

Scheme 10.3 Iron-catalyzed isomerization/DRA reaction.

To the best of our knowledge, there is only one example of efficient cobalt-catalyzed DRA reported in the literature. Indeed, Singh described the use of the cobalt-phthalocyanine complex **62** as the catalyst (1 mol%) to perform the reaction of a broad variety of aliphatic and aromatic aldehydes or ketones with aliphatic and aromatic amines in the presence of 1.5 equiv. of Ph_2SiH_2 in ethanol at 70 °C for 12 hours (13–94% yields) (Scheme 10.4) [68]. Interestingly, the reaction of primary amines selectively leads to the corresponding secondary amines with a huge range of reducible functional groups such as nitro, primary amide, ester, nitrile, halides, lactone, hydroxy, and alkenyl.

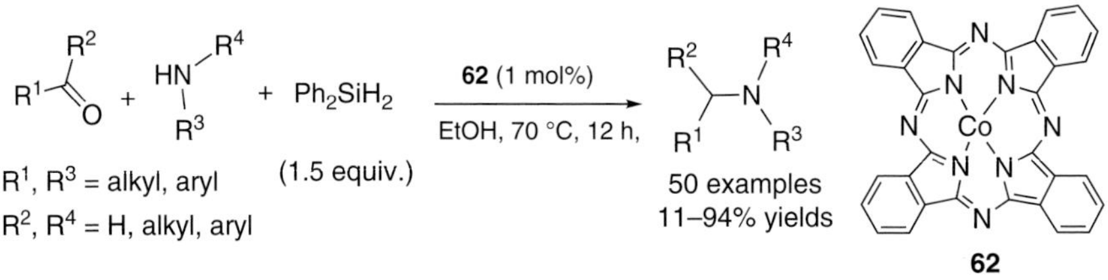

Scheme 10.4 Cobalt-phthalocyanine-catalyzed DRA reaction.

10.4 Reduction of Carboxylic Acid Derivatives

10.4.1 Carboxamides and Ureas

Among the carboxylic acid derivatives, the most difficult ones to reduce are carboxamides mainly because of chemoselectivity issues (C—N vs C=O cleavage) and their catalytic transition metal reductions are well exemplified.

Concomitantly, Beller [69] and Nagashima [70] coworkers reported the first iron-catalyzed hydrosilylation of secondary and tertiary amides leading specifically to the corresponding amines, using 2–10 mol% of $Fe_3(CO)_{12}$ **63** or $Fe(CO)_5$ **64** and 4–10 equiv. of PMHS or 2.2 equiv. of TMDS at 100 °C for 24 hours. The reduction can be conducted at rt. when performing the reaction under light irradiation using a 400 W high-pressure mercury lamp. Employing 5 mol% of well-defined $[CpFe(CO)_2(IMes)][I]$ complex **30** also permits to perform the catalytic reduction of tertiary and secondary carboxamides in the presence of 2 equiv. of phenylsilane in solvent-free conditions at 100 °C for 24 hours under visible light irradiation (Scheme 10.5) [71].

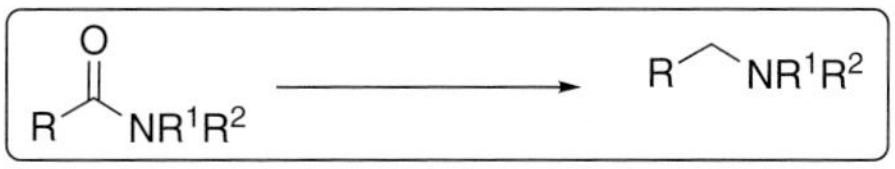

2–10 mol% $Fe_3(CO)_{12}$ **63**, PMHS (4–8 equiv.), n-Bu_2O, 100 °C, 24 h, 28 examples, 50–99% yields 10 mol% $Fe(CO)_5$ **64** or $Fe_3(CO)_{12}$ **63**, TMDS (2.2 equiv.), toluene, 100 °C, 24 h, 8 examples, 21–98% yields

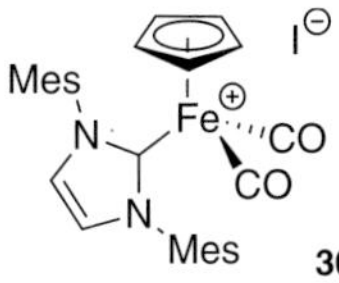

5 mol% **30**, 2 equiv. $PhSiH_3$, 100 °C, visible light, 24 h 21 examples, 77–98% yields

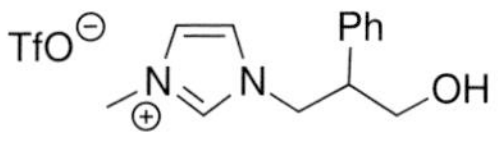

i) 1 mol% $Fe(OAc)_2$, 1.1 mol% **65**, 2.2 mol% n-BuLi, 1 mol% LiCl ii) amide, 3 equiv. PMHS, 65 °C, 5–14 h, 12 examples, 77–92% yields

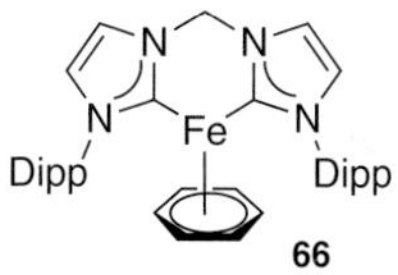

1 mol% **66**, 3 equiv. Ph_2SiH_2, 70 °C, 24 h, 4 examples, 83–99% yields

Scheme 10.5 Iron catalysts for chemoselective hydrosilylation of carboxamides.

Using an *in situ*-prepared Fe/NHC complex from 1 mol% of $Fe(OAc)_2$, 1.1 mol% of ([PhHEMIM][OTf]) **65**, and 1 mol% of LiCl and then treated with 2.2 mol% of *n*BuLi, Adolfsson described the hydrosilylation of tertiary aromatic amides with PMHS performed at 65 °C. In these reactions, the use of LiCl increases both the chemoselectivity and the activity [72]. NHC-Fe(0) complex **66** (1 mol%) was shown to be able to catalyze the hydrosilylation of tertiary amides by reaction with 3 equiv. of Ph_2SiH_2 at 70 °C for 24 hours [73].

On the contrary, with iron catalysts, the hydrosilylation of primary amides is more difficult to perform as the only obtained products are the corresponding nitriles produced by dehydration [74]. To efficiently reduce primary amides to primary amines, Beller and coworkers used two iron species in a sequential manner [75]. Indeed, using 2–5 mol% of the complex $[Et_3NH][HFe_3(CO)_{11}]$ **67** and 3 equiv. of $(EtO)_2MeSiH$ first promotes the dehydration of primary amides to nitriles, which are then reduced to primary amines in the presence of an *in situ*-generated catalyst from 20 mol% of $Fe(OAc)_2$ and 20 mol% of the 3,4,7,8-tetramethyl-1,10-phenanthroline ligand **68** at 100 °C for 28 additional hours (Scheme 10.6).

The selective reduction of ureas is also a difficult task in the synthesis, particularly for the preparation of formamidines without further reduction to aminals, methanol, and amines. Cantat tackled this challenge performing the reduction with 5 mol% of an *in situ*-generated iron catalyst from $Fe(acac)_2$ and the tetraphos

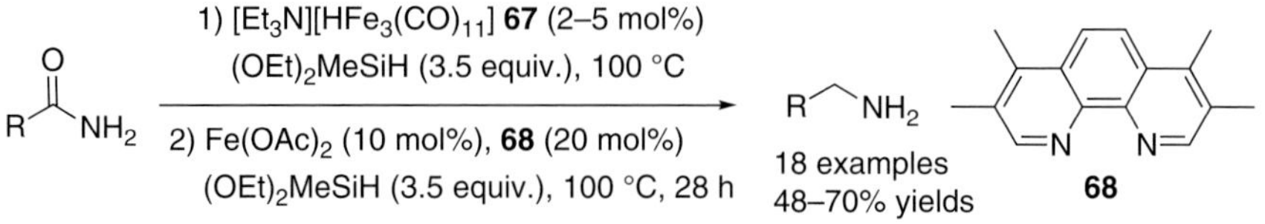

1) $[Et_3N][HFe_3(CO)_{11}]$ **67** (2–5 mol%), $(OEt)_2MeSiH$ (3.5 equiv.), 100 °C
2) $Fe(OAc)_2$ (10 mol%), **68** (20 mol%), $(OEt)_2MeSiH$ (3.5 equiv.), 100 °C, 28 h

R–C(O)NH$_2$ → R–CH$_2$NH$_2$ (18 examples, 48–70% yields)

68

Scheme 10.6 Iron-catalyzed reduction of primary amides to primary amines.

ligand $P(CH_2CH_2PPh_2)_3$ **69** and 1 equiv. of $PhSiH_3$ in THF at 100 °C for 24 hours [76]. A dehydrogenative silylation of the NH bonds is suggested for the formation of carbodiimides. A mixture of formamidines and carbodiimides is then obtained in 69–98% conversions and ratios from 98 : <1 to 29 : 69 (Scheme 10.7).

There are only scarce examples of hydrosilylation of carboxamides catalyzed by cobalt. The first one was reported by our group using a commercially available dicobalt octacarbonyl complex $Co_2(CO)_8$ **70** (0.5 mol%) for the efficient reduction of tertiary (hetero)aromatic amides leading to the corresponding tertiary amines using 2.2 equiv. of TMDS and PMHS as inexpensive hydrosilanes at 100 °C for three hours [77]. Noticeably, reducible functional groups such as esters and alkenyl are tolerated, and the photoassisted reduction can also be performed under UV irradiation (350 nm) at rt. By contrast, the reduction of aliphatic amides is more difficult because of a concomitant hydrogenolysis of the amide.

10.4.2 Carboxylic Esters

The chemoselective reduction of carboxylic esters is still an important and difficult task in multistep methodologies as alcohols, ethers, or aldehydes can be produced. There are only a limited number of reports of selective reduction of carboxylic esters using iron, cobalt, or manganese as the catalysts.

The first example of iron-catalyzed ester reduction via hydrosilylation was reported in 2012. Using 5 mol% of $[Fe(Cp)(CO)_2(PCy_3)][BF_4]$ **27d**, under solvent-free conditions in the presence of 4 equiv. of $PhSiH_3$ under visible light activation at 100 °C for 16 hours, carboxylic esters such as alkanoates and two-substituted acetates lead to the corresponding primary alcohols in 51–88% isolated yields (Scheme 10.8) [78]. Chemoselective reduction of aromatic and aliphatic esters to alcohols can also be conducted with *in situ* catalytic species obtained from 5 mol% of $[Fe(stearate)_2]$ **71** and 10 mol% of ethylenediamine in the presence of 3 equiv. of PMHS at 100 °C for 20 hours (35–86% yields) [79]. The three-coordinate iron(II) *N*-phosphinoamidinate complex **10** has shown superior activity using low catalyst loading (0.25–1 mol%) in the reduction of esters to alcohols performed at rt. for four hours in the presence of 1 equiv. of phenylsilane [80].

The chemoselectivity of these reductions can be finely managed by a careful design of the catalytic system. Indeed, to selectively obtain ethers from esters, the association of the catalyst $Fe_3(CO)_{12}$ **63** (10 mol%) and the hydrosilane TMDS (3 equiv.) was used, and the reduction of aliphatic and alicyclic esters (even steroid esters or benzofuranone) was successfully performed in toluene at 100 °C for two hours (50–85% yields) (Scheme 10.9) [81].

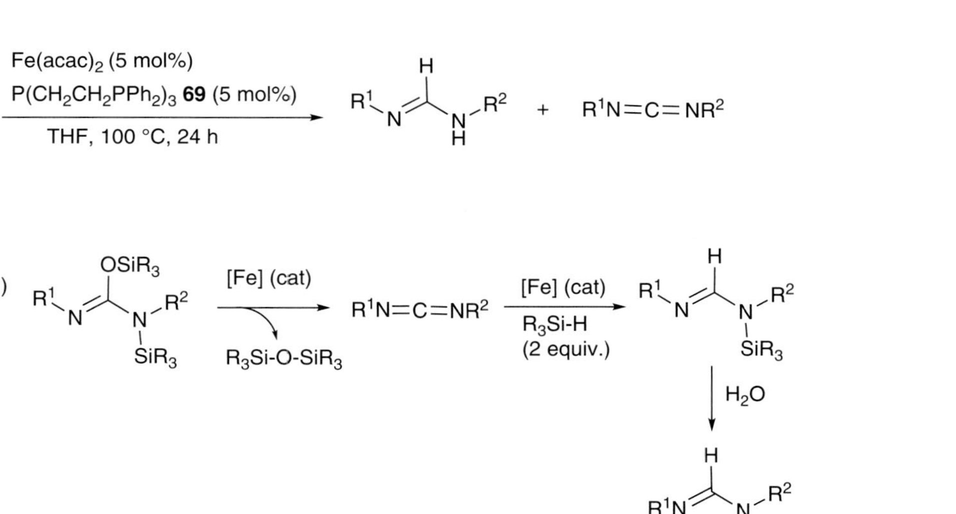

Scheme 10.7 Iron-catalyzed hydrosilylation of ureas.

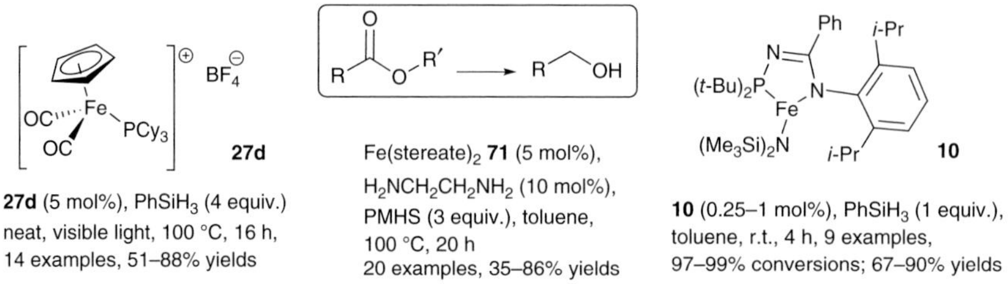

27d (5 mol%), $PhSiH_3$ (4 equiv.)
neat, visible light, 100 °C, 16 h,
14 examples, 51–88% yields

Fe(stereate)$_2$ **71** (5 mol%),
$H_2NCH_2CH_2NH_2$ (10 mol%),
PMHS (3 equiv.), toluene,
100 °C, 20 h
20 examples, 35–86% yields

10 (0.25–1 mol%), $PhSiH_3$ (1 equiv.),
toluene, r.t., 4 h, 9 examples,
97–99% conversions; 67–90% yields

Scheme 10.8 Iron-catalyzed hydrosilylation of esters to alcohols.

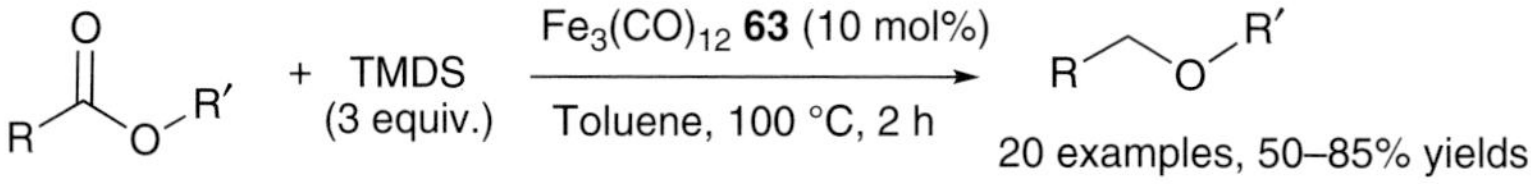

+ TMDS (3 equiv.); $Fe_3(CO)_{12}$ **63** (10 mol%), Toluene, 100 °C, 2 h
20 examples, 50–85% yields

Scheme 10.9 Iron-catalyzed hydrosilylation of esters to ethers.

In organic multistep synthesis, the reduction of esters to give aldehydes is always a tedious step, even if well-known procedures using a stoichiometric alkali reductant can be efficient. This transformation can also be performed starting from (hetero)aromatic and aliphatic esters, using the combination of 5 mol% of [Fe(CO)$_4$(IMes)] **35**, and the silanes R_2SiH_2 (R = Et, Ph) at rt. under UV irradiation (350 nm) [82]. Et_2SiH_2 exhibits the best performance for the reduction of alkanoates, whereas Ph_2SiH_2 is the most efficient for benzoates. Interestingly, the chemoselective reduction of lactones to lactols can also be conducted under similar conditions (Scheme 10.10). It must be noticed that the reaction occurs via an oxidative addition of the dihydrosilane to an unsaturated NHC—Fe intermediate, leading to an iron hydride-silyl complex, which was fully characterized by nuclear magnetic resonance (NMR) and X-ray analysis.

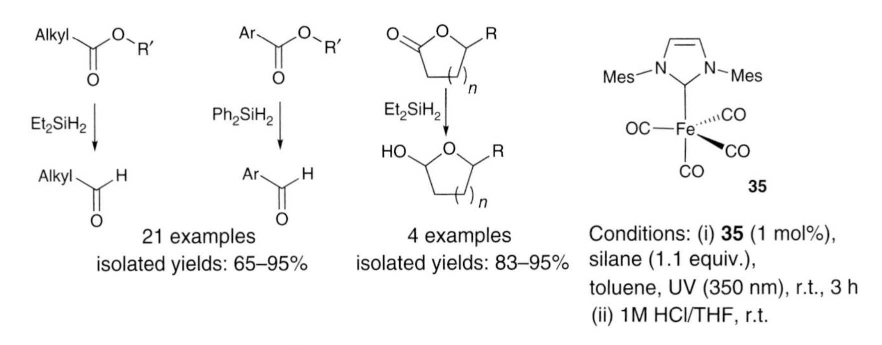

Scheme 10.10 Iron-catalyzed hydrosilylation of esters to aldehydes and lactones to lactols.

Furthermore, more reactive carboxylic compounds such as alkanoyl chlorides selectively lead to aldehydes using 20 mol% of FeO with 5 mol% of tris(2,4,6-trimethoxyphenyl)phosphine (TMPP, **72**) in the presence of 1.12 equiv. of $PhSiH_3$ in toluene at 60–120 °C for 20 hours (38–58% yield) [83].

Although cobalt-catalyzed hydrosilylation of esters is not reported to the best of our knowledge, the manganese version has received some attention. In 1995,

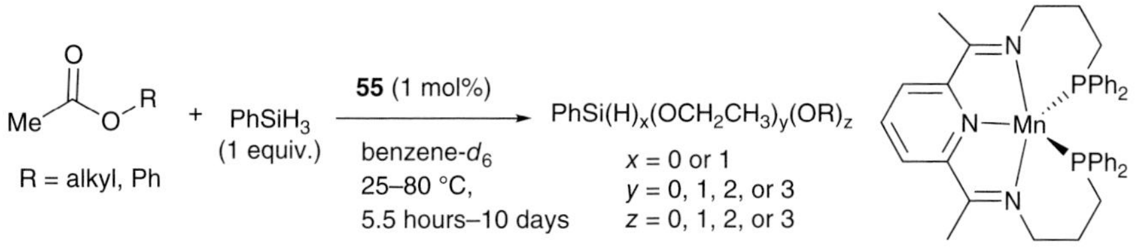

Me–C(=O)–O–R + $PhSiH_3$ (1 equiv.) → [**55** (1 mol%), benzene-d_6, 25–80 °C, 5.5 hours–10 days] $PhSi(H)_x(OCH_2CH_3)_y(OR)_z$

R = alkyl, Ph; x = 0 or 1; y = 0, 1, 2, or 3; z = 0, 1, 2, or 3

55

Scheme 10.11 Manganese-catalyzed reduction of esters.

in a pioneering work, Cutler demonstrated that $(PPh_3)(CO)_5Mn(COMe)$ **73** is able to selectively reduce alkanoic esters to dialkyl ethers via a silyl acetal intermediate in the presence of phenylsilane in benzene at 24 °C for 61–83% yields [84]. By contrast, with benzoic and hindered alkanoic esters, low conversions are observed. Noticeably, lactones can be reduced and moderately yield the corresponding cyclic ethers (35–65%). A decade later, Trovitch investigated pentadentate bis(imino)pyridine bis-diphosphine Mn(0) complex **55** as a catalyst (1 mol%) in the hydrosilylation of alkyl and phenyl acetates in the presence of 1 equiv. of phenylsilane (Scheme 10.11) [61]. A dihydrosilylation process took place, leading to a mixture of tertiary and quaternary silicon derivatives. The hydrosilylation of methyl and ethyl acetate required 24 and 5.5 hours, respectively, at 25 °C to reach a full conversion, whereas the reaction of *i*-propyl, *tert*-butyl, and phenyl acetates had to be performed for several days at rt. or 80 °C.

10.4.3 Carboxylic Acids

The chemoselective reduction of carboxylic acids to either alcohols or aldehydes under mild conditions is a challenging task in multistep synthesis. Even if such transformations can be selectively performed through heterogeneous catalysis using highly drastic conditions, or through homogeneous catalytic hydrogenation, this transformation has been scarcely developed using hydrosilylation with iron or manganese. With iron, using an appropriate combination of iron catalyst and silane, chemoselective reduction of carboxylic acids can be performed [85]. Indeed, performing the reaction with 4 equiv. of phenylsilane and 5 mol% of $[Fe(CO)_3(cod)]$ **61** at rt. under UV irradiation (350 nm) for 24 hours, alkanoic and activated benzoic acids lead to the corresponding alcohols with 67–97% yields after an acidic quench. Alternatively, the use of 5 mol% of $[Fe(t\text{-PBO})(CO)_3]$ (**74**; *t*-PBO, *trans*-4-phenyl-but-3-en-2-one) in combination with 2 equiv. of TMDS at 50 °C for 24 hours permits to reduce alkanoic acids to the corresponding aldehydes in 45–95% isolated yields. To gain insight into the chemoselectivity, a disilylacetal intermediate was isolated and fully characterized by NMR and MS (R = p-Br–C_6H_4–CH_2). This intermediate results from a dehydrogenative silylation of the carboxylic acid initially giving a silyl carboxylate, which is then reduced with a second equivalent of TMDS. Thus, the disilylacetal intermediate is stable in the solution and generates the aldehyde only after the acidic quench, which avoids any over-reduction (Scheme 10.12).

Selective hydrosilylation of aliphatic carboxylic acids to disilylacetals was also reported using $Mn_2(CO)_{10}$ (**75**) as the catalyst (5 mol%) [86]. Once again,

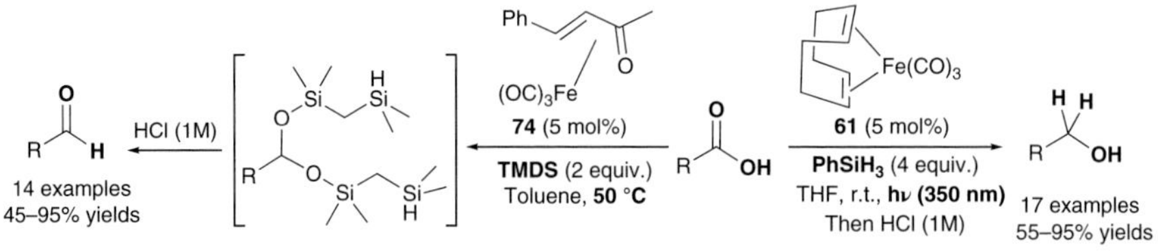

Scheme 10.12 Chemoselective iron-catalyzed hydrosilylation of carboxylic acids.

the careful choice of a tertiary silane (Et_3SiH, 3.3 equiv.) is crucial to gain high chemoselectivity. Working under UV irradiation (350 nm) at rt. for three hours, arylacetic acids, alkenyl, and alkyl carboxylic acids were efficiently reduced, leading to the corresponding disilylacetals $RCH(OSiEt_3)_2$ (60–98% isolated yields), which were then quenched in acidic conditions to produce aldehydes. Noteworthy, functional groups such as hydroxyl, amino, halides, internal C=C, and heteroaromatic rings are tolerated, whereas CF_3 and nitro groups inhibit the reduction and α,β-unsaturated carboxylic acids react with concomitant reduction of the conjugated C=C bond. Benzoic acid derivatives can also be reduced with moderate yields (29–39%).

10.5 Hydroelementation of Carbon Dioxide

10.5.1 Hydrosilylation of Carbon Dioxide

The use of carbon dioxide as a C1 carbon feedstock source is nowadays one of the most important and challenging goals within sustainable chemistry. In this area, transition-metal-catalyzed transformation via hydrosilylation is an efficient tool [87]. Using first-row metals such as iron, cobalt, and manganese, there are few reports on CO_2 hydrosilylation. In 2014, Cantat and coworkers demonstrated that the *in situ* catalyst (5 mol%) generated from $Fe(acac)_2$ and tetraphos ligand $P(CH_2CH_2PPh_2)_3$ **69** can promote the formylation of secondary amines under 1 bar of CO_2 at rt. for 18 hours in the presence of 1 equiv. of $PhSiH_3$, selectively leading to formamides in 24–95% GC-yields [88]. It is noteworthy that this reaction is efficient with aliphatic secondary amines, whereas it is inhibited with diphenylamine. The reaction with primary amines is more challenging as mono- and diformylation can be performed: monoformylated compounds are obtained starting from primary anilines, whereas mixtures of mono- and diformyl compounds are obtained with primary aliphatic amines. Interestingly, ketone and ester functional groups are tolerated but not hydroxyl groups. Using 4 equiv. of $PhSiH_3$ at 100 °C for 18 hours, the catalytic system is able to perform the methylation of methylarylamines to give a mixture of formamides and tertiary methylated amines (Scheme 10.13).

Chirik and coworkers reported the use of pincer cobalt hydride and trimethylsilylmethyl complexes **76** and **77** as catalysts for the hydrosilylation of CO_2 with $PhSiH_3$ and Ph_2SiH_2 (4 equiv.) at rt. (Scheme 10.14) [89]. The reaction with 0.5 mol% of the hydride complex **76** leads to a mixture of silylformate, bis(silyl)acetyl, and silyl ether oligomers. Notably, the use of alkyl complex **77**

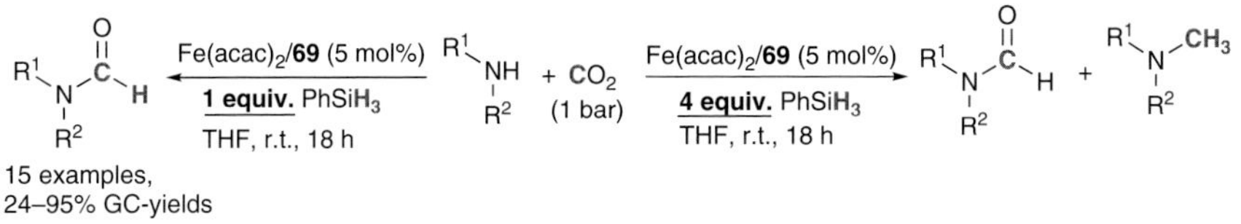

15 examples,
24–95% GC-yields

Scheme 10.13 $Fe(acac)_2/P(CH_2CH_2Ph_2)_3$ catalyst for reductive functionalization of CO_2.

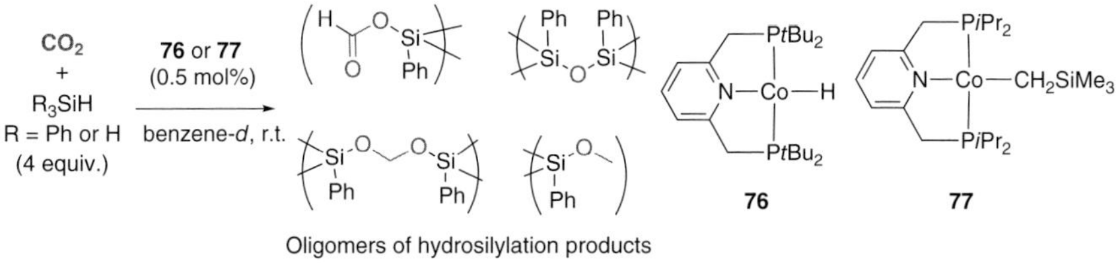

Oligomers of hydrosilylation products

Scheme 10.14 Pincer cobalt complexes for catalyzed hydrosilylation of CO_2.

selectively leads to the silylformate oligomer in two hours with both $PhSiH_3$ and Ph_2SiH_2. Interestingly, it was shown that the cobalt hydride complex is able to perform both the oxidative addition of the hydrosilanes and the insertion of CO_2.

10.5.2 Hydroboration of Carbon Dioxide

Hydroboration is another useful methodology for reducing carbon dioxide, and several systems including transition metals have demonstrated their efficiency in such transformations [90]. With iron, only few examples were recently reported. Sabo-Etienne, Bontemps, and coworkers have shown the way using iron dihydride complex $[(dmpe)_2Fe(H)_2]$ **78** as the catalyst (5 mol%) to perform the selective reduction of carbon dioxide (1 atm) into methoxyborane when using catecholborane (H–Bcat) in C_6D_6 for three hours (59% yield) [91]. When performing the reaction in THF for 47 minutes, with 9-borabicyclo[3.3.1]nonane (9-BBN), bis(boryl)acetal was selectively obtained in 85% with 8% methoxyborane as a side product (Scheme 10.15).

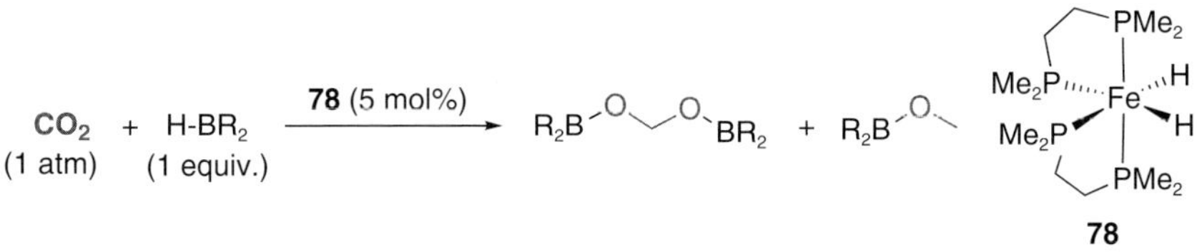

Scheme 10.15 Iron-catalyzed hydroboration of CO_2.

The bis(boryl)acetal derived from 9-BBN can be used as a surrogate of formaldeyde as a methylene source. Indeed, in a one-pot sequence, the *in situ*-obtained bis(boryl)acetal from CO_2 can react with 2,6-diisopropylaniline to give the corresponding methylene imine (83%), with 2-methylaminophenol to give the corresponding hemiaminal (67%), or with *N*,*N*-dimethyl-1,2-benzenediamine to produce the corresponding aminal (77%; Scheme 10.16).

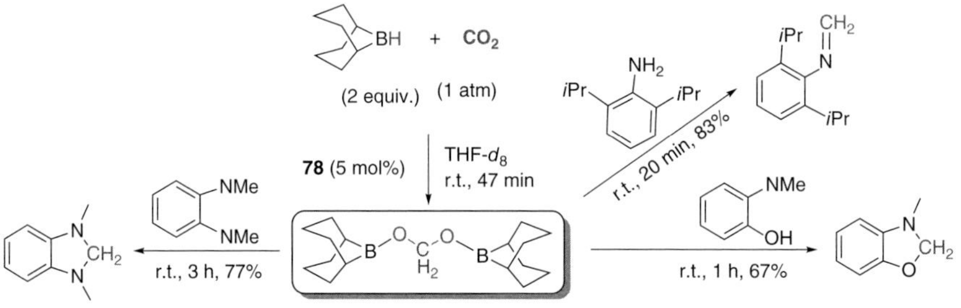

Scheme 10.16 Iron-catalyzed hydroboration/functionalization of CO_2.

Iron triphosphine complex **80** can also be used as a catalyst for the hydroboration of CO_2 (Scheme 10.17) [92]. In the presence of 5 equiv. of 9-BBN in d_3-acetonitrile, **80** (1.5 mol%) gives a mixture of methoxyborane and bis(boryl)acetal (95 : 5) in 75% conversion at 60 °C for 24 hours. After a prolonged reaction time (40 hours), full conversion was observed and the methoxyborane was obtained as the sole product in 99% conversion.

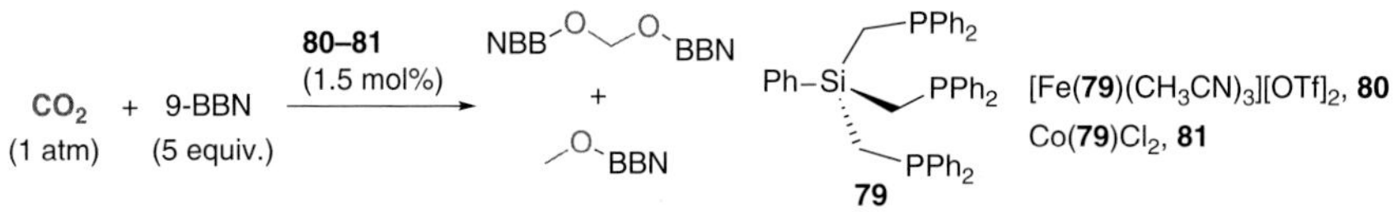

Scheme 10.17 Iron- and cobalt-catalyzed hydroboration of CO_2.

The only report on cobalt-catalyzed hydroboration of carbon dioxide was reported by Berthet and Cantat using tridentate cobalt complex **81** (1.5 mol%) in similar conditions (Scheme 10.17). A mixture of methoxyborane and bis(boryl)acetal (92 : 8) was obtained in full conversion by reaction at 60 °C for 24 hours. After a prolonged reaction time (40 hours), only the methoxyborane is obtained.

10.6 Conclusion

This chapter demonstrates that inexpensive, earth-abundant first-row transition metals such as iron, cobalt, and manganese can be efficient surrogates of noble metals in hydrosilylation and hydroboration of carbonyl and heterocarbonyl derivatives. Without any doubt, the chemoselective hydrosilylation of aldehydes and ketones with impressive functional group tolerances has been the most widely studied among the reactions discussed in this chapter. Importantly, even if iron, cobalt, or manganese do not yet surpass the performances of noble metals, recent pioneering studies on the hydrosilylation and hydroboration of carboxylic derivatives and carbon dioxide pave the way for the near future success in this area of reduction catalysis. Even if it is "difficult to see, always in motion is the future," [93] hydrosilylation and hydroboration reactions using earth-abundant transition metals are not on the dark side.

References

1 (a) For representative recent reviews and books on iron catalysis, see: Bolm, C., Legros, J., Le Paih, J., and Zani, L. (2004). *Chem. Rev.* 104: 6217. (b) Plietker, B. (ed.) (2008). *Iron Catalysis in Organic Chemistry*, 125–143. Weinheim: Wiley-VCH Verlag. (c) Bauer, E.B. (ed.) (2015). Topics in organometallic chemistry. In: *Iron Catalysis II*. (d) Sun, C.L., Li, B.J., and Shi, Z.J. (2011). *Chem. Rev.* 111: 1293. (e) Bézier, D., Sortais, J.-B., and Darcel, C. (2013). *Adv. Synth. Catal.* 355: 19. (f) Gopalaiah, K. (2013). *Chem. Rev.* 113: 3248. (g) Riener, K., Haslinger, S., Raba, A. et al. (2014). *Chem. Rev.* 114: 5215. (h) Bauer, I. and Knölker, H.-J. (2015). *Chem. Rev.* 115: 3170. (i) For representative recent reviews on manganese catalysis, see: Valyaev, D.A., Lavigne, G., and Lugan, N. (2016). *Coord. Chem. Rev.* 308: 191. (j) Khusnutdinov, R.I., Bayguzina, A.R., and Dzhemilev, U.M. (2012). *Russ. J. Org. Chem.* 48: 309. (k) Carney, J.R., Dillon, B.R., and Thomas, S.P. (2016). *Eur. J. Org. Chem.* 3912. (l) For representative reviews on cobalt catalysis, see: Moselage, M., Li, J., and Ackermann, L. (2016). *ACS Catal.* 6: 498. (m) Pellissier, H. and Clavier, H. (2014). *Chem. Rev.* 114: 2775. (n) Gosmini, C. and Moncomble, A. (2010). *Isr. J. Chem.* 50: 568. (o) Crossley, S.W.M., Obradors, C., Martinez, R.M., and Shenvi, R.A. (2016). *Chem. Rev.* 116: 8912.

2 (a) Representative reviews on iron-catalyzed reductions: Junge, K., Schröder, K., and Beller, M. (2011). *Chem. Commun.* 47: 4849. (b) Le Bailly, B.A.F. and Thomas, S.P. (2011). *RSC Adv.* 1: 1435. (c)Zhang, M. and Zhang, A. (2010). *Appl. Organomet. Chem.* 24: 751. (d) Morris, R.H. (2009). *Chem. Soc. Rev.* 38: 2282. (e) Gaillard, S. and Renaud, J.-L. (2008). *ChemSusChem* 1: 505. (f) Representative reviews on manganese-catalyzed reductions: see ref. [1i]; Representative reviews on cobalt-catalyzed reductions: Röse, P. and Hilt, G. (2016). *Synthesis* 463. (g) Quintard, A. and Rodriguez, J. (2016). *ChemSusChem* 9: 28.

3 For a representative example, see: Jacobsen, E.N., Pfaltz, A., and Yamamoto, H. (eds.) (1999). *Comprehensive Asymmetric Catalysis*. Berlin: Springer.

4 Brunner, H. and Fisch, K. (1990). *Angew. Chem. Int. Ed. Engl.* 29: 1131; (1990). *Angew. Chem.* 102: 1189.

5 Shaikh, N.S., Junge, K., and Beller, M. (2007). *Org. Lett.* 9: 5429.

6 Addis, D., Shaikh, N.S., Zhou, S. et al. (2010). *Chem. Asian J.* 5: 1687.

7 Plietker, B. and Dieskau, A. (2009). *Eur. J. Org. Chem.* 775.

8 (a) Dieskau, A., Begouin, J.-M., and Plietker, B. (2011). *Eur. J. Org. Chem.* 5291. (b) Rommel, S., Belger, C., Begouin, J.-M., and Plietker, B. (2015). *ChemCatChem* 7: 1292.

9 Nishiyama, H. and Furuta, A. (2007). *Chem. Commun.* 760.

10 Furuta, A. and Nishiyama, H. (2008). *Tetrahedron Lett.* 49: 110.

11 Muller, K., Schubert, A., Jozak, T. et al. (2011). *ChemCatChem* 3: 887.
12 Shaikh, N.S., Enthaler, S., Junge, K., and Beller, K.M. (2008). *Angew. Chem. Int. Ed.* 47: 2497; (2008). *Angew. Chem.* 120: 2531.
13 Inagaki, T., Phong, L.T., Furuta, A. et al. (2010). *Chem. Eur. J.* 16: 3090.
14 (a) See ref. [9]. (b) Tondreau, A.M., Darmon, J.M., Wile, B.M. et al. (2009). *Organometallics* 28: 3928.
15 Flückiger, M. and Togni, A. (2011). *Eur. J. Org. Chem.* 4353.
16 Yang, J. and Tilley, T.D. (2010). *Angew. Chem. Int. Ed.* 49: 10186; (2010). *Angew. Chem.* 112: 10384.
17 Xue, B., Sun, H., and Li, X. (2015). *RSC Adv.* 5: 52000.
18 (a) Zuo, Z., Sun, H., Wang, L., and Li, X. (2014). *Dalton Trans* 43: 1176. (b) Wang, L., Sun, H., and Li, X. (2015). *Eur. J. Inorg. Chem.* 2732.
19 Tondreau, A.A., Lobkovsky, E., and Chirik, P.J. (2008). *Org. Lett.* 10: 2789.
20 Bhattacharya, P., Krause, J.A., and Guan, H. (2011). *Organometallics* 30: 4720.
21 (a) Bhattacharya, P., Krause, J.A., and Guan, H. (2014). *Organometallics* 33: 6113. (b) Chakraborty, S., Bhattacharya, P., Dai, H., and Guan, H. (2015). *Acc. Chem. Res.* 48: 1995.
22 Wu, S., Li, X., Xiong, Z. et al. (2013). *Organometallics* 32: 3227.
23 Zhao, H., Sun, H., and Li, X. (2014). *Organometallics* 33: 3535.
24 Tondreau, A.M., Darmon, J.M., Wile, B.M. et al. (2009). *Organometallics* 28: 3928.
25 Hosokawa, S., Ito, J.-I., and Nishiyama, H. (2010). *Organometallics* 29: 5773.
26 Zuo, Z., Zhang, L., Leng, X., and Huang, Z. (2015). *Chem. Commun.* 51: 5073.
27 Inagaki, T., Ito, A., Ito, J.;-i., and Nishiyama, H. (2010). *Angew. Chem. Int. Ed.* 49: 9384; (2010). *Angew. Chem.* 122: 9574.
28 Langlotz, B.K., Wadepohl, H., and Gade, L.H. (2008). *Angew. Chem. Int. Ed.* 47: 4670; (2008). *Angew. Chem.* 120: 4748.
29 Bleith, T., Wadepohl, H., and Gade, L.H. (2015). *J. Am. Chem. Soc.* 137: 2456.
30 Bleith, T. and Gade, L.H. (2016). *J. Am. Chem. Soc.* 138: 4972.
31 (a) See ref. [4]. (b) Brunner, H. and Fisch, K. (1991). *J. Organomet. Chem.* 412: C11. (c) Brunner, H. and Rötzer, M. (1992). *J. Organomet. Chem.* 425: 119.
32 Gutsulyak, D.V., Kuzmina, L.G., Howard, J.A.K. et al. (2008). *J. Am. Chem. Soc.* 130: 3732.
33 Zheng, J., Misal Castro, L.C., Roisnel, T. et al. (2012). *Inorg. Chim. Acta* 380: 301.
34 (a) Kandepi, V., V.K.M., Cardoso, J.M.S., Peris, E., and Royo, B. (2010). *Organometallics* 29: 2777. (b) Cardoso, J.M.S., Fernandes, A., Cardoso, B.d.P. et al. (2014). *Organometallics* 33: 5670.
35 Jiang, F., Bézier, D., Sortais, J.-B., and Darcel, C. (2011). *Adv. Synth. Catal.* 353: 239.
36 Bézier, D., Jiang, F., Roisnel, T. et al. (2013). *Eur. J. Inorg. Chem.* 1333.
37 Demir, S., Gökçe, Y., Kaloğlu, N. et al. (2013). *Appl. Organometal. Chem.* 27: 459.
38 César, V., Misal Castro, L.C., Dombray, T. et al. (2013). *Organometallics* 32: 4643.
39 Kumar, D., Prakasham, A.P., Bheeter, L.P. et al. (2014). *J. Organomet. Chem.* 762: 81.

40 (a) Buitrago, E., Zani, L., and Adolfsson, H. (2011). *Appl. Organometal. Chem.* 25: 748. (b) Buitrago, E., Tinnis, F., and Adolfsson, H. (2012). *Adv. Synth. Catal.* 354: 217.
41 Hashimoto, T., Urban, S., Hoshino, R. et al. (2012). *Organometallics* 31: 4474.
42 Warratz, S., Postigo, L., and Royo, B. (2013). *Organometallics* 32: 893.
43 Wekesa, F.S., Arias-Ugarte, R., Kong, L. et al. (2015). *Organometallics* 34: 5051.
44 Zhu, K., Shaver, M.P., and Thomas, S.P. (2015). *Eur. J. Org. Chem.* 2119.
45 Dal Zotto, R.C., Virieux, D., and Campagne, J.M. (2009). *Synlett* 276.
46 (a) Some representatives examples of reviews: Chirik, P.J. (2015). *Acc. Chem. Res.* 48: 1687. (b) Zell, T. and Milstein, D. (2015). *Acc. Chem. Res.* 48: 1979. (c) Foubelo, F., Najera, C., and Yus, M. (2015). *Tetrahedron: Asymmetry* 26: 769. (d) Li, Y.-Y., Yu, S.-L., Shen, W.-Y., and Gao, J.-X. (2015). *Acc. Chem. Res.* 48: 2587.
47 For a recent example, see: Jiang, S. and Liang, D. (2016). *ACS Catal.* 6: 290.
48 Brunner, H. and Amberger, K. (1991). *J. Organomet. Chem.* 417: C63.
49 Niu, Q., Sun, H., Li, X. et al. (2013). *Organometallics* 32: 5235.
50 Zhou, H., Sun, H., Zhang, S., and Li, X. (2015). *Organometallics* 34: 1479.
51 Smith, A.D., Saini, A., Singer, L.M. et al. (2016). *Polyhedron* 114: 286.
52 Inagaki, T., Phong, L.T., Furuta, A. et al. (2010). *Chem. Eur. J.* 16: 3090.
53 Yu, F., Zhang, X.-C., Wu, F.-F. et al. (2012). *Org. Biomol. Chem.* 9: 5652.
54 Sauer, D.C., Wadepohl, H., and Gade, L.H. (2012). *Inorg. Chem.* 51: 12948.
55 Yates, R.L. (1982). *J. Catal.* 78: 111.
56 (a) Gregg, B.T., Hanna, P.K., Crawford, E.J., and Cutler, A.R. (1991). *J. Am. Chem. Soc.* 113: 384. (b) Gregg, B.T. and Cutler, A.R. (1996). *J. Am. Chem. Soc.* 118: 10069.
57 Hanna, P.K., Gregg, B.T., and Cutler, A.R. (1991). *Organometallics* 10: 31.
58 (a) Son, S.U., Paik, S.-J., Lee, I.S. et al. (1999). *Organometallics* 18: 4114. (b) Son, S.U., Paik, S.-J., and Chung, Y.K. (2000). *J. Mol. Catal. A* 151: 87.
59 Zheng, J., Elangovan, S., Valyaev, D.A. et al. (2014). *Adv. Synth. Catal.* 356: 1093.
60 Chidara, V.K. and Du, G. (2013). *Organometallics* 32: 5034.
61 (a) Mukhopadhyay, T.K., Flores, M., Groy, T.L., and Trovitch, R.J. (2014). *J. Am. Chem. Soc.* 136: 882. (b) Ghosh, C., Mukhopadhyay, T.K., Flores, M. et al. (2015). *Inorg. Chem.* 54: 10398.
62 Magnus, P. and Fielding, M.R. (2001). *Tetrahedron Lett.* 42: 6633.
63 Misal Castro, L.C., Sortais, J.-B., and Darcel, C. (2012). *Chem. Commun.* 48: 151.
64 Bhunia, M., Hota, P.K., Vijaykumar, G. et al. (2016). *Organometallics* 35: 2930.
65 Enthaler, S. (2010). *ChemCatChem* 2: 1411.
66 Jaafar, H., Li, H., Misal Castro, L.C. et al. (2012). *Eur. J. Inorg. Chem.* 3546.
67 Li, H., Achard, M., Bruneau, C. et al. (2014). *RSC Adv.* 4: 25892.
68 Kumar, V., Sharma, U., Verma, P.K. et al. (2012). *Adv. Synth. Catal.* 354: 870.
69 Zhou, S., Junge, K., Addis, D. et al. (2009). *Angew. Chem. Int. Ed.* 48: 9507; (2009). *Angew. Chem.* 121: 9671.
70 (a) Sunada, Y., Kawakami, H., Motoyama, Y., and Nagashima, H. (2009). *Angew. Chem. Int. Ed.* 48: 9511; (2009). *Angew. Chem.* 121: 9675.

(b) Tsutsumi, H., Sunada, Y., and Nagashima, H. (2011). *Chem. Commun.* 47: 6581.
71 Bézier, D., Venkanna, G.T., Sortais, J.-B., and Darcel, C. (2011). *ChemCatChem* 3: 1747.
72 Volkov, A., Buitrago, E., and Adolfsson, H. (2013). *Eur. J. Org. Chem.* 2066.
73 Blom, B., Tan, G., Enthaler, S. et al. (2013). *J. Am. Chem. Soc.* 135: 18108.
74 Zhou, S., Addis, D., Das, S. et al. (2009). *Chem. Commun.* 4883.
75 Das, S., Wendt, B., Moeller, K. et al. (2012). *Angew. Chem. Int. Ed.* 51: 1662; (2012). *Angew. Chem.* 124: 1694.
76 Pouessel, J., Jacquet, O., and Cantat, T. (2013). *ChemCatChem* 5: 3552.
77 Dombray, T., Helleu, C., Darcel, C., and Sortais, J.-B. (2013). *Adv. Synth. Catal.* 355: 3358.
78 Bézier, D., Venkanna, G.T., Misal Castro, L.C. et al. (2012). *Adv. Synth. Catal.* 354: 1879.
79 Junge, K., Wendt, B., Zhou, S., and Beller, M. (2013). *Eur. J. Org. Chem.* 2061.
80 Ruddy, A.J., Kelly, C.M., Crawford, S.M. et al. (2013). *Organometallics* 32: 5581.
81 Das, S., Li, Y., Junge, K., and Beller, M. (2012). *Chem. Commun.* 48: 10742.
82 Li, H., Misal Castro, L.C., Zheng, J. et al. (2013). *Angew. Chem. Int. Ed.* 52: 8045; (2013). *Angew. Chem.* 125: 8203.
83 Cong, C., Fujihara, T., Terao, J., and Tsuji, Y. (2014). *Catal. Commun.* 50: 25.
84 Mao, Z., Gregg, B.T., and Cutler, A.R. (1995). *J. Am. Chem. Soc.* 117: 10139.
85 Misal Castro, L.C., Li, H., Sortais, J.-B., and Darcel, C. (2012). *Chem. Commun.* 48: 10514.
86 Zheng, J., Chevance, S., Darcel, C., and Sortais, J.-B. (2013). *Chem. Commun.* 49: 10010.
87 For selective recent reviews, see: (a)Tlili, A., Blondiaux, E., Frogneux, X., and Cantat, T. (2015). *Green Chem.* 17: 157. (b) Fernández-Alvarez, F.J., Aitanib, A.M., and Oro, L.A. (2014). *Catal. Sci. Technol.* 4: 611.
88 Frogneux, X., Jacquet, O., and Cantat, T. (2014). *Catal. Sci. Technol.* 4: 1529.
89 Scheuermann, M.L., Semproni, S.P., Pappas, I., and Chirik, P.J. (2015). *Inorg. Chem.* 53: 9463.
90 (a) For recent reviews on hydroboration of CO_2, see: Chong, C.C. and Kinjo, R. (2015). *ACS Catal.* 5: 3228. (b) Bontemps, S. (2016). *Coord. Chem. Rev.* 308: 117.
91 Guanghua, J., Werncke, C.G., Escudié, Y. et al. (2015). *J. Am. Chem. Soc.* 137: 9563.
92 Aloisi, A., Berthet, J.-C., Genre, C. et al. (2016). *Dalton Trans.* 45: 14774.
93 Yoda to Luke in Star Wars-Episode V – The Empire Strikes Back (1980).

11

Reactive Intermediates and Mechanism in Iron-Catalyzed Cross-coupling

Jared L. Kneebone, Jeffrey D. Sears, and Michael L. Neidig

University of Rochester, Department of Chemistry, Rochester, NY 14627-0216, USA

11.1 Introduction

Applications of iron catalysis in synthetic organic chemistry continue to develop at an accelerated rate, largely motivated by the metal's abundance, sustainability, and redox versatility. Iron-catalyzed carbon–carbon (C–C) cross-coupling continues to be one of the fastest developing subfields with an ever-growing number of methods highlighting the efficient coupling of a variety of activated and unactivated electrophiles with nucleophiles. Methodology development has been the subject of numerous thorough reviews [1], and as the field continues to grow, more emphasis has been placed on obtaining fundamental insight into the mechanistic foundations of these reactions. In contrast to widely used and well-studied palladium- and nickel-catalyzed C–C cross-couplings, the study of iron-based processes at the molecular level has historically been more challenging because of the prevalence of paramagnetic organoiron intermediates that can form during catalysis and the diversity of oxidation and spin states accessible by iron. The continued development of the field heavily depends on discerning mechanism and identifying reactive iron intermediates within these reactions, knowledge that can then facilitate the design of improved cross-coupling systems.

The goal of this chapter is to highlight the important achievements targeting these ends [2], with methods development presented to summarize the historical development of the field and provide context for reported mechanistic studies. Emphasis is placed on experimental approaches that are used historically and more recently to elucidate reactive intermediates and comparisons, and contrasts are drawn within the "ligandless" catalytic systems and those employing supporting ligands or ligand additives. The first section summarizes work done in the field of cross-coupling catalysis using simple iron salts, a field that traverses a wide array of electrophilic and nucleophilic coupling partners. The following sections focus on systems using iron salts combined with *N*,*N*,*N*′,*N*′-tetramethylethylenediamine (TMEDA), *N*-heterocyclic carbenes (NHCs), and phosphines as additives or supporting ligands within well-defined iron precatalysts.

Non-Noble Metal Catalysis: Molecular Approaches and Reactions,
First Edition. Edited by Robertus J. M. Klein Gebbink and Marc-Etienne Moret.

11.2 Cross-coupling Catalyzed by Simple Iron Salts

11.2.1 Methods Overview

Although documentation of the use of iron salts to facilitate C–C homocoupling [3] and cross-coupling [4] dates back over 70 years, the first systematic approaches to thoroughly define the reactivity in cross-coupling catalysis specifically were conducted by Kochi in the 1970s [5]. Before the seminal reports of palladium- and nickel-catalyzed C–C cross-coupling reactions, Kochi and coworkers demonstrated simple ferric salts to be effective in facilitating the coupling of alkyl Grignard reagents with alkenyl electrophiles. Notably, these reactions proceeded with fast reaction times at low temperature using a number of ferric precatalysts including the β-diketonate salts $Fe(acac)_3$ (acac, acetylacetonate) and $Fe(dbm)_3$ (dbm, dibenzoylmethide) (Scheme 11.1). Kochi and coworkers reported facile stereospecific alkylations of alkenyl substrates as well as the facile alkyl disproportionation that can occur when attempting to couple alkyl electrophiles with alkyl Grignard reagents. These results combined with kinetic and spectroscopic mechanistic analyses laid the groundwork for further research in iron-based cross-coupling methods. Despite these seminal reports, the development of synthetically useful iron-catalyzed cross-coupling methods would take another two decades.

Through the 1980s and early 1990s, the work of Molander et al. [10] and Cahiez and coworker [6] generalized Kochi's original protocols and gave birth

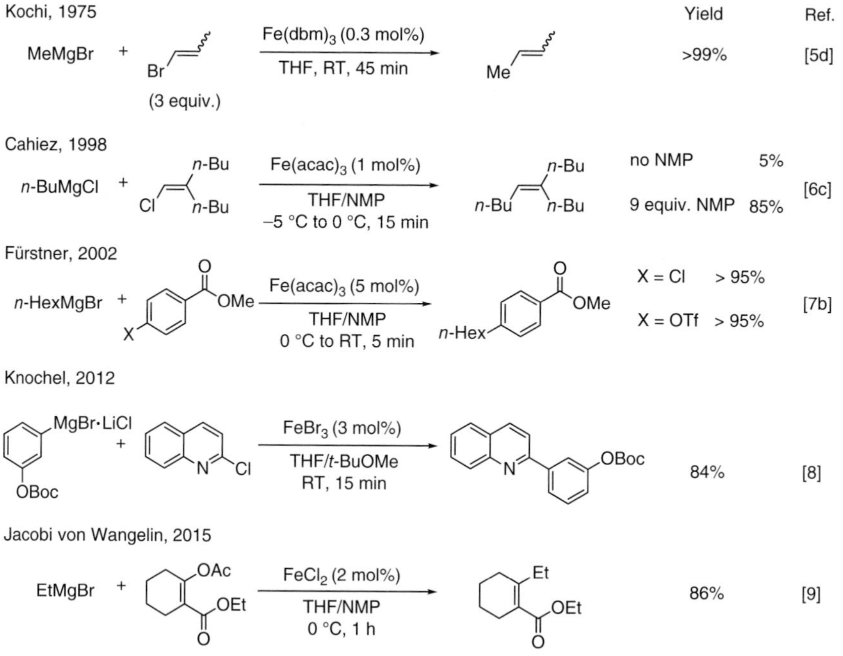

Kochi, 1975: MeMgBr + vinyl bromide (3 equiv.); $Fe(dbm)_3$ (0.3 mol%), THF, RT, 45 min; Yield >99%; Ref. [5d]

Cahiez, 1998: n-BuMgCl + alkenyl chloride; $Fe(acac)_3$ (1 mol%), THF/NMP, −5 °C to 0 °C, 15 min; no NMP 5%, 9 equiv. NMP 85%; [6c]

Fürstner, 2002: n-HexMgBr + aryl X (OMe ester); $Fe(acac)_3$ (5 mol%), THF/NMP, 0 °C to RT, 5 min; X = Cl > 95%, X = OTf > 95%; [7b]

Knochel, 2012: aryl MgBr·LiCl (OBoc) + 2-chloroquinoline; $FeBr_3$ (3 mol%), THF/t-BuOMe, RT, 15 min; 84%; [8]

Jacobi von Wangelin, 2015: EtMgBr + enol acetate (OAc, OEt ester); $FeCl_2$ (2 mol%), THF/NMP, 0 °C, 1 h; 86%; [9]

Scheme 11.1 Representative examples of C–C cross-coupling reactions using simple iron salt precatalysts.

to a renewed interest in developing practical C—C bond formations using iron salts. These studies highlighted the ability of alternative ethereal solvents and cosolvents such as dimethoxyethane (DME) and *N*-methyl-2-pyrrolidone (NMP) to promote efficient alkylation of alkenyl substrates without the need for excess electrophile (Scheme 11.1). Later, Cahiez and coworkers reported facile coupling of functionalized aryl Grignard reagents with alkenyl bromides and iodides under mild conditions using catalytic $Fe(acac)_3$ [11]. The scope of couplings using alkenyl substrates was further expanded by Fürstner and coworkers, Nakamura and coworkers, and Cahiez et al. who respectively highlighted alkenyl triflates as effective coupling partners [12], demonstrated the accessibility of ene-ynes through the coupling of alkenyl electrophiles with alkynyl Grignard reagents [13], and documented the reactivity of vinyl phosphates [14]. Recent work by Jacobi von Wangelin and coworkers further generalized the application of electrophilic pseudohalides. Here, the combination of $FeCl_2$ with NMP cosolvent afforded the coupling of cheap and synthetically relevant alkenyl acetates with primary and secondary alkyl Grignard reagents (Scheme 11.1) [9].

Electrophile scope was expanded to aryl and heteroaryl coupling partners by Fürstner et al. [7] and Figadere and coworkers [15] while obviating the need for directing groups on the electrophilic substrates [16]. Fürstner's work in particular contained ample scope and mild reaction conditions by employing catalytic $Fe(acac)_3$ for coupling aryl and heteroaryl chlorides, tosylates, and triflates (Scheme 11.1) [7, 17]. Biaryl couplings have recently received increased attention by Jacobi von Wangelin and coworker [18] and Knochel and coworkers [8]. The efficient couplings of aryl Grignard reagents with chlorostyrenes using catalytic $Fe(acac)_3$ and pyridyl chlorides using catalytic $FeBr_3$ represent two of these examples (see Scheme 11.1 for an example from Knochel and coworkers).

The number of iron-salt-catalyzed methods using alkyl electrophiles has also increased in the last decade. In 2004, Hayashi and coworker reported a generalized method for coupling aryl Grignard reagents with primary and secondary alkyl halides [19]. Despite the propensity for β-hydrogen elimination within these substrates, good yields were achieved in short reaction times using catalytic $Fe(acac)_3$. The same year, Fürstner and coworker demonstrated the ease of accessing similar aryl–alkyl couplings through the use of a low-valent iron precatalyst [20]. Subsequently, Nakamura and coworker reported efficient couplings of α-bromocarboxylic acid derivatives with aryl Grignard reagents at low temperatures using $Fe(acac)_3$ [21], and Hu and coworkers demonstrated the first examples of "ligandless" C_{sp3}–C_{sp} couplings of a range of secondary alkyl halides and alkynyl Grignard reagents [22].

11.2.2 Mechanistic Investigations

The earliest investigations targeting a broad understanding of the activation of ferric salts during cross-coupling catalysis came in Kochi's seminal reports on the alkylation of alkenyl halides [5b,f]. Evidence for the rapid reduction of iron(III) precatalysts such as $FeCl_3$, $Fe(acac)_3$, and $Fe(dbm)_3$ upon treatment with alkyl Grignard reagents was gained through electron paramagnetic resonance (EPR) analysis of the reaction mixtures. The formation of a $S = 1/2$ species even at low temperatures resulted in the proposal of the generation of an iron(I)-active species following initial treatment of iron with nucleophile. Oxidative addition of electrophile to such an intermediate and subsequent transmetalation of iron by the Grignard reagent was proposed to follow. This organoiron(III) species was then proposed to undergo reductive elimination to release cross-coupled product and reform the active iron(I) resting state (Scheme 11.2a).

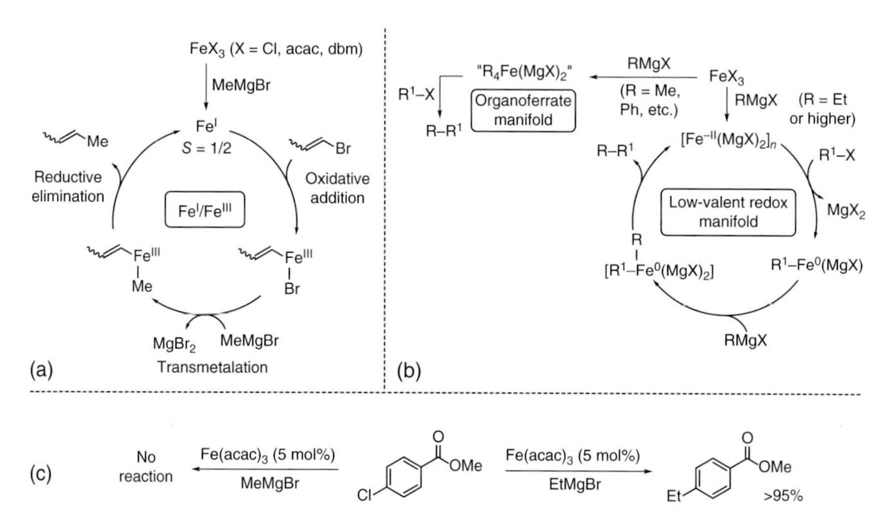

Scheme 11.2 Mechanistic proposals for C–C cross-coupling reactions catalyzed by iron salts from (a) Kochi and (b) Fürstner; (c) distinct reactivity of methyl and ethyl Grignard reagents with aryl chlorides observed by Fürstner.

This canonical mechanism is similar to those conventionally operative in palladium- and nickel-catalyzed cross-couplings. Importantly, at the time of his proposal, Kochi and coworkers was not able to rule out the formation of iron(0) during the initial reduction of iron with Grignard reagent, and the early 2000s saw renewed efforts by Fürstner and coworkers to further interrogate these mechanistic foundations. Inspired by the work of Bogdanovic and coworkers on low-valent inorganic Grignard reagents (IGRs) [23], Fürstner et al. proposed that alkyl Grignard reagents bearing β-hydrogens would favor an Fe(–II)/Fe(0) redox cycle. Reduction to low-valent iron was proposed to result from β-hydrogen elimination from intermediate transmetalated iron species during catalyst activation. Alternatively, catalysis employing nucleophiles lacking β-hydrogens was postulated to proceed through iron(II) organoferrate-active species (Figure 11.2b) [17]. The observation of disparate reactivity of methyl and ethyl Grignard reagents (MeMgBr and EtMgBr) toward aryl chlorides was an important result that, in part, led to this proposal of distinct regimes of reactivity (Scheme 11.2c). In support of the proposed Fe(–II)/Fe(0) manifold, Fürstner et al. demonstrated the effectiveness of low-valent iron model complexes as well-defined precatalysts in the cross-couplings of aryl electrophiles and higher alkyl Grignard reagents [17, 20]. Interestingly, it was observed that the

Table 11.1 Evaluation of low-valent precatalysts **1** and **2** by Fürstner and coworkers.

R–X + R′–MgX → (Precatalyst (mol%), Conditions) R–R′

(TMEDA) Li---Fe^{-II}---Li (TMEDA) **1**

Fe^{0}---Li (TMEDA) **2**

Entry	Precatalyst (mol%)	R–X	R′–MgX	Conditions	Yield of R–R′ (%)
1	$Fe(acac)_3$	4-F_3C-C_6H_4-Cl	*n*-HexMgX	THF/NMP, 0 °C	80
2	**1**				85
3	**1**	cyclohexyl-Br	*p*-tolMgBr	THF, −20 °C	95
4	**2**		PhMgBr	THF, 0 °C to RT	81
5	**1**	Br-CH$_2$CH=CHC(O)CH$_3$	PhMgBr	THF, −30 °C, <10 min	94
6	**2**			THF, −30 °C, 30 min	45

tetrakis(ethylene) iron(−II) complex **1** [24] was just as effective as $Fe(acac)_3$ for these transformations (Table 11.1, entries 1 and 2). The effectiveness of **1** was compared to other low-valent mo del complexes including the iron(0) complex **2** [25, 26] (Table 11.1, entries 3–6), resulting in comparably high yields of arylated cyclohexanes [17].

Fürstner's investigations of the potential role of transmetalated iron(II) ferrates as reactive intermediates in reactions employing nucleophiles lacking β-hydrogens provided the first context which tests Kochi's proposal of an active iron species formed using MeMgBr. Fürstner et al. found that the reaction of $FeCl_3$ with excess methyllithium (MeLi) at low temperature in diethyl ether (Et_2O) resulted in the formation of iron(II) ferrate $[(FeMe_4)(MeLi)][Li(Et_2O)]_2$ (**3**) (Scheme 11.3a) [27]. Similarly, the low temperature treatment of $FeCl_2$ with 4 equiv. of phenyllithium (PhLi) resulted in the isolation of a phenylated iron(II) ferrate species $[FePh_4][Li(Et_2O)_2][Li(dioxane)]$ (**4**) (Scheme 11.3b) [17]. Complex **3** was found to methylate activated electrophiles such as acid chlorides (Scheme 11.3c) and enol triflates. Despite this observed reactivity, the proposal of a mononuclear iron(II)-active species in cross-coupling catalysis contrasted Kochi's original observations of a $S = 1/2$ EPR-active species upon reaction of ferric salts with MeMgBr. Furthermore, an iron(II) center would be incapable of producing such an EPR feature unless it is part of a mixed-valent system. Fürstner et al. also noted that dissolution of red **3** in tetrahydrofuran (THF) results in yellow solutions, an observation that is suggestive of a structural change when exposed to catalytically relevant solvent.

Numerous theoretical investigations probing thermodynamics and transition states within both mechanistic cycles summarized in Scheme 11.2 have also been reported. Norrby and coworkers worked to discern between an Fe(I)/Fe(III)

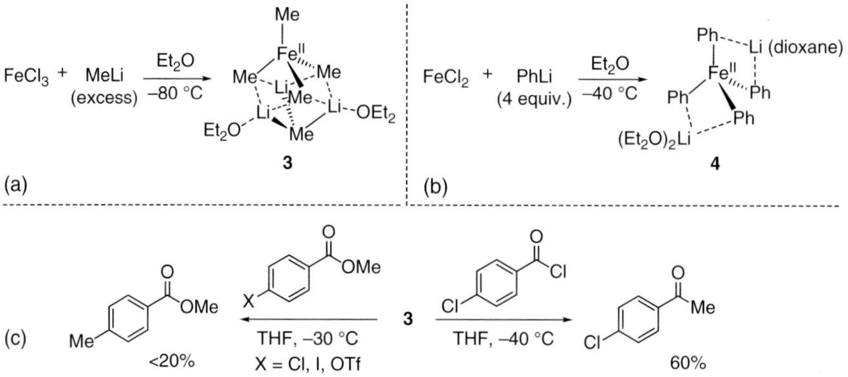

Scheme 11.3 (a and b) Synthesis of homoleptic methylated and phenylated iron(II) ferrates **3** and **4** using organolithium reagents; (c) methylation reactivity of **3** toward aryl and acyl electrophilic positions.

and an Fe(–II)/Fe(0) mechanism through the combination of density functional theory calculations with Hammett reactivity studies in a model reaction employing aryl electrophiles and alkyl Grignard reagents. Hammett studies pointed to sequential oxidative addition/transmetalation steps during catalysis, although the order was not concluded [28]. Similar proposals from Hammett studies were developed in the case of coupling alkyl electrophiles with aryl Grignard reagents. Furthermore, in the case of the reaction of aryl electrophiles with alkyl Grignard reagents, computations of free energies of reductive elimination from iron(–II) centers highlighted unfavorable thermodynamics relative to iron(III) species [29]. Sequential atom transfer processes were suggested to occur in place of a formal two-electron oxidative addition during these catalytic cycles when calculations were performed on an iron alkoxide catalyst [30]. Ren et al. investigated the reaction coordinates calculated for a catalytic cycle involving IGR-type active species of the explicit form $Fe(MgX)_2$, providing insight into the proposed Fe(–II)/Fe(0) cycle [31]. Although insightful, it should be noted that this work heavily relied on proposed active catalyst structures that lack experimental confirmation. Subsequently, Heggen and Thiel conducted a theoretical investigation of the methylation of chlorobenzoyl chloride by Fürstner's complex **3** [32]. The results provided support for the formation of cross-coupled product through a direct substitution mechanism. Experimentally, however, it remains unclear as to whether **3** maintains its solid-state structure in solution under catalytic conditions.

Neidig and coworkers recently reported further discernment of the structural nature of *in situ*-formed iron species when using the solvent and nucleophile employed in Kochi's original catalytic methods. Methylation of $FeCl_3$ with 4 equiv. of MeMgBr at low temperature in THF was observed to produce homoleptic iron(III) ferrate $[MgClTHF_5][FeMe_4]$ (**5**) (Figure 11.1a,b) [33]. This highly temperature-sensitive, distorted square planar species was found to be intermediate spin ($S = 3/2$) in THF solution. Using freeze-trapped solution, EPR **5** was found to be stable at −80 °C in solution, but upon warming, it was

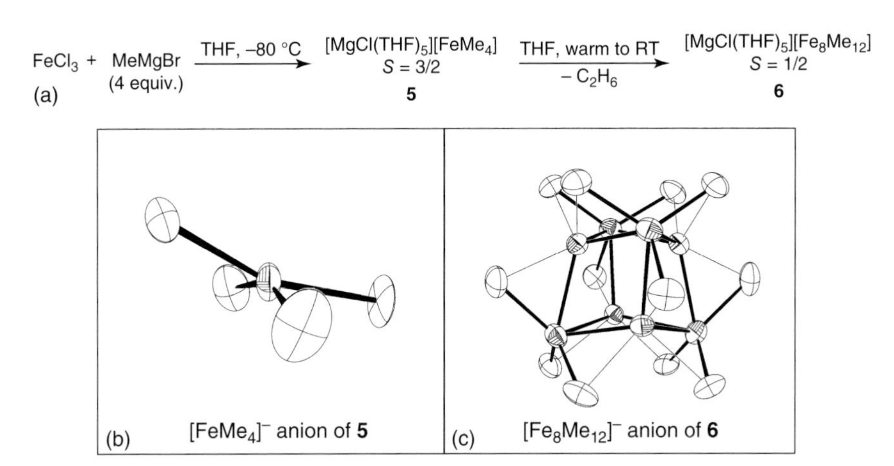

Figure 11.1 (a) Synthesis of homoleptic ferrates **5** and **6** using MeMgBr in THF; solid-state molecular structures of (b) **5** and (c) **6**. Thermal ellipsoids are drawn at the 50% probability level with octant-shaded ellipsoids representing the iron atoms. Cations and hydrogen atoms have been removed for clarity.

observed to convert to a species characterized by a broad $S = 1/2$ EPR signal with concomitant formation of ethane. The $S = 1/2$ species that forms upon thermal decay of **5** is analogous to the spectrum obtained by Kochi, suggesting **5** to be an intermediate in the reduction pathway of iron(III) with MeMgBr.

Neidig and coworkers further demonstrated that recooling mixtures rich in the $S = 1/2$ component could afford the stabilization and isolation of the multinuclear cluster $[MgCl(THF)_5][Fe_8Me_{12}]$ (**6**) (Figure 11.1a,c) [34]. When redissolved in THF, cluster **6**, which is formally mixed valent, is characterized by the same broad $S = 1/2$ EPR signal originally observed by Kochi and coworkers. Assignment of **6** as this originally observed EPR-active species was confirmed through comparison of near-infrared magnetic circular dichroism (NIR MCD) spectra of isolated **6** with samples of *in situ*-generated mixtures from the reaction of $FeCl_3$ or $Fe(acac)_3$ with excess MeMgBr. This approach highlights the power of combining concrete structural characterization with solid state and *in situ* spectroscopic assignments in providing rigorous characterization of the formation, identity, and stability of intermediates such as **6**.

Investigations of **6** as a reactive species in the methylation of β-bromostyrene using $FeCl_3$ were conducted via gas chromatographic (GC) analysis of organic product mixtures combined with spin-quantified, freeze-trapped EPR spectroscopy. As shown in Figure 11.2a,b, limited consumption of **6** and minimal yield of cross-coupled product were detected upon the direct reaction of **6** with β-bromostyrene, but addition of MeMgBr following the initial reaction of **6** with electrophile resulted in rapid and selective formation of cross-coupled product. Notably, Kochi and coworkers hypothesized that the turnover may be dependent on the presence of excess nucleophile, invoking the possibility that an initial intermediate species may form through reaction of **6** with electrophile. Such a species would then require additional nucleophile to facilitate the formation of cross-coupled product [5f]. Importantly, the identification of an iron cluster as

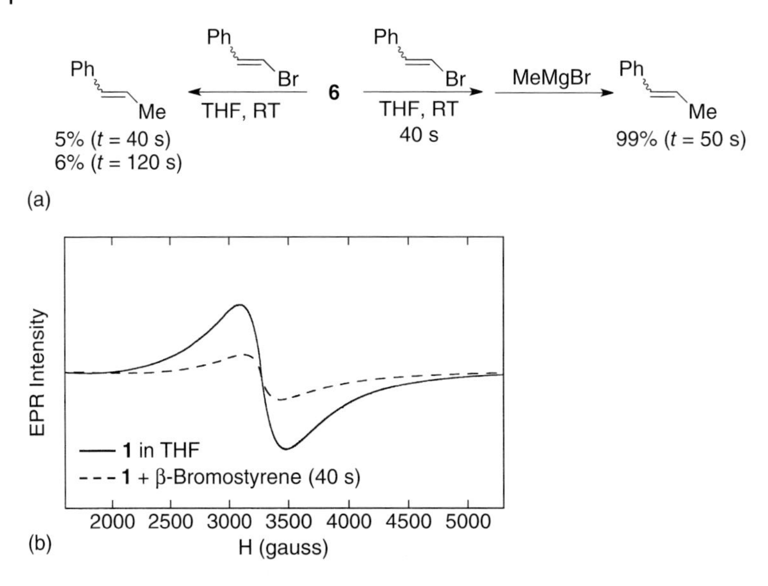

Figure 11.2 (a) Reactivity of **6** with β-bromostyrene in the absence and presence of MeMgBr; (b) 10 K EPR spectrum demonstrating consumption of **6** upon initial reaction with β-bromostyrene in THF. Reprinted with permission from (2016) *J. Am. Chem. Soc.* 138: 7492. Copyright 2016 American Chemical Society.

a reactive species in iron-salt-catalyzed cross-coupling reactions represents a new paradigm in catalyst structure, contrasting previous proposals centered on mononuclear iron species.

Examinations of reactive iron species formed *in situ* upon the treatment of ferric or ferrous salts with phenyl nucleophiles have remain less developed than those with methyl nucleophiles. To date, there exist limited examples of structurally well-characterized homoleptic iron complexes with aryl ligands lacking sterically encumbered alkyl substituents [17, 35]. The combination of Fürstner's proposals of active phenylated ferrates in cross-coupling catalysis with recent work interrogating the reduction pathways of ferric salts with methyl nucleophiles has spawned curiosity as to the structure, reactivity, and pathways to phenylated iron intermediates formed under catalytically relevant conditions. To this end, Bauer and coworkers combined X-ray absorption spectroscopy (XAS) with GC analysis of organic product distributions to provide evidence for reduction to iron(I) following reaction of $Fe(acac)_3$ with excess phenylmagnesium bromide (PhMgBr) [36]. The formation of iron clusters was suggested to be an integral part of the proposed Fe(I)/Fe(III) cycle for the coupling of aryl electrophiles with alkyl Grignard reagents. In addition, electrochemical analyses conducted by Lefevre and Jutand on the same reduction pathway were combined with nuclear magnetic resonance (NMR) and EPR spectroscopic studies to support an Fe(I)/Fe(III) catalytic redox couple. Here, the authors proposed an acac-coordinated monomeric iron(I)–phenyl-active species $[PhFe^{I}(acac)(THF)]^-$ [37].

Very recently, Parchomyk and Koszinowski employed negative-ion mode electrospray ionization mass spectroscopy (ESI-MS) to detect and assign a number of mononuclear ferrates, resulting from phenylation of $Fe(acac)_3$ [38]. A phenylated iron(II) ferrate (or an associated equilibrium species) was assigned as the active catalyst in the cross-coupling of PhMgBr with isopropyl chloride. Reductive elimination to selectively form cross-coupled product was then proposed to occur from an iron(III) ferrate $[Ph_3Fe^{III}(i\text{-}Pr)]^-$. Additional examples of ferrate species involving phenyl nucleophiles were recently provided by the Hu and Lefevre groups who reported the synthesis of several novel iron(0)- and iron(I)-phenyl ferrate species accessible in the presence of organometallic phenyl nucleophiles [39].

11.3 TMEDA in Iron-Catalyzed Cross-coupling

11.3.1 Methods Overview

A number of amine additives and supporting ligands have successfully enhanced the reactivity and selectivity in iron-catalyzed cross-coupling systems [40–44], with TMEDA garnering significant attention over the years [45–50]. Nakamura et al. first reported the cross-coupling of cycloheptyl bromide and PhMgBr catalyzed by $FeCl_3$ with TMEDA as an additive (Table 11.2, entry 1), wherein they observed that slow addition of an equimolar mixture of TMEDA and Grignard reagent decreased undesirable β hydrogen elimination side products with significant increase in cross-coupled product yield [45a]. Bedford et al. corroborated the advantages of using nitrogen-containing additives similar to those reported by Nakamura et al. (Table 11.2, entry 2) [40b]. Unlike in Nakamura's observations, Bedford et al. documented TMEDA as the poorest performing additive while use of 1,4-diazabicyclo[2.2.2]octane (DABCO) influenced more selective product yields for nearly every substrate investigated. It should be noted that the amount of TMEDA used in these two studies differed between the conditions of Nakamura (stoichiometric excess) and Bedford (catalytic loadings).

A catalytic amount of simple ferric salts in combination with excess TMEDA and hexamethyltetramine (HMTA) was later discovered by Cahiez et al. to efficiently promote the cross-coupling of alkenyl and aryl Grignard reagents with alkyl halides [47a]. The cross-coupling of more electron-rich aryl Grignard reagents was found to occur in the presence of a dinuclear $FeCl_3$/TMEDA complex with the formal composition $(FeCl_3)_2(TMEDA)_3$ (Table 11.2, entry 3) [47b]. Cossy and coworkers further demonstrated the broad applicability of combining TMEDA and ferric salts in the cross-coupling of a wide array of alkyl halides while examining the reactivity of various alkenyl Grignard reagents (Table 11.2, entry 4) [48a]. Consistent with Nakamura's observations, Cossy and coworkers observed that slow addition of a premixture of TMEDA and alkenyl Grignard reagent significantly improved the yield of cross-coupled product. More recent examples of the application of TMEDA in iron-catalyzed cross-coupling have demonstrated expansion of alkyl electrophile scope (e.g. tosylates and sulfides) using milder transmetalating reagents such as arylzinc compounds [45d], as well as the applicability of organolithium reagents in aryl–alkyl, alkyl–aryl, and alkyl–alkyl cross-couplings [50].

Table 11.2 Examples of iron-catalyzed cross-coupling reactions with TMEDA as an additive or supporting ligand.

Entry	Catalyst	Electrophile	Grignard reagent	TMEDA (equiv.)	Reaction conditions	Yield (%)	References
1	$FeCl_3$	$C_7H_{13}Br$	PhMgBr	1.2	THF, −78 to 0 °C	71	[45a]
2	$FeCl_3$	$C_6H_{11}Br$	*p*-tolMgBr	0.1	Et_2O, 45 °C	79	[40b]
3	$(FeCl_3)_2(TMEDA)_3$	$C_6H_{11}Br$	PhMgBr	0.1	THF, 0 °C	91	[47b]
4	$FeCl_3$	$C_7H_{13}Br$	$CH_2=C(CH_3)MgBr$	1.9	THF, 0 °C	94	[48a]

11.3.2 Mechanistic Investigations

In 2009, Nakamura and coworkers proposed that slow addition of TMEDA in the cross-coupling of *n*-octyl bromide with mesitylmagnesium bromide (MesMgBr) suppresses unselective reaction pathways [45e]. By isolating mono- and bis-mesitylated iron(II)-TMEDA complexes, they were able to perform single-turnover reactions with excess electrophile. Notably, the reaction of Fe(Mes)Br(TMEDA) with electrophile was significantly slower than with the bis-mesitylated complex $Fe(Mes)_2(TMEDA)$. It was also found that isolated Fe(Mes)Br(TMEDA) reacts with MesMgBr to generate $Fe(Mes)_2(TMEDA)$. Additionally, a radical clock experiment gave evidence to the involvement of a short-lived radical in the reaction of $Fe(Mes)_2(TMEDA)$ with electrophile. Cumulatively, these observations led to the proposal of the mechanistic cycle shown in Scheme 11.4a.

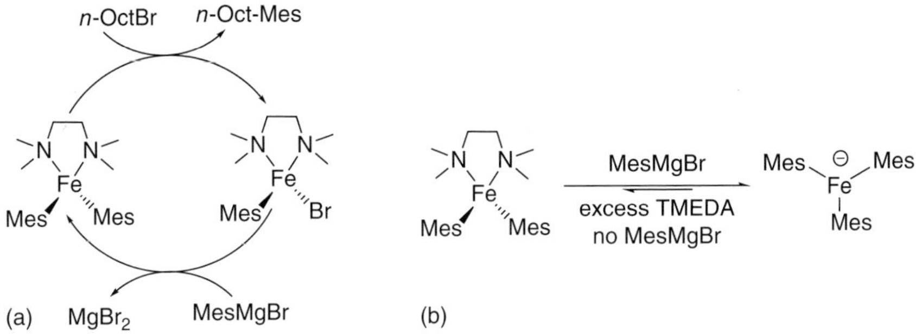

n-OctBr n-Oct-Mes

Mes Fe Mes (TMEDA) Mes Fe Br (TMEDA)

(a) $MgBr_2$ MesMgBr

(b) Mes Fe Mes (TMEDA) — MesMgBr / excess TMEDA, no MesMgBr — Mes Fe⊖ Mes, Mes

Scheme 11.4 (a) Proposed catalytic cycle of the cross-coupling of *n*-octyl bromide and MesMgBr with TMEDA-ligated iron by Nakamura; (b) equilibrium of iron complexes under catalytically relevant equivalents of MesMgBr observed by Bedford.

Although Nakamura et al. proposed that TMEDA may remain coordinated to iron over the course of these reactions, Bedford et al. observed dissociation of TMEDA under catalytically relevant conditions generating homoleptic ferrate complexes that were subsequently characterized by NMR and X-ray crystallography [46]. It was shown that the reaction of either $FeCl_2$ or $FeCl_3$ with MesMgBr in the presence of TMEDA produced the iron(II) ate complex $[Fe(Mes)_3]^-$ (Scheme 11.4b). Further analysis by paramagnetic ^1H NMR supported $[Fe(Mes)_3]^-$ being the major species in solution, whereas $Fe(Mes)_2(TMEDA)$ was not observed. $[Fe(Mes)_3]^-$ was also found to be much more reactive toward electrophile than $Fe(Mes)_2(TMEDA)$, providing full conversion of *n*-octyl bromide to cross-coupled product and octene. Under the same conditions, $FeMes_2(TMEDA)$ gave 25% conversion but with exclusive formation of cross-coupling product.

Additionally, less sterically hindered homoleptic benzylated ferrate complexes $[Fe(Bn)_3]^-$ and $[Fe(Bn)_4]^-$ were synthesized and used to study the effect of TMEDA in the dissociation and formation of TMEDA-supported benzylated iron species. It was found that the TMEDA reversibly reacted with $[Fe(Bn)_3]^-$ but not with $[Fe(Bn)_4]^-$, instead preferring to coordinate with the magnesium-based counterion in the case of the latter, generating $[Mg_3Cl_5(TMEDA)_3]^+$. From these results, Bedford et al. proposed that TMEDA likely plays a role in suppressing competing side reactions by trapping off-cycle iron species.

Maintaining low Grignard reagent concentration relative to TMEDA and the iron catalyst by slowly adding the Grignard reagent during catalysis has consistently been the best approach for minimizing side product formation. One notable difference between the mechanistic studies of Bedford and Nakamura was the concentration of Grignard reagent at which the reaction takes place. Although Nakamura and coworkers used a relative molar ratio of $FeCl_3$, MesMgBr, and TMEDA of 1 : 3 : 8 in THF [45e], the report by Bedford et al. used a higher concentration of MesMgBr, making the ratio 1 : 8 : 8 [46]. Therefore, it remains unclear whether or not the active catalytic species bears TMEDA [51] under catalytic conditions with slow addition of Grignard reagent where a closer relative molar ratio exists at shorter time points or if TMEDA serves an alternative role such as that proposed by Bedford.

11.4 NHCs in Iron-Catalyzed Cross-coupling

11.4.1 Methods Overview

Applications of NHCs as additives and ligands in iron-catalyzed cross-coupling were first reported in the 2000s by Bedford et al. [52]. Using various $FeCl_3$/NHC combinations as well as the well-defined NHC-ligated complex **7**, high yields were observed under mild conditions in the cross-coupling of aryl Grignard reagents with secondary alkyl halides (Table 11.3, entries 1–3). Following Bedford's initial report, Nakamura and coworkers presented an example of biaryl coupling in which the combination of $FeF_3 \cdot 3H_2O$ with SIPr·HCl selectively suppressed homocoupling relative to reactions combining the same iron salt with the NHCs IPr·HCl or SIMes·HCl (Table 11.3, entries 4–6) [53]. Notably, addition of KF to reactions catalyzed by $FeCl_3$ with SIPr·HCl as an additive resulted in increased yields of cross-coupled product (see Table 11.3, entry 7) relative to reactions using only $FeCl_3$ and SIPr·HCl [53a]. Nakamura and coworkers observed this "fluoride effect" in catalysis using combinations of NHCs with cobalt and nickel fluoride salts as well [53b]. To date, the origin of the superior effectiveness of transition metal fluoride precatalysts as well as the differential effectiveness of NHCs such as SIPr·HCl and IPr·HCl remain largely undefined.

Subsequently, the groups of Nakamura, Perry, and Cook expanded the scope of tolerated electrophiles to nonactivated aryl chlorides [58, 59], aryl sulfamates and tosylates [60], and nonactivated chloroalkanes (Table 11.3, entries 8–9) [54]. In this latter example, Nakamura and coworkers reported a deviation from the previously observed efficacy of SIPr·HCl over IPr·HCl as NHC additive. Intriguingly, IPr·HCl outperformed SIPr·HCl in both conversion and selectivity for the *n*-decylbenzene cross-coupled product. Work by the groups of Gaertner [55], Sun [61], Duong [62], Meyer [56], and others [63] have significantly expanded the scope of ionic iron NHC salt precatalysts (Table 11.3, precatalyst **8**, entry 10), iron/NHC additive combinations, and well-defined iron-NHC precatalysts

Table 11.3 Representative examples of iron-catalyzed cross-coupling reactions using NHCs as additives or supporting ligands.

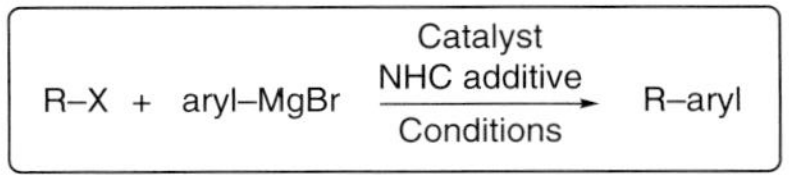

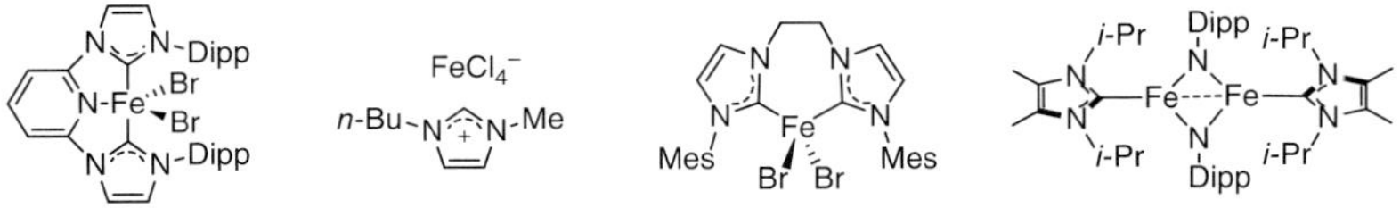

R–X + aryl–MgBr $\xrightarrow[\text{Conditions}]{\text{Catalyst, NHC additive}}$ R–aryl

7 **8** **9** **10**

SItBu·HCl ItBu·HCl SICy·HCl SIPr·HCl IPr·HCl SIMes·HCl

Entry	Catalyst (mol%)	NHC additive (mol%)	R–X	Aryl Grignard	Conditions	Yield (%)	References
1	$FeCl_3$ (5)	SItBu·HCl (10)	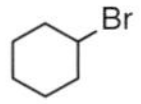	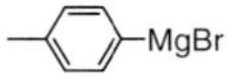	Et_2O, reflux 30 min	97	[52]
2	$FeCl_3$ (5)	SICy·HCl (10)				87	
3	**7**	—				94	
4	$FeF_3 \cdot 3H_2O$	SIPr·HCl (15)	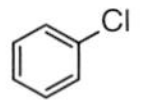	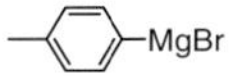	THF, 60 °C 24 h	98	[53]
5	$FeF_3 \cdot 3H_2O$	IPr·HCl (15)				25	
6	$FeF_3 \cdot 3H_2O$	SIMes·HCl (15)				34	
7	$FeCl_3$	SIPr·HCl (15) KF (20)				92	
8	$FeCl_3$ (5)	IPr·HCl (10)	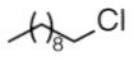	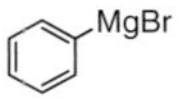	THF, 40 °C 1.5 h	78	[54]
9	$FeCl_3$ (5)	SIPr·HCl (10)				59	

(*Continued*)

Table 11.3 (Continued)

Entry	Catalyst (mol%)	NHC additive (mol%)	R–X	Aryl Grignard	Conditions	Yield (%)	References
10	**8**	—	Br	MgBr	Et_2O, 0 °C 10 min	89	[55]
11	**9**	—			Et_2O, reflux 30 min	75	[56]
12	**10**	—	F	MgBr	THF, RT 48 h	87	[57]

Table 11.4 Alkyl–alkyl cross-coupling with iron salts and NHC additives by Cárdenas and coworkers.

$Fe(OAc)_2$ (2.5 mol%) + IMes·HCl (6 mol%); 30 mol% BrMg(CH2)2-(1,3-dioxan-2-yl), THF, 50–60 °C, 20 min; R–I, slow addition of 3 equiv. Grignard, THF, RT, 8 h → R(CH2)2-(1,3-dioxan-2-yl)

Entry	R–I	Yield (%)
1	1-iodo-3-(4-methoxyphenyl)propane	67
2	iodocyclohexane	88
3	N-PG-4-iodopiperidine	PG = Boc: 67 PG = Bn: 78

(Table 11.3, precatalyst **9**, entry 11) that can facilitate the coupling of alkyl, alkenyl, and aryl electrophiles with aryl Grignard reagents. Furthermore, Deng and coworkers used the NHC-supported iron(II) dimer **10** as a precatalyst in a rare example of iron-catalyzed arylation of fluoride electrophiles (Table 11.3, entry 12) [57].

Further expansion of scope came from Cárdenas and coworkers who in 2013 reported the first example of an alkyl–alkyl cross-coupling method using NHC additives [64]. The group demonstrated that the combination of $Fe(OAc)_2$ with either SIMes·HCl or IMes·HCl led to the coupling of 1-iodo-3-phenylpropane with the acetal-containing (1,3-dioxan-2-ylethyl)magnesium bromide in good yields (Table 11.4). The catalyst system was also observed to tolerate alkyl halides bearing ester and heterocyclic functionalities as well as *N*-Boc and *N*-benzyl protected 4-iodopiperidines. It should be noted that unlike in Nakamura's biaryl coupling method summarized above, variations in NHC backbone saturation were not observed to have an appreciable effect on product yields.

11.4.2 Mechanistic Investigations

The last five years have witnessed the first mechanistic investigations of iron-catalyzed cross-coupling systems employing NHCs as additives and supporting ligands. In a series of studies, Deng and coworkers reported the synthesis of alkylated [65], arylated [65, 66], alkynylated [67], and alkenylated [68] four-coordinate iron(II)-NHC complexes and their stoichiometric reactivity with alkyl, benzyl, and aryl halides, examples of which are shown in Scheme 11.5. Complexes **11–13** exhibit stability toward reductive elimination of their sp^2-

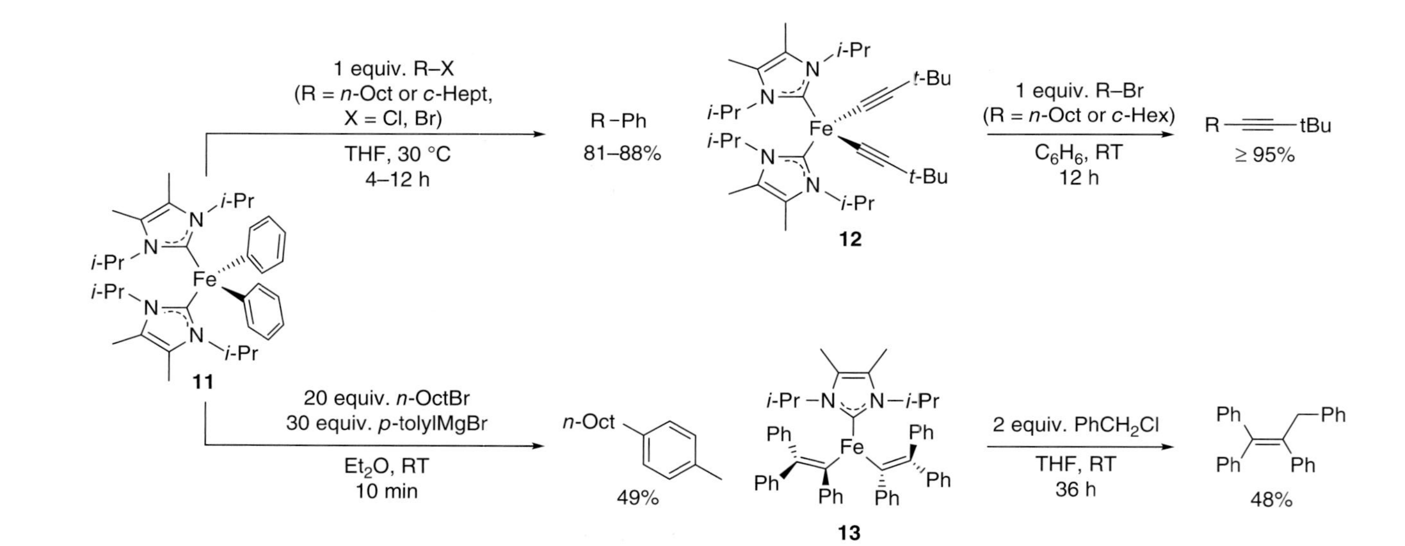

Scheme 11.5 Reactivity of well-defined iron(II)-NHC complexes reported by Deng and coworkers.

and sp-hybridized ligands and were found to possess reactivity toward primary and secondary alkyl halides and benzyl halides to afford phenylated, alkynylated, and alkenylated products. Furthermore, **11** was established to be catalytically competent in the coupling of *p*-tolylmagnesium bromide (*p*-tolMgBr) with *n*-octyl bromide.

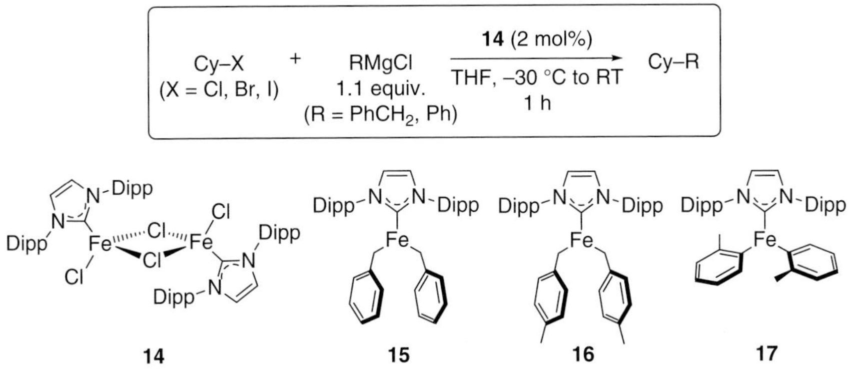

Cy–X (X = Cl, Br, I) + RMgCl 1.1 equiv. (R = $PhCH_2$, Ph) → [**14** (2 mol%), THF, –30 °C to RT, 1 h] → Cy–R

14 **15** **16** **17**

Scheme 11.6 Precatalyst and model intermediates used by Tonzetich and coworkers to study iron-NHC-catalyzed cross-couplings.

Tonzetich and coworkers recently provided a detailed examination of the catalytic effectiveness of three-coordinate iron(II)-NHC complexes in the coupling of primary and secondary alkyl electrophiles with phenyl and benzyl Grignard reagents (Scheme 11.6) [69]. The NHC-supported dimer **14** was found to be a highly selective precatalyst for the cross-coupling of alkyl electrophiles with phenyl and benzyl nucleophiles. Additionally, the group prepared monomeric three-coordinate benzylated and tolylated iron(II)-NHC complexes **15–17** to serve as models for *in situ*-formed transmetalated intermediates. Indeed, each monomeric complex reacted with cyclohexyl bromide cleanly at room temperature in THF. Tonzetich also interrogated the potential role of radical pathways within catalysis using radical traps and radical clock experiments. The addition of the radical scavenger, butylated hydroxytoluene (BHT), to catalytic reaction mixtures prevented the formation of cross-coupled product, and cyclization of 1-bromo-5-hexene was observed during radical clock experiments. Both of these results support radical-based pathways during catalysis. The correlation of organic product distributions with catalyst loading and electrophile concentration demonstrated that in the model catalytic reaction, electrophile radical recombination to iron-bound aryl ligands was more facile than to iron-bound alkyl ligands. Ultimately, the experimental evidence presented in this study resulted in the proposal of an Fe(II)/Fe(III) catalytic cycle involving a doubly transmetalated iron(II)-NHC-active species (Scheme 11.7a).

Cárdenas' investigation of alkyl–alkyl couplings with iron salts and NHC additives represents the first and, to our knowledge, only attempt to elucidate active catalyst oxidation state within an iron-NHC cross-coupling system without using a well-defined NHC-supported iron precatalyst [64]. The group demonstrated that an effective precatalyst can be generated *in situ* by heating a

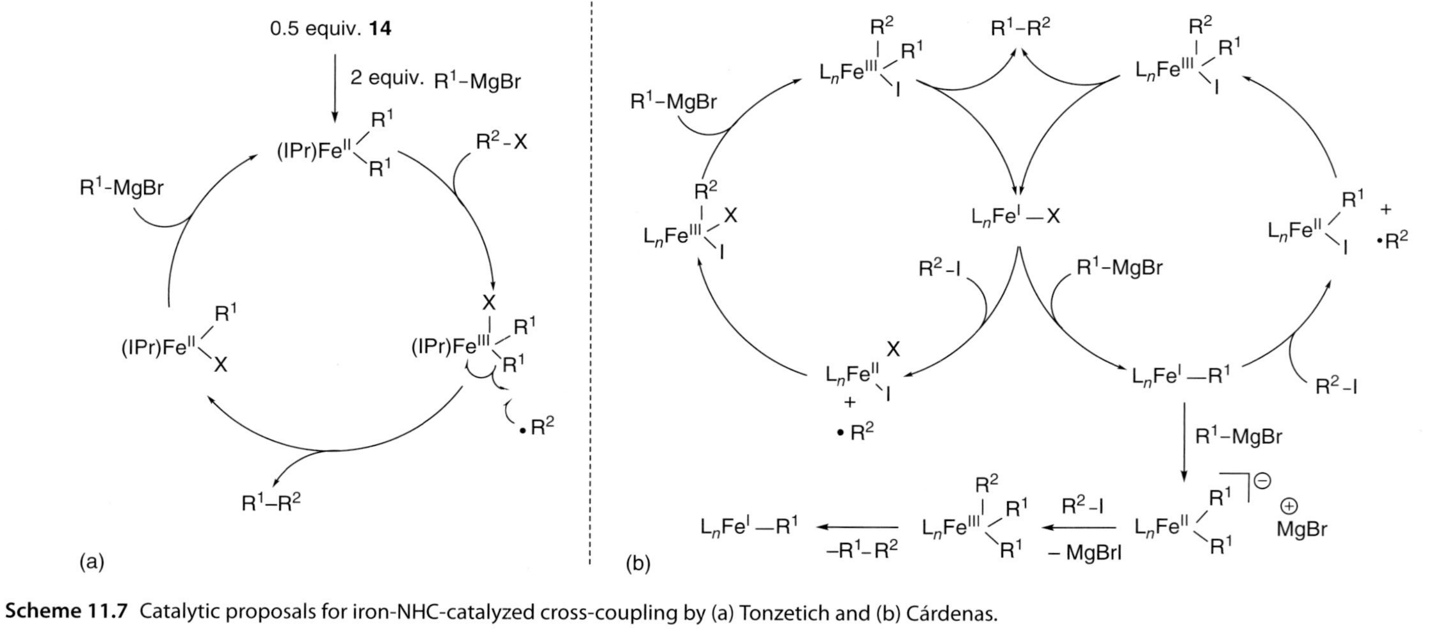

Scheme 11.7 Catalytic proposals for iron-NHC-catalyzed cross-coupling by (a) Tonzetich and (b) Cárdenas.

mixture of $Fe(OAc)_2$, NHC, and 12 equiv. of Grignard reagent (see Table 11.4) Subsequent addition of electrophile and slow addition of Grignard reagent over approximately 24 hours resulted in good yields of cross-coupled product. GC monitoring of the formation of homocoupled nucleophile as a function of Grignard reagent equivalents during the precatalyst activation process was used to argue for a formal reduction to iron(I) before addition of electrophile. Further analysis of this reaction mixture by EPR spectroscopy revealed the formation of a $S = 1/2$ species during the activation, again leading to the proposal of an iron(I) species being formed before reaction with electrophile. Observations of cyclopropylmethyl ring opening and 5-hexenyl ring closure during radical clock experiments were used to support halogen abstraction and subsequent electrophile radical recombination, which led to the proposal of a radical-based mechanism in which an NHC-supported iron(III) intermediate releases cross-coupled product through reductive elimination, regenerating the proposed active iron(I)-NHC catalyst (Scheme 11.7b). This proposed mechanistic cycle highlighted two possible pathways, differing in the order of transmetalation and oxidative addition of electrophile. It was suggested that anionic alkylated iron-NHC species are likely not operative based on the lack of homocoupled electrophile present in the product mixtures.

Despite this important contribution to mechanistic interrogation of iron-NHC-mediated cross-coupling reactions, a number of questions remain regarding the active catalytic species and mechanism. Notably, Cárdenas and coworkers assigned an iron(I)-active catalyst based on the observation of a $S = 1/2$ EPR signal during the precatalyst activation procedure, with no further spectroscopic or synthetic investigations reported. EPR analysis of samples extracted from catalytic reaction mixtures was also not reported. Furthermore, the EPR spectra of the precatalyst activation mixture lacked quantitative analysis. The quantitation of EPR-active species in a mixture via spin quantitation has been well described both theoretically and practically in the literature [70, 71], and the absence of this type of analysis leaves the extent of reduction upon pretreatment undefined. Quantitation of homocoupled nucleophile was also used by the authors to judge the extent of iron reduction during pretreatment. However, care must be taken when interpreting results of such an experiment, as it has been recently demonstrated that in some cases, inflated yields of homocoupled nucleophile can result from hydrolysis quenching of transmetalated iron species [71].

11.5 Phosphines in Iron-Catalyzed Cross-coupling

11.5.1 Methods Overview

Over the last 10 years, phosphines have gained increased interest as additives and supporting ligands in iron-catalyzed cross-coupling. The use of mono- and bidentate phosphines as reaction additives was first investigated by Bedford et al. in 2006 for the coupling of aryl Grignard reagents with secondary alkyl halides [52], where both mono- and bidentate phosphines were used in conjunction with $FeCl_3$

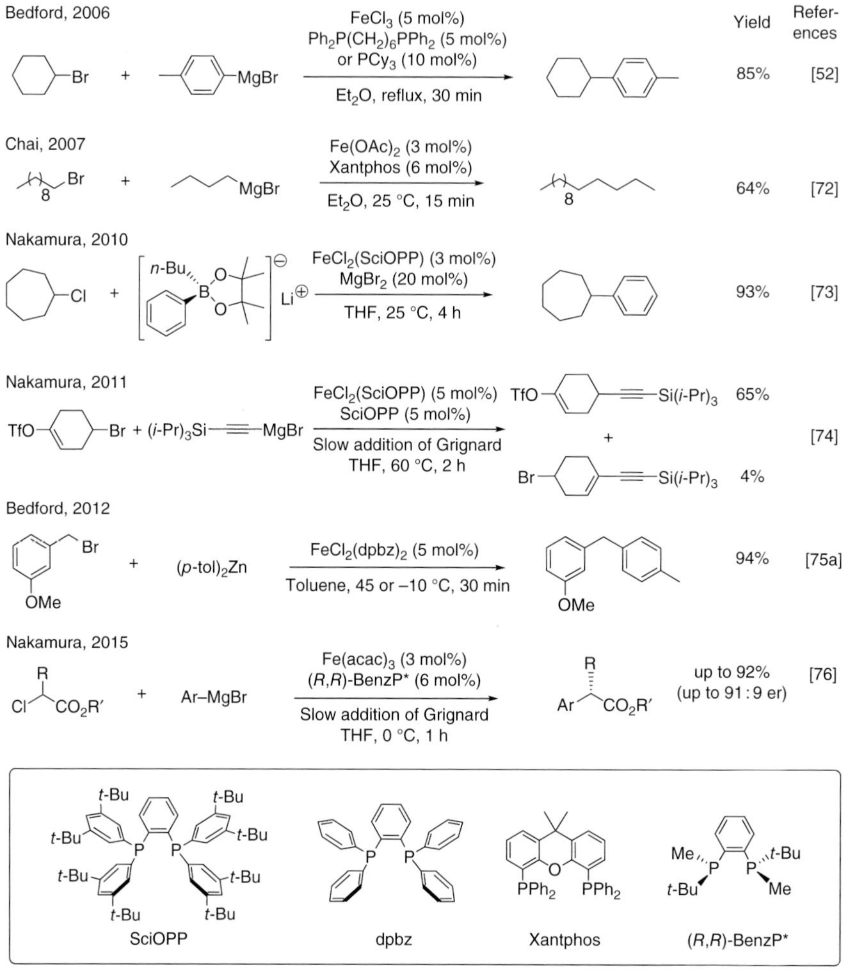

Scheme 11.8 Representative examples of C–C cross-coupling reactions using phosphine additives or phosphine-supported iron(II) precatalysts.

to provide high cross-coupling yields (Scheme 11.8). Subsequent studies by Bedford and others have used bidentate phosphines as reaction additives to expand substrate scope and generalize protocols to using softer nucleophilic sources such as organozinc and organoboron reagents, and the groups of Chai [72] and Nakamura [77] have reported the use of the chelating phosphine Xantphos to promote effective alkyl–alkyl Kumada and Suzuki–Miyaura cross-coupling methods (see Scheme 11.8 for an example from Chai and coworkers).

A growing body of work has also incorporated bidentate phosphines as supporting ligands within well-defined precatalysts. Commercially available phosphines such as dpbz (1,2-bis(diphenylphosphino)benzene) [78], dppe (1,2-bis(diphenylphosphino)ethane) [79], dppp (1,2-bis(diphenylphosphino) propane) [80], and others [81] have been used to this end in Kumada, Suzuki–Miyaura, and Negishi coupling reactions. Six-coordinate $FeCl_2(dpbz)_2$

has been used frequently by Bedford and coworkers in the Negishi coupling of arylzinc reagents with benzylic halides [75] (Scheme 11.8) and phosphates [82] as well as 2-halopyridine and pyrimidine substrates [83]. Bedford has also shown that dpbz and dppe can stabilize five-coordinate bis-chelated iron(I) species and that these complexes function as effective precatalysts in aryl–benzyl Negishi couplings [75a, 79]. Recently, Baran and coworkers have employed dpbz as an additive in conjunction with $Fe(acac)_3$ to catalyze the arylation of redox-active esters using both arylzinc and aryl Grignard reagents [84].

Steric tuning of the aryl periphery of dpbz was performed by the Nakamura group in attempts to maintain low-coordination and high spin states at the iron center, factors that were judged to be essential to active catalyst reactivity in systems using chelating amine ligands [45e]. The resulting sterically bulky SciOPP ligand (see Scheme 11.8) has been used as both a supporting ligand and additive in the Kumada and Suzuki–Miyaura couplings of alkyl halides [73], benzyl halides [85], and halohydrins [86] with alkenyl, aryl, and alkynyl [74, 87] nucleophiles (see Scheme 11.8 for selected examples). Interestingly, Nakamura and coworkers have reported protocols in which there exists a disparity in reactivity when using $FeCl_2$(SciOPP) or $FeCl_2(dpbz)_2$ as precatalysts [73, 88]. For example, a particularly large reactivity gap was observed in Kumada and Suzuki–Miyaura couplings of secondary alkyl halides with activated phenylborates (93% yield using $FeCl_2$(SciOPP) vs 14% yield using $FeCl_2(dpbz)_2$, see Ref. [82]). Recently, a chiral SciOPP-type additive (*R*,*R*)-BenzP* has also been reported by the group, enabling the first example of an iron-catalyzed enantioselective cross-coupling reaction. Here, α-haloesters and arylmagnesium reagents were coupled in high yields and enantiomeric ratios [76]. The group also reported the synthesis of aryl *C*-glycosides using an iron-catalyzed cross-coupling of halosugars [89].

11.5.2 Mechanistic Investigations

To date, the most detailed mechanistic investigations of C–C cross-coupling reactions performed in the presence of phosphines have focused on reactions using well-defined iron(II) precatalysts supported by chelating phosphine ligands. Numerous studies have been carried out by the groups of Bedford, Nakamura, Fürstner, and Neidig, together employing a diverse array of experimental methods to elucidate reactive intermediates and elementary reaction steps. In 2010, Nakamura and coworkers presented a mechanistic interrogation of the $FeCl_2$(SciOPP)-catalyzed Suzuki–Miyaura coupling of nonactivated primary and secondary alkyl halides with activated arylborates using radical clock experiments [73]. Use of (bromomethyl)cyclopropane as a coupling partner resulted in high yield of the ring-opened cross-coupled product. The combination of this result with the lack of observed homocoupled biaryls in product mixtures resulted in the mechanistic proposal summarized in Scheme 11.9. In this cycle, $Fe(Ar)_2$(SciOPP)-active species were proposed to react with electrophile through homolytic cleavage of the R—X bond of the electrophile, generating a haloiron(III)-SciOPP intermediate that could then recombine with the alkyl radical to form cross-coupled product. The resulting

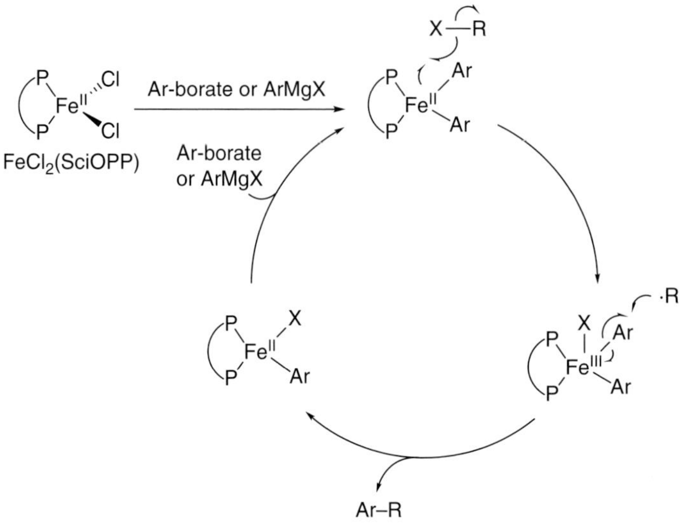

Scheme 11.9 Mechanistic proposal by Nakamura and coworkers for iron-SciOPP-catalyzed cross-couplings of aryl borates with alkyl halides.

monoarylated iron(II)-SciOPP species could again be transmetalated by nucleophile to regenerate $Fe(Ar)_2(SciOPP)$. Analogous Fe(II)/Fe(III) mechanisms have subsequently been proposed by Nakamura and coworkers to be operative in iron-SciOPP-catalyzed Kumada cross-couplings using aryl [88] and alkynyl nucleophiles [74, 87].

The proposal of SciOPP-supported iron(II)-active species in aryl–alkyl cross-coupling reactions was rigorously examined by Neidig and coworkers using a multitechnique spectroscopic approach, first for the coupling of primary alkyl halides with MesMgBr (Figure 11.3a) [90]. The combination of freeze-trapped ^{57}Fe Mössbauer, EPR, and magnetic circular dichroism (MCD) spectroscopy with independent synthetic efforts and reaction monitoring via GC allowed the group to correlate *in situ*-generated iron speciation with organic

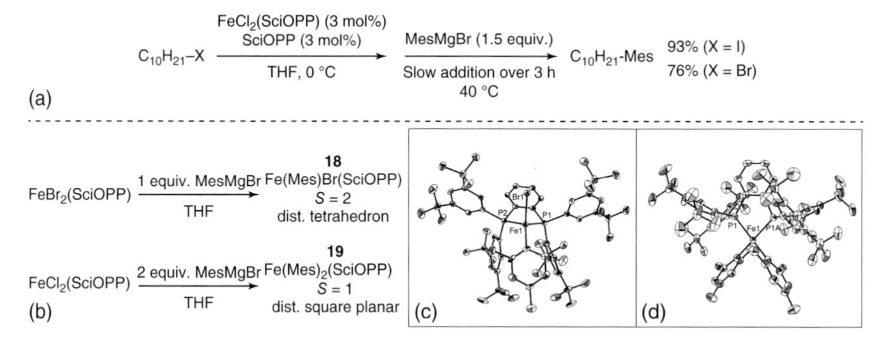

Figure 11.3 (a) Iron-SciOPP-catalyzed cross-coupling of MesMgBr with primary alkyl halides developed by Nakamura and coworkers; (b) synthesis and characterization of mesitylated iron(II)-SciOPP complexes by Neidig and coworkers; solid-state molecular structures of (c) **18** and (d) **19**. Thermal ellipsoids are drawn at the 50% probability level with hydrogen atoms removed for clarity.

product distributions, ultimately deciphering the identity of on- and off-cycle intermediates within catalysis. Stoichiometric treatment of $FeCl_2$(SciOPP) with MesMgBr (1 and 2 equiv.) resulted in very clean conversion to mono- and bis-mesitylated iron(II)-SciOPP species Fe(Mes)Br(SciOPP) (**18**) and $Fe(Mes)_2$(SciOPP) (**19**) (Figure 11.3b–d). Variable-temperature, variable-field (VTVH) MCD spectroscopy was used to rigorously characterize the distorted tetrahedral **18** as high-spin iron(II) ($S = 2$), whereas the distorted square planar bis-mesitylated complex **19** was found to be intermediate spin ($S = 1$). Furthermore, mesitylation with excess Grignard reagent was found to result in dissociation of SciOPP from the iron center, affording the stable ferrate $[Fe(Mes)_3]^-$. Freeze-trapped ^{57}Fe Mössbauer spectroscopy was subsequently used to follow the reaction profile of *in situ*-generated **19** with 1-bromodecane under pseudo-first-order conditions. Consumption of **19** was observed along with formation of cross-coupled product (analyzed by GC) and **18** (observed by 80 K frozen solution Mössbauer analysis, see Figure 11.4). Similar experiments probing the reaction of *in situ*-generated $[Fe(Mes)_3]^-$ with electrophiles show that the ferrate reacts at rates relevant to that found in catalysis, but with poor selectivity for cross-coupled product. The excess SciOPP ligand used in the catalytic protocol combined with the slow addition of Grignard reagent during catalysis was found to disfavor the formation of $[FeMes_3]^-$ during catalytic turnover. Notably, no formation of reduced iron below iron(II) was observed to occur during stoichiometric treatment of $FeCl_2$(SciOPP) with nucleophile or during catalytic turnover.

The extension of this experimental approach to iron-SciOPP-catalyzed systems using phenyl nucleophiles again demonstrated that SciOPP-supported iron(II)-active species are responsible for catalytic turnover. The reduction of the steric demand of the nucleophile was also observed to complicate the

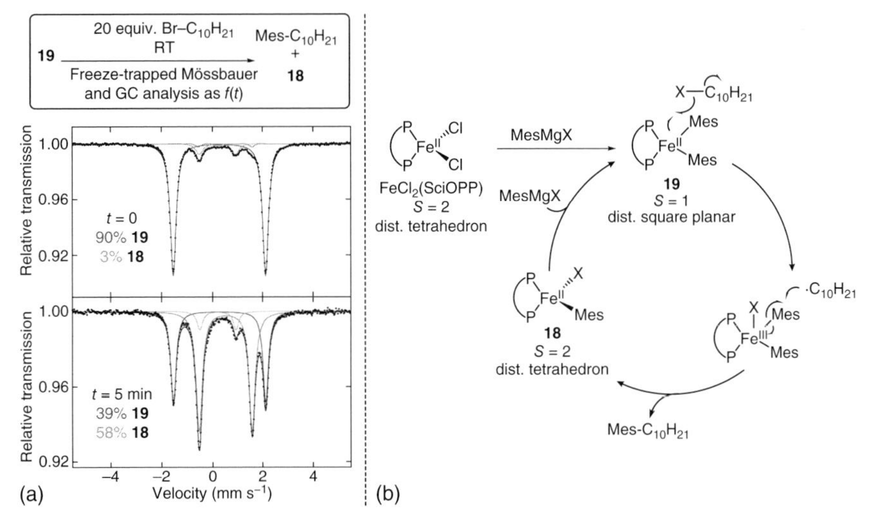

Figure 11.4 (a) Reaction of **19** with *n*-decyl bromide to form cross-coupled product and complex **18**; (b) catalytic cycle experimentally confirmed by Neidig and coworkers. Reprinted with permission from (2014) *J. Am. Chem. Soc.* 136: 9132. Copyright 2014 American Chemical Society.

iron distributions formed upon transmetalation [71]. Although complexes **18** and **19** exhibited stability in THF solution at ambient conditions, reactions of $FeCl_2$(SciOPP) with stoichiometric PhMgBr resulted in a distribution of iron species even at short time points at room temperature when using both 1 or 2 equiv. of Grignard reagent to iron. Instability of *in situ*-formed mono-phenylated Fe(Ph)X(SciOPP) (**20**; X = Cl or Br) and bis-phenylated species $Fe(Ph)_2$(SciOPP) (**21**) and $Fe(Ph)_2$(THF)(SciOPP) (**22**) to phenyl ligand redistribution and, in the case of **21** and **22**, reductive elimination to form the iron(0) complex Fe(η^6-biphenyl)(SciOPP) (**23**) was defined using freeze-trapped solution Mössbauer spectroscopy (Figure 11.5a–c). These observations sharply contrast the reactivity and stability of mesitylated complexes **18** and **19**.

GC analyses of organic product distributions were combined with freeze-trapped Mössbauer spectroscopy to evaluate the reactivity and selectivity of these phenylated iron(II)-SciOPP species. In contrast to the system using mesityl Grignard reagent, both **20** and mixtures of bisphenylated iron(II)-SciOPP species **21** and **22** were found to be reactive toward electrophile at rates relevant to that of catalysis. Complex **20** was found to be more selective in forming cross-coupled product (Figure 11.5d–e), with the bisphenylated species responsible for more selectivity toward cycloheptene by-product. Although the iron(0) complex **23** was found to display reactivity toward electrophile, coupling occurs at a rate approximately 4 orders of magnitude slower than the phenylated iron(II) species. Lastly, despite the observed formation of minor amounts of iron(I) during stoichiometric phenylation, this species does not react further over long time points. In the end, the conclusions of these studies again support an Fe(II)/Fe(III) mechanism and demonstrate that low-valent iron is not required for the generation of highly reactive species and effective arylation of alkyl substrates in systems using bidentate phosphine ligands.

Furthermore, Neidig and coworkers found that an Fe(II)/Fe(III) mechanism was also operative in the iron-SciOPP-catalyzed cross-coupling of alkynyl Grignard reagents with alkyl electrophiles [91]. In contrast to iron-SciOPP-catalyzed aryl–alkyl cross-coupling systems, neutral mono- and bis-alkynylated iron (II)-SciOPP complexes displayed inherent instability toward alkynyl ligand redistribution in the solution resulting in the formation of SciOPP-coordinated ferrate species in the form of iron(I)-SciOPP. Although it was demonstrated that iron(I)-SciOPP species are accessed during catalytic turnover, these species were found to be off-cycle and lacked selective reactivity with electrophile.

Studies by Bedford and coworkers utilizing less sterically demanding dpbz and dppe ligands as supporting ligands in complexes such as $FeCl_2(dpbz)_2$ [75a,b] and $FeCl_2$(dppe) [79] precatalysts have led to the suggestion of catalytic cycles involving iron(I)-active species. The group has synthesized and characterized a number of halide-bound and arylated five-coordinate iron(I) species supported by dpbz [75a] and dppe [75b, 79], ligation (complexes **24**–**30**, Scheme 11.10a). These species have been characterized by EPR spectroscopy to be low-spin ($S = 1/2$), and, notably, the group has provided examples in which these complexes can serve as functional precatalysts in the cross-coupling of arylzinc reagents and benzyl electrophiles, resulting product yields comparable to reactions using the iron(II) precatalysts $FeCl_2(dpbz)_2$ and $FeCl_2$(dpbz) (Scheme 11.10b) [75a].

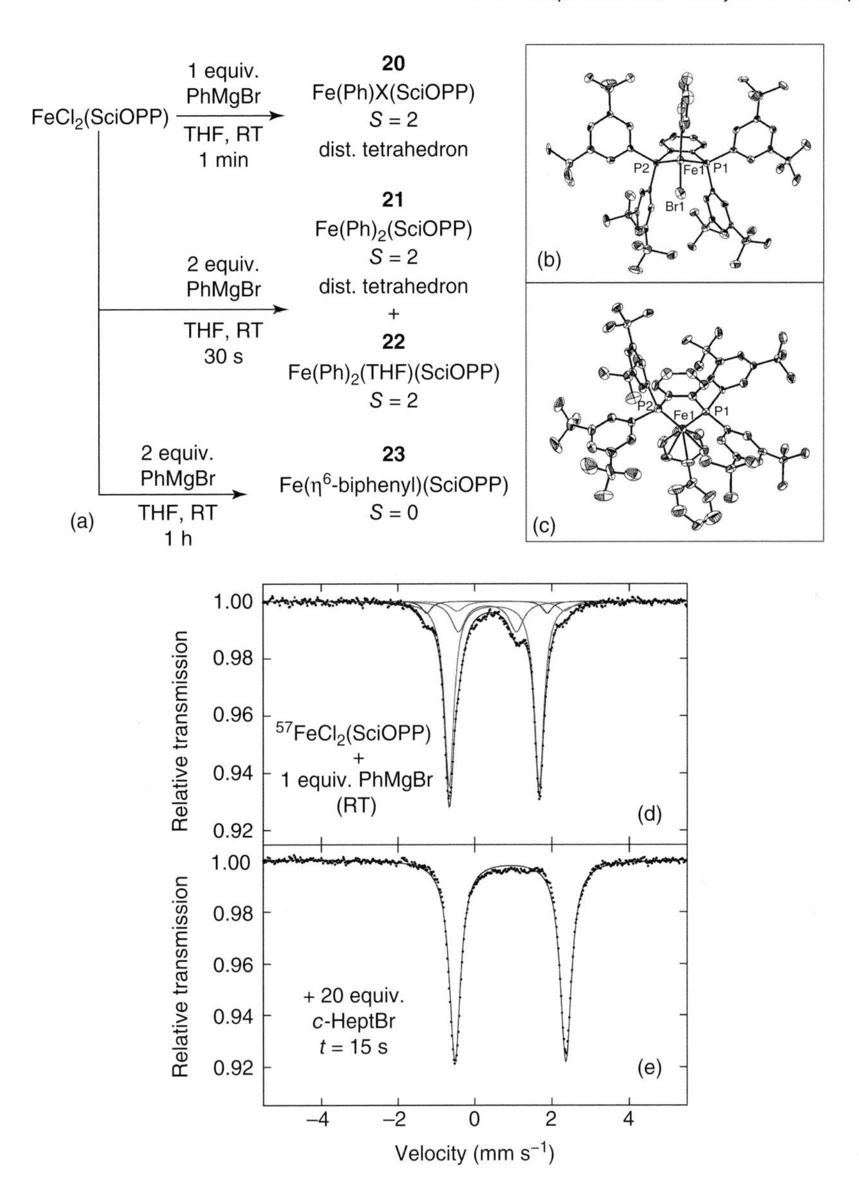

Figure 11.5 (a) Generation of phenylated iron(II)-SciOPP species using PhMgBr; solid-state molecular structures of (b) **20** and (c) **23** with thermal ellipsoids drawn at the 50% probability level and hydrogen atoms removed from clarity; 80 K Mössbauer spectra demonstrating (d) *in situ* generation of **20** and (e) its consumption through reaction with cycloheptyl bromide to give FeX_2(SciOPP) (X = Cl or Br). Reprinted with permission from (2015) *J. Am. Chem. Soc.* 137: 11432. Copyright 2015 American Chemical Society.

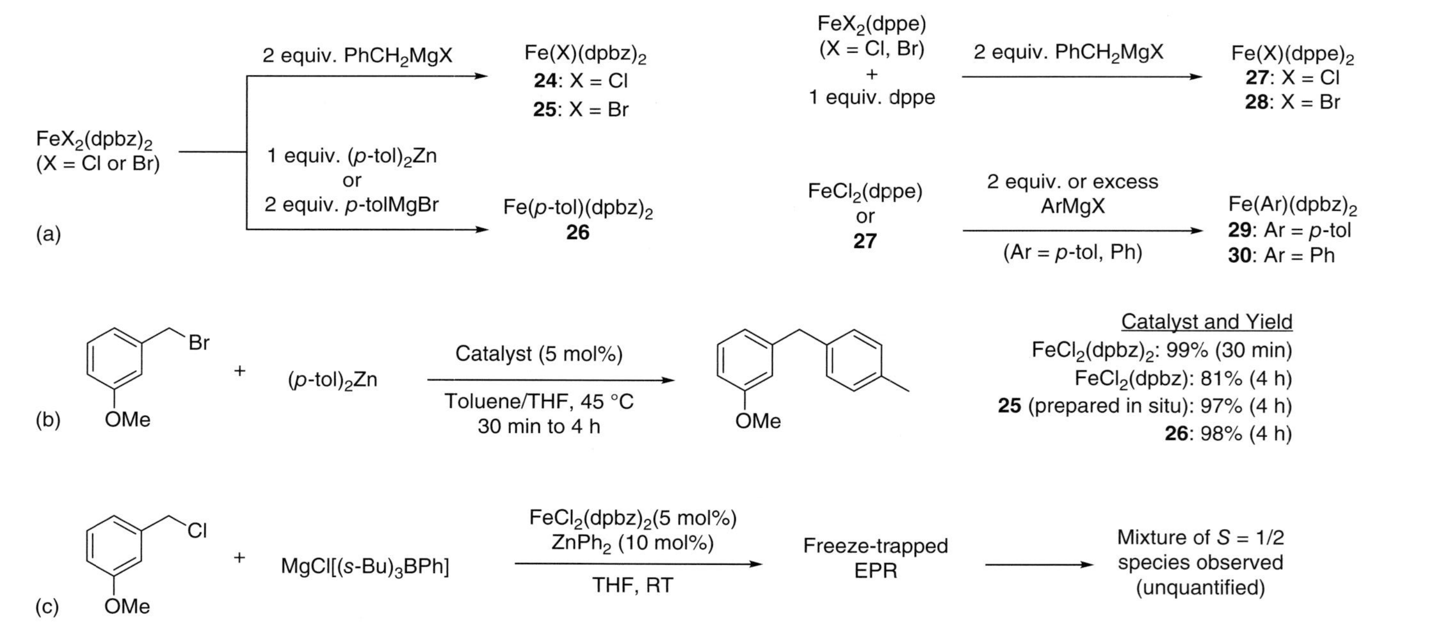

Scheme 11.10 (a) Synthesis of dpbz- and dppe-supported iron(I) species by Bedford and coworkers; (b) catalytic activity of iron(II) and iron(I) phosphine complexes within the Negishi cross-coupling reactions; (c) observation of iron(I) species in Suzuki–Miyaura couplings.

Furthermore, freeze-trapped EPR spectroscopy samples of the catalytic mixture in the Suzuki–Miyaura coupling of 3-methoxybenzyl chloride with tri(*sec*-butyl)phenylborate (Scheme 11.10c) demonstrated the formation of $S = 1/2$ iron species in the solution, the spectral profiles strongly resembling those of the complexes in the series **24**–**30** [75b] It is important to note that the iron(II) precatalyst $FeCl_2(dpbz)_2$ was used in this catalytic reaction. Thus, the observation of iron(I) in solution provided the evidence that lower valent iron could be accessed during the reaction. However, the lack of quantitation of the EPR-active iron present in the mixture renders assessment of the extent of its formation ambiguous. Moving forward, studies targeting the quantitation of all *in situ*-formed iron species in these systems will be required to more completely elucidate mechanistic foundations of these reactions and draw conclusions as to how comparable their reactivity pathways are relative to systems using more sterically demanding chelating phosphine ligands.

11.6 Future Outlook

In recent years, iron catalysis has been the focus of substantial methodology development in efforts to discover more sustainable reaction protocols and novel reactivity in C–C cross-coupling. This chapter has summarized the growing body of research aimed at elucidating reactive intermediates within these catalytic processes. This research targets a more fundamental understanding of mechanism and reactivity and thus aims to support further development of improved catalytic systems. In particular, multitechnique approaches combining spectroscopic methods (e.g. MCD, EPR, and ^{57}Fe Mössbauer) with synthetic studies and correlating these experiments with organic product distributions have to date provided some of the most detailed insight into the identity and reactivity of *in situ*-generated iron intermediates. Moving forward, the continued complementation of such inorganic spectroscopic methods with conventional organic analytical techniques will continue to prove powerful in determining catalyst structure and reactivity profiles in iron-catalyzed C—C cross-coupling.

Acknowledgments

M.L.N. gratefully acknowledges the National Institutes of Health (R01GM111480) for financial support.

References

1 For selected reviews focusing on methods development, see (a) Kochi, J.K. (2002). *J. Organomet. Chem.* 653: 11. (b) Bolm, C., Legros, J., Le Paih, J., and Zani, L. (2004). *Chem. Rev.* 104: 6217. (c) Fürstner, A. and Martin, R. (2005). *Chem. Lett.* 34: 624. (d) Correa, A., Garcia Mancheno, O., and Bolm, C. (2008). *Chem. Soc. Rev.* 37: 1108. (e) Enthaler, S., Junge, K., and Beller, M. (2008). *Angew. Chem. Int. Ed.* 47: 3317. (f) Leitner, A. (2008).

Iron-catalyzed cross-coupling reactions. In: *Iron Catalysis in Organic Chemistry* (ed. B. Plietker), 147. Wiley-VCH. (g) Sherry, B.D. and Fürstner, A. (2008). *Acc. Chem. Res.* 41: 1500. (h) Czaplik, W.M., Mayer, M., Cvengros, J., and Jacobi von Wangelin, A. (2009). *ChemSusChem* 2: 396. (i) Fürstner, A. (2009). *Angew. Chem. Int. Ed.* 48: 1364. (j) Knochel, P., Thaler, T., and Diene, C. (2010). *Isr. J. Chem.* 50: 547. (k) Jana, R., Pathak, T.P., and Sigman, M.S. (2011). *Chem. Rev.* 111: 1417. (l) Ilies, L. and Nakamura, E. (2014). *The Chemistry of Organoiron Compounds*, 539. (m) Nakamura, E., Hatakeyama, T., Ito, S. et al. (eds.) (2014). Iron-catalyzed cross-coupling reactions. In: *Organic Reactions*, vol. 83, 1. Wiley. (n) Bauer, I. and Knölker, H.-J. (2015). *Chem. Rev.* 115: 3170. (o) Bedford, R.B. and Brenner, P.B. (2015). *Top. Organomet. Chem.* 50: 19. (p) Cahiez, G., Moyeux, A., and Cossy, J. (2015). *Adv. Synth. Catal.* 357: 1983. (q) Kuzmina, O.M., Steib, A.K., Moyeux, A. et al. (2015). *Synthesis* 47: 1696. (r) Legros, J. and Figadere, B. (2015). *Nat. Prod. Rep.* 32: 1541. (s) Haas, D., Hammann, J.M., Greiner, R., and Knochel, P. (2016). *ACS Catal.* 6: 1540.

2 For recent reviews placing emphasis on mechanistic studies, see (a) Bedford, R.B. (2015). *Acc. Chem. Res.* 48: 1485. (b) Cassani, C., Bergonzini, G., and Wallentin, C.-J. (2016). *ACS Catal.* 6: 1640. (c) Guérinot, A. and Cossy, J. (2016). *Top. Curr. Chem.* 374: 1. (d) Mako, T.L. and Byers, J.A. (2016). *Inorg. Chem. Front.* 3: 766.

3 (a) Gilman, H. and Lichtenwalter, M. (1939). *J. Am. Chem. Soc.* 61: 957. (b) Kharasch, M.S. and Fields, E.K. (1941). *J. Am. Chem. Soc.* 63: 2316.

4 Vavon, G. and Mottez, P. (1944). *C. R. Acad. Sci.* 218: 557.

5 (a) Tamura, M. and Kochi, J. (1971). *J. Organomet. Chem.* 31: 289. (b) Tamura, M. and Kochi, J.K. (1971). *J. Am. Chem. Soc.* 93: 1487. (c) Tamura, M. and Kochi, J.K. (1971). *Bull. Chem. Soc. Jpn.* 44: 3063. (d) Neumann, S.M. and Kochi, J.K. (1975). *J. Org. Chem.* 40: 599. (e) Kwan, C.L. and Kochi, J.K. (1976). *J. Am. Chem. Soc.* 98: 4903. (f) Smith, R.S. and Kochi, J.K. (1976). *J. Org. Chem.* 41: 502.

6 (a) Cahiez, G. and Marquais, S. (1996). *Pure Appl. Chem.* 68: 53. (b) Cahiez, G. and Marquais, S. (1996). *Tetrahedron Lett.* 37: 1773. (c) Cahiez, G. and Avedissian, H. (1998). *Synthesis* 1998: 1199.

7 (a) Fürstner, A., Leitner, A., Mendez, M., and Krause, H. (2002). *J. Am. Chem. Soc.* 124: 13856. (b) Fürstner, A. and Leitner, A. (2002). *Angew. Chem. Int. Ed.* 41: 609.

8 Kuzmina, O.M., Steib, A.K., Flubacher, D., and Knochel, P. (2012). *Org. Lett.* 14: 4818.

9 Gärtner, D., Stein, A.L., Grupe, S. et al. (2015). *Angew. Chem. Int. Ed.* 54: 10545.

10 Molander, G.A., Rahn, B.J., Shubert, D.C., and Bonde, S.E. (1983). *Tetrahedron Lett.* 24: 5449.

11 Dohle, W., Kopp, F., Cahiez, G., and Knochel, P. (2001). *Synlett* 2001: 1901.

12 Scheiper, B., Bonnekessel, M., Krause, H., and Fürstner, A. (2004). *J. Org. Chem.* 69: 3943.

13 Hatakeyama, T., Yoshimoto, Y., Gabriel, T., and Nakamura, M. (2008). *Org. Lett.* 10: 5341.

14 Cahiez, G., Gager, O., and Habiak, V. (2008). *Synthesis* 2008: 2636.

15 Quintin, J., Franck, X., Hocquemiller, R., and Figadere, B. (2002). *Tetrahedron Lett.* 43: 3547.

16 Pridgen, L.N., Snyder, L., and Prol, J. (1989). *J. Org. Chem.* 54: 1523.

17 Fürstner, A., Martin, R., Krause, H. et al. (2008). *J. Am. Chem. Soc.* 130: 8773.

18 Gülak, S. and Jacobi von Wangelin, A. (2012). *Angew. Chem. Int. Ed.* 51: 1357.

19 Nagano, T. and Hayashi, T. (2004). *Org. Lett.* 6: 1297.

20 Martin, R. and Fürstner, A. (2004). *Angew. Chem. Int. Ed.* 43: 3955.

21 Jin, M. and Nakamura, M. (2011). *Chem. Lett.* 40: 1012.

22 Cheung, C.W., Ren, P., and Hu, X. (2014). *Org. Lett.* 16: 2566.

23 (a) Aleandri, L.E., Bogdanovic, B., Bons, P. et al. (1995). *Chem. Mater.* 7: 1153. (b) Bogdanovic, B. and Schwickardi, M. (2000). *Angew. Chem. Int. Ed.* 39: 4610.

24 Jonas, K., Schieferstein, L., Krüger, C., and Tsay, Y.H. (1979). *Angew. Chem. Int. Ed. Engl.* 18: 550.

25 Fürstner, A., Martin, R., and Majima, K. (2005). *J. Am. Chem. Soc.* 127: 12236.

26 Fürstner, A., Majima, K., Martín, R. et al. (2008). *J. Am. Chem. Soc.* 130: 1992.

27 Furstner, A., Krause, H., and Lehmann, C.W. (2006). *Angew. Chem. Int. Ed.* 45: 440.

28 (a) Kleimark, J., Hedström, A., Larsson, P.-F. et al. (2009). *ChemCatChem* 1: 152. (b) Hedström, A., Bollman, U., and Norrby, P.-O. (2011). *Chem. Eur. J.* 17: 11991. (c) Kleimark, J., Larsson, P.-F., Emamy, P. et al. (2012). *Adv. Synth. Catal.* 354: 448.

29 Hedström, A., Lindstedt, E., and Norrby, P.-O. (2013). *J. Organomet. Chem.* 748: 51.

30 Bekhradnia, A. and Norrby, P.-O. (2015). *Dalton Trans.* 3959.

31 Ren, Q., Guan, S., Jiang, F., and Fang, J. (2013). *J. Phys. Chem. A* 117: 756.

32 Heggen, B. and Thiel, W. (2016). *J. Organomet. Chem.* 804: 42.

33 Al-Afyouni, M.H., Fillman, K.L., Brennessel, W.W., and Neidig, M.L. (2014). *J. Am. Chem. Soc.* 136: 15457.

34 Muñoz, S.B. III,, Daifuku, S.L., Brennessel, W.W., and Neidig, M.L. (2016). *J. Am. Chem. Soc.* 138: 7492.

35 Alonso, P.J., Arauzo, A.B., Forniés, J. et al. (2006). *Angew. Chem. Int. Ed.* 45: 6707.

36 Schoch, R., Desens, W., Werner, T., and Bauer, M. (2013). *Chem. Eur. J.* 19: 15816.

37 Lefevre, G. and Jutand, A. (2014). *Chem. Eur. J.* 20: 4796.

38 Parchomyk, T. and Koszinowski, K. (2016). *Chem. Eur. J.* 22: 15609.

39 (a) Zhurkin, F.E., Wodrich, M.D., and Hu, X. (2017). *Organometallics* 36: 499. (b) Clémancey, M., Cantat, T., Blondin, G. et al. (2017). *Inorg. Chem.* 56: 3834.

40 (a) Bedford, R.B., Bruce, D.W., Frost, R.M. et al. (2004). *Chem. Commun.* 2822. (b) Bedford, R.B., Bruce, D.W., Frost, R.M., and Hird, M. (2005). *Chem. Commun.* 4161.

41 (a) Chowdhury, R.R., Crane, A.K., Fowler, C. et al. (2008). *Chem. Commun.* 94. (b) Hasan, K., Dawe, L.N., and Kozak, C.M. (2011). *Eur. J. Inorg. Chem.* 2011: 4610. (c) Qian, X., Dawe, L.N., and Kozak, C.M. (2011). *Dalton Trans.* 40: 933. (d) Reckling, A.M., Martin, D., Dawe, L.N. et al. (2011).

J. Organomet. Chem. 696: 787. (e) Chard, E.F., Dawe, L.N., and Kozak, C.M. (2013). *J. Organomet. Chem.* 737: 32.

42 Yamaguchi, Y., Ando, H., Nagaya, M. et al. (2011). *Chem. Lett.* 40: 983.

43 Barré, B., Gonnard, L., Campagne, R. et al. (2014). *Org. Lett.* 16: 6160.

44 (a) Bauer, G., Cheung, C.W., and Hu, X. (2015). *Synthesis* 47: 1726. (b) Bauer, G., Wodrich, M.D., Scopelliti, R., and Hu, X. (2015). *Organometallics* 34: 289. (c) Bauer, G. and Hu, X. (2016). *Inorg. Chem. Front.* 3: 741.

45 (a) Nakamura, M., Matsuo, K., Ito, S., and Nakamura, E. (2004). *J. Am. Chem. Soc.* 126: 3686. (b) Nakamura, M., Ito, S., Matsuo, K., and Nakamura, E. (2005). *Synlett* 2005: 1794. (c) Hatakeyama, T., Nakagawa, N., and Nakamura, M. (2009). *Org. Lett.* 11: 4496. (d) Ito, S., Fujiwara, Y.-I., Nakamura, E., and Nakamura, M. (2009). *Org. Lett.* 11: 4306. (e) Noda, D., Sunada, Y., Hatakeyama, T. et al. (2009). *J. Am. Chem. Soc.* 131: 6078.

46 Bedford, R.B., Brenner, P.B., Carter, E. et al. (2014). *Angew. Chem. Int. Ed.* 53: 1804.

47 (a) Cahiez, G., Duplais, C., and Moyeux, A. (2007). *Org. Lett.* 9: 3253. (b) Cahiez, G., Habiak, V., Duplais, C., and Moyeux, A. (2007). *Angew. Chem. Int. Ed.* 46: 4364.

48 (a) Guérinot, A., Reymond, S., and Cossy, J. (2007). *Angew. Chem. Int. Ed.* 46: 6521. (b) Bensoussan, C., Rival, N., Hanquet, G. et al. (2013). *Tetrahedron* 69: 7759.

49 (a) Czaplik, W.M., Mayer, M., and Jacobi von Wangelin, A. (2009). *Angew. Chem. Int. Ed.* 48: 607. (b) Czaplik, W.M., Mayer, M., Grupe, S., and Jacobi von Wangelin, A. (2010). *Pure Appl. Chem.* 82: 1545. (c) Czaplik, W.M., Mayer, M., and Jacobi von Wangelin, A. (2011). *ChemCatChem* 3: 135.

50 Jia, Z., Liu, Q., Peng, X.-S., and Wong, H.N.C. (2016). *Nat. Commun.* 7: 11955.

51 Bedford, R.B., Brenner, P.B., Elorriaga, D. et al. (2016). *Dalton Trans.* 45: 15811.

52 (a) Bedford, R.B., Betham, M., Bruce, D.W. et al. (2006). *J. Org. Chem.* 71: 1104. For the original report of complex 7(b)Danopoulos, A.A., Tsoureas, N., Wright, J.A., and Light, M.E. (2004). *Organometallics* 23: 166.

53 (a) Hatakeyama, T. and Nakamura, M. (2007). *J. Am. Chem. Soc.* 129: 9844. (b) Hatakeyama, T., Hashimoto, S., Ishizuka, K., and Nakamura, M. (2009). *J. Am. Chem. Soc.* 131: 11949.

54 Ghorai, S.K., Jin, M., Hatakeyama, T., and Nakamura, M. (2012). *Org. Lett.* 14: 1066.

55 Bica, K. and Gaertner, P. (2006). *Org. Lett.* 8: 733.

56 Meyer, S., Orben, C.M., Demeshko, S. et al. (2011). *Organometallics* 30: 6692.

57 Mo, Z., Zhang, Q., and Deng, L. (2012). *Organometallics* 31: 6518.

58 Perry, M.C., Gillett, A.N., and Law, T.C. (2012). *Tetrahedron Lett.* 53: 4436.

59 Agata, R., Iwamoto, T., Nakagawa, N. et al. (2015). *Synthesis* 47: 1733.

60 Agrawal, T. and Cook, S.P. (2012). *Org. Lett.* 15: 96.

61 (a) Gao, H., Yan, C., Tao, X. et al. (2010). *Organometallics* 29: 4189. (b) Deng, H., Xing, Y., Xia, C. et al. (2012). *Dalton Trans.* 41: 11597. (c) Xia, C., Xie, C., Wu, Y. et al. (2013). *Org. Biomol. Chem.* 11: 8135.

62 (a) Chua, Y.-Y. and Duong, H.A. (2014). *Chem. Commun.* 50: 8424. (b) Chua, Y.-Y. and Duong, H.A. (2016). *Chem. Commun.* 52: 1466.
63 Li, B.-J., Xu, L., Wu, Z.-H. et al. (2009). *J. Am. Chem. Soc.* 131: 14656.
64 Guisan-Ceinos, M., Tato, F., Bunuel, E. et al. (2013). *Chem. Sci.* 4: 1098.
65 Xiang, L., Xiao, J., and Deng, L. (2011). *Organometallics* 30: 2018.
66 Liu, Y., Xiao, J., Wang, L. et al. (2015). *Organometallics* 34: 599.
67 Wang, X., Zhang, J., Wang, L., and Deng, L. (2015). *Organometallics* 34: 2775–2782.
68 Liu, Y., Wang, L., and Deng, L. (2015). *Organometallics* 34: 4401.
69 Przyojski, J.A., Veggeberg, K.P., Arman, H.D., and Tonzetich, Z.J. (2015). *ACS Catal.* 5: 5938.
70 Aasa, R. and Vanngard, T. (1975). *J. Magn. Reson.* 19: 308.
71 Daifuku, S.L., Kneebone, J.L., Snyder, B.E.R., and Neidig, M.L. (2015). *J. Am. Chem. Soc.* 137: 11432.
72 Dongol, K.G., Koh, H., Sau, M., and Chai, C.L.L. (2007). *Adv. Synth. Catal.* 349: 1015.
73 Hatakeyama, T., Hashimoto, T., Kondo, Y. et al. (2010). *J. Am. Chem. Soc.* 132: 10674.
74 Hatakeyama, T., Okada, Y., Yoshimoto, Y., and Nakamura, M. (2011). *Angew. Chem. Int. Ed.* 50: 10973.
75 (a) Adams, C.J., Bedford, R.B., Carter, E. et al. (2012). *J. Am. Chem. Soc.* 134: 10333. (b) Bedford, R.B., Brenner, P.B., Carter, E. et al. (2014). *Organometallics* 33: 5767.
76 Jin, M., Adak, L., and Nakamura, M. (2015). *J. Am. Chem. Soc.* 137: 7128.
77 Hatakeyama, T., Hashimoto, T., Kathriarachchi, K.K.A.D.S. et al. (2012). *Angew. Chem. Int. Ed.* 51: 8834.
78 Hatakeyama, T., Kondo, Y., Fujiwara, Y.-I. et al. (2009). *Chem. Commun.* 1216.
79 Bedford, R.B., Carter, E., Cogswell, P.M. et al. (2013). *Angew. Chem. Int. Ed.* 52: 1285.
80 Lin, X., Zheng, F., and Qing, F.-L. (2012). *Organometallics* 31: 1578.
81 Sun, C.-L., Krause, H., and Fürstner, A. (2014). *Adv. Synth. Catal.* 356: 1281.
82 Bedford, R.B., Huwe, M., and Wilkinson, M.C. (2009). *Chem. Commun.* 600.
83 Bedford, R.B., Hall, M.A., Hodges, G.R. et al. (2009). *Chem. Commun.* 6430.
84 Toriyama, F., Cornella, J., Wimmer, L. et al. (2016). *J. Am. Chem. Soc.* 138: 11132.
85 Kawamura, S. and Nakamura, M. (2013). *Chem. Lett.* 42: 183.
86 Kawamura, S., Kawabata, T., Ishizuka, K., and Nakamura, M. (2012). *Chem. Commun.* 48: 9376.
87 Nakagawa, N., Hatakeyama, T., and Nakamura, M. (2015). *Chem. Lett.* 44: 486.
88 Hatakeyama, T., Fujiwara, Y.-I., Okada, Y. et al. (2011). *Chem. Lett.* 40: 1030.
89 Adak, L., Kawamura, S., Toma, G. et al. (2017). *J. Am. Chem. Soc.* 139: 10693.
90 Daifuku, S.L., Al-Afyouni, M.H., Snyder, B.E.R. et al. (2014). *J. Am. Chem. Soc.* 136: 9132.
91 Kneebone, J.L., Brennessel, W.W., and Neidig, M.L. (2017). *J. Am. Chem. Soc.* 139: 6988.

12

Recent Advances in Cobalt-Catalyzed Cross-coupling Reactions

Oriol Planas[1], Christopher J. Whiteoak[2], and Xavi Ribas[1]

[1] Universitat de Girona, Institut de Química Computacional i Catàlisi (IQCC) and Departament de Química, Campus Montilivi, 17071 Girona, Catalonia, Spain
[2] Sheffield Hallam University, Biomolecular Sciences Research Centre, Faculty of Health and Wellbeing, City Campus, Sheffield S1 1WB, UK

12.1 Introduction

Cross-coupling reactions have revolutionized synthetic chemistry over the past few decades, with the most successful cross-coupling protocols having been based on palladium, where the importance and relevance of these advancements were recognized with the awarding of the 2010 Nobel Prize to Heck, Negishi, and Suzuki [1]. These now well-established protocols require the use of preactivated substrates, such as aryl halides or triflates, that generate unnecessary waste and require prefunctionalization of the substrate. As a result, more recently, the field of palladium cross-coupling has come to include C–H activation rather than solely the use of these aforementioned preactivated substrates, thus further expanding the practical application of this approach to preparing new C—C and C—X bonds [2]. The major drawback of C–H activation approaches toward cross-coupling remains the selective activation of ubiquitous C—H bonds, which can be addressed by the use of chelating groups, and indeed, most C–H functionalization protocols rely on the chelate direction in order to overcome this challenge (also see Chapter 15). Meanwhile, there has also been a drive to replace expensive palladium with more abundant and cheaper first-row transition metal complexes, particularly in direct C–H functionalization [3]. Herein, this chapter will exemplify the key examples of recent advances of cobalt in cross-coupling reactions using both C–H activation and prefunctionalized approaches.

The first report of the use of cobalt in cross-coupling was as far back as 1939, where Gilman and Lichtenwalter described the homocoupling of phenylmagnesium iodide using stoichiometric amounts of cobalt chloride as the mediator [4]. Later, in 1941, Kharasch and Fields developed a catalytic variant of this work, providing a Co(II)-catalyzed protocol for the coupling of Grignard reagents and organic halides [5]. A decade later, in 1955, Murahashi provided the first example of a low-valent, cobalt-catalyzed cross-coupling protocol involving a C(sp^2)–H activation and carbonylation [6], further widening the mechanistic variety of

Non-Noble Metal Catalysis: Molecular Approaches and Reactions,
First Edition. Edited by Robertus J. M. Klein Gebbink and Marc-Etienne Moret.

cobalt-catalyzed transformations. This report demonstrated how $Co_2(CO)_8$ can be used to catalyze the coupling of Schiff bases and carbon monoxide to readily form isoindolone derivatives. This protocol clearly operates through a different mechanism than that of the previous report by Kharasch and Fields and provided the first example of cobalt-catalyzed C–H functionalization (Scheme 12.1a).

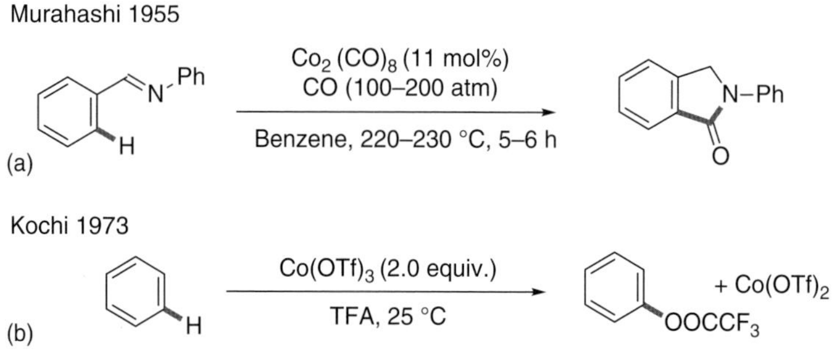

Murahashi 1955

$Co_2(CO)_8$ (11 mol%)
CO (100–200 atm)
Benzene, 220–230 °C, 5–6 h
(a)

Kochi 1973

$Co(OTf)_3$ (2.0 equiv.)
TFA, 25 °C
+ $Co(OTf)_2$
$OOCCF_3$
(b)

Scheme 12.1 (a) Early Co(0)-catalyzed annulation of *N*-benzylideneaniline with carbon monoxide through a C–H activation approach forming isoindolone derivatives [6]. (b) Co(III)-mediated trifluoroacetylation of aromatic compounds found to operate through a SET mechanism [7].

Since this time, important advances have been made in homogeneous cobalt catalysis, particularly in the field of hydroformylation [8] and development of the Pauson–Khand reaction [9]. In terms of cross-coupling operating through C–H activation/functionalization, little attention was paid to this field until 1973 when Kochi et al. reported a Co(III)-catalyzed cross-coupling of benzene with trifluoroacetate operating through a novel single-electron transfer (SET) mechanism (Scheme 12.1b) [7]. Despite these exciting breakthroughs, the field of cobalt-catalyzed cross-couplings returned to be relatively dormant until 1994, when Kisch and coworkers reported the ortho-alkenylation of aromatic azo compounds using a Co(I)-catalyzed protocol, with the product dependent on the substituents of the aromatics [10].

More recently, the trigger for the new rapid expansion of this field came in the report in 2010 from Yoshikai and coworkers who devised, likely inspired by the protocol of Kisch, a simple ternary catalyst system consisting of $CoBr_2$, a phosphine ligand, and a Grignard reagent to efficiently couple alkynes and aryl pyridines (Scheme 12.2) [11]. In this protocol, the Co(II) is reduced to the reactive Co(I) species *in situ* by the Grignard reagent. It was proposed that the Co(I) species is oxidatively inserted into the C—H bond, with a final reductive elimination step providing the cross-coupled product. Besides these successful low-valent approaches, discoveries have also been made using higher valent catalyst systems, which are proposed to operate through a concerted metalation–deprotonation (CMD) step. The use of Co(II) salts first described by Daugulis and coworker [12] and cationic Cp*Co(III) catalysts pioneered by Matsunaga and coworkers [13] has provided significant advancements and inspiration for many researchers.

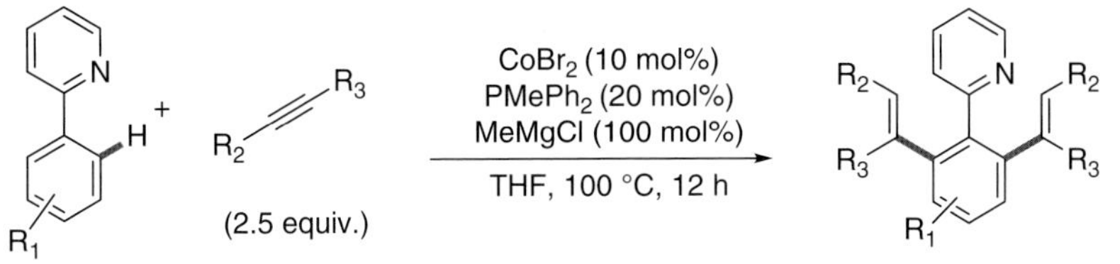

$CoBr_2$ (10 mol%)
$PMePh_2$ (20 mol%)
MeMgCl (100 mol%)
THF, 100 °C, 12 h
(2.5 equiv.)

Scheme 12.2 Ternary catalyst system reported by Yoshikai and coworkers for coupling of alkynes to aryl pyridine substrates [11].

The field of cobalt-catalyzed cross-coupling through both prefunctionalized substrates and C—H functionalization protocols has been extensively reviewed by Cahiez and Moyeux [14], Ackermann and coworkers [15], Niu and coworkers [16], and Whiteoak and coworker [17]. These contributions provide a broad account of developments in this field, including descriptions of the proposed mechanisms. In this chapter, we will provide a complimentary overview of this rapidly growing field using selected examples to highlight the potential of cobalt for aiding the expansion of the synthetic chemistry toolbox. In this context, we will describe cross-couplings that involve a C—H activation of the substrate (including aldehydes and hydroacylation) and also more traditional cross-coupling procedures starting from aryl halides and pseudohalides as substrates.

12.2 Cobalt-Catalyzed C—C Couplings Through a C—H Activation Approach

In the following sections of this chapter, we will discuss and provide examples of C—C bond forming reactions catalyzed by cobalt. These protocols are based on a range of mechanisms, and as a result, we will consider each of these individually.

12.2.1 Low-Valent Cobalt Catalysis

After the initial report [11], Yoshikai and coworkers has since reported on the application of the ternary catalyst system for the coupling of alkynes and alkenes to a number of different substrates and supporting ligands [18]. In one of these examples, the coupling of alkynes to the C2-position of indoles, the authors focused on the optimization of the phosphine ligand, and it was found that the reaction was sensitive to the added ligand and that 2-[2-(diphenylphosphanyl)ethyl]pyridine (Pyphos) was optimal [18a]. The authors noted that the enhanced performance when using the Pyphos ligand may have relevance to the acceleration effect observed when additional pyridine was added to the ternary system for the ortho-alkenylation of aromatic ketimines [18b]. A different report from the same authors also demonstrated how changing the ligand from a phosphine to an *N*-heterocyclic carbene could affect the regioselective outcome during the hydroarylation of styrenes (Scheme 12.3) [18c].

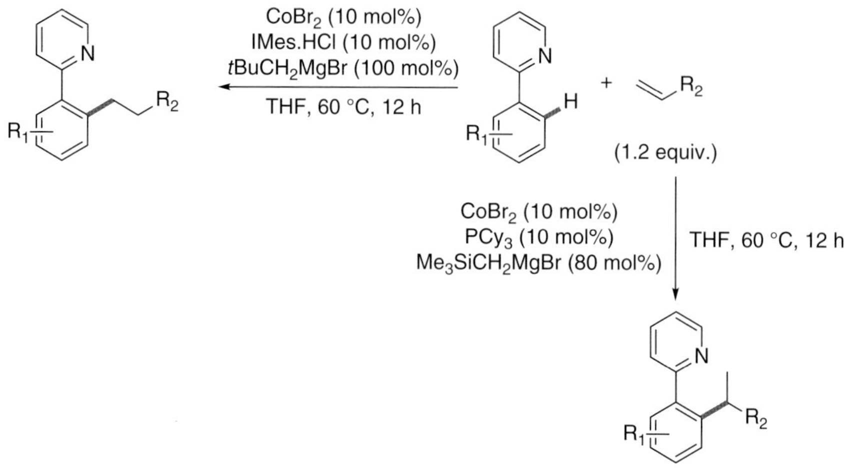

$CoBr_2$ (10 mol%)
IMes.HCl (10 mol%)
tBuCH$_2$MgBr (100 mol%)
THF, 60 °C, 12 h

(1.2 equiv.)

$CoBr_2$ (10 mol%)
PCy_3 (10 mol%)
Me_3SiCH_2MgBr (80 mol%)
THF, 60 °C, 12 h

Scheme 12.3 Regioselectivity difference when coupling styrenes to aryl pyridines using phosphines or *N*-heterocyclic carbenes as ligands [18c]. (IMes = 1,3-dimesitylimidazol-2-ylidene).

Using this same mechanistic approach, Nakamura and coworkers described the cobalt-catalyzed ortho-alkylation of secondary benzamides using alkyl chlorides as coupling partners [19]. This procedure does not require the addition of an external ligand and provides a rare example of the coupling of aromatic compounds with saturated hydrocarbons. Meanwhile, in 2012, Ackermann and coworker reported on the cross-coupling of aryl pyridines/indoles with sulfamates, carbamates, and phosphates through a C—H activation of the aryl pyridine/indole and C—O bond cleavage of the coupling partner [20]. More recently, Ackermann and coworkers has further extended the scope of this protocol to provide a new alkylation protocol employing alkyl acetates, carbonates, carbamates, and phosphates as coupling partners [21].

Acylation reactions have also attracted interest with Brookhart and coworker first reporting a cobalt-catalyzed example in 1997 [22]. This report details the inter- and intramolecular hydroacylation of vinyltrimethylsilane with aromatic aldehydes using a Cp*Co(I)-based catalyst system (Scheme 12.4).

Thereafter, less attention was paid to the potential of cobalt-catalyzed acylation reactions despite this early success, until 2014 when Dong and coworkers, inspired by the work of Brookhart, provided a significant breakthrough. In this work, the authors described the coupling of aldehydes and 1,3-dienes, observing that aromatic aldehydes favor 1,4-hydroacylation products and aliphatic aldehydes favor 1,2-hydroacylation products (Scheme 12.5a) [23]. In 2017, Dong and coworkers reported the synthesis of cyclobutanones with excellent enantiocontrol under mild conditions using cobalt catalysis (Scheme 12.5b) [24]. Mechanistic studies suggest a pathway where Co(0)/Co(II) catalytic cycle is involved, in which Co(0) is responsible of the C—H cleavage step.

Soon after, Yoshikai and coworker presented an intramolecular hydroacylation in compounds containing ketone or olefin functionalities with aldehyde functionalities, providing phthalide and indanone derivatives depending on the external ligand [25]. The protocol not only provided the desired products but

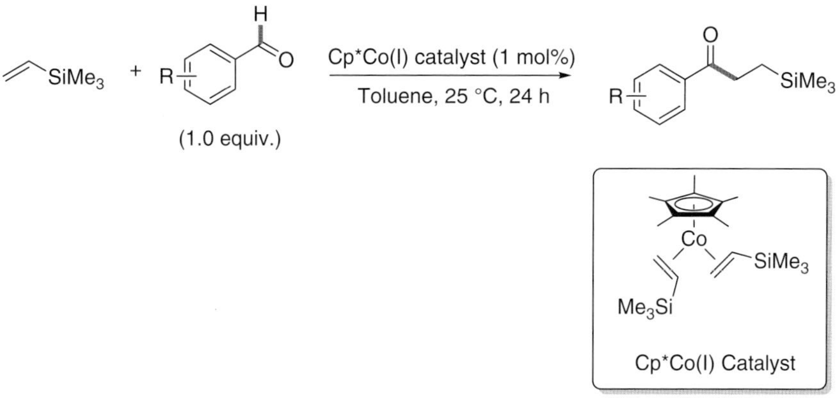

Scheme 12.4 Hydroacylation of olefins with benzaldehydes reported by Brookhart and coworker [22].

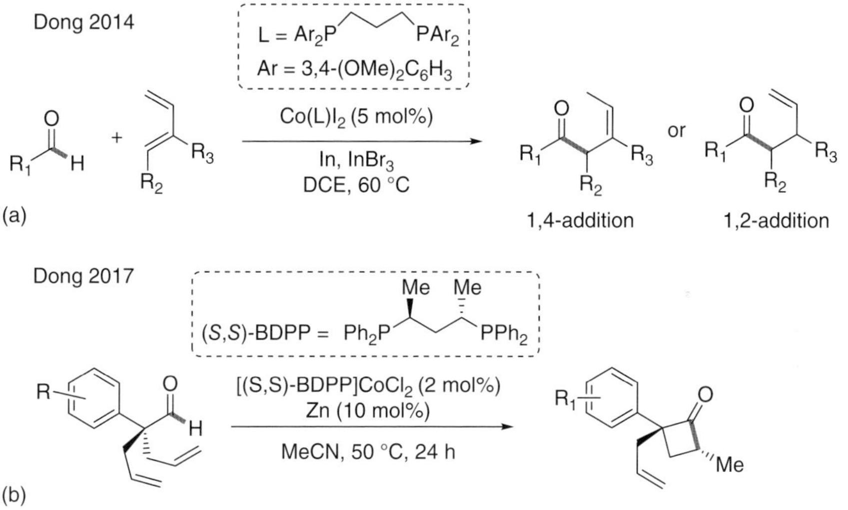

Scheme 12.5 Protocols reported by Dong and coworkers for (a) hydroacylation of 1,3-dienes [23] and (b) the synthesis of chiral cyclobutanones using cobalt as a catalyst [24].

also furnished product distributions with high enantioselectivities through the use of chiral ligands.

Although many of these low-valent transformations are proposed to proceed via the *in situ* formation of a catalytic Co(I) species, Petit and coworkers has recently applied the Co(0) compound $Co(PMe_3)_4$ in low-valent cobalt catalysis (Scheme 12.6) [26]. This Co(0) compound obviates the need for a stoichiometric reductant, thus making the catalyst system much simpler. The same authors have also recently expanded the substrate scope, permitting the facile preparation of substituted hydropyridine compounds [27].

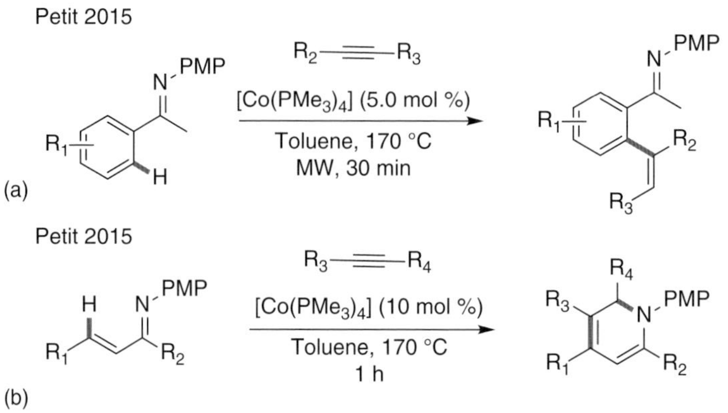

Petit 2015

R_2—≡—R_3

$[Co(PMe_3)_4]$ (5.0 mol %)

Toluene, 170 °C
MW, 30 min

(a)

Petit 2015

R_3—≡—R_4

$[Co(PMe_3)_4]$ (10 mol %)

Toluene, 170 °C
1 h

(b)

Scheme 12.6 Use of $[Co(PMe_3)_4]$ as a catalyst reported by Petit and coworkers (PMP = *p*-methoxyphenyl) [26, 27].

12.2.2 High-Valent Cobalt Catalysis

In comparison to the low-valent cobalt-catalyzed methodologies, the high-valent cobalt catalysis has only recently started to receive significant attention for C—C bond forming reactions. These examples can be separated into Co(II)- and Co(III)-catalyzed routes; although when starting from Co(II) species, the hypothesized intermediate is usually proposed to be a Co(III) organometallic intermediate, resulting from *in situ* oxidation of the Co(II) species [12, 28a, b].

It was not until 2013 when Matsunga and coworkers reported on the use of a readily accessible Cp*Co(III) complex, which was able to couple aryl pyridines and imines in a selective manner [13a, b], that the field of high-valent, cobalt-catalyzed cross-coupling started to demonstrate extended application potential. Later, the same authors performed the same coupling, but using indoles as the substrate, showing impressive turnover numbers (TONs) of up to 180 [29]. This latter contribution provided a unique C2-selective route by simple Lewis-acid-catalyzed Friedel–Crafts addition, which occurs at the C3 position, thus highlighting the potential of Cp*Co(III) catalysts to be added to synthetic chemist's toolbox. These initial reports have inspired significant growth in the field of high-valent, cobalt-catalyzed, and cross-coupling reactions.

In 2014, Glorius and coworkers reported on the allylation of indole substrates using allyl carbonates as coupling partners, again obtaining high TON's (up to 2200) using the Cp*Co(III) catalyst [30]. Since this initial report, the group of Glorius and coworkers has provided several other examples utilizing the same catalyst system, preparing a variety of heterocyclic compounds using a cobalt-catalyzed annulation methodology [31]. One of these reports demonstrated how, by controlling the proposed organometallic intermediate, it is possible to switch between the preparation of indoles and quinolines (Scheme 12.7a) [31c]. A similar regioselectivity switch was observed by Niu and coworkers using the bidentate pyridine-*N*-oxide benzamide substrate depicted in Scheme 12.7b [32].

Another example has been reported by Ellman and coworker, who described a facile route toward indazoles and furans using aldehydes as the coupling

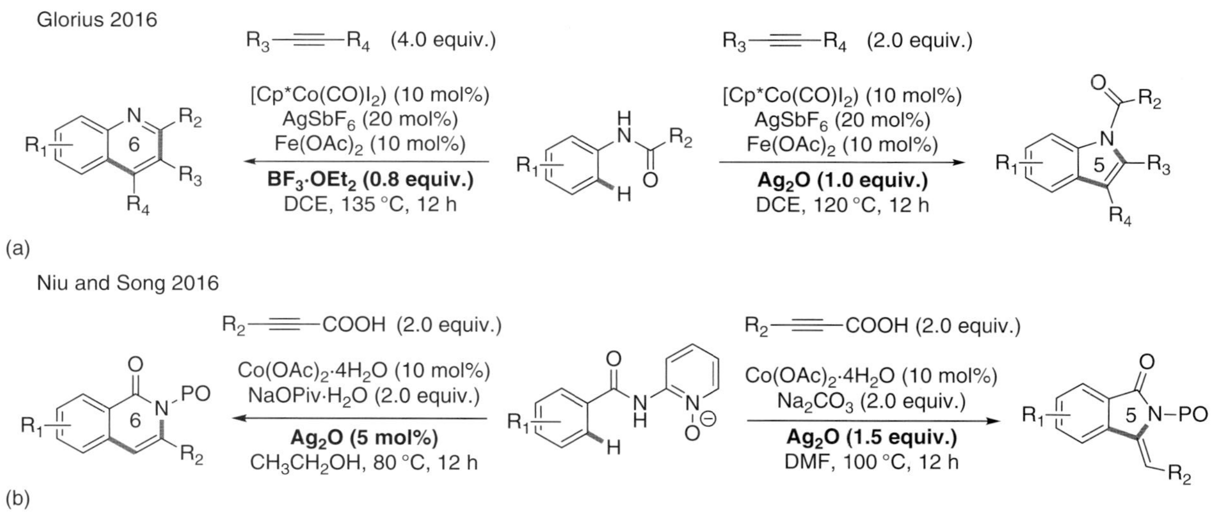

Glorius 2016

R_3—≡—R_4 (4.0 equiv.)

[Cp*Co(CO)I_2) (10 mol%)
$AgSbF_6$ (20 mol%)
$Fe(OAc)_2$ (10 mol%)
$BF_3 \cdot OEt_2$ (0.8 equiv.)
DCE, 135 °C, 12 h

R_3—≡—R_4 (2.0 equiv.)

[Cp*Co(CO)I_2) (10 mol%)
$AgSbF_6$ (20 mol%)
$Fe(OAc)_2$ (10 mol%)
Ag_2O (1.0 equiv.)
DCE, 120 °C, 12 h

(a)

Niu and Song 2016

R_2—≡—COOH (2.0 equiv.)

$Co(OAc)_2 \cdot 4H_2O$ (10 mol%)
$NaOPiv \cdot H_2O$ (2.0 equiv.)
Ag_2O (5 mol%)
CH_3CH_2OH, 80 °C, 12 h

R_2—≡—COOH (2.0 equiv.)

$Co(OAc)_2 \cdot 4H_2O$ (10 mol%)
Na_2CO_3 (2.0 equiv.)
Ag_2O (1.5 equiv.)
DMF, 100 °C, 12 h

(b)

Scheme 12.7 (a) Switchable cyclization for the synthesis of quinolines and indoles reported by Glorius and coworkers [31c]. (b) Decarboxylative C–H activation/annulation reaction to furnish isoquinolones and isoindolinones by Niu and coworkers [32].

partner [33]. More recently, the same authors have reported on a highly stereoselective three-component cascade reaction involving aldehydes and ethyl vinyl ketones, where cobalt was shown to be a more effective catalyst than rhodium [34]. Meanwhile, Ackermann and coworker has used isocyanates and acyl azides as coupling partners in aryl and alkenyl aminocarbonylation reactions [35]. The same group has also reported the coupling of bromoalkynes to indoles [36] and a route toward highly functionalized quinolines [37].

Pyrroles are very important building blocks in synthetic chemistry, and a new route to access multisubstituted pyrroles has been provided by Zhang and coworkers through the facile cobalt-catalyzed coupling of enamides and alkynes [38]. Meanwhile, Sundararaju and coworkers has provided a route toward substituted quinolines through the cross-coupling of alkynes and aromatic oximes [39]. Finally, although alkynes and olefins are the most common unsaturated coupling partners, Cheng and coworkers has provided a novel route toward 2*H*-chromenes through the coupling of allenes and a vinylic phenol substrate [40]. Further examples in which cobalt catalysis showed an improved efficient toward other group nine catalysts were reported by Kanai and coworkers in allylation reactions [41].

In order to further expand the potential of cobalt catalysis, the use of diazo compounds as coupling partners has also been studied, as a methodology to provide highly reactive carbene intermediates through the release of benign N_2. In 2015, Glorius and coworkers provided a novel cobalt-catalyzed route toward a family of extended π-systems showing intense absorption and emission bands using the Cp*Co(III) catalyst (Scheme 12.8a) [31a]. The same group later published a protocol for the preparation of isoquinolin-3-ones using this diazo compounds' approach, with an unprotected imine as the directing group [31d]. Through the same approach, Ackermann and coworkers has provided a route toward isoquinolines using aryl amidines as substrates [43]. Similarly, Wang and

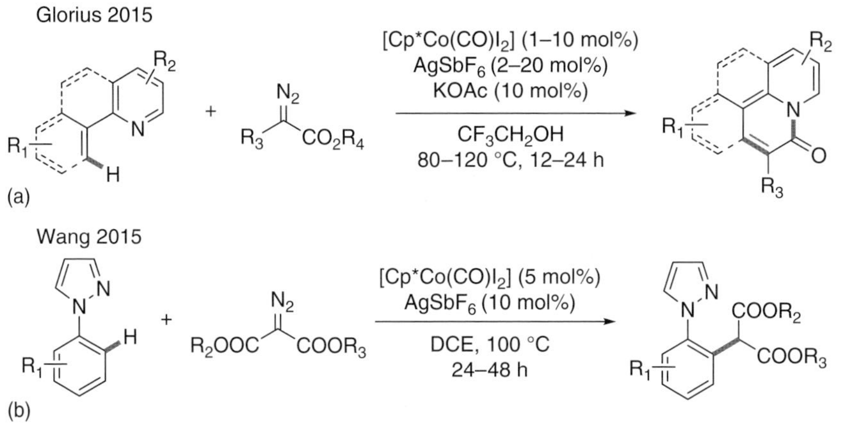

Glorius 2015

[Cp*Co(CO)I_2] (1–10 mol%)
$AgSbF_6$ (2–20 mol%)
KOAc (10 mol%)
CF_3CH_2OH
80–120 °C, 12–24 h

(a)

Wang 2015

[Cp*Co(CO)I_2] (5 mol%)
$AgSbF_6$ (10 mol%)
DCE, 100 °C
24–48 h

(b)

Scheme 12.8 Use of diazo compounds as coupling partners for the synthesis of (a) extended π-systems reported by Glorius and coworkers [31a] and (b) alkylated arenes by Wang and coworkers [42].

coworkers described the alkylation of arenes using diazo malonates and cobalt as catalyst, thus constructing C(sp^2)—C(sp^3) bonds (Scheme 12.8b) [42].

One of the biggest challenges in cross-coupling remains the coupling of inert C(sp^3)—H bonds because of their inert nature. In this context, Sundararaju and coworkers has provided a key example through the alkenylation of 8-methylquinoline substrates (Scheme 12.9a) [44]. This extremely challenging conversion indicates the excellent potential for cobalt catalysis in the future development of new cross-coupling protocols. Furthermore, recently, Shi and coworkers used diazo esters for the functionalization of C(sp^3)—H bonds using 8-methylquinoline as the substrate (Scheme 12.9b) [45].

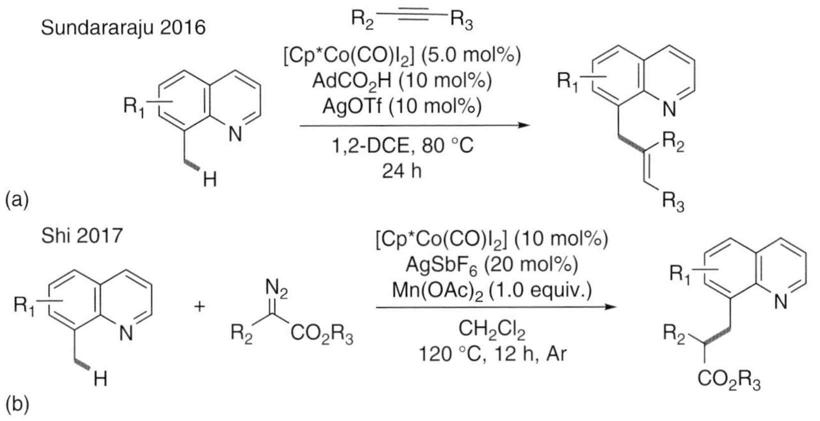

Sundararaju 2016

R_2—≡—R_3
[Cp*Co(CO)I_2] (5.0 mol%)
$AdCO_2H$ (10 mol%)
AgOTf (10 mol%)
1,2-DCE, 80 °C
24 h

(a)

Shi 2017

[Cp*Co(CO)I_2] (10 mol%)
$AgSbF_6$ (20 mol%)
$Mn(OAc)_2$ (1.0 equiv.)
CH_2Cl_2
120 °C, 12 h, Ar

(b)

Scheme 12.9 Functionalization of 8-methylquinoline with (a) alkynes and (b) diazomalonates reported by Sundararaju and coworkers [44] and Shi and coworkers [45].

At the same time the Cp*Co(III) catalysis was flourishing, Daugulis and coworker provided several examples of cobalt-salt-catalyzed annulation reactions using alkynes (Scheme 12.10a) [12], alkenes (Scheme 12.10h) [51], and carbon monoxide (Scheme 12.10d) [48] as coupling partners with a substrate containing an 8-aminoquinoline motif as the directing group (Scheme 12.10). The authors were able to use simple Co(II) salts as catalysts and manganese acetate salts as oxidants to form *in situ* Co(III) organometallic intermediates. Indeed, in one of these reports, the authors were able to isolate a Co(III) organometallic intermediate, which was characterized by nuclear magnetic resonance (NMR) [12]. More recently, Maiti and coworkers have provided crystallographic evidence of this intermediate, which was found to be operative in a C—H activation protocol using alkenes as coupling partners (Scheme 12.10g) [28b]. In contrast to the six-membered rings obtained through the coupling of alkenes reported by Daugulis, later work by Ackermann and coworker has shown that five-membered rings are also accessible using alkene derivatives [52]. Also building on the work by Daugulis, a novel simple procedure for the preparation of sultam motifs has been independently provided by both Ribas and coworkers (Scheme 12.10b) [46] and Sundararaju and coworker (Scheme 12.10c) [47].

Most recently, Daugulis and coworkers extended the substrate scope still further, using phosphinic acid amides providing a new route toward

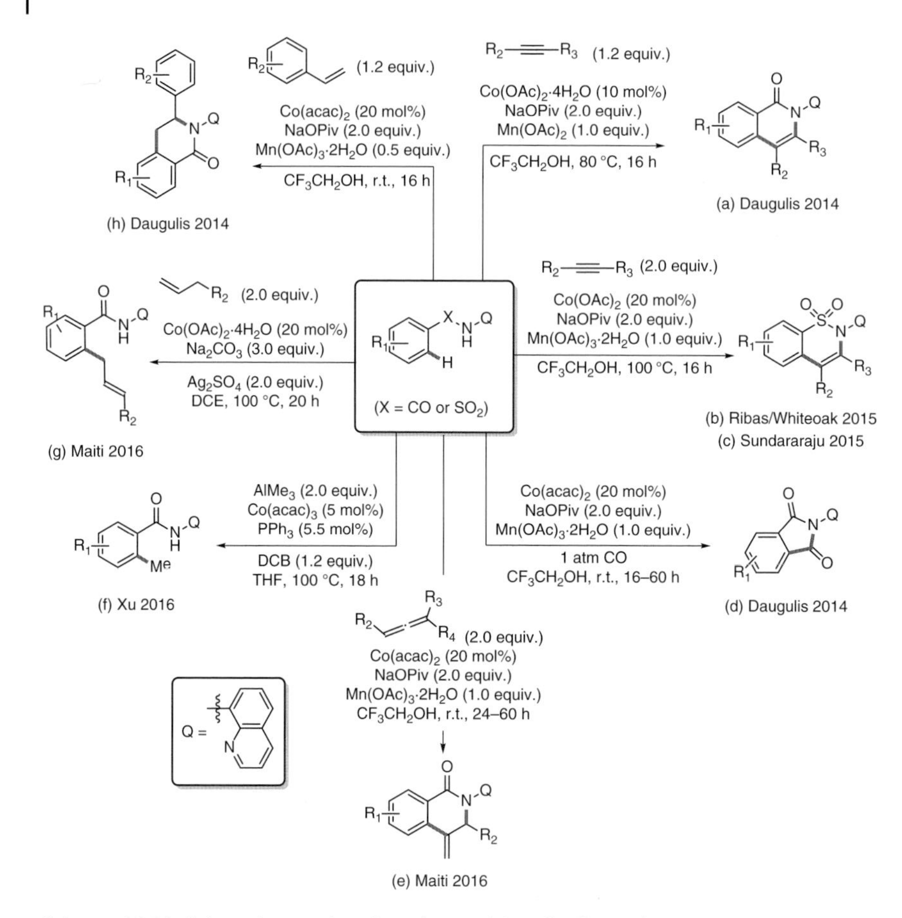

Scheme 12.10 Selected examples of products achieved utilizing the 8-aminoquinoline directing group approach pioneered by (a) Daugulis and coworker [12], (b) Ribas/Whiteoak coworkers [46], (c) Sundararaju and coworker [47], (d) Daugulis and co-worker [48], (e) Thrimurtulu and coworkers [49], (f) Wang and coworkers [50], (g) Maiti and coworkers [28b], and (h) Daugulis and coworker [51].

P,*P*-diarylphosphinic acids [53]. Maiti and coworkers have extended the use of 8-aminoquinoline as the directing group in cobalt-catalyzed C—H activation using allenes as coupling partners (Scheme 12.10e) [49]. In addition, alkylation reactions have also been reported by Xu and coworkers, using trimethylaluminum as coupling partner (Scheme 12.10f) [50]. Since these works, others have also contributed to this field and to this end, recently, Chatani and coworker have provided a very insightful overview of the use of bidentate directing groups in the field of Co(II)-catalyzed C—H functionalization [54]. Further reading on this topic is provided in Chapter 15 of this book.

In attempts to fully elucidate the mechanism of cobalt-catalyzed C—H functionalization protocols, Ribas and coworker reported a rare example of the synthesis and characterization of a benchtop stable family of aryl–Co(III) compounds through a C(sp^2)—H activation using a macrocyclic arene model substrate (Scheme 12.11) [55]. Regioselective formation of five- and

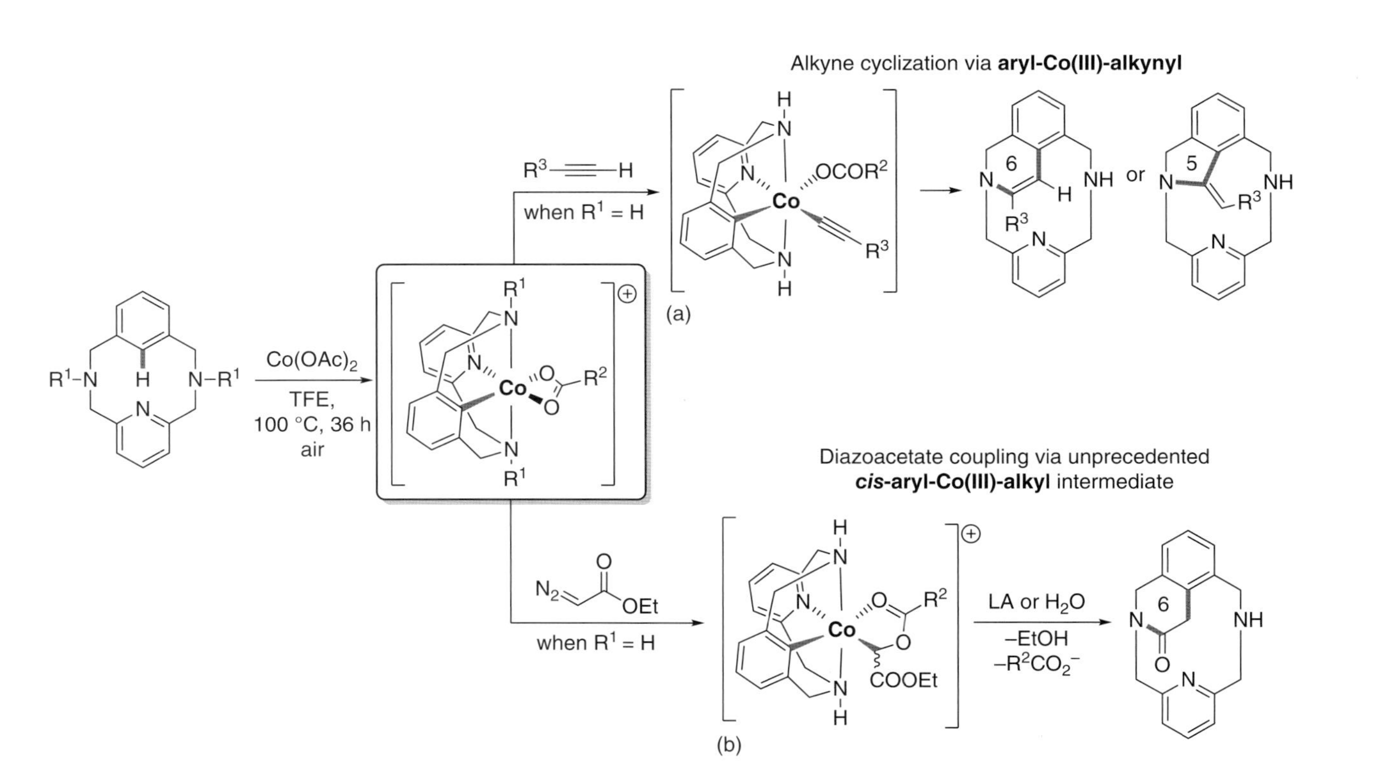

Scheme 12.11 Synthesis and reactivity of organometallic aryl–Co(III) complexes reported by Ribas and coworkers [55, 56] (LA, Lewis acid).

six-membered ring products was achieved when the organometallic intermediates were reacted with terminal alkynes as coupling partners, depending on both the electronic properties of the alkynes and the temperature (Scheme 12.11a). Furthermore, reaction with internal alkynes also furnished six-membered ring isoquinolines in a regioselective manner.

The aryl–Co(III) intermediates were demonstrated to be catalytically competent intermediates in cobalt-catalyzed annulations using alkynes as coupling partners. Evidence obtained from the resulting regioselectivity of the reactions and density functional theory (DFT) studies indicated that a mechanism involving an organometallic aryl–Co(III)-alkynyl intermediate is preferred with terminal alkynes, in contrast to the generally accepted β-migratory insertion pathway. One year later, Ribas and coworkers have reported the synthesis of a rare C-metalated aryl–Co(III) enolate by reacting the previously reported aryl–Co(III) complex with ethyl diazoacetate (Scheme 12.11b) [56]. Interestingly, the aryl–Co(III) enolate intermediate was able to evolve to the macrocyclic amide depicted in Scheme 12.11, which the authors rationalized through an S_N2-type mechanism in which the aryl moiety acts as a nucleophile. The work from Ribas and coworkers represents an unprecedented organometallic mechanistic study in cobalt-catalyzed C—H activation that describes fundamental steps and elucidates several reaction intermediates previously unknown.

12.3 Cobalt-Catalyzed C—C Couplings Using a Preactivated Substrate Approach (Aryl Halides and Pseudohalides)

The vast majority of the cobalt-catalyzed C—C cross-coupling reactions is dominated by the use of Grignard reagents (R–Mg–X) as nucleophilic transmetalating agents, building on the original work of Kharasch and Fields. This section includes examples of arylation, alkenylation, or alkylation of aryl, alkenyl, or alkyl halides or pseudohalides (such as triflates). However, in few cases, the coupling reaction is performed using *in situ*-formed organozinc nucleophilic partners, or more scarcely two aryl halide partners.

12.3.1 Aryl or Alkenyl Halides, C(sp^2)–X

The use of C(sp^2)–X as electrophile partners and a range of organomagnesium compounds as nucleophiles in cobalt catalysis is a highly developed area, and several examples have been reported [14]. In this section, a brief selection of these methodologies will be discussed.

Early examples demonstrating the ability of cobalt to catalyze C(sp^2)–C(sp^2) cross-coupling reactions were provided by Kharasch and Fuchs back in 1943 [57], when alkenyl chlorides and bromides were coupled with Ar–Mg–X compounds, using $CoCl_2$ as the catalyst as mentioned already in this chapter. Cahiez and Avedissian further optimized this reaction in the late 1990s [58] and Hayashi and coworkers expanded the scope of this methodology by using alkenyl triflates and $Co(acac)_3$ as the catalyst (Scheme 12.12a) [59]. Nakamura and coworkers have since extended this synthetic methodology to provide biaryl products from aryl halides and ArMgBr in good yields under mild reaction conditions (Scheme 12.12b) [60].

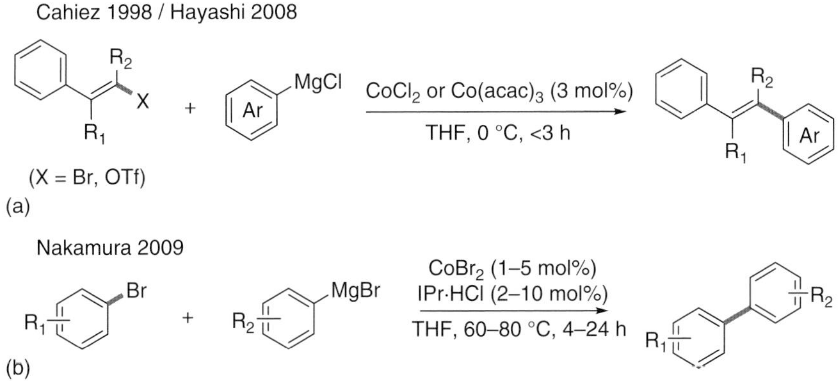

Scheme 12.12 Cobalt-catalyzed C(sp^2)–C(sp^2) cross-coupling using Grignard reagents, described by (a) Cahiez and Avedissian [58] and Hayashi and coworkers [59], and (b) by Nakamura and coworkers [60] (IPr, 1,3-Diisopropylimidazol-2-ylidene).

In a further advancement of this methodology, alkenyl halides have been used as coupling counterparts in Co(II)-catalyzed reactions using alkyl–MgBr or alkyl–ZnBr, providing efficient routes to C(sp^2)–C(sp^3)-coupled products (Scheme 12.13a) [61]. In addition to alkenyl halides, Oshima has utilized alkyl halides and alkenyl or alkynyl Grignard reagents [62], and also aromatic halides together with alkyl Grignard reagents for the preparation of alkyl-substituted aromatic compounds (Scheme 12.13b) [63].

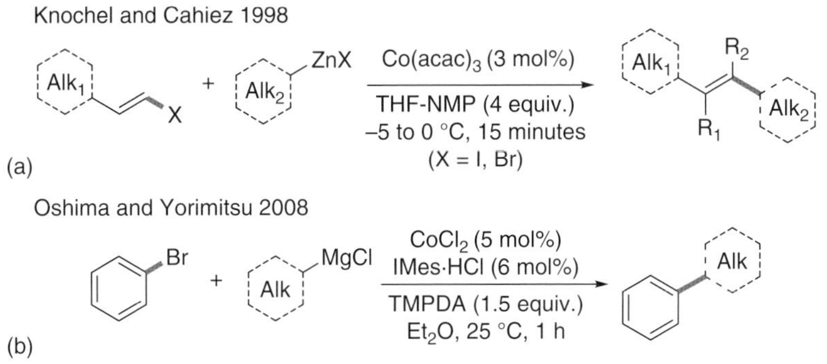

Scheme 12.13 Cobalt-catalyzed C(sp^2)–C(sp^3) cross-coupling using Grignard or organozinc reagents [61, 63].

12.3.2 Alkyl Halides, C(sp^3)–X

Alkyl halides have successfully been used as electrophiles in cobalt-catalyzed cross-coupling reactions with alkenyl [62] and aryl Grignard reagents (Scheme 12.14) [64]. Thus, these methodologies represent good examples for the construction of C(sp^2)–C(sp^3) bonds in mild conditions and using readily available reagents [65a, b].

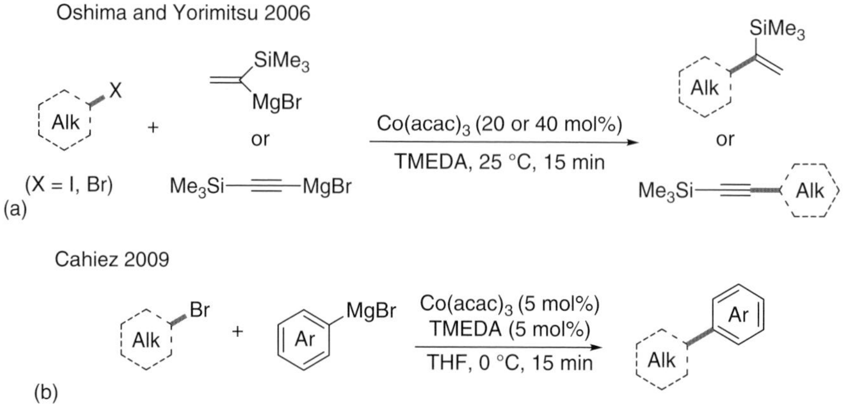

Oshima and Yorimitsu 2006

(a) Alk–X (X = I, Br) + CH$_2$=C(SiMe$_3$)MgBr or Me$_3$Si–≡–MgBr → [Co(acac)$_3$ (20 or 40 mol%), TMEDA, 25 °C, 15 min] → Alk–C(SiMe$_3$)=CH$_2$ or Me$_3$Si–≡–Alk

Cahiez 2009

(b) Alk–Br + Ar–MgBr → [Co(acac)$_3$ (5 mol%), TMEDA (5 mol%), THF, 0 °C, 15 min] → Alk–Ar

Scheme 12.14 Cobalt-catalyzed C(sp^3)–C(sp^2) cross-coupling using (a) alkene and (b) aryl Grignard reagents [62, 64] (TMEDA, tetramethylethylenediamine).

In the former methodology, Oshima and coworkers described a cobalt-mediated cross-coupling of alkyl halides with alkenyl and alkynyl Grignard compounds (Scheme 12.14a). In this example, a SET mechanism is proposed to govern the transformation, enabling sequential radical alkenylation and alkynylation reactions. Interestingly, this methodology was further extended by Cahiez and coworkers who employed Co(III) catalysts with aryl–Grignard reagents as coupling partners (Scheme 12.14b) [64]. Furthermore, benzyl halides have also been successfully used together with alkynyl–MgBr [66] or alkyl–ZnI (Scheme 12.15) [67].

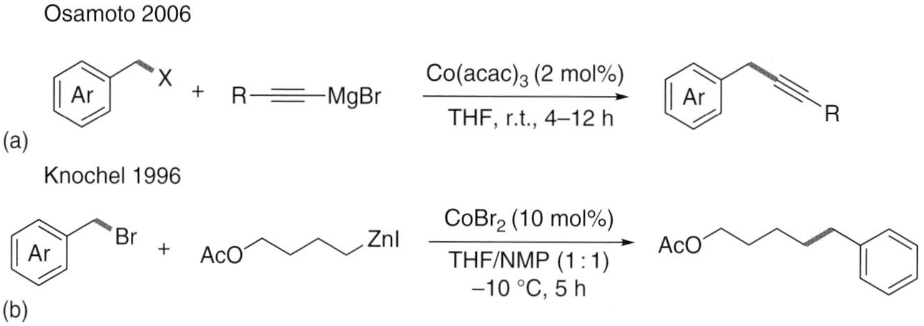

Osamoto 2006

(a) ArCH$_2$X + R–≡–MgBr → [Co(acac)$_3$ (2 mol%), THF, r.t., 4–12 h] → ArCH$_2$–≡–R

Knochel 1996

(b) ArCH$_2$Br + AcO(CH$_2$)$_5$ZnI → [CoBr$_2$ (10 mol%), THF/NMP (1 : 1), −10 °C, 5 h] → AcO(CH$_2$)$_5$CH$_2$Ph

Scheme 12.15 Cobalt-catalyzed C(sp^3)–C(sp^2) cross-coupling using benzyl halides and (a) alkynyl Grignard reagents [66] and (b) organozinc compounds [67].

Cobalt-catalyzed, Heck-type reactions, i.e. alkylation of alkenes using alkyl halides, were first reported by Branchaud and Detlefsen in 1991 using bis(dimethylglyoximato)cobalt(III) (cobaloxime) as the catalyst. This reaction, which was reported to proceed through a radical pathway, afforded higher yields compared to the corresponding Pd-catalyzed reaction, probably because of the enhanced stability of alkyl–cobalt species toward β-hydride elimination [68]. The Heck-type, cobalt-catalyzed cross-coupling reaction was further optimized by Oshima and coworkers [69]. In their work, the authors optimized the reaction by adding a reducing agent (Me_3SiCH_2MgCl), which promotes a radical pathway that leads to the regioselective formation of *E*-alkenes (Scheme 12.16).

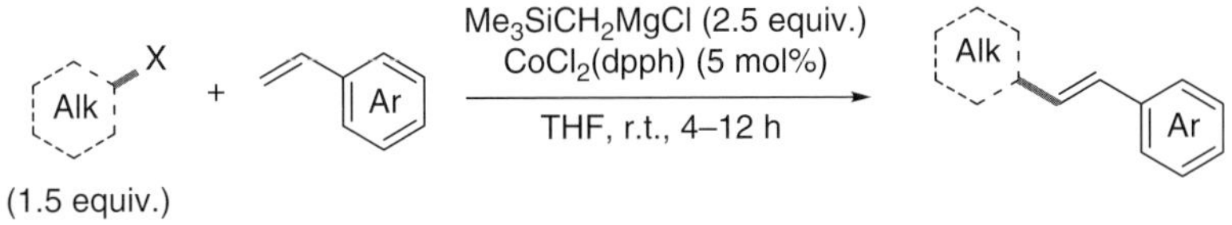

Alk–X (1.5 equiv.) + Ar–CH=CH$_2$ → [Me_3SiCH_2MgCl (2.5 equiv.), $CoCl_2$(dpph) (5 mol%), THF, r.t., 4–12 h] → Alk–CH=CH–Ar

Scheme 12.16 Cobalt-catalyzed C(sp^3)–C(sp^2) Heck-type coupling described by Oshima and coworkers [69] (dpph = 1,6-bis(diphenylphosphino)hexane).

12.3.3 Alkynyl Halides, C(sp)–X

The use of alkynyl halides as coupling partners in cobalt-catalyzed cross-coupling transformations was implemented several years ago. As mentioned before, Kharasch and Fuchs demonstrated the practicality of cobalt catalysis in C(sp^2)–C(sp^2) cross-coupling reactions in 1943 [57]. Some years later, the same authors extended this work and pioneered the cobalt-catalyzed coupling of alkynyl halides with alkynyl Grignard reagents as coupling partners [70]. Weedon and coworkers, in 1954 [71], applied a similar protocol using a variety of alkynyl halides and alkyl and aryl Grignard reagents, thus expanding the scope of both the organomagnesium reagents and the alkynyl halide substrates (Scheme 12.17).

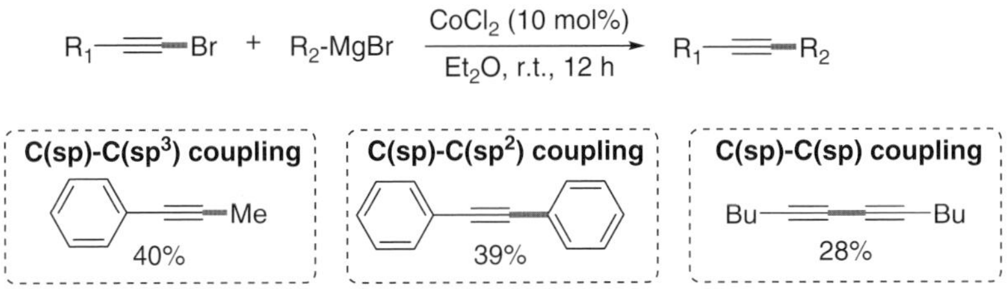

R_1–≡–Br + R_2-MgBr → [$CoCl_2$ (10 mol%), Et_2O, r.t., 12 h] → R_1–≡–R_2

C(sp)-C(sp^3) coupling: Ph–≡–Me, 40%
C(sp)-C(sp^2) coupling: Ph–≡–Ph, 39%
C(sp)-C(sp) coupling: Bu–≡–≡–Bu, 28%

Scheme 12.17 Cobalt-catalyzed C(sp)–C cross-couplings using alkynyl halides [71].

12.3.4 Aryl Halides Without Organomagnesium

Fewer reagents other than organomagnesiums have been found to efficiently deliver the nucleophilic alkenyl reaction partner. Périchon and coworkers reported in 2003 on the use of allyl acetates as nucleophiles in the cobalt-catalyzed electrochemical cross-coupling with aryl halides to obtain the aryl–allylated products (Scheme 12.18a) [72a, b]. A few years later, the same authors expanded the cross-coupling to the use of alkenyl acetates to afford the aryl alkenylated compounds (Scheme 12.18b) [73]. In this case, the reaction was

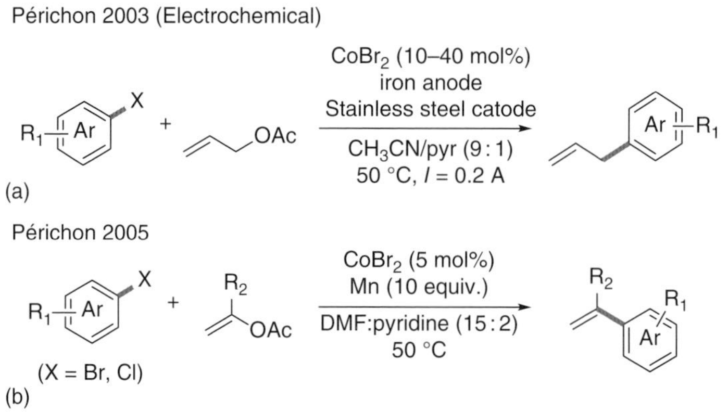

Périchon 2003 (Electrochemical)

$CoBr_2$ (10–40 mol%)
iron anode
Stainless steel catode
CH_3CN/pyr (9 : 1)
50 °C, I = 0.2 A

(a)

Périchon 2005

$CoBr_2$ (5 mol%)
Mn (10 equiv.)
DMF:pyridine (15 : 2)
50 °C

(X = Br, Cl)

(b)

Scheme 12.18 Cobalt-catalyzed C(sp^2)–C(sp^2) cross-couplings using aryl halides and (a) allyl acetates or (b) alkenyl acetates [72, 73].

carried out without the need of electrochemical conditions but using manganese as the reductant.

Recently, Gosmini and coworkers reported the cobalt-catalyzed cross-coupling reaction using two different aryl halides/triflates to obtain the corresponding biaryl products in good yields (Scheme 12.19) [74]. Interestingly, the transformation is proposed to go through a Co(I)/Co(III) catalytic cycle, although a radical pathway could not be ruled out. This catalytic process involves a simple, inexpensive, and environmentally friendly cobalt catalyst. This fact, combined with the wide range of aryl halides and triflates tolerated, makes this protocol a good alternative to the use of Grignard reagents.

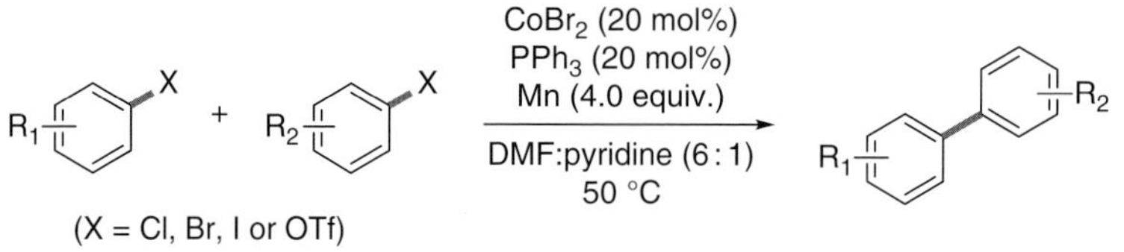

$CoBr_2$ (20 mol%)
PPh_3 (20 mol%)
Mn (4.0 equiv.)
DMF:pyridine (6 : 1)
50 °C

(X = Cl, Br, I or OTf)

Scheme 12.19 Cobalt-catalyzed C(sp^2)–C(sp^2) cross-couplings using two different aryl halides [74].

12.4 Cobalt-Catalyzed C—X Couplings Using C—H Activation Approaches

During the last five years, significant steps toward catalytic systems that allow performing the construction of C—X bonds (X = N, O, S, etc.) using cobalt as the catalyst have been reported. Although cobalt-catalyzed hydrosilylation [75], hydrophosphination [76], hydroboration [77], and borylation [78] reactions are known, direct C—H bond functionalization to furnish C—X bonds is relatively underdeveloped [14, 79]. Nevertheless, the explosion of high-valent cobalt catalysis has allowed for the development of a variety of C—X couplings starting from inert C—H bonds [13b, 15, 16]. In this section, recent cobalt-catalyzed C—H functionalization in the context of C—X bond formation will be highlighted.

12.4.1 C—N Bond Formation

High-valent, cobalt-catalyzed C—N bond formation via C—H cleavage is still rare. Alkyl and aryl azides have been recognized as important reagents for amination reactions [80], but cobalt-mediated coupling of arenes with secondary amines is still in its infancy. Recently, Niu and coworkers described a new method for the amination of arylamides with secondary alkylamines through $C(sp^2)$—H bond functionalization (Scheme 12.20a) [81]. Based on previous reports, experiments with radical traps and DFT studies, a SET was suggested as the key step in a tentative reaction mechanism. Independently, Zhang and coworkers reported a similar amination protocol utilizing the 8-aminoquinoline directing group [83]. Mechanistic observations, such as a KIE value of 1.3, suggest that the reaction proceeds by a SET mechanism, although a Co^{II}/Co^{IV} catalytic amination pathway could not be excluded. An interesting example of C—N bond formation through organometallic C—H activation has been recently reported by Niu and coworkers (Scheme 12.20b) [82]. The authors developed an oxidative C—H/N—H cross-dehydrogenative coupling (CDC) reaction between unactivated arenes and simple aniline substrates to accomplish the synthesis of triarylamines. These CDC aminations seem to open the door to other cobalt-catalyzed C—H bond cross-coupling reactions.

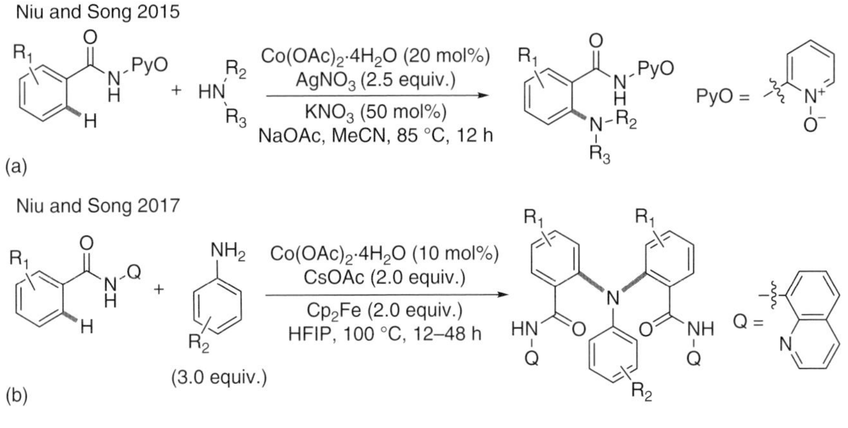

Niu and Song 2015

$Co(OAc)_2·4H_2O$ (20 mol%), $AgNO_3$ (2.5 equiv.), KNO_3 (50 mol%), NaOAc, MeCN, 85 °C, 12 h

PyO =

(a)

Niu and Song 2017

$Co(OAc)_2·4H_2O$ (10 mol%), CsOAc (2.0 equiv.), Cp_2Fe (2.0 equiv.), HFIP, 100 °C, 12–48 h

(3.0 equiv.)

Q =

(b)

Scheme 12.20 Cobalt-catalyzed amination using (a) pyridine oxide (PyO) [81] and (b) 8-aminoquinoline [82] as the directing group reported by Niu and coworkers.

A SET mechanism was also proposed by Whiteoak and Ribas who described a novel room temperature cobalt-catalyzed remote nitration protocol of 8-aminoquinoline carboxyamides with *tert*-butyl nitrite (TBN) for the preparation of 5- and 7-nitroaminoquinolines (Scheme 12.21a) [84]. The inhibition observed when electron-withdrawing groups were present in the substrate together with a KIE value of 0.97 suggested a unique remote C—H functionalization SET mechanism previously unknown for cobalt, similar to previous work using copper as a catalyst [87]. Following this protocol, Song and coworkers disclosed highly chemo- and regioselective nitration of C(sp^3)—H bonds using $Co(OAc)_2$ as the catalyst and TBN as the nitro source (Scheme 12.21b) [85]. Although the protocol was limited to the use of pyrimidine derivatives, it showed a broad spectrum of functional group tolerance. Based on previous reports and radical scavenging experiments using (2,2,6,6-tetramethylpiperidin-1-yl)oxyl (TEMPO), the authors proposed a plausible SET mechanism for this cobalt-catalyzed C(sp^3)–H nitration. In the same field, Das and coworkers has recently reported a similar cobalt-catalyzed regioselective *ortho*-C(sp^2)—H bond nitration of *N*-phenylpyridine-2-amines using $AgNO_2$ as nitro source proposed to proceed through a proton-coupled electron transfer (PCET) mechanism (Scheme 12.21c) [86].

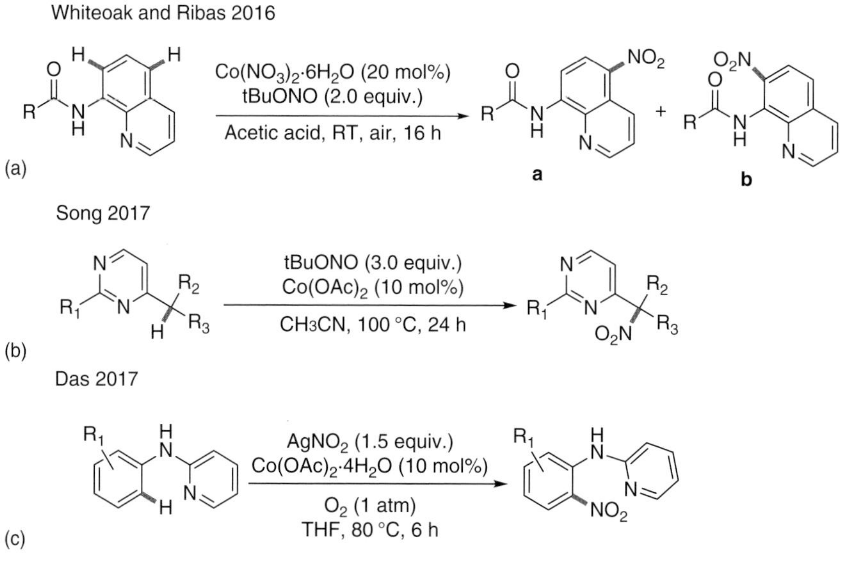

Whiteoak and Ribas 2016

$Co(NO_3)_2·6H_2O$ (20 mol%), tBuONO (2.0 equiv.), Acetic acid, RT, air, 16 h → a + b

(a)

Song 2017

tBuONO (3.0 equiv.), $Co(OAc)_2$ (10 mol%), CH3CN, 100 °C, 24 h

(b)

Das 2017

$AgNO_2$ (1.5 equiv.), $Co(OAc)_2·4H_2O$ (10 mol%), O_2 (1 atm), THF, 80 °C, 6 h

(c)

Scheme 12.21 Cobalt-catalyzed C—H bond nitration reactions reported by (a) Whiteoak/Ribas and coworkers [84], (b) Song and coworkers [85], and (c) Das and coworkers [86].

New advances in cobalt-catalyzed amidation of C(sp^2)–H and C(sp^3)–H have also been reported. In this context, Ge and coworkers developed a highly selective intramolecular amination of propionamide derivatives via a cobalt-catalyzed C(sp^3)–H functionalization process (Scheme 12.22, right) [88]. The reaction favors the β-C—H bonds, providing β-lactam derivatives in a highly site- and diastereoselective manner (dr > 20 : 1). Moreover, this methodology represents the first cobalt-catalyzed intermolecular amination of C(sp^3)—H bonds using electron-poor amides (Scheme 12.22, left). Thus, perfluorinated amides, together with $Co(acac)_3$ as the catalyst, furnished C—H bond amidation products of β-C—H bonds in a selective manner and good yields.

Amidation of C(sp^2)—H and C(sp^3)—H bonds has also been studied using powerful Cp^*Co^{III} catalysts. Kanai and coworkers described the pioneering

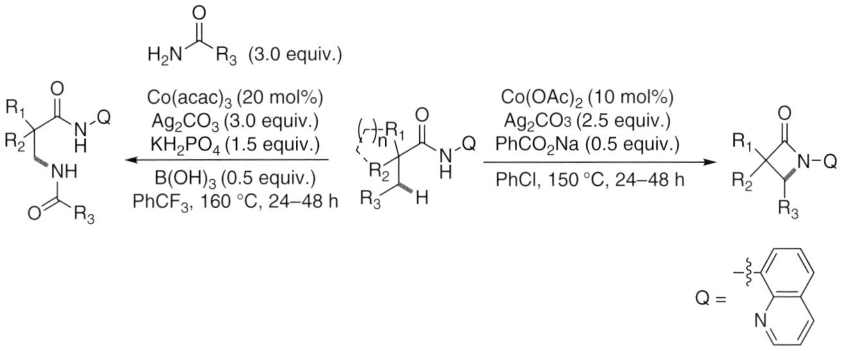

Scheme 12.22 Preparation of β-lactams (right) and β-amido products (left) using 8-aminoquinoline as the directing group reported by Ge and coworkers [88].

C2-selective sulfamidation of indoles using sulfonyl azides as coupling partners, obtaining a variety of sulfonamide products (Scheme 12.23, right) [89]. Co^{III}-catalyzed C—H activation of the C2-position is proposed to undergo via either an S_EAr or a CMD mechanism, furnishing an intermediate organometallic Co^{III} species. Similar reactivity was also achieved using phosphoryl azides, further expanding the scope of this coupling protocol (Scheme 12.23, left) [90].

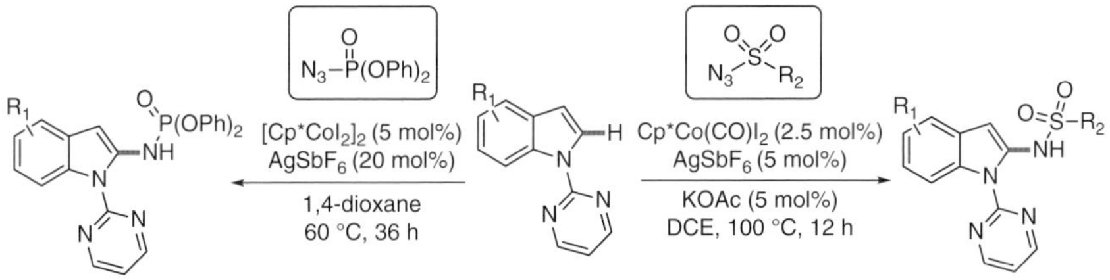

Scheme 12.23 Cp*Co(III)-catalyzed sulfamidation and phosphoramidation of C(sp^2)—H bonds reported by Matsunaga and coworkers [89, 90].

While the previously described amidation methods with Cp*Co(III) catalysts were restricted to the synthesis of phosphonamides and sulfonamides, the use of dioxazolone derivatives as coupling partners allowed for the preparation of aryl as well as alkyl amides with both C(sp^2)—H and C(sp^3)—H bonds (Scheme 12.24).

The first report of dioxazolones in cobalt-catalyzed, cross-coupling reactions was described by Chang and coworkers in 2014 for the conversion of aryl pyridine substrates (Scheme 12.24a) [91]. In addition to the use of dioxazolones, Chang and coworkers has also reported on the use of *O*-acylcarbamates as the amido source [100]. Dioxazolone derivatives have also proved to be efficient amidation reagents in the C—H functionalization of indole motifs, a protocol that was independently described by Jiao and coworkers [92] and Ackermann and coworkers [93] (Scheme 12.24b,c). This amidating reagent has also found application in annulation reactions, furnishing a variety of heterocyclic compounds. Li, Glorius, and coworkers independently reported the synthesis of

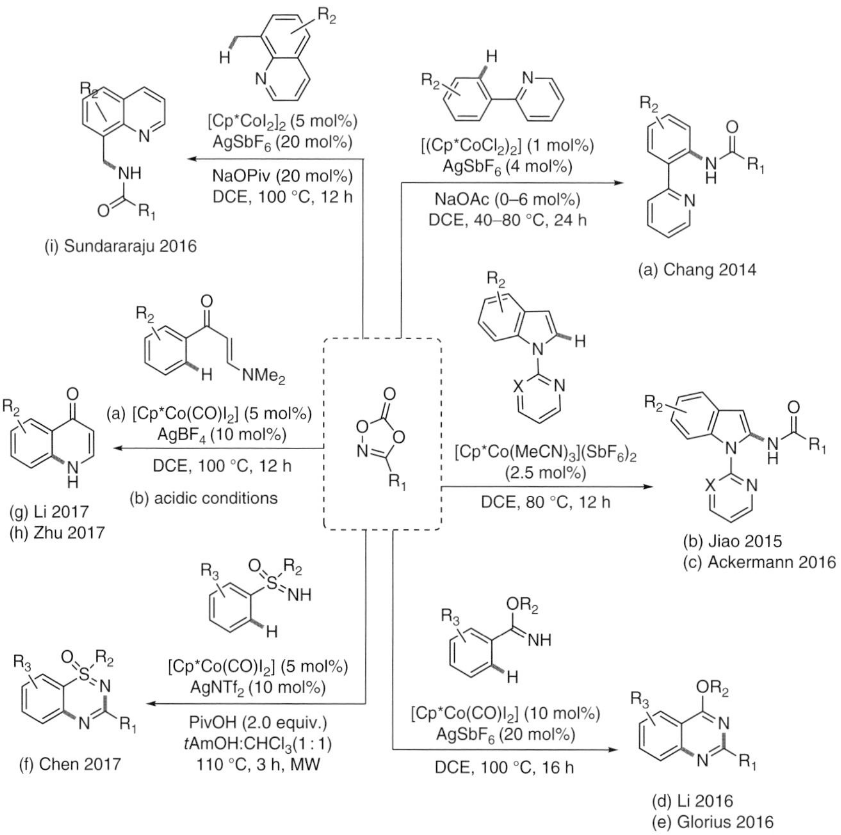

Scheme 12.24 Cp*Co(III)-catalyzed amidation of C(sp^2)—H bonds using dioxazolones as amidating reagents reported by (a) Chang and coworker [91], (b) Jiao and coworkers [92], (c) Ackermann and coworkers [93], (d) Li and coworkers [94], (e) Glorius and coworkers [95], (f) Chen and coworkers [96], (g) Li and coworkers [97], (h) Zhu and coworkers [98], and (i) Sundararaju and coworkers [99].

quinazolines from a formal [4+2] cycloaddition of arenes with free imines and dioxazolones (Scheme 12.24d,e) [94, 95]. In 2017, Chen and coworkers reported on a microwave-assisted preparation of thiadiazine-1-oxide compounds [96], which have important properties as biologically active compounds (Scheme 12.24f). In the same year, Li and coworkers [97] and Zhu and coworkers [98] independently disclosed the synthesis of quinolone compounds starting from aryl enaminones (Scheme 12.24g,h). In this case, the synthesis of the final annulated product was achieved through a one-pot reaction including a first amination step and a second acid-assisted cyclization reaction. Further applications of dioxazolones were developed by Sundararaju, who used them in the cobalt-catalyzed C(sp^3)–H activation and functionalization using a quinoline directing group (Scheme 12.24i) [99].

Recently, Li and coworkers have developed new strategies taking advantage of the properties of anthranils as aminating agents [101]. When Cp*Co(III) was used as a catalyst together with $Cu(OAc)_2$ as the cocatalyst, 1*H*-indazoles containing different functional groups were obtained in excellent yields from readily available imidates and anthranils (Scheme 12.25). Mechanistic experiments showed a KIE value of 1.4, indicating that C—H bond cleavage is not involved in the rate-determining step. Furthermore, the use of radical scavengers such as *tert*-butyl-4-methylphenol (BHT) partially inhibited the reaction, suggesting that a radical pathway is likely involved. This latter methodology, together with the protocols described above, constitute a proof of the future potential of cobalt catalysis for application in practical C—N forming protocols.

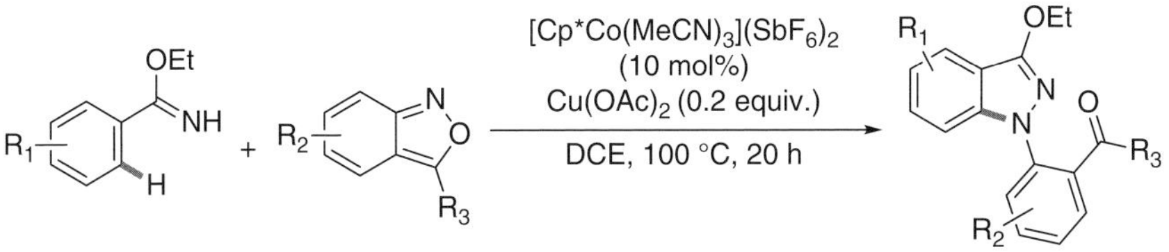

Scheme 12.25 Co(III)/Cu(II)-catalyzed synthesis of 1*H*-indazoles reported by Li and coworkers [101].

12.4.2 C—O and C—S Bond Formation

The synthesis of C—O and C—S bonds is the fundamental process in organic chemistry, and ether- and thioether-containing groups are widely employed in the synthesis of pharmaceuticals and functional materials [102]. Cobalt-mediated C—O bond formation was pioneered by Kochi et al. [7, 101, 103, 104], but just one recent example of cobalt-catalyzed alkoxylation of C—H bonds can be found in the literature [105]. In this protocol developed by Niu and coworkers, alkoxylation of $C(sp^2)$—H bonds in aromatic and olefinic benzamides assisted by a 2-aminopyridine-1-oxide group (PyO) was achieved (Scheme 12.26a). A wide variety of alcohols were coupled using $Co(OAc)_2 \cdot 4H_2O$ as the catalyst and Ag_2O as the oxidant, furnishing the corresponding aromatic and olefinic ethers in moderate-to-excellent yields (34–88%). Electron paramagnetic resonance (EPR) spectroscopy studies as well as experiments with radical quenchers suggested that a SET pathway was likely to be operative. Furthermore, DFT calculations performed by Wei and coworkers provided further support to this hypothesis as well as valuable insights into rational prediction of catalyst design and reactivity using cobalt [107]. Very recently, Zheng and coworkers reported a cobalt-catalyzed CDC of arenes with carboxylic acids to furnish a variety of aryl esters (Scheme 12.26b) [106]. The applicability of this methodology relies in the conversion of the ester moieties into free-hydroxyl groups, furnishing an entirely new library of substituted phenol derivatives. Furthermore, several control experiments revealed that the transformation proceeds through an organometallic mechanism in which an aryl–Co(III) species is involved.

The direct thiolation of C—H bonds represents an attractive strategy for the construction of thioethers, which are common in bioactive compounds and in a variety of intermediates in organic synthesis [102e]. Despite advances toward

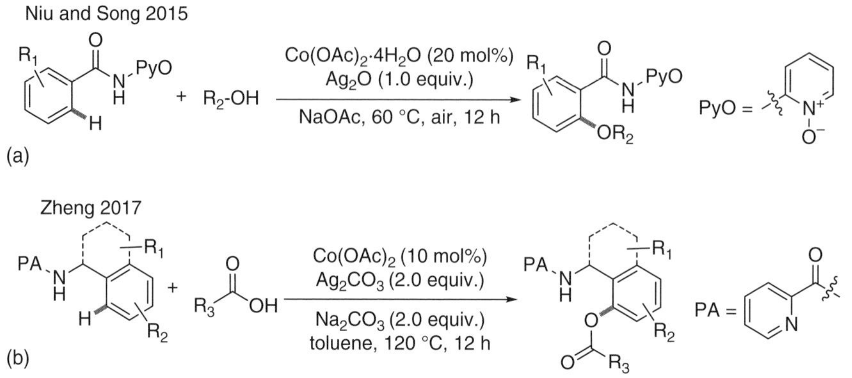

Scheme 12.26 Niu and Song's alkoxylation of aromatic and olefinic substrates using 2-aminopyridine-1-oxide (PyO) as the directing group [105] and Zheng's acyloxylation reaction of C—H bonds using picolinamide (PA) as the directing group [106].

an efficient technology to achieve these transformations, there is still a need for an improvement in terms of catalyst efficiency and cost. In this context, Glorius and coworkers reported a novel Cp*Co(III)-catalyzed C—H thiolation through a CDC pathway using indole substrates bearing 2-pyrimidine as the directing group, furnishing C2-thiolated indoles employing a cooperative system with Cu(II) (Scheme 12.27a) [108]. Reactivity of Co(III) was demonstrated to be unique in this transformation, as employing Rh(III) resulted in trace amounts of thioester products. Furthermore, mass spectrometric analysis of crude reaction mixtures after five minutes revealed the presence of a C—H-metalated intermediate, as well as copper thiolates. These observations suggested that copper plays an essential role as a coupling partner activating agent. Wang and coworkers recently described the formation of C—S bonds through trifluoromethylthiolation of C(sp^2)—H bonds using silver(I) trifluoromethylthiolate (Scheme 12.27b) [109]. In this protocol, C—H functionalization proceeded efficiently with 2-aryl pyridines and 2-arylpyrimidines, but it failed with a range of other common substrates. The scope of this reaction was further expanded by Yoshino and coworkers, who presented the direct trifluoromethylthiolating of C(sp^2)—H bonds employing *N*-trifluoromethylthiobenzenesulfonimide as the trifluoromethylthiolate source (Scheme 12.27c) [110]. In the latter protocol, the reaction scope included substrates such as 6-arylpurines, and it showed higher efficiency with single-component catalysts.

12.4.3 C—X Bond Formation (X = Cl, Br, I, and CN)

Molecules containing C—X bonds represent an important class of organic compounds because of their high synthetic utility [111]. Although C—H bond halogenation is well known using precious metals [2, 112], examples of C—X bond construction using cobalt as the catalyst are limited. A first Cp*Co(III)-mediated halogenation was reported by Glorius and coworkers (Scheme 12.28) [30]. Cyano groups, considered as pseudohalogens, were coupled to the C2-position

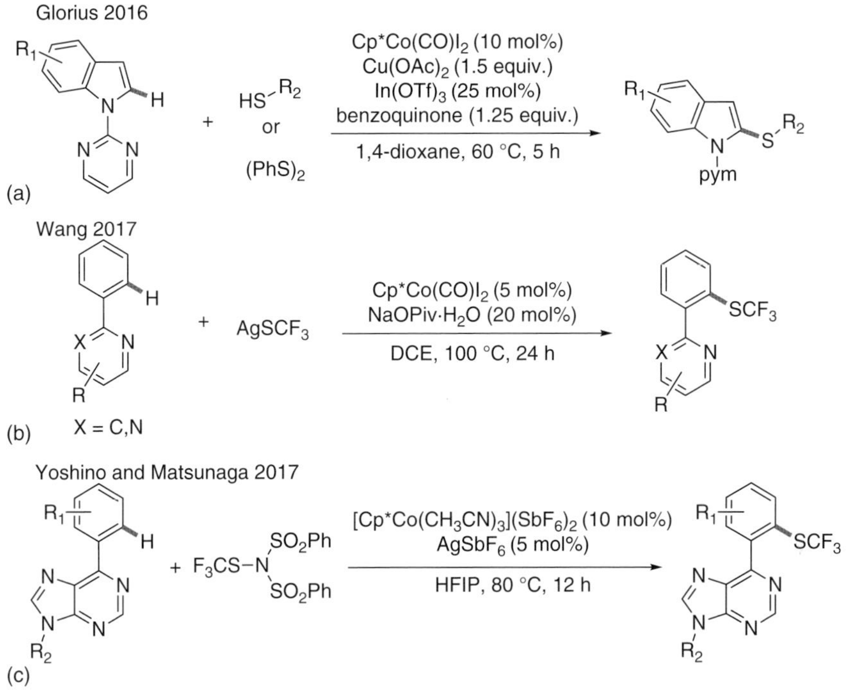

Scheme 12.27 Cp*Co(III)-catalyzed C—S bond formation reactions reported by (a) Glorius and coworkers [108], (b) Wang and coworkers [109], and (c) Yoshino and coworkers [110].

of indole in excellent yields using *N*-cyano-*N*-phenyl-*p*-toluenesulfonamide (NCTS). Furthermore, halogenation of aromatic and olefinic C—H bonds was also achieved using easily available halogen sources such as *N*-iodosuccinimide (NIS) and *N*-bromophthalimide (NBS). This protocol, as well as work reported by Lade and coworker [113], represents potentially valuable methods for obtaining organonitriles, –iodides, and –bromides from readily available starting materials.

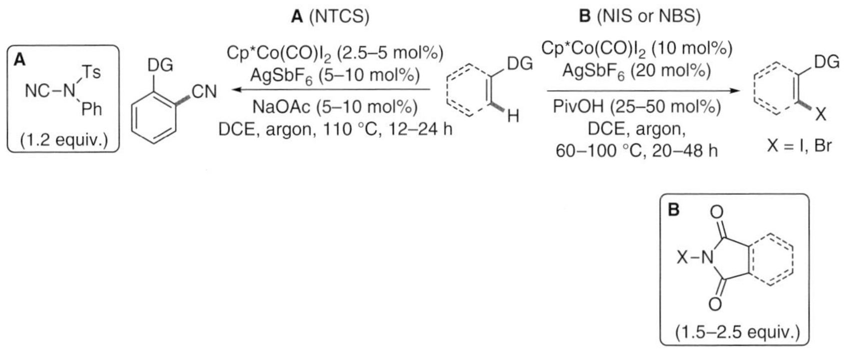

Scheme 12.28 Cp*Co(III)-catalyzed halogenation and cyanation of C—H bonds reported by Glorius and coworkers [30] (DG = directing group).

12.5 Cobalt-Catalyzed C—X Couplings Using a Preactivated Substrate Approach (Aryl Halides and Pseudohalides)

The recent development in cross-coupling reactions using transition metal catalysis afforded effective methods for the construction of C—X bonds (X = N, O, S). In this field, cobalt-catalyzed, cross-coupling transformations are tremendously useful for C—C bond forming reactions [14, 79]. However, methodologies involving the construction of C—heteroatom bonds are underdeveloped, and only a few examples can be found in the literature. In this section, the known examples of cobalt-catalyzed, cross-coupling reactions to construct C—X bonds will be highlighted.

12.5.1 C(sp^2)–S Coupling

The first cobalt-catalyzed C—X cross-couplings reported were achieved using thiols as nucleophiles. In 2006, Cheng and coworkers described the efficient synthesis of thioethers using aryl iodides or bromides and aryl and alkyl thiols under mild conditions and low (1–2 mol%) catalyst loadings (Scheme 12.29a) [114]. The authors also described the C(sp^2)–S coupling using iodoacrylates using the same protocol. This methodology highlighted the potential of cobalt to perform C—S bond constructions through an inexpensive and efficient protocol. Following the work of Cheng and coworkers, Tsai and coworkers reported a similar reaction, expanding the scope to aryl halides and alkyl thiols [115]. Interestingly, this protocol tolerates aryl chlorides as coupling partners, albeit they need harsh conditions and longer reaction times. Furthermore, yields were moderate, but the reaction can be carried out in water as a solvent under an oxygenated atmosphere, which is an uncommon feature in these types of transformations (Scheme 12.29b).

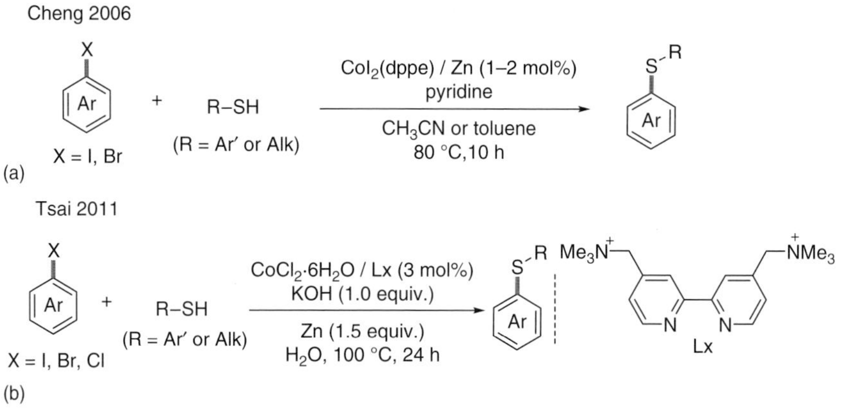

Cheng 2006
X
Ar
+ R–SH
(R = Ar′ or Alk)
X = I, Br
CoI_2(dppe) / Zn (1–2 mol%)
pyridine
CH_3CN or toluene
80 °C,10 h
S–R
Ar
(a)

Tsai 2011
X
Ar
+ R–SH
(R = Ar′ or Alk)
X = I, Br, Cl
$CoCl_2 \cdot 6H_2O$ / Lx (3 mol%)
KOH (1.0 equiv.)
Zn (1.5 equiv.)
H_2O, 100 °C, 24 h
S–R
Ar
$Me_3\overset{+}{N}$
$\overset{+}{N}Me_3$
N N
Lx
(b)

Scheme 12.29 Cobalt-catalyzed C–S formation disclosed by (a) Cheng and coworkers [114] and (b) Tsai and coworkers [115] (dppe = 1,2-bis(diphenylphosphino)ethane).

12.5.2 C(sp^2)–N Coupling

In 2009, Teo and Chua reported the first example of cobalt-catalyzed N-arylation of cyclic secondary amides and amides. The yields were found to be moderate, but catalysis was conducted in water as a solvent [116]. In this unprecedented example, the catalytic system is composed of $CoCl_2{\cdot}6H_2O$ as a catalyst, $K_3PO_4{\cdot}H_2O$ as a base, and dimethylethylenediamine (DMEDA) as the ligand. A variety of nitrogen nucleophiles, including amines and heterocyclic compounds such as indoles and pyrrolidones, could be used (Scheme 12.30a). In 2014, the same authors optimized the reaction for the arylation of benzylamide derivatives obtaining excellent yields [117]. In these examples, the authors tentatively propose an unprecedented two-electron Co(II)/Co(IV) redox cycle. Based on Teo's protocol, in 2010, Punniyamurthy and coworkers developed a cyclization of 2-haloarylbenzamidines and 2-haloarylbenzamides via intramolecular C—N cross-couplings using Co(II) as the catalyst to furnish benzimidazoles (Scheme 12.30b) [118]. In this protocol, the authors suggested an oxidative addition of Co(II) to the C—X bond and excluded the presence of trace amounts of copper through atomic absorption analysis.

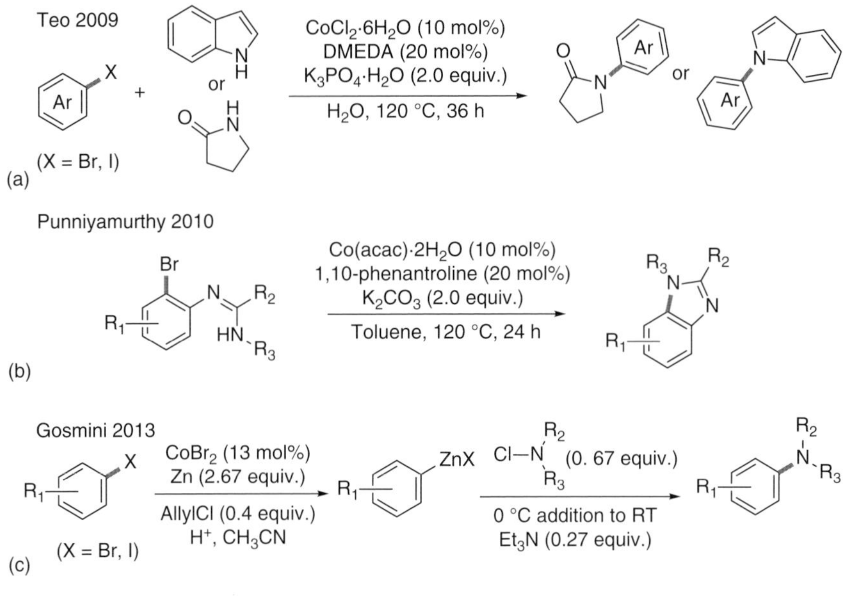

Scheme 12.30 Cobalt-catalyzed C(sp^2)–N cross-couplings reported by (a) Teo and coworker [117], (b) Punniyamurthy and coworkers [118], and (c) Gosmini and coworkers [119].

In 2013, Gosmini and coworkers reported on the cobalt-catalyzed electrophilic arylation of *N*-chloroamines by *in situ* formation of the nucleophilic aryl–Zn species from aryl halides (Scheme 12.30c) [119]. The *in situ* generation of aryl–ZnX from the corresponding halides or triflates was developed by the same group years before [120]. Furthermore, this simple and convenient protocol displays a wide substrate scope and a tolerance to a large number of functional groups, including esters, ketones, and sulfides.

12.5.3 C(sp^2)–O Coupling

The development of cobalt-catalyzed C—O bond formation through cross-coupling transformations is still rare, and only one example can be found in the literature. In 2015, Ranu and coworkers pioneered the bimetallic Co/Cu-catalyzed cross-coupling for the synthesis of biaryl and aryl–vinyl ethers, using copper as an efficient transmetalating agent in a proposed Co(I)/Co(III) cycle (Scheme 12.31) [121]. The two metals work in a cooperative manner, and the absence of any of them totally quenches the reaction. This methodology provides easy access to a library of aryl–vinyl, heteroaryl–styryl, and aryl–aryl ethers, which are difficult to obtain by other methods.

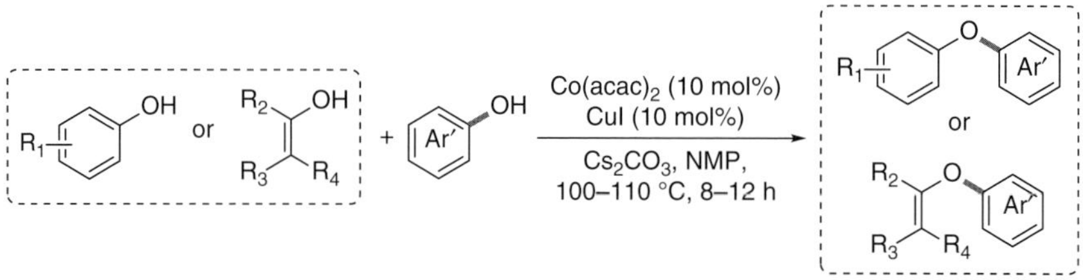

Co(acac)$_2$ (10 mol%)
CuI (10 mol%)
Cs$_2$CO$_3$, NMP,
100–110 °C, 8–12 h

Scheme 12.31 Cobalt-catalyzed C(sp^2)–O cross-couplings with phenols [121].

12.6 Miscellaneous

Two alternative cobalt-catalyzed methodologies deserve to be mentioned: the Michael addition and the allylation of 1,3-dicarbonyl compounds. Périchon and coworkers reported in 2006 the cobalt-catalyzed Michael addition between aryl halide and triflates and activated alkenes in good yields, where metallic Mn was used as a reducing reagent [122]. On the other hand, the allylation of 1,3-dicarbonyl compounds was reported by Iqbal and coworkers using either allylic acetates or allylic alcohols [123]. Regio- and stereoselectivities were modest, and mixtures of isomers were obtained.

Photocatalytic cobalt-based methodologies have also been recently introduced for C—H functionalization to construct both C—C and C—heteroatom bonds, specifically for C—H thiolation. In the first case, a $CoCl_2$/dimethylglyoxime (dmgH) catalyst was used to form C(sp^3)–C(sp^2) coupling products at room temperature and in water, under the irradiation of 450 nm LEDs [124]. Tetrahydroisoquinolines and indoles were synthesized in good yields via this environmentally benign cross-dehydrogenative coupling methodology (Scheme 12.32a). In the second example, C(sp^2)—S bonds are formed in a cobalt-catalyzed intramolecular cyclization of thiobenzoanilides to afford benzothiazoles using visible light irradiation and Ru(bpy)$_3$ (bpy, 2,2′-bipyridine) as a photosensitizer (Scheme 12.32b) [125]. The application of photocatalysis in cobalt catalysis is a promising strategy to develop new mild and benign synthetic methodologies.

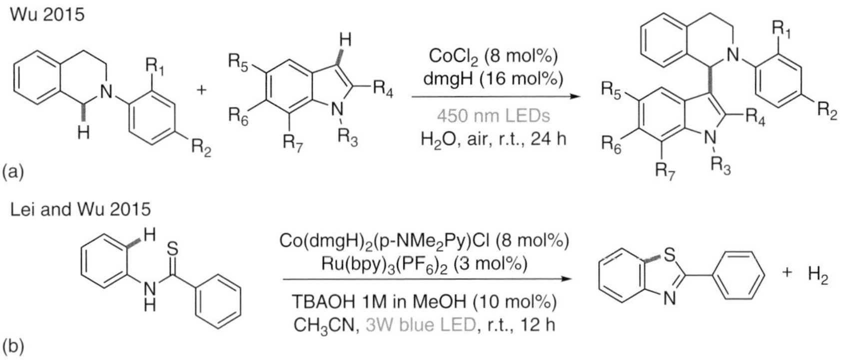

Wu 2015

$CoCl_2$ (8 mol%)
dmgH (16 mol%)
450 nm LEDs
H_2O, air, r.t., 24 h

(a)

Lei and Wu 2015

$Co(dmgH)_2(p\text{-}NMe_2Py)Cl$ (8 mol%)
$Ru(bpy)_3(PF_6)_2$ (3 mol%)
TBAOH 1M in MeOH (10 mol%)
CH_3CN, 3W blue LED, r.t., 12 h
+ H_2

(b)

Scheme 12.32 Photocatalyzed formation of (a) $C(sp^3)–C(sp^2)$ by Wu et al. and (b) $C(sp^2)—S$ bonds by Lei and coworkers [124, 125].

12.7 Conclusions and Future Prospects

In conclusion, cobalt catalysis appears to have significant potential for cross-coupling transformations, especially C—H functionalization, as demonstrated by the number of publications reported over the last few years [13–16]. Strikingly, cobalt catalysts can emulate the known reaction patterns of analogous group nine catalysts, but they are also able to exhibit higher catalytic activities or unique reactivities. Cp*-derived ligands for Co(I) and Co(III) catalysts designed by Brookhart and coworker [22], and Matsunaga and coworkers [13] mimic the previously used, more expensive, rhodium and iridium catalysts. However, based on the different properties of these metals, this strategy enabled to improve selectivities and provide novel chemical transformations, which are mainly based on the higher nucleophilicity of the carbon—cobalt(III) bond. Thus, amidation reactions have proven to be more efficient when using cobalt as catalysts [91], as well as regioselectivities were improved in allylation [41] and annulation reactions [12, 28, 31, 46, 51]. Furthermore, recent studies were carried out toward the mechanistic understanding of these transformations [28], pointing toward a cobalt(III)-mediated CMD C—H activation pathway [55]. Despite this huge advance, future efforts should be directed toward the design of more effective and robust catalysts, which will allow lower catalyst loadings, as well as the stereoselective construction of organic molecules through mild and efficient cobalt-catalyzed transformations. The full potential of cobalt catalysis, especially high-valent cobalt C—H functionalization, remains to be discovered, and new developments are expected in this rapidly evolving research area.

Acknowledgments

We acknowledge the financial support from the European Research Council for the Starting Grant Project ERC-2011-StG-277801 to X.R. and from MINECO of Spain for project CTQ2016-77989-P to X.R. We also thank Generalitat de Catalunya for project 2014SGR862. We thank the MECD of Spain for a FPU PhD grant to O.P. Finally, X.R. also thanks ICREA for an ICREA-Acadèmia award. CW thanks funding from the Biomedical Sciences Research Centre at Sheffield Hallam University.

References

1 Johansson Seechurn, C.C.C., Kitching, M.O., Colacot, T.J., and Snieckus, V. (2012). *Angew. Chem. Int. Ed.* 51: 5062.
2 Chen, X., Engle, K.M., Wang, D.-H., and Yu, J.-Q. (2009). *Angew. Chem. Int. Ed.* 48: 5094.
3 Pototschnig, G., Maulide, N., and Schnürch, M. (2017). *Chem. Eur. J.* 23: 9206.
4 Gilman, H. and Lichtenwalter, M. (1939). *J. Am. Chem. Soc.* 61: 957.
5 Kharasch, M.S. and Fields, E.K. (1941). *J. Am. Chem. Soc.* 63: 2316.
6 Murahashi, S. (1955). *J. Am. Chem. Soc.* 77: 6403.
7 Kochi, J.K., Tang, R.T., and Bernath, T. (1973). *J. Am. Chem. Soc.* 95: 7114.
8 Hebrard, F. and Kalck, P. (2009). *Chem. Rev.* 109: 4272.
9 Blanco-Urgoiti, J., Anorbe, L., Perez-Serrano, L. et al. (2004). *Chem. Soc. Rev.* 33: 32.
10 Halbritter, G., Knoch, F., Wolski, A., and Kisch, H. (1994). *Angew. Chem. Int. Ed. Engl.* 33: 1603; (1994). *Angew. Chem.* 106: 1676.
11 Gao, K., Lee, P.-S., Fujita, T., and Yoshikai, N. (2010). *J. Am. Chem. Soc.* 132: 12249.
12 Grigorjeva, L. and Daugulis, O. (2014). *Angew. Chem. Int. Ed.* 53: 10209; (2014). *Angew. Chem.* 126: 10373.
13 (a) Yoshino, T., Ikemoto, H., Matsunaga, S., and Kanai, M. (2013). *Angew. Chem. Int. Ed.* 52: 2207; (2013). *Angew. Chem.* 125: 2263. (b) Yoshikai, N. (2015). *ChemCatChem* 7: 732.
14 Cahiez, G. and Moyeux, A. (2010). *Chem. Rev.* 110: 1435.
15 Moselage, M., Li, J., and Ackermann, L. (2016). *ACS Catal.* 6: 498.
16 Wei, D., Zhu, X., Niu, J.-L., and Song, M.-P. (2016). *ChemCatChem* 8: 1242.
17 Chirila, P.G. and Whiteoak, C.J. (2017). *Dalton Trans.* 46: 9721.
18 (a) Ding, Z. and Yoshikai, N. (2012). *Angew. Chem. Int. Ed.* 51: 4698; (2012). *Angew. Chem.* 124: 4776. (b) Lee, P.-S., Fujita, T., and Yoshikai, N. (2011). *J. Am. Chem. Soc.* 133: 17283. (c) Gao, K. and Yoshikai, N. (2011). *J. Am. Chem. Soc.* 133: 400. (d) Ding, Z. and Yoshikai, N. (2010). *Org. Lett.* 12: 4180. (e) Gao, K. and Yoshikai, N. (2011). *Angew. Chem. Int. Ed.* 50: 6888; (2011). *Angew. Chem.* 123: 7020. (f) Ding, Z. and Yoshikai, N. (2011). *Synthesis* 16: 2561. (g) Yamakawa, T. and Yoshikai, N. (2013). *Org. Lett.* 15: 196. (h) Yamakawa, T. and Yoshikai, N. (2013). *Tetrahedron* 69: 4459. (i) Jinghua, D., Pin-Sheng, L., and Naohiko, Y. (2013). *Chem. Lett.* 42: 1140. (j) Gao, K. and Yoshikai, N. (2014). *Acc. Chem. Res.* 47: 1208.
19 Chen, Q., Ilies, L., and Nakamura, E. (2011). *J. Am. Chem. Soc.* 133: 428.

20 Song, W. and Ackermann, L. (2012). *Angew. Chem. Int. Ed.* 51: 8251; (2012). *Angew. Chem.* 124: 8376.
21 Moselage, M., Sauermann, N., Richter, S.C., and Ackermann, L. (2015). *Angew. Chem. Int. Ed.* 54: 6352; (2015). *Angew. Chem.* 127: 6450.
22 Lenges, C.P. and Brookhart, M. (1997). *J. Am. Chem. Soc.* 119: 3165.
23 Chen, Q.-A., Kim, D.K., and Dong, V.M. (2014). *J. Am. Chem. Soc.* 136: 3772.
24 Kim, D.K., Riedel, J., Kim, R.S., and Dong, V.M. (2017). *J. Am. Chem. Soc.* 139: 10208.
25 Yang, J. and Yoshikai, N. (2014). *J. Am. Chem. Soc.* 136: 16748.
26 Fallon, B.J., Derat, E., Amatore, M. et al. (2015). *J. Am. Chem. Soc.* 137: 2448.
27 Fallon, B.J., Garsi, J.-B., Derat, E. et al. (2015). *ACS Catal.* 5: 7493.
28 (a) Zhang, J., Chen, H., Lin, C. et al. (2015). *J. Am. Chem. Soc.* 137: 12990. (b) Maity, S., Kancherla, R., Dhawa, U. et al. (2016). *ACS Catal.* 6: 5493.
29 Yoshino, T., Ikemoto, H., Matsunaga, S., and Kanai, M. (2013). *Chem. Eur. J.* 19: 9142.
30 Yu, D.-G., Gensch, T., de Azambuja, F. et al. (2014). *J. Am. Chem. Soc.* 136: 17722.
31 (a) Zhao, D., Kim, J.H., Stegemann, L. et al. (2015). *Angew. Chem. Int. Ed.* 54: 4508; (2015). *Angew. Chem.* 127: 4591. (b) Lerchen, A., Vásquez-Céspedes, S., and Glorius, F. (2016). *Angew. Chem. Int. Ed.* 55: 3208; (2016). *Angew. Chem.* 128: 3261. (c) Lu, Q., Vásquez-Céspedes, S., Gensch, T., and Glorius, F. (2016). *ACS Catal.* 6: 2352. (d) Kim, J.H., Greßies, S., and Glorius, F. (2016). *Angew. Chem. Int. Ed.* 55: 5577; (2016). *Angew. Chem.* 128: 5667.
32 Hao, X.-Q., Du, C., Zhu, X. et al. (2016). *Org. Lett.* 18: 3610.
33 Hummel, J.R. and Ellman, J.A. (2015). *J. Am. Chem. Soc.* 137: 490.
34 Boerth, J.A., Hummel, J.R., and Ellman, J.A. (2016). *Angew. Chem. Int. Ed.* 55: 12650; (2016). *Angew. Chem.* 128: 12840.
35 Li, J. and Ackermann, L. (2015). *Angew. Chem. Int. Ed.* 54: 8551; (2015). *Angew. Chem.* 127: 8671.
36 Sauermann, N., González, M.J., and Ackermann, L. (2015). *Org. Lett.* 17: 5316.
37 Wang, H., Koeller, J., Liu, W., and Ackermann, L. (2015). *Chem. Eur. J.* 21: 15525.
38 Yu, W., Zhang, W., Liu, Y. et al. (2016). *RSC Adv.* 6: 24768.
39 Sen, M., Kalsi, D., and Sundararaju, B. (2015). *Chem. Eur. J.* 21: 15529.
40 Kuppusamy, R., Muralirajan, K., and Cheng, C.-H. (2016). *ACS Catal.* 6: 3909.
41 Suzuki, Y., Sun, B., Sakata, K. et al. (2015). *Angew. Chem. Int. Ed.* 54: 9944; (2015). *Angew. Chem.* 127: 10082.
42 Liu, X.-G., Zhang, S.-S., Wu, J.-Q. et al. (2015). *Tetrahedron Lett.* 56: 4093.
43 Li, J., Tang, M., Zang, L. et al. (2016). *Org. Lett.* 18: 2742.
44 Sen, M., Emayavaramban, B., Barsu, N. et al. (2016). *ACS Catal.* 6: 2792.
45 Yan, S.-Y., Ling, P.-X., and Shi, B.-F. (2017). *Adv. Synth. Catal.* 359: 2912.
46 Planas, O., Whiteoak, C.J., Company, A., and Ribas, X. (2015). *Adv. Synth. Catal.* 357: 4003.
47 Kalsi, D. and Sundararaju, B. (2015). *Org. Lett.* 17: 6118.

48 Grigorjeva, L. and Daugulis, O. (2014). *Org. Lett.* 16: 4688.

49 Thrimurtulu, N., Dey, A., Maiti, D., and Volla, C.M.R. (2016). *Angew. Chem. Int. Ed.* 55: 12361; (2016). *Angew. Chem.* 128: 12549.

50 Wang, H., Zhang, S., Wang, Z. et al. (2016). *Org. Lett.* 18: 5628.

51 Grigorjeva, L. and Daugulis, O. (2014). *Org. Lett.* 16: 4684.

52 Ma, W. and Ackermann, L. (2015). *ACS Catal.* 5: 2822.

53 Nguyen, T.T., Grigorjeva, L., and Daugulis, O. (2016). *ACS Catal.* 6: 551.

54 Kommagalla, Y. and Chatani, N. (2017). *Coord. Chem. Rev.* 350: 117.

55 Planas, O., Whiteoak, C.J., Martin-Diaconescu, V. et al. (2016). *J. Am. Chem. Soc.* 138: 14388.

56 Planas, O., Roldán-Gómez, S., Martin-Diaconescu, V. et al. (2017). *J. Am. Chem. Soc.* doi: 10.1021/jacs.7b07880.

57 Kharasch, M.S. and Fuchs, C.F. (1943). *J. Am. Chem. Soc.* 65: 504.

58 Cahiez, G. and Avedissian, H. (1998). *Tetrahedron Lett.* 39: 6159.

59 Shirakawa, E., Imazaki, Y., and Hayashi, T. (2008). *Chem. Lett.* 37: 654.

60 Hatakeyama, T., Hashimoto, S., Ishizuka, K., and Nakamura, M. (2009). *J. Am. Chem. Soc.* 131: 11949.

61 Avedissian, H., Bérillon, L., Cahiez, G., and Knochel, P. (1998). *Tetrahedron Lett.* 39: 6163.

62 Ohmiya, H., Yorimitsu, H., and Oshima, K. (2006). *Org. Lett.* 8: 3093.

63 Hamaguchi, H., Uemura, M., Yasui, H., Yorimitsu, H., and Oshima. K. (2008). *Chem. Lett.* 37: 1178.

64 Cahiez, G., Chaboche, C., Duplais, C., and Moyeux, A. (2009). *Org. Lett.* 11: 277.

65 (a) Ohmiya, H., Wakabayashi, K., Yorimitsu, H., and Oshima, K. (2006). *Tetrahedron* 62: 2207. (b) Ohmiya, H., Yorimitsu, H., and Oshima, K. (2006). *J. Am. Chem. Soc.* 128: 1886.

66 Kuno, A., Saino, N., Kamachi, T., and Okamoto, S. (2006). *Tetrahedron Lett.* 47: 2591.

67 Reddy, C.K. and Knochel, P. (1996). *Angew. Chem. Int. Ed. Engl.* 35: 1700; (1996). *Angew. Chem.* 108: 1812.

68 Branchaud, B.P. and Detlefsen, W.D. (1991). *Tetrahedron Lett.* 32: 6273.

69 Ikeda, Y., Nakamura, T., Yorimitsu, H., and Oshima, K. (2002). *J. Am. Chem. Soc.* 124: 6514.

70 Kharasch, M.S., Lambert, F.L., and Urry, W.H. (1945). *J. Org. Chem.* 10: 298.

71 Black, H.K., Horn, D.H.S., and Weedon, B.C.L. (1704). *J. Chem. Soc.* 1954.

72 (a) Durandetti, M., Nédélec, J.-Y., and Périchon, J. (1996). *J. Org. Chem.* 61: 1748. (b) Gomes, P., Gosmini, C., and Périchon, J. (2003). *J. Org. Chem.* 68: 1142.

73 Amatore, M., Gosmini, C., and Périchon, J. (2005). *Eur. J. Org. Chem.* 2005: 989.

74 Amatore, M. and Gosmini, C. (2008). *Angew. Chem. Int. Ed.* 47: 2089; (2008). *Angew. Chem.* 120: 2119.

75 Sun, J. and Deng, L. (2016). *ACS Catal.* 6: 290.

76 Ohmiya, H., Yorimitsu, H., and Oshima, K. (2005). *Angew. Chem. Int. Ed.* 44: 2368; (2005). *Angew. Chem.* 117: 2420.

77 (a) Obligacion, J.V., Neely, J.M., Yazdani, A.N. et al. (2015). *J. Am. Chem. Soc.* 137: 5855. (b) Zhang, L., Zuo, Z., Wan, X., and Huang, Z. (2014). *J. Am. Chem. Soc.* 136: 15501. (c) Zuo, Z. and Huang, Z. (2016). *Org. Chem. Front.* 3: 434. (d) Zhang, L., Zuo, Z., Leng, X., and Huang, Z. (2014). *Angew. Chem. Int. Ed.* 53: 2696; (2014). *Angew. Chem.* 126: 2734.
78 (a) Obligacion, J.V., Semproni, S.P., and Chirik, P.J. (2014). *J. Am. Chem. Soc.* 136: 4133. (b) Palmer, W.N., Obligacion, J.V., Pappas, I., and Chirik, P.J. (2016). *J. Am. Chem. Soc.* 138: 766.
79 Gandeepan, P. and Cheng, C.-H. (2015). *Acc. Chem. Res.* 48: 1194.
80 (a) Cenini, S., Gallo, E., Penoni, A. et al. (2000). *Chem. Commun.* 2265. (b) Ragaini, F., Penoni, A., Gallo, E. et al. (2003). *Chem. Eur. J.* 9: 249. (c) Lyaskovskyy, V., Suarez, A.I.O., Lu, H. et al. (2011). *J. Am. Chem. Soc.* 133: 12264. (d) Lu, H., Hu, Y., Jiang, H. et al. (2012). *Org. Lett.* 14: 5158. (e) Villanueva, O., Weldy, N.M., Blakey, S.B., and MacBeth, C.E. (2015). *Chem. Sci.* 6: 6672.
81 Zhang, L.-B., Zhang, S.-K., Wei, D. et al. (2016). *Org. Lett.* 18: 1318.
82 Du, C., Li, P.-X., Zhu, X. et al. (2017). *ACS Catal.* 7: 2810.
83 Yan, Q., Xiao, T., Liu, Z., and Zhang, Y. (2016). *Adv. Synth. Catal.* 358: 2707.
84 Whiteoak, C.J., Planas, O., Company, A., and Ribas, X. (2016). *Adv. Synth. Catal.* 358: 1679.
85 Zhou, Y., Tang, Z., and Song, Q. (2017). *Chem. Commun.* 53: 8972.
86 Nageswar Rao, D., Rasheed, S., Raina, G. et al. (2017). *J. Org. Chem.* 82: 7234.
87 Suess, A.M., Ertem, M.Z., Cramer, C.J., and Stahl, S.S. (2013). *J. Am. Chem. Soc.* 135: 9797.
88 Wu, X., Yang, K., Zhao, Y. et al. (2015). *Nat. Commun.* 6: 6462.
89 Sun, B., Yoshino, T., Matsunaga, S., and Kanai, M. (2014). *Adv. Synth. Catal.* 356: 1491.
90 Sun, B., Yoshino, T., Matsunaga, S., and Kanai, M. (2015). *Chem. Commun.* 51: 4659.
91 Park, J. and Chang, S. (2015). *Angew. Chem. Int. Ed.* 54: 14103; (2015). *Angew. Chem.* 127: 14309.
92 Liang, Y., Liang, Y.-F., Tang, C. et al. (2015). *Chem. Eur. J.* 21: 16395.
93 Mei, R., Loup, J., and Ackermann, L. (2016). *ACS Catal.* 6: 793.
94 Wang, F., Wang, H., Wang, Q. et al. (2016). *Org. Lett.* 18: 1306.
95 Wang, X., Lerchen, A., and Glorius, F. (2016). *Org. Lett.* 18: 2090.
96 Huang, J., Huang, Y., Wang, T. et al. (2017). *Org. Lett.* 19: 1128.
97 Wang, F., Jin, L., Kong, L., and Li, X. (2017). *Org. Lett.* 19: 1812.
98 Shi, P., Wang, L., Chen, K. et al. (2017). *Org. Lett.* 19: 2418.
99 Barsu, N., Rahman, M.A., Sen, M., and Sundararaju, B. (2016). *Chem. Eur. J.* 22: 9135.
100 Patel, P. and Chang, S. (2015). *ACS Catal.* 5: 853.
101 (a) Li, L., Wang, H., Yu, S. et al. (2016). *Org. Lett.* 18: 3662. (b) Yu, S., Tang, G., Li, Y. et al. (2016). *Angew. Chem. Int. Ed.* 55: 8696; (2016). *Angew. Chem.* 128: 8838.
102 (a) Enthaler, S. and Company, A. (2011). *Chem. Soc. Rev.* 40: 4912. (b) Li, J.J. and Johnson, D.S. (2010). *Modern Drug Synthesis*. Hoboken, NJ:

Wiley. (c) Li, J.J., Johnson, D.S., Sliskovic, D.R., and Roth, B.D. (2004). *Contemporary Drug Synthesis*. Hoboken, NJ: Wiley. (d) Müller, F. (1999). *Agrochemicals*. Weinheim: Wiley-VCH. (e) Shen, C., Zhang, P., Sun, Q. et al. (2015). *Chem. Soc. Rev.* 44: 291.

103 Lande, S.S. and Kochi, J.K. (1968). *J. Am. Chem. Soc.* 90: 5196.

104 Tang, R. and Kochi, J.K. (1973). *J. Inorg. Nucl. Chem.* 35: 3845.

105 Zhang, L.-B., Hao, X.-Q., Zhang, S.-K. et al. (2015). *Angew. Chem. Int. Ed.* 54: 272; (2015). *Angew. Chem.* 127: 274.

106 Lan, J., Xie, H., Lu, X. et al. (2017). *Org. Lett.* 19: 4279.

107 Guo, X.-K., Zhang, L.-B., Wei, D., and Niu, J.-L. (2015). *Chem. Sci.* 6: 7059.

108 Gensch, T., Klauck, F.J.R., and Glorius, F. (2016). *Angew. Chem. Int. Ed.* 55: 11287; (2016). *Angew. Chem.* 128: 11457.

109 Liu, X.G., Li, Q., and Wang, H. (2017). *Adv. Synth. Catal.* 359: 1942.

110 Yoshida, M., Kawai, K., Tanaka, R. et al. (2017). *Chem. Commun.* 53: 5974.

111 Ribas, X. (2013). *C–H and C–X Bond Functionalization*. Cambridge: RSC.

112 (a) Chen, X., Engle, K.M., Wang, D.-H., and Yu, J.-Q. (2009). *Angew. Chem.* 121: 5196. (b) Wang, X.-C., Hu, Y., Bonacorsi, S. et al. (2013). *J. Am. Chem. Soc.* 135: 10326. (c) Schröder, N., Lied, F., and Glorius, F. (2015). *J. Am. Chem. Soc.* 137: 1448. (d) Wang, L. and Ackermann, L. (2014). *Chem. Commun.* 50: 1083. (e) Yu, Q., Hu, L.A., Wang, Y. et al. (2015). *Angew. Chem. Int. Ed.* 54: 15284; (2015). *Angew. Chem.* 127: 15499.

113 Pawar, A.B. and Lade, D.M. (2016). *Org. Biomol. Chem.* 14: 3275.

114 Wong, Y.-C., Jayanth, T.T., and Cheng, C.-H. (2006). *Org. Lett.* 8: 5613.

115 Lan, M.-T., Wu, W.-Y., Huang, S.-H. et al. (2011). *RSC Adv.* 1: 1751.

116 Teo, Y.-C. and Chua, G.-L. (2009). *Chem. Eur. J.* 15: 3072.

117 Tan, B.Y.-H. and Teo, Y.-C. (2014). *Org. Biomol. Chem.* 12: 7478.

118 Saha, P., Ali, M.A., Ghosh, P., and Punniyamurthy, T. (2010). *Org. Biomol. Chem.* 8: 5692.

119 Qian, X., Yu, Z., Auffrant, A., and Gosmini, C. (2013). *Chem. Eur. J.* 19: 6225.

120 Fillon, H., Gosmini, C., and Périchon, J. (2003). *J. Am. Chem. Soc.* 125: 3867.

121 Kundu, D., Tripathy, M., Maity, P., and Ranu, B.C. (2015). *Chem. Eur. J.* 21: 8727.

122 Amatore, M., Gosmini, C., and Périchon, J. (2006). *J. Org. Chem.* 71: 6130.

123 (a) Bhatia, B., Reddy, M.M., and Iqbal, J. (1993). *Tetrahedron Lett.* 34: 6301. (b) Maikap, G.C., Madhava Reddy, M., Mukhopadhyay, M. et al. (1994). *Tetrahedron* 50: 9145. (c) Mukhopadhyay, M. and Iqbal, J. (1995). *Tetrahedron Lett.* 36: 6761.

124 Wu, C.-J., Zhong, J.-J., Meng, Q.-Y. et al. (2015). *Org. Lett.* 17: 884.

125 Zhang, G., Liu, C., Yi, H. et al. (2015). *J. Am. Chem. Soc.* 137: 9273.

13

Trifluoromethylation and Related Reactions

Jérémy Jacquet[1], Louis Fensterbank[1], and Marine Desage-El Murr[2]

[1] *Institut Parisien de Chimie Moléculaire, UMR CNRS 8232, Sorbonne Universités, Université Paris 6, 4, place Jussieu, 75005 Paris, France*
[2] *Institut de Chimie, UMR CNRS 7177, Université de Strasbourg, 1, rue Blaise Pascal, 67000 Strasbourg, France*

Introduction of trifluoromethyl groups in molecular scaffolds is of prime interest. Indeed, this group endows the overall chemical structure with wide-ranging properties such as enhanced lipophilicity and tuned electronics that can aptly be put to use in medicinal chemistry or materials science [1].

However, these exciting prospects come at a price, which is directly linked to the difficulty in introducing the CF_3 group. Homogeneous catalysis using transition metals has provided interesting forays in this field. Initial pioneering results were obtained with noble metals, but earth-abundant, first-row transition metals have since been excellent contributors. One of the specific features of trifluoromethylating agents is their wide range of electronic nature as these include electrophilic, nucleophilic, and radical sources. Another currently developing synthetic application deals with the introduction of the $-SCF_3$ group. This group is chemically close to the CF_3 group but also incorporates a heteroatom that confers enhanced properties such as very high lipophilicity.

In this chapter, we will focus on the non-noble, metal-catalyzed, or mediated introduction of the aforementioned groups, starting with trifluoromethylation (13.1), trifluoromethylthiolation (13.2), and perfluoroalkylation (13.3). This field of catalysis applied to synthesis is very lively and exhaustivity is not possible. Therefore, in each section, we have focused on the more recent and representative contributions.

13.1 Trifluoromethylation Reactions

13.1.1 Copper(I) Salts with Nucleophilic Trifluoromethyl Sources

The combination of copper(I) salts with trifluoromethyltrialkylsilanes has proven to be an efficient strategy for the construction of $C(sp^2)-CF_3$ bonds. The reaction between these two partners via a transmetallation step of the nucleophilic CF_3 moiety to copper generates a Cu^I-CF_3 intermediate, which is reactive toward electrophiles and nucleophiles.

Non-Noble Metal Catalysis: Molecular Approaches and Reactions,
First Edition. Edited by Robertus J. M. Klein Gebbink and Marc-Etienne Moret.

13.1.1.1 Reactions with Electrophiles

A catalytic system was described by Amii in 2009 for the trifluoromethylation of aryl iodides, using 10 mol% of copper(I) iodide, 10 mol% of 1,10-phenanthroline as a ligand, in the presence of an excess of TES–CF_3 (TES, triethylsilyl) and KF [2]. The catalytic cycle includes the transmetallation of the nucleophilic CF_3 to the copper center, via an activation of TES–CF_3 by KF, followed by the reaction of the resulting Cu^I–CF_3 species with an aryl iodide (Scheme 13.1). The donor character of the 1,10-phenanthroline ligand increases the nucleophilic character of the coordinated CF_3 group and therefore accelerates the reaction with the electrophile and the subsequent regeneration of the catalyst. In the absence of the ligand, decomposition of TES–CF_3 becomes faster than the regeneration of the initial species, and a stoichiometric amount of copper is then required for the reaction.

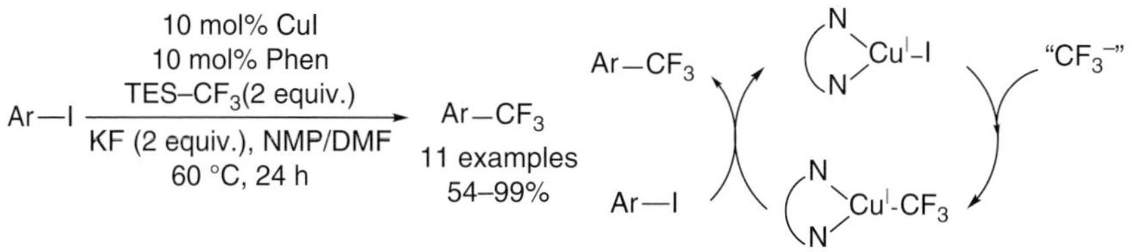

Ar—I → Ar—CF_3 (10 mol% CuI, 10 mol% Phen, TES–CF_3(2 equiv.); KF (2 equiv.), NMP/DMF, 60 °C, 24 h); 11 examples, 54–99%

Ar—CF_3; Ar—I; (N,N)Cu^I–I; (N,N)Cu^I-CF_3; "CF_3^-"

Scheme 13.1 Trifluoromethylation of aryl iodides.

Over the last few years, several well-defined trifluoromethylated copper(I) complexes were prepared by transmetallation of a nucleophilic CF_3 group ($TMSCF_3$, CHF_3; TMS, trimethylsilyl) and evaluated in trifluoromethylation of aryl halides [3]. For example, Hartwig's group synthesized in 2011 the trifluoromethylated copper complex [(Phen)Cu^I–CF_3] starting from [(Phen)Cu^I–I] and TMS–CF_3. This complex, stable and isolable, efficiently converts a wide diversity of aryl iodides and bromides into trifluoromethylated compounds under mild conditions [4].

The poor stability of trifluoromethylsilanes has led chemists to develop more stable and practical nucleophilic CF_3 sources. Although silylated hemiaminal **1** was initially developed for the trifluoromethylation of aldehydes and ketones [5], this compound was shown to be an efficient source in the copper-catalyzed trifluoromethylation of aryl iodides [6]. Activated by a fluoride ion, the resulting hemiaminaloate coordinates with the copper complex and enables the formation of a Cu–CF_3 species via a CF_3 group migration (Scheme 13.2).

The reaction of trimethoxyborate with TMS–CF_3, in the presence of a fluoride source, generates (trifluoromethyl)trimethoxyborate **2**. This isolable and air-stable nucleophilic CF_3 source is efficient in the copper-catalyzed trifluoromethylation of aryl iodides and does not require any additives (Scheme 13.3) [7].

Regarding the electrophile, trifluoromethylation of an aryldiazonium salt has recently been studied [8]. Although a quasi-stoichiometric amount of copper salt is needed, this elegant Sandmeyer trifluoromethylation offers access to trifluoromethylarenes under mild conditions. Radical clock experiments suggest a monoelectronic reduction of diazonium by the Cu^I–CF_3 species. The loss of N_2

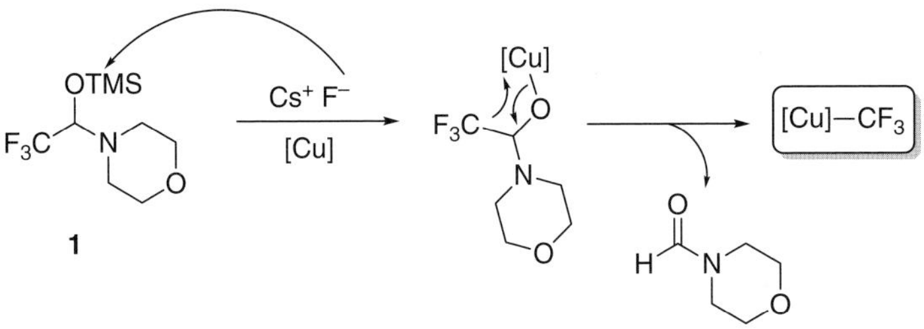

Scheme 13.2 Mechanism for the copper-mediated trifluoromethylation involving a silylated hemiaminal.

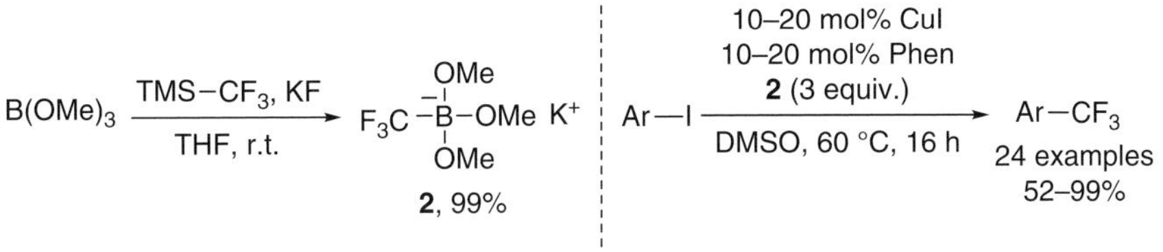

Scheme 13.3 Copper-catalyzed trifluoromethylation with (trifluoromethyl)trimethoxyborate as the CF_3 source.

results in the formation of an aryl radical, which attacks the oxidized $Cu^{II}–CF_3$ species to generate trifluoromethylarene (Scheme 13.4).

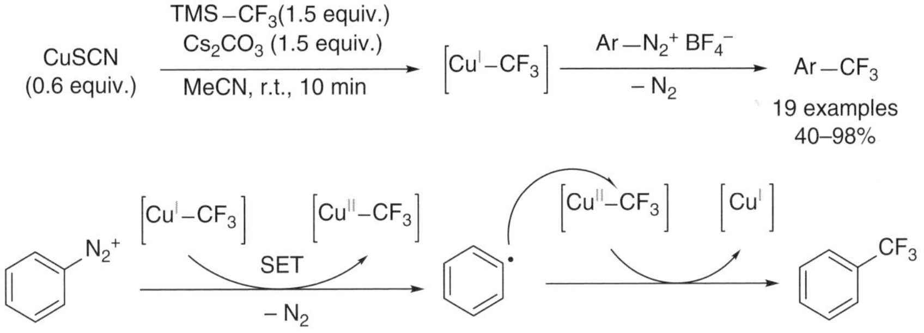

Scheme 13.4 Copper-catalyzed trifluoromethylation of diazonium salts.

13.1.1.2 Reactions with Nucleophiles: Oxidative Coupling

In trifluoromethylation, the first oxidative coupling was applied to alkynes for the formation of C(sp)–CF_3 bonds (Scheme 13.5) [9]. In this methodology, a large excess of TMS–CF_3 is needed in order to avoid the formation of homo-coupling products via a dialkynylcopper complex. The presence of an oxidant is also essential for the reaction: air appears to be the best oxidant, whereas a high concentration of dioxygen was found to inhibit the reaction.

The Qing group also studied the copper-catalyzed oxidative trifluoromethylation of 1,3,4-oxadiazoles and 1,3-azoles, containing C—H bonds whose acidity is close to that of alkynes [10]. In the presence of a base (*t*-BuONa), a co-base (NaOAc), and an oxidant (air or *tert*-butyl peroxide), the reaction offers access to

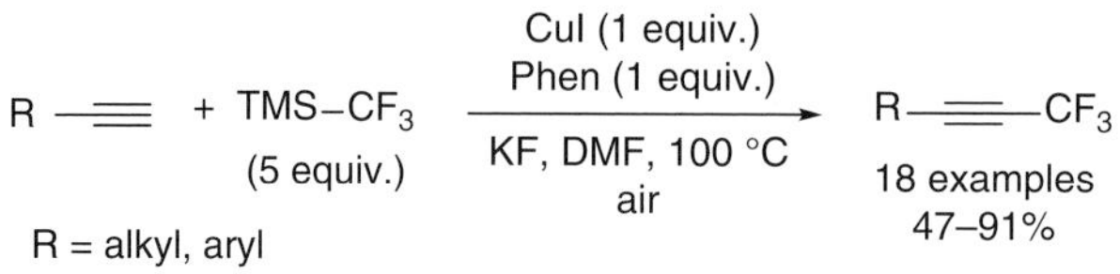

Scheme 13.5 Copper-mediated oxidative trifluoromethylation of alkynes.

a wide diversity of trifluoromethylated heteroarenes, commonly encountered in medicinal chemistry. This methodology has also been applied to indoles, which require a lower catalytic loading (10 mol% of $Cu(OH)_2$). Although the authors suggest the formation of a Cu^I–CF_3 species with 1,3,4-oxadiazoles and 1,3-azoles, $CF_3^\bullet$ radicals might be involved in the case of indoles (Scheme 13.6).

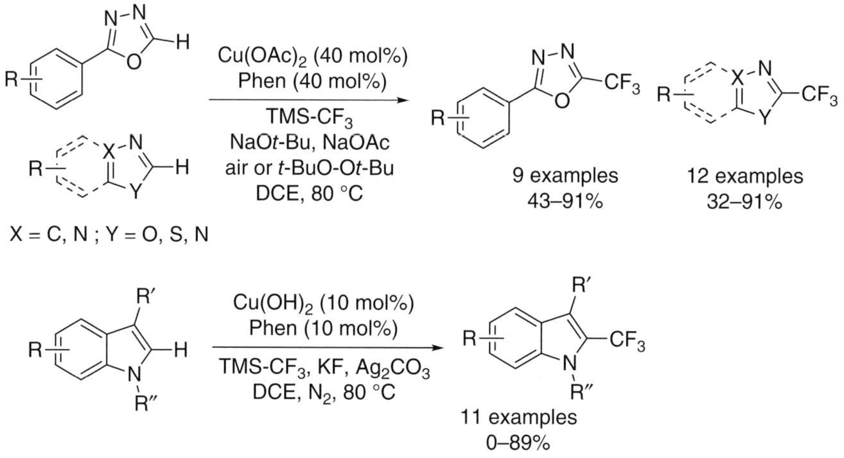

Scheme 13.6 Copper-catalyzed trifluoromethylation of heteroaromatics.

Boronic acids and boronates are easy-to-access and stable carbon nucleophiles frequently used in organic synthesis [11]. Several systems have been developed for the oxidative trifluoromethylation of these nucleophiles, using copper(I) or (II) complexes with TMS-CF_3 (or derivatives) [12], or pregenerated complexes [13], in the presence of an oxidant. Interestingly, catalytic conditions were achieved by controlling addition rates of the two nucleophiles with a syringe pump [14]. A rare example of alkylboronic acids coupling was reported in 2012: under oxidative conditions, using TMS–CF_3 and tetramethylated 1,10-phenanthroline as the ligand, the reaction enables the formation of C(sp^3)–CF_3 bonds [15]. Good yields were obtained with primary substrates and also with secondary substrates, opening the way toward an asymmetric version (Scheme 13.7).

13.1.2 Generation of $CF_3^\bullet$ Radicals Using Langlois' Reagent

Sodium trifluoromethanesulfinate was first introduced in the early 1990s by Langlois, and it was noticed that $CF_3^\bullet$ radicals could be generated after oxidation

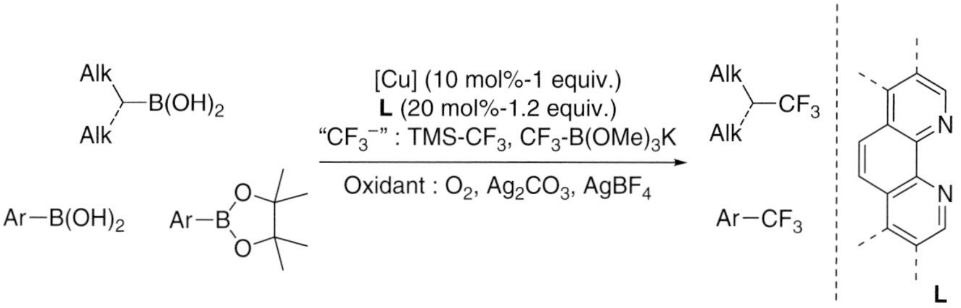

Scheme 13.7 Copper-catalyzed trifluoromethylation of boronic derivatives.

of the reagent [16]. This process, which has been applied in many methodologies, is generally enabled by the association of the reagent CF_3SO_2Na with a catalytic copper salt and an excess of TBHP (*tert*-butylhydroperoxide), or more recently with photoredox systems (Scheme 13.8) [17]. Mechanisms are not clearly established, but a pathway for the generation of $CF_3^\bullet$ radicals is frequently proposed and involves a reduction of TBHP by copper into *tert*-butoxy radical, which oxidizes the trifluoromethanesulfinate anion. Immediate decomposition of the resulting radical generates sulfur dioxide and $CF_3^\bullet$ radical.

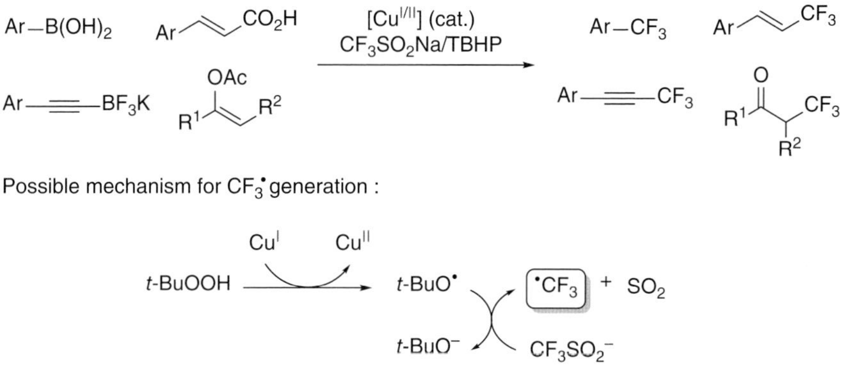

Scheme 13.8 Copper-catalyzed radical trifluoromethylation.

13.1.3 Copper and Electrophilic CF_3^+ Sources

Copper, as a non-noble metal, is known to favor monoelectronic processes and is also able to exist as Cu^0, Cu^I, Cu^{II}, and Cu^{III} (more rarely Cu^{IV}) [18] species. These properties have led chemists to associate copper salts with electrophilic CF_3 sources, in particular, Togni's [19] and Umemoto's [20] reagents, for the development of new methodologies in trifluoromethylation.

The interaction between these hypervalent iodine derivatives and copper can lead to different processes, depending on several parameters (reaction conditions, substrate, and ligands) (Scheme 13.9) [21]. Determination of an exact mechanism seems difficult; however, radical clock experiments, spectroscopic studies, and density functional theory (DFT) calculations bring key information supporting radical, cationic, or organometallic intermediates.

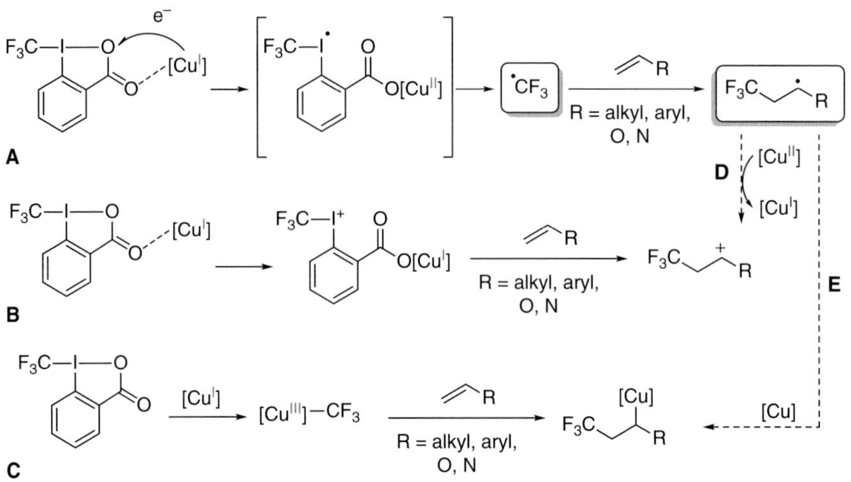

Scheme 13.9 Mechanisms for copper-catalyzed radical trifluoromethylation.

First, an intramolecular single-electron transfer (SET) from the copper center to the activated Togni's reagent generates a Cu^{II}–carboxylate intermediate. Homolytic cleavage of the destabilized I–CF_3 bond releases a $CF_3^{\bullet}$ radical, which can add across unsaturated moieties (**A**). Depending on its stability, the resulting alkyl radical can undergo monoelectronic oxidation (**D**) or an alkylcopper can be formed (**E**) [22]. This radical pathway competes with an electrophilic activation of the trifluoromethylating agent by the copper complex, which acts as a Lewis acid in this process. A subsequent heterolytic cleavage of the I—O bond results in the formation of a highly reactive CF_3^+ source, particularly toward activated unsaturated moieties, which are able to stabilize carbocationic intermediates (**B**). Finally, the reaction between a copper complex and Togni's reagent can also lead to a Cu–CF_3 species. An alkylcopper complex is then formed, after 1,2-insertion of a double bond (**C**).

The interaction of a copper complex with Umemoto's reagent is also associated with two possible pathways (Scheme 13.10): an organometallic pathway, which involves a Cu^{III}–CF_3 intermediate stemming from an oxidative addition (**A**) and a radical pathway characterized by a monoelectronic reduction of the trifluoromethylsulfonium into a $CF_3^{\bullet}$ radical (**B**).

Although the association of an electrophilic trifluoromethylating agent with copper was successfully applied to the trifluoromethylation of some electrophiles (aryl iodides and diazoniums) [23] and nucleophiles (boronic acids) [24], recent strategies are based on the addition of a trifluoromethyl group across an insaturation (activated or unactivated double, triple bonds, (hetero)arenes), thus giving access to a wide diversity of trifluoromethylated structures [1c, 25]. Selected examples of this profuse chemistry are reported here.

The electrophilic trifluoromethylation of silyl enol ethers and enamines has proven to be an elegant way to access α-trifluoromethylated aldehydes and ketones. Previously developed with photoredox catalysis [26], trifluoromethylation of silyl enol ethers was studied by Guo in 2014. Using a catalytic amount of

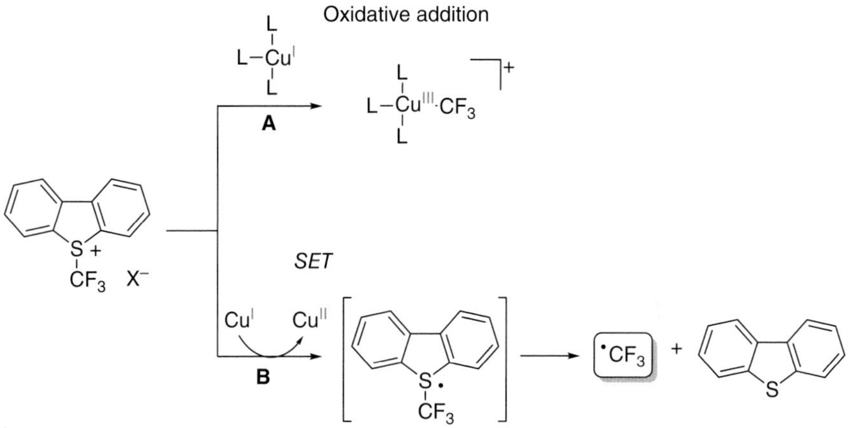

Scheme 13.10 Possible reaction pathways for activation of Umemoto's reagent by copper complexes.

copper isocyanate and Togni's reagent II, a practical and nonexpensive method is revealed, enabling access to a wide diversity of α-trifluoromethylated ketones (Scheme 13.11) [27].

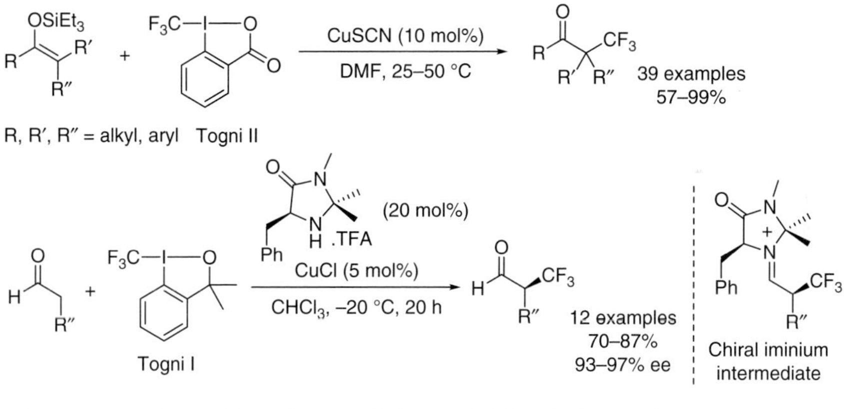

Scheme 13.11 Electrophilic trifluoromethylation of silyl enol ethers and enamines.

Merging organocatalysis and copper-activated iodonium salts, MacMillan achieved in 2010 the enantioselective trifluoromethylation of aldehydes [28]. Condensation of the amine catalyst with an aldehyde generates a chiral π-electron-rich enamine, which is reactive toward the copper-activated Togni's reagent I. Excellent yields and enantioselectivites are reached. In this methodology, MacMillan uses copper chloride as a Lewis acid, whereas Guo suggests that $CF_3^{\bullet}$ radicals are involved, based on radical clock and trapping experiments.

Baudoin has shown that $CF_3^{\bullet}$ radicals could be directly added on an aryl-substituted *N,N′*-dialkylhydrazone [29]. In the presence of catalytic copper chloride and Togni's reagent II, addition of a $CF_3^{\bullet}$ radical on the carbon of the C=N bond generates a trifluoromethylated aminyl radical. Subsequent copper oxidation of the N-centered radical followed by a proton loss delivers the trifluoromethylated hydrazone. This methodology was extended to α,β-unsaturated hydrazones [30]. It was noticed that the $CF_3^{\bullet}$ radical was added in β-position with high regioselectivity, in particular with *N*,*N*-dibenzylhydrazones. Regarding the stereochemistry of the trifluoromethylated double bond, *E*-configuration is mainly obtained with α-unsubstituted hydrazones, whereas *Z*-configuration is preferred with α-substituted hydrazones (Scheme 13.12).

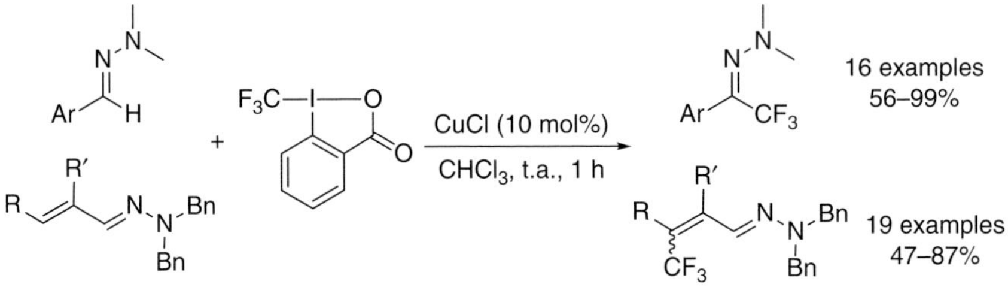

CuCl (10 mol%)
$CHCl_3$, t.a., 1 h
16 examples
56–99%
19 examples
47–87%

Scheme 13.12 Electrophilic trifluoromethylation of hydrazones.

Three similar systems were published in 2011 for the copper-catalyzed allylic trifluoromethylation of unactivated alkenes, providing an easy access to trifluoromethylated unsaturated structures under mild conditions. Systems reported by Buchwald [31] and Wang [32] are based on the combination of a copper salt with Togni's reagent II in methanol. Mechanisms involved with these systems remain unclear as studies indicate that $CF_3^{\bullet}$ radicals are likely involved, but electrophilic activation or Cu–CF_3 intermediates cannot be excluded.

Liu uses Umemoto's reagent as a trifluoromethylating agent, and collidine as a base and ligand [33]. DFT calculations support a Heck-like four-membered-ring transition state after oxidative addition of the CF_3^+ and coordination of the double bond to the copper center (Scheme 13.13).

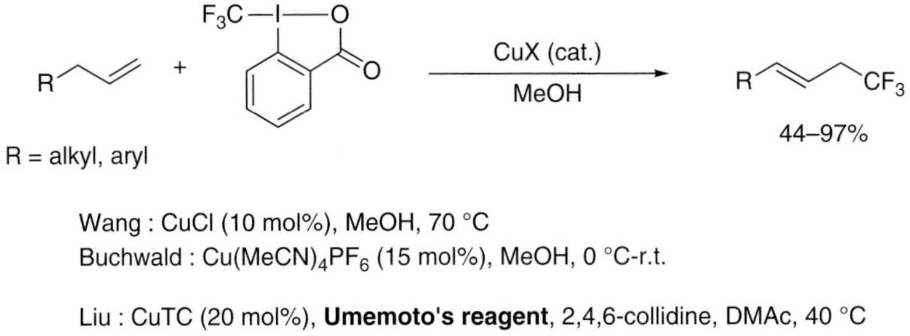

CuX (cat.)
MeOH
44–97%
R = alkyl, aryl

Wang : CuCl (10 mol%), MeOH, 70 °C
Buchwald : $Cu(MeCN)_4PF_6$ (15 mol%), MeOH, 0 °C-r.t.

Liu : CuTC (20 mol%), **Umemoto's reagent**, 2,4,6-collidine, DMAc, 40 °C

Scheme 13.13 Copper-catalyzed allylic trifluoromethylation of unactivated alkenes.

Under similar conditions, allylsilanes can also be used to target trifluoromethylated allylic structures [34]. Mechanistic aspects are identical to those previously discussed except that the formation of the insaturation at the final step is obtained after the loss of trimethylsilyl group from a cationic intermediate (Scheme 13.14a) rather than deprotonation.

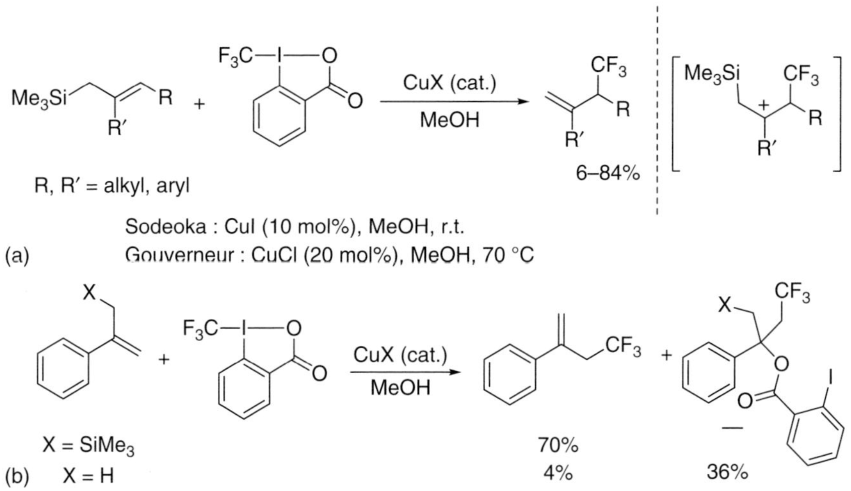

Scheme 13.14 Copper-catalyzed allylic trifluoromethylation of allyl silanes and alkenes.

Such methodology is of great interest because it enables the application of allylic trifluoromethylation to di- and tri-substituted alkenes. In the absence of a trimethylsilyl group, ionic recombination of the stabilized carbocation with benzoate is favored (Scheme 13.14b).

Recent strategies have focused on the nucleophilic trapping of a cationic intermediate, stemming from a formal electrophilic addition of CF_3^+ on the insaturation (Scheme 13.15). By varying the nature of the nucleophile and the substrate, inter-, intra- [35], molecular carbo- [36], oxy- [37], amino- [35b, 38] trifluoromethylation reactions were developed and applied to alkenes, but also alkynes [39] and allenes [40].

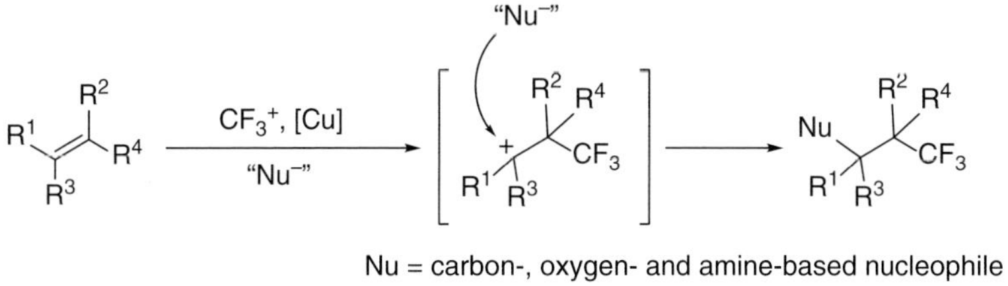

Scheme 13.15 Copper-catalyzed trifluoromethylation of alkenes with nucleophilic trapping.

Addition of an electrophilic/radical CF_3 group across unsaturated moieties can initiate a cascade reaction [41]. Nevado reported a copper-catalyzed radical trifluoromethylation/aryl migration/desulfonylation cascade starting from *N*-tosyl acrylamide (Scheme 13.16) [42]. It was noticed that the nature of the N-substituent controlled the last event of the cascade reaction, leading to two possible structures: α-aryl-β-trifluoromethylamides (**A**) or oxindoles (**B**). Despite its electrophilic character, a $CF_3^{\bullet}$ radical can add on the electron-poor double bond to form a stabilized radical. A 5-*ipso*-cyclization involving the

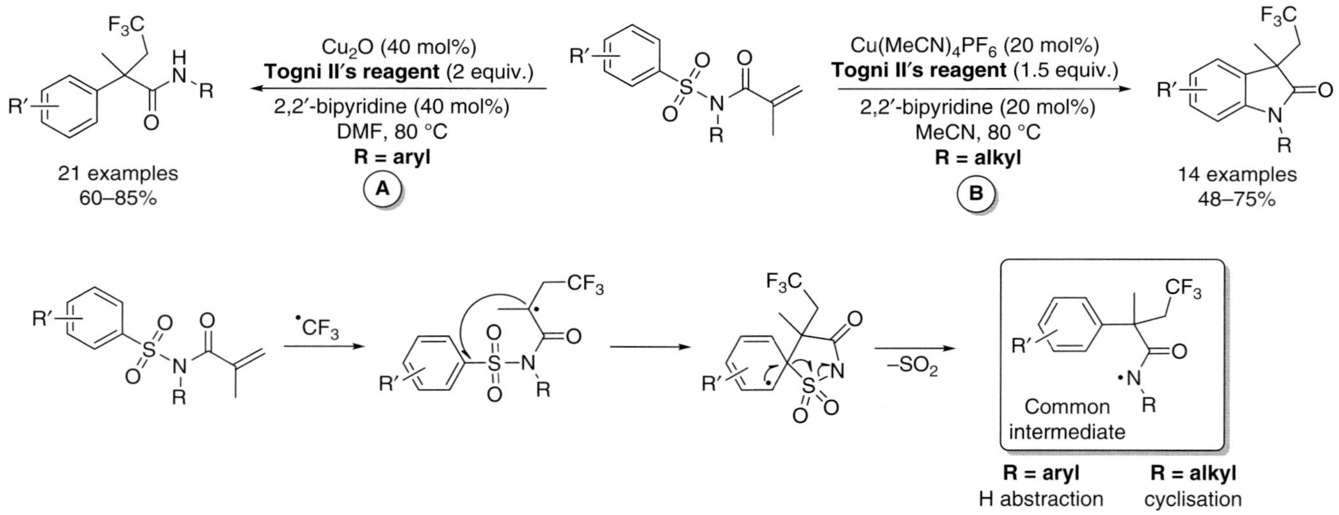

Cu_2O (40 mol%)
Togni II's reagent (2 equiv.)
2,2′-bipyridine (40 mol%)
DMF, 80 °C
R = aryl
A

21 examples
60–85%

$Cu(MeCN)_4PF_6$ (20 mol%)
Togni II's reagent (1.5 equiv.)
2,2′-bipyridine (20 mol%)
MeCN, 80 °C
R = alkyl
B

14 examples
48–75%

$^{\bullet}CF_3$

$-SO_2$

Common intermediate

R = aryl
H abstraction

R = alkyl
cyclisation

Scheme 13.16 Copper-catalyzed radical trifluoromethylation/aryl migration/desulfonylation cascade.

aromatic ring then generates an aryl radical and subsequent rearomatization triggers a desulfonylation step. The resulting amidyl radical evolves differently depending on the nature of the N-substituent: if R is an aryl substituent, hydrogen abstraction leads to an α-aryl-β-trifluoromethylamide, whereas, when R is an alkyl group, cyclization occurs and an oxindole is obtained. Mechanistic studies suggest that this cyclization is not a radical process but rather involves a copper enolate.

Based on single-electron transfers promoted by a photoexcited catalyst, photoredox catalysis enables the controlled generation of radicals under mild conditions [43] and has proved to be an elegant and powerful tool for the development of methodologies in trifluoromethylation [44]. Polypyridyl ruthenium and iridium complexes are commonly used in this chemistry but tend to be replaced by cheaper non-noble metals, such as copper [45].

A recent example was reported by Reiser, in which trifluoromethylchlorosulfonylation of alkenes catalyzed by [Cu(dap)$_2$Cl] is achieved (Scheme 13.17) [46]. This reaction, which is not possible with ruthenium and iridium photocatalysts, demonstrates the ability of copper to stabilize a SO_2Cl^- anion, via the formation of a [Cu–SO_2Cl] species. The reaction suffers, however, from a strong substrate dependence: the authors suggest that a destabilization of [Cu–SO_2Cl] intermediate by atom donors on the substrate can occur and favors SO_2Cl^- decomposition: the chlorotrifluoromethylation product is therefore observed.

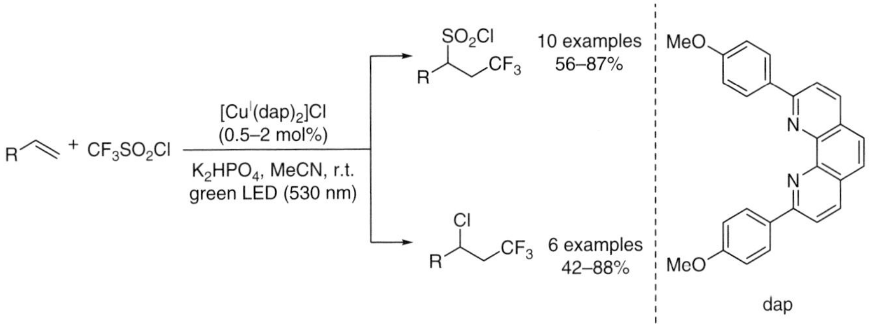

Scheme 13.17 Copper-catalyzed trifluoromethylchlorosulfonylation of alkenes.

The group of Vincent developed a photoreducible copper(II) complex, coordinated by two DMEDA ligands (DMEDA, *N*,*N*′-dimethylethylenediamine), and counterions are carboxylates linked to a benzophenone chromophore [47]. Under irradiation, this chromophore enables an electron transfer (PET, photoinduced electron transfer) between a protic solvent, such as methanol, and copper(II), thereby reduced to copper(I). This species has proved to be an efficient catalyst for cycloaddition reactions [47], but also for the radical trifluoromethylation of unactivated alkenes (Scheme 13.18) [48]. In the presence of Togni's reagent II and under a simple UV lamp (365 nm), the reaction requires only 0.5 mol% of the catalyst and is very fast (one to two hours). Under sunlight, the reaction is

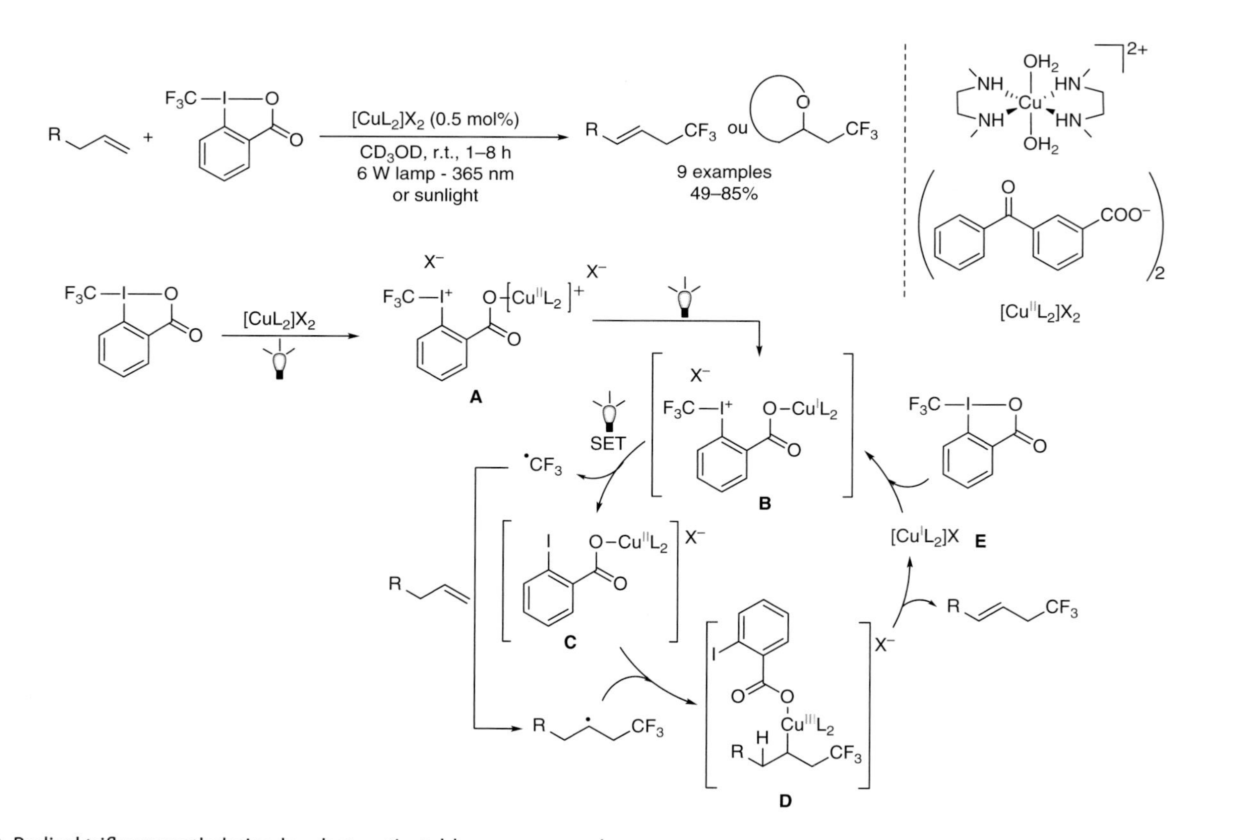

Scheme 13.18 Radical trifluoromethylation by photoactivatable copper complex.

slower (five to eight hours), but better yields are obtained. Based on EPR studies, the proposed mechanism starts with a copper-induced heterolytic cleavage of the I—O bond, leading to intermediate **A**. The photoinduced process enables the reduction of this copper(II) complex into a copper(I) complex **B**, and photoactivation of **B** triggers an electron transfer, which generates a $CF_3^•$ radical. Addition of the $CF_3^•$ radical on the double bond and recombination of the resulting alkyl radical with copper complex **C** leads to an alkylcopper(III) intermediate **D**. Subsequent β-H-elimination frees the trifluoromethylated compound, and species **B** is regenerated by reaction of **E** with Togni's reagent.

Cooperative catalysis with redox noninnocent ligands has recently emerged as an elegant alternative to photoredox catalysis, involving a well-defined system for the controlled generation of $CF_3^•$ radicals in mild conditions [49]. The bis(*o*-iminobenzosemiquinonato)copper(II) complex is able to reduce electrophilic trifluoromethylating agents into $CF_3^•$ radicals, and the corresponding system was successfully applied to the trifluoromethylation of silyl enol ethers, heteroarenes, and to the hydrotrifluoromethylation of alkynes. Spectroscopic studies demonstrate that redox ligands, acting as an electron reservoir, are involved in SET processes, whereas the copper oxidation state is preserved (Scheme 13.19).

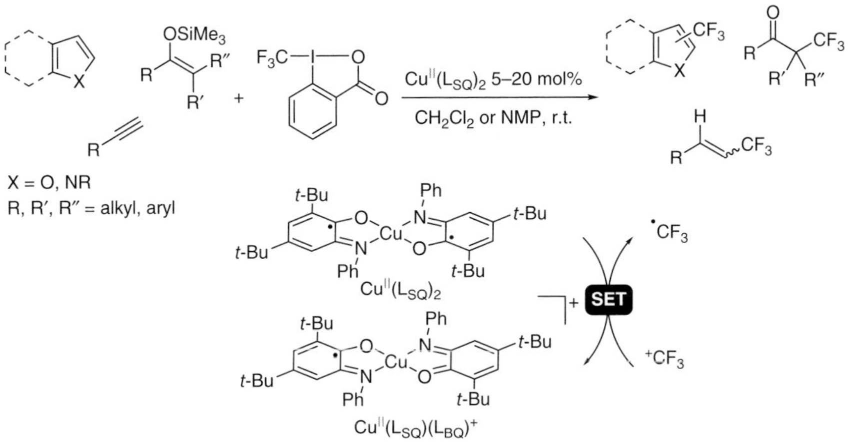

X = O, NR
R, R′, R″ = alkyl, aryl

$Cu^{II}(L_{SQ})_2$ 5–20 mol%
CH_2Cl_2 or NMP, r.t.

$Cu^{II}(L_{SQ})_2$
$Cu^{II}(L_{SQ})(L_{BQ})^+$
SET
$^•CF_3$
$^+CF_3$

Scheme 13.19 Radical trifluoromethylation with copper complex bearing redox-active ligands.

13.2 Trifluoromethylthiolation Reactions

Among CF_3 derivatives, the SCF_3 group has elicited great interest and is now present in many agrochemical and pharmaceutical products. The use of practical nucleophilic and electrophilic trifluoromethylthiolation reagents in copper, but also nickel catalysis has given birth to versatile and straightforward methodologies, enabling the incorporation of SCF_3 group starting from different types of substrates.

13.2.1 Nucleophilic Trifluoromethylthiolation

13.2.1.1 Copper-Catalyzed Nucleophilic Trifluoromethylthiolation

Although trifluoromethylated copper complexes are stable and efficient agents for CF_3 group transfers, attempts for the preparation of $CuSCF_3$ species are proved to be arduous and their application is limited in scope [50]. In 2013, Huang achieved the synthesis of (bpy)$CuSCF_3$ (bpy, 2,2′-bipyridine) and used this nucleophilic SCF_3 transfer agent with both electron-poor and electron-rich electrophiles [51]. Recently, the same complex was used for the synthesis of trifluoromethylthiolated pyridinones starting from iodopyridinones [52]. These methodologies being limited by stoichiometric amounts of copper, strong efforts were made for the development of copper-catalyzed trifluoromethylthiolations using stoichiometric amounts of nucleophilic SCF_3 source. In this field, $AgSCF_3$ appeared, in several systems, as a relevant reagent.

Liu reported a cross-coupling-like reaction, for the formation of $C(sp^2)–SCF_3$ bonds starting from aryl halides and using $AgSCF_3$ as a transmetallation agent [53]. The presence of *ortho*-directing groups, such as pyridyl, imine, ester, amide, or triazole, is essential for the reaction and might enable a precoordination of copper species, facilitating the oxidative addition step. Throughout the optimization of the catalytic system, it was also noticed that a strongly coordinating pyridyl moiety as a directing group (DG) or even DMF as a solvent is used to perform the reaction at room temperature (Scheme 13.20).

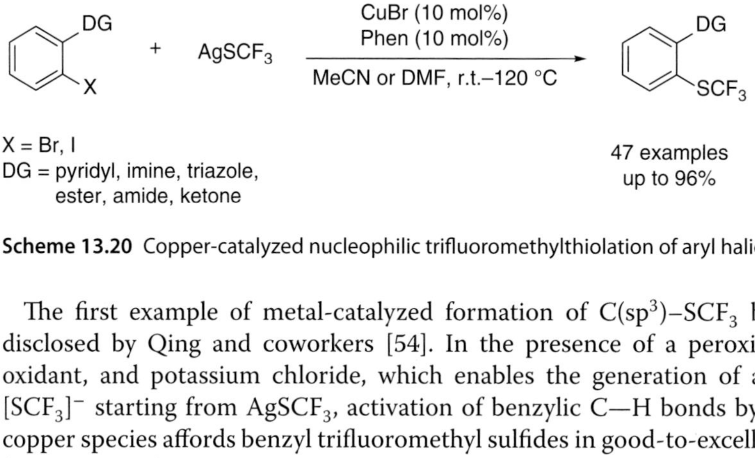

DG
X
+ $AgSCF_3$
CuBr (10 mol%)
Phen (10 mol%)
MeCN or DMF, r.t.–120 °C
DG
SCF_3
X = Br, I
DG = pyridyl, imine, triazole, ester, amide, ketone
47 examples
up to 96%

Scheme 13.20 Copper-catalyzed nucleophilic trifluoromethylthiolation of aryl halides.

The first example of metal-catalyzed formation of $C(sp^3)–SCF_3$ bond was disclosed by Qing and coworkers [54]. In the presence of a peroxide, as an oxidant, and potassium chloride, which enables the generation of a reactive $[SCF_3]^-$ starting from $AgSCF_3$, activation of benzylic C—H bonds by catalytic copper species affords benzyl trifluoromethyl sulfides in good-to-excellent yields (Scheme 13.21).

Because the use of $AgSCF_3$ implies stoichiometric amounts of silver, several groups turned their attention to simple nonmetallic nucleophilic SCF_3 sources, such as tetramethylammonium trifluoromethylthiolate. This reagent, which is readily prepared starting from tetrabutylammonium fluoride (TBAF), elemental sulfur S_8, and TMS–CF_3 [55], was notably used by Vicic for copper-mediated oxidative trifluoromethylthiolation of aryl boronic acids [56]. Recently, Gooßen and coworkers showed that a catalytic amount of CuSCN in the presence of Me_4NSCF_3 could convert aryl diazonium [57] and α-diazo esters [58] into the corresponding trifluoromethyl thioethers under mild conditions. A series of

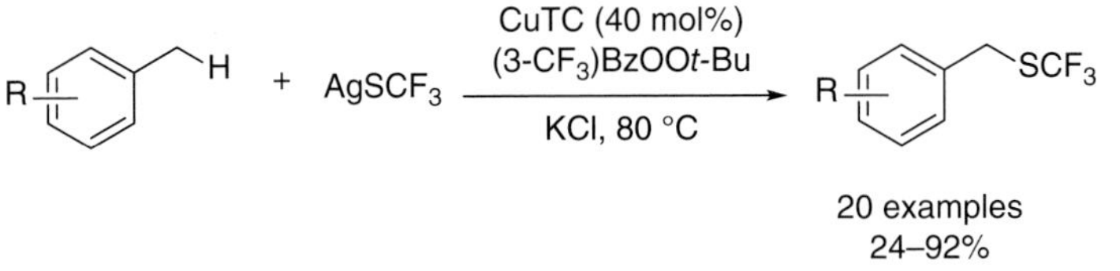

Scheme 13.21 Copper-catalyzed nucleophilic trifluoromethylthiolation of benzylic C—H bonds.

experiments support the formation of a $CuSCF_3$ species, but the mechanism seems to be different depending on the substrate: whereas a Sandmeyer-type SET mechanism involving aryl radicals is suggested for aryl diazonium, a radical pathway is excluded with α-diazo esters (Scheme 13.22).

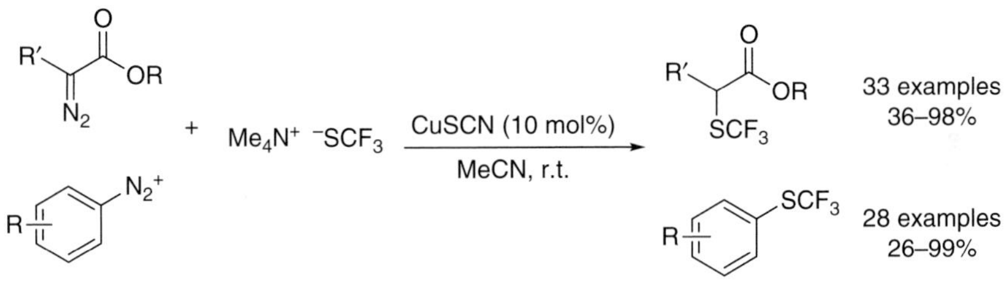

Scheme 13.22 Copper-catalyzed nucleophilic trifluoromethylthiolation with tetramethylammonium trifluoromethylthiolate.

Another strategy consists in the combination of elemental sulfur S_8 and $TMSCF_3$, which was first applied to oxidative trifluoromethylthiolation of aryl boronic acids [59]. Using a catalytic amount of copper thiocyanate, this transformation offers a simple and inexpensive access to aryl trifluoromethyl thioethers. Mechanistic studies, associated with Karlin's work [60], suggest the formation of a copper arylthiolate stemming from the reaction of an aryl boronic acid with a Cu^I disulfide complex.

This strategy was extended to activated electrophiles such as α-bromoketones [61] and allylic and propargylic halides [62]. Although Cu–SCF_3 species might be involved for the latter, TEMPO experiments conducted with α-bromoketones suggest the formation of $CF_3^\bullet$ radicals. Despite incomplete mechanistic elucidations, this methodology is a convenient way for the formation of $C(sp^3)–SCF_3$ bonds and many valuable synthons (Scheme 13.23).

In 2014, the Gooßen group developed a methodology for trifluoromethylthiolation of aryl diazonium salts, in which the elemental sulfur source is replaced by sodium thiocyanate [63]. This transformation can be decomposed in a Sandmeyer thiocyanation step of aryl diazonium, involving a CuSCN species, followed by the reaction of the resulting aryl thiocyanate with the nucleophilic trifluoromethylation reagent $TMSCF_3$. As evidenced by a series of controlled experiments, no $CuSCF_3$ species is formed during the process. Unlike previous methods involving a direct transfer of SCF_3 from a preformed $CuSCF_3$ complex and limited to a few electron-poor electrophiles [50], this straightforward

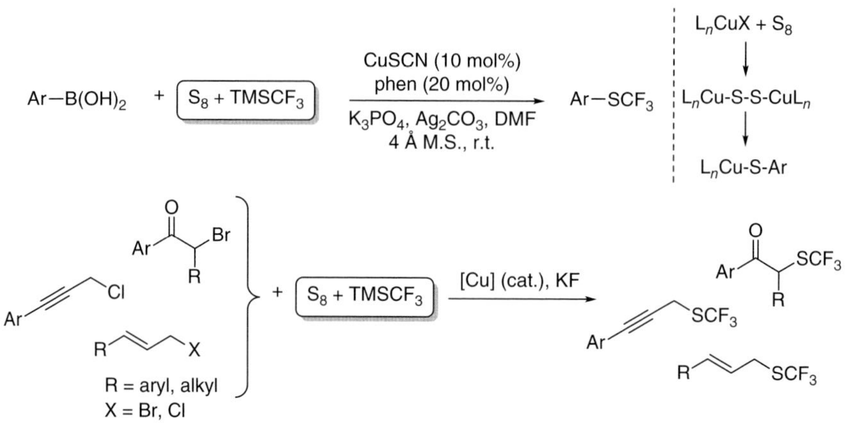

Scheme 13.23 Copper-catalyzed nucleophilic trifluoromethylthiolation with elemental sulfur S_8 and $TMSCF_3$.

methodology is broadly applicable both to electron-poor and electron-rich diazoniums (Scheme 13.24).

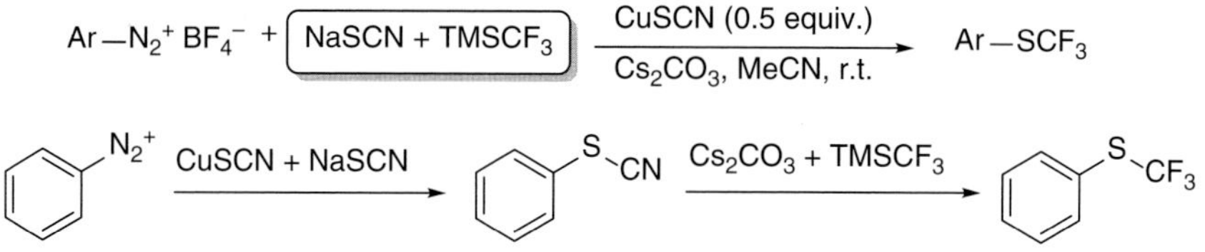

Scheme 13.24 Copper-catalyzed nucleophilic trifluoromethylthiolation of aryl diazonium salts.

13.2.1.2 Nickel-Catalyzed Nucleophilic Trifluoromethylthiolation

To date, copper remains the most common non-noble metal for nucleophilic trifluoromethylthiolation; however, more and more nickel-based catalytic systems are reported, which offer a specific reactivity and opens the way toward new developments in this field.

Vicic and coworker described a methodology for the nucleophilic trifluoromethylthiolation of aryl iodides and bromides [64]. The reaction is catalyzed by $Ni(COD)_2$ (COD, 1,4-cyclooctadiene) in the presence of the 4,4′-dimethoxybipyridine ligand dmbpy and uses Me_4NSCF_3. Unlike the aforementioned copper-based system [53], no *ortho*-directing group is needed here to activate $C(sp^2)$–I and $C(sp^2)$–Br bonds (Scheme 13.25).

As evidenced by Schoenebeck's work, nickel-based catalytic systems are also able to activate $C(sp^2)$–Cl bonds and introduce a SCF_3 moiety to yield $C(sp^2)–SCF_3$ coupling products [65]. The reaction is conducted with 10 mol% of $Ni(COD)_2$ with diphenylphosphinoferrocene (dppf) as the ligand and Me_4NSCF_3 as the coupling reagent. Computational and experimental studies suggest a Ni^0/Ni^{II} cycle involving oxidative addition of the $C(sp^2)$–Cl bond and $C(sp^2)–SCF_3$ reductive elimination. Interestingly, addition of acetonitrile

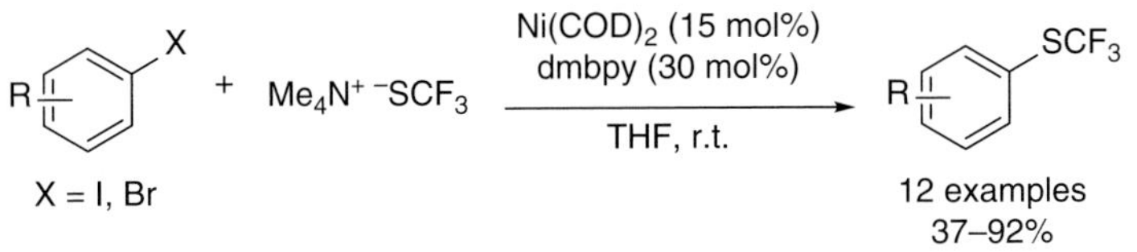

$$\text{R-C}_6\text{H}_4\text{-X} + \text{Me}_4\text{N}^+ \ {}^-\text{SCF}_3 \xrightarrow[\text{THF, r.t.}]{\text{Ni(COD)}_2\text{ (15 mol\%)},\ \text{dmbpy (30 mol\%)}} \text{R-C}_6\text{H}_4\text{-SCF}_3$$

X = I, Br

12 examples
37–92%

Scheme 13.25 Nickel-catalyzed nucleophilic trifluoromethylthiolation of aryl iodides and bromides.

resulted in improved conversions, notably for challenging substrates. This phenomenon could be due to the weaker coordinating ability of acetonitrile, compared to COD, thereby enabling the generation of a more reactive Ni^0 species. The $Ni(COD)_2$/dppf system was then extended to the activation of $C(sp^2)$–O bonds [66]. Aryl and vinyl triflates and nonaflates are, under similar conditions, converted into their corresponding aryl and vinyl trifluoromethyl thioethers. A ligand-less system was also recently reported for the trifluoromethylthiolation of aryl chlorides, using $Ni(COD)_2$ and $AgSCF_3$, and requiring *ortho*-directing groups for the activation of the $C(sp^2)$–Cl bond [67] (Scheme 13.26).

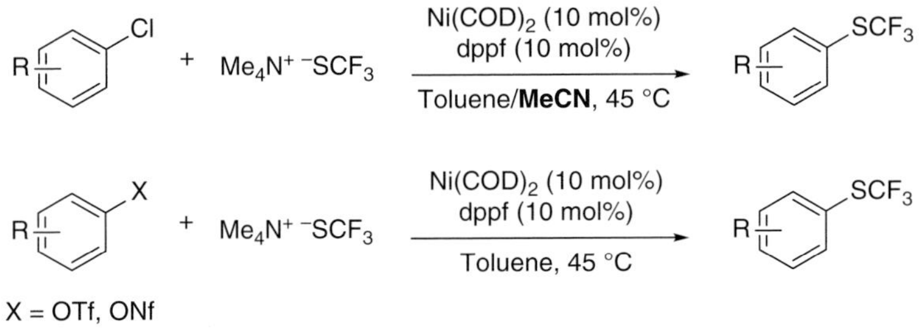

$$\text{R-C}_6\text{H}_4\text{-Cl} + \text{Me}_4\text{N}^+ \ {}^-\text{SCF}_3 \xrightarrow[\text{Toluene/}\mathbf{MeCN}\text{, 45 °C}]{\text{Ni(COD)}_2\text{ (10 mol\%)},\ \text{dppf (10 mol\%)}} \text{R-C}_6\text{H}_4\text{-SCF}_3$$

$$\text{R-C}_6\text{H}_4\text{-X} + \text{Me}_4\text{N}^+ \ {}^-\text{SCF}_3 \xrightarrow[\text{Toluene, 45 °C}]{\text{Ni(COD)}_2\text{ (10 mol\%)},\ \text{dppf (10 mol\%)}} \text{R-C}_6\text{H}_4\text{-SCF}_3$$

X = OTf, ONf

Scheme 13.26 Nickel-catalyzed nucleophilic trifluoromethylthiolation of aryl chlorides and aryl triflates.

13.2.2 Electrophilic Trifluoromethylthiolation

Drawing inspiration from Togni's reagent, Shen's group prepared dimethyl-tri fluoromethylsulfanyl-benziodoxole, starting from the chloroiodane precursor and $AgSCF_3$ [68]. This compound appeared as a reagent of choice for electrophilic trifluoromethylthiolation [69]. Interestingly, the initially proposed λ^3-iodane structure was recently revised by Buchwald, proving existence of a O–SCF_3 bond (**4**) instead of hypervalent iodine structure **3** [70]. From this result, the iodine-less reagent **5** was also prepared, and the poor influence of the iodine atom (or even any other substituent) on the reactivity was revealed (Scheme 13.27) [71].

Using these SCF_3 sources (**4** or **5**), copper-based catalytic systems were developed for electrophilic trifluoromethylthiolation of aryl, vinyl, alkylboronic acids, and alkynes (Scheme 13.28) [68a, 71, 72]. A wide range of functional groups were tolerated and trifluoromethylthioethers were obtained in good-to-excellent yields. It was also shown that such SCF_3 sources combined with catalytic copper

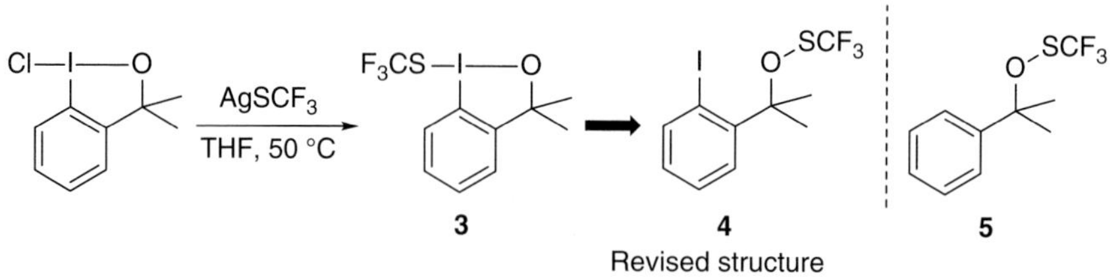

Scheme 13.27 Revised structure for dimethyl-trifluoromethylsulfanyl-benziodoxole.

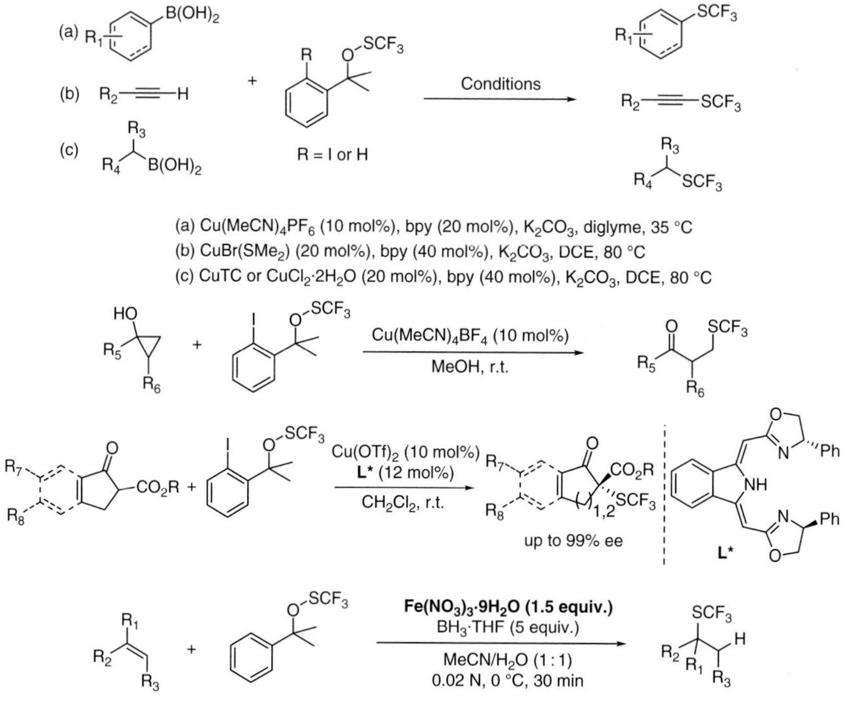

Scheme 13.28 Copper-based catalytic systems for electrophilic trifluoromethylthiolation of aryl, vinyl, alkylboronic acids, and alkynes.

could be involved in a ring-opening cross-coupling, converting cyclopropanols into β-SCF_3-substituted ketones and aldehydes [73]. Based on his previous work on copper-catalyzed enantioselective alkylation of β-keto-esters using copper–boxmi complexes (boxmi, bis(oxazolinylmethylidene)isoindolines) [74], Gade and coworkers developed a methodology for the enantioselective trifluoromethylthiolation of the latter [75]. The copper–boxmi complex was suggested to act as a Lewis acid, which could chelate the substrate, control its ester–enolate forms, and create a chiral environment close to the site of reaction with the SCF_3^+ source. The valuable SCF_3 products were obtained with high yields and excellent enantioselectivities. An original strategy to access alkyl-SCF_3 compounds, based on hydrotrifluoromethylthiolation of unactivated alkenes, was disclosed by Shen and coworkers [76]. A stoichiometric amount of Fe^{III} salt, associated with BH_3·THF as hydrogen source, enables to perform the reaction under mild conditions and to get addition products with excellent Markovnikov selectivities, particularly with terminal olefins. Results obtained from radical clock experiments strongly support a free radical mechanism.

The potential of *N*-(trifluoromethylthio)phtalimide as a SCF_3^+ source was assessed by Shen and coworkers in 2014, and the study was facilitated by an enhanced method to access the reagent [77] The synthesis, initially implying the toxic and corrosive gas CF_3SCl, was achieved using bromophtalimide and $AgSCF_3$, under mild conditions and in very good yield. *N*-(Trifluoromethylthio) phtalimide has shown good performances as a SCF_3^+ source for the trifluoromethylthiolation of boronic acids, catalyzed by a bipyridine–copper complex. Belonging to the same family of stable N-SCF_3 reagents, trifluorosulfenamides have demonstrated their efficiency in many electrophilic trifluoromethylthiolation protocols, notably in copper-catalyzed cross-coupling reactions with boronic acids [78]. This reaction, studied by Billard, was found to occur under very mild conditions, without any base addition or large excess of reagent. Interestingly, the presence of an optimal amount of water (7 equiv.) was found to be crucial to reach high yields of aryl trifluoromethylthioethers; however, the role of water was not clearly established. Recently, the same group focused on the mono- and bis-trifluoromethylthiolation of alkynes, which are controlled by adjusting catalyst loading [79] (Scheme 13.29).

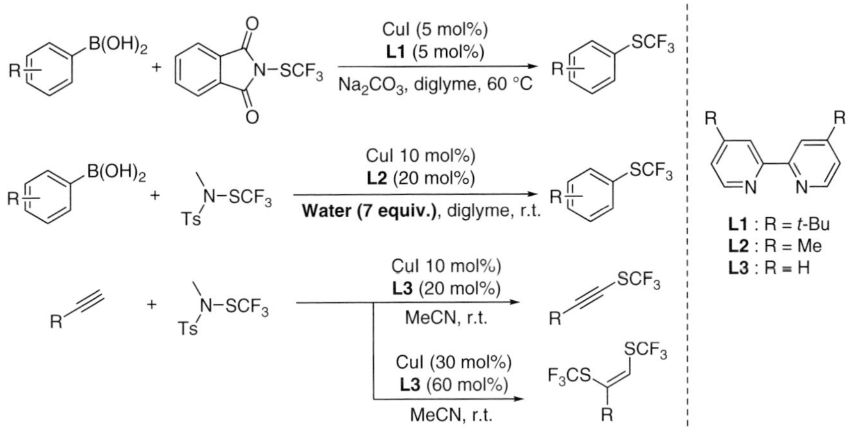

Scheme 13.29 Copper-catalyzed trifluoromethylthiolation of boronic acids and alkynes.

To avoid the use of SCF_3 sources prepared from CF_3SCl or $AgSCF_3$ precursors, Shibata's group worked on the development of a reagent bearing a stable CF_3SO_2 unit as a "hidden" CF_3S source (Scheme 13.30) [80]. The reactivity of a hypervalent iodonium ylide and its interaction with copper were studied in the context of electrophile-type trifluoromethylthiolation of nucleophiles. Indeed, in the presence of catalytic CuCl, enamines, β-ketoesters, and indoles were efficiently converted into the corresponding trifluoromethylthiolated products. The mechanism was not totally elucidated; however, copper seems to play

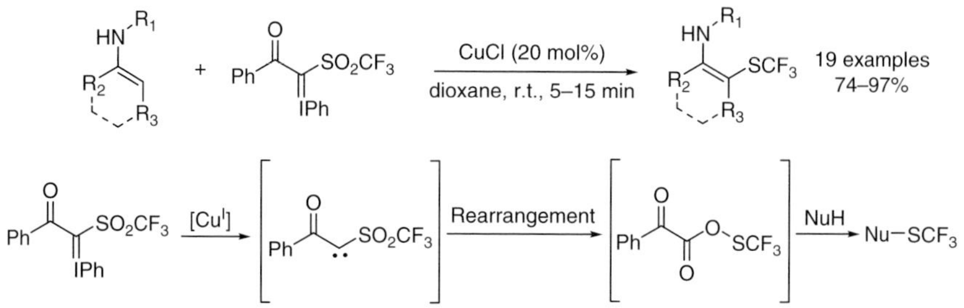

Scheme 13.30 An alternative SCF_3 source for copper-catalyzed electrophilic trifluoromethylthiolation.

a crucial role in the activation of the ylide, which evolves to form a sulfonyl carbene, detected by mass spectroscopy (HR-ESI-MS). This resulting carbene then rearranges to give a thioperoxoate. This reactive species would act as the electrophilic source of SCF_3, thus reacting with the nucleophile via a SET process or an ionic pathway. A wide diversity of nucleophiles was successively studied: pyrroles [81], silyl enol ethers, allylsilanes [82], allylic alcohols, and boronic acids [83], demonstrating the high versatility of this methodology.

13.3 Perfluoroalkylation Reactions

Because of their interesting properties including high electronegativity, stability, lipophilicity, and water and oil repellency, perfluoroalkylated compounds have also elicited important synthetic developments. Among several methodologies, stoichiometric copper complexes have been utilized. Thus, copper-mediated perfluoralkylation has been accomplished based on the use of perfluoro analogs of Hartwig's $PhenCuCF_3$ either from aryl iodide precursors [4] or from nonhalogenated aromatic derivatives in a reaction involving an initial iridium(I)-catalyzed C—H borylation [13a, 84].

Following Amii's pioneering copper-catalyzed coupling of electron-deficient aryliodides with trifluoromethylsilanes [2], the group of Daugulis in 2011 devised a copper-catalyzed arylation of 1*H*-perfluoroalkanes. Using a mixture of 10 mol% CuCl, 20 mol% phenanthroline in the presence of TMP_2Zn as a base (TMP, bis-2,2,6,6-tetramethylpiperidide), various aryliodides including heterocyclic systems and 1*H*-perfluoroalkanes could efficiently be coupled. Mechanistic investigations suggest a preliminary deprotonation of the 1*H*-perfluoroalkane reagent by the zinc amide. Copper transmetallation generates a mixture of copper species that react with the aryliodide component (Scheme 13.31).

Also using aryliodide precursors, the group of Mikami has developed some stable but reactive bis(perfluoroalkyl)zinc reagents of the general formula $Zn(R_F)_2(DMPU)_2$ (DMPU, 1,3-dimethyl-3,4,5,6-tetrahydro-2(1*H*)-pyrimidone) [85]. These reagents are easily prepared from the corresponding perfluoroalkyl

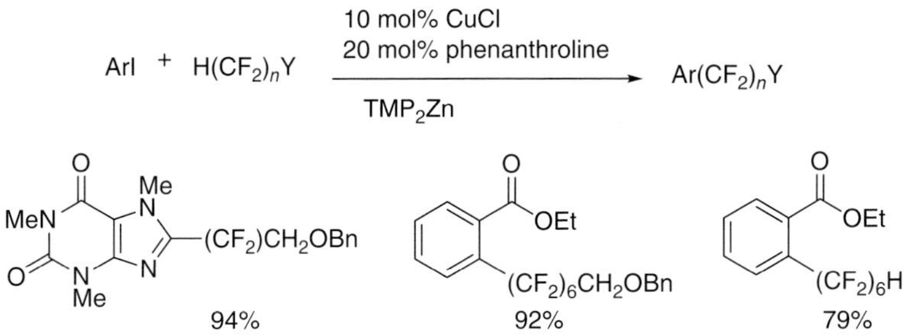

ArI + $H(CF_2)_nY$ —(10 mol% CuCl, 20 mol% phenanthroline, TMP_2Zn)→ $Ar(CF_2)_nY$

$(CF_2)CH_2OBn$, 94%; $(CF_2)_6CH_2OBn$, 92%; $(CF_2)_6H$, 79%

Scheme 13.31 Cu(I)-catalyzed perfluoroalkylation of aryliodides with 1*H*-perfluoroalkanes.

iodides and diethylzinc and isolated as white solids. Upon CuI catalysis, and without external ligand or activator, they can be used to perfluoroalkylate aryl or vinyl iodides. Different classes of perfluorated derivatives including heterocycles are accessible. It should be noted that the arylation reaction also works with $Zn(CF_3)_2(DMPU)_2$ (Scheme 13.32).

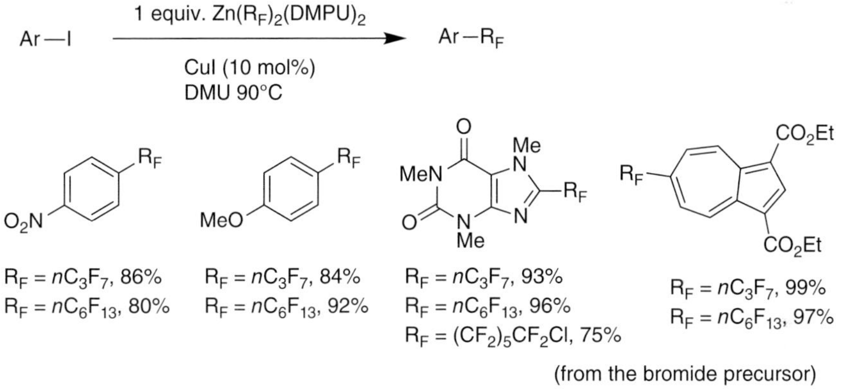

Ar—I —(1 equiv. $Zn(R_F)_2(DMPU)_2$, CuI (10 mol%), DMU 90°C)→ Ar—R_F

$R_F = nC_3F_7$, 86%; $R_F = nC_6F_{13}$, 80%
$R_F = nC_3F_7$, 84%; $R_F = nC_6F_{13}$, 92%
$R_F = nC_3F_7$, 93%; $R_F = nC_6F_{13}$, 96%; $R_F = (CF_2)_5CF_2Cl$, 75% (from the bromide precursor)
$R_F = nC_3F_7$, 99%; $R_F = nC_6F_{13}$, 97%

Scheme 13.32 Cu(I)-catalyzed perfluoroalkylation of aryliodides using zinc reagents.

As mentioned in Ref. [79], Tlili and Billard have devised the copper-catalyzed perfluoroalkylthiolation of alkynes using perfluoroalkanesulfenamides precursors. Using the same type of catalytic system as before consisting of CuI (10 mol%) and a bpy ligand (20 mol%) and in the presence of pentafluoroethyl- or heptafluoropropyl *N*-methyl-tosylamides, satisfactory yields of *S*-perfluoro alkyne adducts were obtained. The reaction showed a high tolerance of functionality on the alkyne moiety. *Z*-diastereoselective bis-pentafluoroethylthiolation could also be achieved. Presumably, the reaction is not radical in nature, neither does it involve a copper acetylide species. The mechanism remains under speculation, presumably involving a Cu(III) species that would result from an oxidative addition into the N—S bond and which would activate the alkyne to generate an intermediate vinyl copper species (Scheme 13.33).

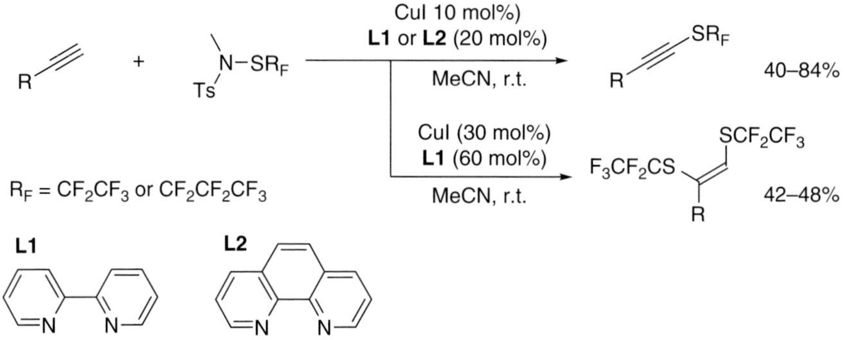

R + N–SR_F (Ts) → CuI 10 mol%), **L1** or **L2** (20 mol%), MeCN, r.t. → R–SR_F 40–84%

→ CuI (30 mol%), **L1** (60 mol%), MeCN, r.t. → F_3CF_2CS / SCF_2CF_3 / R 42–48%

R_F = CF_2CF_3 or $CF_2CF_2CF_3$

L1 **L2**

Scheme 13.33 Cu(I)-catalyzed perfluoroalkylthiolation of alkynes using perfluoroalkanesulfenamides precursors.

13.4 Conclusion

Introduction of trifluoromethyl and related groups by means of earth-abundant metal catalysis has witnessed an intense growth in the last decade. Efficient methodologies have been developed that allow performing such reactions with a host of chemical partners including both electrophiles and nucleophiles. Bespoke reagents have been devised and successfully applied to the synthesis of advanced molecular scaffolds, thus providing promising forays into the ubiquity and widespread implementation of such methods. One can anticipate that this field has only reached its maturity and important findings are still to come.

References

1 For selected recent reviews: (a) Purser, S., Moore, P.R., Swallow, S., and Gouverneur, V. (2008). *Chem. Soc. Rev.* 37: 320–330. (b) Ma, J.-A. and Cahard, D. (2008). *Chem. Rev.* 108: PR1. (c) Furuya, T., Kamlet, A.S., and Ritter, T. (2011). *Nature* 473: 470–477. (d) Tomashenko, O.A. and Grushin, V.V. (2011). *Chem. Rev.* 111: 4475–4521. (e) Wu, X.-F., Neumann, H., and Beller, M. (2012). *Chem. Asian J.* 7: 1744–1754. (f) Studer, A. (2012). *Angew. Chem. Int. Ed.* 51: 8950–8958; (2012). *Angew. Chem.* 124: 9082–9090. (g) Liang, T., Neumann, C.N., and Ritter, T. (2013). *Angew. Chem. Int. Ed.* 52: 8214–8264; (2013). *Angew. Chem.* 125: 8372–8423.

2 Oishi, M., Kondo, H., and Amii, H. (2009). *Chem. Commun.* 1909–1911.

3 (a) Dubinina, G.G., Furutachi, H., and Vicic, D.A. (2008). *J. Am. Chem. Soc.* 130: 8600–8601. (b) Tomashenko, O.A., Escudero-Adán, E.C., Martínez Belmonte, M., and Grushin, V.V. (2011). *Angew. Chem. Int. Ed.* 50: 7655–7659; (2011). *Angew. Chem.* 123: 7797–7801. (c) Zanardi, A., Novikov, M.A., Martin, E. et al. (2011). *J. Am. Chem. Soc.* 133: 20901–20913.

4 Morimoto, H., Tsubogo, T., Litvinas, N.D., and Hartwig, J.F. (2011). *Angew. Chem. Int. Ed.* 50: 3793–3798; (2011). *Angew. Chem.* 123: 3877–3882.

5 Billard, T., Bruns, S., and Langlois, B.R. (2000). *Org. Lett.* 2: 2101–2103.
6 Kondo, H., Oishi, M., Fujikawa, K., and Amii, H. (2011). *Adv. Synth. Catal.* 353: 1247–1252.
7 Knauber, T., Arikan, F., Röschenthaler, G.-V., and Gooßen, L.J. (2011). *Chem. Eur. J.* 17: 2689–2697.
8 Danoun, G., Bayarmagnai, B., Grünberg, M.F., and Gooßen, L.J. (2013). *Angew. Chem. Int. Ed.* 52: 7972–7975; (2013). *Angew. Chem.* 125: 8130–8133.
9 Chu, L. and Qing, F.-L. (2010). *J. Am. Chem. Soc.* 132: 7262–7263.
10 Chu, L. and Qing, F.-L. (2012). *J. Am. Chem. Soc.* 134: 1298–1304.
11 Hall, D.G. (2005). *Boronic Acids: Preparation and Applications in Organic Synthesis and Medicine.* Weinheim: Wiley-VCH.
12 (a) Chu, L. and Qing, F.-L. (2010). *Org. Lett.* 12: 5060–5063. (b) Senecal, T.D., Parsons, A.T., and Buchwald, S.L. (2010). *J. Org. Chem.* 76: 1174–1176. (c) Khan, B.A., Buba, A.E., and Gooßen, L.J. (2012). *Chem. Eur. J.* 18: 1577–1581.
13 (a) Litvinas, N.D., Fier, P.S., and Hartwig, J.F. (2012). *Angew. Chem. Int. Ed.* 51: 536–539; (2012). *Angew. Chem.* 51: 8950–8958. (b) Novák, P., Lishchynskyi, A., and Grushin, V.V. (2012). *Angew. Chem. Int. Ed.* 51: 7767–7770; (2012). *Angew. Chem.* 124: 7887–7890.
14 Jiang, X., Chu, L., and Qing, F.-L. (2012). *J. Org. Chem.* 77: 1251–1257.
15 Xu, J., Xiao, B., Xie, C.-Q. et al. (2012). *Angew. Chem. Int. Ed.* 51: 12551–12554; (2012). *Angew. Chem.* 124: 12719–12722.
16 (a) Tordeux, M., Langlois, B.R., and Wakselman, C. (1989). *J. Org. Chem.* 54: 2452–2453. (b) Langlois, B.R., Laurent, E., and Roidot, N. (1991). *Tetrahedron Lett.* 32: 7525–7528.
17 Zhang, C. (2014). *Adv. Synth. Catal.* 356: 2895–2906.
18 Sinha, W., Sommer, M.G., Deibel, N. et al. (2015). *Angew. Chem. Int. Ed.* 54: 13769–13774; (2015). *Angew. Chem.* 127: 13973–13978.
19 Charpentier, J., Früh, N., and Togni, A. (2015). *Chem. Rev.* 115: 650–682.
20 Zhang, C. (2014). *Org. Biomol. Chem.* 12: 6580–6589.
21 (a) Ling, L., Liu, K., Li, X., and Li, Y. (2015). *ACS Catal.* 5: 2458–2468. (b) Kawamura, S., Egami, H., and Sodeoka, M. (2015). *J. Am. Chem. Soc.* 137: 4865–4873.
22 (a) Kochi, J.K., Bemis, A., and Jenkins, C.L. (1968). *J. Am. Chem. Soc.* 90: 4616–4625. (b) Jenkins, C.L. and Kochi, J.K. (1972). *J. Am. Chem. Soc.* 94: 843–855.
23 (a) Zhang, C.-P., Wang, Z.-L., Chen, Q.-Y. et al. (2011). *Angew. Chem. Int. Ed.* 50: 1896–1900; (2011). *Angew. Chem.* 123: 1936–1940. (b) Dai, J.-J., Fang, C., Xiao, B. et al. (2013). *J. Am. Chem. Soc.* 135: 8436–8439.
24 (a) Xu, J., Luo, D.-F., Xiao, B. et al. (2011). *Chem. Commun.* 47: 4300–4302. (b) Liu, T. and Shen, Q. (2011). *Org. Lett.* 13: 2342–2345.
25 (a) Merino, E. and Nevado, C. (2014). *Chem. Soc. Rev.* 43: 6598–6608. (b) Alonso, C., Martínez de Marigorta, E., Rubiales, G., and Palacios, F. (2015). *Chem. Rev.* 115: 1847–1935. (c) Prieto, A., Baudoin, O., Bouyssi, D., and Monteiro, N. (2016). *Chem. Commun.* 52: 869–881.
26 Pham, P.V., Nagib, D.A., and MacMillan, D.W.C. (2011). *Angew. Chem. Int. Ed.* 50: 6119–6122; (2012). *Angew. Chem.* 123: 6243–6246.

27 Li, L., Chen, Q.-Y., and Guo, Y. (2014). *J. Org. Chem.* 79: 5145–5152.
28 Allen, A.E. and MacMillan, D.W.C. (2010). *J. Am. Chem. Soc.* 132: 4986–4987.
29 Pair, E., Monteiro, N., Bouyssi, D., and Baudoin, O. (2013). *Angew. Chem. Int. Ed.* 52: 5346–5349; (2013). *Angew. Chem.* 125: 5454–5457.
30 Prieto, A., Jeamet, E., Monteiro, N. et al. (2014). *Org. Lett.* 16: 4770–4773.
31 Parsons, A.T. and Buchwald, S.L. (2011). *Angew. Chem. Int. Ed.* 50: 9120–9123; (2011). *Angew. Chem.* 123: 9286–9289.
32 Wang, X., Ye, Y., Zhang, S. et al. (2011). *J. Am. Chem. Soc.* 133: 16410–16413.
33 Xu, J., Fu, Y., Luo, D.-F. et al. (2011). *J. Am. Chem. Soc.* 133: 15300–15303.
34 (a) Shimizu, R., Egami, H., Hamashima, Y., and Sodeoka, M. (2012). *Angew. Chem. Int. Ed.* 51: 4577–4580; (2012). *Angew. Chem.* 124: 4655–4658. (b) Mizuta, S., Galicia-López, O., Engle, K.M. et al. (2012). *Chem. Eur. J.* 18: 8583–8587.
35 Selected examples:(a)Zhu, R. and Buchwald, S.L. (2012). *J. Am. Chem. Soc.* 134: 12462–12465. (b) Egami, H., Kawamura, S., Miyazaki, A., and Sodeoka, M. (2013). *Angew. Chem. Int. Ed.* 52: 7841–7844; (2013). *Angew. Chem.* 125: 7995–7998. (c) Bai, X., Lv, L., and Li, Z. (2016). *Org. Chem. Front.* 3: 804–808.
36 Selected example:He, Y.-T., Li, L.-H., Yang, Y.-F. et al. (2013). *Org. Lett.* 16: 270–273.
37 Selected example:Janson, P.G., Ghoneim, I., Ilchenko, N.O., and Szabo, J. (2012). *Org. Lett.* 14: 2882–2885.
38 Selected examples:Wang, F., Qi, X., Liang, Z. et al. (2014). *Angew. Chem. Int. Ed.* 53: 1881–1886; (2014). *Angew. Chem.* 126: 1912–1917.
39 Selected examples: (a) Wang, Y., Jiang, M., and Liu, J.-T. (2014). *Chem. Eur. J.* 20: 15315–15319. (b) Xu, J., Wang, Y.-L., Gong, T.-J. et al. (2014). *Chem. Commun.* 50: 12915–12918.
40 Selected example:Yu, Q. and Ma, S. (2013). *Chem. Eur. J.* 19: 13304–13308.
41 (a) Han, G., Liu, Y., and Wang, Q. (2014). *Org. Lett.* 16: 3188–3191. (b) Cheng, C., Liu, S., Lu, D., and Zhu, G. (2016). *Org. Lett.* 18: 2852–2855.
42 Kong, W., Casimiro, M., Merino, E., and Nevado, C. (2013). *J. Am. Chem. Soc.* 135: 14480–14483.
43 (a) Ravelli, D., Dondi, D., Fagnoni, M., and Albini, A. (2009). *Chem. Soc. Rev.* 38: 1999–2011. (b) Nicewicz, D.A. and MacMillan, D.W.C. (2008). *Science* 322: 77–80. (c) Levin, M.D., Kim, S., and Toste, F.D. (2016). *ACS Cent. Sci.* 2: 293–301.
44 Koike, T. and Akita, M. (2014). *Top. Catal.* 57: 967–974.
45 Paria, S. and Reiser, O. (2014). *ChemCatChem* 6: 2477–2483.
46 Bagal, D.B., Kachkovskyi, G., Knorn, M. et al. (2015). *Angew. Chem. Int. Ed.* 54: 6999–7002; (2015). *Angew. Chem.* 127: 7105–7108.
47 Beniazza, R., Lambert, R., Harmand, L. et al. (2014). *Chem. Eur. J.* 20: 13181–13187.
48 Beniazza, R., Molton, F., Duboc, C. et al. (2015). *Chem. Commun.* 51: 9571–9574.
49 Jacquet, J., Blanchard, S., Derat, E. et al. (2016). *Chem. Sci.* 7: 2030–2036.
50 Adams, D.J., Goddard, A., Clark, J.H., and Macquarrie, D.J. (2000). *Chem. Commun.* 987–988.

51 Weng, Z., He, W., Chen, C. et al. (2013). *Angew. Chem. Int. Ed.* 52: 1548–1552; (2013). *Angew. Chem.* 125: 1588–1592.
52 Luo, B., Zhang, Y., You, Y., and Weng, Z. (2016). *Org. Biomol. Chem.* 14: 8615–8622.
53 Xu, J., Mu, X., Chen, P. et al. (2014). *Org. Lett.* 16: 3942–3945.
54 Chen, C., Xu, X.-H., Yang, B., and Qing, F.-L. (2014). *Org. Lett.* 16: 3372–3375.
55 (a) P. Kirsch, C. V. Roeschenthaler, B. Bissky, A. Kolomeitsev, (Merck GmbH), Procedure for the production of perfluoroalkylthiolates and arylperfluoroalkylthiolates. DE-A1 10254597, 2003. (b) Tyrra, W., Naumann, D., Hoge, B., and Yagupolskii, Y.L. (2003). *J. Fluorine Chem.* 119: 101–107.
56 Zhang, C.-P. and Vicic, D.A. (2012). *Chem. Asian J.* 7: 1756–1758.
57 Matheis, C., Wagner, V., and Gooßen, L.J. (2016). *Chem. Eur. J.* 22: 79–82.
58 Matheis, C., Krause, T., Bragoni, V., and Gooßen, L.J. (2016). *Chem. Eur. J.* 22: 12270–12273.
59 Chen, C., Xie, Y., Chu, L. et al. (2012). *Angew. Chem. Int. Ed.* 51: 2492–2495; (2012). *Angew. Chem.* 124: 2542–2545.
60 Helton, M.E., Chen, P., Paul, P.P. et al. (2003). *J. Am. Chem. Soc.* 125: 1160–1161.
61 Huang, Y., He, X., Lin, X. et al. (2014). *Org. Lett.* 16: 3284–3287.
62 (a) Rong, M., Li, D., Huang, R. et al. (2014). *Eur. J. Org. Chem.* 5010–5016. (b) Li, J., Wang, P., Xie, F.-F. et al. (2015). *Eur. J. Org. Chem.* 3568–3571.
63 Danoun, G., Bayarmagnai, B., Gruenberg, M.F., and Gooßen, L.J. (2014). *Chem. Sci.* 5: 1312–1316.
64 Zhang, C.-P. and Vicic, D.A. (2011). *J. Am. Chem. Soc.* 134: 183–185.
65 Yin, G., Kalvet, I., Englert, U., and Schoenebeck, F. (2015). *J. Am. Chem. Soc.* 137: 4164–4172.
66 Dürr, A.B., Yin, G., Kalvet, I. et al. (2016). *Chem. Sci.* 7: 1076–1081.
67 Nguyen, T., Chiu, W., Wang, X. et al. (2016). *Org. Lett.* 18: 5492–5495.
68 (a) Shao, X., Wang, X.-Q., Yang, T. et al. (2013). *Angew. Chem. Int. Ed.* 52: 3457–3460; (2013). *Angew. Chem.* 125: 3541–3544. (b) Ma, B.-Q., Shao, X., and Shen, Q. (2015). *J. Fluorine Chem.* 171: 73–77.
69 Shao, X., Xu, C., Lu, L., and Shen, Q. (2015). *Acc. Chem. Res.* 48: 1227–1236.
70 Vinogradova, E.V., Müller, P., and Buchwald, S.L. (2014). *Angew. Chem. Int. Ed.* 53: 3125–3128; (2014). *Angew. Chem.* 126: 3189–3192.
71 Shao, X., Xu, C., Lu, L., and Shen, Q. (2015). *J. Org. Chem.* 80: 3012–3021.
72 Shao, X., Liu, T., Liu, L., and Shen, Q. (2014). *Org. Lett.* 16: 4738–4741.
73 Li, Y., Ye, Z., Bellman, T.M. et al. (2015). *Org. Lett.* 17: 2186–2189.
74 Deng, Q.-H., Wadepohl, H., and Gade, L.H. (2012). *J. Am. Chem. Soc.* 134: 2946–2949.
75 Deng, Q.-H., Rettenmeier, C., Wadepohl, H., and Gade, L.H. (2014). *Chem. Eur. J.* 20: 93–97.
76 Yang, T., Lu, L., and Shen, Q. (2015). *Chem. Commun.* 51: 5479–5481.
77 Kang, K., Xu, C., and Shen, Q. (2014). *Org. Chem. Front.* 1: 294–297.
78 Glenadel, Q., Alazet, S., Tlili, A., and Billard, T. (2015). *Chem. Eur. J.* 21: 14694–14698.

79 Tlili, A., Alazet, S., Glenadel, Q., and Billard, T. (2016). *Chem. Eur. J.* 22: 10230–10234.
80 Yang, Y.-D., Azuma, A., Tokunaga, E. et al. (2013). *J. Am. Chem. Soc.* 135: 8782–8785.
81 Huang, Z., Yang, Y.-D., Tokunaga, E., and Shibata, N. (2015). *Org. Lett.* 17: 1094–1097.
82 Arimori, S., Takada, M., and Shibata, N. (2015). *Org. Lett.* 17: 1063–1065.
83 Arimori, S., Takada, M., and Shibata, N. (2015). *Dalton Trans.* 44: 19456–19459.
84 For a minireview, see:Liu, T. and Shen, Q. (2012). *Eur. J. Org. Chem.* 6679–6687.
85 Aikawa, K., Nakamura, Y., Yokota, Y. et al. (2015). *Chem. Eur. J.* 21: 96–100.

14

Catalytic Oxygenation of C=C and C—H Bonds

Pradip Ghosh, Marc-Etienne Moret, and Robertus J. M. Klein Gebbink

Utrecht University, Debye Institute for Nanomaterials Science, Organic Chemistry and Catalysis, Universiteitsweg 99, 3584 CG, Utrecht, The Netherlands

14.1 Introduction

The oxygenation of C=C and C—H bonds is an important process in both the chemical industry and academia (Scheme 14.1) [1, 2]. The conversion of these moieties to useful fine chemicals, viz. epoxides and diols, and alcohols and carbonyl compounds has been the subject of fundamental research activities over many years. In practice, millions of tons of raw materials obtained from petroleum-based industries are converted through oxidative processes into functionalized chemicals, often involving harsh reaction conditions, stoichiometric amounts of oxidant, and the formation of undesired by-products. In some cases, noble metals are used as a catalyst, e.g. silver salts are used to catalyze the oxidation of ethylene to ethylene oxide [3]. The desire to replace prototypical catalysts based on heavy metals by catalysts based on base metals represents one of the biggest challenges in the fields of oxidation catalysis and oxidative synthesis. This is a rapidly expanding frontier in academic and industrial research that is of importance for practical reasons (cost and availability, environmental impact, and toxicity) and offers numerous opportunities for the discovery of fundamentally new reactivity using non-noble metals.

This chapter aims to present recent achievements in the catalytic oxygenation of C=C and C—H bonds by homogeneous catalysts based on late first-row transition metals (Mn, Fe, Co, Ni, and Cu). In view of the numerous recent developments in the field, the chapter focuses on the so-called nonheme catalysts and will not discuss the use of catalysts based on porphyrins and related ligands. Special emphasis is given to the use of H_2O_2 as sacrificial oxidant, whereas some examples using other oxidants will also be discussed. Among the commonly used stoichiometric oxidants, hydrogen peroxide has a number of desirable features such as high atom efficiency, high active oxygen content, and generation of water as the only by-product. In addition, hydrogen peroxide, being a liquid reactant easily and safely handled, presents significant advantages over molecular oxygen and is relative inexpensive and commercially available on a large scale.

Non-Noble Metal Catalysis: Molecular Approaches and Reactions,
First Edition. Edited by Robertus J. M. Klein Gebbink and Marc-Etienne Moret.

Scheme 14.1 Oxidation of C=C and C—H bonds.

14.2 Oxygenation of C=C Bonds

Olefins are widely used as starting materials in organic synthesis and the fine chemicals industry. Epoxidation of olefins is among the most straightforward strategies for the (enantioselective) functionalization of olefins and the incorporation of carbon–hetero atom bonds in petrochemical feedstocks. Because of their intrinsic reactivity of the three-membered ring, epoxides are valuable intermediates for the synthesis of high added-value chemicals, including chiral ones [4, 5]. Epoxides are attractive for further derivatization because they readily react with a variety of nucleophiles. Olefin cis-dihydroxylation, on the other hand, is an important reaction for organic synthesis because enantiopure syn-diols are very useful synthons for the construction of chiral building blocks. The development of generally applicable and environmentally benign catalysts for the enantioselective cis-dihydroxylation of alkenes remains a challenge. In the subsequent sections, recent advances in the oxygenation of C=C bonds using late first-row transition metals are discussed.

14.2.1 Manganese Catalysts

Manganese is the third most abundant transition metal in the earth crust and an essential trace element for life on earth. Manganese complexes are known to catalyze the epoxidation of functionalized alkenes in the presence of terminal oxidants such as H_2O_2, O_2, PhIO, CH_3CO_3H, or NaOCl. In some cases, the epoxidations are performed in the presence of a coreductant.

The first report on the epoxidation of olefins, using cationic Mn–salen complexes **1a-h**, was published by Kochi and coworkers (Figure 14.1) [6]. Mn(III) complexes of the salen ligand (salen = *N*,*N*′-ethylenebis(salicylideneaminato)) are effective catalysts for the epoxidation of various olefins including styrenes, stilbenes, and cyclic and acyclic alkenes using iodosylbenzene as the terminal oxidant. These epoxidations were found to take place in a stereospecific manner and in high yields (50–75%) within 15 minutes at ambient temperatures in acetonitrile. Enhanced catalytic activities were noted upon introduction of electron-withdrawing groups in the R_2 position of the salen framework (R_2 = Cl, NO_2).

Later on, the groups of Jacobsen and Katsuki published at about the same time on asymmetric olefin epoxidations using chiral Mn–salen complexes [7, 8]. Jacobsen and coworkers reported on the use of the C_2-symmetric manganese–salen complexes **2** and **3** wherein the diamine moiety is part of (*R*,*R*) or (*S*,*S*)-1,2-diamino-1,2-diphenylethane (Figure 14.2). Katsuki and coworkers used chiral complexes **4** and **5** containing a bulky chiral group *ortho* to the phenoxide oxygen atom to scrutinize the effect of asymmetric centers

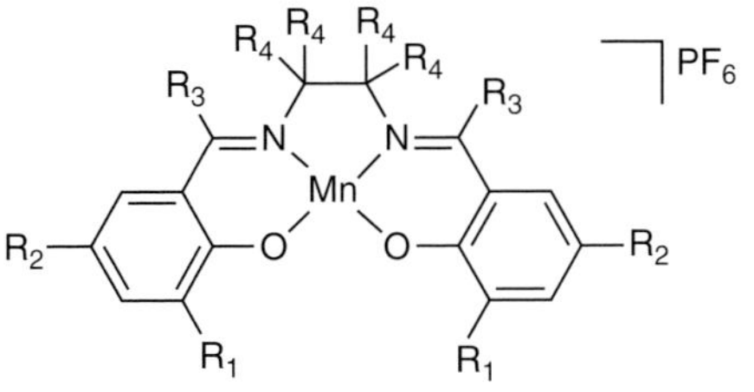

Complex	R_1	R_2	R_3	R_4
1a	H	H	H	H
1b	OMe	H	H	H
1c	H	OMe	H	H
1d	H	H	Ph	H
1e	H	Cl	H	H
1f	H	NO_2	H	H
1g	H	NO_2	H	CH_3
1h	NO_2	NO_2	H	H

Figure 14.1 Structure of the (salen)Mn catalysts **1a-h** used by Kochi. Source: Srinivasan et al. 1986 [6]. Reproduced with permission of ACS.

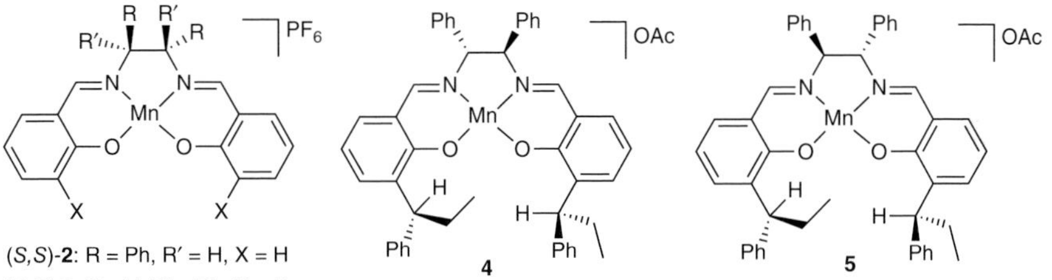

(*S*,*S*)-**2**: R = Ph, R′ = H, X = H
(*R*,*R*)-**2**: R = H, R′ = Ph, X = H
3: R = H, R′ = Ph, X = ^{t}Bu

Figure 14.2 Chiral Mn–salen complexes **2–5** used for the asymmetric epoxidation of olefins by Jacobsen and Katsuki.

located closer to the metal center. In the asymmetric epoxidation of (*E*)- and (*Z*)-1-phenyl-l-propene and dihydronaphthalene using **4** and **5**, it was noted that the sense of asymmetric induction by these complexes is opposite, pointing at the importance of the chirality of the ligand backbone. The introduction of the additional chiral groups does not influence the overall ee of the reactions; in the epoxidation of (*E*)-1-phenyl-l-propene, complex **2** yielded the (1*S*,2*S*)-epoxide product in 20% enantiomeric excess (ee), whereas complex **5** gave the same product in 21% ee.

Since then, epoxidation chemistry using chiral Mn–salen complexes has been widely explored and has proven to be a valuable tool in asymmetric synthesis, as the steric and electronic properties of the catalysts can be tuned in a very straightforward manner by choosing the appropriate diamine and salicylaldehyde precursors [9, 10].

A landmark study on olefin epoxidations catalyzed by aminopyridine manganese complexes was reported in 2003 by Stack and coworkers [11]. They reported on highly selective and efficient manganese complexes containing the tetradentate mep or mcp ligands (mep = *N*,*N*′-dimethyl-*N*,*N*′-bis(2-pyridylmethyl)ethane-1,2-diamine; mcp = (*R*,*R*)-*N*,*N*′-dimethyl-*N*,*N*′-bis(2-pyridylmethyl)cyclohexane-1,2-diamine) (Figure 14.3). $[Mn^{II}(mep)(CF_3SO_3)_2]$ (**6**) and $[Mn^{II}(R,R\text{-mcp})(CF_3SO_3)_2]$ (**7**) can rapidly (less than five minutes) epoxidize

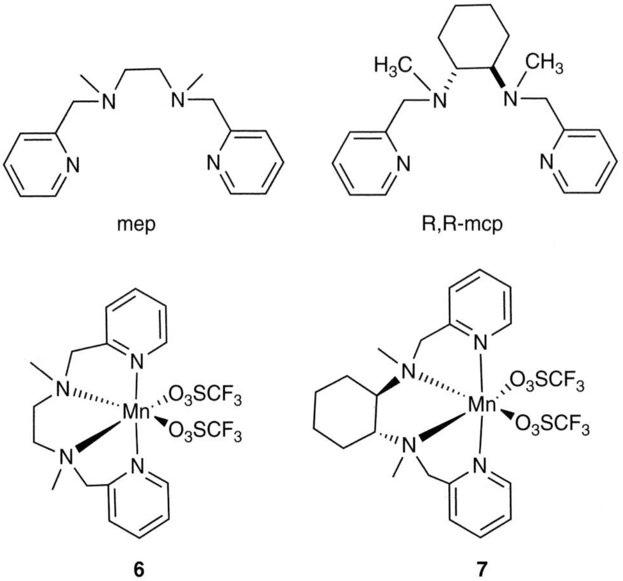

Figure 14.3 The mep and mcp ligands and the corresponding Mn complexes **6** and **7** used by Stack et al. Source: Murphy et al. 2003 [11]. Reproduced with permission of ACS.

electron-deficient olefins using peracetic acid (CH_3CO_3H, PAA) as the oxidant with high turnover numbers (TONs). Complex **6** rapidly and efficiently epoxidizes terminal olefins with 1.2 equiv. of peracetic acid (PAA) in MeCN at ambient temperatures and at 5 mol% catalyst loading. The kinetically and thermodynamically stable complex **7** is even more efficient: only 0.1 mol% catalyst is required to epoxidize vinyl cyclohexane at a rate greater than 250 turnovers per minute. Complex **7** can rapidly epoxidize both electron-rich and electron-poor olefins. The latter substrates require a slight increase in either the catalyst loading or the amount of PAA to achieve high conversion and yields, which indicates the involvement of an electrophilic oxidant. The corresponding zinc complex is unable to catalyze the epoxidations, which suggests the involvement of high-valent Mn species.

The influence of the ligand scaffold in the manganese complex and the pH of the reaction medium on the epoxidation of terminal olefins using PAA has been scrutinized [12]. For a number of Mn complexes containing polyamine ligands, a significant increase in epoxidation reactivity was noted under less acidic conditions, whereas strongly acidic conditions resulted in reduced catalytic activities because of catalyst decomposition. In general, highly preorganized, neutral tetradentate aminopyridine ligands such as mep and mcp allow the octahedral manganese center to accommodate two labile ligands in a cis manner in thermodynamically and kinetically stable complexes, which are more resistant to metal decomplexation. Such complexes generally provide potent catalysts for terminal olefin oxidation. In contrast, octahedral manganese complexes in which tetradentate ligands bind within the equatorial plane, creating two trans labile sites, show lower reactivity toward terminal olefins.

Recently, Stack and coworkers have reported a simplified and highly effective Mn-based system for olefin epoxidation [13]. This system can produce sizable quantities of a broad range of epoxides within minutes. The system consists of a mixture of $Mn(CF_3SO_3)_2$, picolinic acid (2-$PyCO_2H$), and base-modified commercial peracetic acid (PAA_M). Under optimized conditions (0.4 mol% $Mn(CF_3SO_3)_2$, 2 mol% 2-$PyCO_2H$, and slow addition of 1.1 equiv. PAA_M over five minutes at 0 °C), the system is able to transform olefins of varied geometric and electronic nature to their epoxides with good selectivity and high oxidant efficiency (Scheme 14.2). Aliphatic olefins are converted to their epoxides with retention of the geometry at the olefin center. The 1 : 5 ratio of Mn(II) salt to 2-$PyCO_2H$ has been rationalized to maximize the formation of bis-ligated Mn species in the solution. The role of "picolinic acid-like" ligands, i.e. an aromatic nitrogen donor and a carboxylate donor, has been scrutinized, and the study established that the presence of 2-$PyCO_2H$ binding motifs are required for catalysis. The anionic picolinate ligand forms a five-membered chelate that is assumed to stabilize high-valent metal-oxo complexes [14]. These studies also scrutinize the integrity of aminopyridine ligands in epoxidation reactions using PAA and complement the report of Browne and coworkers, which demonstrates the dihydroxylation of electron-deficient olefins using a similar catalyst system [15].

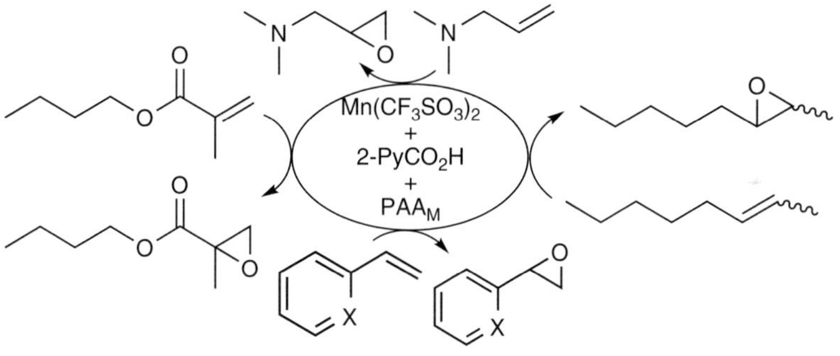

Scheme 14.2 Epoxidation of olefins with the Mn/2-$PyCO_2H$ catalyst system.

Following these initial studies, several manganese complexes containing modified electron-rich aminopyridine N_4 ligands were investigated for the catalytic epoxidation of olefins [16]. Costas and coworkers reported on $[Mn(^{H,Me}PyTACN)(CF_3SO_3)_2]$ (**8**) (PyTACN = 1-(2-pyridylmethyl)-4,7-dimethyl -1,4,7-triazacyclononane; Figure 14.4.) as a robust, efficient, and selective catalyst for the epoxidation of a broad range of olefins using PAA as the oxidant [17]. However, the use of PAA does bear some important general drawbacks: it has a poorer atom economy and is more expensive than H_2O_2, and commercial solutions of PAA are very acidic, which limits its use in the epoxidation of acid-sensitive substrates and has implications on catalyst stability (see above). Later, the authors extended this chemistry to the use of H_2O_2 in combination with acetic acid because PAA is assumed to form *in situ* under these conditions [18]. Initial experiments to epoxidize 1-octene

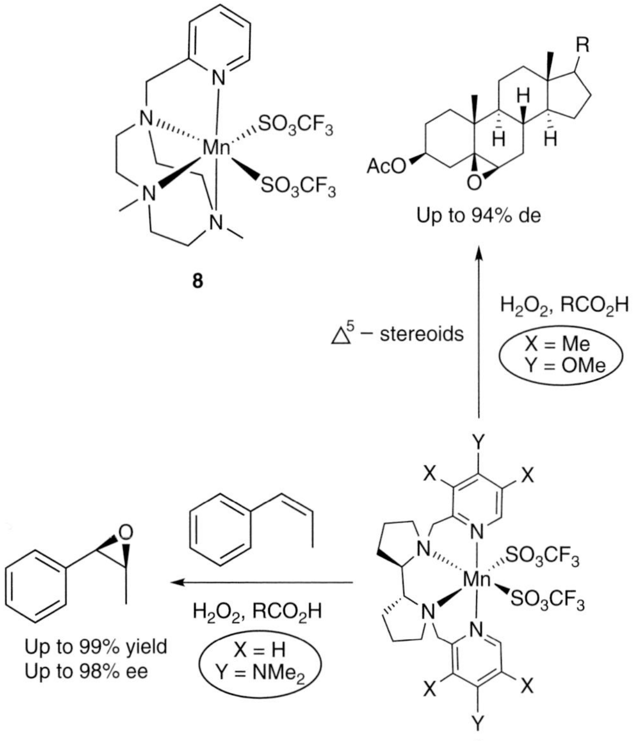

Figure 14.4 Mn catalysts used for olefin epoxidation by Cusso et al. 2013 [17]. Reproduced with permission of ACS.

using H_2O_2 (1.2 equiv.) in the presence of 0.1 mol% catalyst **8** resulted in <1% epoxide formation because of the immediate disproportionation of the peroxide. Upon addition of increasing amounts of acetic acid, the yield of the epoxide product increased, and 90% yield was achieved using 14 equiv. of acetic acid. Later, the same group described how the electronic properties of a series of manganese catalysts of general formula [Mn((*S*,*S*)-^{X}PDP)(CF_3SO_3)$_2$] (**9**) (^{H}PDP = 2-[[2-(1-(pyrid-2-ylmethyl)pyrrolidin-2-yl)pyrrolidin-1-yl]methyl]pyridine) exert a profound and systematic impact on epoxidation activity and product stereoselectivity [19]. Upon changing the substituent at the para position of the pyridine rings from –H to –NMe_2, the yield (38–75%) and enantioselectivity (43–82%) increased in the epoxidation of *cis*-β-methylstyrene (Figure 14.4). Variation of the substituents on the pyridine rings of the ligand turned out to be a key factor affecting yield and enantioselectivity. β-Selectivity in the epoxidation of Δ^5-unsaturated steroids with moderate-to-good yields was achieved using more electron-rich ligands (X = Me, Y = OMe) in the presence of 2 equiv. of H_2O_2 and 15 equiv. of pivalic acid (Figure 14.4).

Asymmetric olefin epoxidation using Mn catalyst [Mn((*S*,*S*)-PDP)(CF_3SO_3)$_2$] in the presence of different carboxylic acid additives was investigated by Bryliakov, Talsi, and coworkers [20]. The enantioselectivities increase with increasing steric bulk of the carboxylic acid, and hence, they proposed that the carboxylic acid plays a dual role in catalysis: (i) it assists the activation of hydrogen peroxide, and (ii) it controls enantioselectivity by the incorporation of the carboxylic acid into the active species of the enantioselectivity-determining step. Later, the effect of steric and electronic properties of the aminopyridine moiety in the (*S*,*S*)-PDP ligand was also reported [21]. The incorporation of electron-donating groups on the pyridine moiety leads to increased enantioselectivities in epoxidations of electron-deficient olefins, as was also observed by Costas and coworkers [19].

In 2009, Sun and coworkers reported on the epoxidations of unfunctionalized olefins and α,β-enones with H_2O_2 (6 equiv.) using a series of C_2-symmetric mcp-based manganese complexes $[LMn^{II}(CF_3SO_3)_2]$ (L = (*R*,*R*)-mcp; (*R*,*R*)-pmcp; (*R*,*R*)-nmcp; (*R*,*R*)-bpmcp) (Figure 14.5) in which aromatic groups were introduced at the 2-pyridylmethyl positions of the (*R*,*R*)-mcp ligand. Using 1 mol% of catalyst and 5 equiv. acetic acid, they obtained epoxides with enantioselectivities up to 89% ee for electron-deficient olefins, in particular for substituted chalcones [16b]. Upon the introduction of strongly donating dimethylamino groups as in complex **10** (Figure 14.5), activities in the asymmetric epoxidation of styrene derivatives with H_2O_2 as the oxidant improved, even when only a catalytic amount of carboxylic acid was used as an additive [22]. Changing the carboxylic acid was found to modulate the stereoselectivity in these reactions, with dimethylbutanoic acid (DMBA) giving the best results.

The effect of Brønsted acids in epoxidation reactions catalyzed by Mn-mcp catalysts was also studied by the Sun group [23]. Using the complex

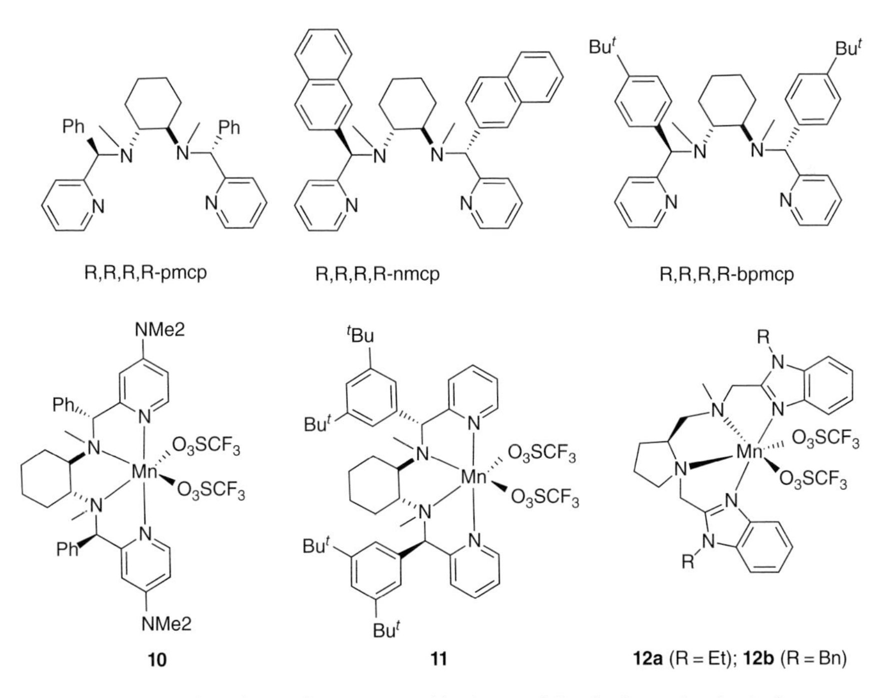

Figure 14.5 Ligands and complexes reported by Sun et al. for the (enantioselective) epoxidation of olefins. Source: Miao et al. 2016 [22]. Reproduced with permission of ACS.

$[Mn^{II}(dbp\text{-}mcp)(CF_3SO_3)_2]$ (**11**) (dbp-mcp = (1*R*,2*R*)-*N*,*N*′-dimethyl-*N*,*N*′-bis((R)-(3,5-di-*tert*-butyl-phenyl)-2-pyridinylmethyl)cyclohexane-1,2-diamine), the yields of epoxide products as well as the chemo- and enantioselectivities were found to increase in the presence of H_2SO_4. In the presence of a small amount of H_2SO_4 (3 mol%), the epoxidation of cyclooctene proceeded in 91% yield, whereas the formation of only small amounts of epoxides was noted in the presence of other Brønsted acids such as H_3PO_4, $HClO_4$, CF_3SO_3H, HCl, and acetic acid. No epoxidation occurred when Na_2SO_4 was used instead of H_2SO_4, whereas 73% epoxide was obtained in the presence of 3 mol% $NaHSO_4$. It was therefore proposed that H_2SO_4 plays a crucial role in the generation of the effective oxidant in the reaction of **11** and H_2O_2. The epoxidation of chalcone derivatives by **11** (0.20 mol%) and H_2O_2 (1.5 equiv.) in the presence of H_2SO_4 (1 mol%) affords high yields (up to 93%) of the epoxide products with >97% enantioselectivities.

The same group has also designed C_1-symmetric N_4 ligands featuring a more rigid chiral diamine derived from proline and two benzimidazole donor moieties [24]. The manganese triflate complexes **12a,b** are able to catalyze the asymmetric epoxidation of olefins with high enantioselectivities (up to 94% ee) along with high yields (60% to >90%) at only 0.01–0.2 mol% catalyst loading. This system is applicable for the synthesis of epoxides on a gram scale without any loss in enantioselectivity.

The stoichiometric cis-dihydroxylation of olefins using permanganate (MnO_4^-) has been known for a long time. However, considerable progress has been made in the development of atom-efficient and environmentally friendly catalytic methods for these reactions using Mn complexes in combination with H_2O_2. In 2005, Feringa and coworkers reported that the dimeric complex $[Mn_2O_3(tmtacn)_2]^{2+}$ (**13**) (tmtacn = 1,4,7-trimethyl-1,4,7-triazacyclononane) catalyzes the cis-dihydroxylation and epoxidation of olefins employing H_2O_2 as the oxidant [25]. The authors proposed that the catalytically active bis-μ-carboxylato Mn^{III}_2 complex $[Mn^{III}_2(\mu\text{-}O)(\mu\text{-}RCO_2)_2(tmtacn)_2]^{2+}$ (**14**) is formed *in situ* by the reaction of **13** with H_2O_2 in the presence of a carboxylic acid (Scheme 14.3). Cyclooctene can be converted to the cis-diol with over 2000 turnovers using 0.1 mol% of **13**. The catalytic activity increases when an electron-deficient carboxylic acid such as CCl_3CO_2H or 2,6-Cl_2PhCO_2H is used and the selectivity toward cis-dihydroxylation is highest in the presence of sterically hindered carboxylato ligands. Complex **13** in combination with

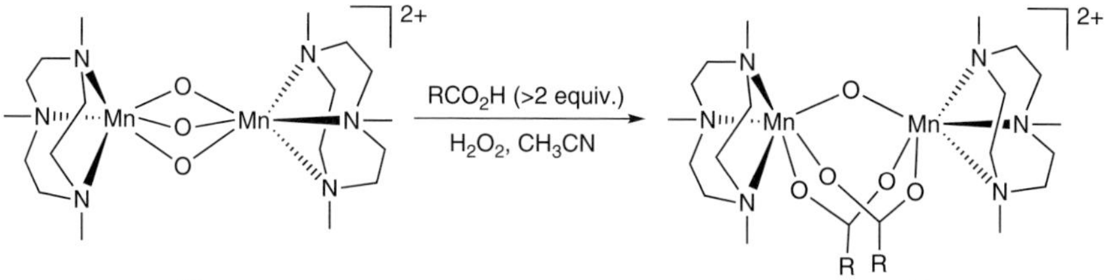

Scheme 14.3 Dinuclear Mn complexes explored by Feringa for olefin cis-dihydroxylation.

2,6-dichlorobenzoic acid is the most active, and the ratio of cis-diol/epoxide is 7 for the oxidation of cyclooctene.

In 2008, the same group reported the first manganese-based catalyst system for the asymmetric cis-dihydroxylation of olefins with H_2O_2. They used chiral carboxylato ligands in combination with **13** to achieve large turnovers in enantioselective olefin cis-dihydroxylation [26]. The prochiral cis-alkene 2,2-dimethylchromene was selected as a substrate to identify chiral carboxylic acids suitable for inducing enantioselectivity (Scheme 14.4, top). A series of 24 chiral carboxylic acids containing a stereogenic center α to the carboxylic acid was screened in this study. Some selected carboxylic acids used in this study are shown in Scheme 14.4. The cis-diol product was the major product in all cases, and the cis-/trans-diol ratio typically varied between 2.7 and 4.3. The highest cis-/trans-diol ratio of 7.0 (conversion 53%) was observed for (1*S*)-(+)-ketopinic acid, and this indicates that steric bulk close to the carboxylic acid functionality favors cis-dihydroxylation over the epoxidation. Chiral carboxylic acids containing a phenyl or naphthyl group attached directly to the stereogenic carbon atom showed similar ee's of 13–20% with substrate conversions above 60%, whereas *N*-carbamate-protected amino acids showed ee's of c. 30%. Carboxylic acids lacking an aromatic group at the stereogenic center provided good substrate conversion with low ee for the cis-diol product. The use of commercial chiral carboxylic acids holds a distinct advantage over the synthetically challenging development of chiral versions of the tmtacn ligand. The highest selectivity and activity in these cis-dihydroxylation reactions were obtained with electron-rich cis-olefins.

Later on, Browne and coworkers introduced a manganese-catalyzed cis-dihydroxylation system for electron-deficient olefins [27]. This system is based on the manganese salt $Mn(ClO_4)_2{\cdot}6H_2O$ in combination with 2-$PyCO_2H$, a ketone, and a base (NaOAc) and uses H_2O_2 as the oxidant. Generally, a 1 : 6 : 10 mixture of a Mn^{II} salt, picolinic acid, and NaOAc showed good substrate conversion and yield of the cis-diol products. Substituted pyridinecarboxylic acids, e.g. pyridine-2,5-dicarboxylic acid, showed full activity, whereas picolinic acid *N*-oxide or pyridine 2,6-dicarboxylic acid did not show any catalytic activity, indicating that the cocatalyst is acting as a ligand to manganese. This system is relatively insensitive in terms of activity and selectivity to the nature of the base used during catalysis. NaOAc, KOAc, $NaHCO_3$, Na_2CO_3, or NaOH provided comparable results, and the number of equivalents of base employed did not affect the reaction significantly, provided that it was in excess with respect to the 2-$PyCO_2H$ and Mn^{II}. This system can convert electron-deficient olefins such as diethyl fumarate and *N*-alkyl-maleimides to the corresponding cis-dihydroxylation products in quantitative yields and at >1000 turnovers per manganese (Scheme 14.4 bottom).

14.2.2 Iron Catalysts

Iron salts and complexes have been known for long for their use as catalysts in the epoxidation of olefins with hydrogen peroxide [28–31]. Que and coworkers reported the first example of a synthetic iron complex capable of catalyzing

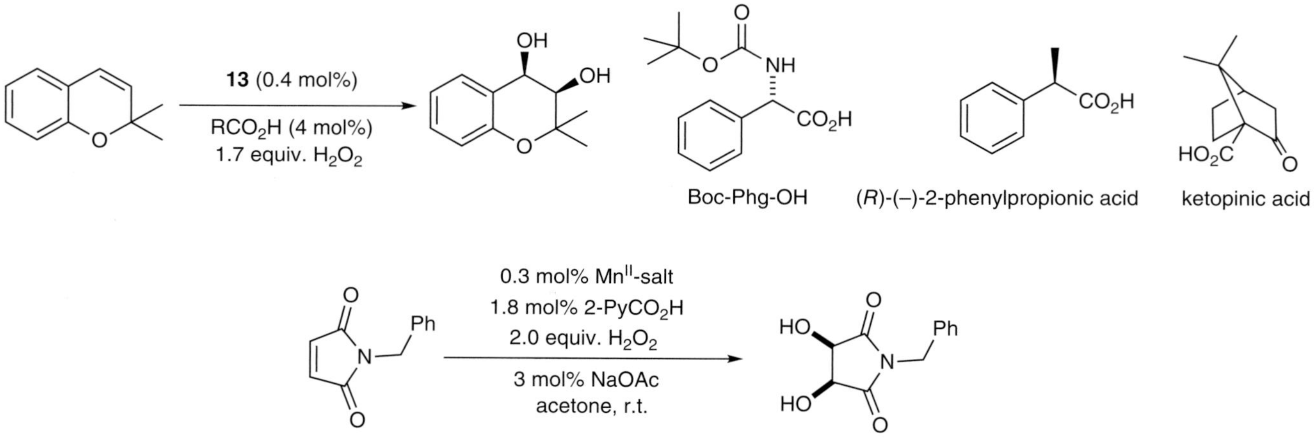

13 (0.4 mol%)

RCO_2H (4 mol%)

1.7 equiv. H_2O_2

Boc-Phg-OH

(*R*)-(−)-2-phenylpropionic acid

ketopinic acid

0.3 mol% Mn^{II}-salt

1.8 mol% 2-$PyCO_2H$

2.0 equiv. H_2O_2

3 mol% NaOAc

acetone, r.t.

Scheme 14.4 (Enantioselective) cis-hydroxylations reported by Feringa and Browne.

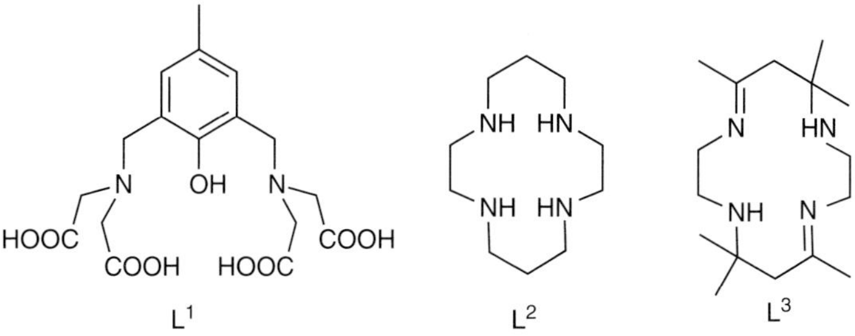

Figure 14.6 Ligand used by Que and Valentine to generate Fe-based epoxidation catalysts.

olefin epoxidations with H_2O_2 in 1986 [32]. This iron complex, $(Me_4N)[Fe_2(L^1)(OAc)_2]$ (L^1 = *N*,*N*′-(2-hydroxy-5-methyl-1,3-xylylene)bis(*N*-carboxymethylglycin; Figure 14.6), was reported to catalyze the epoxidation of several olefins, including cyclohexene, styrene, and stilbene. The epoxidation of cis-stilbene by this complex resulted in 95% trans- and 5% cis-epoxide. In 1991, Valentine and coworkers reported on several iron(II) cyclam complexes; some of them, $[LFe(CF_3SO_3)_2]$, (L = L^2 and L^3), are capable of catalyzing the epoxidation of cyclohexene with up to 20 turnovers (Figure 14.6) [33].

Beller and coworkers developed a practical iron-catalyzed alkene epoxidation procedure applicable at room temperature and under aerobic conditions. They reported that a combination of commercially available $FeCl_3 \cdot 6H_2O$, pyridine-2,6-dicarboxylic acid, and an organic nitrogen-containing base (such as pyrrolidine, benzylamines, or methylimidazole derivatives) leads to the *in situ* formation of an iron catalyst that activates H_2O_2 and catalyzes the epoxidation of a variety of substrates, e.g. 1,2-disubstituted aromatic olefins and 1,3-dienes, in a chemo- and stereoselective manner (Scheme 14.5) [34]. By replacing the organic base with a chiral 1,2-diphenyl-ethylene-1,2-diamine (DPEN) derivative, the asymmetric version of this reaction was achieved, which allows for the epoxidation of trans-stilbene derivatives with 40–94% yield and 10–97% ee (Scheme 14.5) [35].

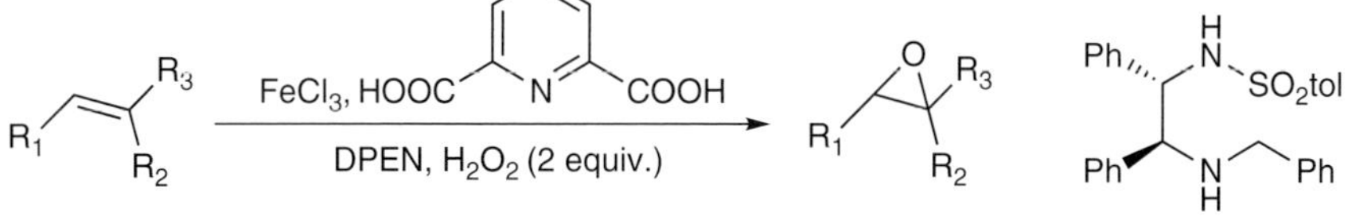

Scheme 14.5 Fe-based epoxidation catalyst developed by Beller (left) and chiral DPEN derivative used as the ligand (right).

The most successful family of iron catalysts recently developed for asymmetric epoxidation is based on tetradentate ligands with a bis-amine-bis-pyridine (or related heterocycle) structure. Upon binding to the metal, these ligands form three five-membered chelate rings, which provide high stability to the complexes. The cis-α topology of this type of complex has been proven to be highly suitable for asymmetric epoxidation catalysts [36, 37]. The development of these catalysts parallels the development of the corresponding Mn-based catalysts.

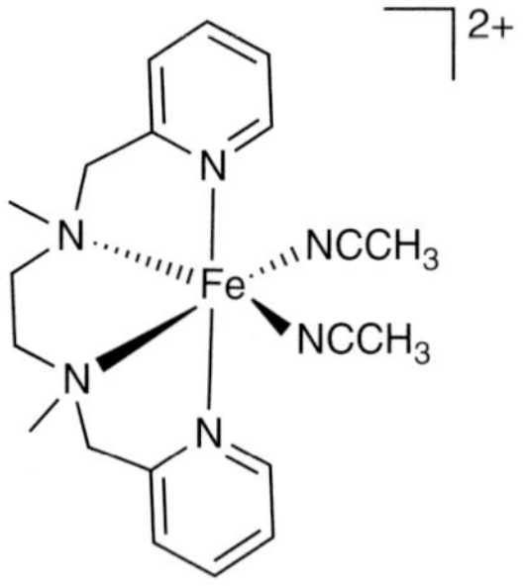

15

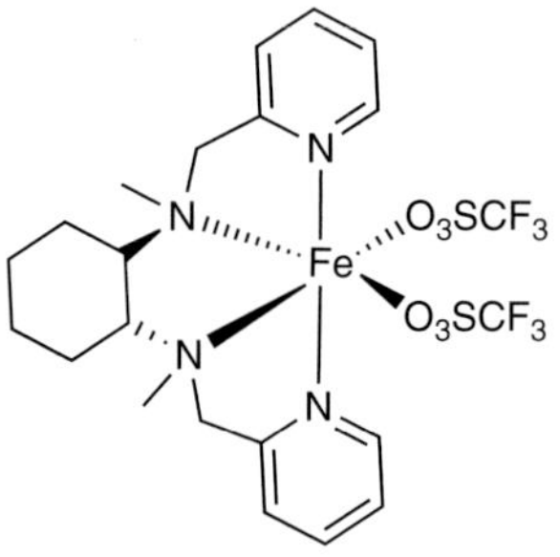

16

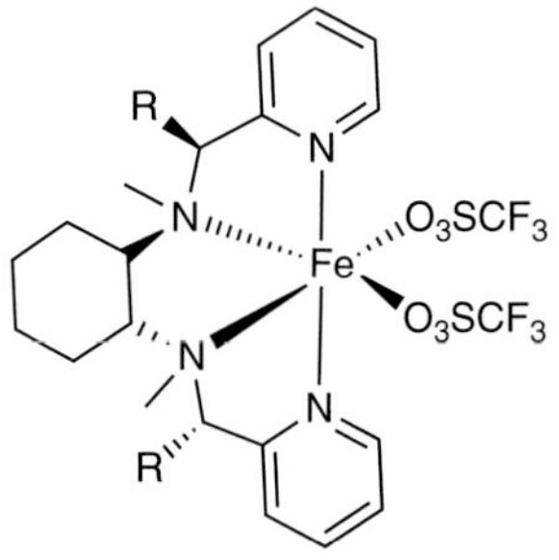

17a (R = Ph)
17b (R = 4-^{t}Bu-C_6H_4)

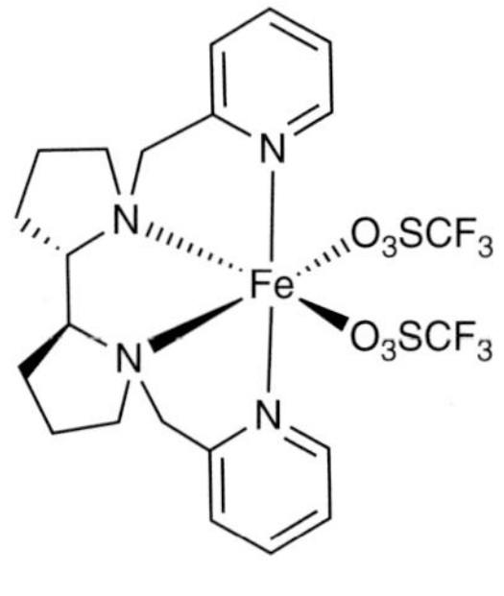

18

Figure 14.7 Mep- and mcp-based iron epoxidation catalysts reported by Jacobsen (**15**), Que (**16**), Sun (**17**), and Talsi (**18**).

Jacobsen and coworkers reported on aliphatic olefin epoxidations on a preparative scale using the $[Fe(mep)(CH_3CN)_2](ClO_4)_2$ catalyst **15** in combination with H_2O_2 (Figure 14.7) [38]. This iron catalyst (5 mol%) allows for the epoxidation of 1-decene using various amounts of H_2O_2; 4 equiv. of oxidant was found to be sufficient for the complete conversion of olefin. However, only 40% of epoxide product formation was reported along with a variety of over-oxidized compounds as by-products. Upon changing the counter anion from ClO_4^- to SbF_6^-, the catalyst became more efficient; in this case, 1.5 equiv. of H_2O_2 was required for complete conversion of the olefin, and the selectivity toward the formation of 1,2-epoxydecane improved to 71%. Addition of 1 equiv. of acetic acid with respect to the catalyst improved the selectivity of epoxide formation to 82%. A wide range of aliphatically substituted alkenes proved to be excellent substrates for epoxidation with catalyst **15**. Terminal olefins, which normally show a low reactivity to electrophilic oxidants and typically require long reaction times with any of the known epoxidation methods, were found to undergo epoxidation within five minutes and in 60–90% isolated yields.

Initial studies by Que and coworkers using (*R,R′*)-$[Fe(mcp)(CF_3SO_3)_2]$ (**16**) in asymmetric epoxidations with H_2O_2 showed a modest 12% ee in the epoxidation of *trans*-2-heptene (Figure 14.7) [39]. Sun and coworkers reported the modification of the mcp ligand through the introduction of aryl groups (Ph and 4-^{t}Bu-C_6H_4) at the 2-pyridinylmethyl positions in the ligand backbone. The modified iron-mcp complexes **17a,b** showed an improved reactivity in the epoxidation of trans-chalcones [40]. Compared to the reactivity of **16** in the epoxidation of trans-chalcone using hydrogen peroxide and acetic acid, catalyst **17** showed an increase in both yield and selectivity (54% yield and 77% ee for **17b** vs 47% yield and 54% ee for **16**). However, the reported substrate scope of these catalysts is limited to aromatic *trans*-α,β-enones.

Bryliakov and Talsi reported the reactivity of [Fe(PDP)(CF_3SO_3)$_2$] complex **18**, derived from the bis-pyrrolidine-based PDP ligand, in the enantioselective epoxidation of a diverse set of olefins with aqueous H_2O_2 [20]. These studies showed that the replacement of cyclohexyldiamine by a bis-pyrrolidine backbone led to a catalyst with improved enantioselectivities in the epoxidation of chalcones. For example, the epoxidation of trans-chalcone improves from 54% ee with 2 mol% of **16** to 71% ee when 1 mol% of **18** is used.

A somewhat less rigid proline-derived, mono-pyrrolidine ligand backbone was later introduced by Sun and coworkers (see **12a** in Figure 14.5) [41]. Complexes derived from this ligand, such as **19**, differ from those derived from cyclohexanediamine- and bipyrrolidine-based ligands in that they lack C_2-symmetry (Figure 14.8). Yields (up to 68%) and enantioselectivities (up to 56% ee) for bis-pyridyl complex **19** remain moderate for trans-chalcone. Interestingly, replacement of pyridines by benzimidazoles in catalyst **20** provided excellent yields and enantioselectivities: trans-chalcone and tetralone derivatives were epoxidized in up to 99% yield with 97% ee, and in 99% yield with 98% ee, respectively. Later, the same authors also developed complex **21**, the benzimidazole analog of Fe-mcp complex **16**, which proved to be an efficient catalyst for H_2O_2-mediated enantioselective epoxidation of chalcone derivatives furnishing high yields (up to 93%) and ee's up to 91% [42].

Costas and coworkers investigated the electronic effects on the catalytic epoxidation activities of the iron-PDP family of complexes with the general formula [Fe((*S*,*S*)-^{X}PDP)(CF_3SO_3)$_2$]. Using *cis*-β-methylstyrene as a model substrate, they concluded that the enantioselectivity of the epoxidation reactions improves with an increase of the electron-donating properties of the PDP ligand [43]. In this

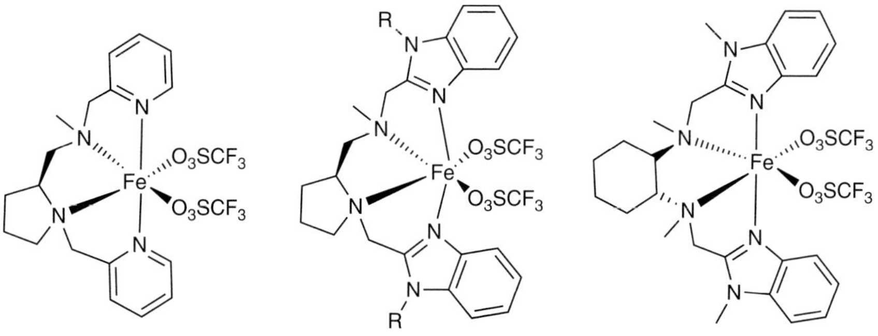

Figure 14.8 Fe-based epoxidation catalysts developed by Sun. Source: Miao et al. 2016 [22]. Reproduced with permission of ACS.

case, chiral carboxylic acids were also introduced to improve the enantioselectivity. Using chiral amino acids, they were able to extend the substrate scope of these systems to more challenging terminal olefins, obtaining the desired epoxides with high enantioselectivity [44]. Very recently, they have also reported on iron complex **22** with a C_1-symmetric tetradentate PDP-based ligand bearing a bulky TIPS substituent (TIPS = trisisopropylsilyl), which catalyzes the asymmetric epoxidation of cyclic enones and cyclohexene ketones with aqueous hydrogen peroxide (Scheme 14.6). Good-to-excellent yields and enantioselectivities (up to 99% yield, and 95% ee) were obtained under mild conditions and in short reaction times. The electrophilic oxidant that is generated under these conditions is able to distinguish between different olefinic moieties with distinct electronic properties, and epoxidation occurs at the more electron-rich site in excellent yields and stereoselectivities (Scheme 14.6 right) [45].

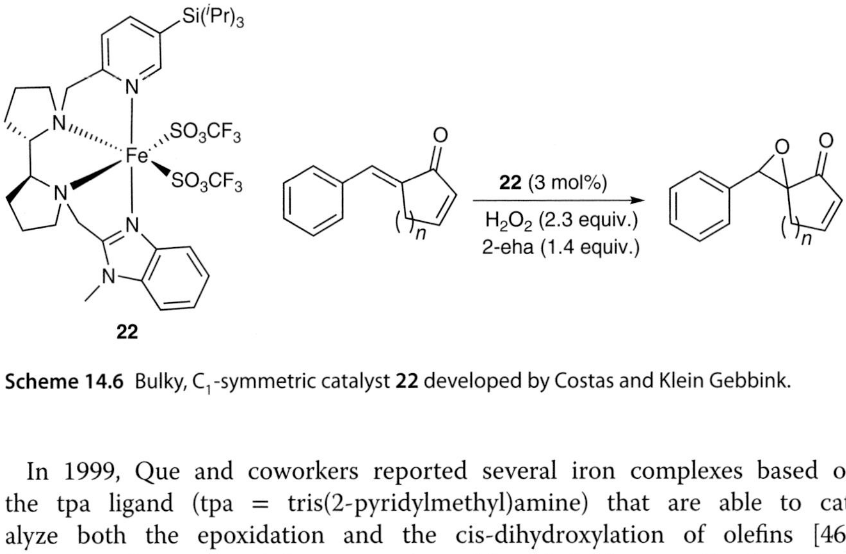

$Si(^iPr)_3$ N N Fe SO_3CF_3 SO_3CF_3 N N N **22**

O ( )n **22** (3 mol%) H_2O_2 (2.3 equiv.) 2-eha (1.4 equiv.) O O ( )n

Scheme 14.6 Bulky, C_1-symmetric catalyst **22** developed by Costas and Klein Gebbink.

In 1999, Que and coworkers reported several iron complexes based on the tpa ligand (tpa = tris(2-pyridylmethyl)amine) that are able to catalyze both the epoxidation and the cis-dihydroxylation of olefins [46]. Iron complexes containing two cis-positioned labile coordination sites, such as $[Fe(tpa)(CH_3CN)_2](ClO_4)_2$ and $[Fe(6\text{-}Me_3\text{-}tpa)(CH_3CN)_2](ClO_4)_2$ (6-Me_3-tpa = tris-(6-methyl-2-pyridylmethyl)amine), yield a mixture of cis-diol and epoxide products with preferential formation of the diol. Subsequently, the oxidation of several olefins with hydrogen peroxide in the presence of a series of iron complexes was systematically examined, wherein the effect of the ligand on the diol vs epoxide product ratio was monitored (Figure 14.9) [47]. For example, Fe(mep) complex **15** converts cyclooctene to cyclooctene oxide and the cis-diol in respective yields of 75% and 9% relative to H_2O_2, whereas methyl-substituted complex $[Fe(6\text{-}Me_2\text{-}mep)(CF_3SO_3)_2]$ (**23**) under the same conditions affords the epoxide and cis-diol in respective yields of 15% and 64%. Chiral mcp-complex **16** catalyzes the oxidation of *trans*-2-heptene at modest product ee's; 29% ee for the diol and 12% for the epoxide. More promising were the results for the same reaction catalyzed by the corresponding, methyl-substituted complex *S,S*-$[Fe(6\text{-}Me_2\text{-}mcp)(CF_3SO_3)_2]$ (**24**), which affords the cis-diol with an ee of 79%,

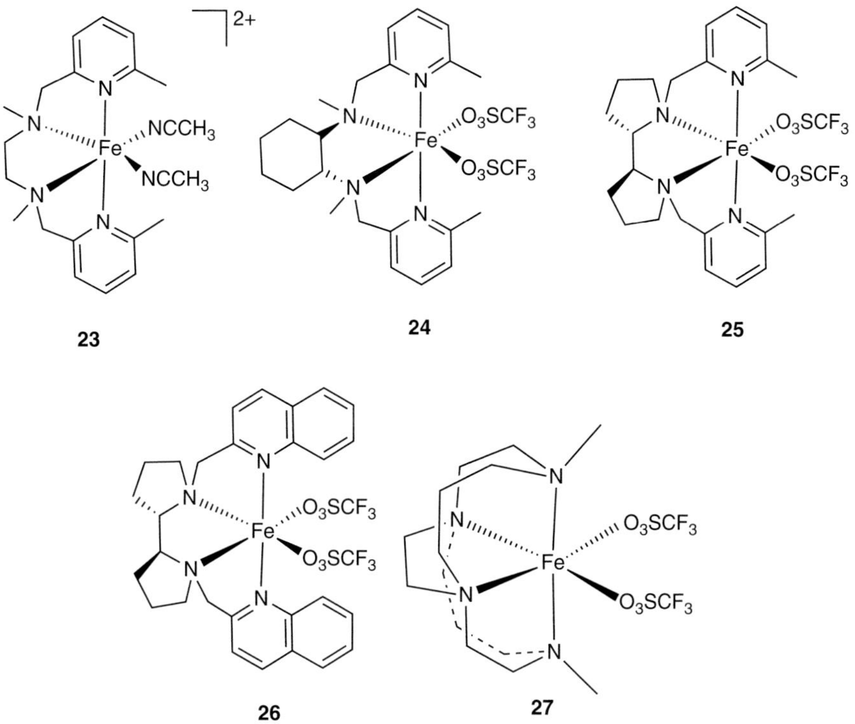

23 24 25

26 27

Figure 14.9 Fe complexes studied by Que et al. in olefin oxidations. Reproduced with permission of ACS.

along with some racemic epoxide. The use of the 1*R*,2*R* enantiomer affords the same ee values but with the opposite configuration for the major diol product. The major cis-diol products obtained in the oxidation of *trans*-2-heptene by *R*,*R*-[Fe(mcp)$(CF_3SO_3)_2$] and *R*,*R*-[Fe(6-Me_2-mcp)$(CF_3SO_3)_2$] have the same chirality. Therefore, the configuration of the 1,2-cyclohexanediamine moiety determines the chirality of the product.

In 2008, Que and coworkers reported on iron(II) complexes **25** and **26** derived from PDP-based ligands, bearing 6-methyl-pyridine (6-Me_2-pdp) or 2-quinoline (bqbp) donor moieties (Figure 14.9) [48]. These complexes showed excellent enantioselectivity for cis-dihydroxylation of olefins with H_2O_2 as an oxidant. The asymmetric induction results obtained with **25** are the best along the series of ligands shown in Figure 14.9; in particular, these are significantly greater than those obtained with **24**, e.g. for the cis-dihydroxylation of *trans*-2-heptene, 97% of cis-diol was noted for **25** against 79% for **24**. In 2011, the same group also investigated the catalytic activity of the iron(II) complex [Fe(Me_2EBC)$(CF_3SO_3)_2$] (**27**; Me_2EBC = 4,11-dimethyl-1,4,8,11-tetraazabicyclo[6.6.2]hexadecane) for the cis-hydroxylation of olefins with H_2O_2 (Figure 14.9) [49]. With this catalyst, the oxidation of more electron-rich olefins is favored, with the order of oxidation preference found to be cis-cyclooctene > 1-octene > *t*-butyl acrylate > dimethyl fumarate.

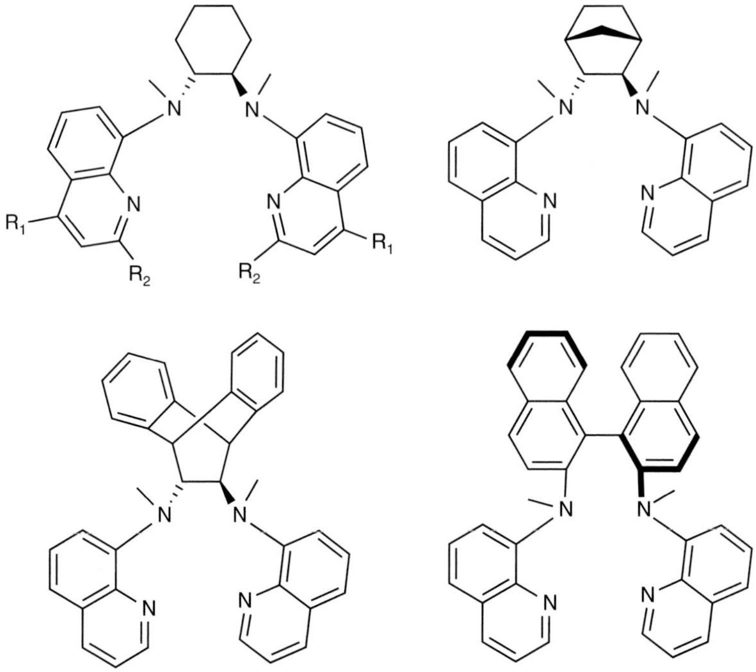

Figure 14.10 Ligands studied by Che et al. for Fe-catalyzed olefin dihydroxylation.

More recently, highly enantioselective iron-catalyzed cis-dihydroxylations of olefins with hydrogen peroxide were reported by Che and coworkers [50]. Iron complexes containing tetradentate N_4 ligands featuring different chiral diamine backbones and quinoline donor moieties were used as catalysts in this study (Figure 14.10). A series of different electron-rich/poor olefins were tested with these catalysts, showing that these catalysts yield cis-diols up to 85% isolated yield and up to 99% ee. The N_4 ligand containing a binaphthyl backbone turned out to be a promising chiral ligand scaffold, and its iron complex catalyzes the cis-hydroxylation of (*Z*)-alkenes such as coumarin, indene, and (*Z*)-β-methylstyrene giving the cis-diols with higher ee values (83%, 67%, and 56%, respectively) than those obtained with catalyst **24**. The gram-scale formation of diols has been achieved with methyl (*E*)-cinnamates and dimethyl fumarate as substrates.

Next to the prototypical tetradentate aminopyridine-type ligands, a number of other ligand topologies have also been explored for iron-mediated olefin oxidation. In 2005, Que and coworkers reported the bis-ligated iron(II) complex $[(ph\text{-}dpah)_2Fe](CF_3SO_3)_2$ (ph-dpah = (di-(2-pyridyl)methyl)benzamide), containing a bio-inspired facial N,N,O-donor ligand, which catalyzes the cis-dihydroxylation of olefins with a high diol selectivity (Figure 14.11) [51]. For both cis- and *trans*-2-heptene, 99% of diol products were obtained with the retention of configuration. Later, several other facial N,N,O-donor ligands have also been investigated for the catalytic cis-hydroxyaltion of olefins using iron and hydrogen peroxide [52]. Klein Gebbink and

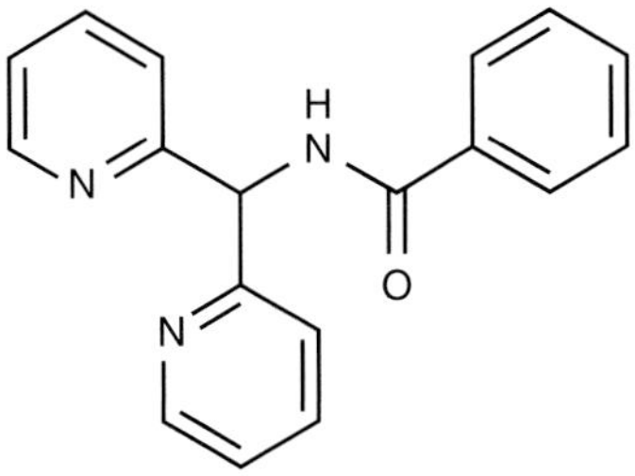

ph-dpah

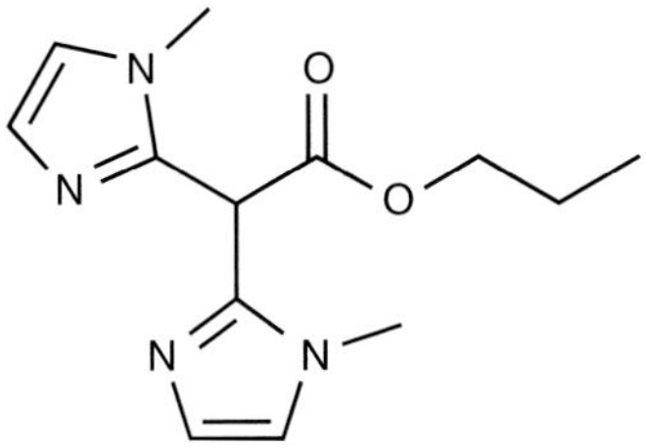

$BMIP^{Pr}$

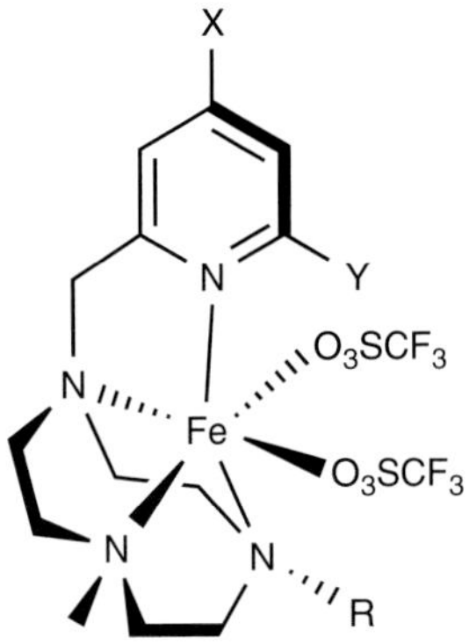

R	Y	X
CH_3	H	H
CH_3	H	CH_3
CH_3	H	Cl
CH_3	H	NO_2
CH_3	H	$N(CH_3)_2$
CH_3	F	H
CH_3	CH_3	H
CH_3	Cl	H
CH_3	CH_3	CH_3
i-Pr	H	H
i-Pr	CH_3	H

28

Figure 14.11 Alternative ligand topologies used for Fe-catalyzed olefin oxidation.

coworkers reported the synthesis of iron complexes derived from propyl 3,3-bis(1-methylimidazol-2-yl)propionate ($BMIP^{Pr}$) and related ligands (Figure 14.11). The triflate complex $[Fe(BMIP^{Pr})_2](CF_3SO_3)_2$ was found to be an active catalyst for the epoxidation and cis-dihydroxylation of olefinic substrates with the conversion of H_2O_2 ranging from 39% to 51% [53]. For complexes derived from the ph-dpah and $BMIP^{Pr}$ ligands, it is assumed that partial ligand dissociation has to occur to open up the coordination sphere around iron and enable catalysis.

In a 2008 study, Costas and coworkers tested iron complexes based on the pyridyl-functionalized triazacyclononane (PyTACN) backbone in the oxidation of cyclooctene and cis-2-heptene with H_2O_2 (Figure 14.11) [54]. These complexes convert olefins to the corresponding epoxide and cis-diol with high turnover numbers ranging from 4.1 to 141 for the diol product and 0.5 to 29 for the epoxide. The yields of the epoxide and cis-diol are highly dependent on the structure of the complexes employed as the catalyst. In the presence of excess cyclooctene, all the complexes convert the peroxide into epoxide and diol products, with good-to-excellent yields (50–81%). Later, Beller and Costas reported on modified PyTACN ligands and tested the corresponding iron complexes $[Fe(^{R,Y,X}PyTACN)(CF_3SO_3)_2]$ (**28**) for the cis-dihydroxylation of olefins with hydrogen peroxide [55]. Competition experiments showed that

28 exhibits preferential selectivity toward the oxidation of cis-olefins over the trans analogs and also affords better yields and high cis-diol/epoxide ratios when cis-olefins are oxidized. The catalytic activity of this series of complexes toward epoxidation vs cis-dihydroxylation is governed to a minor extent by electronic effects imposed by the substituted pyridine ring.

14.2.3 Cobalt, Nickel, and Copper Catalysts

Although iron and manganese complexes have been studied extensively for the catalytic epoxidation and cis-hydroxyaltion of olefins, cobalt, nickel, and copper complexes have been less intensively studied for this purpose in recent years. A selected number of examples of olefin oxidation catalysts based on the latter metals is discussed in this section.

Cobalt(II) and cobalt(III) complexes are well-known catalysts for the epoxidation of olefins using molecular oxygen as the terminal oxidant and aldehydes, acetals, and β-ketoesters as coreductants [56–58]. Journaux and coworkers reported on square-planar cobalt(III) complexes (**29**) containing bis-*N,N′*-disubstituted oxamides and related ligands (Figure 14.12) [59]. These complexes catalyze the epoxidation of tri- and disubstituted olefins in moderate-to-high yields in the presence of pivalaldehyde and molecular oxygen. Iqbal and coworkers described the Schiff base cobalt(II) complexes **30**–**32**, which are effective catalysts for the aerobic epoxidation of alkenes including steroids and terpenoids in the presence of either aliphatic aldehydes or β-ketoesters at ambient temperature [60–64].

Kim and coworkers reported on a robust, trimeric cobalt complex (**33**) as an effective catalyst for olefin epoxidations using *meta*-chloroperoxybenzoic acid (mCPBA) (Figure 14.12) [65]. Cyclic olefins such as cyclopentene, cycloheptene, and cyclooctene as well the terminal olefin 1-hexene were oxidized to the corresponding epoxides in good yields (60–90%). The same group has described a cobalt(III) complex (**34**) derived from a tetradentate ligand containing two deprotonated amide moieties that catalyzes olefin epoxidation upon treatment with mCPBA as the terminal oxidant [66].

The first examples of olefin epoxidation by nickel(II) catalysts derived from different tetraazamacrocycles (both neutral and anionic), Schiff bases, porphyrins, and bidentate phosphines, using iodosylbenzene as the oxidant, were reported by Koola and Kochi [67]. Later, Burrows and coworkers reported on the catalytic activity of several nickel(II) salen and cyclam complexes in the epoxidation of a wide variety of olefins using iodosylbenzene or hypochlorite as the terminal oxidant [68–70]. In 1991, Yamada et al. reported the epoxidation of a variety of olefins using $Ni(dmp)_2$ (dmp = 1,3-bis(p-methoxyphenyl)-1,3-propanedionato, Figure 14.13) in the presence of molecular oxygen as the oxidant and aldehydes as the reductant [71]. They screened several aldehydes as reductant in the epoxidation of 2-methyl-2-decene as a model substrate. When a primary aldehyde was employed as the reductant, the yield and conversion were 10% and 7%, respectively, whereas the use of aldehydes bearing a secondary or tertiary carbon next to the carbonyl group as the reductant gave full conversion to the epoxide product. Next, they tested several other 1,3-diketone-type ligands in the epoxidation of

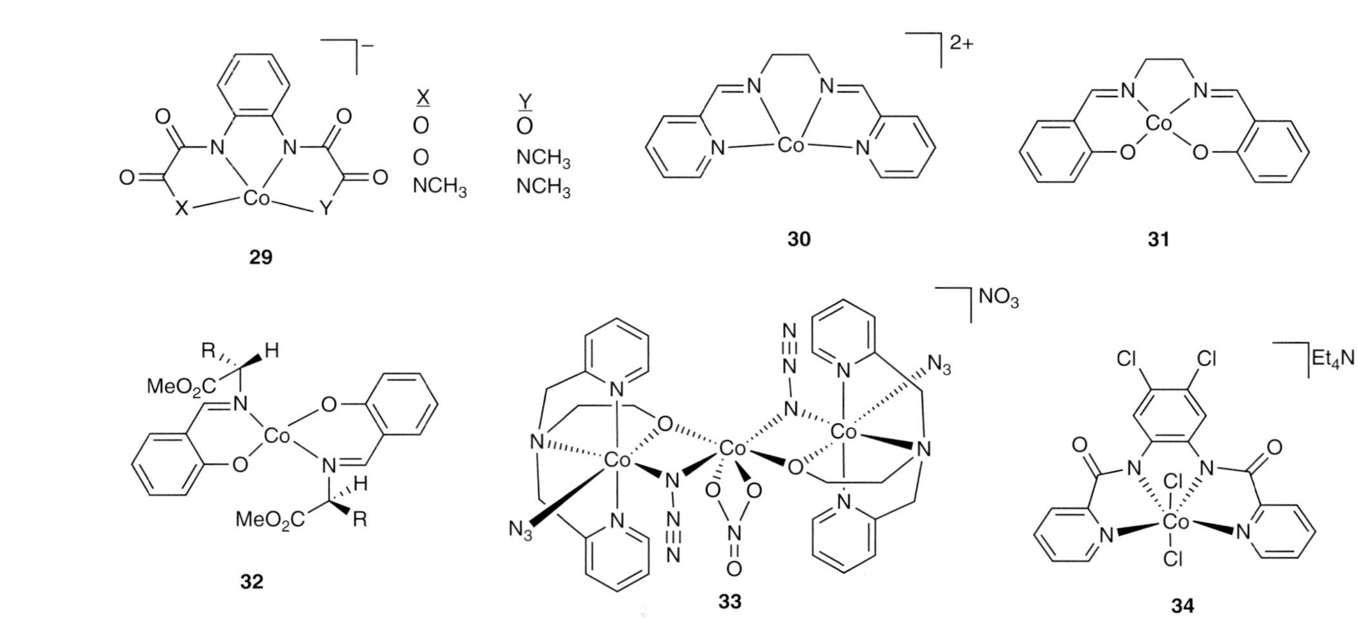

Figure 14.12 Cobalt-based catalysts for the oxidation of olefins.

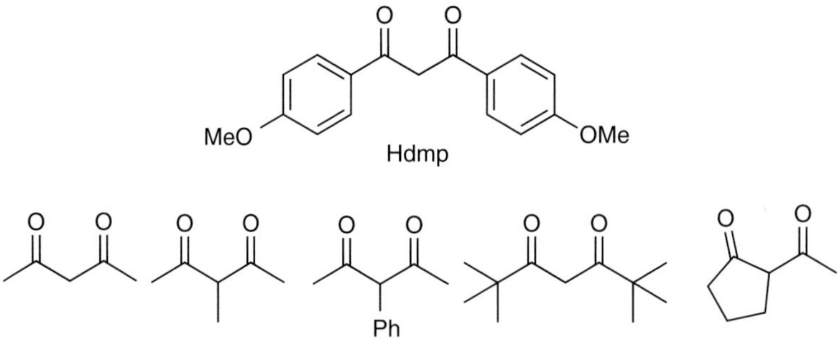

Figure 14.13 1,3-Diketone-type ligands used in Ni-catalyzed aerobic epoxidations by Yamada et al. [71].

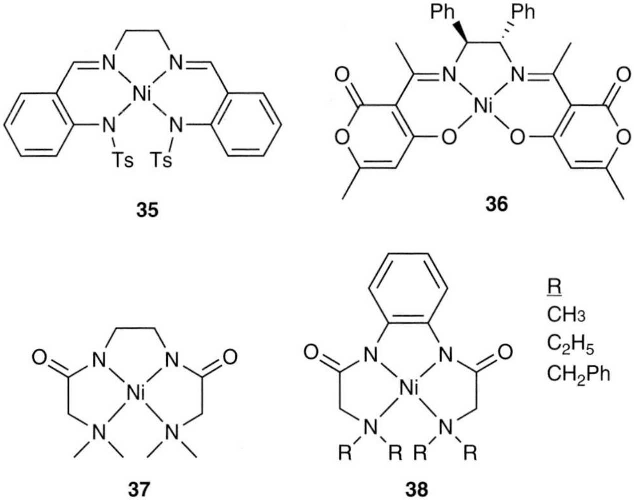

Figure 14.14 Ni catalysts used for the aerobic epoxidation of olefins (Ts = para-toluenesulfonate).

2-methyl-2-decene using iso-butyraldehyde and molecular oxygen (Figure 14.13). These 1,3-diketone-type ligands were found to be an excellent ligand scaffold for the synthesis of Ni catalysts for the epoxidation of trisubstituted olefins.

In the same year, Katsuki and coworkers also described the epoxidation of linear olefins in the presence of 2-methylpropanal and molecular oxygen, using salen-type complexes **35** and **36** (Figure 14.14) [72]. The chiral nickel complex **36** catalyzes the epoxidation of 1-octene, indene, 1-hexene, and *tert*-4-octene with 18–41% ee and 66–85% yield. Gupta and coworkers reported the synthesis of square-planar, bis-amido nickel(II) complexes **37** and **38** [73]. Also, these Ni^{II} complexes are efficient catalysts for the epoxidation of olefins in the presence of an aldehyde as the coreductant and dioxygen as the oxidizing agent.

Valentine and coworkers reported on the use of several mono- and binuclear copper complexes based on the PY2 chelating moiety (PY2 = bis(2-ethylpyridine)

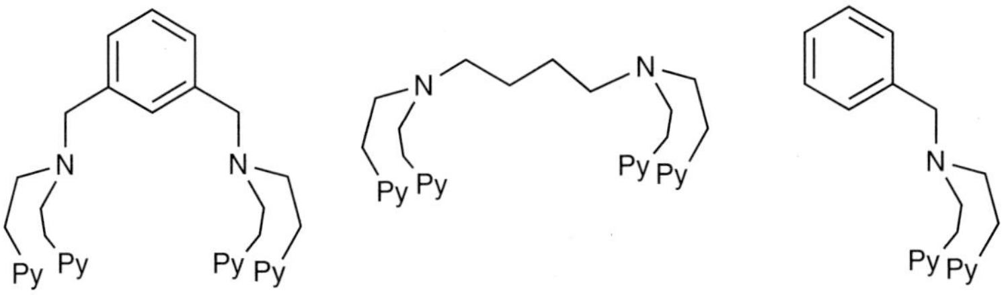

Figure 14.15 Ligand systems based on the PY2 chelating moiety used by Valentine (Py = 2-pyridyl).

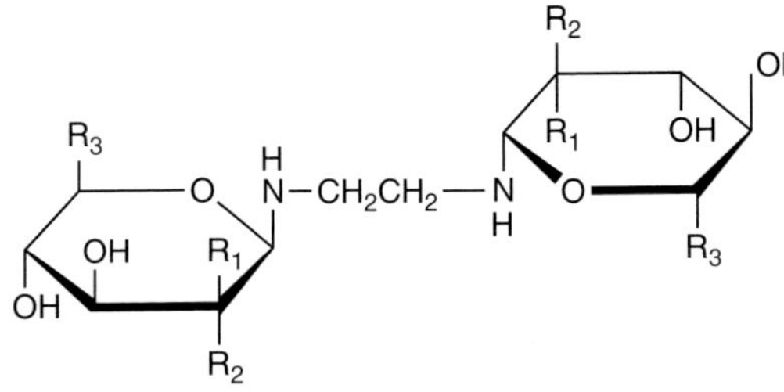

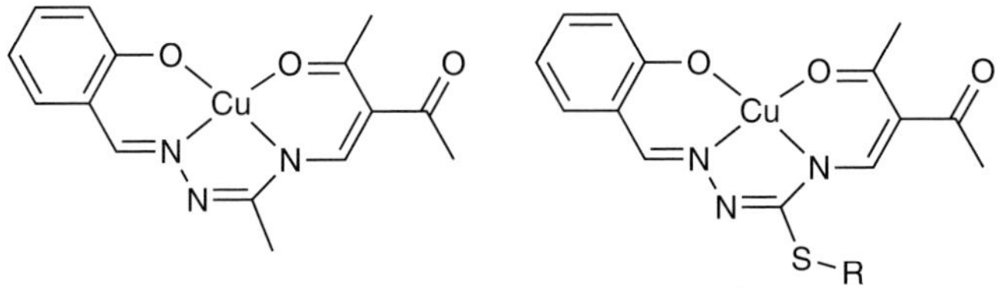

Figure 14.16 Sugar-derived ligand (top) and Cu-based epoxidation catalysts (bottom) reported by Yamamoto and Arion, respectively. Source: (top) Tanase et al. 1993 [76]. Reproduced with permission of ACS; (bottom) Gerbeleu et al. 1995 [75]. Reproduced with permission of Elsevier.

amine) as epoxidation catalysts using PhIO as the oxidant (Figure 14.15) [74]. They found that binuclear PY2 complexes are more active catalysts than their mononuclear counterparts.

Yamamoto and coworkers described the synthesis of copper complexes derived from *N*-glycoside-based ligands (Figure 14.16) [76]. Interestingly, they noted that these sugar-derived complexes catalyze the epoxidation of a number of unfunctionalized olefins, e.g. (*E*)-stilbene, (*Z*)-stilbene, and (*E*)-β-methylstyrene, with *tert*-butyl hydroperoxide (TBHP) as the oxidant, albeit with low enantioselectivity. In 1994, Arion and coworkers reported copper amidrazone complex **39**, which is an efficient catalyst for the epoxidation of norbornene in tetrahydrofuran (THF) with molecular oxygen (Figure 14.16) [75]. The presence of an additional thioether donor moiety in the ligand backbone resulted in a diminished catalytic activity of the corresponding complexes **40** because of the coordination of the –SR group in the axial position of the Cu center.

14.3 Oxygenation of C—H Bonds

The selective oxidation of C—H bonds is a reaction of fundamental importance in synthetic chemistry and biology. An example of a C—H oxidation reaction carried out at an industrial, multiton scale is the cobalt-catalyzed aerobic oxidation of cyclohexane to a mixture of cyclohexanol and cyclohexanone, which serve as intermediates for the production of Nylon-6,6′. Cyclohexane conversion is typically kept at a low 3–8% in view of the low selectivity of this radical chain reaction. Most biological C—H oxidation reactions are carried out by dedicated metalloenzymes. In the past decades, the development of bioinspired mononuclear first-row transition metal complexes for the selective oxidation of C—H bonds has received wide interest. However, selective oxidation of aliphatic C—H bonds in an environmentally friendly manner still remains as a challenge for synthetic chemists [77]. In this section, recent progress on the development of efficient mononuclear nonheme C—H oxidation catalysts based on base metal complexes is presented.

14.3.1 Manganese Catalysts

Manganese-containing enzymes involved in reactions with O_2 and H_2O_2, such as hydrogen peroxide catalases, superoxide dismutases, the oxygen-evolving complex of photosystem II, and oxidoreductases, have been widely studied, among others to decipher their reaction mechanism and to identify oxygenated Mn intermediates. In contrast, the development of biomimetic manganese-based catalysts for C—H bond oxidations has remained relatively underinvestigated as compared to their iron-based counterparts [78]. Recent investigations on the use of Mn complexes for selective C—H bond oxidation have demonstrated that Mn catalysts are able to oxidize relatively weak C—H bonds. This section briefly discusses the recent progress made in the development of manganese-catalyzed C—H oxidation catalysts.

In 1998, Shul'pin and Lindsay Smith reported catalytic alkane oxygenations with H_2O_2 in the presence of Wieghardt's [79] dinuclear Mn^{IV} complex $[Mn_2O_3(tmtacn)_2]^{2+}$, **13** (Scheme 14.3) [80]. The catalyst is very efficient for the C—H oxygenation of simple alkanes such as hexane, cyclohexane, and cyclopentane and the reactions proceed with PAA or with H_2O_2 in the presence of acetic acid. Turnover numbers up to 1350 were noted for the oxidation of hexane with alcohol/ketone ratios of about 1.3–1.4 [81]. Subsequently, the same catalyst was reported to catalyze the oxidation of other alkanes, e.g. methane, ethane, and higher normal and branched alkanes at room temperature, in some cases providing up to 3100 turnovers [82–84]. A set of Mn complexes were synthesized using tacn ligand derivatives, including chiral ones, by Süss-Fink and coworkers. Their ability to oxidize linear and branched alkanes with H_2O_2 was tested in the presence of either oxalate or ascorbic acid as cocatalysts [85]. The oxidation of cyclohexane yielded cyclohexyl hydroperoxide as the major product, and after reduction using triphenyl phosphine, a TON of 68 was achieved for the alcohol product after five hours.

Later, Costas with coworkers reported on the mononuclear, tacn-based complex $[(^{H,Me}Pytacn)Mn(CF_3SO_3)_2]$ (**8**, Figure 14.4), which is able to carry out eight catalytic turnovers in the oxidation of *cis*-1,2-dimethylcyclohexane [86]. Building on these and other findings on the development of aminopyridine ligand frameworks, Bryliakov and coworkers reported excellent catalytic efficiencies of manganese-aminopyridine complex **41**, derived from the C_1-symmetric aminopyridine ligand developed by Sun and coworkers [41], for the oxidation of a variety of secondary and tertiary alkanes in the presence of acetic acid additives, providing up to 970 catalytic turnovers (Scheme 14.7) [87]. In a typical example, the oxidation of 2,6-dimethyloctane yields an equimolar mixture of "remote" and "proximal" oxidation products, whereas introduction of an electron-withdrawing substituent substantially deactivates the proximal tertiary C—H group. For the substrate 1-bromo-3,7-dimethyloctane, the observed remote/proximal ratio was found to be 97 : 1.

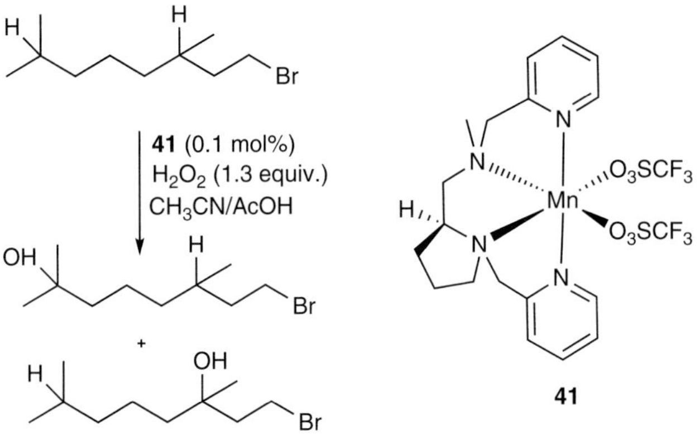

41 (0.1 mol%)
H_2O_2 (1.3 equiv.)
CH_3CN/AcOH

Remote/proximal = 97:1

41

Scheme 14.7 Selective Mn-catalyzed C—H oxidations reported by Bryliakov.

Subsequently, Sun and coworkers reported on a related benzimidazole-based manganese complex that efficiently catalyzes benzylic, aliphatic, and tertiary C—H oxidation with hydrogen peroxide as the oxidant in the presence of acetic acid as an additive [88]. The oxidation of adamantane was examined using 1.3 equiv. of H_2O_2, which gave 1-adamantanol as the main product (40% yield) along with 3.5% of 2-adamantanol and 2-adamantanone.

More recent examples of selective Mn-catalyzed C—H oxidations include the oxidation of protected aliphatic amines by a Mn-PDP-type catalyst reported by Costas and coworkers [89] The enantioselective oxidation of nonactivated aliphatic substrates by a bulky Mn-mcp-type catalyst was also reported by Costas and coworkers [90], and the enantioselective oxidation of benzylic substrates by a Mn-PDP-type catalyst was reported by Bryliakov and coworkers [91].

14.3.2 Iron Catalysts

Iron complexes containing multidentate N-donor ligands have been studied extensively over the past 25 years and by now have been shown to be able to oxidize aliphatic C—H bonds in a highly predictable and selective manner. In 1997, the first example of highly stereospecific oxidations of *cis*- and *trans*-1,2-dimethylcyclohexane with >99% retention of stereochemistry was described by Que and coworkers using the iron complex $[Fe(tpa)(CH_3CN)_2](ClO_4)_2$ as the catalyst along with hydrogen peroxide as the oxidant [92]. Later, they extended their reactivity studies to tpa-based iron(II) complexes with substituted pyridine moieties, concluding that the 5-Me_3-tpa-based iron complex is a more efficient catalyst than $[Fe(tpa)(CH_3CN)_2](ClO_4)_2$. On the other hand, ligands bearing two or three 6-Me substituents lead to iron complexes that are less reactive [93, 94]. Also in 1997, μ-oxo bridged diiron(III) complexes bearing either mep or tpa ligands were reported by Nishida and coworkers. The structural features of the two complexes are very similar, although these two complexes exhibit very different reactivity toward the oxidation of cyclohexane with H_2O_2 with only negligible oxidation taking place with the tpa-based complex [95].

In 2007, White and coworkers introduced a highly predictable and selective C—H oxidation protocol for the conversion of complex substrates using H_2O_2 as an oxidant and an iron catalyst [96]. This landmark study introduced the use of the bipyrrolidine ligand backbone in the field of oxidation catalysis. The selectivity of C—H bond oxidations by $[Fe(PDB)(CH_3CN)_2](SbF_6)_2$ (**42**) is dictated by the electronic and steric properties of the different C—H bonds present within the organic substrate and proceed without the need of directing groups (Scheme 14.8). The hydroxylation of C—H bonds preferentially occurs at the most electron-rich or the least hindered tertiary C—H bond. Very interestingly, although considering the electronic and steric factors in a single substrate, steric factors can override electronic factors in determining site selectivity. Moreover, selective hydroxylations of natural products can be achieved utilizing this catalyst, which provided further evidence for its functional group tolerance. Later, the same group reported on the electronic, steric, and stereoelectronic factors that determine highly selective methylene C—H oxidations in a number of natural products using the same iron catalyst [97], and on further use of this and related catalysts in selective C—H oxidations [98, 99]. The reliability and activity of this and related catalyst systems are such that the C—H bond can itself be put forward as a functional group in organic synthesis [100, 101].

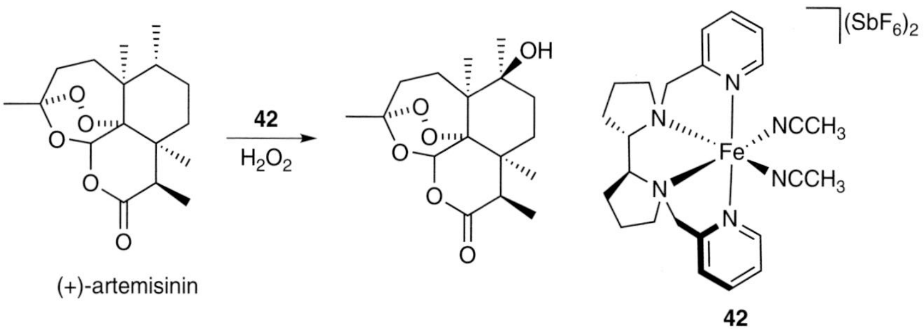

(+)-artemisinin

42

H_2O_2

OH

$(SbF_6)_2$

N Fe NCCH$_3$ NCCH$_3$

42

Scheme 14.8 Predictable and selective aliphatic C—H oxidation reported by White.

Further modifications of the pyridyl moieties in the PDP ligand by White, Costas, and others have led to a further optimization of the Fe-PDP system. Variation of the bipyrrolidine backbone has been reported by the groups of Rybak-Akimova and Klein Gebbink through the use of a bipiperidine and a bis-isoindoline backbone, respectively [102, 103].

Kodera and coworkers reported on an iron complex derived from the pentadentate H-dpaq ligand (dpaq = 2-[bis-(pyrid-2-ylmethyl)amino]-*N*-quinol-8-yl-acetamidate), which can catalyze the selective hydroxylation of C—H bonds using H_2O_2 as the oxidant. The catalyst $[Fe^{III}(dpaq)(H_2O)]^{2+}$ showed a virtually similar activity to that of Fe-PDP catalyst **42** in the oxidation of *cis*-4-methylcyclohexyl-1-pivalate with >90% retention of stereochemistry (Scheme 14.9) [104]. Regioselective hydroxylation of 1-substituted 3,7-dimethyloctane was also tested and complex $[Fe^{III}(dpaq)(H_2O)]^{2+}$ showed greater selectivity in comparison to the Fe-PDP complex.

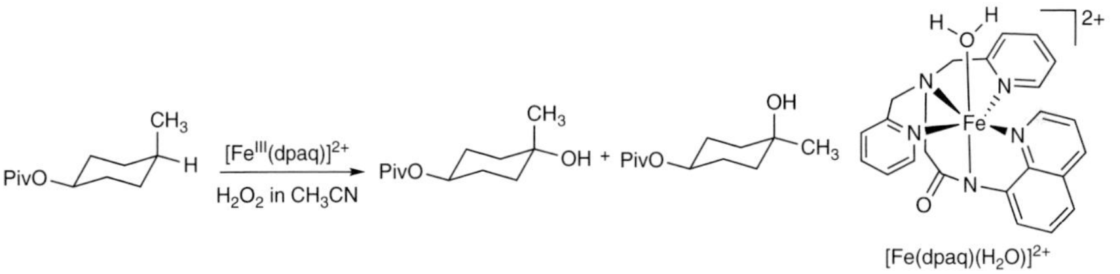

PivO–C₆H₁₀(CH₃)(H) $\xrightarrow[H_2O_2 \text{ in } CH_3CN]{[Fe^{III}(dpaq)]^{2+}}$ PivO–C₆H₁₀(CH₃)(OH) + PivO–C₆H₁₀(OH)(CH₃)

$[Fe(dpaq)(H_2O)]^{2+}$

Scheme 14.9 Selective C—H oxidation reported by Kodera et al.

Another catalyst platform, reported by Costas and coworkers, that has been extensively studied for the oxidation of alkanes is the iron catalyst $[Fe(^{Me,H}Pytacn)(CF_3SO_3)_2]$ [54, 105–107]. This catalyst is capable of oxidizing C—H bonds with the retention of configuration (>95%) and in yields up to 54% at a catalyst loading of 3 mol%. The same group also described the effect of steric bulk on catalytic oxidation through substitution on the *N*-atoms of the triazamacrocycle and on the 6-position of the pyridine moiety. The 6-methylpyridine-substituted iron complex $[Fe(^{Me,Me}Pytacn)(CF_3SO_3)_2]$ not only showed unusually high efficiency in stereospecific oxidations but also exhibits an enhanced selectivity toward methylene sites.

The iron catalyst $[Fe(mcp)(CF_3SO_3)_2]$ has been studied for quite some time. A recent report by Costas and coworkers shows the ability of this catalyst to perform the selective oxidation of methylene sites in alkanes [108]. Similar to **42**, this selectivity is dependent on the electronic and stereoelectronic environments of the C—H bonds. In some cases, though, $[Fe(mcp)(CF_3SO_3)_2]$ appears to be an even more sterically sensitive catalyst than the PDP-based catalyst in distinguishing between multiple methylene C—H bonds within a single substrate. A remarkable example is the oxidation of trans-decalin; the mcp catalyst preferentially oxidizes at the least hindered methylene site C-3 (the ratio of C-3 oxidation to C-2 oxidation is 1.5 : 1), whereas the PDP catalyst yields a 1 : 1 ratio of products (Scheme 14.10).

The effect of steric bulk of the catalyst on product selectivity was further studied by Costas and Klein Gebbink, by appending trialkylsilyl groups at

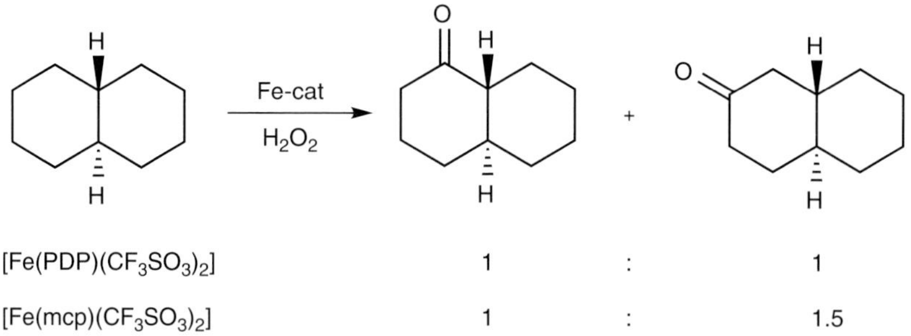

$[Fe(PDP)(CF_3SO_3)_2]$	1	:	1
$[Fe(mcp)(CF_3SO_3)_2]$	1	:	1.5

Scheme 14.10 Methylene oxidation by $[Fe(PDP)(CF_3SO_3)_2]$ (**42**) and $[Fe(mcp)(CF_3SO_3)_2]$.

the *meta*-position of the pyridine rings in tetradentate aminopyridine ligands based on mcp and PDP [109]. Preferential oxidation of secondary over tertiary C—H bonds and site-selective oxidation of methylenic sites in terpenoid and steroidal substrates was achieved by using iron catalysts **43** and **44** with H_2O_2 (Scheme 14.11). Unprecedented site-selective oxidation at C6 and C12 methylenic sites in steroidal substrates is governed by the chirality of the catalysts, i.e. changing the stereogeneity of the ligand backbone reverses the C6/C12 oxidation selectivity.

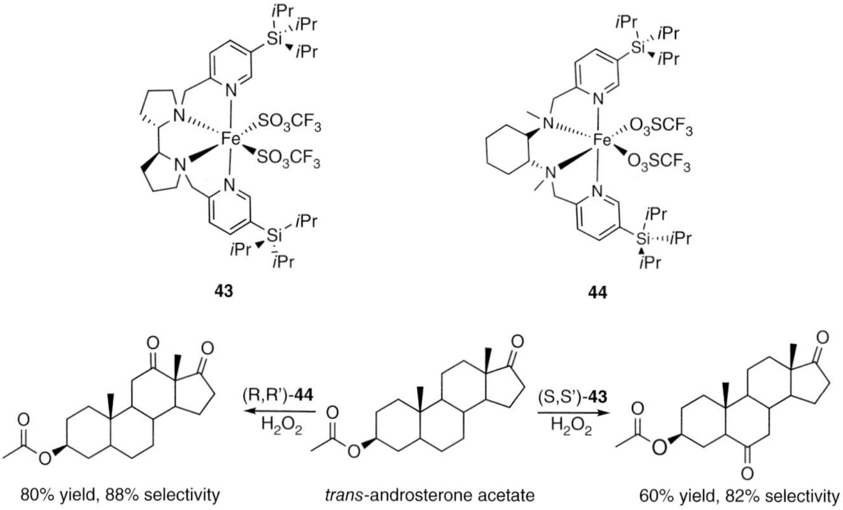

Scheme 14.11 Regioselective oxidation of *trans*-androsterone acetate governed by catalyst chirality.

14.3.3 Cobalt Catalysts

The use of cobalt-based catalysts for the oxidation of C—H bonds is known for a few industrial applications, e.g. for the synthesis of terephthalic acid from para-xylene [110] and adipic acid from cyclohexane [111] In these processes, molecular O_2 is used as the oxidant, and the cobalt catalyst is involved in

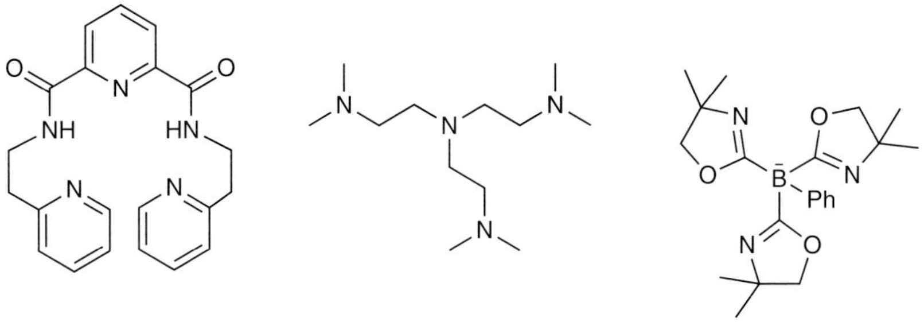

Figure 14.17 Ligands used for catalytic alkane oxidations with cobalt.

radical chain auto-oxidation initiated by decomposition of alkylhydroperoxides generated in the reaction mixture. These systems bear drawbacks as elevated pressures and temperatures are required for oxidation, and alternative catalyst systems and protocols are therefore of interest.

Analogous to the industrial processes, cobalt(III) alkylperoxy complexes have been studied for the catalytic hydroxylation of alkanes with alkylhydroperoxides by Mascharak and coworkers [112] A series of Co^{III}-alkylperoxo complexes derived from dianionic ligands combining three *N*-heterocyclic groups and two diamido units were isolated and structurally characterized (Figure 14.17, left). These cobalt complexes are capable of oxidizing alkanes in the presence of TBHP. Turnover numbers over 100 and ~10% conversion of cyclohexane to cyclohexanol and cyclohexanonone within four hours have been reported. Cobalt-catalyzed hydroxylations of alkanes such as cyclohexane and adamantane with good alcohol selectivity using mCPBA as the oxidant were reported by Knör and coworkers [113]. In this case, the tripodal tris[2-(dimethylamino)ethyl]amine ligand (Me_6TREN; Figure 14.17, middle) was employed and the corresponding cobalt(II) complex converts adamantane to 1-adamantanol (TON = 68), 2-adamantanol (TON = 8), and 2-adamantanone (TON = 26), but the selectivity of tertiary over secondary C—H bond oxidation is quite low (3°/2° = 6). Sadow and coworkers reported on a [tris(oxazolinyl)borato]cobalt(II)acetate complex that catalyzes the oxidation of cyclohexane to cyclohexanol with *m*CPBA as the terminal oxidant to afford 536 equiv. of cyclohexanol per 1 equiv. of catalyst after seven hours [114].

14.3.4 Nickel Catalysts

Few examples are known in which Ni complexes are used to catalyze the oxygenation of C—H bonds. Selected examples of studies that have succeeded in using nickel-based complexes in such oxidative transformations are highlighted in this section. In some of these examples, it was reported that aromatic C—H bonds can selectively be oxidized in the presence of aliphatic C—H bonds.

Itoh and coworkers reported on nickel complex $[Ni^{II}(OAc)(tpa)(H_2O)](BPh_4)$ as a very efficient catalyst for alkane hydroxylation with m-CPBA. The complex is robust enough to achieve high catalytic turnover (656) and shows a moderate alcohol selectivity (A/K = 8.5) for the oxidation of cyclohexane [115]. In 2007,

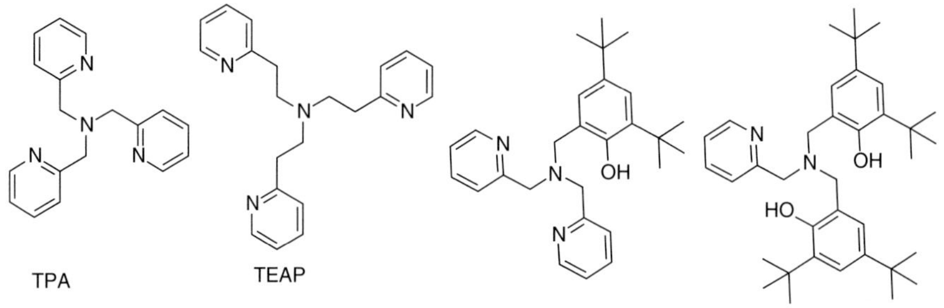

TPA TEAP

Figure 14.18 Ligands used by Itoh for Ni-catalyzed C—H oxidations.

the same group reported Ni^{II} complexes supported by tetradentate ligands containing two or three 2,4-di-*tert*-butylphenol groups (Figure 14.18). The Ni^{II} complex of the diphenol ligand has been found to act as a very efficient catalyst in alkane hydroxylations with mCPBA [116, 117]. In the oxygenation of cyclohexane, the reaction proceeded catalytically to give cyclohexanol as the major product (93%) along with cyclohexanone as the minor product (7%), representing an alcohol/ketone selectivity up to 13.3. In 2015, these studies were extended to the direct hydroxylation of benzene using H_2O_2 as the oxidant [118]. In this case, the nickel complex supported by the tepa ligand (tepa = tris-[2-(pyrid-2-yl)ethyl]amine) exhibited a prominent ability to produce phenol in 21% yield based on benzene, without the formation of any overoxidation products. The Ni-tepa complex reached a total turnover number of 749, which is the highest value ever reported for a molecular catalyst in the direct hydroxylation of benzene. This aromatic hydroxylation method has been shown to also be applicable to the oxygenation of alkylbenzene derivatives. The selectivity for the cresol product in the direct hydroxylation of toluene reached an impressive 90%.

In 2011, Palaniandavar and coworkers reported nickel(II) complexes of tripodal N_4 ligands as catalysts for alkane oxidation using mCPBA as an oxidant (Figure 14.19). These complexes efficiently catalyze the hydroxylation of alkanes with TONs in the range of 340–620 and good alcohol selectivity for cyclohexane. The design of these ligands includes the replacing of one of the pyridyl donors in the prototypical tpa ligand by a weakly binding NMe_2 or NEt_2 group, and further pyridine replacement by imidazole or sterically demanding quinoline and benzimidazole moieties. The comparison of the hydroxylation reaction of cyclohexane with those of related nickel(II) catalysts has been

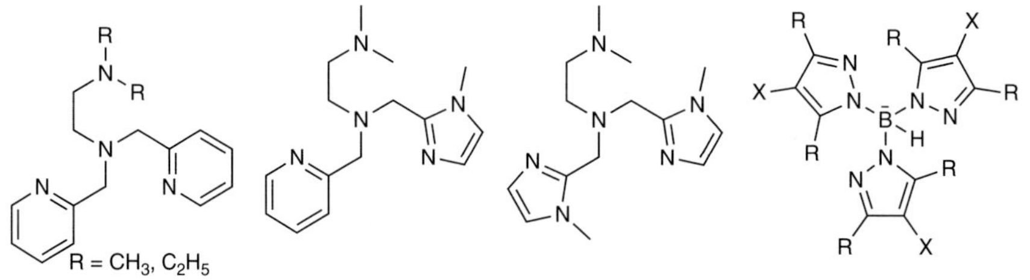

$R = CH_3, C_2H_5$

Figure 14.19 Ligands used by Palaniandavar and Hikichi in C—H oxidations with nickel.

performed under identical conditions. Replacement of one of the pyridyl donors in tris(2-pyridylmethyl)amine (TPA) by $-NMe_2$ leads to a slight increase in the catalytic activity with no change in the selectivity. Under identical conditions, the Ni-tpa catalyst converted cyclohexane to cyclohexanol with TON = 505 (A/K = 8.1), whereas the $-NMe_2$-substituted catalyst showed TON = 622 (A/K = 8.7). Upon replacing a next pyridyl group by an imidazolylmethyl arm or by another *N*-heterocycle, the alcohol/ketone ratio decreased very significantly from 8.7 to 5.7. The catalytic activities of the Ni^{II} complexes were further explored for the aliphatic oxidation of ethylbenzene and cumene [119]. Interestingly, the oxidation of cumene selectively gave 2-phenyl-2-propanol without any side product formation.

Hikichi et al. reported nickel complexes with structurally and electronically tunable hydrotris(3,5-dialkyl-4-X-pyrazolyl)borate ligands ($Tp^{R2,X}$, R = Me or ^{i}Pr and X = H or Br) that catalyze alkane oxidations with mCPBA [120]. In the oxidation of cyclohexane, the $Tp^{Me2,Br}$ nickel complex showed a higher alcohol yield (A/K = 53) than the complex derived from the non-brominated ligand (A/K = 32) when the reaction was performed in $CF_3C_6H_5$. The electronic and steric properties of the $Tp^{R2,X}$ ligands affect the catalytic properties, with less hindered ligands containing electron-withdrawing substituents exhibiting a higher alcohol selectivity.

14.3.5 Copper Catalysts

In the past few decades, much attention has focused on biomimetic copper complexes in order to achieve high selectivities and reactivities in C—H oxidations [121]. Copper-catalyzed hydroxylations of alkanes have not yet been developed to efficient synthetic procedures for a range of substrates, though, because of the lack of regioselectivity and the formation of over-oxidized or desaturation products (olefins). A large number of mono- and multinuclear copper complexes have been investigated for the catalytic activation of aliphatic and aromatic C—H bonds, and some selected examples of successful alkane hydroxylations with copper will be presented here [122–124].

Sir Derek Barton investigated the oxidation of alkanes using $Cu(fod)_2$ (fod = 2,2-dimethyl-6,6,7,7,8,8,8-heptafluoro-3,5-octanedionate) with TBHP under an oxygen atmosphere [125]. The oxidation of cyclooctane afforded the corresponding alcohol and ketone, along with the olefin desaturation product. In the oxidation of cyclooctane in air, approximately equal proportions of olefin and ketone were formed, whereas under an atmosphere of oxygen, the amount of ketone increased substantially giving an olefin to ketone ratio of 1 : 19. The turnover number for this particular reaction was calculated to be 11.3 per catalyst molecule, with an efficiency of 77% and an overall conversion of 11.3%.

Hydroxylation of alkanes with H_2O_2 catalyzed by β-diketiminatocopper(II) complexes **45** were reported by Itoh and coworkers. Complexes containing an electron-withdrawing substituent on the α-carbon (R=NO_2 or CN) of the diketiminate ligand exhibited the highest catalytic activities among this series of complexes (Figure 14.20) [126]. Overall, the system showed limited activity with turnover numbers around 26 in the oxidation of cyclohexane. Pérez and

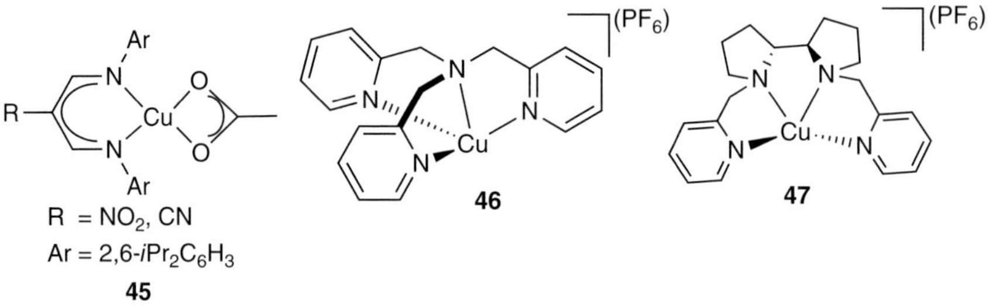

R = NO_2, CN
Ar = 2,6-$i$$Pr_2C_6H_3$
45

46 (PF_6)

47 (PF_6)

Figure 14.20 Mononuclear copper complexes used in catalytic alkane oxidations.

coworkers also reported several copper complexes of the general formula $Tp^xCu(CH_3CN)$ (Tp^x = hydrotrispyrazolylborate) as potential catalysts for the conversion of benzene to phenol with H_2O_2 under acid-free conditions [127]. In the presence of 0.01 mol% catalyst, 14–30% conversion of benzene into phenol was noted with selectivity toward phenol of 67–85%. Garcia-Bosch and Siegler have reported copper(I) complexes **46** and **47** bearing a tpa or PDP ligand for the oxidation of alkanes with H_2O_2 as the oxidant (Figure 14.20) [128]. Cyclohexane, cyclopentane, cycloheptane, and cyclooctane were oxidized to the corresponding secondary alcohols and ketones in 36–47% yield with good selectivity (A/K ratio in between 4.1 and 12.2) and without the formation of overoxidized products. The linear substrate n-hexane can be oxidized by these systems to the corresponding alcohol and ketone products (20–30% yield, A/K ratio = 3.9–9.1), with oxidation preferentially taking place at the internal, methylene positions.

14.4 Conclusions and Outlook

In this chapter, we have tried to provide the reader with a representative overview of the development of molecular oxidation catalysts based on late first-row transition metals during the past two decades. Although Co, Ni, and Cu catalysts have been studied to some extent for the oxygenation of C=C and C—H bonds, the development of selective oxidations by means of mononuclear nonheme Mn and Fe complexes has blossomed over this time period. From the early observations on the involvement of metal-based oxidants that are able to convert substrates with the retention of substrate configuration and the first reports on enantioselective catalytic oxidation protocols, the fields of Mn and Fe oxidation catalysis have advanced to its applicability in the organic synthesis of complex molecules. Future developments in the field are foreseen in its application in manufacturing. This will likely include the development of new ligand platforms that allow for high selectivity and efficiency and new ligand platforms that rely on cheap building blocks and are easily synthetized. Further developments are expected in terms of process engineering, including process optimization and automatization, the use of alternative reaction media, and the use of flow chemistry and process miniaturization.

Acknowledgment

The authors acknowledge the European Union for Marie Curie Postdoctoral Fellowship support (H2020-MSCA-IF-2014–657765), the Sectorplan Natuur- en Scheikunde (Tenure-track grant at Utrecht University), and Utrecht University for financial support.

References

1 Backvall, J.E. (ed.) (2004). *Modern Oxidation Methods*. Weinheim: Wiley.
2 Sheldon, R. (2012). *Metal-Catalyzed Oxidations of Organic Compounds: Mechanistic Principles and Synthetic Methodology Including Biochemical Processes*. Amsterdam: Elsevier Science.
3 Lefort, T.E. (1935). French Patent 729,952, 1931; U.S. Patent 1,998,878.
4 Cavani, F. and Teles, J.H. (2009). *ChemSusChem* 2: 508.
5 Roberts, S.M. and Whittall, J. (eds.) (2007). *Catalysts for Fine Chemical Synthesis Regio- and Stereocontrolled Oxidations and Reductions*, vol. 5. England: Wiley.
6 Srinivasan, K., Michaud, P., and Kochi, J.K. (1986). *J. Am. Chem. Soc.* 108: 2309.
7 Zhang, W., Loebach, J.L., Wilson, S.R., and Jacobsen, E.N. (1990). *J. Am. Chem. Soc.* 112: 2801.
8 Irie, R., Nodda, K., Ito, Y., and Katsuki, T. (1990). *Tetrahedron Lett.* 31: 7345.
9 Kausuki, T. (1995). *Coord. Chem. Rev.* 140: 189.
10 McGarrigle, E.M. and Gilheany, D.G. (2005). *Chem. Rev.* 105: 1563.
11 Murphy, A., Dubois, G., and Stack, T.D.P. (2003). *J. Am. Chem. Soc.* 125: 5250.
12 Murphy, A., Pace, A., and Stack, T.D.P. (2004). *Org. Lett.* 18: 3119.
13 Moretti, R.A., Du Bois, J., and Stack, T.D.P. (2016). *Org. Lett.* 18: 2528.
14 Wu, X., Seo, M.S., Davis, K.M. et al. (2011). *J. Am. Chem. Soc.* 133: 20088.
15 Dong, J.J., Saisaha, P., Meinds, T.G. et al. (2012). *ACS Catal.* 2: 1087.
16 (a) For selected examples of manganese catalysts with MCP-derived ligands, see: Guillemot, G., Neuburger, M., and Pfaltz, A. (2007). *Chem. Eur. J.* 13: 8960. (b) Wu, M., Wang, B., Wang, S. et al. (2009). *Org. Lett.* 11: 3622. (c) Ottenbacher, R.V., Bryliakov, K.P., and Talsi, E.P. (2010). *Inorg. Chem.* 49: 8620. (d) Ottenbacher, R.V., Bryliakov, K.P., and Talsi, E.P. (2011). *Adv. Synth. Catal.* 353: 885. (e) Maity, N.C., Bera, P.K., Ghosh, D. et al. (2014). *Catal. Sci. Technol.* 4: 208.
17 Garcia-Bosch, I., Company, A., Fontrodona, X. et al. (2008). *Org. Lett.* 10: 2095.
18 Garcia-Bosch, I., Ribas, X., and Costas, M. (2009). *Adv. Synth. Catal.* 351: 348.
19 Cusso, O., Garcia-Bosch, I., Font, D. et al. (2013). *Org. Lett.* 15: 6158.
20 Lyakin, O.Y., Ottenbacher, R.V., Bryliakov, K.P., and Talsi, E.P. (2012). *ACS Catal.* 2: 1196.

21 Ottenbacher, R.V., Samsonenko, D.G., Talsi, E.P., and Bryliakov, K.P. (2016). *ACS Catal.* 6: 979.
22 Shen, D., Qiu, B., Xu, D. et al. (2016). *Org. Lett.* 18: 372.
23 Miao, C., Wang, B., Wang, Y. et al. (2016). *J. Am. Chem. Soc.* 138: 936.
24 Wang, B., Miao, C., Wang, S. et al. (2012). *Chem. Eur. J.* 18: 6750.
25 de Boer, J.W., Brinksma, J., Browne, W.R. et al. (2005). *J. Am. Chem. Soc.* 127: 7990.
26 de Boer, J.W., Browne, W.R., Harutyunyan, S.R. et al. (2008). *Chem. Commun.* 3747.
27 Saisaha, P., Pijper, D., van Summeren, R.P. et al. (2010). *Org. Biomol. Chem.* 8: 4444.
28 Sugimoto, H. and Sawyer, D.T. (1985). *J. Org. Chem.* 50: 1784.
29 Sugimoto, H., Spencer, L., and Sawyer, D.T. (1987). *Proc. Natl. Acad. Sci. U.S.A.* 84: 1731.
30 Tohma, M., Tomita, T., and Kimura, M. (1973). *Tetrahedron Lett.* 4359.
31 Yamamoto, T. and Kimura, M. (1977). *J. Chem. Soc. Chem. Commun.* 948.
32 Murch, B.P., Bradley, F.C., and Que, L. Jr., (1986). *J. Am. Chem. Soc.* 108: 5027.
33 Nam, W., Ho, R., and Valentine, J.S. (1991). *J. Am. Chem. Soc.* 113: 7052.
34 Anilkumar, G., Bitterlich, B., Gelalcha, F.G. et al. (2007). *Chem. Commun.* 289.
35 Gelalcha, F.G., Bitterlich, B., Anilkumar, G. et al. (2007). *Angew. Chem. Int. Ed.* 47: 7293.
36 Cussó, O., Ribas, X., and Costas, M. (2015). *Chem. Commun.* 51: 14285.
37 Mas-Ballesté, R., Costas, M., van den Berg, T., and Que, L. Jr., (2006). *Chem. Eur. J.* 12: 7489.
38 White, M.C., Doyle, A.G., and Jacobsen, E.N. (2001). *J. Am. Chem. Soc.* 123: 7194.
39 Costas, M., Tipton, A.K., Chen, K. et al. (2001). *J. Am. Chem. Soc.* 123: 6722.
40 Wu, M., Miao, C.-X., Wang, S. et al. (2011). *Adv. Synth. Catal.* 353: 3014.
41 Wang, B., Wang, S., Xia, C., and Sun, W. (2012). *Chem. Eur. J.* 18: 7332.
42 Wang, X., Miao, C., Wang, S. et al. (2013). *ChemCatChem* 5: 2489.
43 Cussó, O., Garcia-Bosch, I., Ribas, X. et al. (2013). *J. Am. Chem. Soc.* 135: 14871.
44 Cussó, O., Ribas, X., Loret-Fillol, J., and Costas, M. (2015). *Angew. Chem. Int. Ed.* 54: 2729.
45 Cussó, O., Cianfanelli, M., Ribas, X. et al. (2016). *J. Am. Chem. Soc.* 138: 2732.
46 Chen, K. and Que, L. Jr., (1999). *Angew. Chem. Int. Ed.* 38: 2227.
47 Chen, K., Costas, M., Kim, J. et al. (2002). *J. Am. Chem. Soc.* 124: 3026.
48 Suzuki, K., Oldenburg, P.D., and Que, L. Jr., (2008). *Angew. Chem. Int. Ed.* 47: 1887.
49 Feng, Y., England, J., and Que, L. Jr., (2011). *ACS Catal.* 1: 1035.
50 Zang, C., Liu, Y., Xu, Z.-J. et al. (2016). *Angew. Chem. Int. Ed.* 55: 10253; (2016). *Angew. Chem.* 128: 10409.
51 Oldenburg, P.D., Shteinman, A.A., and Que, L. Jr., (2005). *J. Am. Chem. Soc.* 127: 15672.

52 Oldenburg, P.D., Feng, Y., Pryjomska-Ray, I. et al. (2010). *J. Am. Chem. Soc.* 132: 17713.
53 Bruijnincx, P.C.A., Buurmans, I.L.C., Gosiewska, S. et al. (2008). *Chem. Eur. J.* 14: 1228.
54 Company, A., Gómez, L., Fontrodona, X. et al. (2008). *Chem. Eur. J.* 14: 5727.
55 Prat, I., Font, D., Company, A. et al. (2013). *Adv. Synth. Catal.* 355: 947.
56 Koola, J.D. and Kochi, J.K. (1987). *J. Org. Chem.* 52: 4545.
57 Guilmet, E. and Meunier, B. (1980). *Tetrahedron Lett.* 21: 4449.
58 Takai, T., Hata, E., Yorozu, K., and Mukaiyama, T. (1992). *Chem. Lett.* 2077.
59 Estrada, J., Fernandez, I., Pedro, J.R. et al. (1997). *Tetrahedron Lett.* 38: 2377.
60 Punniyamurthy, T., Bhatia, B., and Iqbal, J. (1993). *Tetrahedron Lett.* 34: 4657.
61 Punniyamurthy, T. and Iqbal, J. (1994). *Tetrahedron Lett.* 35: 4003.
62 Punniyamurthy, T., Bhatia, B., and Iqbal, J. (1994). *J. Org. Chem.* 59: 850.
63 Punniyamurthy, T., Reddy, M.M., Kalra, S.S., and Iqbal, J. (1996). *J. Pure Appl. Chem.* 619.
64 Punniyamurthy, T., Bhatia, B., Reddy, M.M. et al. (1997). *Tetrahedron* 53: 7649.
65 Shin, J.W., Rowthu, S.R., Hyun, M.Y. et al. (2011). *Dalton Trans.* 40: 5762.
66 Joo Song, Y., Hyun, M.Y., Lee, J.H. et al. (2012). *Chem. Eur. J.* 18: 6094.
67 Koola, J.D. and Kochi, J.K. (1987). *Inorg. Chem.* 26: 908.
68 Kinneary, J.F., Wagler, T.R., and Burrows, C.J. (1988). *Tetrahedron Lett.* 29: 877.
69 Yoon, H. and Burrows, C.J. (1988). *J. Am. Chem. Soc.* 110: 4087.
70 Kinneary, J.F., Albert, J.S., and Burrows, C.J. (1988). *J. Am. Chem. Soc.* 110: 6124.
71 Yamada, T., Takai, T., Rhode, O., and Mukaiyama, T. (1991). *Bull. Chem. Soc. Jpn.* 64: 2109.
72 Irie, R., Ito, Y., and Katsuki, T. (1991). *Tetrahedron Lett.* 32: 6891.
73 Singh, J., Hundal, G., and Gupta, R. (2008). *Eur. J. Inorg. Chem.* 2052.
74 Tai, A.F., Margerum, L.D., and Valentine, J.S. (1986). *J. Am. Chem. Soc.* 108: 5006.
75 Gerbeleu, N.V., Palanciuc, S.S., Simonov, Y.A. et al. (1995). *Polyhedron* 14: 521.
76 Tanase, T., Mano, K., and Yamamoto, Y. (1993). *Inorg. Chem.* 32: 3995.
77 Codola, Z., Lloret-Fillol, J., and Costas, M. (2014). *Progress in Inorganic Chemistry*, vol. 59 (ed. K.D. Karlin), 447–531. Hoboken, NJ: Wiley.
78 Talsi, E.P. and Bryliakov, K.P. (2012). *Coord. Chem. Rev.* 256: 1418.
79 Wieghardt, K., Bossek, U., Nuber, B. et al. (1988). *J. Am. Chem. Soc.* 110: 1398.
80 Lindsay Smith, J.R. and Shul'pin, G.B. (1998). *Tetrahedron Lett.* 39: 4909.
81 Shul'pin, G.B. and Lindsay Smith, J.R. (1998). *Russ. Chem. Bull.* 47: 2379.
82 Shul'pin, G.B., Süss-Fink, G., and Lindsay Smith, J.R. (1999). *Tetrahedron* 55: 5345.
83 Shul'pin, G.B., Nizova, G.V., Kozlov, Y.N., and Pechenkina, I.G. (2002). *New J. Chem.* 26: 1238.

84 Shul'pin, G.B., Nizova, G.V., Kozlov, Y.N. et al. (2005). *J. Organomet. Chem.* 690: 4498.
85 Romakh, V.B., Therrien, B., Süss-Fink, G., and Shul'pin, G.B. (2007). *Inorg. Chem.* 46: 1315.
86 Gómez, L., Garcia-Bosch, I., Company, A. et al. (2009). *Angew. Chem. Int. Ed.* 48: 5720.
87 Ottenbacher, R.V., Samsonenko, D.G., Talsi, E.P., and Bryliakov, K.P. (2012). *Org. Lett.* 14: 4310.
88 Shen, D., Miao, C., Wang, S. et al. (2014). *Org. Lett.* 16: 1108.
89 Milan, M., Carboni, G., Salamone, M. et al. (2017). *ACS Catal.* 7: 5903.
90 Milan, M., Bietti, M., and Costas, M. (2017). *ACS Cent. Sci.* 3: 196.
91 Talsi, E.P., Samsonenko, D.G., Ottenbacher, R.V., and Bryliakov, K.P. (2017). *ChemCatChem* 9: 4580.
92 Kim, C., Chen, K., Kim, J., and Que, L. Jr. (1997). *J. Am. Chem. Soc.* 119: 5964.
93 Chen, K. and Que, L. Jr. (2001). *J. Am. Chem. Soc.* 123: 6327.
94 Chen, K., Costas, M., and Que, L. Jr. (2002). *J. Chem. Soc. Dalton Trans.* 672.
95 Okuno, T., Sayo, I., Ohba, S., and Nishida, Y. (1997). *J. Chem. Soc. Dalton Trans.* 3547.
96 Chen, M.S. and White, M.C. (2007). *Science* 318: 783.
97 Chen, M.S. and White, M.C. (2010). *Science* 327: 566.
98 Bigi, M.A., Reed, S.A., and White, M.C. (2012). *J. Am. Chem. Soc.* 134: 9721.
99 Gormisky, P.E. and White, M.C. (2013). *J. Am. Chem. Soc.* 135: 14052.
100 Howell, J.M., Feng, K., Clark, J.R. et al. (2015). *J. Am. Chem. Soc.* 137: 14590.
101 White, M.C. (2012). *Science* 335: 807.
102 Mikhalyova, E.A., Makhlynets, O.V., Palluccio, T.D. et al. (2012). *Chem. Commun.* 48: 687.
103 Chen, J., Lutz, M., Milan, M. et al. (2017). *Adv. Synth. Catal.* 359: 2590.
104 Hitomi, Y., Arakawa, K., Funabiki, T., and Kodera, M. (2012). *Angew. Chem. Int. Ed.* 51: 3448; (2012). *Angew. Chem.* 124: 3504.
105 Company, A., Gómez, L., Güell, M. et al. (2007). *J. Am. Chem. Soc.* 129: 15766.
106 Prat, I., Company, A., Postils, V. et al. (2013). *Chem. Eur. J.* 19: 6724.
107 Prat, I., Gómez, L., Canta, M. et al. (2013). *Chem. Eur. J.* 19: 1908.
108 Canta, M., Font, D., Gómez, L. et al. (2014). *Adv. Synth. Catal.* 356: 818.
109 Font, D., Canta, M., Milan, M. et al. (2016). *Angew. Chem. Int. Ed.* 55: 5776; (2016). *Angew. Chem.* 128: 5870.
110 Saffer, A. and Babyside, N.Y. (1958). US Patent 2,833,816.
111 Srinivas, D., Chavan, S.A., and Ratnasamy, P. (2001). US Patent 6,521,789 B1.
112 Chavez, F.A. and Mascharak, P.K. (2000). *Acc. Chem. Res.* 33: 539.
113 Tordin, E., List, M., Monkowius, U. et al. (2013). *Inorg. Chim. Acta* 402: 90.
114 Reinig, R.R., Mukherjee, D., Weinstein, Z.B. et al. (2016). *Eur. J. Inorg. Chem.* 2486.
115 Nagataki, T., Tachi, Y., and Itoh, S. (2006). *Chem. Commun.* 4016.
116 Nagataki, T., Ishii, K., Tachi, Y., and Itoh, S. (2007). *Dalton Trans.* 36: 1120.
117 Nagataki, T. and Itoh, S. (2007). *Chem. Lett.* 36: 748.

118 Morimoto, Y., Bunno, S., Fujieda, N. et al. (2015). *J. Am. Chem. Soc.* 137: 5867.
119 Balamurugan, M., Mayilmurugan, R.B., Sureshan, E., and Palaniandavar, M. (2011). *Dalton Trans.* 40: 9413.
120 Hikichi, S., Hanaue, K., Fujimura, T. et al. (2013). *Dalton Trans.* 42: 3346.
121 Allen, S.E., Walvoord, R.R., Padilla-Salinas, R., and Kozlowski, M.C. (2013). *Chem. Rev.* 113: 6234.
122 Kirillov, A.M., Kirillova, M.V., and Pombeiro, A.J.L. (2012). *Coord. Chem. Rev.* 256: 2741.
123 Kirillov, A.M., Kirillova, M.V., and Pombeiro, A.J.L. (2013). *Adv. Inorg. Chem.* 65: 1–31. (Ed.: R. VanEldik and C. D. Hubbard).
124 Wendlandt, A.E., Suess, A.M., and Stahl, S.S. (2011). *Angew. Chem. Int. Ed.* 50: 11062.
125 Barton, D.H.R., Beviere, S.D., and Hill, D.R. (1994). *Tetrahedron* 50: 2665.
126 Shimokawa, C., Teraoka, J., Tachi, Y., and Itoh, S. (2006). *J. Inorg. Biochem.* 100: 1118.
127 Conde, A., Díaz-Requejo, M.M., and Pérez, P.J. (2011). *Chem. Commun.* 47: 8154.
128 Garcia-Bosch, I. and Siegler, M.A. (2016). *Angew.Chem. Int. Ed.* 55: 12873; (2016). *Angew. Chem.* 128: 13065.

15

Organometallic Chelation-Assisted C—H Functionalization

Parthasarathy Gandeepan and Lutz Ackermann

Georg-August-Universität, Institut für Organische und Biomolekulare Chemie, Tammannstraße 2, 37077 Göttingen, Germany

15.1 Introduction

The development of efficient methods for organic synthesis continues to be an important challenge in academic and industrial research. During the past few decades, transition metal catalysis has dramatically improved the efficacy of synthetic strategies [1]. Particularly, functionalizations of unactivated carbon–hydrogen (C—H) bonds have emerged as powerful tools in organic synthesis, allowing high atom and step economy, reduced amounts of undesired wastes, and improved efficiency [2]. Because the organic substrates of interest usually possess many C—H bonds with similar reactivity, one of the major challenges is associated with achieving positional selectivity [3]. Generally, the site selectivity of the C–H functionalization can be controlled by the use of chelation assistance with directing groups (DGs). The Lewis base DGs coordinate with the metal complex, which assist the metal to approach entropically favored C—H bonds followed by cyclometalation through proximity-induced C—H cleavage (Scheme 15.1) [4].

Although the direct functionalization of C—H bonds by transition metal complexes is a formidable tool for effective organic syntheses, the currently most commonly used noble metals, such as ruthenium, rhodium, palladium, iridium, platinum, or gold, suffer from low natural abundance, high costs, and significant environmental impact by high toxicities [5]. In the past few years, the direct functionalization of unactivated C—H bonds using non-noble metals, including manganese, iron, cobalt, nickel, and copper, is surfacing as an appealing alternative to precious metal catalysis because of their lower costs and considerably lower toxicity [6–10]. This review aims at highlighting the most important advances in chelation-assisted aromatic C–H functionalization by base metal complexes.

Non-Noble Metal Catalysis: Molecular Approaches and Reactions,
First Edition. Edited by Robertus J. M. Klein Gebbink and Marc-Etienne Moret.

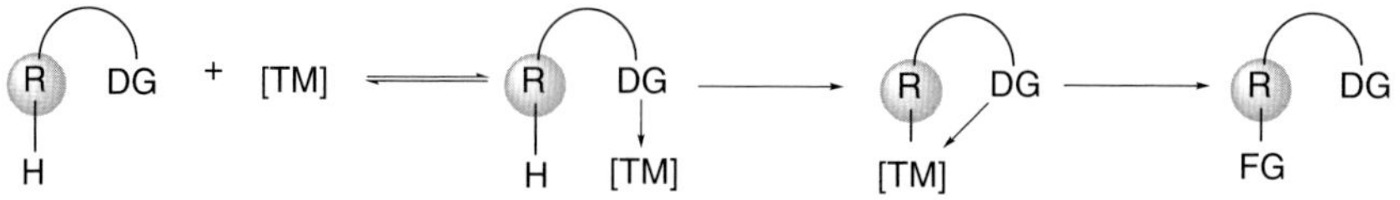

Scheme 15.1 Chelation-assisted C—H functionalization.

15.2 C—C Bond Formation via C—H Activation

15.2.1 Reaction with Unsaturated Substrates

15.2.1.1 Addition to C—C Multiple Bonds

The addition of C—H bonds to alkenes or alkynes represents an important class of reactions that enable the synthesis of substituted alkanes or alkenes, respectively, in a highly atom- and step-economical manner. These transformations largely proceed through chelation-assisted C—H cyclometalation followed by *cis*-addition to C—C π-bonds. Based on the pioneering work by Kisch [11a], Yoshikai and coworkers demonstrated the regiodivergent hydroarylation of styrenes **1** with 2-arylpyridines **2** [11b]. The catalytic system consisting of $CoBr_2$, PCy_3, and Me_3SiCH_2MgCl afforded the branched alkane **3**, whereas the use of the NHC ligand 1,3-bis(2,4,6-trimethylphenyl)imidazolium chloride (IMesHCl) in place of PCy_3 preferentially delivered the linear alkane **4** (Scheme 15.2a). In an independent study, the group of Nakamura reported a similar addition reaction with aromatic amides **5** and alkenes **6** (Scheme 15.2b). In this process, a variety of unactivated alkenes could be used to furnish linear alkanes **7** by anti-Markovnikov addition [11c]. By the action of Grignard reagents, *in situ*-generated, low-valent, cobalt catalyst-enabled hydroarylation with alkenes was also successfully applied to aromatic aldimines, ketimines, and heteroarenes [11d–h]. Further development in this area was achieved by Matsunaga and Kanai and coworkers using a well-defined cobalt(III) complex as the catalyst [12]. Thus, catalytic amounts of $[Cp^*Co(C_6H_6)](PF_6)_2$ in the absence of any additives affected the addition onto α,β-unsaturated carbonyl compounds **8** with 2-arylpyridines **2** (Scheme 15.2c) [12a]. Subsequently, Wang showed that $[MnBr(CO)_5]$ was a viable catalyst for the C—H hydroarylation of 2-arylpyridines **2** onto unsaturated carbonyl compounds **10** (Scheme 15.2d) [13].

The organometallic chelation-assisted addition of C—H bonds to alkynes has been achieved with a variety of base metal catalysts (Scheme 15.3). In 1994, Kisch and coworkers reported the first cobalt(I)-catalyzed hydroarylation of alkynes **13** with aromatic azoarenes **12** (Scheme 15.3a) [11a]. Inspired by these results, many reactions using a catalytic amount of cobalt complex along with stoichiometric Grignard reagents were exploited for the hydroarylation of alkynes with 2-arylpyridines **2** [14a], aromatic aldimines [14b], and ketimines **15** (Scheme 15.3b) [14c, d]. Heteroaromatics, such as indoles **17** and benzimidazoles, were also successfully employed in the alkyne hydroarylation regime by means of low-valent cobalt catalysis (Scheme 15.3c) [15]. A C2-selective, redox-neutral C—H alkenylation of indoles with both terminal and internal alkynes using a high-valent cobalt complex was achieved by the group of Matsunaga and Kanai (Scheme 15.3d) [12a]. In addition to cobalt catalysis, hydroarylations of terminal alkynes **19** also proved viable via manganese catalysis

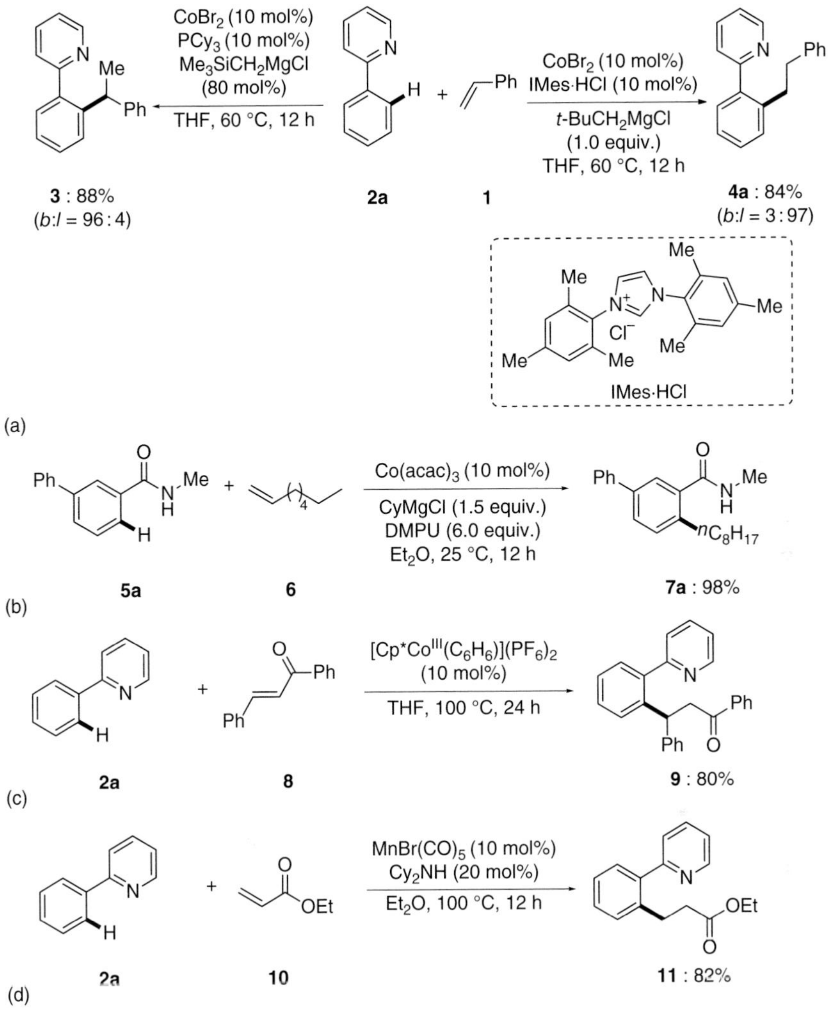

Scheme 15.2 Hydroarylation of alkenes.

[16]. Thus, Wang and coworkers reported the pyridine-directed addition of aromatic C—H bonds to a variety of terminal alkynes **19** in the presence of catalytic amounts of [$MnBr(CO)_5$] and Cy_2NH (Scheme 15.3e) [16a]. A related work by Li and coworkers expanded the scope of manganese-catalyzed hydroarylation reactions to heteroaromatic substrates [16b]. Here, the 2-pyrimidyl (pym) group [16d] was used as the DG for the C2-selective functionalization of indoles **17** and pyrroles.

15.2.1.2 Addition to C—Heteroatom Multiple Bonds

Addition to Imines First-row transition metal catalysis is not restricted to additions to C—C multiple bonds. Indeed, the addition manifold could be achieved with a variety of C—heteroatom multiple bonds, such as imines, carbonyls, and isocyanates as well [17]. For instance, Yoshikai and coworker employed

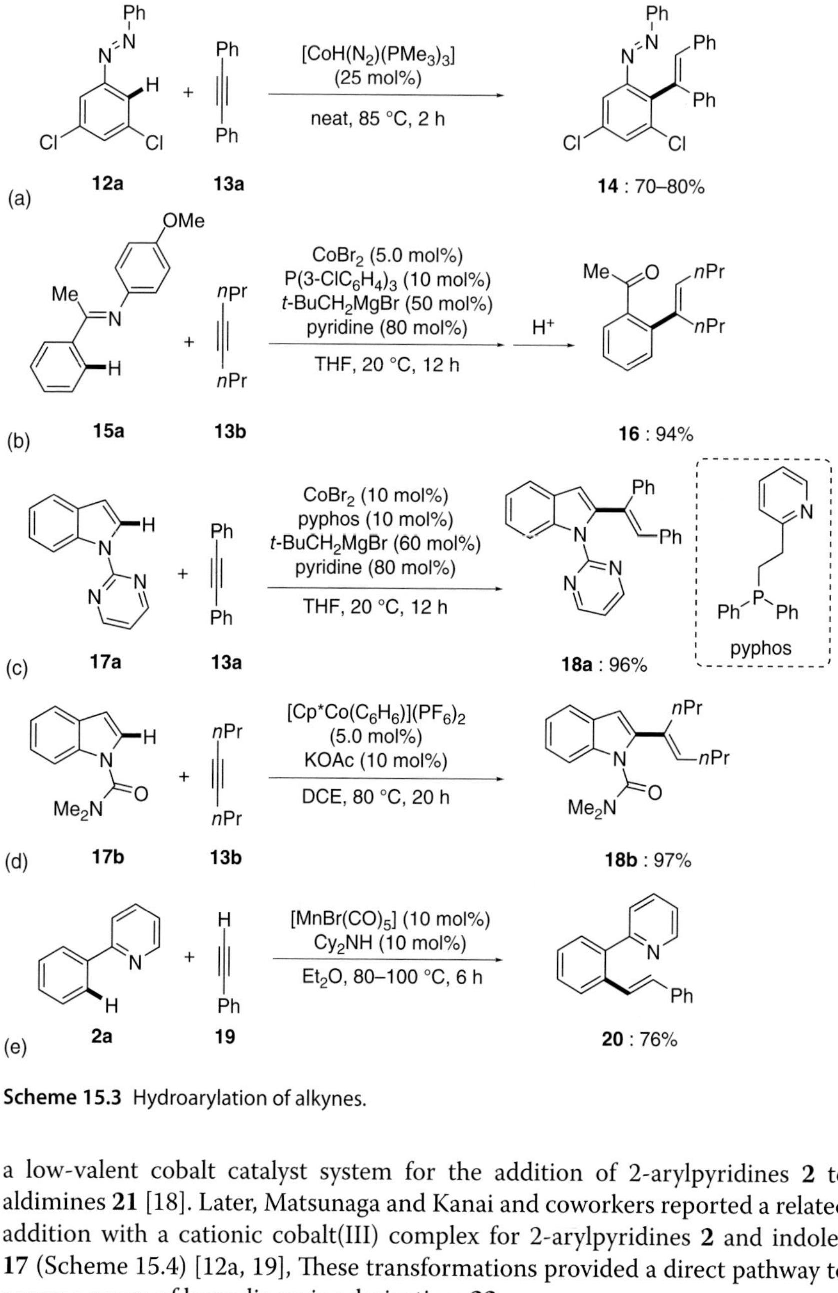

Scheme 15.3 Hydroarylation of alkynes.

a low-valent cobalt catalyst system for the addition of 2-arylpyridines **2** to aldimines **21** [18]. Later, Matsunaga and Kanai and coworkers reported a related addition with a cationic cobalt(III) complex for 2-arylpyridines **2** and indoles **17** (Scheme 15.4) [12a, 19], These transformations provided a direct pathway to access a range of benzylic amine derivatives **22**.

Addition to Aldehydes A pioneering manganese-catalyzed C—H activation and addition to the C=O bond of aldehydes **24** was reported by Kuninobo and Takai and coworkers [20a]. The reaction proceeded in the presence of catalytic amounts of [MnBr(CO)$_5$], along with stoichiometric quantities of Et$_3$SiH (**25**), to afford

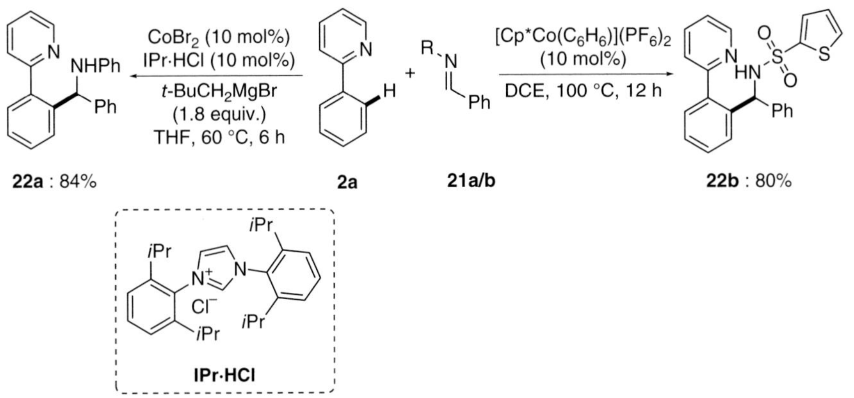

Scheme 15.4 Cobalt-catalyzed C–H activation with addition to imines.

the silyl-protected alcohols **26**. Mechanistic studies suggested that the cyclometalation occurs by oxidative addition (OA) of the C—H bond to the active manganese(I) catalyst. It is worth noting that the manganese(I)-catalyzed addition reaction sets the stage for a diastereoselective process through a chiral imidazoline auxiliary (Scheme 15.5a). Later, Wang significantly improved the atom and step economy of the addition reaction to aldehydes to access benzylic alcohols **27** directly instead of the silyl ethers **26**, by employing Me_2Zn and $ZnBr_2$ as stoichiometric additives with $[MnBr(CO)_5]$ as the catalyst (Scheme 15.5b) [20b]. Importantly, the catalytic system also enabled the challenging addition to the difficult nitrile group. Recently, Ellman and coworkers developed a convenient

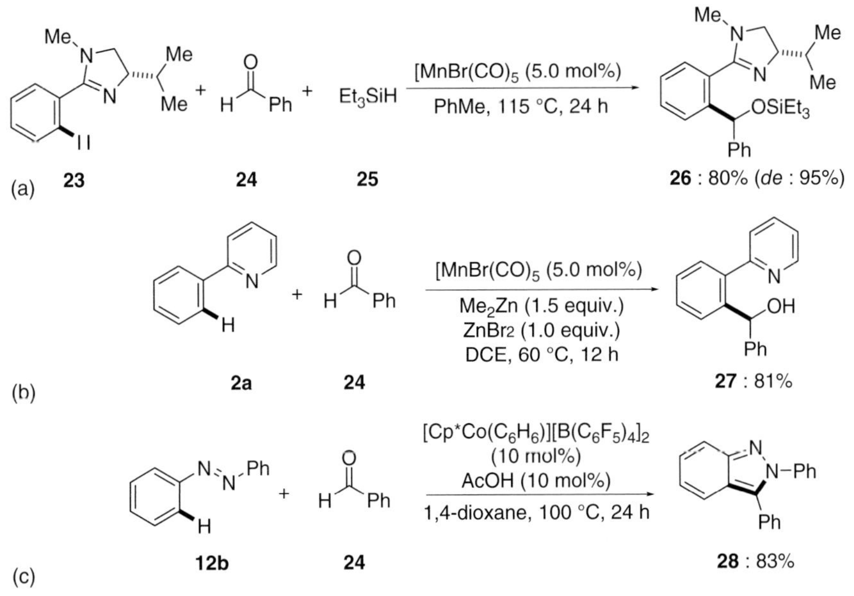

Scheme 15.5 C–H functionalization by addition to aldehydes **24**.

method to access *N*-aryl-2*H*-indazoles **28** from azoarenes **12** and aldehydes **24** through a cobalt(III)-catalyzed C—H activation proceeding through additions to aldehydes and trapping of the thus-formed alcohols by the DG (Scheme 15.5c) [20c]. Under similar reaction conditions, unsaturated oxime ethers underwent the reaction with aldehydes to provide substituted furans.

Addition to Isocyanates The Ackermann group reported the synthesis of aryl amides **31** by means of cobalt(III)-catalyzed C—H addition to isocyanates **30** (Scheme 15.6a) [21a]. In this process, both 1-aryl and 1-vinylpyrazoles **29** were effectively functionalized with ample substrate scope. Moreover, the same research group also demonstrated a versatile aminocarbonylation reaction with acyl azides. Concurrently, the Ellman group also described the amidation of aryl pyrazoles **29** with isocyanates **30** using a cationic cobalt(III) complex [21b]. The first manganese-catalyzed C—H activation and addition to isocyanates **30** was accomplished by the Ackermann group (Scheme 15.6b) [22]. The reaction permitted the selective preparation of various heteroaromatic amides **32** in the presence of catalytic amounts of [MnBr(CO)$_5$].

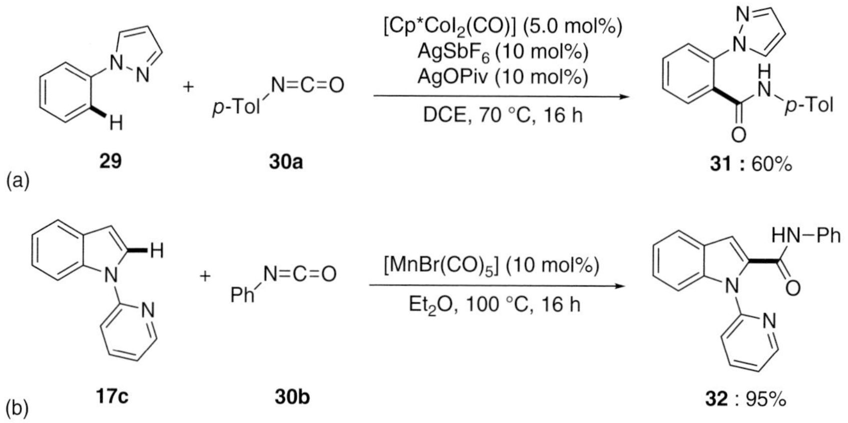

(a) **29** + *p*-Tol–N=C=O **30a** → [Cp*CoI$_2$(CO)] (5.0 mol%), AgSbF$_6$ (10 mol%), AgOPiv (10 mol%), DCE, 70 °C, 16 h → **31** : 60%

(b) **17c** + Ph–N=C=O **30b** → [MnBr(CO)$_5$] (10 mol%), Et$_2$O, 100 °C, 16 h → **32** : 95%

Scheme 15.6 Organometallic C–H functionalization through addition to isocyanates.

15.2.1.3 Oxidative C—H Olefination

Chelation-assisted oxidative olefination reactions of C—H bonds by base metal catalysis are rather scarce [23]. A notable report by Matsunaga and Kanai and coworkers using a cobalt(III) complex enabled the oxidative C—H alkenylation of benzamides **5** with acrylates **10** (Scheme 15.7a) [23a]. The reaction tolerated a wide range of substituents on the benzamide substrates, whereas the protocol proved limited to activated olefins. A similar catalytic system could be applied to the alkenylation of anilides as well.

In a recent study, the Ackerman group revealed a general strategy for the direct alkenylation of (hetero)aromatic C—H bonds with a range of cyclic and acyclic enol esters **34** (Scheme 15.7b) [23b]. The catalytic system consisting of CoI$_2$, 1,3-bis(2,6-diisopropylphenyl)imidazolium chloride (IPrHCl), and CyMgCl gave optimal results and allowed C—H activation at ambient temperature. In addition

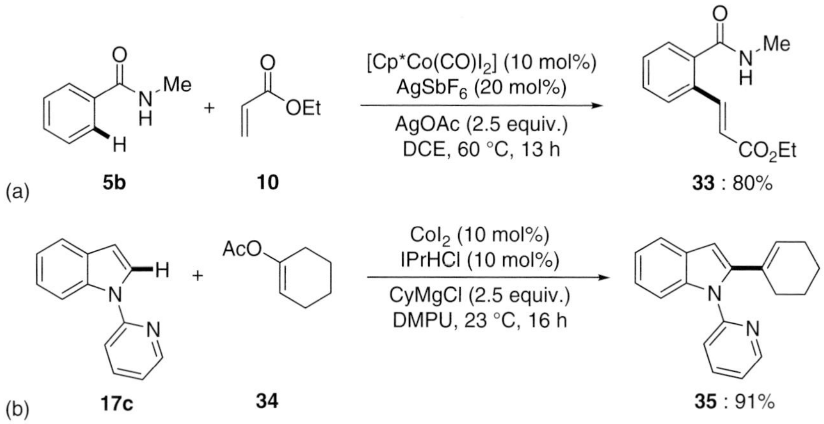

Scheme 15.7 Cobalt-catalyzed C–H alkenylation.

to enol esters, the low-valent cobalt catalyst proved applicable to enol phosphates, carbamates, and carbonates.

15.2.1.4 C—H Allylation

Recently, the direct allylation of (hetero)aromatic C—H bonds with allyl acetates, allyl alcohols, and allyl carbonates by cobalt(III) complexes were independently explored by the groups of Ackermann [24], Glorious [25], and Matsunaga and Kanai [26]. Thus, the Ackermann group developed a remarkable C—H functionalization with vinylcyclopropanes **36** for the synthesis of substituted allylic olefins **37** (Scheme 15.8a) [24b]. This unique process proceeded under mild reaction conditions to give *Z*-alkenes via an unusual C—H/C—C activation. Moreover, the Ackermann group also realized the C—H allylation by manganese catalysis, in which a range of aromatic kitimines **15** were effectively coupled with allyl carbonates **38** in the presence of $[Mn_2(CO)_{10}]$ to give *ortho*-allylated aromatic ketones **39** upon hydrolysis (Scheme 15.8b) [24c]. Related to the concept of cobalt-catalyzed C—H allylation, Sundararaju revealed the use of nickel catalysis in the C—H allylation of aromatic amides with allyl bromides **41** (Scheme 15.8c) [27c]. Further progress in this field was constituted by the Ackermann group developing an iron-catalyzed, triazole-assisted C—H allylation of arenes, heteroarenes, and alkenes with readily available allyl chlorides **44** (Scheme 15.8d) [24d].

15.2.1.5 Oxidative C—H Functionalization and Annulations

Chelation-assisted C—H functionalization through the annulation of alkenes by means of 3d transition metal catalysis were rarely reported as of yet (Scheme 15.9) [28]. In 2015, the Daugulis group described the annulation of alkenes by aromatic amides **40** via cobalt-catalyzed *ortho*-C—H activation [28a]. A variety of vinyl arenes as well as cyclic and acyclic aliphatic alkenes effectively participated in the reaction. During the same time, the Ackermann group reported a methodology for the synthesis of isoindolinones **47** from aromatic amides **40** and electron-deficient alkenes **10** in a solvent mixture consisting of

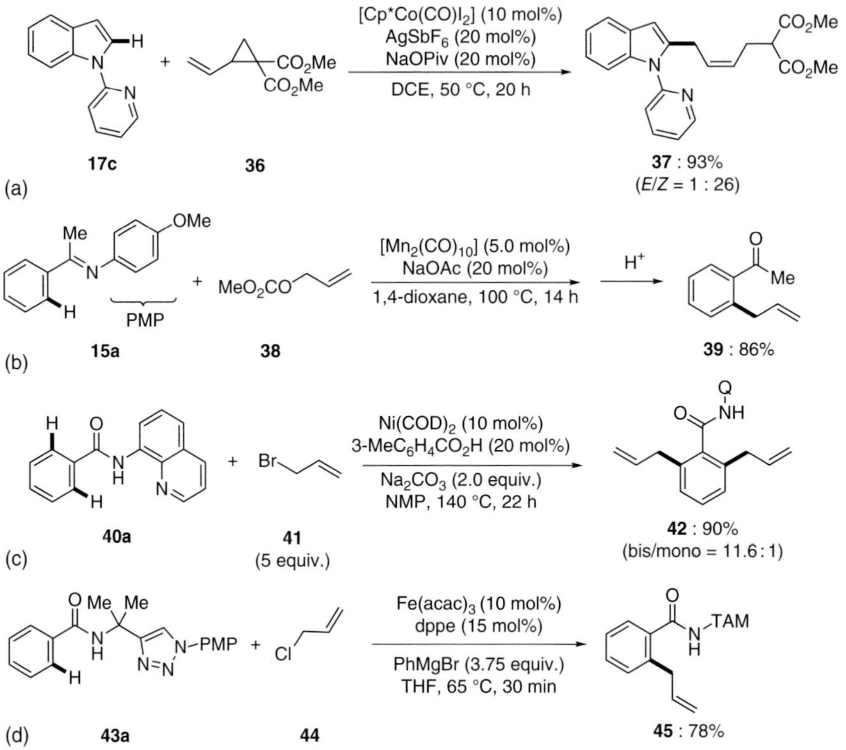

Scheme 15.8 Chelation-assisted C–H allylation.

polyethylene glycol (PEG) and 2,2,2-trifluoroethanol (TFE) [28b]. The reaction proceeded through the *N,N*-bidentate chelation-assisted C—H cobaltation followed by migratory insertion of the olefin into the C—Co bond and subsequent β-hydride elimination, delivering the *ortho*-alkenylation products. The intramolecular hydroamination of the thus-formed olefinated product furnished the final isoindolinone products **47**. A similar type of oxidative cyclization of aromatic amides with maleimides **48** was achieved using a catalytic system consisting of $Cu(OAc)_2$ and Cy_2NMe [28c]. An interesting cobalt-catalyzed annulation reaction by the Cheng group allowed the diastereoselective synthesis of dihydroepoxybenzofluorenone derivatives **51** from aromatic/vinylic amides and bicyclic alkenes **50** via C—H/C—N activation [28d].

Organometallic C—H functionalization and annulation with alkynes were identified as a powerful method in heterocycle synthesis. In particular, the synthesis of isoquinoline derivatives **53** was extensively studied owing to their high natural abundance in a large number of natural, bioactive compounds [29a]. In 2015, the Ackermann [29b], Kanai/Matsunaga [29c], and Sundararaju [29d] groups concurrently reported the synthesis of substituted isoquinolines **53** by cobalt(III)-catalyzed C—H/N—O functionalization of (hetero)aromatic ketoxime derivatives **52** with alkynes **13** (Scheme 15.10a) [29]. In this process, the N—O bond of the oxime derivatives acted as an internal oxidant, and the

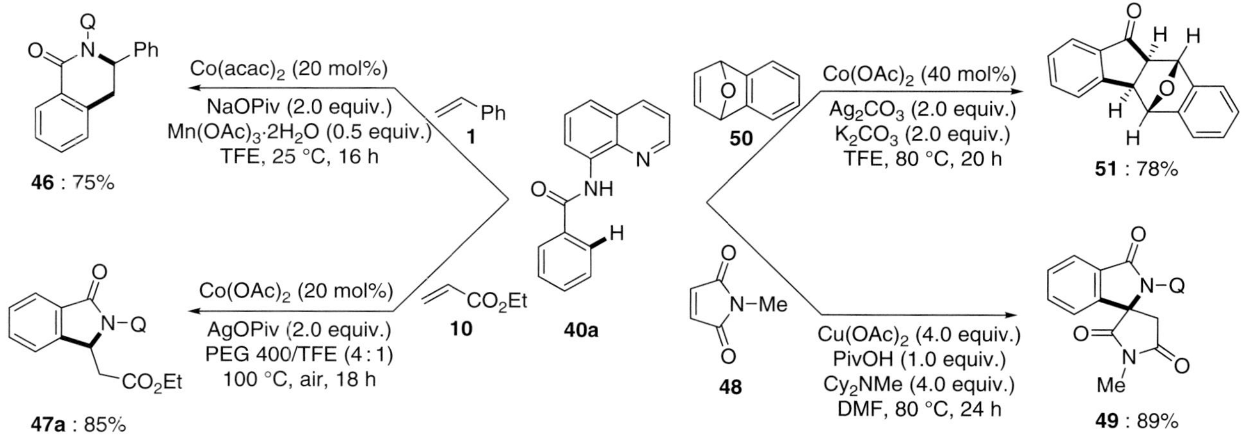

40a

1: Co(acac)$_2$ (20 mol%), NaOPiv (2.0 equiv.), Mn(OAc)$_3$·2H$_2$O (0.5 equiv.), TFE, 25 °C, 16 h → **46** : 75%

10: Co(OAc)$_2$ (20 mol%), AgOPiv (2.0 equiv.), PEG 400/TFE (4 : 1), 100 °C, air, 18 h → **47a** : 85%

50: Co(OAc)$_2$ (40 mol%), Ag$_2$CO$_3$ (2.0 equiv.), K$_2$CO$_3$ (2.0 equiv.), TFE, 80 °C, 20 h → **51** : 78%

48: Cu(OAc)$_2$ (4.0 equiv.), PivOH (1.0 equiv.), Cy$_2$NMe (4.0 equiv.), DMF, 80 °C, 24 h → **49** : 89%

Scheme 15.9 Chelation-guided C–H functionalization and annulation with alkenes.

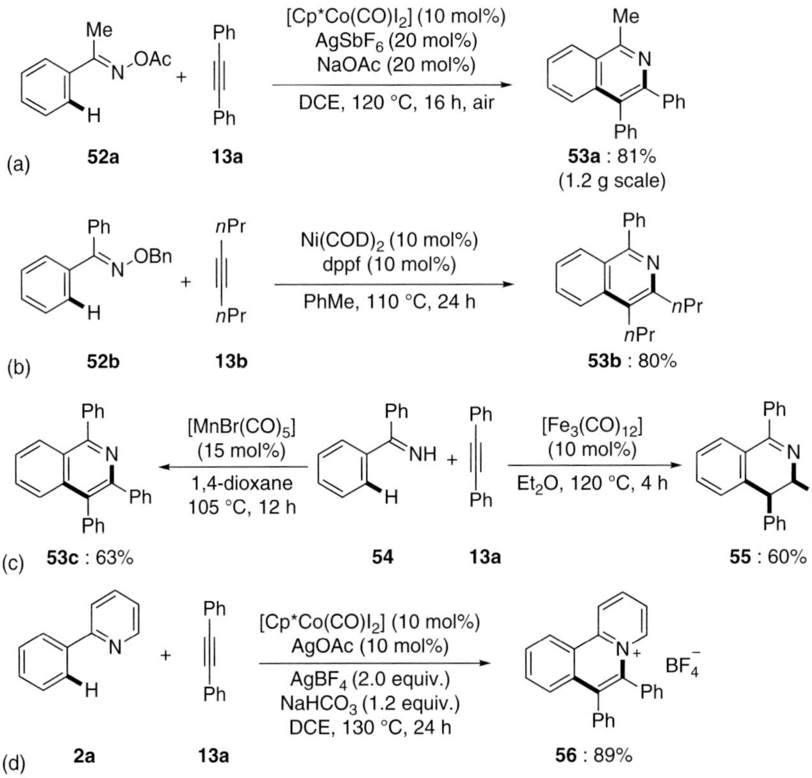

Scheme 15.10 Synthesis of isoquinoline derivatives via C–H activation.

catalyst thus did not require any additional external oxidant. Further, Kurahashi and Matsubara employed $Ni(COD)_2$ and 1,1′-bis(diphenylphosphino)ferrocene (dppf) as the catalytic system for isoquinoline synthesis from aromatic ketoxime ethers **52** with alkynes **13** (Scheme 15.10b) [30]. Furthermore, substituted isoquinolines **53** were conveniently obtained through the dehydrogenative [4+2] annulation of alkynes with aromatic N–H ketimines **54** by means of manganese(I) catalysis (Scheme 15.10c) [31a]. This manganese-catalyzed, redox-neutral process produced H_2 as evidenced by mass spectrometry as the only by-product without any oxidants or additives. Recently, the Wang group also demonstrated the efficiency of the iron complex $[Fe_3(CO)_{12}]$ in the redox-neutral annulation of internal alkynes **13** by aromatic N–H ketimines **54** to provide *cis*-3,4-dihydroisoquinolines **55** via C—H cleavage (Scheme 15.10c) [31b]. In their recent studies, the Cheng group realized an efficient synthetic route for various *N*-heterocyclic cations **56** from 2-arylpyridiens **2** through a cobalt(III)-catalyzed C—H activation and annulation with alkynes **13** (Scheme 15.10d) [32].

As discussed above, base metal complexes were shown to participate in C—H activation followed by alkyne insertion, which enabled the synthesis of complex heterocyclic compounds [28–32]. Based on this general strategy, a variety of aromatic amides were cyclized with alkynes in a formal [4+2] [33] or [4+1] [34] manner by means of first-row transition metal catalysis (Scheme 15.11).

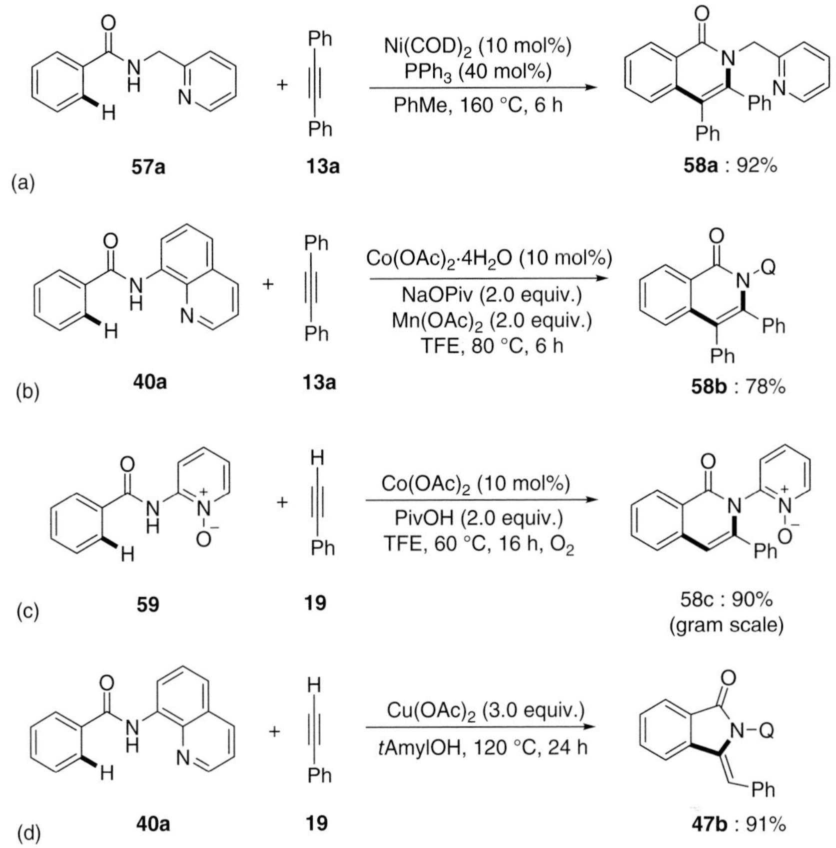

(a) 57a + 13a → 58a : 92%
$Ni(COD)_2$ (10 mol%)
PPh_3 (40 mol%)
PhMe, 160 °C, 6 h

(b) 40a + 13a → 58b : 78%
$Co(OAc)_2{\cdot}4H_2O$ (10 mol%)
NaOPiv (2.0 equiv.)
$Mn(OAc)_2$ (2.0 equiv.)
TFE, 80 °C, 6 h

(c) 59 + 19 → 58c : 90% (gram scale)
$Co(OAc)_2$ (10 mol%)
PivOH (2.0 equiv.)
TFE, 60 °C, 16 h, O_2

(d) 40a + 19 → 47b : 91%
$Cu(OAc)_2$ (3.0 equiv.)
*t*AmylOH, 120 °C, 24 h

Scheme 15.11 Chelation-assisted C–H functionalization of aromatic amides with alkynes.

In 2011, the Chatani group demonstrated the nickel-catalyzed annulation of internal alkynes **13** by aromatic amides **57** containing a 2-pyridinylmethylamine group (Scheme 15.11a) [33a]. Later in 2014, similar transformations using the simple cobalt salt $Co(OAc)_2{\cdot}4H_2O$ were developed by Daugulis and coworkers (Scheme 15.11b) [33e]. In this reaction, amides **40** with an 8-aminoquinoline (AQ) moiety acted as the substrates for selective C—H cleavages. Significant progress in the formation of isoquinolones **58** from aromatic amides bearing a bidentate 2-pyridyl-*N*-oxide (PyO) group and alkynes using cobalt catalysis was made by the Ackermann group (Scheme 15.11c) [33f]. In this transformation, O_2 was used as the sole oxidant under mild reaction conditions. Interestingly, the reaction of quinoline-appended aromatic amides with terminal alkynes catalyzed by copper [34a, b], cobalt [34c], and nickel [34d] complexes provided the 3-methyleneisoindolin-1-one **47** derivatives via a formal [4+1] cycloaddition instead of the [4+2] pathway (Scheme 15.11d).

The versatile cobalt(III) catalysis regime enabled the step-economical synthesis of substituted indoles through directed C—H activation followed by alkyne annulations (Scheme 15.12) [35]. The redox-neutral [3+2] annulation process of aromatic nitrones **73** and alkynes **13** was developed by the Ackermann group,

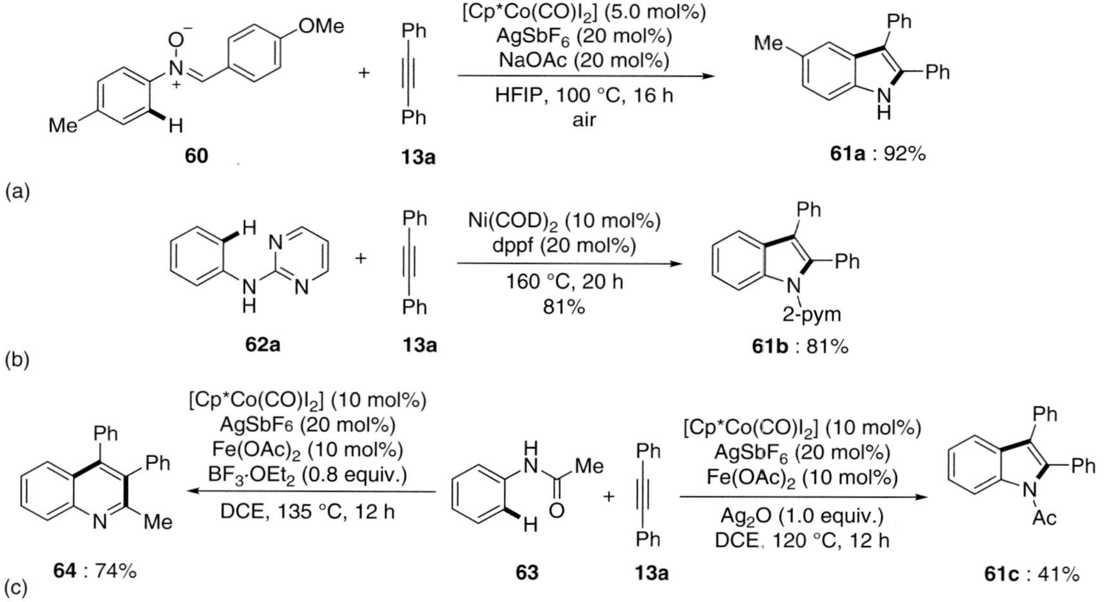

(a)

60 + 13a → 61a : 92%

$[Cp^*Co(CO)I_2]$ (5.0 mol%)
$AgSbF_6$ (20 mol%)
NaOAc (20 mol%)
HFIP, 100 °C, 16 h
air

(b)

62a + 13a → 61b : 81%

$Ni(COD)_2$ (10 mol%)
dppf (20 mol%)
160 °C, 20 h
81%

(c)

63 + 13a → 61c : 41%

$[Cp^*Co(CO)I_2]$ (10 mol%)
$AgSbF_6$ (20 mol%)
$Fe(OAc)_2$ (10 mol%)
Ag_2O (1.0 equiv.)
DCE, 120 °C, 12 h

63 → 64 : 74%

$[Cp^*Co(CO)I_2]$ (10 mol%)
$AgSbF_6$ (20 mol%)
$Fe(OAc)_2$ (10 mol%)
$BF_3{\cdot}OEt_2$ (0.8 equiv.)
DCE, 135 °C, 12 h

Scheme 15.12 Synthesis of indoles by organometallic C–H functionalization.

providing a wide range of unprotected indoles **61** in a single step with ample substrate scope and excellent yields (Scheme 15.12a) [35b]. Likewise, the same research group devised a nickel catalyst to access indoles from anilines **62** and alkynes **13** [35a]. The reaction proceeded smoothly with catalytic amounts of $Ni(COD)_2$ and dppf under solvent-free reaction conditions without any additional additives (Scheme 15.12b). Recently, the Glorius group employed Boc-protected hydrazines and alkynes as the substrates for the indole synthesis [35c]. The control of the chemoselectivity to form quinolines **64** or indoles **61** from anilides **63** and alkynes **13** proved viable here (Scheme 15.12c) [35d].

15.2.1.6 C—H Alkynylations

Chelation-assisted organometallic C—H functionalization by means of copper [36], nickel [37], and cobalt catalysis [38] significantly improved the synthesis of alkynylated arenes with respect to the classical Sonogashira–Hagihara coupling reaction (Scheme 15.13). After the Yu group's copper-mediated C—H alkynylation of amides (Scheme 15.13a) [16a], a variety of C—H alkynylations of aromatic amides with terminal alkynes **19** and alkynylbromides **67** were reported. The first nickel-catalyzed C—H alkynylation of anilines with alkynyl bromide **67** was achieved by the Ackermann group (Scheme 15.13b) [37e]. Very recently, the same researchers disclosed a catalytic high-valent cobalt system for the C—H alkynylation of indoles **17** under exceedingly mild reaction conditions, which is at an ambient temperature of 23 °C (Scheme 15.13c) [38a].

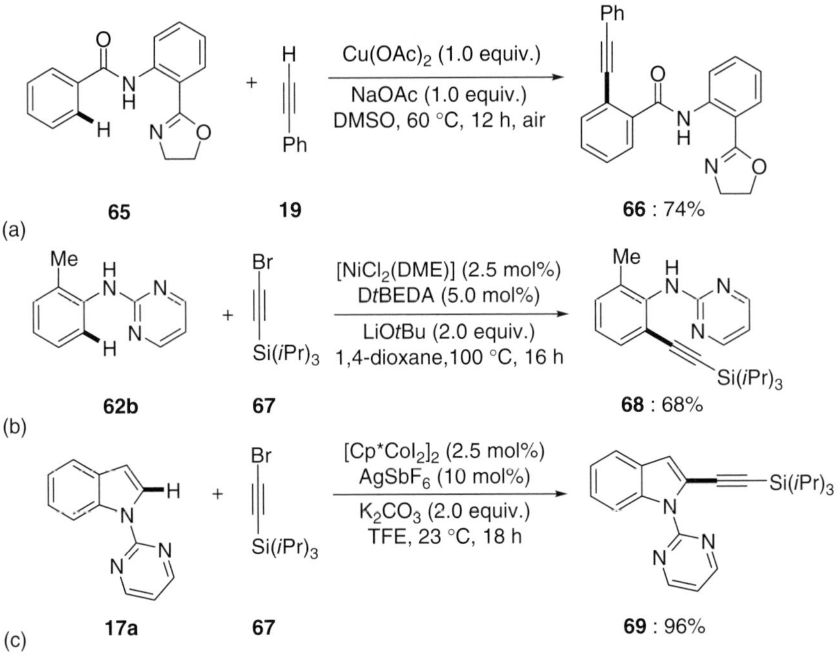

(a) **65** + **19** → **66**: 74%; $Cu(OAc)_2$ (1.0 equiv.), NaOAc (1.0 equiv.), DMSO, 60 °C, 12 h, air

(b) **62b** + **67** → **68**: 68%; $[NiCl_2(DME)]$ (2.5 mol%), D*t*BEDA (5.0 mol%), LiO*t*Bu (2.0 equiv.), 1,4-dioxane, 100 °C, 16 h

(c) **17a** + **67** → **69**: 96%; $[Cp^*CoI_2]_2$ (2.5 mol%), $AgSbF_6$ (10 mol%), K_2CO_3 (2.0 equiv.), TFE, 23 °C, 18 h

Scheme 15.13 Directed C–H alkynylation of (hetero)arenes.

15.2.2 C—H Cyanation

The development of new methodologies for the synthesis of organic nitriles continues to be of great interest because of their numerous applications in functional group transformations and bioactive compound syntheses [39]. Recently, great developments were made in their synthesis through the use of nontoxic cyanation reagents for step-economical C—H cyanation routes (Scheme 15.14) [39c]. Over a decade, copper catalysis was effectively employed for the C—H cyanation with various cyanating reagents, including nitromethane, acetonitrile, benzyl nitrile, and azobisisobutyronitrile (AIBN) [40]. In contrast, the Ackermann group developed a cobalt(III)-catalyzed C—H activation strategy for the cyanation of (hetero)aromatic C—H bonds with *N*-cyano-*N*-phenyl-*p*-toluenesulfonamide (NCTS) (**70**) (Scheme 15.14a) [41a]. In an independent report, Glorius disclosed an analogous catalytic system for the C—H cyanation of (hetero)arenes with NCTS [25a]. The concept of cobalt(III)-catalyzed C—H cyanation was also realized on 6-arylpurines **72a**, in which *N*-cyanosuccinimide (**73**) was exploited as the effective cyanating source (Scheme 15.14b) [41b].

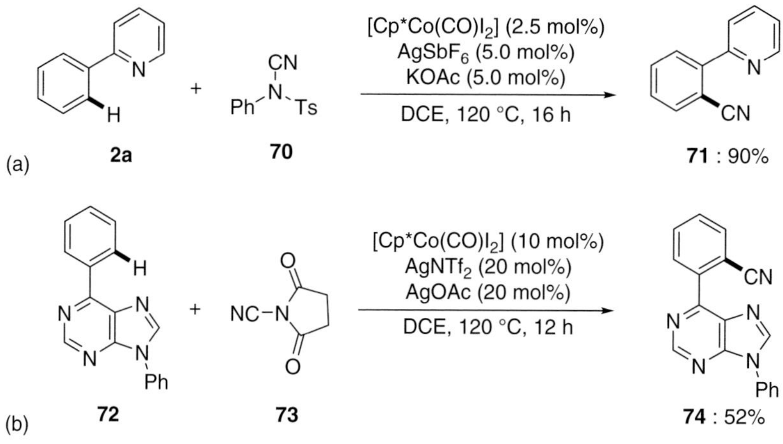

(a) **2a** + **70** → [Cp*Co(CO)I_2] (2.5 mol%), $AgSbF_6$ (5.0 mol%), KOAc (5.0 mol%), DCE, 120 °C, 16 h → **71** : 90%

(b) **72** + **73** → [Cp*Co(CO)I_2] (10 mol%), $AgNTf_2$ (20 mol%), AgOAc (20 mol%), DCE, 120 °C, 12 h → **74** : 52%

Scheme 15.14 Cobalt-catalyzed C–H cyanation.

15.2.3 C—H Arylation

The catalytic C—H arylation with low-valent cobalt systems is effective with a variety of electrophiles, including inexpensive but usually difficult to activate aryl chlorides, carbamates, and sulfamates [42]. In this regard, Ackermann and Song reported a catalytic system consisting of $Co(acac)_2$ (acac, acetylacetonate) and IMesHCl for the arylation of (hetero)aromatic compounds via *N*-heterocycle-assisted C—H/C—O activation (Scheme 15.15a) [42a]. Furthermore, the Ackermann group expanded the scope of their catalytic system to aromatic amides and aryl imidates [42b–d]. Meanwhile, Yoshikai also employed a low-valent cobalt system for the synthesis of biaryl ketones **78** starting from ketimines **15** and aryl chlorides **77** (Scheme 15.15b) [42e].

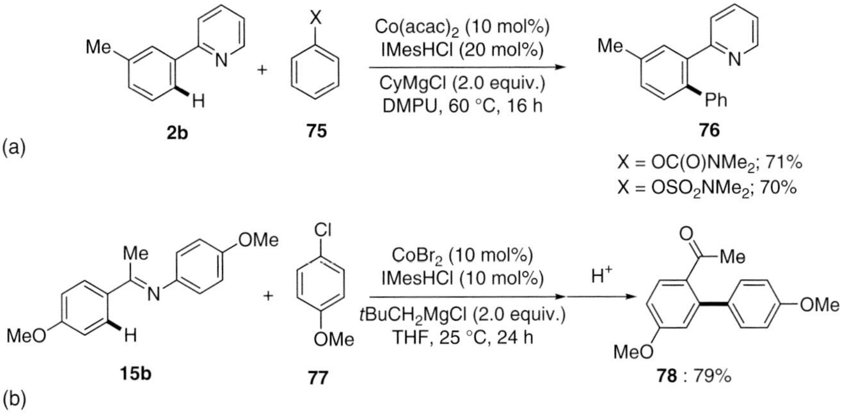

Scheme 15.15 Cobalt-catalyzed C–H arylation.

Organic transformations by means of iron catalysis are highly attractive because of the earth abundancy and low toxicity of iron [43]. The studies by the Nakamura group made inexpensive iron catalysis also viable for the chelation-controlled C—H arylation [44]. In 2008, they reported a remarkable arylation of arenes assisted by *N*-heterocycles with *in situ*-generated Ph_2Zn from PhMgBr and $ZnCl_2$·TMEDA (TMEDA, *N*,*N*,*N*′,*N*′-tetramethylethylenediamine) (Scheme 15.16a). The reaction proceeded at 0 °C with 10 mol% of $Fe(acac)_3$ and 1,10-phenanthroline as well as 2.0 equiv. of dichloroisobutane (DCIB) as the sacrificial oxidant [44a]. Later, the concept was also applied to the C—H arylation of aromatic amides and imines [44b–f]. Recently, Ackermann and coworkers developed a catalytic system consisting of $FeCl_3$, 1,2-bis(diphenylphosphino)ethane (dppe), and DCIB for the C—H arylation of aromatic and aliphatic amides by the assistance of an easily removable triazole-based bidentate chelation system (Scheme 15.16b) [45].

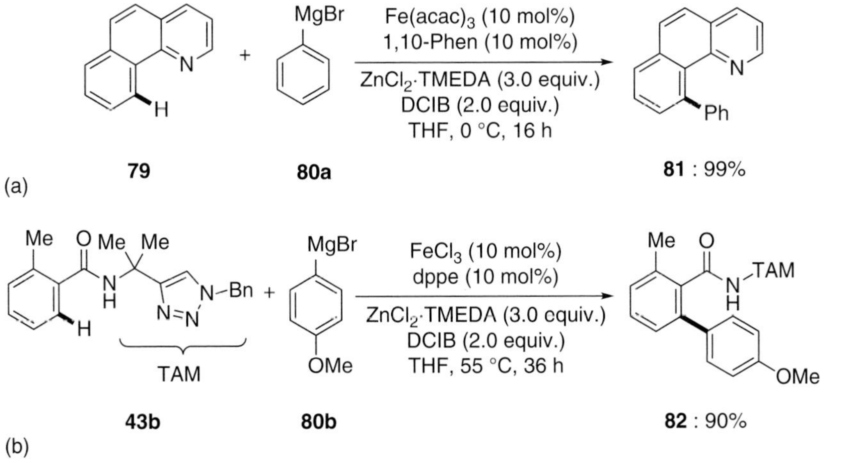

Scheme 15.16 Iron-catalyzed C–H arylation.

Copper catalysis was effectively employed in directed C—H arylation reactions in a stoichiometric or catalytic manner [10]. Intermolecular dehydrogenative C—C bond formation through double C—H activation is one of the most step-economical ways for the synthesis of biaryls. The Hirano/Miura group extensively studied the copper-mediated C—H/C—H coupling for biaryl formation [46]. Hence, they found that the direct coupling of 2-phenylpyridines **2** was viable with 1,3-azoles **83** in the presence of 5.0 equiv. of $Cu(OAc)_2$ and 1.0 equiv. of pivalic acid (PivOH) (Scheme 15.17a) [46a]. A subsequent process by the Dai/Yu group allowed for the use of arylboronates **85** as the coupling partners in the copper-catalyzed C—H arylation of aromatic amides **65** containing a oxazoline-based bidentate auxiliary (Scheme 15.17b) [47a].

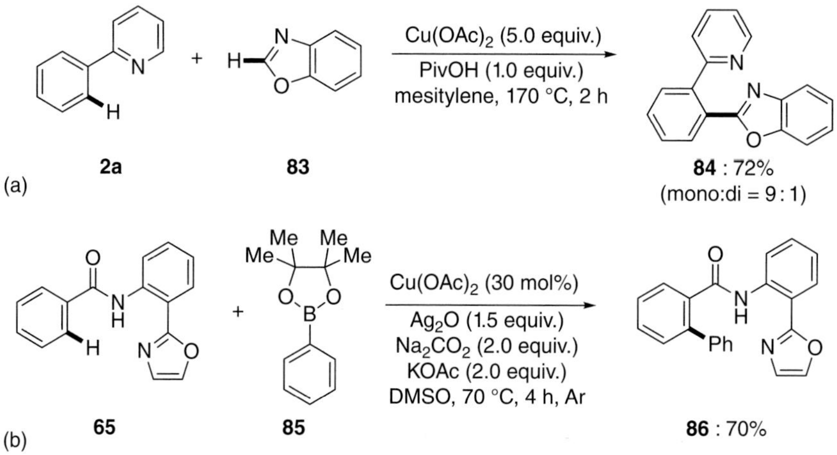

(a) **2a** + **83** → **84**: 72% (mono:di = 9 : 1)
$Cu(OAc)_2$ (5.0 equiv.), PivOH (1.0 equiv.), mesitylene, 170 °C, 2 h

(b) **65** + **85** → **86**: 70%
$Cu(OAc)_2$ (30 mol%), Ag_2O (1.5 equiv.), Na_2CO_2 (2.0 equiv.), KOAc (2.0 equiv.), DMSO, 70 °C, 4 h, Ar

Scheme 15.17 Copper-mediated/catalyzed C–H arylation.

During the past decade, nickel complexes were frequently employed in the synthesis of biaryl compounds through cross-couplings and heteroarene C—H functionalization, whereas similar processes of unactivated aromatic C—H bonds continue to be scarce [8]. Recently, Chatani and coworkers accomplished the *ortho*-selective C—H arylation of aromatic amides **40** with aryl iodides **87** employing the AQ-based *N,N*-bidentate ligand at an elevated reaction temperature of 160 °C (Scheme 15.18) [48].

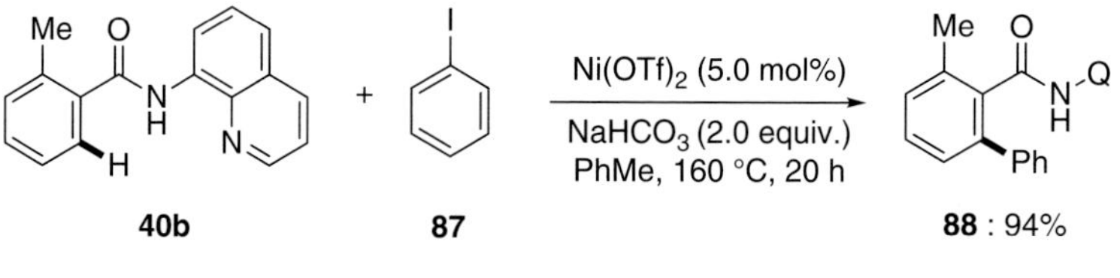

40b + **87** → **88**: 94%
$Ni(OTf)_2$ (5.0 mol%), $NaHCO_3$ (2.0 equiv.), PhMe, 160 °C, 20 h

Scheme 15.18 Nickel-catalyzed chelation-assisted C–H arylation.

15.2.4 C—H Alkylation

Alkylation of aromatic C—H bonds by non-noble metals has been performed with both electrophilic (alkyl halides/pseudo halides) and nucleophilic reagents (organometallic reagents) [49]. The competence of cobalt catalysis in C—H alkylations was established by the groups of Nakamura, Ackermann, and Yoshikai. For instance, Nakamura reported on the *ortho*-C—H alkylation of secondary amides **5** by alkyl chlorides **89** at ambient temperature using the $Co(acac)_2$/CyMgCl/1,3-dimethyl-3,4,5,6-tetrahydro-2-pyrimidinone (DMPU) catalytic system (Scheme 15.19a) [50]. The Ackermann group found that the NHC precursor IPrHCl significantly accelerated the efficiency of cobalt-catalyzed C—H alkylation processes (Scheme 15.19b) [42c]. Furthermore, Ackermann's protocol was also applicable to perform C—H benzylations of indoles **17** with benzylic phosphates [42b]. The cobalt-catalyzed C—H alkylation concept was further applied to the alkylation and benzylation of aromatic ketimines **15** by Yoshikai (Scheme 15.19c) [51].

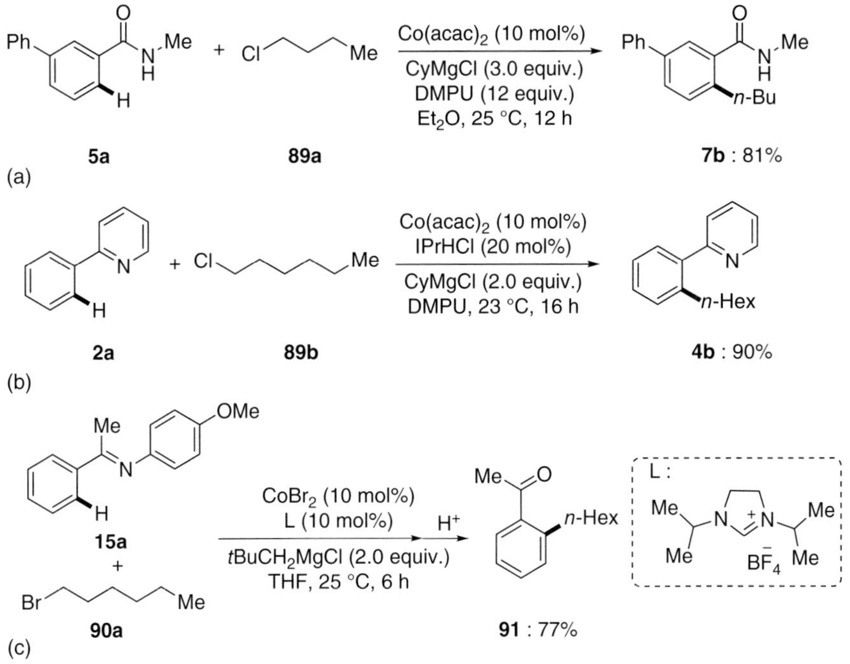

Scheme 15.19 Cobalt-catalyzed C–H alkylation.

The catalytic C—H alkylations were also performed with inexpensive iron complexes (Scheme 15.20) [52–54]. Thus, Ilies and Nakamura demonstrated the iron-catalyzed C—H alkylation of aromatic and olefinic carboxamides **92** with alkyl chlorides by the assistance of AQ-based bidentate auxiliaries

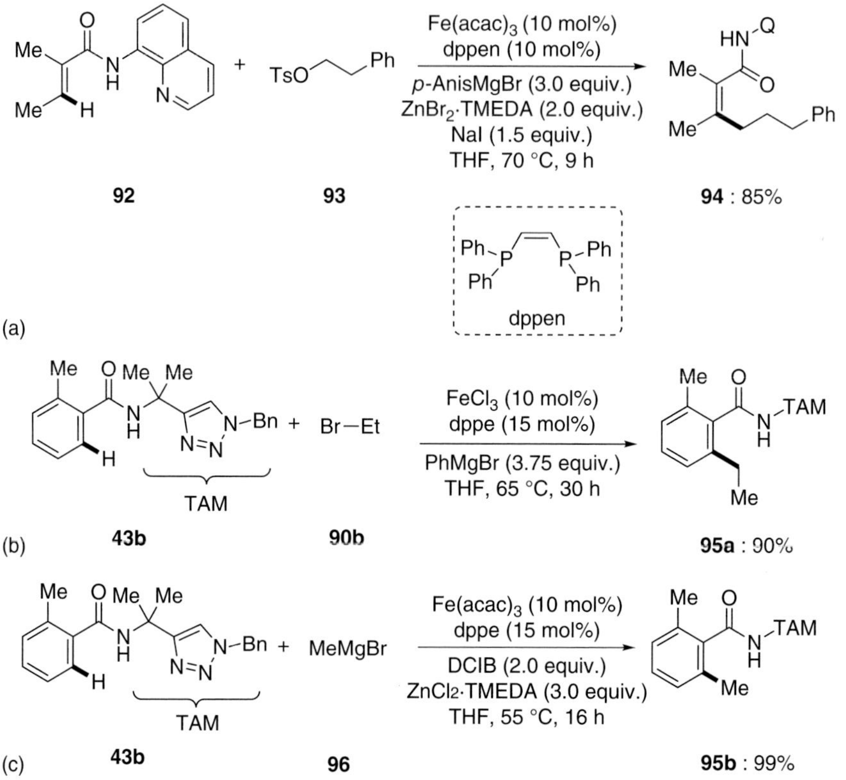

Scheme 15.20 Iron-catalyzed C–H alkylation.

(Scheme 15.20a) [52]. In this process, a variety of alkyl electrophiles, including primary and secondary alkyl tosylates **93**, mesylates, and halides, were suitable to give satisfactory product yields. A report by the Ackermann group established the utilization of the easily removable triazole-based triazolyldimethyl (TAM) amide group for the expedient C—H alkylation of (hetero)aromatic amides **43** with alkyl bromides **90** (Scheme 15.20b) [24d]. The reaction displayed great functional group tolerance and ample scope. Very recently, iron catalysis was also successfully employed for the methylation of unactivated C—H bonds [54]. In their independent studies, Ackermann [54a], and Ilies and Nakamura [54b, c] employed MeMgBr or $AlMe_3$ as the methylating reagents, respectively, for the C—H alkylations (Scheme 15.20c).

Intermolecular *ortho*-selective C—H alkylation of aryl and vinyl amides with alkyl halides was succeeded with the catalytic system consisting of $Ni(OTf)_2$ and PPh_3 by the Chatani group (Scheme 15.21a) [55]. These reaction conditions were also suitable for the C—H methylation of aromatic amides using TsOMe and NaI or $PhMe_3NI$ as the methylating reagents [55c]. Recently, the Ackermann group discovered that bis(2-dimethylaminoethyl)ether (BDMAE) can be an effective ligand for $NiCl_2$(DME) in the C—H alkylation of aromatic amides **40** [56a]. In their studies, the combination of $NiCl_2$(DME) and

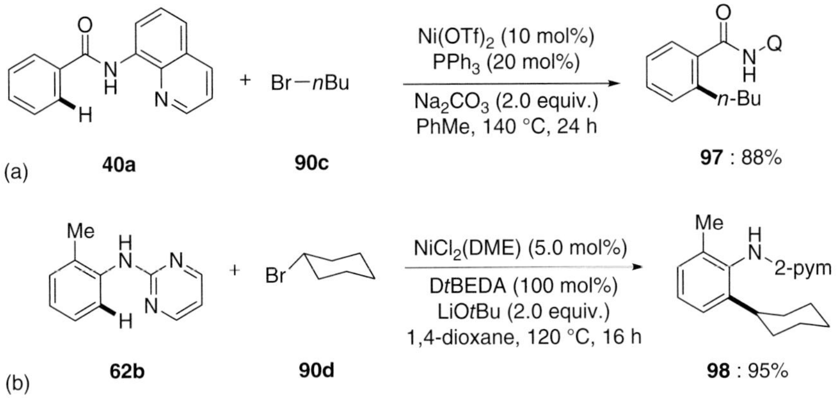

Scheme 15.21 Chelation-assisted nickel-catalyzed C–H alkylation.

di-*tert*-butylethane-1,2-diamine (D*t*BEDA) was found to be a successful system for the C—H alkylation of various anilines **62** (Scheme 15.21b) [56b].

15.3 C—Heteroatom Formation via C—H Activation

15.3.1 C—N Formation via C—H Activation

The amino group is an important functionality widely present in natural and bioactive compounds [57]. Because of the great applications of amines in molecular synthesis, developing step- and cost-effective methods for their synthesis is of outmost importance in academic and industrial research [57a]. During the past decade, transition-metal-catalyzed C—H functionalization strategies for the construction of C—N bonds became highly popular because of the atom and step economy and broad scope [58].

15.3.1.1 C—H Amination with Unactivated Amines

In 2006, Yu [40a] and Chatani [59a] independently reported the C—H amination of 2-phenylpyridines **2** with tosylamine ($TsNH_2$, **99**) and aniline, respectively, using a stoichiometric amount of $Cu(OAc)_2$. Despite the relatively narrow scope, these results inspired many research groups to develop C—H amination with amines using catalytic amounts of copper(II) salts (Scheme 15.22) [59–61]. Recent progress in this arena showed that nickel [62] and cobalt complexes [63] were also suitable catalysts for the direct C—H amination with unactivated amines.

15.3.1.2 C—H Amination with Activated Amine Sources

Organometallic C—H aminations with unactivated amines usually require a stoichiometric external oxidant [59–63]. Recently, the utilization of organic azides **104** as the nitrogenation agents set the stage for transformations under oxidant-free and mild reaction conditions [64, 65]. In 2014, Zhu and coworkers developed a copper-promoted *ortho*-C—H amination protocol for the direct access to primary amino groups using $TMSN_3$ (**104a**) (Scheme 15.23a) [64].

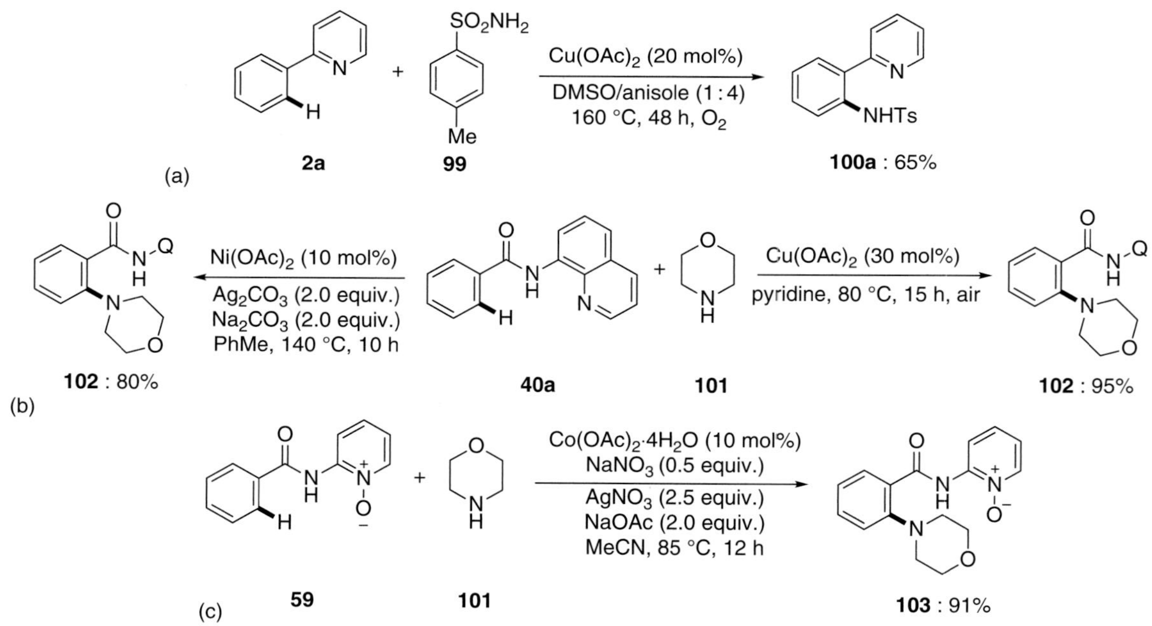

(a) **2a** + **99** (SO_2NH_2, Me) → $Cu(OAc)_2$ (20 mol%), DMSO/anisole (1 : 4), 160 °C, 48 h, O_2 → **100a** (NHTs) : 65%

(b) **40a** + **101** → $Cu(OAc)_2$ (30 mol%), pyridine, 80 °C, 15 h, air → **102** : 95%

40a + **101** → $Ni(OAc)_2$ (10 mol%), Ag_2CO_3 (2.0 equiv.), Na_2CO_3 (2.0 equiv.), PhMe, 140 °C, 10 h → **102** : 80%

(c) **59** + **101** → $Co(OAc)_2{\cdot}4H_2O$ (10 mol%), $NaNO_3$ (0.5 equiv.), $AgNO_3$ (2.5 equiv.), NaOAc (2.0 equiv.), MeCN, 85 °C, 12 h → **103** : 91%

Scheme 15.22 Chelation-assisted C–H amination using amines.

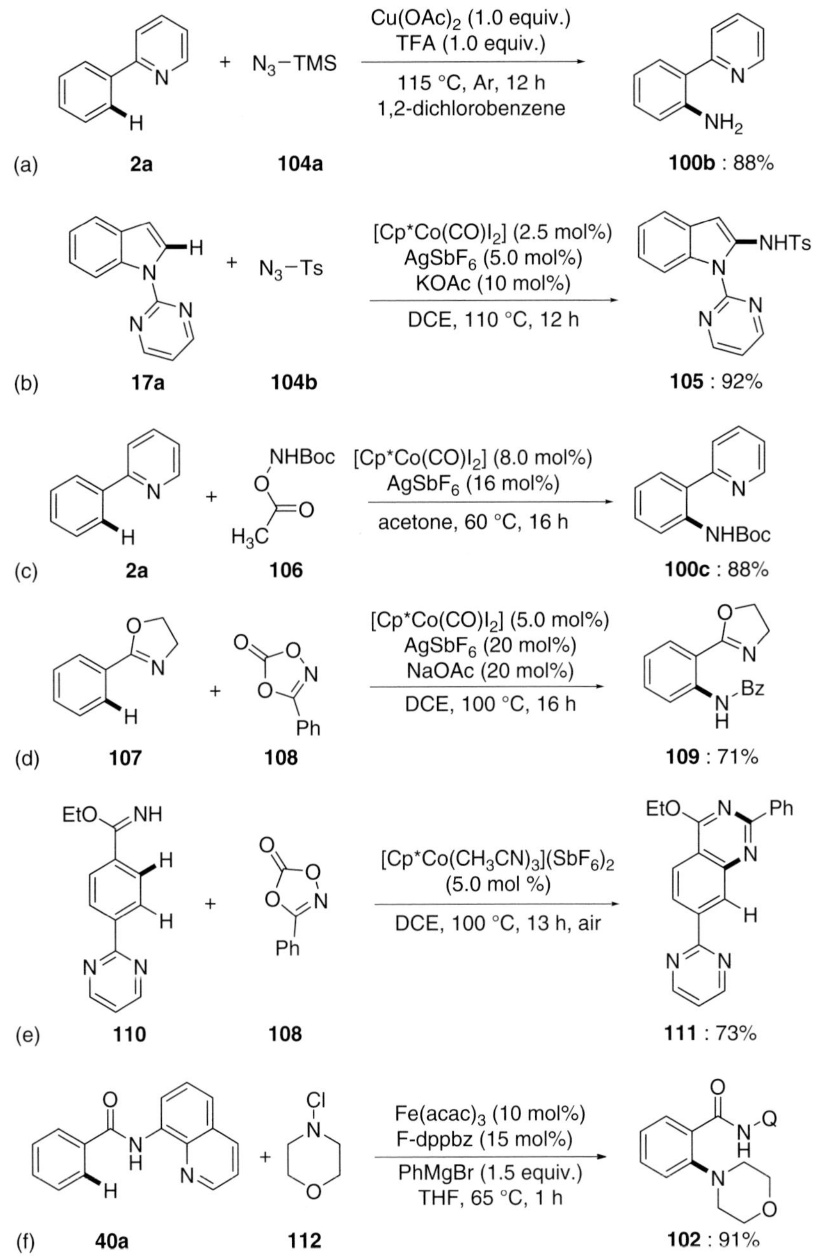

(a) **2a** + N_3–TMS (**104a**) → $Cu(OAc)_2$ (1.0 equiv.), TFA (1.0 equiv.); 115 °C, Ar, 12 h, 1,2-dichlorobenzene → NH_2 **100b** : 88%

(b) **17a** + N_3–Ts (**104b**) → $[Cp^*Co(CO)I_2]$ (2.5 mol%), $AgSbF_6$ (5.0 mol%), KOAc (10 mol%); DCE, 110 °C, 12 h → NHTs **105** : 92%

(c) **2a** + NHBoc-O-C(=O)-H_3C (**106**) → $[Cp^*Co(CO)I_2]$ (8.0 mol%), $AgSbF_6$ (16 mol%); acetone, 60 °C, 16 h → NHBoc **100c** : 88%

(d) **107** + **108** (Ph) → $[Cp^*Co(CO)I_2]$ (5.0 mol%), $AgSbF_6$ (20 mol%), NaOAc (20 mol%); DCE, 100 °C, 16 h → N(H)-Bz **109** : 71%

(e) **110** (EtO, NH) + **108** (Ph) → $[Cp^*Co(CH_3CN)_3](SbF_6)_2$ (5.0 mol %); DCE, 100 °C, 13 h, air → EtO, Ph **111** : 73%

(f) **40a** + Cl–N morpholine (**112**) → $Fe(acac)_3$ (10 mol%), F-dppbz (15 mol%); PhMgBr (1.5 equiv.), THF, 65 °C, 1 h → N(H)-Q **102** : 91%

Scheme 15.23 Chelation-assisted C–H amination with functionalized amine sources.

Independently, the Matsunaga/Kanai group explored cobalt(III) catalysis for the C2-selective amination of indoles **17** using sulfonyl and phosphoryl azides **104** (Scheme 15.23b) [65]. Recent studies by the Chang group demonstrated that the cobalt(III)-catalyzed C—H amination can be performed with *O*-acylcarbamates **106** as the aminating reagents (Scheme 15.23c) [66]. The concept of cobalt(III)-catalyzed C—H amination was further improved by employing 1,4,2-dioxazol-5-ones **108** as the amidation reagents for a broad range of substrates, including anilides, phenylpyridines, benzamides, and purine derivatives (Scheme 15.23d) [67]. In their recent studies, Ackermann and coworkers developed a convenient method to access quinazolines **111** from aryl imidates **110** and dioxazolones **108** via a cobalt(III)-catalyzed C—N formation through C—H activation (Scheme 15.23e) [68]. It is worth noting that the reaction proceeded selectively through imidate assistance in the presence of other strongly coordinating *N*-heterocycles, such as pyrimidines, oxazolines, pyrazoles, and pyridines, as the substituent on the aromatic imidates. Further progress was achieved by Ilies and Nakamura using iron catalysis [69]. This iron-catalyzed process was performed with *N*-chloroamines **112** as the aminating reagents. A variety of carboxamides **40** derived from AQ were aminated with both cyclic and acyclic *N*-chloroamines in good yields (Scheme 15.23f). The choice of the ligand 1,2-bis[bis(4-fluorophenyl)phosphine]benzene (F-dppbz) was crucial for the success of this transformation.

15.3.2 C—O Formation via C—H Activation

Transition-metal-catalyzed C—O bond formation through C—H functionalization is one of the most straightforward routes to install the hydroxyl group into organic molecules [70]. Among different base metals, stoichiometric copper was extensively studied in C—H oxygenations [70–74]. Recently, many direct C—H hydroxylation reactions with a stoichiometric amount of copper salt were accomplished by employing bidentate DGs [71]. Notably, the Shi group realized the convenient installation of the hydroxyl group on amides via a copper-promoted C—H functionalization (Scheme 15.24a) [71b]. In addition to the direct hydroxylation, a number of alkoxylation [72] and benzoxylation [73] transformations were also accomplished with the aid of copper catalysis. Recent research by the Daugulis group showcased the efficiency of copper complexes in the direct phenoxylation of aromatic amides **40** (Scheme 15.24b) [74]. Recently, simple $Co(OAc)_2$ was successfully employed as the catalyst in the *ortho*-alkoxylation of aromatic and alkenyl amides **59** bearing the pyridine-1-oxide moiety as the coordinating group with a wide range of alcohols (Scheme 15.24c) [75]. Similar alkoxylations were also observed by the Cheng group with AQ-derived carboxamides **40** [28d].

15.3.3 C—Halogen Formation via C—H Activation

Site-selective halogenation of aromatic compounds via C—H activation has been identified as a powerful tool in organic synthesis [76a–c]. In 2006, Yu found that the reaction of 2-arylpyridines **2** in 1,2-dichloroethane (DCE) in the presence of $Cu(OAc)_2$ under an atmosphere of oxygen gave the *ortho*-selective

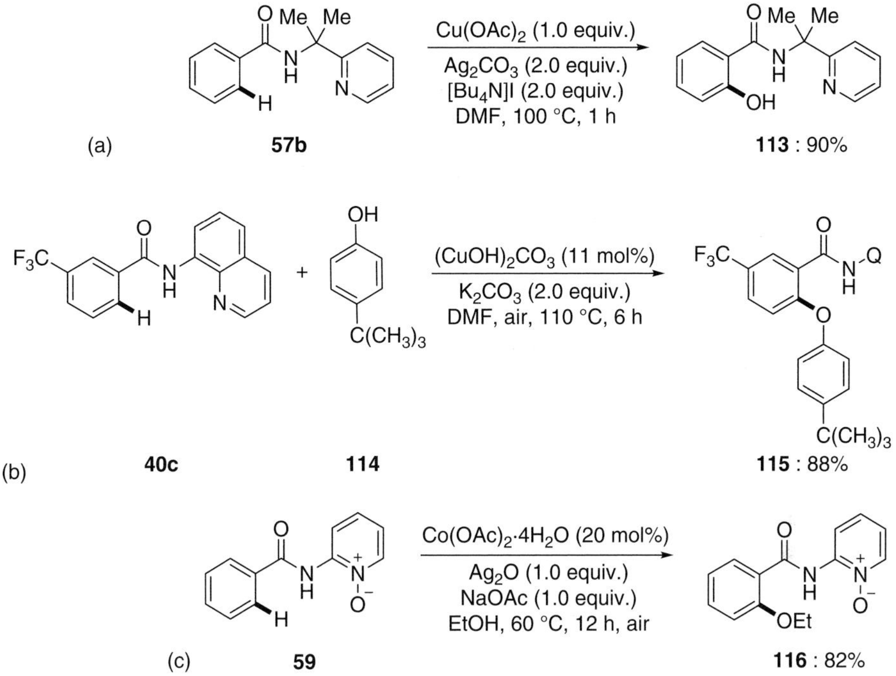

Scheme 15.24 Chelation-controlled C—O bond formation via C—H activation.

dichlorinated 2-aryl pyridines [40a]. In this transformation, an excess of DCE acted as the chlorinating reagent. The similar chlorination can also be performed using LiCl and $CuCl_2$ as the chlorinating sources [76d, e]. In contrast, Carretero and coworkers employed *N*-halosuccinimides for the halogenation of anilines **117** in the presence of 10 mol% of a copper(II) salt and molecular oxygen as the oxidant (Scheme 15.25a) [76f]. Recently, molecular iodine was used as the iodide source in the nickel-catalyzed *ortho*-C—H iodination of aromatic amides **40** (Scheme 15.25b) [77a]. Concurrently, the Shi group used the lithium salts LiBr and LiI for the nickel-catalyzed C—H bromination and iodination, respectively, of (hetero)aromatic amides **57** bearing the 2-pyridinylisopropyl (PIP) auxiliary [77b]. The halogenation reaction of aromatic and alkenylic C—H bonds using *N*-iodosuccinimide (NIS, **120**) and *N*-bromophthalimide (NBP) by cobalt(III) complexes was realized as well (Scheme 15.25c) [25a]. Here, 2-arylpyridines **2** and amides were selectively halogenated by chelation assistance. Recently, the concept was further extended to the halogenation of biologically active 6-arylpurines with NIS [78].

Fluorine is an important substituent present in many pharmaceuticals, agrochemicals, and imaging materials [79a]. In spite of their importance, the construction of the C—F bond in a cost-effective manner continues to be challenging and most of the reactions were catalyzed by noble metals [79b–g]. In 2013, Daugulis and coworkers disclosed a copper-catalyzed AQ auxiliary-assisted C—H fluorination of aromatic amides using AgF as the fluoride source (Scheme 15.25d) [80]. By controlling the amount of the AgF reagent, mono- or difluorinated amides were selectively achieved.

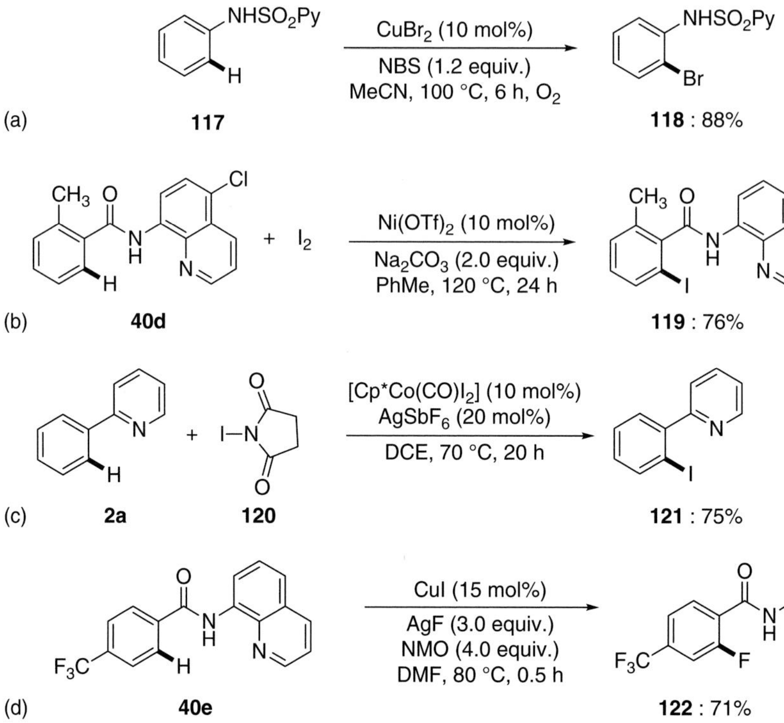

(a) **117** → $CuBr_2$ (10 mol%), NBS (1.2 equiv.), MeCN, 100 °C, 6 h, O_2 → **118** : 88%

(b) **40d** + I_2 → $Ni(OTf)_2$ (10 mol%), Na_2CO_3 (2.0 equiv.), PhMe, 120 °C, 24 h → **119** : 76%

(c) **2a** + **120** → $[Cp^*Co(CO)I_2]$ (10 mol%), $AgSbF_6$ (20 mol%), DCE, 70 °C, 20 h → **121** : 75%

(d) **40e** → CuI (15 mol%), AgF (3.0 equiv.), NMO (4.0 equiv.), DMF, 80 °C, 0.5 h → **122** : 71%

Scheme 15.25 Chelation-controlled C–H halogenation.

15.3.4 C—Chalcogen Formation via C—H Activation

Organometallic directed C—H functionalization to construct C—C, C—N, and C—O bonds has received great attention, whereas the corresponding C—S and C—Se bond forming reactions were less explored [81]. Very recently, the Ackermann group employed an inexpensive Cu catalyst for the C—H selenylation of triazoles **123** with the assistance of a weakly coordinating DG (Scheme 15.26a) [82a]. The group also disclosed a versatile catalytic nickel system for the C—H

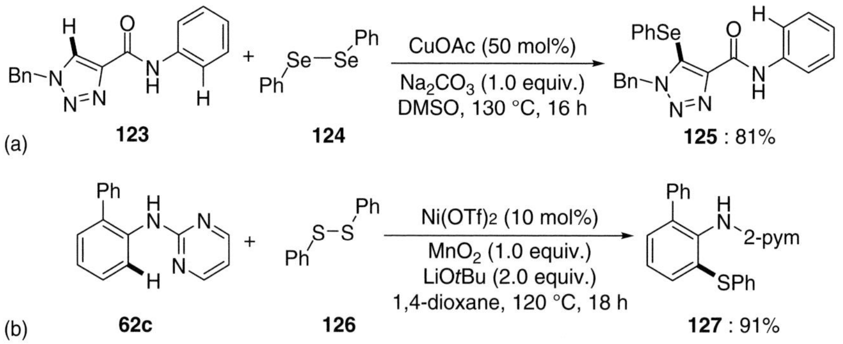

(a) **123** + **124** → CuOAc (50 mol%), Na_2CO_3 (1.0 equiv.), DMSO, 130 °C, 16 h → **125** : 81%

(b) **62c** + **126** → $Ni(OTf)_2$ (10 mol%), MnO_2 (1.0 equiv.), LiO*t*Bu (2.0 equiv.), 1,4-dioxane, 120 °C, 18 h → **127** : 91%

Scheme 15.26 Chelation-assisted C–H chalcogenation.

thiolation and selenylation of anilines **62** with the easily removable pyrimidyl DG in a positional selective manner (Scheme 15.26b) [82b]. Detailed mechanistic studies suggested that the C—H cleavage occurred as the rate-determining step.

15.4 Conclusions

During the past few years significant attention has been devoted toward developing efficient chelation-controlled C—H functionalizations with inexpensive and nontoxic 3d transition metals. Through the appropriate choice of ligands, a number of mild catalytic systems have been developed by means of low-valent cobalt and iron catalysis. Recently, high-valent cobalt catalysis has become a rising star in directed C—H functionalization, which however often resembles the selectivities observed with rhodium(III) catalysis. Nickel and copper complexes were found to be successful in C—H functionalization, particularly with bidentate DGs. Very recently, the concept of catalytic C—H functionalization with manganese(I) complexes came to the limelight. A variety of C—C forming reactions were developed with this versatile manganese catalysis manifold.

Indeed, considerable progress has been reported to solve the challenges associated with the utilization of base metal complexes in the C—H functionalization reactions. However, this research area demands further major attention to address the following issues: (i) often high catalyst loadings (0.1–1.0 equiv.) were used, (ii) high reaction temperatures proved mandatory, (iii) bidentate DGs are frequently required, and (iv) the substrate scope was limited. Nevertheless, given the recent remarkable advances, further promising and exciting developments are expected in the topical area of base-metal-catalyzed C—H functionalization.

Acknowledgments

Generous support by the Alexander von Humboldt foundation (fellowship to P. G.) and the European Research Council under the Seventh Framework Program of the European Community (FP720072013; ERC Grant Agreement No. 307535) are gratefully acknowledged.

References

1 (a) Andersson, P.G. (2012). *Innovative Catalysis in Organic Synthesis: Oxidation, Hydrogenation, and C–X Bond Forming Reactions*. Weinheim, Germany: Wiley-VCH. (b) Ackermann, L. (2009). *Modern Arylation Methods*. Weinheim, Germany: Wiley-VCH. (c) Beller, M. and Bolm, C. (2004). *Transition Metals for Organic Synthesis: Building Blocks and Fine Chemicals*, 2e. Weinheim, Chichester: Wiley-VCH. (d) Tsuji, J. (2000). *Transition Metal Reagents and Catalysts: Innovations in Organic Synthesis*. Chichester, New York: Wiley. (e) Diederich, F.O. and Stang, P.J. (1998). *Metal-Catalyzed Cross-Coupling Reactions*. Weinheim, Chichester: Wiley-VCH. (f) Colquhoun, H.M. (1984).

New Pathways for Organic Synthesis: Practical Applications of Transition Metals. New York, NY: Plenum Press.

2 (a) Li, C.-J. (2014). *From C–H to C–C Bonds: Cross-Dehydrogenative-Coupling*. Royal Society of Chemistry. (b) Yu, J.-Q. and Shi, Z. (2010). *C–H Activation*. Berlin: Springer. (c) Ackermann, L. (2007). *Top. Organomet. Chem.* 24: 35–60. (d) Dyker, G. (2005). *Handbook of C–H Transformations: Applications in Organic Synthesis*. Weinheim, Germany: Wiley-VCH. (e) Shilov, A.E. and Shulpin, G.B. (2000). *Activation and Catalytic Reactions of Saturated Hydrocarbons in the Presence of Metal Complexes*. Dordrecht, London: Kluwer Academic Publishers.

3 (a) Ma, W., Gandeepan, P., Li, J., and Ackermann, L. (2017). *Org. Chem. Front.* 4: 1435–1467. (b) Bandara, H.M.D., Jin, D., Mantell, M.A. et al. (2016). *Cat. Sci. Technol.* 6: 5304–5310. (c) Yoshida, S., Shimomori, K., Nonaka, T., and Hosoya, T. (2015). *Chem. Lett.* 44: 1324–1326. (d) Takaya, J., Ito, S., Nomoto, H. et al. (2015). *Chem. Commun.* 51: 17662–17665. (e) Furukawa, T., Tobisu, M., and Chatani, N. (2015). *J. Am. Chem. Soc.* 137: 12211–12214. (f) Gigant, N. and Bäckvall, J.-E. (2014). *Org. Lett.* 16: 4432–4435. (g) Zheng, L. and Wang, J. (2012). *Chem. Eur. J.* 18: 9699–9704. (h) Kubota, A., Emmert, M.H., and Sanford, M.S. (2012). *Org. Lett.* 14: 1760–1763. (i) Emmert, M.H., Cook, A.K., Xie, Y.J., and Sanford, M.S. (2011). *Angew. Chem. Int. Ed.* 50: 9409–9412. (j) Stuart, D.R. and Fagnou, K. (2007). *Science* 316: 1172–1175. (k) Becker, J. and Hölderich, W.F. (1998). *Catal. Lett.* 54: 125–128.

4 (a) Gandeepan, P. and Cheng, C.-H. (2015). *Chem. Asian J.* 10: 824–838. (b) Zheng, Q.-Z. and Jiao, N. (2014). *Tetrahedron Lett.* 55: 1121–1126. (c) Ros, A., Fernández, R., and Lassaletta, J.M. (2014). *Chem. Soc. Rev.* 43: 3229–3243. (d) Kakiuchi, F., Kochi, T., and Murai, S. (2014). *Synlett* 25: 2390–2414. (e) De Sarkar, S., Liu, W., Kozhushkov, S.I., and Ackermann, L. (2014). *Adv. Synth. Catal.* 356: 1461–1479. (f) Rousseau, G. and Breit, B. (2011). *Angew. Chem. Int. Ed.* 50: 2450–2494. (g) Ackermann, L. (2011). *Chem. Rev.* 111: 1315–1345. (h) Ackermann, L., Vicente, R., and Kapdi, A.R. (2009). *Angew. Chem. Int. Ed.* 48: 9792–9826.

5 (a) Ackermann, L. (2015). *Org. Process. Res. Dev.* 19: 260–269. (b) Kuhl, N., Schröder, N., and Glorius, F. (2014). *Adv. Synth. Catal.* 356: 1443–1460. (c) Ackermann, L. (2014). *Acc. Chem. Res.* 47: 281–295. (d) Kozhushkov, S.I. and Ackermann, L. (2013). *Chem. Sci.* 4: 886–896. (e) Satoh, T. and Miura, M. (2010). *Chem. Eur. J.* 16: 11212–11222. (f) Lyons, T.W. and Sanford, M.S. (2010). *Chem. Rev.* 110: 1147–1169. (g) Chen, X., Engle, K.M., Wang, D.-H., and Yu, J.-Q. (2009). *Angew. Chem. Int. Ed.* 48: 5094–5115.

6 Liu, W. and Ackermann, L. (2016). *ACS Catal.* 6: 3743–3752.

7 (a) Cera, G. and Ackermann, L. (2016). *Top. Curr. Chem.* 374: 57. (b) Lindhorst, A.C., Haslinger, S., and Kühn, F.E. (2015). *Chem. Commun.* 51: 17193–17212. (c) Bauer, I. and Knölker, H.-J. (2015). *Chem. Rev.* 115: 3170–3387. (d) Jia, F. and Li, Z. (2014). *Org. Chem. Front.* 1: 194–214. (e) Sun, X., Li, J., Huang, X., and Sun, C. (2012). *Curr. Inorg. Chem.* 2: 64–85. (f) Sherry, B.D. and Fürstner, A. (2008). *Acc. Chem. Res.* 41: 1500–1511.

8 (a) Moselage, M., Li, J., and Ackermann, L. (2016). *ACS Catal.* 6: 498–525. (b) Gandeepan, P. and Cheng, C.-H. (2015). *Acc. Chem. Res.* 48: 1194–1206.

(c) Yoshikai, N., Gao, K., and Yamakawa, T. (2014). *Synthesis* 46: 2024–2039. (d) Tilly, D., Dayaker, G., and Bachu, P. (2014). *Cat. Sci. Technol.* 4: 2756–2777. (e) Hyster, T.K. (2014). *Catal. Lett.* 145: 458–467. (f) Gao, K. and Yoshikai, N. (2014). *Acc. Chem. Res.* 47: 1208–1219. (g) Ackermann, L. (2014). *J. Org. Chem.* 79: 8948–8654.

9 (a) Yamaguchi, J., Muto, K., and Itami, K. (2016). *Top. Curr. Chem.* 374: 55. (b) Ritleng, V., Henrion, M., and Chetcuti, M.J. (2016). *ACS Catal.* 6: 890–906. (c) Johnson, S.A. (2015). *Dalton Trans.* 44: 10905–10913. (d) Castro, L.C.M. and Chatani, N. (2015). *Chem. Lett.* 44: 410–421. (e) Tasker, S.Z., Standley, E.A., and Jamison, T.F. (2014). *Nature* 509: 299–309. (f) Nakao, Y. (2011). *Chem. Rec.* 11: 242–251.

10 (a) Subramanian, P., Rudolf, G.C., and Kaliappan, K.P. (2016). *Chem. Asian J.* 11: 168–192. (b) Liu, J., Chen, G., and Tan, Z. (2016). *Adv. Synth. Catal.* 358: 1174–1194. (c) Jadhav, A.P., Ray, D., Rao, V.U.B., and Singh, R.P. (2016). *Eur. J. Org. Chem.* 2016: 2369–2382. (d) Hirano, K. and Miura, M. (2015). *Chem. Lett.* 44: 868–873. (e) Hao, W. and Liu, Y. (2015). *Beilstein J. Org. Chem.* 11: 2132–2144. e) Guo, X.-X., Gu, D.-W., Wu, Z., and Zhang, W. (2015). *Chem. Rev.* 115: 1622–1651. (f) Hirano, K. and Miura, M. (2012). *Chem. Commun.* 48: 10704–10714. (g) Gephart, R.T. and Warren, T.H. (2012). *Organometallics* 31: 7728–7752. (h) Wendlandt, A.E., Suess, A.M., and Stahl, S.S. (2011). *Angew. Chem. Int. Ed.* 50: 11062–11087.

11 (a) Halbritter, G., Knoch, F., Wolski, A., and Kisch, H. (1994). *Angew. Chem. Int. Ed. Engl.* 33: 1603–1605. (b) Gao, K. and Yoshikai, N. (2011). *J. Am. Chem. Soc.* 133: 400–402. (c) Ilies, L., Chen, Q., Zeng, X., and Nakamura, E. (2011). *J. Am. Chem. Soc.* 133: 5221–5223. (d) Lee, P.-S. and Yoshikai, N. (2013). *Angew. Chem. Int. Ed.* 52: 1240–1244. (e) Dong, J., Lee, P.-S., and Yoshikai, N. (2013). *Chem. Lett.* 42: 1140–1142. (f) Xu, W., Pek, J.H., and Yoshikai, N. (2016). *Adv. Synth. Catal.* 358: 2564–2568. (g) Lee, P.-S. and Yoshikai, N. (2015). *Org. Lett.* 17: 22–25. (h) Ding, Z. and Yoshikai, N. (2013). *Angew. Chem. Int. Ed.* 52: 8574–8578.

12 (a) Yoshino, T., Ikemoto, H., Matsunaga, S., and Kanai, M. (2013). *Angew. Chem. Int. Ed.* 52: 2207–2211. (b) Ikemoto, H., Yoshino, T., Sakata, K. et al. (2014). *J. Am. Chem. Soc.* 136: 5424–5431.

13 Zhou, B., Ma, P., Chen, H., and Wang, C. (2014). *Chem. Commun.* 50: 14558–14561.

14 (a) Gao, K., Lee, P.-S., Fujita, T., and Yoshikai, N. (2010). *J. Am. Chem. Soc.* 132: 12249–12251. (b) Yamakawa, T. and Yoshikai, N. (2013). *Tetrahedron* 69: 4459–4465. (c) Lee, P.-S., Fujita, T., and Yoshikai, N. (2011). *J. Am. Chem. Soc.* 133: 17283–17295. (d) Fallon, B.J., Derat, E., Amatore, M. et al. (2015). *J. Am. Chem. Soc.* 137: 2448–2451.

15 Ding, Z. and Yoshikai, N. (2012). *Angew. Chem. Int. Ed.* 51: 4698–4701.

16 (a) Zhou, B., Chen, H., and Wang, C. (2013). *J. Am. Chem. Soc.* 135: 1264–1267. (b) Shi, L., Zhong, X., She, H. et al. (2015). *Chem. Commun.* 51: 7136–7139. (c) Yang, X., Jin, X., and Wang, C. (2016). *Adv. Synth. Catal.* 358: 2436–2442. (d) Ackermann, L. and Lygin, A.V. (2011). *Org. Lett.* 13: 3332–3335.

17 (a) Yang, L. and Huang, H. (2015). *Chem. Rev.* 115: 3468–3517. (b) Boyarskiy, V.P., Ryabukhin, D.S., Bokach, N.A., and Vasilyev, A.V. (2016). *Chem. Rev.* 116: 5894–5986. (c) Crisenza, G.E.M. and Bower, J.F. (2016). *Chem. Lett.* 45: 2–9. (d) Shi, X.-Y., Han, W.-J., and Li, C.-J. (2016). *Chem. Rec.* 16: 1178–1190.

18 Gao, K. and Yoshikai, N. (2012). *Chem. Commun.* 48: 4305–4307.

19 Yoshino, T., Ikemoto, H., Matsunaga, S., and Kanai, M. (2013). *Chem. Eur. J.* 19: 9142–9146.

20 (a) Kuninobu, Y., Nishina, Y., Takeuchi, T., and Takai, K. (2007). *Angew. Chem. Int. Ed.* 46: 6518–6520. (b) Zhou, B., Hu, Y., and Wang, C. (2015). *Angew. Chem. Int. Ed.* 54: 13659–13663. (c) Hummel, J.R. and Ellman, J.A. (2015). *J. Am. Chem. Soc.* 137: 490–498.

21 (a) Li, J. and Ackermann, L. (2015). *Angew. Chem. Int. Ed.* 54: 8551–8554. (b) Hummel, J.R. and Ellman, J.A. (2015). *Org. Lett.* 17: 2400–2403.

22 Liu, W., Bang, J., Zhang, Y., and Ackermann, L. (2015). *Angew. Chem. Int. Ed.* 54: 14137–14140.

23 (a) Suzuki, Y., Sun, B., Yoshino, T. et al. (2015). *Tetrahedron* 71: 4552–4556. (b) Moselage, M., Sauermann, N., Richter, S.C., and Ackermann, L. (2015). *Angew. Chem. Int. Ed.* 54: 6352–6355.

24 (a) Ackermann, L., Moselage, M., Sauermann, N. et al. (2015). *Synlett* 26: 1596–1600. (b) Zell, D., Bu, Q., Feldt, M., and Ackermann, L. (2016). *Angew. Chem. Int. Ed.* 55: 7408–7412. (c) Liu, W., Richter, S.C., Zhang, Y., and Ackermann, L. (2016). *Angew. Chem. Int. Ed.* 55: 7747–7750. (d) Cera, G., Haven, T., and Ackermann, L. (2016). *Angew. Chem. Int. Ed.* 55: 1484–1488.

25 (a) Yu, D.-G., Gensch, T., de Azambuja, F. et al. (2014). *J. Am. Chem. Soc.* 136: 17722–17725. (b) Gensch, T., Vásquez-Céspedes, S., Yu, D.-G., and Glorius, F. (2015). *Org. Lett.* 17: 3714–3717.

26 (a) Suzuki, Y., Sun, B., Sakata, K. et al. (2015). *Angew. Chem. Int. Ed.* 54: 9944–9947. (b) Bunno, Y., Murakami, N., Suzuki, Y. et al. (2016). *Org. Lett.* 18: 2216–2219.

27 (a) Kong, L., Yu, S., Tang, G. et al. (2016). *Org. Lett.* 18: 3802–3805. (b) Maity, S., Kancherla, R., Dhawa, U. et al. (2016). *ACS Catal.* 6: 5493–5499. (c) Barsu, N., Kalsi, D., and Sundararaju, B. (2015). *Chem. Eur. J.* 21: 9364–9368. (d) Asako, S., Ilies, L., and Nakamura, E. (2013). *J. Am. Chem. Soc.* 135: 17755–17757. (e) Asako, S., Norinder, J., Ilies, L. et al. (2014). *Adv. Synth. Catal.* 356: 1481–1485.

28 (a) Grigorjeva, L. and Daugulis, O. (2014). *Org. Lett.* 16: 4684–4687. (b) Ma, W. and Ackermann, L. (2015). *ACS Catal.* 5: 2822–2825. (c) Miura, W., Hirano, K., and Miura, M. (2015). *Org. Lett.* 17: 4034–4037. (d) Gandeepan, P., Rajamalli, P., and Cheng, C.-H. (2016). *Angew. Chem. Int. Ed.* 55: 4308–4311.

29 (a) Bentley, K.W. (1992). *Nat. Prod. Rep.* 9: 365–391. (b) Wang, H., Koeller, J., Liu, W., and Ackermann, L. (2015). *Chem. Eur. J.* 21: 15525–15528. (c) Sun, B., Yoshino, T., Kanai, M., and Matsunaga, S. (2015). *Angew. Chem. Int. Ed.* 54: 12968–12972. (d) Sen, M., Kalsi, D., and Sundararaju, B. (2015). *Chem. Eur. J.* 21: 15529–15533. (e) Muralirajan, K., Kuppusamy, R., Prakash, S., and Cheng, C.-H. (2016). *Adv. Synth. Catal.* 358: 774–783. (f) Zhang, S.-S., Liu,

X.-G., Chen, S.-Y. et al. (2016). *Adv. Synth. Catal.* 358: 1705–1710. (g) Wang, F., Wang, Q., Bao, M., and Li, X. (2016). *Chin. J. Catal.* 37: 1423–1430.

30 Yoshida, Y., Kurahashi, T., and Matsubara, S. (2011). *Chem. Lett.* 40: 1140–1142.

31 (a) He, R., Huang, Z.-T., Zheng, Q.-Y., and Wang, C. (2014). *Angew. Chem. Int. Ed.* 53: 4950–4953. (b) Jia, T., Zhao, C., He, R. et al. (2016). *Angew. Chem. Int. Ed.* 55: 5268–5271.

32 (a) Gandeepan, P. and Cheng, C.-H. (2016). *Chem. Asian J.* 11: 448–460. (b) Prakash, S., Muralirajan, K., and Cheng, C.-H. (2016). *Angew. Chem. Int. Ed.* 55: 1844–1848. (c) Lao, Y.-X., Zhang, S.-S., Liu, X.-G. et al. (2016). *Adv. Synth. Catal.* 358: 2186–2191.

33 (a) Shiota, H., Ano, Y., Aihara, Y. et al. (2011). *J. Am. Chem. Soc.* 133: 14952–15955. (b) Planas, O., Whiteoak, C.J., Company, A., and Ribas, X. (2015). *Adv. Synth. Catal.* 357: 4003–4012. (c) Hao, X.-Q., Du, C., Zhu, X. et al. (2016). *Org. Lett.* 18: 3610–3613. (d) Matsubara, T., Ilies, L., and Nakamura, E. (2016). *Chem. Asian J.* 11: 380–384. (e) Nguyen, T.T., Grigorjeva, L., and Daugulis, O. (2016). *ACS Catal.* 6: 551–554. (f) Mei, R., Wang, H., Warratz, S. et al. (2016). *Chem. Eur. J.* 22: 6759–6763. (g) Sivakumar, G., Vijeta, A., and Jeganmohan, M. (2016). *Chem. Eur. J.* 22: 5899–5903.

34 (a) Dong, J., Wang, F., and You, J. (2014). *Org. Lett.* 16: 2884–2887. (b) Zhang, Y., Wang, Q., Yu, H., and Huang, Y. (2014). *Org. Biomol. Chem.* 12: 8844–8850. (c) Zhang, L.-B., Hao, X.-Q., Liu, Z.-J. et al. (2015). *Angew. Chem. Int. Ed.* 54: 10012–10015. (d) Zheng, X.-X., Du, C., Zhao, X.-M. et al. (2016). *J. Org. Chem.* 81: 4002–4011.

35 (a) Song, W. and Ackermann, L. (2013). *Chem. Commun.* 49: 6638–6640. (b) Wang, H., Moselage, M., González, M.J., and Ackermann, L. (2016). *ACS Catal.* 6: 2705–2709. (c) Lerchen, A., Vásquez-Céspedes, S., and Glorius, F. (2016). *Angew. Chem. Int. Ed.* 55: 3208–3211. (d) Lu, Q., Vásquez-Céspedes, S., Gensch, T., and Glorius, F. (2016). *ACS Catal.* 6: 2352–2356. (e) Zhou, S., Wang, J., Wang, L. et al. (2016). *Org. Lett.* 18: 3806–3809. (f) Zhang, Z.-Z., Liu, B., Xu, J.-W. et al. (2016). *Org. Lett.* 18: 1776–1779. (g) Liang, Y. and Jiao, N. (2016). *Angew. Chem. Int. Ed.* 55: 4035–4039. (h) Yu, W., Zhang, W., Liu, Y. et al. (2016). *RSC Adv.* 6: 24768–24772.

36 (a) Shang, M., Wang, H.-L., Sun, S.-Z. et al. (2014). *J. Am. Chem. Soc.* 136: 11590–11593. (b) Liu, Y.-J., Liu, Y.-H., Yin, X.-S. et al. (2015). *Chem. Eur. J.* 21: 205–209.

37 (a) Liu, Y.-H., Liu, Y.-J., Yan, S.-Y., and Shi, B.-F. (2015). *Chem. Commun.* 51: 11650–11653. (b) Liu, Y.-J., Liu, Y.-H., Yan, S.-Y., and Shi, B.-F. (2015). *Chem. Commun.* 51: 6388–6391. (c) Yi, J., Yang, L., Xia, C., and Li, F. (2015). *J. Org. Chem.* 80: 6213–6221. (d) Landge, V.G., Shewale, C.H., Jaiswal, G. et al. (2016). *Cat. Sci. Technol.* 6: 1946–1951. (e) Ruan, Z., Lackner, S., and Ackermann, L. (2016). *ACS Catal.* 6: 4690–4693.

38 (a) Sauermann, N., González, M.J., and Ackermann, L. (2015). *Org. Lett.* 17: 5316–5319. (b) Zhang, Z.-Z., Liu, B., Wang, C.-Y., and Shi, B.-F. (2015). *Org. Lett.* 17: 4094–4097. (c) Landge, V.G., Jaiswal, G., and Balaraman, E. (2016). *Org. Lett.* 18: 812–815.

39 (a) Ellis, G.P. and Romney-Alexander, T.M. (1987). *Chem. Rev.* 87: 779–794. (b) Wen, Q., Jin, J., Zhang, L. et al. (2014). *Tetrahedron Lett.* 55: 1271–1280. (c) Ping, Y., Ding, Q., and Peng, Y. (2016). *ACS Catal.* 5989–6005.

40 (a) Chen, X., Hao, X.-S., Goodhue, C.E., and Yu, J.-Q. (2006). *J. Am. Chem. Soc.* 128: 6790–6791. (b) Jin, J., Wen, Q., Lu, P., and Wang, Y. (2012). *Chem. Commun.* 48: 9933–9935. (c) Kou, X., Zhao, M., Qiao, X. et al. (2013). *Chem. Eur. J.* 19: 16880–16886. (d) Pan, C., Jin, H., Xu, P. et al. (2013). *J. Org. Chem.* 78: 9494–9498. (e) Xu, H., Liu, P.-T., Li, Y.-H., and Han, F.-S. (2013). *Org. Lett.* 15: 3354–3357.

41 (a) Li, J. and Ackermann, L. (2015). *Angew. Chem. Int. Ed.* 54: 3635–3638. (b) Pawar, A.B. and Chang, S. (2015). *Org. Lett.* 17: 660–663.

42 (a) Song, W. and Ackermann, L. (2012). *Angew. Chem. Int. Ed.* 51: 8251–8254. (b) Punji, B., Song, W., Shevchenko, G.A., and Ackermann, L. (2013). *Chem. Eur. J.* 19: 10605–10610. (c) Li, J. and Ackermann, L. (2015). *Chem. Eur. J.* 21: 5718–5722. (d) Mei, R. and Ackermann, L. (2016). *Adv. Synth. Catal.* 358: 2443–2448. (e) Gao, K., Lee, P.-S., Long, C., and Yoshikai, N. (2012). *Org. Lett.* 14: 4234–4237.

43 (a) Bolm, C., Legros, J., Le Paih, J., and Zani, L. (2004). *Chem. Rev.* 104: 6217–6254. (b) Fürstner, A. (2009). *Angew. Chem. Int. Ed.* 48: 1364–1367. (c) Cassani, C., Bergonzini, G., and Wallentin, C.-J. (2016). *ACS Catal.* 6: 1640–1648. (d) Guérinot, A. and Cossy, J. (2016). *Top. Curr. Chem.* 374: 49.

44 (a) Norinder, J., Matsumoto, A., Yoshikai, N., and Nakamura, E. (2008). *J. Am. Chem. Soc.* 130: 5858–5859. (b) Yoshikai, N., Matsumoto, A., Norinder, J., and Nakamura, E. (2009). *Angew. Chem. Int. Ed.* 48: 2925–2928. (c) Ilies, L., Asako, S., and Nakamura, E. (2011). *J. Am. Chem. Soc.* 133: 7672–7675. (d) Ilies, L., Konno, E., Chen, Q., and Nakamura, E. (2012). *Asian J. Org. Chem.* 1: 142–145. (e) Sirois, J.J., Davis, R., and DeBoef, B. (2014). *Org. Lett.* 16: 868–871. (f) Shang, R., Ilies, L., Asako, S., and Nakamura, E. (2014). *J. Am. Chem. Soc.* 136: 14349–14352.

45 Gu, Q., Al Mamari, H.H., Graczyk, K. et al. (2014). *Angew. Chem. Int. Ed.* 53: 3868–3871.

46 (a) Kitahara, M., Umeda, N., Hirano, K. et al. (2011). *J. Am. Chem. Soc.* 133: 2160–2162. (b) Nishino, M., Hirano, K., Satoh, T., and Miura, M. (2012). *Angew. Chem. Int. Ed.* 51: 6993–6997. (c) Nishino, M., Hirano, K., Satoh, T., and Miura, M. (2013). *Angew. Chem. Int. Ed.* 52: 4457–4461. (d) Odani, R., Hirano, K., Satoh, T., and Miura, M. (2013). *J. Org. Chem.* 78: 11045–11052. (e) Odani, R., Hirano, K., Satoh, T., and Miura, M. (2014). *Angew. Chem. Int. Ed.* 53: 10784–10988. (f) Odani, R., Hirano, K., Satoh, T., and Miura, M. (2015). *J. Org. Chem.* 80: 2384–2391. (g) Zhao, S., Yuan, J., Li, Y.-C., and Shi, B.-F. (2015). *Chem. Commun.* 51: 12823–12826. (h) Wang, M., Hu, Y., Jiang, Z. et al. (2016). *Org. Biomol. Chem.* 14: 4239–4246.

47 (a) Shang, M., Sun, S.-Z., Dai, H.-X., and Yu, J.-Q. (2014). *Org. Lett.* 16: 5666–5669. (b) Gui, Q., Chen, X., Hu, L. et al. (2016). *Adv. Synth. Catal.* 358: 509–514.

48 Yokota, A., Aihara, Y., and Chatani, N. (2014). *J. Org. Chem.* 79: 11922–11932.

49 Ackermann, L. (2010). *Chem. Commun.* 46: 4866–4877.

50 Chen, Q., Ilies, L., and Nakamura, E. (2011). *J. Am. Chem. Soc.* 133: 428–429.

51 (a) Gao, K. and Yoshikai, N. (2013). *J. Am. Chem. Soc.* 135: 9279–9282. (b) Xu, W., Paira, R., and Yoshikai, N. (2015). *Org. Lett.* 17: 4192–4195.

52 Ilies, L., Matsubara, T., Ichikawa, S. et al. (2014). *J. Am. Chem. Soc.* 136: 13126–13129.

53 Fruchey, E.R., Monks, B.M., and Cook, S.P. (2014). *J. Am. Chem. Soc.* 136: 13130–13133.

54 (a) Graczyk, K., Haven, T., and Ackermann, L. (2015). *Chem. Eur. J.* 21: 8812–8815. (b) Shang, R., Ilies, L., and Nakamura, E. (2015). *J. Am. Chem. Soc.* 137: 7660–7663. (c) Shang, R., Ilies, L., and Nakamura, E. (2016). *J. Am. Chem. Soc.* 138: 10132–10135.

55 (a) Aihara, Y. and Chatani, N. (2013). *J. Am. Chem. Soc.* 135: 5308–5311. (b) Aihara, Y., Wuelbern, J., and Chatani, N. (2015). *Bull. Chem. Soc. Jpn.* 88: 438–446. (c) Uemura, T., Yamaguchi, M., and Chatani, N. (2016). *Angew. Chem. Int. Ed.* 55: 3162–3165. (d) Kubo, T. and Chatani, N. (2016). *Org. Lett.* 18: 1698–1701.

56 (a) Song, W., Lackner, S., and Ackermann, L. (2014). *Angew. Chem. Int. Ed.* 53: 2477–2480. (b) Ruan, Z., Lackner, S., and Ackermann, L. (2016). *Angew. Chem. Int. Ed.* 55: 3153–3157.

57 (a) Lawrence, S.A. (2004). *Amines: Synthesis, Properties and Applications.* Cambridge: Cambridge University Press. (b) Rappoport, Z. (2007). *The Chemistry of Anilines.* Weinheim, Germany: Wiley-VCH. (c) Ricci, A. (2008). *Amino Group Chemistry: From Synthesis to the Life Sciences.* Weinheim, Germany: Wiley-VCH.

58 (a) Collet, F., Dodd, R.H., and Dauban, P. (2009). *Chem. Commun.* 5061–5074. (b) Cho, S.H., Kim, J.Y., Kwak, J., and Chang, S. (2011). *Chem. Soc. Rev.* 40: 5068–5083. (c) Bariwal, J. and Van der Eycken, E. (2013). *Chem. Soc. Rev.* 42: 9283–9303. d) Louillat, M.-L. and Patureau, F.W. (2014). *Chem. Soc. Rev.* 43: 901–910. (e) Shin, K., Kim, H., and Chang, S. (2015). *Acc. Chem. Res.* 48: 1040–1052. (f) Jiao, J., Murakami, K., and Itami, K. (2016). *ACS Catal.* 6: 610–633. (g) Zhou, Y., Yuan, J., Yang, Q. et al. (2016). *ChemCatChem* 8: 2178–2192.

59 (a) Uemura, T., Imoto, S., and Chatani, N. (2006). *Chem. Lett.* 35: 842–843. (b) Shuai, Q., Deng, G., Chua, Z. et al. (2010). *Adv. Synth. Catal.* 352: 632–636. (c) John, A. and Nicholas, K.M. (2011). *J. Org. Chem.* 76: 4158–4162. (d) Wang, L., Priebbenow, D.L., Dong, W., and Bolm, C. (2014). *Org. Lett.* 16: 2661–2663. (e) Xu, H., Qiao, X., Yang, S., and Shen, Z. (2014). *J. Org. Chem.* 79: 4414–4422. (f) Li, G., Jia, C., Chen, Q. et al. (2015). *Adv. Synth. Catal.* 357: 1311–1315.

60 (a) Tran, L.D., Roane, J., and Daugulis, O. (2013). *Angew. Chem. Int. Ed.* 52: 6043–6046. (b) Roane, J. and Daugulis, O. (2016). *J. Am. Chem. Soc.* 138: 4601–4607. (c) Sadhu, P. and Punniyamurthy, T. (2016). *Chem. Commun.* 52: 2803–2806.

61 Shang, M., Sun, S.-Z., Dai, H.-X., and Yu, J.-Q. (2014). *J. Am. Chem. Soc.* 136: 3354–3357.

62 Yan, Q., Chen, Z., Yu, W. et al. (2015). *Org. Lett.* 17: 2482–2485.

63 Zhang, L.-B., Zhang, S.-K., Wei, D. et al. (2016). *Org. Lett.* 18: 1318–1321.

64 (a) Peng, J., Xie, Z., Chen, M. et al. (2014). *Org. Lett.* 16: 4702–4705. (b) Peng, J., Chen, M., Xie, Z. et al. (2014). *Org. Chem. Front.* 1: 777–781.

65 (a) Sun, B., Yoshino, T., Matsunaga, S., and Kanai, M. (2014). *Adv. Synth. Catal.* 356: 1491–1495. (b) Sun, B., Yoshino, T., Matsunaga, S., and Kanai, M. (2015). *Chem. Commun.* 51: 4659–4661.

66 Patel, P. and Chang, S. (2015). *ACS Catal.* 5: 853–858.

67 (a) Park, J. and Chang, S. (2015). *Angew. Chem. Int. Ed.* 54: 14103–14107. (b) Liang, Y., Liang, Y.-F., Tang, C. et al. (2015). *Chem. Eur. J.* 21: 16395–16399. (c) Mei, R., Loup, J., and Ackermann, L. (2016). *ACS Catal.* 6: 793–797.

68 Wang, H., Lorion, M.M., and Ackermann, L. (2016). *Angew. Chem. Int. Ed.* 55: 10386–10390.

69 Matsubara, T., Asako, S., Ilies, L., and Nakamura, E. (2014). *J. Am. Chem. Soc.* 136: 646–649.

70 (a) Alonso, D.A., Nájera, C., Pastor, I.M., and Yus, M. (2010). *Chem. Eur. J.* 16: 5274–5284. (b) Thirunavukkarasu, V.S., Kozhushkov, S.I., and Ackermann, L. (2014). *Chem. Commun.* 50: 29–39. (c) Moghimi, S., Mahdavi, M., Shafiee, A., and Foroumadi, A. (2016). *Eur. J. Org. Chem.* 2016: 3282–3299.

71 (a) Gallardo-Donaire, J. and Martin, R. (2013). *J. Am. Chem. Soc.* 135: 9350–9353. (b) Li, X., Liu, Y.-H., Gu, W.-J. et al. (2014). *Org. Lett.* 16: 3904–3907. (c) Sun, S.-Z., Shang, M., Wang, H.-L. et al. (2015). *J. Org. Chem.* 80: 8843–8848.

72 (a) Bhadra, S., Dzik, W.I., and Gooßen, L.J. (2013). *Angew. Chem. Int. Ed.* 52: 2959–2962. (b) Bhadra, S., Matheis, C., Katayev, D., and Gooßen, L.J. (2013). *Angew. Chem. Int. Ed.* 52: 9279–9283. (c) Zhang, L.-B., Hao, X.-Q., Zhang, S.-K. et al. (2014). *J. Org. Chem.* 79: 10399–10409. (d) Yin, X.-S., Li, Y.-C., Yuan, J. et al. (2015). *Org. Chem. Front.* 2: 119–123.

73 (a) Wang, W., Luo, F., Zhang, S., and Cheng, J. (2010). *J. Org. Chem.* 75: 2415–2418. (b) Huang, Z.-Z., Bian, Y.-J., Xiang, C.-B., and Chen, Z.-M. (2011). *Synlett* 2011: 2407–2409. (c) Li, L., Yu, P., Cheng, J. et al. (2012). *Chem. Lett.* 41: 600–602. (d) Chen, X., Zhu, C., Cui, X., and Wu, Y. (2013). *Chem. Commun.* 49: 6900–6902. (e) Khemnar, A.B. and Bhanage, B.M. (2014). *Org. Biomol. Chem.* 12: 9631–9637. (f) Rout, S.K., Guin, S., Gogoi, A. et al. (2014). *Org. Lett.* 16: 1614–1617.

74 Roane, J. and Daugulis, O. (2013). *Org. Lett.* 15: 5842–5845.

75 (a) Zhang, L.-B., Hao, X.-Q., Zhang, S.-K. et al. (2015). *Angew. Chem. Int. Ed.* 54: 272–275. (b) Guo, X.-K., Zhang, L.-B., Wei, D., and Niu, J.-L. (2015). *Chem. Sci.* 6: 7059–7071.

76 (a) Cavallo, G., Metrangolo, P., Milani, R. et al. (2016). *Chem. Rev.* 116: 2478–2601. (b) Petrone, D.A., Ye, J., and Lautens, M. (2016). *Chem. Rev.* 116: 8003–8104. (c) Voskressensky, L., Golantsov, N., and Maharramov, A. (2016). *Synthesis* 48: 615–643. (d) Mo, S., Zhu, Y., and Shen, Z. (2013). *Org. Biomol. Chem.* 11: 2756–2760. (e) Zhao, J., Cheng, X., Le, J. et al. (2015). *Org. Biomol. Chem.* 13: 9000–9004. (f) Urones, B., Martínez, Á.M., Rodríguez, N. et al. (2013). *Chem. Commun.* 49: 11044–11046.

77 (a) Aihara, Y. and Chatani, N. (2016). *ACS Catal.* 6: 4323–4329. (b) Zhan, B.-B., Liu, Y.-H., Hu, F., and Shi, B.-F. (2016). *Chem. Commun.* 52: 4934–4937.

78 Pawar, A.B. and Lade, D.M. (2016). *Org. Biomol. Chem.* 14: 3275–3283.

79 (a) Zhou, Y., Wang, J., Gu, Z. et al. (2016). *Chem. Rev.* 116: 422–518. (b) Hull, K.L., Anani, W.Q., and Sanford, M.S. (2006). *J. Am. Chem. Soc.* 128: 7134–7135. (c) Ball, N.D. and Sanford, M.S. (2009). *J. Am. Chem. Soc.* 131: 3796–3797. (d) Wang, X., Mei, T.-S., and Yu, J.-Q. (2009). *J. Am. Chem. Soc.* 131: 7520–7521. (e) Engle, K.M., Mei, T.-S., Wang, X., and Yu, J.-Q. (2011). *Angew. Chem. Int. Ed.* 50: 1478–1491. (f) Sibi, M.P. and Landais, Y. (2013). *Angew. Chem. Int. Ed.* 52: 3570–3572. (g) Mu, X. and Liu, G. (2014). *Org. Chem. Front.* 1: 430–433.

80 Truong, T., Klimovica, K., and Daugulis, O. (2013). *J. Am. Chem. Soc.* 135: 9342–9345.

81 (a) Chu, L., Yue, X., and Qing, F.-L. (2010). *Org. Lett.* 12: 1644–1647. (b) Tran, L.D., Popov, I., and Daugulis, O. (2012). *J. Am. Chem. Soc.* 134: 18237–18240. (c) Lin, C., Li, D., Wang, B. et al. (2015). *Org. Lett.* 17: 1328–1331. (d) Sharma, P., Rohilla, S., and Jain, N. (2015). *J. Org. Chem.* 80: 4116–4122. (e) Yan, S.-Y., Liu, Y.-J., Liu, B. et al. (2015). *Chem. Commun.* 51: 4069–4072. (f) Yang, K., Wang, Y., Chen, X. et al. (2015). *Chem. Commun.* 51: 3582–3585. (g) Zhu, L., Cao, X., Qiu, R. et al. (2015). *RSC Adv.* 5: 39358–39365. (h) Gensch, T., Klauck, F.J.R., and Glorius, F. (2016). *Angew. Chem. Int. Ed.* 55: 11287–11291. (i) Mandal, A., Sahoo, H., and Baidya, M. (2016). *Org. Lett.* 18: 3202–3205. (j) Peng, P., Wang, J., Li, C. et al. (2016). *RSC Adv.* 6: 57441–57445.

82 (a) Cera, G. and Ackermann, L. (2016). *Chem. Eur. J.* 22: 8475–8478. (b) Müller, T. and Ackermann, L. (2016). *Chem. Eur. J.* 22: 14151–14154. (c) Gandeepan, P., Koeller, J., and Ackermann, L. (2017). *ACS Catal.* 7: 1030–1034.

16

Catalytic Water Oxidation: Water Oxidation to O_2 Mediated by 3d Transition Metal Complexes

Zoel Codolá[1], *Julio Lloret-Fillol*[2,3], *and Miquel Costas*[1]

[1] *Universitat de Girona, Institut de Química Computacional I Catàlisi (IQCC), Departament de Química, Campus Montilivi, E17071 Girona, Catalonia, Spain*
[2] *The Barcelona Institute of Science and Technology, Institute of Chemical Research of Catalonia (ICIQ), Avinguda Paisos Catalans 16, 43007, Tarragona, Catalonia, Spain*
[3] *Catalan Institution for Research and Advanced Studies (ICREA), Passeig Lluïs Companys, 23, 08010, Barcelona, Spain*

16.1 Water Oxidation – From Insights into Fundamental Chemical Concepts to Future Solar Fuels

Sunlight is the only energy source that can meet the growing energy requirements of the current society [1], but its permanent supply is a difficult task because of the day/night cycles [2]. To overcome the out-of-phase sunlight energy consumption/production, a convenient energy storage system is required. Chemical bonding is envisioned as one of the most promising alternatives for energy storage because of its high energy density, with natural photosynthesis being a mechanism to reflect on [3].

Artificial photosynthesis is one of the most ambitious scientific and technological goals of our society [1–4]. This is indeed a very complex process integrating multielectronic and multiprotonic reactions, for which fundamental understanding is required. On the one hand, the reduction of protons (Scheme 16.1; Eq. (2)) and CO_2 (Eqs. (3)–(7)) is necessary for generating high-energy-content molecules than can be used as fuels, the so-called solar fuels. On the other hand, water oxidation (Eq. (1)) is required to provide the electrons for these reductions. Water oxidation and CO_2 reduction constitute *per se* very difficult chemical problems and remain as bottlenecks for the development of efficient artificial photosynthesis.

16.1.1 The Oxygen-Evolving Complex. A Well-Defined Tetramanganese Calcium Cluster

Water oxidation in nature takes place at the oxygen-evolving complex (OEC) [5]. The OEC is an impressive water oxidation catalyst; it produces oxygen at outstanding rates in the range of 100–300 O_2 molecules·s^{-1}, with an estimated turnover number (TON) of about half a million (a value estimated based on the

Non-Noble Metal Catalysis: Molecular Approaches and Reactions,
First Edition. Edited by Robertus J. M. Klein Gebbink and Marc-Etienne Moret.

Water oxidation half reaction: an energetic "uphill" reaction

$$2H_2O \longrightarrow O_2 + 4H^+ + 4e^- \qquad \Delta G^\circ = 113.4\ \text{kcal·mol}^{-1},\ \Delta H^\circ = 136.4\ \text{kcal·mol}^{-1},\ E^\circ = 1.23\ \text{V} \qquad (1)$$

Proton and selected CO_2 reduction half reactions

$$4H^+ + 4e^- \longrightarrow 2H_2 \qquad E^\circ = 0.0\ \text{V} \qquad (2)$$

$$CO_2 + 1e^- \longrightarrow CO_2^{\cdot -} \qquad E^\circ = -1.9\ \text{V} \qquad (3)$$

$$CO_2 + 2H^+ + 2e^- \longrightarrow CO + H_2O \qquad E^\circ = -0.53\ \text{V} \qquad (4)$$

$$CO_2 + 2H^+ + 2e^- \longrightarrow HCO_2H \qquad E^\circ = -0.61\ \text{V} \qquad (5)$$

$$CO_2 + 6H^+ + 6e^- \longrightarrow CH_3OH + H_2O \qquad E^\circ = -0.38\ \text{V} \qquad (6)$$

$$CO_2 + 8H^+ + 8e^- \longrightarrow CH_4 + 2H_2O \qquad E^\circ = -0.24\ \text{V} \qquad (7)$$

Scheme 16.1 Selected standard reduction potentials for the water electrolysis and CO_2 reduction (potentials vs SHE at pH 0).

half lifetime of the protein in which the OEC resides, this protein is replaced every ~30 minutes by a new copy) [6] and an overpotential lower than 200 mV [7, 8]. The TON, turnover frequency (TOF), and overpotential values of the OEC together are unbeatable by artificial water oxidation catalysts discovered so far.

The Mn_4O_4Ca complex located in the light-harvesting complex 1 (LH1) of the protein complex photosystem II (PSII) is directly responsible for this catalytic activity. Umena et al. reported the most accurate crystal structure determination ever for PSII, with a resolution of 1.9 Å. In this structure, a distorted cubane-like Mn_3O_4Ca cluster appears as the most plausible conformation of the OEC, which is bound to an additional manganese atom by an oxo bridge completing the oxygen-evolving cluster (Mn_4O_5Ca, Figure 16.1) [9]. Despite the high resolution, the authors admit that the data are still insufficient in order to reveal the detailed structure of the water oxidation catalytic center. Indeed, distances obtained by X-ray crystallography and extended X-ray absorption fine spectroscopy (EXAFS) are slightly different, suggesting that the cluster may also suffer from photoreduction by radiation and the data do not exclusively represent the structure of the S_1 state, but a superposition of reduced states [10]. More recently, a "radiation-damage-free" structure of PSII in the S1 state at a resolution of 1.95 Å using femtosecond X-ray pulses evidences Mn· · ·Mn distances that are shorter by 0.1–0.2 Å than in the X-ray diffraction (XRD) structure [11].

The groups of Chabaud [12] and Kok [13] reported the first key insights into the OEC function establishing five sequential stages (S states) along the reaction. The so-called Kok cycle consists of the accumulation of four oxidizing equivalents (from S_0 to S_4) and a light-independent O_2 release step ($S_4 \rightarrow S_0$), with S_1 (2 Mn^{III}, 2 Mn^{IV}) being the resting stage under dark conditions. The characterization of the different S states is not straightforward. S_0 and S_2 have been characterized by ^{55}Mn-electron nuclear double resonance (ENDOR) spectroscopy, restricting the possible oxidation states to $Mn^{III}{}_3Mn^{IV}$ and $Mn^{III}Mn^{IV}{}_3$. By combination of density functional theory (DFT) and spectroscopy, two possible conformations of the

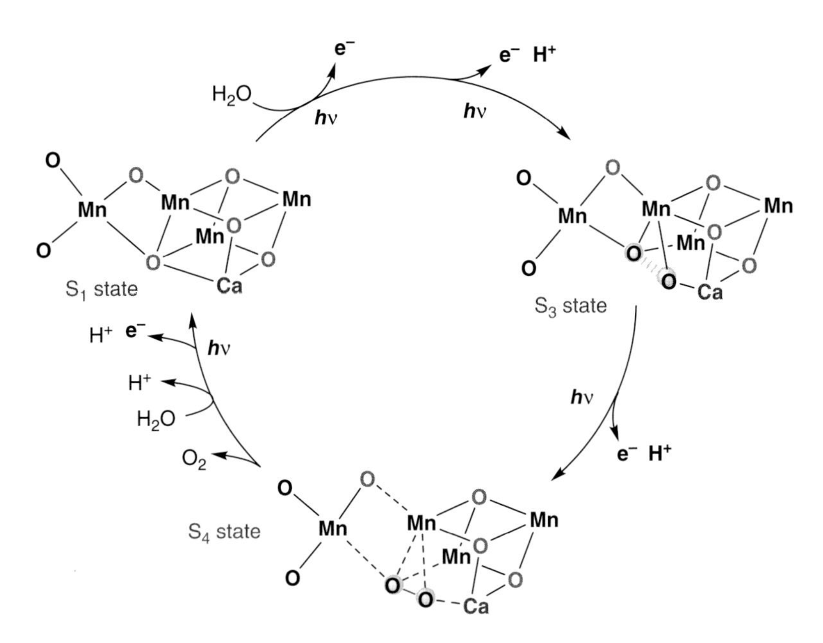

Figure 16.1 Possible mechanism of O=O bond formation in the OEC. Source: Suga et al. 2017 [16]. Reproduced with permission of Springer Nature.

S_2 state, electronically and structurally distinct, were established. The "open" and the "closed" forms, with spin ground states of 1/2 and 5/2, respectively, lead to the formation of different conformations of the subsequent S_3 state. The current consensus is that the S_4 state is responsible for the O—O bond formation event. Indirect evidences of the mechanism have been obtained by monitoring the rate exchange of $H_2O/H_2{}^{18}O$ at the substrate sites of the cluster by time-resolved mass spectrometry, identifying Ca^{2+} as an essential water-binding site. The two water molecules involved in the O—O bond formation reaction were also found to bind at different sites and redox states. Although the first H_2O (W_s) molecule is present from the S_0 state onward and has a slow exchange rate, the second H_2O (W_f) molecule is coordinated during the S_2–S_3 transition. The existence of the W_s molecule limits the number of possible O—O bond formation pathways to (i) the Ca^{2+}-assisted nucleophilic attack of a water molecule on the electrophilic μ-oxo bridge or (ii) the coupling reaction between two metal-oxyl radicals, which requires the binding of an extra water molecule to the cluster, triggering a structural expansion at the S_3 state [14, 15]. This additional water substrate forms the oxyl radical in S_4 (Siegbahn model) [10].

Insight into the structural changes between S_1 and S_3 at a resolution of 2.35 Å of PSII has been obtained by time-resolved femtosecond crystallography with an X-ray free-electron laser (XFEL) after two-flash illumination at room temperature. Importantly, the two-flash-minus-dark isomorphous difference Fourier map suggests the insertion of a new oxygen atom (O6) close to O5, providing an O=O distance of 1.5 Å between these two oxygen atoms (Figure 16.1). This provides a mechanism for the O=O bond formation [16].

In the OEC surrounding, molecules of water were found coordinated with the metal centers and linking the Mn cluster to tyrosine Yz and the last one to a nearby histidine (D1-His 190). D1-His190 seems to play a crucial role in photosynthesis, providing an exit pathway for protons and rendering the water oxidation process thermodynamically favorable ($\Delta G = -8.4\,kcal\,mol^{-1}$) [17].

In summary, PSII orchestrates the whole catalytic water oxidation reaction. It is hypothesized that the surrounding protein residues of the OEC also play a role by finely modulating the redox potentials and by providing specific channels for access of substrate (H_2O), energy, and products (O_2, H^+, and electrons) of the reaction [18], while dealing with the inevitable side reactions. The inherent complexity associated with the multiproton and multielectron nature of the water oxidation reaction, in combination with the unique architecture and reactivity of the intermediates implicated, represents major issues for the mimicking and developing of synthetic functional models [19].

16.1.2 Synthetic Models for the Natural Water Oxidation Reaction

Despite the complexity of the chemistry of the OEC, progress in the fields of synthetic inorganic chemistry and catalysis over the past two decades has shown that simple coordination compounds can act as catalysts for the water oxidation reaction, highlighting the possibility to develop simple and sustainable artificial water oxidation catalysts. Some of these synthetic compounds have also provided structural, spectroscopic, and reactivity models that have helped understand the water oxidation reaction taking place in the OEC. In addition, water oxidation has become a topic of major interest in the inorganic chemistry community and has been a major contributor to the development of the chemistry of high oxidation states.

16.1.3 Oxidants in Water Oxidation Reactions

Water oxidation is a thermodynamically uphill process and requires an external driving force. In nature, the required energy is provided by light. Model catalytic systems for the water oxidation reaction operating with light are still rare, and their mechanistic analysis is often very complex. Instead, model catalytic systems are often studied by using a chemical sacrificial oxidant (SO) [20] or an electric current as an external driving force. The main advantage of using SOs is that the mechanistic analysis of the reactions is often simpler than in electrochemically or photochemically driven reactions. This is best evidenced in (i) the straightforward analysis of bulk solutions, (ii) the facile measurements of TON and TOF (TONs and TOFs), (iii) the fast screenings, (iv) the simple methodologies for establishing kinetics, and (v) also the study of the intermediates [20]. All these characteristics facilitate the determination of the reaction mechanisms. Oxo-transfer agents such as hypochlorite (HClO, 1.39 V vs SCE [sodium calomel electrode]), oxone (SO_5^-, 1.85 V vs SCE), and periodate (at low pH H_5IO_6, 1.60 V vs SCE) and electron transfer complexes such as $[Ce(NO_3)_6](NH_4)_2$ (cerium ammonium nitrate, CAN, 1.72 V vs SCE), $[Co(OH_2)_6]^{3+}$ (1.92 V vs SCE), and $[Ru(bpy)_3]^{3+}$ (bpy = 2,2′-bipyridine, 1.29 V vs SCE) are common SOs used in water oxidation (WO) because their oxidation potential is high enough to oxidize water, but only at very slow rates [20, 21]. Because SOs are usually employed in high concentrations, precaution has to be taken with the oxo transfer agents as they could be the real source of oxygen and, generally, single-electron oxidants (CAN and $[Ru(bpy)_3]^{3+}$) are preferred. On the other hand, $[Co(OH_2)_6]^{3+}$ can also be problematic because it can directly oxidize water at equivalent rates as in some catalyzed processes. These SOs do not mimic the conditions desired for an artificial photosystem, and therefore, data obtained must be considered with caution [20]. It has been described that Ce(IV) could also be involved in O—O bond formation [22]. Nevertheless, the mechanistic details obtained with this oxidant are often very valuable and are very difficult to obtain by other means.

Electrochemical water oxidation studies also provide very valuable information about the mechanism and complementary to information obtained by chemical oxidants. When possible, Pourbaix diagrams provide information about speciation at given pHs and redox potentials and their transformation, which have important implications on the mechanism and catalyst design. On the other hand, electrochemistry can provide information about overpotentials needed for yielding the electrocatalytic water oxidation species, which is otherwise difficult to obtain by other techniques, and it is also an essential parameter for catalyst optimization. It is although more problematic to obtain accurate TON and TOF values because of the difficult quantification of the real amount of catalyst contributing to the observed catalytic activity [23]. After every water oxidation catalytic cycle, 4 equiv. of protons per molecule of oxygen are released, lowering the pH of the solution. To maintain a constant pH and avoid the depletion of catalytic activity, it is common to use buffered solutions. Nevertheless, the possible influence of these buffers on the mechanism or even on the formation of the real catalytic species must be taken into account, especially in electrocatalytic studies.

Very interesting are the systems that produce O_2 when using a photosensitizer (PS) and an electron acceptor (Scheme 16.2). Ruthenium(II) polypyridine complexes are the most commonly used photosensitizers with redox potentials in the 1.2–1.9 V vs SCE range [20]. As an ultimate oxidant, $Na_2S_2O_8$ is generally used. The oxidative quenching of the Ru(II) excited state by $Na_2S_2O_8$ oxidizes the PS with concomitant formation of SO_4^{2-} and SO_4^{-} radical. This methodology is not free of technical difficulties though. In this regard, some authors suggested that

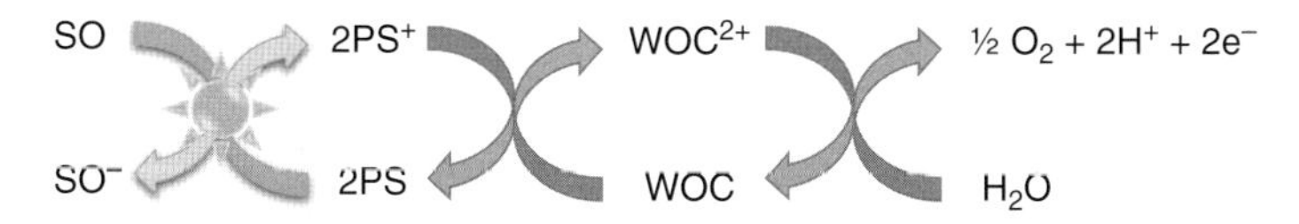

Scheme 16.2 Schematic light-driven water oxidation mechanism. SO stands for the sacrificial oxidant (usually $Na_2S_2O_8$) and PS for the photosensitizer (usually $[Ru(bpy)_3]^{2+}$). A first electron transfer is photoinduced by the excitation and oxidation of the PS (PS → PS* → PS⁺), reducing the SO ($S_2O_8^{2-}$ to SO_4^{2-} and $SO_4{\cdot}^{-}$). The PS⁺ oxidizes the catalyst, which accumulates four oxidizing equivalents to finally oxidize a water molecule.

the SO_4^{-} radical (2.1 V vs SCE) may have an active role in the generation of high oxidation states at the metal centers [24]. For instance, it has been found that it can trigger the formation of active water oxidation nanoparticles masking the real activity of molecular complexes [25].

16.2 Model Well-Defined Water Oxidation Catalysts

This chapter describes relevant work in the study of water oxidation catalysts based on the first-row transition metals. Some relevant examples reported until 2016 of homogeneous water oxidation catalysts based on the first-row transition metals are displayed in Figure 16.2, most of which are discussed in this chapter. Focus is placed on studies with coordination compounds, aiming at providing fundamental understanding of the reactions. The chapter does not intend to be comprehensive, but instead aims to discuss some relevant and representative work that could provide the reader with a general overview. For more extensive discussions, the reader is directed toward some recent reviews [26].

16.2.1 Manganese Water Oxidation Catalysts

As discussed above, the OEC is constituted by three manganese atoms and a calcium atom bridged by four oxygen atoms in a cubane-distorted cluster ($CaMn_3O_4$). An external Mn atom bound through an oxo bridge to this cluster consummates the Mn_4O_5Ca structure. Building this structure with a synthetic compound and implementation of functionality into these synthetic structures constitutes a very challenging yet unmet goal. Only a limited number of structural models, structurally simpler than the OEC but containing Mn oxides arranged as a cubane-like cluster, have been found to be active catalysts for water oxidation. Selected examples are detailed below.

16.2.1.1 Bioinspired Mn_4O_4 Models

Tetranuclear manganese complexes with the general formula $L_6Mn_4O_4$ (L^- = diarylphosphinate ligand, (p-R-$C_6H_4)_2PO_2^-$ (R = H, alkyl, OMe), Scheme 16.3) can be regarded as one of the first models of the OEC. By supporting these clusters in Nafion, photoelectrochemical-driven water oxidation was observed at 1.0–1.4 V vs SHE. During photoelectrolysis for 65 hours, the membrane passes a net charge equivalent to 1000 turnovers per cluster [27]. It was proposed that upon absorption of light, O_2 is produced, generating a "pinned butterfly" structure that will bind two water molecules and extrude four protons, regenerating the cubane structure.

However, more recently, Spiccia and coworkers have proven that $L_6Mn_4O_4$ clusters supported on Nafion are transformed into Mn^{2+} oxides (birnessite), which, in turn, are electro-oxidized into $Mn^{III}/^{IV}$-O_x nanoparticles (NPs), which were proven to be the species responsible for water oxidation [28].

This example is actually representative of a rather common situation in the field. A number of molecular compounds, initially considered to be catalysts,

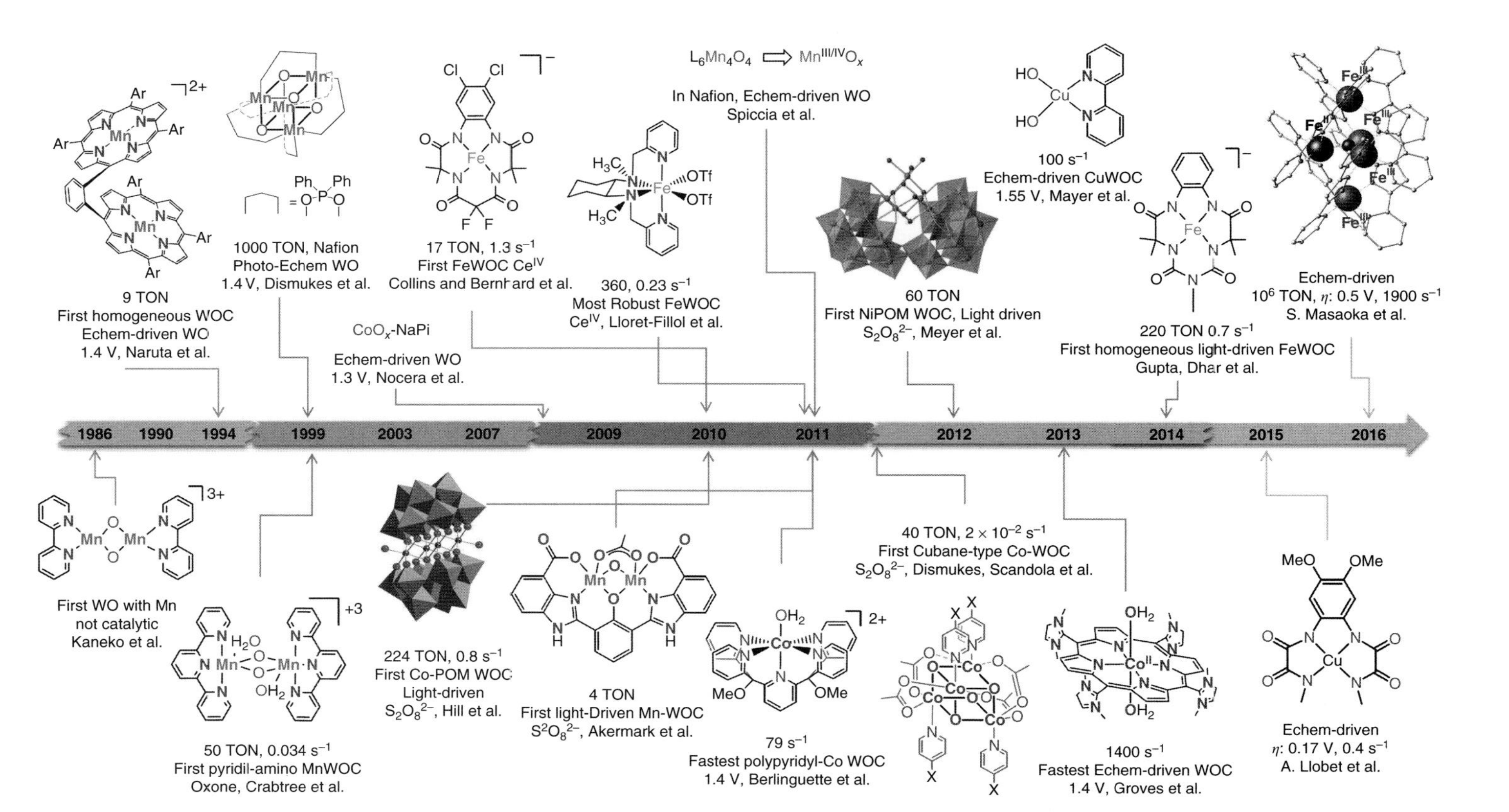

Figure 16.2 Time line with the most emblematic first-row transition metal WOCs. TON, TOF (s^{-1}), breakthrough, oxidation method, and the corresponding authors are included. E-chem-driven stands for electrochemical-driven water oxidation, and photo/e-chem is used for a system powered by a combination of light and electrical current.

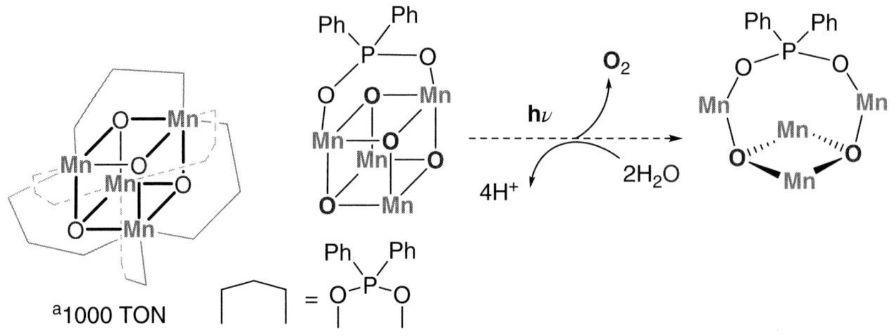

Scheme 16.3 Proposed structural rearrangement of the $L_6Mn_4O_4$ cluster during water oxidation. Only one phosphinate ligand is shown in the catalytic cycle to emphasize the coordination/decoordination process. [a] Photoelectrochemical TON obtained for the $L_6Mn_4O_4$ cluster supported on Nafion after 65 hours at 1.4 V.

are not stable under the oxidizing conditions required for water oxidation and decompose into heterogeneous species that are the real water oxidation catalysts. Recognizing molecular water oxidation catalysts may be straightforward when decomposition products (generally metal oxides) are not active or their activity is substantially smaller than that of the molecular catalyst precursor. However, in the case of catalytically active metal oxides, identification of the activity of the molecular catalysts can be quite difficult.

16.2.1.2 Biomimetic Models Including a Lewis Acid

Calcium is an essential cofactor for biological water oxidation, but its role in the OEC is still under debate. The preparation of structural analogs of the OEC containing a calcium atom may help to clarify its role, but it has been found synthetically challenging. Probably, the closest rational synthetic approach to the real structure of the OEC, a $[Mn_3CaO_4]^{6+}$ model, was described by Agapie and coworkers (Figure 16.3) [29]. Different Lewis acid ions could be installed in the cluster structure, and the electrochemical properties of the cluster were evaluated. Reduction potentials were found dependent on the Lewis acidity of the redox-inactive metal incorporated, supporting the idea that Ca^{2+} has a redox modulator role in the OEC [30]. The lower the pK_a of the M(aqua)$^+$ ion (proportional to the Lewis acidity of the redox-inactive metal), the higher the redox potential of the Mn centers because of the lower electron density on the metal [30b]. However, a deviation from the trend (Sr^{2+} and Ca^{2+} exhibit the same redox) and the intriguing observation that only the replacement of Ca^{2+} for Sr^{2+} partially maintains the function of the enzyme [31] remain puzzling observations awaiting explanation. The incorporation of an external Mn atom in synthetic models to fully reproduce the OEC structure also remains a challenge for synthetic inorganic chemists [32].

Water oxidation activity with Mn clusters is very difficult to attain. Toward this goal, the work of the Scandola, Kortz, and Bonchio groups is particularly remarkable, reporting polyoxometalate $[Mn^{III}_3Mn^{IV}O_3(CH_3COO)_3(A\text{-}\alpha\text{-}SiW_9O_{34})]^{6-}$ (Mn_4POM) as a multiredox Mn^{III}/Mn^{IV} water oxidation catalyst that evolves through five electronic states (S_i, where $i = 0$–4) like PSII [4]. Mn_4POM forms

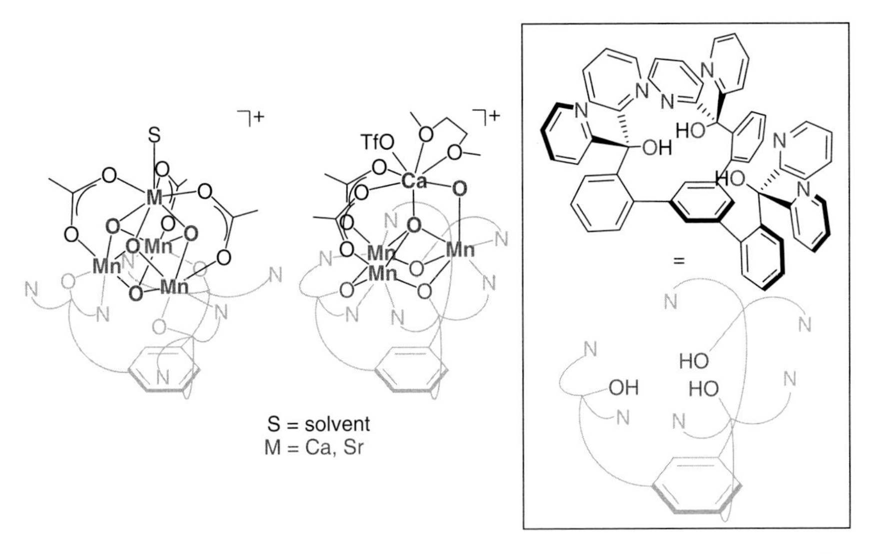

Figure 16.3 $[Mn_3CaO_4]^{6+}$ models described by T. Agapie and coworkers. Source: Kanady et al. 2011 [29]. Reproduced with permission of AAAS.

up to 5.2 tons of O_2 upon illumination of a solution containing $[Ru(bpy)_3]^{2+}$ and $Na_2S_2O_8$ in $NaHCO_3/Na_2SiF_6$ buffer (pH 5.2) with a quantum efficiency of 1.7%. Nanosecond laser flash photolysis showed an $S_0 \rightarrow S_1$ transition within milliseconds and the possibility of the cluster to accumulate multiple $1e^-$ oxidations, resembling the OEC.

16.2.1.3 Catalytic Water Oxidation with Manganese Coordination Complexes

The ability of manganese complexes to catalyze the water oxidation reaction has been extensively explored, and a full description of the topic is beyond the scope of this work. In the following section, we describe some representative examples.

In a landmark study, Brudvig and coworkers have reported that $[Mn_2^{III,IV}(\mu\text{-}O)_2(terpy)_2](NO_3)_3$ (terpy = 2,2′ : 6′,2″-terpyridine) produces O_2 by reaction with OCl^- or oxone as SOs (4 TON and >50 TON, respectively) but not when using CAN [33]. Mechanistic studies suggest a very complex reaction picture (Scheme 16.4). Oxone can bind at the Mn^{III} and at the Mn^{IV} sites, and this initiates different reaction paths. The productive path entails binding of the oxidant to the Mn^{III} site to form $[(terpy)(HSO_5)\text{-}Mn^{III}(\mu\text{-}O)_2Mn^{IV}(terpy)]^{3+}$, which is catalytically competent and evolves to $[(terpy)Mn^{V}(O)(\mu\text{-}O)_2Mn^{IV}(terpy)]^{3+}$ (best described as a $Mn_2^{IV/IV}$-oxyl radical). This species reacts with water, forming $[(terpy)Mn^{III}(OOH)(\mu\text{-}O)_2Mn^{IV}(terpy)]^{2+}$, which eventually evolves O_2. O—O bond cleavage leading to the formation of the formal Mn^V intermediate is rate determining under excess of oxone [34]. This mechanistic picture is supported by DFT calculations and by isotope labeling studies, which shows incorporation of ^{18}O from $H_2{}^{18}O$ into the evolved O_2 [34].

The use of CAN as an outer sphere, single-electron oxidant is preferred to avoid parallel sources of O_2 not arising from H_2O. However, the low pH of Ce^{IV} solutions is an important issue for the stability of the complexes, and few

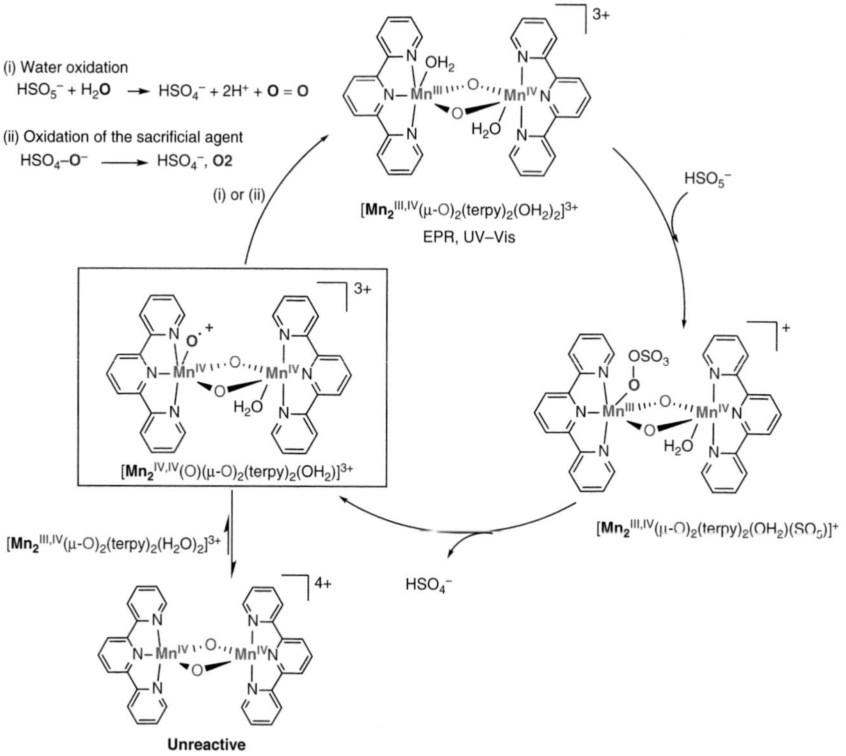

Scheme 16.4 Mechanism proposed for the oxidation of water with $[Mn_2(\mu\text{-}O)_2(terpy)_2(H_2O)_2]^{3+}$ (terpy = 2,2′ : 6,2″-terpyridine) when using oxone as an oxidant. The proposed intermediate responsible for the oxygen release has been highlighted [34].

ligands are stable under these harsh conditions. The original $[Mn_2^{III,IV}(\mu\text{-}O)_2(terpy)_2(H_2O)_2]^{3+}$ complex is unfortunately unstable under these conditions. To address this problem, the manganese complex can be adsorbed onto different layered materials (e.g. kaolin or mica). Remarkably, a maximum TON of 14 is observed, and the stability of the entire $[Mn_2^{III,IV}(\mu\text{-}O)_2(terpy)_2]^{3+}$/clay material after 30 minutes of treatment with CAN is confirmed by EXAFS [35].

Particularly interesting are also the family of dimeric Mn–porphyrin complexes described by Naruta (Scheme 16.5). These complexes contain two tetraarylporphyrin moieties (aryl moieties = 4-tBuC$_6$H$_4$, 2,4,6-Me$_3$C$_6$H$_2$, and C_6F_5) linked by a 1,2-phenylene bridge. The complexes produce oxygen electrochemically when the potential is set up above 1.4 V vs SHE. The highest activity reported, 9.2 TON, is observed for the C_6F_5 aryl-substituted Mn–porphyrin at 2 V. The origin of the activity is related to the spatial disposition of the two Mn–porphyrin units, which facilitates a face-to-face intramolecular interaction between the two Mn=O units [36].

The first homogeneous water oxidation catalyst working in the presence of an outer sphere, single-electron transfer oxidant ($[Ru(bpy)_3]^{3+}$) is the dimeric $[Mn_2(dCIP)(OMe)(CH_3CO_2)]$ (dCIP = 2-(3-(7-carboxy-1H-3λ^4-benzol

Scheme 16.5 (Left) Schematic diagram of dimeric Mn–porphyrin complexes that act as catalysts for electrochemical oxidation of water. The active species for the water oxidation is suggested to be Mn^V; Ar = C_6F_5. (Right). Proposed mechanism for the water oxidation [36].

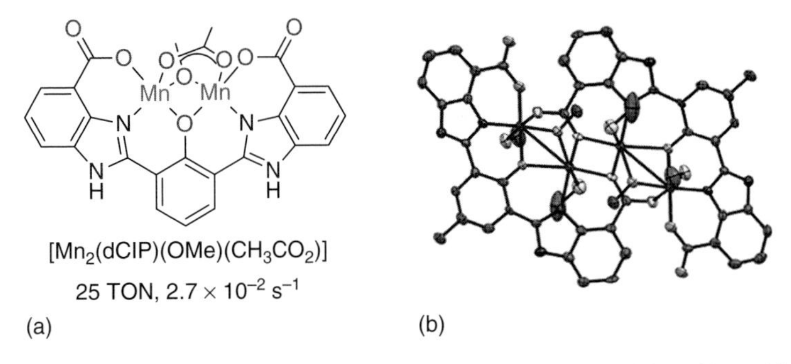

Figure 16.4 (a) Line drawing of the first homogeneous Mn-WOC working with single-electron oxidants. (b) X-ray crystal structure at 50% probability level of the biomimetic complex formed upon refluxing in MeOH. Karlsson et al. 2011 [37]. Reproduced with permission of John Wiley & Sons.

[*d*]imidazol-2-yl)-2-hydroxyphenyl)-1*H*-benzo[*d*]imidazole-4-carboxylic acid) [37]. [Mn_2(dCIP)(OMe)(CH_3CO_2)] carries out the WO with a ton of 25 and TOF of 0.027 s^{-1}. Interestingly, crystallization in hot methanol yields a tetrameric structure of the compound, reminiscent of the OEC cluster (Figure 16.4). Labeling studies with 5.8% ^{18}O-enriched water are in agreement with a water oxidation process. The catalyst is competent for the light-driven process with a TON of 4 ($Na_2S_2O_8$ as SO and $[Ru(bpy)_3]^{2+}$ or $[Ru(bpy)_2(4,4'\text{-}CO_2Et\text{-}bpy)](PF_6)_2$ as photosensitizers).

16.2.2 Water Oxidation with Molecular Iron Catalysts

Because of its availability and lack of environmental impact, iron represents a particularly interesting metal in terms of catalyst development. First evidences that iron complexes can catalyze the oxidation of water were reported by Elizarova et al. in the early 1980s [38] and by Kaneko and coworkers in 1998 [39], but the subject experienced a major interest since the initial report by Bernhard, Collins, and coworkers in 2009 describing fast catalytic oxidation with a molecular iron complex [40]. Since then, different families of homogeneous water oxidation catalysts based on iron have been described.

16.2.2.1 Iron Catalysts with Tetra-Anionic Tetra-Amido Macrocyclic Ligands

Tetra-anionic tetra-amido macrocyclic ligands (TAML's) have been extensively explored as privileged ligand scaffolds to support transition metals in high oxidation states [41]. The tetra-anionic nature of this class of ligands provides an effective charge stabilization, which in combination with their robust nature against oxidative conditions translates into unusually stable complexes where the metal adopts high oxidation states. The unusual stability of Fe–TAML complexes does not compromise their reactivity in oxidation reactions with external molecules, and they are actually very powerful peroxidase mimics. This reactivity has found utility in water cleaning–detoxification processes.

Water oxidation with Fe–TAML complexes (Scheme 16.6) can be achieved upon addition of CAN or $NaIO_4$ as SOs to aqueous solutions of the complexes, producing oxygen with a remarkably high TOF value of $1.3\,s^{-1}$, but with low TONs **16** [40]. Kinetic studies indicate that the O_2 evolution rate is first order in concentration of the catalyst, suggesting that a mono-iron site is responsible for the water oxidation reaction. Spectroscopic data shows that $Fe^V(O)$ species form under these experimental conditions, although kinetic evidence for their catalytic competence has not been delivered so far. DFT calculations performed by Cramer and coworkers suggest that $[TAML^{+\cdot}\text{-}Fe^V=O]$, where the ligand is one electron oxidized, is responsible for the O—O bond formation via a water nucleophilic attack (WNA) [42].

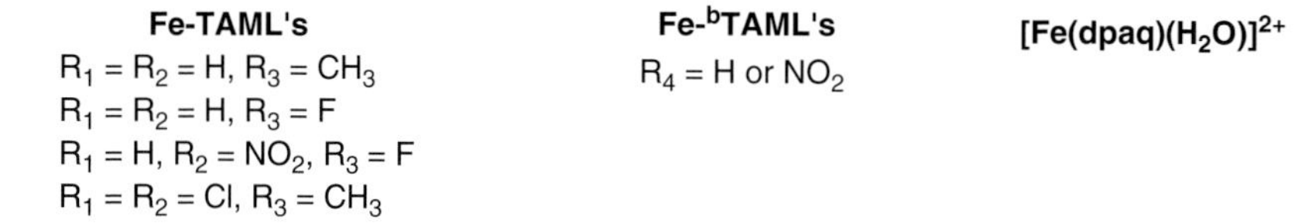

Fe-TAML's
$R_1 = R_2 = H$, $R_3 = CH_3$
$R_1 = R_2 = H$, $R_3 = F$
$R_1 = H$, $R_2 = NO_2$, $R_3 = F$
$R_1 = R_2 = Cl$, $R_3 = CH_3$

Fe-^{b}TAML's
$R_4 = H$ or NO_2

$[Fe(dpaq)(H_2O)]^{2+}$

Scheme 16.6 Schematic representation of the Fe–TAML, biuret Fe–^{b}TAML, and $[Fe(dpaq)(H_2O)]^{2+}$ complexes studied in water oxidation.

TAML–Fe complexes do not show electrocatalytic WO activity in a homogeneous phase, presumably because of rapid decomposition. However, immobilization in Nafion provides materials that show electrocatalytic WO [43]. Oxygen was produced when applying a static current of 5 mA with a faradaic yield of 45% at an estimate TOF-O_2 of $0.081\,s^{-1}$ with a parasitic carbon oxidation of the supporting material.

Biuret-modified Fe–TAML complexes (Fe–^{b}TAML, Scheme 16.6) show enhanced stability against hydrolysis and catalyze photochemical WO under homogeneous conditions [44]. Upon irradiation at 440 nm, a remarkably fast photochemical WO (TON = 220, TOF 0.76 s^{-1}) can be accomplished employing $[Ru(bpy)_3]^{2+}$ as the photosensitizer and $Na_2S_2O_8$ as the terminal oxidant. Most interestingly, because of their improved stability, Fe–^{b}TAML complexes permit spectroscopic characterization of reaction intermediates during catalysis. This analysis reveals the rapid formation of an μ-oxo-Fe^{IV} dimer, which then evolves into a Fe^{V}(O) species. Neither the μ-oxo-Fe^{IV} dimer nor the Fe^{V}(O) monomer is kinetically competent to oxidize water and, thus, further oxidation of the latter appears necessary for generating water-oxidizing species, in a scenario that appears reminiscent to that occurring for Fe–TAML complexes.

16.2.2.2 Mononuclear Complexes with Monoanionic Polyamine Ligands

The prototypical example of this type of catalysts is $[Fe(dpaq)(H_2O)]^{2+}$ (dpaq = 2-[bis(pyridine-2-ylmethyl)]amino-*N*-quinolin-8-yl-acetamido, Scheme 16.6), which acts as a water oxidation electrocatalyst in propylene carbonate–water mixtures. Kinetic analysis of the electrocatalytic behavior suggests formation of a Fe^{V}(O) species that then undergoes bimolecular water oxidation (first order in catalyst and first order in water). Sustained water oxidation catalysis yields 29 turnovers of O_2 over a 15-hour electrolysis experiment with a 45% faradaic yield. However, no significant electrocatalytic activity in aqueous solution is observed over an extended pH range. In a parallel study, the same complex was inactive in water oxidation when using CAN as an oxidant [45].

16.2.2.3 Iron Catalysts with Neutral Ligands

Selected iron complexes with aminopyridine ligands show high activity in water oxidation reactions using chemical oxidants [45, 46]. The activity of this class of catalysts is highly dependent on the nature of the ligand. The best activity is obtained with complexes that bear tetradentate ligands and leave two cis-labile sites, whereas complexes with bidentate, tridentate, pentadentate, or tetradentate ligands that leave trans-labile sites are virtually inactive. The complexes $[Fe(OTf)_2(Pytacn)]$ (Pytacn = 1,4-dimethyl-7-(2-pyridylmethyl)-1,4,7-triazacyclononane, OTf = trifluoromethanesulfonate anion) and $[Fe(OTf)_2(mcp)]$ (mcp = *N*,*N*′-dimethyl-*N*,*N*′-bis(2-pyridylmethyl)cyclohexane-*trans*-1,2-diamine) (Scheme 16.7) are particularly interesting because they exhibit remarkable stability against hydrolytic and oxidative decomposition paths in chemically driven water oxidation reactions (at low pHs), and high-valent reaction intermediates accumulate during their catalytic reactions. $[Fe(OTf)_2(mcp)]$ is among the most efficient water oxidation catalysts based on the first-row transition metals described to date, reaching TON values up to 360 (TOF_{max} = 0.23 s^{-1}) and >1000 (TOF_{max} = 0.06 s^{-1}) when using CAN (pH 1) and $NaIO_4$ (pH 2) as the chemical oxidant, respectively [46a]. Unfortunately, these complexes are not competent in electrocatalytically or photochemically driven water oxidation processes. In the latter case, they undergo rapid decomposition forming iron oxide nanoparticles that are then catalytically competent [25, 46f, 47].

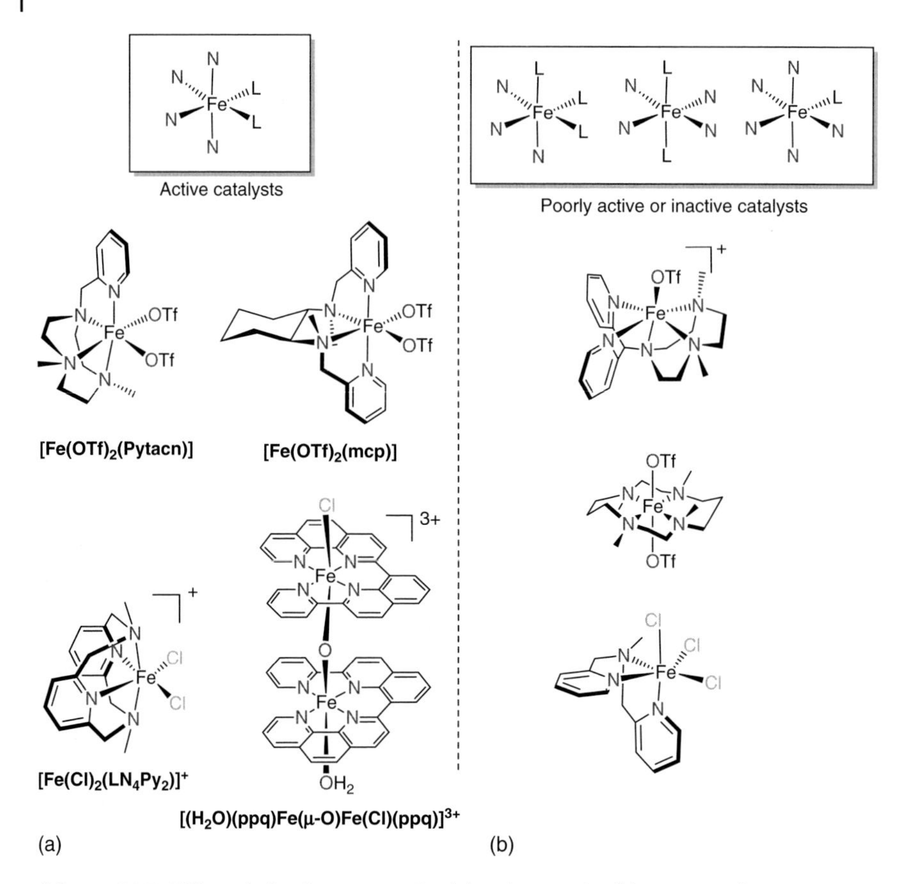

Scheme 16.7 Differentiation between active (a) and nonactive (b) structures for water oxidation with iron complexes bearing neutral aminopyridine ligands.

Isotopic labeling experiments performed on reactions catalyzed with the $[Fe(OTf)_2(L)]$, L = Pytacn, and mcp catalysts indicate that oxygen atoms of the evolved O_2 originate exclusively from water. Negligible CO_2 production is measured, and dynamic light scattering (DLS) experiments show no evidence for NP formation. These two experimental observations argue against complete ligand oxidative degradation and production of iron oxides and instead are indicative of a homogeneous process. Most significantly, $[Fe^{IV}(O)(S)(L)]^{2+}$ (S being presumably water) can be identified as the resting state during catalytic reactions, and O_2 evolution occurs only when these species are present in the solution. Once these species are consumed, O_2 evolution ceases despite the presence of excess of oxidant. However, kinetic studies in CAN-driven oxidations show that $[Fe^{IV}(O)(S)(L)]^{2+}$ do not react with water. Instead, detailed kinetic studies following O_2 evolution and CAN consumption produce congruent rates that exhibit first-order dependence on $[Fe^{IV}{=}O]$ and on CAN when the latter is used in slight excess. However, using large amounts of oxidant (CAN > 20 equiv.) leads to saturation of the reaction rates and formation of a new UV–Vis chromophore [46b]. This new species can be spectroscopically characterized in catalytic reactions performed with the [Fe(OTf)$_2$(mcp)] complex. High resolution mass spectrometry (HRMS) and resonance Raman experiments lead to the formulation of the intermediate as a heterometallic [Fe^{IV}(O)(μ-O)Ce^{IV}] species. This heterometallic species can be understood as an inner sphere intermediate in the single-electron oxidation of the Fe^{IV}(O) species by Ce^{IV} to form a Fe^{V}(O)(OH) intermediate where the O—O bond is formed by undergoing a WNA (Scheme 16.8). The later scenario is validated by a DFT study, which also points out the importance of the hydroxide ligand in order to bind and orient the water substrate toward nucleophilic attack to the Fe=O unit. Computations also show that the oxidation of Fe^{IV}(O)(OH$_2$) by Ce^{IV} to form Fe^{V}(O)(OH) is possible because the reaction entails a proton-coupled electron transfer (PCET) process. However, single-electron oxidation of Fe^{IV}(O) complexes with ligands that do not permit the PCET process, such as pentadentate amines, is not possible.

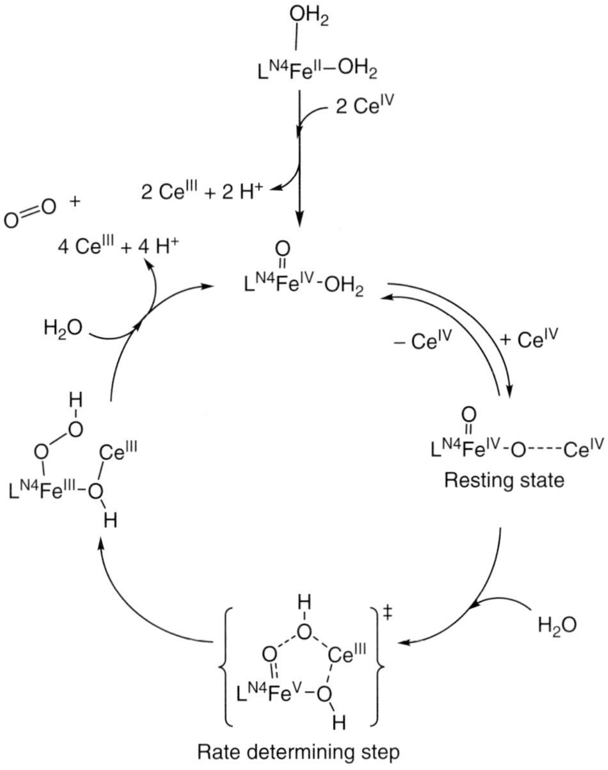

Scheme 16.8 Mechanism postulated for water oxidation with iron complexes bearing tetradentate aminopyridine ligands.

Interestingly, Che and coworkers propose a different succession of events for the O—O bond formation of the iron complex [Fe(LN$_4$Py$_2$)Cl$_2$]$^+$ (LN$_4$Py$_2$ = *N,N′*-dimethyl-2,11-diaza[3, 3](2,6)pyridinophane) (Scheme 16.7). No evidence of an iron–cerium complex was observed, and it was proposed that an $Fe^{IV}(O)(OH)$ evolves to $Fe^{V}(O)(O)$ ($S = 3/2$), which engages in O—O bond formation through a WNA mechanism [46g].

An oxo-bridged diiron complex, where each center contains a planar tetradentate polypyridyl ligand, showing remarkable water oxidation activity has been recently described (Scheme 16.7) [48]. The complex $[(H_2O)(ppq)Fe(\mu\text{-}O)Fe(Cl)(ppq)]^{3+}$ (ppq = 2-(pyrid-2′-yl)-8-(1″,10″-phenanthrolin-2″-yl)-quinoline) reacts with CAN, producing O_2 at a remarkably fast rate (TOF = 7920 h^{-1}). Interestingly, the complex also exhibits electrocatalytic behavior. Cyclic voltammetry (CV) indicates that the complex undergoes a two-electron oxidation from $H_2O\text{—}Fe^{III}Fe^{III}$ to $H_2O\text{—}Fe^{IV}Fe^{IV}$, which is then proposed to isomerize to an $O{=}Fe^{V}Fe^{III}$ species, finally responsible for the water oxidation.

16.2.2.4 Water Oxidation by a Multi-iron Catalyst

Multimetallic complexes with electronically coupled metal centers are very interesting because they are susceptible to delocalize and accumulate multiple charges, making them particularly suitable for engaging in multielectronic reactions. The OEC actually constitutes a paradigmatic example. The design of such type of complexes is not only very interesting but also very difficult because control over the molecular architecture, stability, and electronic structure of multimetallic complexes is not obvious, most often unpredictable. However, the outcomes of this approach can be truly outstanding, as recently illustrated by Masaoka, who has described a pentairon complex $[Fe^{II}_4Fe^{III}(\mu_3\text{-}O)(\mu\text{-}L)_6]^{3+}$ in which the metal ions are strongly electronically coupled via the oxo bridge and that shows outstanding electrocatalytic water oxidation ability. In this complex, five iron centers arrange in the vertexes of a trigonal bipyramid, with an oxo ligand in the center, which connects the five iron atoms. Each of the two iron atoms at the apical positions is connected to the three atoms of the trigonal plane by dinucleating 3,5-bis(2-pyridyl)pyrazole ligands. In acetonitrile, the complex exhibits a rich electrochemical behavior, supporting up to four chemically and electrochemically reversible one-electron oxidation reduction steps corresponding to Fe^{II}/Fe^{III} couples. In the presence of water, the last oxidation results in an electrocatalytic wave. Kinetic analysis of the electrocatalytic behavior reveals a large TON (10^6–10^7 for 120 minutes) at remarkably high rates (TOF = 140–1400 s^{-1}) [49]. The authors propose that O—O formation takes place via reaction of two spatially properly disposed terminal iron-oxo units. A single negative aspect of this catalyst is that oxidation takes place at a large overpotential, a factor that may be addressed by proper ligand design.

16.2.3 Cobalt Water Oxidation Catalysts

Cobalt compounds have gained particular attention as water oxidation catalysts since the pioneering report of Nocera's group, showing that precipitates from cobalt complexes (or cobalt oxophosphate salts "CoPi") in water show water oxidation activity [50]. Modified electrodes obtained by deposition of these precipitates exhibit a highly improved long-term water oxidation activity. In light of this discovery, cobalt complexes have been investigated with the aim of developing well-defined homogeneous catalysts, which could help in mechanistic understanding. Currently, a large quantity of cobalt complexes based on polyoxometalates [51], porphyrins, phthalocyanins, corroles, TAML [44b], polypyridines [52], polyamines [53], polypyridinamines [54], and salen [55] ligands have been shown to possess catalytic WO activity. In addition, multinuclear cobalt clusters have been used as models to shed light on the water oxidation mechanism operating on heterogeneous materials. A major complication in these studies arises from the fact that simple Co^{2+} ions and cobalt oxides are also active WO catalysts, and a great effort is usually needed to distinguish between homogeneous and heterogeneous systems. The work on cobalt compounds is quite extensive, and a proper review goes beyond the scope of this chapter. We have chosen representative cases of the field.

Studies on $[Co(PY5)(OH_2)](ClO_4)_2$ (Figure 16.5) are illustrative of the difficulties of ascertaining the nature of the water oxidation catalysts with this class of compounds. Berlinguette first showed that the complex shows an irreversible oxidation at ~1.4 V (at pH 9.2). This process is assigned to a catalytic behavior occurring upon the formation of a $[Co^{IV}OH]^{3+}$ species. The reaction rate (k_{cat}) is 79 s^{-1}. In addition, the complex is stable over a pH range of 7.6–10.3, without leading to film deposition under electrochemical conditions, but contribution of cobalt oxides cannot be completely ruled out [56]. The pH-dependent electrochemical features of the complex differ from Co^{2+} (aq) ions, suggesting that a part of the current may arise from a molecular compound. However, diluted Co^{2+} (aq) shows a similar catalytic behavior. First-order kinetics in cobalt and OH^- concentration suggests that the rate-determining step (RDS) is the electrophilic attack of a high valent $[Co^{IV}\text{-}OH]^{3+}$ to OH^-, with an observed kinetic isotope effect (KIE) of 4.7 [57].

Later on, the Ott and Thapper groups have explored the electro- and photocatalytic activity of the related $[Co(PY5OH)(Cl)](BF_4)$ complex (Figure 16.5).

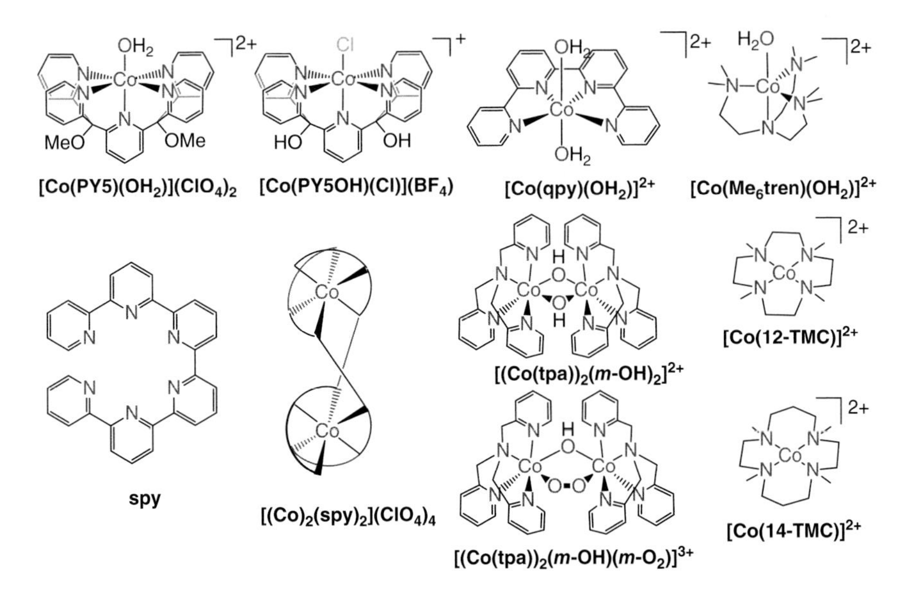

Figure 16.5 Polypyridyl, polypyridylamino, and polyamine cobalt complexes.

A similar WO overpotential was found with a TON of 51 for the photocatalytic activity ($[Ru(bpy)_3]^{2+}/Na_2S_2O_8$ at pH 8, borate buffer). Nanoparticle formation in the 7.5–10 pH range is discarded based on DLS experiments; nevertheless, at pH > 10, a clear formation of nanoparticles was observed. When using $[Ru(bpy)_3]^{3+}$ as an oxidant in a Co/Ru molar ration of 1 : 100, a TON value of 15 was found [52a].

Lau and coworkers introduced the $[Co(qpy)(OH_2)]^{2+}$ complex in 2012 as the first WO and water reduction (WR) catalyst (Figure 16.5, qpy = 2,2′:6′,2″:6″,2‴-quaterpyridine). The catalytic WO activity under photochemical conditions ($[Ru(bpy)_3]^{2+}$ and $Na_2S_2O_8$ at pH 8, borate buffer) leads to a TON of 335 after 1.5 hours of irradiation at 457 nm [58]. The entire complex was detected by electrospray ionization mass spectroscopy (ESI-MS), and DLS analysis does not support the presence of NPs. The O_2 yield for the thermal WO is 67%, with an initial TOF of $4\,s^{-1}$. The same research group has also reported the WO catalytic activity of the helical cobalt dimer $[(Co)_2(spy)_2](ClO_4)_4$ (Figure 16.5, spy = 2,2′:6′,2″:6″,2‴:6‴,2⁗:6⁗,2′′′′′-sexipyridine) [52b]. Interestingly, O_2 production increases with the catalyst concentration up to 2 μM, leading to a maximum TON of 442 after three hours of irradiation and a maximum TOF of $1.9\,s^{-1}$. Several lines of evidence indicate that the complex behaves as a molecular catalyst. The complex is stable for more than 10 days at room temperature at pH 5–9, decomposing at low pH. No nanoparticles are detected by DLS after catalysis, whereas when using the same concentration of Co^{2+} ions, nanoparticles of 350 nm are clearly formed.

Closely related systems are the cobalt complexes based on aminopyridine ligands. Cobalt complexes derived from the tris(2-pyridylmethyl)amine (TPA) ligand type have been found to lead to oxygen evolution under photo- and electrochemical conditions [54a–c]. The catalytic activity is related to the formation of cobalt oxide nanoparticles [54d]. Electrodeposition, titration experiments (bpy), energy-dispersive X-ray (EDX), and X-ray photoelectron spectroscopy (XPS) show that cobalt oxides deposited on the electrode are responsible for WO. Importantly, DLS is not sensitive enough to show the presence of nanoparticles but transmission electron microscopy (TEM) is. The authors also claim that the formed CoOx nanoparticles can get adhered to the surface of magnetic stirring bars, which cause false negatives. Extreme care should be taken to ensure that the catalytic species are well-defined complexes [52c].

Nam, Fukuzumi, and coworkers have reported a clear-cut example where a heterogeneous material was the real catalyst (Figure 16.5) [59]. Under photocatalytic conditions at pH 7–10, complexes $[Co(Me_6tren)(OH_2)]^{2+}$, $[Co(Cp^*)(bpy)(OH_2)]^{2+}$, $[Co(12\text{-}TMC)]^{2+}$, and $[Co(14\text{-}TMC)]^{2+}$ (Me_6tren = tris(*N*,*N*′-dimethylaminoethyl)amine, Cp* = η^5-pentamethylcyclopentadientyl, 12-TMC = 1,4,7,10-tetramethyl-1,4,7,10-tetraazacyclododecane, 13-TMC = 1,4, 7,10-tetramethyl-1,4,7,10-tetraazacyclotriecane) degrade after three minutes of irradiation. DLS and TEM show the formation of NPs of different sizes (20–200 nm, depending on the precatalyst). The water-soluble complex $[Co(cyclam)](ClO_4)$ ClO_4 (cyclam = 1,4,8,11-tetraazacyclotetradecane) also serves as a precursor for the electrodeposition of an active cobalt-based material on carbon electrodes. The modified electrode shows a water oxidation activity of 6.5 mA cm^{-2} at an overpotential of 580 mV at pH 12 with 98% faradaic yield [53]. Along a similar vein, other hydrolytically instable cobalt complexes have served as precatalysts for electro- and photo-water oxidation [60].

16.2.4 Nickel-Based Water Oxidation Catalysts

Nickel-based complexes have been explored to a very limited extent as catalysts in water oxidation reactions. However, this situation is changing since recent discoveries of nickel-based materials that efficiently catalyze the water oxidation reaction [61]. NiOx is indeed an excellent water oxidation catalyst and represents a potential contributor to water oxidation activity detected with molecular compounds. The interest for elucidating the details of these reactions has fueled the study of well-defined coordination compounds as models that permit an understanding of the reactions at a molecular scale.

Strong evidence in favor of a single-site nickel-based water oxidation catalyst operating in solution was first provided for the macrocyclic complex $[Ni(\textit{meso}\text{-}L)]^{2+}$ (Scheme 16.9) [62]. The complex was isolated and crystallographically characterized as a four-coordinate nickel complex with the macrocyclic ligand enforcing a square-planar coordination geometry. In a water solution, the complex is in equilibrium with the octahedrally coordinated $[Ni(\textit{meso}\text{-}L)(H_2O)_2)]^{2+}$, where the two water molecules occupy the axial positions. In phosphate buffer (but also in carbonate and acetate buffers) at neutral pH, $[Ni(\textit{meso}\text{-}L)(H_2O)_2](ClO_4)_2$ electrocatalyzes the oxidation of water at a relatively low overpotential ($E_{p,a} = 1.41$ V vs NHE) using a glassy carbon (GC) or a indium tin oxide (ITO) electrode. The CV of the complex exhibits two oxidation waves; the first one corresponds to oxidation from $[Ni^{II}(\textit{meso}\text{-}L)(H_2O)_2]^{2+}$ to $[Ni^{III}(\textit{meso}\text{-}L)(OH)(H_2O)]^{2+}$, and the second one that corresponds to the electrocatalytic wave is assigned to oxidation from $[Ni^{III}(\textit{meso}\text{-}L)(OH)(H_2O)]^{2+}$ to $[Ni^{IV}(\textit{meso}\text{-}L)(O)(H_2O)]^{2+}$ or $[Ni^{III}(\textit{meso}\text{-}L)(O^{\cdot})(H_2O)]^{2+}$, which is then

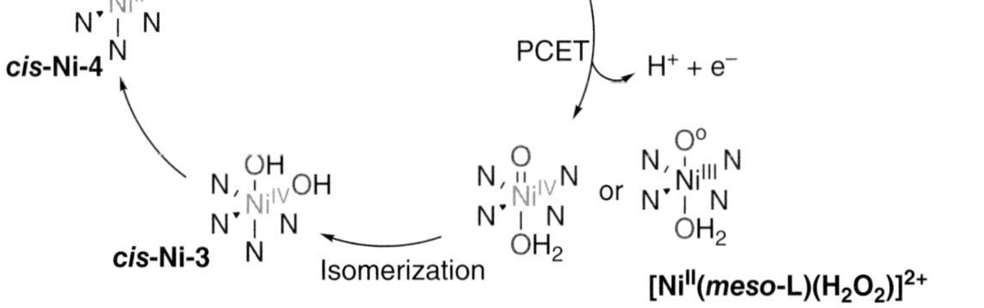

Scheme 16.9 Mechanism of electrocatalytic water oxidation proposed for $[Ni(\textit{meso}\text{-}L)(H_2O)_2](ClO_4)_2$.

responsible for the water oxidation reaction, presumably via a Ni(IV)-Ni(II) cycle. Insights into the reaction mechanism were elucidated by DFT methods. Two main points deserve a special consideration: (i) isomerization from a *trans*-$(H_2O)Ni(O)$ to a *cis*-$Ni(OH)_2$ is facile (they differ in <2 kcal mol^{-1}) and (ii) O—O bond formation occurs via HO—OH coupling, resulting in a $[Ni^{II}(meso\text{-}L)(H_2O_2)]^{2+}$ intermediate that is further oxidized to produce O_2.

Because the presence of two cis-labile sites appears to be a crucial element in the O—O bond formation step, the same authors have studied electrocatalytic water oxidation with the nickel complexes $[Ni(men)(H_2O)_2]^{2+}$ (men = *N,N′*-dimethyl-*N,N′*-bis(2-pyridylmethyl)ethane-1,2-diamine and $[Ni(mcp)(H_2O)_2]^{2+}$ based on tetradentate aminopyridine ligands (Scheme 16.10) [63]. Both complexes exhibit electrocatalytic behavior at a low overpotential, very much resembling $[Ni(meso\text{-}L)(H_2O)_2)](ClO_4)_2$, although they proved somewhat less active.

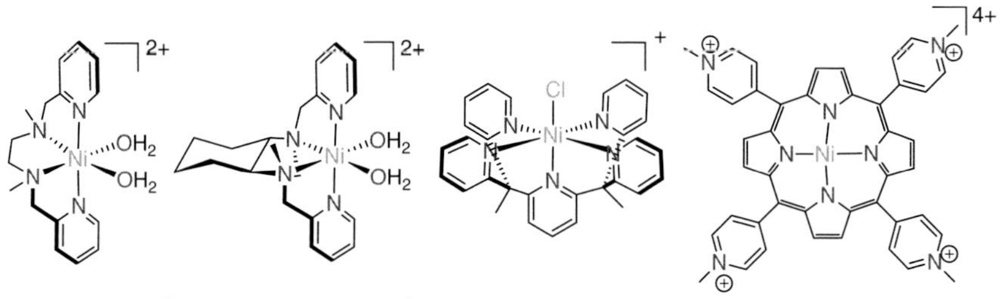

$[Ni(men)(H_2O)_2]^{2+}$ $[Ni(mcp)(H_2O)_2]^{2+}$ $[Ni(Py5)(Cl)]^{+}$ $[Ni(PorphPyMe)]^{4+}$

Scheme 16.10 Ni complexes employed as water oxidation catalysts.

A nickel complex with a pentapyridine ligand ($[Ni(Py5)(Cl)]^{+}$) has also been described by Sun and coworkers (Scheme 16.10) [64]. This compound also acts as an electrocatalyst in phosphate buffer, and in this case, water oxidation is presumed to take place via a nucleophilic attack of the water molecule on a high-valent Ni^{V}-oxo species, the O—O bond formation being assisted by proton transfer to HPO_4^{2-}. Interestingly, the reaction rate of this complex is highly enhanced by the proton acceptor base HPO_4^{2-}, reaching remarkably high rates (1820 s^{-1}).

Structurally distinct from these compounds is the Ni–porphyrin complex $[Ni(PorphPyMe)]^{4+}$ (Scheme 16.10) [65], which also acts as a water oxidation electrocatalyst, operating in a pH range of 2.0–8.0 and at low overpotentials (onset at 1.0 V at neural pH). Inspection of the electrode by scanning electron microscope and EDX spectroscopy demonstrated that NiOx are not deposited on the electrode and suggested a molecular nature of the catalyst. The mechanism of O—O bond formation was investigated by DFT methods, CV, and by determining KIEs. The crucial O—O bond formation is proposed to occur via the reaction of a Ni(III)—O·species that reacts with a water molecule via an oxygen atom transfer, coupled with a hydrogen atom transfer from the water molecule to a second water molecule or a base (acetate of phosphate) [65].

16.2.5 Copper-Based Water Oxidation Catalysts

The first evidence for copper-catalyzed water oxidation was described by Elizarova et al. [38, 66]. In these reports, $CuCl_2$, $Cu(bpy)_2Cl_2$, and $Cu(bpy)_3Cl_2$ were used as catalysts and $[Ru(bpy)_3]^{3+}$ as an oxidant. More recently, Mayer and coworkers described that the simple $[(bpy)Cu(\mu\text{-}OH)]_2{}^{2+}$ complex is a catalyst for electrochemical water oxidation. CVs of the complex exhibit a large irreversible current with 750 mV overpotential at high pHs (pH > 12). The monomeric $[(bpy)Cu(OH)_2]$ is found dominant in the solution (electron paramagnetic resonance [EPR], CV), and a TOF_{max} of $100\,s^{-1}$ is observed. The authors discarded the electrodeposition of the catalyst on the electrode and formation of NPs [67]. Furthermore, kinetic analysis of the electrochemical data pinpoints toward a water oxidation reaction that is first order in copper (Scheme 16.11). The 6,6′-OH substituted bipyridyl ligand was later designed to facilitate PCET during catalytic reaction and also to facilitate oxidation of the copper center (lowering redox potentials) by electron-donating groups [68]. Effectively, the overpotential is reduced by 200 mV compared to the unsubstituted system. Catalyst stability is supported by UV–Vis and nuclear magnetic resonance (NMR) (>90% of the ligand was recovered after acidification), and a TON of 400 is determined during bulk electrolysis at 1.1 V vs SHE (pH 12.4, 0.1 M NaOAc/NaOH) in a fritted cell (the counter electrode is separated with a membrane to avoid the reduction of the catalyst). An apparent TOF of $0.4\,s^{-1}$ was calculated.

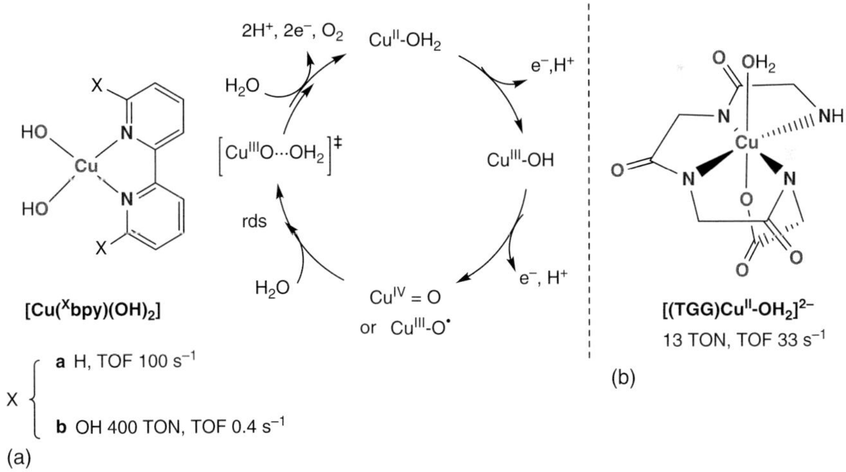

Scheme 16.11 Schematic representation of Cu-WOCs, $[(^{X}bpy)Cu(\mu\text{-}OH)]_2{}^{2+}$ (a), and $[(TGG)Cu^{II}\text{-}OH_2]^{2-}$ (b) and the mechanistic proposal for the electrochemical acid–base water oxidation [68].

Under appropriate conditions, the use of simple Cu^{2+} salts also provides a sustained WO [69]. At high concentrations of CO_3^- or PO_4^{3-} (~1 M, pH = 10.8), kinetics agree with a pH-dependent direct coupling mechanism (second order in $[Cu^{2+}]$ and the activity decreases at lower pH values). An electrodeposited material is found to contribute to the catalysis, and the spectroscopic data (UV–Vis and pulsed EPR) point toward a [buffer–Cu] complex as the active species. However, a first order on copper concentration is observed in the presence of $NaHCO_3$ (CO_2-saturated, 0.1 M, pH 6.7) or acetate buffer (0.1 M, pH 6), and thus, both mechanisms, direct coupling and WNA, are compatible for water oxidation with copper salts.

An electrocatalytic oxidation wave with an onset at 1.1 V (vs SHE in NaPi, pH 11) is also observed for a triglycylglycine Cu^{II} complex $[(TGG)Cu^{II}\text{-}OH_2]^{2-}$ (Scheme 16.11) [70]. The current is almost stable for longtime electrolysis, and no changes are observed because of electrodeposition or catalyst decomposition, favoring a robust homogeneous system. Kinetics agree with a single-site process, being a formal $Cu^{IV}{=}O$ active species involved in a rate-determining O—O bond formation process, taking place with a $k_{O—O}$ of 33 s^{-1}, which in fact corresponds to the TOF of the catalytic reaction.

Recently, Llobet and coworkers have described electrocatalytic water oxidation with a family of mono-copper complexes bearing tetra-anionic tetradentate amidate ligands (Figure 16.6) [71, 72]. The series of mononuclear square-planar Cu(II) complexes undergo a first metal-based Cu^{III}/Cu^{II} oxidation followed by a pH-dependent, ligand-based, single-electron oxidation that is associated with an electrocatalytic wave. Foot-of-the-wave analysis gave a catalytic rate constant of 3.6 s^{-1} at pH 11.5 and 12 s^{-1} at pH 12.5. Interestingly, as the electron-donating capacity at the aromatic ring in the ligand increases, the overpotential is drastically reduced, reaching a remarkably low value of 170 mV for the dimethoxy-substituted complex. Computational analyses suggest that this family of complexes follows a singular mechanistic path entailing two consecutive intramolecular single-electron transfer steps; initial attack of a hydroxyl moiety at the $[L{\cdot}Cu^{III}\text{-}OH]$ species forms a $(HO—OH){\cdot}^{-}$ radical anion fragment with a partial O—O bond (with an O—O distance of 2.3 Å, Figure 16.6), which is hydrogen bonded to the $[(L)Cu^{III}]$ complex. Then, a second electron transfer leads to a HO—OH species that is hydrogen bonded to the $[(L)Cu^{II}]$ complex. The $[(L)Cu^{II}(HOOH)]$ species is then oxidized to $[(L)Cu^{III}(HOOH)]$ and a subsequent PCET leads to [(L1)Cu(III)(HOO·)], which evolves producing a proton and O_2.

16.3 Conclusion and Outlook

Although biological water oxidation takes place at a manganese cluster, the reaction has been traditionally difficult to accomplish with synthetic catalysts based on first-row transition metals. Indeed, studies on the oxidation of the water molecule with well-defined coordination complexes have traditionally been focused on precious metals. However, under the pressure of finding more sustainable catalytic methodologies, catalysts based on earth-abundant metals have become common in the past decade. Although manganese was basically the only one explored a decade ago, earth-abundant metals such as copper, iron, cobalt, and nickel have become common constituents of water oxidation

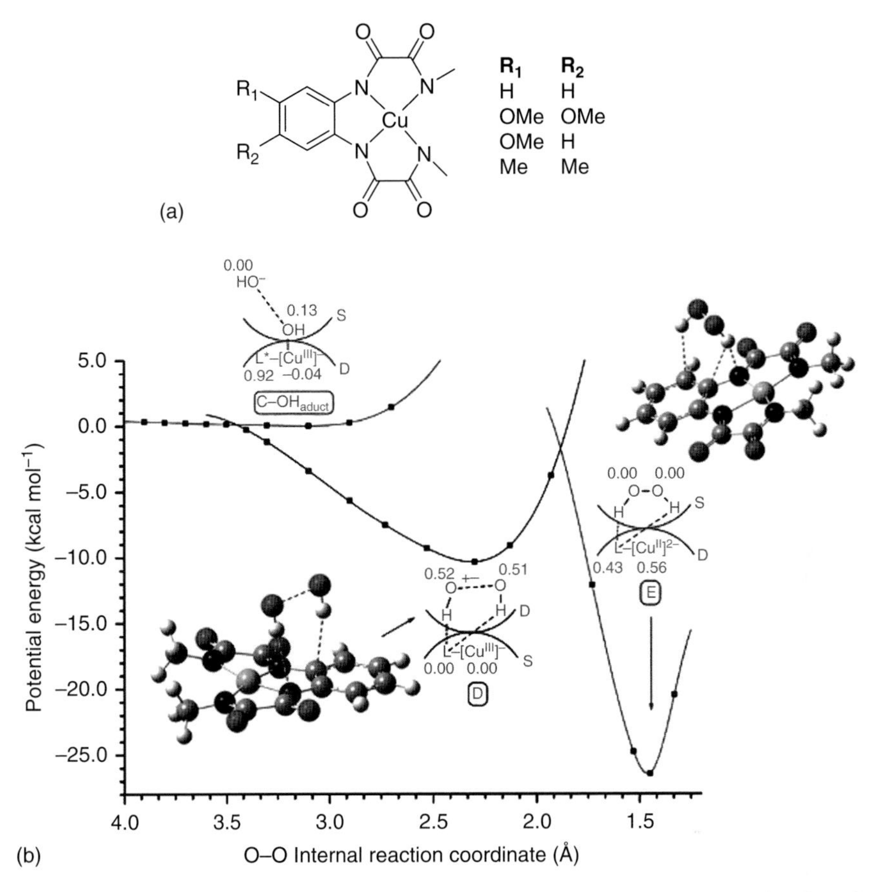

Figure 16.6 (a) Schematic diagram of the mono-copper complexes with tetraamidate ligands. (b) Potential energy-relaxed scan for the O—O bond formation. Source: Funes-Ardoiz et al. 2017 [72]. Reprinted (adapted) with permission from ACS. Copyright (2017) American Chemical Society. The energy barrier for the second I_{SET} is estimated from changes of potential energy in the coordinate scan.

catalysts explored nowadays. The design of these catalysts, their mode of action (as homogeneous catalysts or as precursors of catalytically active nanoparticles), and the elucidation of their reaction mechanisms have become a topic of major interest for the coordination chemistry and bioinorganic chemistry fields. It is obvious that useful water-spitting devices could not be based on coordination complexes acting as catalysts in homogeneous water solutions, but instead, these coordination complexes should be viewed as excellent platforms for studying the mechanisms at the molecular scale, providing valuable information that help creating efficient heterogeneous catalysts susceptible to be incorporated into useful devices. Although most of the catalysts that have been described up to date are based on monometallic centers, operating via highly electrophilic metal-oxo species, recent studies point toward novel directions that entail metal–ligand cooperation and multimetallic centers as tools to accumulate multielectronic oxidation events required to form the O—O bond from the oxidation of water. It is envisioned that such multimetallic compounds, which may be viewed as synthetic analogs of the Mn cluster of the OEC, will become a topic for further development of the field in the upcoming years. Furthermore, it is likely that the lessons learned in the design of catalysts for water oxidation will provide useful directions in catalyst design for other challenging multielectronic transformations of current interest such as CO_2 or N_2 reduction.

References

1 Ronge, J., Bosserez, T., Martel, D. et al. (2014). *Chem. Soc. Rev.* 43: 7963.
2 Lewis, N.S. and Nocera, D.G. (2006). *Proc. Natl. Acad. Sci. U.S.A.* 103: 15729–15735.
3 (a) Kalyanasundaram, K. and Graetzel, M. (2010). *Curr. Opin. Biotechnol.* 21: 298–310. (b) Bard, A.J. and Fox, M.A. (1995). *Acc. Chem. Res.* 28: 141–145.
4 Al-Oweini, R., Sartorel, A., Bassil, B.S. et al. (2014). *Angew. Chem.* 126: 11364–11367. (2014). *Angew. Chem. Int. Ed.* 53: 11182–11185.
5 Biesiadka, J., Loll, B., Kern, J. et al. (2004). *Phys. Chem. Chem. Phys.* 6: 4733–4736.
6 Nocera, D.G. (2009). *Inorg. Chem.* 48: 10001–10017.
7 (a) Duan, L., Bozoglian, F., Mandal, S. et al. (2012). *Nat. Chem.* 4: 418. (b) Duan, L., Araujo, C.M., Ahlquist, M.S.G., and Sun, L. (2012). *Proc. Natl. Acad. Sci. U.S.A.* 109: 15584.
8 (a) Shevela, D. and Messinger, J. (2012). *Biochim. Biophys. Acta, Bioenerg.* 1817: 1208–1212. (b) Duan, L., Tong, L., Xu, Y., and Sun, L. (2011). *Energy Environ. Sci.* 4: 3296–3313.
9 Umena, Y., Kawakami, K., Shen, J.R., and Kamiya, N. (2011). *Nature* 473: 55–60.
10 Cox, N., Pantazis, D.A., Neese, F., and Lubitz, W. (2013). *Acc. Chem. Res.* 46: 1588–1596.
11 Suga, M., Akita, F., Hirata, K. et al. (2014). *Nature* 517: 99–103.
12 Joliot, P., Barbieri, G., and Chabaud, R. (1969). *Photochem. Photobiol.* 10: 309–329.
13 Kok, B., Forbush, B., and McGloin, M. (1970). *Photochem. Photobiol.* 11: 457–475.
14 Betley, T.A., Surendranath, Y., Childress, M.V. et al. (2008). *Philos. Trans. R. Soc. London, Ser. B* 363: 1293–1303.
15 Cox, N., Retegan, M., Neese, F. et al. (2014). *Science* 345: 804–808.
16 Suga, M., Akita, F., Sugahara, M. et al. (2017). *Nature* 543: 131–135.
17 Sartorel, A., Bonchio, M., Campagna, S., and Scandola, F. (2013). *Chem. Soc. Rev.* 42: 2262–2280.
18 Yano, J. and Yachandra, V. (2014). *Chem. Rev.* 114: 4175–4205.
19 (a) Lutterman, D.A., Surendranath, Y., and Nocera, D.G. (2009). *J. Am. Chem. Soc.* 131: 3838–3839. (b) Reece, S.Y., Hamel, J.A., Sung, K. et al. (2011). *Science* 334: 645–648.
20 Parent, A.R., Crabtree, R.H., and Brudvig, G.W. (2013). *Chem. Soc. Rev.* 42: 2247–2252.

21 Haynes, W.M. (2014). *CRC Handbook of Chemistry and Physics*, 95e. CRC Press.
22 Yoshida, M., Masaoka, S., Abe, J., and Sakai, K. (2010). *Chem. Asian J.* 5: 2369–2378.
23 (a) Costentin, C., Passard, G., and Savéant, J.-M. (2015). *J. Am. Chem. Soc.* 137: 5461–5467. (b) Rountree, E.S., McCarthy, B.D., Eisenhart, T.T., and Dempsey, J.L. (2014). *Inorg. Chem.* 53: 9983–10002.
24 Akhtar, U.S., Tae, E.L., Chun, Y.S. et al. (2016). *ACS Catal.* 6: 8361–8369.
25 Chen, G., Chen, L., Ng, S.-M. et al. (2013). *Angew. Chem.* 125: 1833–1835. (2013). *Angew. Chem. Int. Ed.* 52: 1789–1791.
26 (a) Karkas, M.D., Verho, O., Johnston, E.V., and Akermark, B. (2014). *Chem. Rev.* 114: 11863–12001. (b) Blakemore, J.D., Crabtree, R.H., and Brudvig, G.W. (2015). *Chem. Rev.* 115: 12974–13005. (c) Najafpour, M.M., Safdari, R., Ebrahimi, F. et al. (2016). *Dalton Trans.* 45: 2618–2623.
27 Brimblecombe, R., Swiegers, G.F., Dismukes, G.C., and Spiccia, L. (2008). *Angew. Chem.* 120: 7445–7448. (2008). *Angew. Chem. Int. Ed.* 47: 7335–7338.
28 Hocking, R.K., Brimblecombe, R., Chang, L.-Y. et al. (2011). *Nat. Chem.* 3: 461–466.
29 Kanady, J.S., Tsui, E.Y., Day, M.W., and Agapie, T. (2011). *Science* 333: 733–736.
30 (a) Kanady, J.S., Mendoza-Cortes, J.L., Tsui, E.Y. et al. (2012). *J. Am. Chem. Soc.* 135: 1073–1082. (b) Tsui, E.Y., Tran, R., Yano, J., and Agapie, T. (2013). *Nat. Chem.* 5: 293–299. (c) Siegbahn, P.E.M. (2014). *Phys. Chem. Chem. Phys.* 16: 11893–11900.
31 Brudvig, G.W. (2008). *Philos. Trans. R. Soc. London, Ser. B* 363: 1211–1219.
32 Kanady, J.S., Lin, P.-H., Carsch, K.M. et al. (2014). *J. Am. Chem. Soc.* 136: 14373–14376.
33 (a) Limburg, J., Brudvig, G.W., and Crabtree, R.H. (1997). *J. Am. Chem. Soc.* 119: 2761–2762. (b) Limburg, J., Vrettos, J.S., Liable-Sands, L.M. et al. (1999). *Science* 283: 1524–1527.
34 (a) Limburg, J., Vrettos, J.S., Chen, H. et al. (2000). *J. Am. Chem. Soc.* 123: 423–430. (b) Tagore, R., Crabtree, R.H., and Brudvig, G.W. (2008). *Inorg. Chem.* 47: 1815–1823.
35 (a) Yagi, M. and Narita, K. (2004). *J. Am. Chem. Soc.* 126: 8084–8085. (b) Narita, K., Kuwabara, T., Sone, K. et al. (2006). *J. Phys. Chem. B* 110: 23107–23114.
36 Naruta, Y., Sasayama, M., and Sasaki, T. (1994). *Angew. Chem. Int. Ed.* 33: 1839–1841.
37 Karlsson, E.A., Lee, B.-L., Åkermark, T. et al. (2011). *Angew. Chem.* 123: 11919–11922. (2011). *Angew. Chem. Int. Ed.* 50: 11715–11718.
38 Elizarova, G.L., Matvienko, L.G., Lozhkina, N.V. et al. (1981). *React. Kinet. Catal. Lett.* 16: 191–194.
39 Abe, T., Shiroishi, H., Kinoshita, K., and Kaneko, M. (1998). *Macromol. Symp.* 131: 81–86.
40 Ellis, W.C., McDaniel, N.D., Bernhard, S., and Collins, T.J. (2010). *J. Am. Chem. Soc.* 132: 10990–10991.

41 (a) Collins, T.J. (2002). *Acc. Chem. Res.* 35: 782. (b) Ghosh, A., de Oliveira, F.T., Yano, T. et al. (2005). *J. Am. Chem. Soc.* 127: 2505. (c) Oliveira, F.T.D., Chanda, A., Banerjee, D. et al. (2007). *Science* 315: 835–838. (d) Chanda, A., Shan, X., Chakrabarti, M. et al. (2008). *Inorg. Chem.* 47: 3669.
42 Ertem, M.Z., Gagliardi, L., and Cramer, C.J. (2012). *Chem. Sci.* 3: 1293–1299.
43 Demeter, E.L., Hilburg, S.L., Washburn, N.R. et al. (2014). *J. Am. Chem. Soc.* 136: 5603–5606.
44 (a) Panda, C., Debgupta, J., Díaz Díaz, D. et al. (2014). *J. Am. Chem. Soc.* 136: 12273–12282. (b) Das, D., Pattanayak, S., Singh, K.K. et al. (2016). *Chem. Commun.* 52: 11787–11790. (c) Pattanayak, S., Chowdhury, D.R., Garai, B. et al. (2017). *Chem. Eur. J.* 23: 3414–3424.
45 Zhang, B., Li, F., Yu, F. et al. (2014). *Chem. Asian J.* 9: 1515–1518.
46 (a) Fillol, J.L., Codolà, Z., Garcia-Bosch, I. et al. (2011). *Nat. Chem.* 3: 807–813. (b) Codolà, Z., Garcia-Bosch, I., Acuña-Parés, F. et al. (2013). *Chem. Eur. J.* 19: 8042–8047. (c) Codolà, Z., Gómez, L., Kleespies, S.T. et al. (2015). *Nat. Commun.* 6: 5865. (d) Hoffert, W.A., Mock, M.T., Appel, A.M., and Yang, J.Y. (2013). *Eur. J. Inorg. Chem.* 2013: 3846–3857. (e) Detz, R.J., Abiri, Z., Kluwer, A.M., and Reek, J.N.H. (2015). *ChemSusChem* 8: 3057–3061. (f) Parent, A.R., Nakazono, T., Lin, S. et al. (2014). *Dalton Trans.* 43: 12501–12513. (g) To, W.-P., Wai-Shan Chow, T., Tse, C.-W. et al. (2015). *Chem. Sci.* 6: 5891–5903. (h) Wang, D. and Que, L. Jr. (2013). *Chem. Commun.* 49: 10682–10684.
47 (a) Hong, D., Mandal, S., Yamada, Y. et al. (2013). *Inorg. Chem.* 52: 9522–9531. (b) Parent, A.R. and Sakai, K. (2014). *ChemSusChem* 7: 2070.
48 Wickramasinghe, L.D., Zhou, R., Zong, R. et al. (2015). *J. Am. Chem. Soc.* 137: 13260–13263.
49 Okamura, M., Kondo, M., Kuga, R. et al. (2016). *Nature* 530: 465–468.
50 Kanan, M.W. and Nocera, D.G. (2008). *Science* 321: 1072–1075.
51 (a) Yin, Q., Tan, J.M., Besson, C. et al. (2010). *Science* 328: 342–345. (b) Lieb, D., Zahl, A., Wilson, E.F. et al. (2011). *Inorg. Chem.* 50: 9053–9058. (c) Stracke, J.J. and Finke, R.G. (2011). *J. Am. Chem. Soc.* 133: 14872–14875. (d) Car, P.-E., Guttentag, M., Baldridge, K.K. et al. (2012). *Green Chem.* 14: 1680–1688. (e) Goberna-Ferrón, S., Vigara, L., Soriano-López, J., and Galán-Mascarós, J.R. (2012). *Inorg. Chem.* 51: 11707–11715. (f) La Ganga, G., Puntoriero, F., Campagna, S. et al. (2012). *Faraday Discuss.* 155: 177–190. (g) Lv, H., Geletii, Y.V., Zhao, C. et al. (2012). *Chem. Soc. Rev.* 41: 7572–7589. (h) Schiwon, R., Klingan, K., Dau, H., and Limberg, C. (2013). *Chem. Commun.* 50: 100–102. (i) Song, F., Ding, Y., Ma, B. et al. (2013). *Energy Environ. Sci.* 6: 1170–1184. (j) Soriano-López, J., Goberna-Ferrón, S., Vigara, L. et al. (2013). *Inorg. Chem.* 52: 4753–4755. (k) Stracke, J.J. and Finke, R.G. (2013). *ACS Catal.* 3: 1209–1219. (l) Stracke, J.J. and Finke, R.G. (2013). *ACS Catal.* 79–89. (m) Vickers, J.W., Lv, H., Sumliner, J.M. et al. (2013). *J. Am. Chem. Soc.* 135: 14110–14118.
52 (a) Das, B., Orthaber, A., Ott, S., and Thapper, A. (2015). *Chem. Commun.* 51: 13074–13077. (b) Chen, M., Ng, S.M., Yiu, S.M. et al. (2014). *Chem. Commun.* 50: 14956–14959. (c) Ullman, A.M., Brodsky, C.N., Li, N. et al. (2016). *J. Am. Chem. Soc.* 138: 4229–4236.

53 Duraisamy, S., Kim, H., and Shin, W. (2017). *Int. J. Hydrogen Energy* 42: 7908–7916.
54 (a) Wang, H.Y., Mijangos, E., Ott, S., and Thapper, A. (2014). *Angew. Chem. Int. Ed.* 53: 14499–14502. (b) Ishizuka, T., Watanabe, A., Kotani, H. et al. (2016). *Inorg. Chem.* 55: 1154–1164. (c) Wang, H., Xin, Z., Xiang, R. et al. (2016). *Chin. J. Chem.* 34: 757–762. (d) Wang, J.-W., Sahoo, P., and Lu, T.-B. (2016). *ACS Catal.* 6: 5062–5068.
55 (a) Pizzolato, E., Natali, M., Posocco, B. et al. (2013). *Chem. Commun.* 49: 9941–9943. (b) Lopez, A.M., Natali, M., Pizzolato, E. et al. (2014). *Phys. Chem. Chem. Phys.* 16: 12000–12007. (c) Gonawala, S., Baydoun, H., Wickramasinghe, L., and Verani, C.N. (2016). *Chem. Commun.* 52: 8440–8443.
56 Wasylenko, D.J., Ganesamoorthy, C., Borau-Garcia, J., and Berlinguette, C.P. (2011). *Chem. Commun.* 47: 4249–4251.
57 Wasylenko, D.J., Palmer, R.D., Schott, E., and Berlinguette, C.P. (2012). *Chem. Commun.* 48: 2107–2109.
58 Leung, C.-F., Ng, S.-M., Ko, C.-C. et al. (2012). *Energy Environ. Sci.* 5: 7903–7907.
59 Hong, D., Jung, J., Park, J. et al. (2012). *Energy Environ. Sci.* 5: 7606–7616.
60 (a) Zhao, Y., Lin, J., Liu, Y. et al. (2015). *Chem. Commun.* 51: 17309–17312. (b) Younus, H.A., Ahmad, N., Chughtai, A.H. et al. (2017). *ChemSusChem* 10: 862–875.
61 (a) Gong, M., Li, Y.G., Wang, H.L. et al. (2013). *J. Am. Chem. Soc.* 135: 8452–8455. (b) Subbaraman, R., Tripkovic, D., Chang, K.C. et al. (2012). *Nat. Mater.* 11: 550–557.
62 Zhang, M., Zhang, M.T., Hou, C. et al. (2014). *Angew. Chem.* 126: 13258–13264. (2014). *Angew. Chem. Int. Ed.* 53: 13042–13048.
63 (a) Wang, J.W., Zhang, X.Q., Huang, H.H., and Lu, T.B. (2016). *ChemCatChem* 8: 3287–3293. (b) Luo, G.Y., Huang, H.H., Wang, J.W., and Lu, T.B. (2016). *ChemSusChem* 9: 485–491.
64 Wang, L., Duan, L.L., Ambre, R.B. et al. (2016). *J. Catal.* 335: 72–78.
65 Han, Y.Z., Wu, Y.Z., Lai, W.Z., and Cao, R. (2015). *Inorg. Chem.* 54: 5604–5613.
66 Elizarova, G.L., Matvienko, L.G., Lozhkina, N.V. et al. (1981). *React. Kinet. Catal. Lett.* 16: 285–288.
67 Barnett, S.M., Goldberg, K.I., and Mayer, J.M. (2012). *Nat. Chem.* 4: 498–502.
68 Zhang, T., Wang, C., Liu, S. et al. (2013). *J. Am. Chem. Soc.* 135: 15314–15317.
69 Chen, Z. and Meyer, T.J. (2012). *Angew. Chem.* 125: 728–731. (2013). *Angew. Chem. Int. Ed.* 52: 700–703.
70 Zhang, M.-T., Chen, Z., Kang, P., and Meyer, T.J. (2013). *J. Am. Chem. Soc.* 135: 2048–2051.
71 Garrido-Barros, P., Funes-Ardoiz, I., Drouet, S. et al. (2015). *J. Am. Chem. Soc.* 137: 6758–6761.
72 Funes-Ardoiz, I., Garrido-Barros, P., Llobet, A., and Maseras, F. (2017). *ACS Catal.* 7: 1712–1719.

17

Base-Metal-Catalyzed Hydrogen Generation from Carbon- and Boron Nitrogen-Based Substrates

Elisabetta Alberico[1], Lydia K. Vogt[2], Nils Rockstroh[2], and Henrik Junge[2]

[1] *Consiglio Nazionale delle Ricerche, Istituto di Chimica Biomolecolare, Traversa La Crucca 3, 07100 Sassari, Italy*
[2] *Leibniz Institute for Catalysis at the University of Rostock, Albert Einstein-Straße 29a, 18059 Rostock, Germany*

17.1 Introduction

17.1.1 State of the Art of Hydrogen Generation from Carbon- and Boron Nitrogen-Based Substrates

Wind and sunlight are progressively contributing to a sufficient and sustainable energy supply. For example, already in 2013, in the northeastern German state Mecklenburg and Western Pomerania, more electric energy was harvested from renewable sources (120%) than annually consumed. Because of the intermittent nature of renewable electric energy, the development of efficient, viable, and reversible ways to store it has become increasingly important. Approaches for the interconversion of electric and chemical energy appear as a promising solution as the necessary storage capacity is expected to increase rapidly during the next years. Such methods include, for instance, power-to-gas techniques where hydrogen generation from water as a suitable starting material is an important first step [1]. This is already achieved by coupling either wind power or photovoltaics with PEM (proton exchange membrane, also known as polymer electrolyte membrane) or alkaline electrolysis cells in various pilot plants. The resulting hydrogen can be either stored or subsequently applied for further synthesis of hydrogen and energy carriers as well as intermediates for chemical industry. Because every additional conversion results in a loss of overall efficiency, the first priority is always the direct usage of (renewable) electric energy.

In the field of hydrogen storage, so far, mainly physical methods such as compression and liquefaction as well as chemical storage in metal hydrides are applied [2]. However, these methods have drawbacks such as energy losses up to 40% and in part low gravimetric energy densities. Alternatively, carbon compounds such as CO_2 and aromatic cyclic compounds such as toluene and naphthalene represent another promising option for hydrogen storage. In every case, a cycle comprising a dehydrogenated and a hydrogenated form of the storage material can be envisioned (Scheme 17.1).

Non-Noble Metal Catalysis: Molecular Approaches and Reactions,
First Edition. Edited by Robertus J. M. Klein Gebbink and Marc-Etienne Moret.

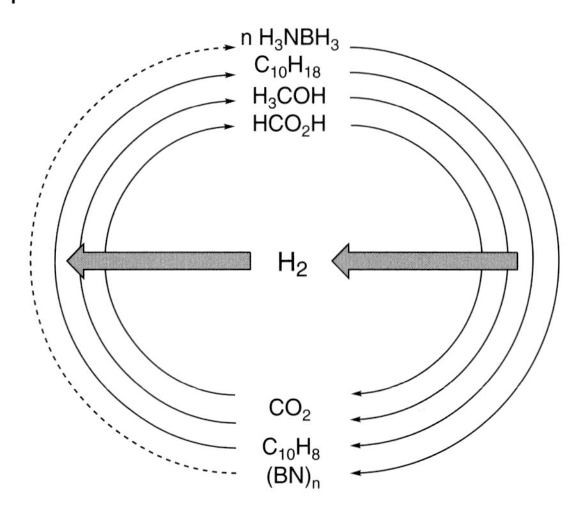

Scheme 17.1 Carbon- and BN-based hydrogen storage cycles.

As the first part, the cycle involves the loading of different substrates with hydrogen. Although the hydrogenation of C-based compounds has been extensively investigated and outstanding activities were achieved, difficulties for the BN-based materials still exist (*vide infra*). As an example, for the CO_2 hydrogenation to formic acid (FA) or formate (Figure 17.1) activities (turnover frequency, TOF) of up to 1 100 000 and 73 000 as well as productivities (turnover number, TON) of up to 250 000 and 3 500 000, respectively, were achieved applying homogeneous catalysts [3] based on noble metals such as iridium (**1**) [4] or ruthenium (**2**) [5] PNP pincer-type complexes. Homogeneous catalysts in CO_2 hydrogenation to methanol are not yet as productive. So far, >2000 turnovers were obtained with the molecular defined Ru-PNP pincer complex **3** (Figure 17.1) [6], whereas up to 3500 tons per year are produced applying heterogeneous catalysts in the "George Olah CO_2 to Renewable Methanol Plant" in Iceland [7]. As demonstrated by the mentioned CO_2 hydrogenation reactions to FA or formate, pincer-type complexes enabled a quantum jump toward much higher catalyst activities and productivities. In part, this is due to the participation of the ligand in the catalytic cycle, which is described as "non-innocent" behavior. In addition, these complexes are also able to catalyze the reverse reaction of the storage cycle, the hydrogen release from formic acid, formate, methanol, and other, in part biomass-derived, alcohols.

For example, **4** was shown to be able to perform the aqueous-phase reforming of methanol to hydrogen and carbon dioxide at temperatures below 100 °C with a TOF of 2687 (after two hours) and a TON of >350 000 [8]. Noteworthy, applying **3** or **4**, (bio)ethanol, glycerol, and 2-propanol can also be used for the generation of hydrogen and valuable additional products such as acetyl acetate and acetic acid, lactic acid, and acetone, respectively [9]. Other recent examples of methanol dehydrogenation under mild conditions include catalysts $[K(dme)_2][RuH(trop_2dad)]$ ($trop_2dad$,

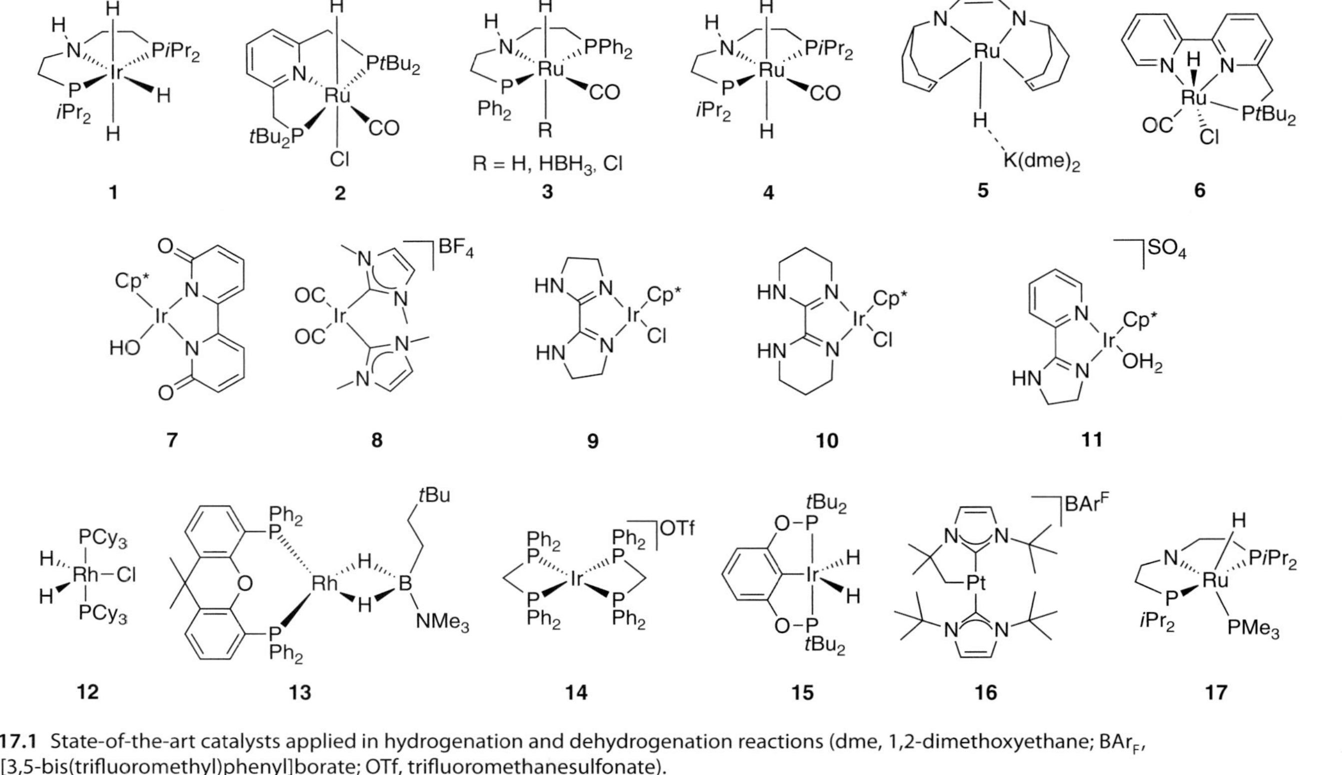

Figure 17.1 State-of-the-art catalysts applied in hydrogenation and dehydrogenation reactions (dme, 1,2-dimethoxyethane; BAr_F, tetrakis[3,5-bis(trifluoromethyl)phenyl]borate; OTf, trifluoromethanesulfonate).

1,4-bis(5*H*-dibenzo[*a,d*]cyclohepten-5-yl)-1,4-diazabuta-1,3-diene) (**5**) [10a], **6** [10b], **7** [10c], and **8** [10d].

A direct comparison among different catalysts is often hampered by the different experimental conditions implemented. Nevertheless, for formic acid dehydrogenation to date, pentamethylcyclopentadienyl (Cp*) iridium complexes with *N,N*-bidentate ligands (e.g. consisting of pyridine, pyrimidine, pyrazole, or 4,5-dihydro-1*H*-imidazole moieties) possess outstanding activity and durability, with a TOF as high as $487\,000\,h^{-1}$ at 90 °C, measured after two minutes (Figure 17.1, catalyst **9**) and an unprecedented TON of 2 400 000 at 80 °C (Figure 17.1, catalyst **10**) [11]. In view of a possible practical application, the Cp*Ir complex bearing a bidentate pyridyl-imidazoline ligand **11** (Figure 17.1) deserves to be mentioned. In the presence of 10 μmol of this catalyst, $1.02\,m^3$ of a H_2/CO_2 (1/1) gas mixture was produced by dehydrogenation of formic acid in water at 50 °C over a period of 15 days. No additives or pH adjustment were required [12]. This sets the benchmark against which catalysts should be evaluated. A more detailed discussion of the results of hydrogen generation from formic acid or formate and alcohols including noble metal-based catalysts is provided elsewhere [13]. Several relevant examples of either CO_2 hydrogenation to formic acid or formate salts and hydrogen release from the same compounds have yet been reported. However, even more important is the reversible combination of both reactions, which constitutes a step forward to the development of storage devices for renewable energy. Besides the already mentioned pincer-type catalysts **1** and **2** (Figure 17.1), non-pincer-type complexes such as [RuCl($m$tppms)$_2$]$_2$ (*m*tppms, sodium diphenylphosphinobenzene-3-sulfonate) [14], [$RuCl_2$(benzene)]$_2$/dppm (dppm, 1,2-bis(diphenylphosphino)methane) [15], [{Ir(Cp*)(OH_2)}$_2$(thbpm)] (thbpm, 4,4′,6,6′-tetrahydroxybipyrimidine) [3c], and [Ir^{III}(Cp*)(4-(1*H*-pyrazol-1-yl-κN^2)benzoic acid-κC^3)(OH_2)]$_2SO_4$ [16] were also successfully applied in a hydrogen storage cycle as illustrated in Scheme 17.1.

A further possibility for hydrogen storage is based on liquid organic hydrogen carriers (LOHC). LOHC are organic compounds with extended π-conjugated electron systems, such as polycyclic aromatic hydrocarbons and heterocycles or π-conjugated organic polymers. The idea to hydrogenate aromatic compounds and its application as a seasonal energy storage medium were discussed almost 20 years ago for the first time [17]. The underlying hydrogenation of aromatic compounds, e.g. naphthalene to decaline, was already developed in the 1940s [18]. Recently, compounds such as *N*-ethyl carbazole and dibenzyltoluene have been considered as alternatives for toluene or decaline because of their advantageous boiling and hazard properties [19]. However, the release of hydrogen from LOHC is highly endothermic, and high temperatures are required. Thus, 15.3–16.5 kcal (Scheme 17.2, Eqs. (5) and (6)) are necessary per mole of released H_2 [25]. Therefore, especially nitrogen-containing heterocyclic compounds gained significant attention as they can be dehydrogenated at milder conditions because of their favorable thermodynamics of hydrogen release [26]. However, very often, only the heteroatom-containing ring undergoes the (de)hydrogenation at those mild conditions [27], which limits the hydrogen storage capacity. The majority of successfully applied catalysts comprises heterogeneous systems with supported Pt or Pd particles being the metal of

Reaction	$\Delta G°$ [kcal mol^{-1}]	$\Delta H°$ [kcal mol^{-1}]	$\Delta S°$ [cal mol^{-1}]	
HCO_2H (l) → CO_2 (g) + H_2 (g)	−7.83	+7.52	+51.59	(1)
HCO_2H (l) → CO_2 (aq.) + H_2 (g)	+0.96	n.a.	n.a.	(2)
NH_4^+ (aq.) + HCO_2^- (aq.) → CO_2 (g) + H_2 (g) + NH_3 (aq.)	+2.27	+20.13	+59.71	(3)
H_3COH (l) + H_2O (l) → 3 H_2 (g) + CO_2 (g)	+2.13	+31.22	n.a.	(4)
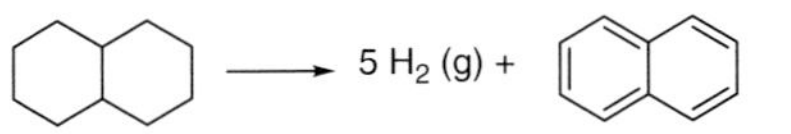	+28.90	+76.43/ +71.01	+156.21	(5)
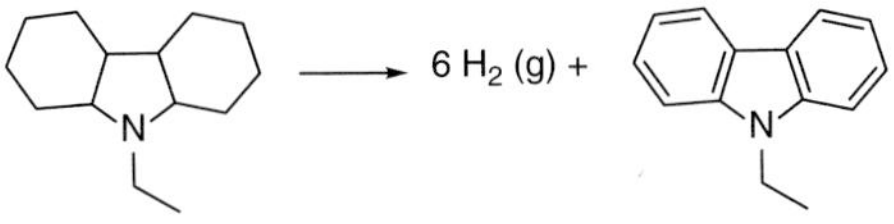	n.a.	+75.95	n.a.	(6)
H_3NBH_3 (s) → 3 H_2 (g) + $(BN)_n$ (s)	n.a.	−5.18	n.a.	(7)

Scheme 17.2 Thermodynamic data for the dehydrogenation reaction of selected substrates [20b, 21–24].

choice [19b, 28]. The dehydrogenation temperature can be further decreased by the usage of homogeneous catalysts as well as by light irradiation [29]. The results concerning precious metal-based catalysts are discussed in more detail elsewhere [2, 25–27, 30].

Ammonia–borane and amine–boranes constitute another class of materials possessing a comparably high hydrogen content of up to 19.6 wt% [2, 31]. Thus, especially, ammonia borane (AB) theoretically outperforms the gravimetric hydrogen storage capacities of formic acid (4.4 wt%), methanol (12.6 wt%), and decaline (7.29 wt%) [32]. Thermal dehydrogenation of BN-based hydrogen carriers occurs in three distinct steps: at approximately 107–117 °C, at approximately 150 °C, and above 500 °C [20]. The enthalpy for the first dehydrogenation is reported to be between −5.2 [20b, 21] and −5.1 kcal mol^{-1} [20d, 33] (Scheme 17.2, Eq. (7)). Moreover, the mode of dehydrogenation greatly influences the types of its products. Thus, several BN(H) compounds may be yielded under thermal conditions as well as borates in the presence of solvents such as water [31a]. From thermodynamic considerations, the regeneration of AB from borates is highly energy demanding and also the regeneration from BN(H) compounds poses severe problems, which still have to be solved satisfyingly [34]. Moreover, the regeneration of the dehydrogenated products of AB is also of environmental and economic interest as boron is a quite rare element [26]. Numerous reports describe the application of heterogeneous catalysts based on precious metals such as Ru, Rh, Pt, and Pd [35], but also based on non-noble metals such as Co, Ni, and Cu [36]. In addition, homogeneous catalysts on the basis of noble and non-noble metals have also been developed. In the case of rhodium as the central metal, dimethyl AB could be dehydrogenated at 25 °C. Although some complexes formed Rh^0 particles during the course of the reaction [37], others remained stable (see Figure 17.1, **12** and **13**) [38]. With [Ir(dppm)$_2$][OTf] (**14**, Figure 17.1), dehydrogenation was even possible at −30 °C [39]. There have been several other complexes (see Figure 17.1, **15–17**), which rapidly liberate H_2 from AB or amine boranes at room temperature based on Pt [40], Ir [41], and Ru [42–44].

In general, hydrogen liberation from all carriers includes two main demands with respect to practical applications: Firstly, H_2 should be provided instantaneously at a constant rate, and secondly, it should be liberated under appropriate conditions. However, for the mentioned substrates such as formic acid, formate, methanol, as well as other, in particular, biomass-derived alcohols, and LOHC, the hydrogen release is an endothermic process, and only for AB, an exothermic process (Scheme 17.2). Nevertheless, for formic acid, formates, alcohols, ammonia, and amine boranes, the dehydrogenation starts to take place at rather low temperatures (e.g. −30 °C for AB, 0 °C for formic acid amine adduct, and 65 °C for MeOH/water). In contrast, for the dehydrogenation of LOHC, much harsher conditions are necessary because of the enthalpy of ca. 71.7 kcal mol^{-1}. Among catalysts promoting these reactions, noble-metal-containing systems clearly dominate so far. Nevertheless, a few non-noble metal-based catalysts have already been shown to be a viable alternative. These will be discussed in more detail in the following paragraphs.

17.1.2 Development of Base Metal Catalysts for Catalytic Hydrogen Generation

First, general mechanistic features of homogeneously catalyzed dehydrogenation reactions for hydrogen generation can be discussed. The metal-mediated acceptorless dehydrogenation comprises the transfer of two hydrogens, a proton, and a hydride from a suitable substrate to a metal complex (Scheme 17.3, Eq. (8)), from which they are subsequently released as a hydrogen molecule (Scheme 17.3, Eq. (9)) [45]. A perusal of the literature shows that homogeneous catalysts competent for such processes can be classified into two classes, depending on whether such a transfer requires the substrate to be covalently bound to the metal in its inner coordination sphere or not. This, in turn, is mainly determined by the type of ancillary ligands present in the catalyst.

$$\overset{\delta^+}{\mathrm{H}}\ \ \overset{\delta^-}{\mathrm{H}}\ \ \mathrm{X{-}Y} + [\mathrm{cat}] \longrightarrow \mathrm{X{=}Y} + [\mathrm{cat}]\cdot\mathrm{H_2} \qquad (8)$$

$$[\mathrm{cat}]\cdot\mathrm{H_2} \longrightarrow [\mathrm{cat}] + \mathrm{H_2} \qquad (9)$$

Scheme 17.3 Sequence of events defining the acceptorless dehydrogenation of a suitable substrate. The hydrogen termini abstracted from the substrate by interaction with the catalyst generate a metal–dihydrogen complex and are subsequently released as hydrogen gas.

Although the sequence of steps within the catalytic cycle may vary depending on the substrate and/or the catalyst, the coordination of the substrate to the metal in the inner coordination sphere activates the former for the transfer of a hydride to the metal (Scheme 17.4). This might require prior dissociation of an ancillary ligand to provide a vacant coordination site for the incoming substrate. Subsequent protonation of the metal hydride, either at the metal or the hydride, generates a metal dihydrogen complex, from which hydrogen is eventually released.

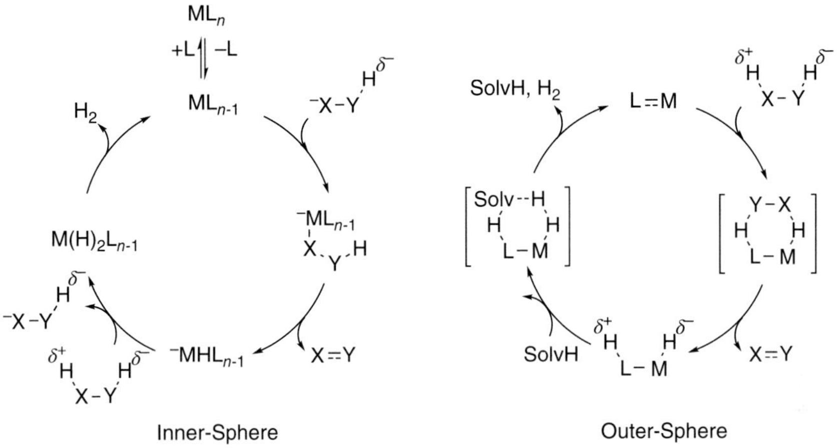

Scheme 17.4 Schematic representation of the two general mechanisms, inner and outer sphere, for dehydrogenation of a substrate and subsequent release of hydrogen gas from the catalyst.

Alternatively, the transfer of the hydrogen termini may occur in the second or outer coordination sphere of the metal (Scheme 17.4). In this case, the ancillary ligand possesses a basic site, most often an amide nitrogen, which can accept the protic hydrogen of the substrate, thus activating the latter for direct hydride transfer to the metal. The delivery of the two hydrogen atoms from the substrate may be concerted or stepwise. Noyori et al. has coined the term "metal-ligand bifunctional catalysis" [46] to describe this kind of mechanism, in which the so-called "non-innocent" ligand [47] is directly involved in substrate activation and bond breaking/formation. The generation of the hydrogen molecule at the metal and its subsequent release are indeed aided by protic solvents or the substrate itself: in fact, they act as proton shuttles, by protonating the metal hydride to form a dihydrogen complex while being hydrogen bonded to the protonated basic site of the ligand [48]. In some cases, experimental evidence has been provided for the operation of such a mechanism [48a], and density functional theory (DFT) calculations have proven the dehydrogenation process to be energetically less demanding, even in the absence of suitable hydrogen acceptors [48b]. Besides, proper reactor design helps shifting the equilibrium of these reactions by easing hydrogen gas release.

In some cases, coordination of the deprotonated substrate to the metal rather than direct hydride transfer occurs after protonation of the basic site of the ligand. In this instance, hydride delivery proceeds through β-hydride transfer from the coordinated substrate and requires generation of a vacant cis-coordination site. However, examples have also been reported when this is not strictly necessary. Despite the presence of a cooperative ligand, such

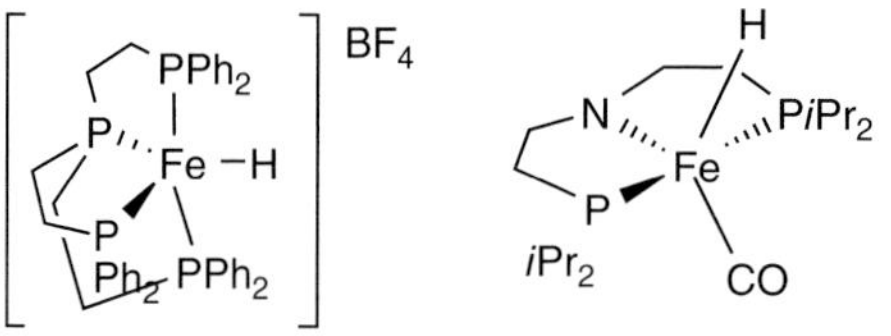

Figure 17.2 Iron-based catalysts promoting the acceptorless dehydrogenation of various substrates through an inner (**18**) or an outer sphere/blended (**19**) mechanism, respectively.

a blended, inner sphere mechanism is often observed for catalysts bearing polydentate ligands with one labile donor, which is able to temporarily dissociate from the metal.

Iron complexes **18** and **19** serve to showcase the type of catalysts, which are capable of promoting the acceptorless dehydrogenation of suitable substrates (Figure 17.2) [49]. Catalyst **18** has proven to be extremely efficient for the decomposition of formic acid proceeding through an inner-sphere mechanism, as discussed in Section 17.2.

Complex **18** may be regarded as the third generation of a series of iron-based catalysts competent for formic acid decomposition. On the way to catalyst improvement, the replacement of monodentate triphenylphosphine for tribenzylphosphine allowed for the chelation of the iron center through ortho-metalation of one of the substituents in the latter ligand. Furthermore, the introduction of the tetradentate tris[2-(diphenylphosphino)ethyl]phosphine (*tetraphos*, PP_3) (as in **18**) led to further improvement of catalyst efficiency both in terms of activity and stability. Indeed, a strongly chelating ligand is highly desirable in order to make up for the substitutional lability of first-row transition metals as compared to second- and third-row ones. Complex **19** is noteworthy as it is able to catalyze the efficient dehydrogenation of both aqueous methanol and formic acid to H_2 and CO_2, as well as the dehydrogenation of amine borane. The complex contains an aliphatic pincer ligand whose nitrogen, in its catalytic active amido form, is able to accept a protic hydrogen atom from the substrate, thus cooperating with the metal in the dehydrogenation of the latter. Evidence of both outer sphere and blended mechanisms has been put forward depending on the substrate being dehydrogenated (*vide infra*). In either case, the formal oxidation state of the metal does not change, as observed instead for catalytic reactions which proceed, for example, through oxidative addition at the metal and subsequent reductive elimination. Therefore, the use of non-innocent ligands has paved the way to the application of first-row transition metals, which would otherwise prefer one-electron redox changes. The pincer coordination and the presence of strong-field ligands such as phosphines, hydride, and CO serve to stabilize the metal in its low-spin state and avoid the formation of unpaired electron species.

17.2 Hydrogen Generation from Formic Acid

In the presence of suitable metal catalysts, formic acid can be decomposed to afford hydrogen and carbon dioxide according to Eqs. (1)–(3) in Scheme 17.2.

Despite its low hydrogen content, 4.4 wt%, formic acid is regarded as an appealing hydrogen vector. Several aspects are taken into account when assessing its role as storage and carrier medium: It is liquid at room temperature and is nontoxic, although corrosive, which implies easy storage, handling, and transportation. It is available on large scale, mainly from the hydrolysis of methylformate [50], and can be even derived from biomass [51]. The by-product of dehydrogenation, CO_2, can be trapped [52] and hydrogenated back to HCO_2H [3d], thus setting the stage for the possible implementation of a hydrogen battery [5, 15, 53] or usage as a C1 synthon for the synthesis of value-added chemicals [3e, 54]. Alternatively, the decomposition of formic acid can also proceed through dehydration, which is favored at high temperatures. In this case, CO is produced, which prevents the direct use of the liberated hydrogen in fuel cells as it poisons the precious electrocatalyst at the cell anode already at concentrations of a few ppm. Competent catalysts should therefore promote the decarboxylation reaction selectively (Eqs. (1)–(3); Scheme 17.2).

State-of-the-art systems have been discussed above. These contain noble metals such as iridium and ruthenium. However, the high cost and limited availability of precious metals has spurred researchers to design homogeneous catalysts based on cheaper and more abundant non-noble metals. Thus, chromium [55], molybdenum [56], manganese [57], iron [58–63], cobalt [58, 59e, 64], nickel [58, 65], copper [66], and aluminum [67] salts and/or ligand-supported complexes were reported to promote the dehydrogenation of formic acid. As expected, activities are lower than those recorded with the best noble-metal-based catalysts. Iron catalysts are an exception, and very high efficiencies have been achieved. Therefore, a special focus will be laid on the latter.

17.2.1 Iron

In 2009, as part of an investigation of the decomposition of the HCO_2H/Et_3N azeotrope (5 : 2, triethylammoniumfluoride (TEAF)) to H_2 and CO_2 promoted by Ru(II) and Ru(III) complexes, a few first-row transition metal salts including $FeCl_2$ and $FeCl_3$ were tested as catalysts [58]. Indeed, hydrogen evolution was observed with the iron salts, but the activity recorded at 120 °C was negligible. Activity and selectivity could be noticeably enhanced by combining a suitable iron precursor with phosphorus and nitrogen donors. The first successful results were reported by Beller and coworkers in 2010. A catalyst generated *in situ* by combining $[Fe_3(CO)_{12}]$, triphenylphosphine, and either terpyridine or phenanthroline allowed for the decomposition of TEAF in dimethylformamide (DMF) at 40 °C under visible light irradiation with only traces of CO being detected [59a]. A TON >100 and a TOF of up to 200 h^{-1} were achieved by proper choice of the chelating nitrogen ligand, which proved to be important to extend the catalyst lifetime. Light was required both to generate the iron hydride active species and for catalytic turnover. The same authors were able to improve the productivity of this catalytic system by 1 order of magnitude (TON = 1266 over 51 hours at 60 °C) by replacing triphenylphosphine with tribenzylphosphine (PBn_3) [59b]. Mechanistic investigations by NMR spectroscopy, corroborated by DFT calculations, showed that under light irradiation, one of the coordinated PBn_3 ligands within the *in situ*-generated Fe(0) complex underwent ortho-metalation of one of its aryl substituents. The improved efficiency of the catalyst system was therefore ascribed to the consequences of this event. Thus, a switch in the oxidation state from Fe(0) to Fe(II) as well as the presence of a Fe—C(phenyl) bond and the ability of the ligand to act as a chelator through the formation of a stable five-membered ring with the metal were discussed.

A major breakthrough was later reported by the same authors. A third-generation catalyst, obtained by combining $Fe(BF_4)_2{\cdot}(H_2O)_6$ with 1 equiv. of the tetradentate ligand PP_3, was able to catalyze the selective decomposition of HCO_2H in propylene carbonate at 40 °C, achieving a TON of more than 825 over three hours [59c,d]. No base additive and light irradiation were required. The activity of the catalyst was strongly affected by the solvent (no activity was observed in pure HCO_2H), the presence of halide ions, the water content, and the ligand to metal ratio. By using an excess of the ligand (4 equiv. to the metal), the thermal stability of the catalyst was improved to afford a maximum TOF of 9425 h^{-1} at 80 °C (Scheme 17.5) [59c,d].

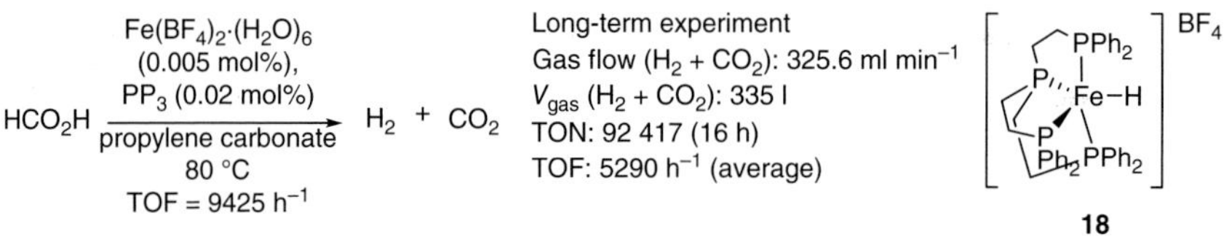

HCO_2H —[$Fe(BF_4)_2{\cdot}(H_2O)_6$ (0.005 mol%), PP_3 (0.02 mol%), propylene carbonate, 80 °C, TOF = 9425 h^{-1}]→ H_2 + CO_2

Long-term experiment
Gas flow (H_2 + CO_2): 325.6 ml min^{-1}
V_{gas} (H_2 + CO_2): 335 l
TON: 92 417 (16 h)
TOF: 5290 h^{-1} (average)

$[Fe(H)PP_3]BF_4$ **18**

Scheme 17.5 Dehydrogenation of formic acid promoted by catalyst **18** generated *in situ* from $Fe(BF_4)_2{\cdot}(H_2O)_6$ and the PP_3 ligand.

To assess the productivity of the catalyst in a long-term experiment, a device was constructed to allow for the continuous replenishment of HCO_2H and thus to secure its constant concentration in solution. At 80 °C, a constant gas flow of 325.6 ml min^{-1} ($H_2 + CO_2$) was released over 16 hours to deliver a final TON of 92 417 (Scheme 17.5). The catalyst proved to be robust as solvents and HCO_2H did not have to be dried or distilled before use. However, the accumulation of water and chloride ions in the batch reactor, which are both present as impurities in the used formic acid, was deemed responsible for catalyst deactivation and activity drop after 16 hours. Indeed, the addition of 1 equiv. of NaCl, with respect to Fe, was sufficient to drastically reduce the activity; with 15 equiv., the activity was almost nullified. Based on the results of an in-depth mechanistic investigation, which combined the kinetic studies and *in situ* experimental techniques (NMR, IR, Raman, UV–Vis, XANES, and EXAFS) with theoretical calculations, evidence was provided that the dehydrogenation of HCO_2H occurs through two competing catalytic cycles (Scheme 17.6, Cycles **A** and **B**) [59e]. The iron monohydride species $[Fe(II)(H)PP_3]^+$ **18**, arising from β-hydride elimination of the coordinated formate in $[Fe(II)(\eta^2\text{-}O_2CH)PP_3]^+$ **20**, is a key to both.

Catalytic turnover may then proceed through protonation of the monohydride species by HCO_2H, release of a dihydrogen molecule, and regeneration of the $[Fe(II)(\eta^2\text{-}O_2CH)PP_3]^+$ species **20** (Scheme 17.6, Cycle **A**). Alternatively, formate might coordinate to $[Fe(II)(H)PP_3]^+$ **18** to generate a neutral species

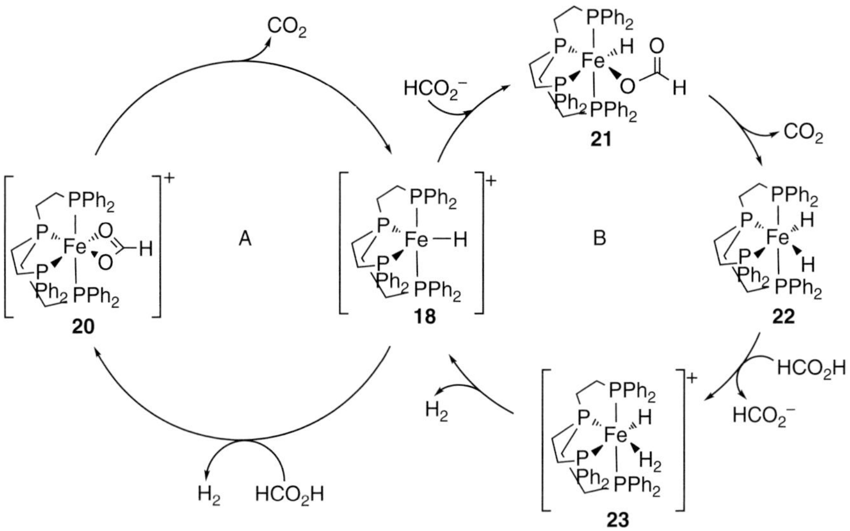

Scheme 17.6 Proposed catalytic cycles for the dehydrogenation of formic acid promoted by catalyst **18** generated *in situ* from $Fe(BF_4)_2{\cdot}(H_2O)_6$ and the PP_3 ligand.

$[Fe(II)(H)(\eta^1\text{-}O_2CH)PP_3]$ **21**, which generates a neutral dihydride species $[Fe(II)(H)_2PP_3]$ **22** (Scheme 17.6, Cycle **B**) after β-hydride elimination from the coordinated formate. Protonation of **22** affords the complex $[Fe(II)(H)(H_2)PP_3]^+$ **23** [68], from which dihydrogen is released. The presence of this species under catalytic conditions nicely explains why activity is negatively affected by an increase of hydrogen pressure. $[Fe(II)(H)(H_2)PP_3]^+$ **23** becomes the catalyst resting state at high conversion when HCO_2H concentration is low [59e]. $[Fe(II)(H)(\eta^1\text{-}O_2CH)PP_3]$ **21** is a coordinatively saturated 18-electron species; therefore, the possibility of β-hydride elimination in the absence of a cis vacant coordination site was questioned by Yang, who found by DFT that the direct hydride transfer from a formate ion to the metal in $[Fe(II)(H)PP_3]^+$ **18** is energetically more favorable [69a]. The solid-state structure of $[Fe(II)(H)(H_2)PP_3]^+$ **23** indicates that protonation of the neutral dihydride species **22** takes place at the hydride trans to the bridgehead phosphorus, a preference corroborated by the theoretical calculations performed by Ahlquist and coworkers [69b]. The other hydride in $[Fe(II)(H)_2PP_3]$ **22** acts as a spectator ligand that affects reactivity at the metal center through its electronic properties but is not directly involved in catalysis. Therefore, other ligands might replace this "dummy" hydride in order to tune the reactivity at the metal center and thus improve the catalyst efficiency.

In view of the development of a viable hydrogen storage and release system based on a reversible carbon dioxide formic acid cycle (Scheme 17.1), it is worth mentioning that the $Fe(BF_4)_2{\cdot}(H_2O)_6/PP_3$ catalyst can also promote the hydrogenation of sodium bicarbonate to sodium formate. At 80 °C in methanol under 60 bars of hydrogen and in the presence of 0.14% of the catalyst, sodium formate was formed in 88% yield over 20 hours, with a TON of 610 [70].

Gonsalvi and coworkers explored the efficiency of the iron catalyst modified by linear tetraphosphine 1,1,4,7,10,10-hexaphenyl-1,4,7,10-tetraphosphadecane (tetraphos-1, P_4). As there are two potential isomers, *rac* and *meso*, *rac* (S,S;R,R) P_4 in combination with $Fe(BF_4)_2{\cdot}(H_2O)_6$ provided the most active catalytic system both for HCO_2H dehydrogenation and sodium bicarbonate hydrogenation [60]. An initial TOF of up to 1737 h^{-1} and a TON of up to 6061 have been recorded for the dehydrogenation of HCO_2H in propylene carbonate at 60 °C with a substrate/catalyst ratio of 10 000 (Scheme 17.7).

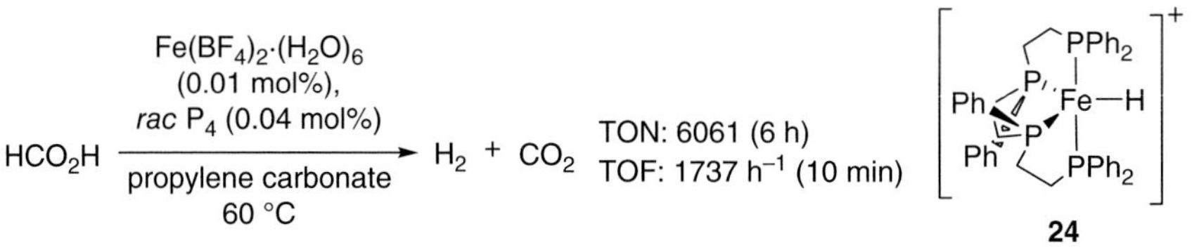

HCO_2H —[$Fe(BF_4)_2{\cdot}(H_2O)_6$ (0.01 mol%), *rac* P_4 (0.04 mol%); propylene carbonate, 60 °C]→ $H_2 + CO_2$ TON: 6061 (6 h) TOF: 1737 h^{-1} (10 min)

$[\text{Fe(H)}(P_4)]^+$ **24** (Ph, Ph, P, P, PPh_2, PPh_2, Fe—H)

Scheme 17.7 Dehydrogenation of formic acid promoted by catalyst **24** generated *in situ* from $Fe(BF_4)_2{\cdot}(H_2O)_6$ and excess *rac* P_4 ligand.

Similar to the PP_3 case, catalyst productivity is greatly enhanced in the presence of an excess of the ligand up to a ligand-to-metal ratio of 4. NMR investigations under operando conditions clearly showed that the reaction follows a cationic mechanism analogous to **A** in Scheme 17.6. The superior efficiency provided by the *rac*-ligand as to the *meso* one stems from its selective coordination mode at the iron center, which secures the availability of two cis-disposed coordination sites to allow for formate coordination and subsequent β-hydride elimination.

By sulfonation of the aryl substituents in the polyphosphine ligand, Laurenczy and coworkers were able to prepare a water-soluble version of the $Fe(BF_4)_2{\cdot}(H_2O)_6/PP_3$ catalytic system [61]. When 2 equiv. of the ligand were used, complete dehydrogenation of HCO_2H in a 5 M aqueous solution at 80 °C at a substrate/catalyst ratio of 200 was achieved within 50 minutes, corresponding to a TOF of 240. The catalyst proved to be thermally stable and oxidation resistant if exposed to air. Noteworthy, no activity loss was observed after eight weeks in solution at 80 °C, attaining a total TON exceeding 13 000 under the condition of periodic replenishment of the converted HCO_2H.

Milstein and coworkers reported the first example of an iron pincer complex [(*t*Bu-PNP)Fe(H)$_2$(CO)] **25** (*t*Bu-PNP, 2,6-bis(di-*tert*-butylphosphinomethyl) pyridine) being able to promote the selective dehydrogenation of HCO_2H. Under optimized conditions in tetrahydrofuran (THF) at 40 °C in the presence of 50 mol% of NEt_3 (NEt_3, triethylamine) and 0.05 mol% of the catalyst, a TOF of 836 h^{-1} in the first hour was achieved (Scheme 17.8) [62].

In a long-term experiment with a catalyst loading of 0.001%, 1 mol of HCO_2H in dioxane was fully converted after 10 days, corresponding to a total TON of 100 000. No hydrogen evolution was recorded in the absence of NEt_3 and activity increased at increasing base concentrations. The solvent was mandatory, and better results were observed in THF, 1,4-dioxane, and dimethylsulfoxide (DMSO). The proposed catalytic cycle for the decomposition of HCO_2H with **25** is outlined

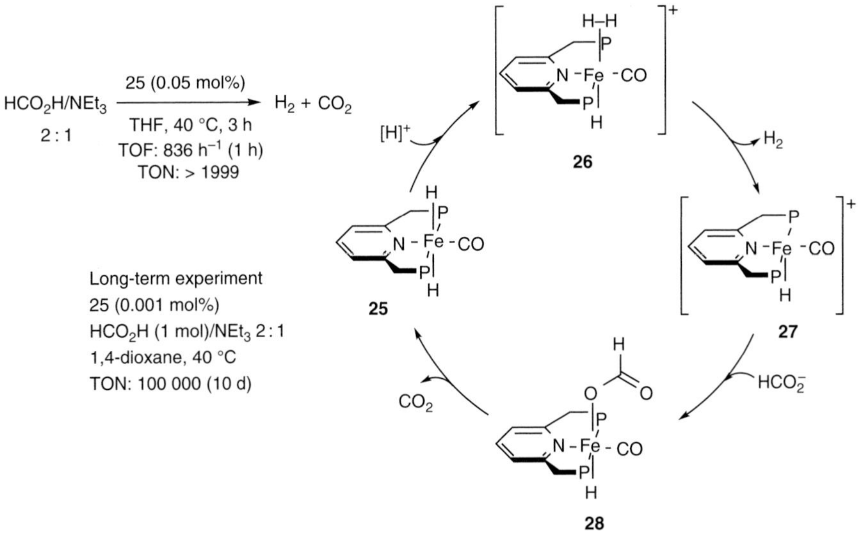

HCO_2H/NEt_3 2 : 1 → $H_2 + CO_2$ (25 (0.05 mol%); THF, 40 °C, 3 h; TOF: 836 h^{-1} (1 h); TON: > 1999)

Long-term experiment
25 (0.001 mol%)
HCO_2H (1 mol)/NEt_3 2 : 1
1,4-dioxane, 40 °C
TON: 100 000 (10 d)

Scheme 17.8 Dehydrogenation of formic acid/triethylamine adduct promoted by catalyst **25** and proposed mechanism (P = P*t*Bu_2).

in Scheme 17.8: Protonation of the dihydride **25** generates a dihydrogen complex **26**, from which hydrogen evolves to give the cationic coordinatively unsaturated monohydride species **27**. Then, the formate anion can coordinate to the iron center to form species **28**, from which CO_2 can be released regenerating the iron dihydride **25**. This step cannot proceed through β-hydride elimination unless one arm of the ligand dissociates to provide the required cis-vacant coordination site. Instead, DFT predicts that intramolecular rearrangement of the *O*-bound formate to directly transfer the hydride to the metal in **27** is energetically more favorable. According to such mechanism, the potentially cooperative ligand is not directly involved in the decomposition of HCO_2H. The same catalyst is able to promote the hydrogenation of CO_2 (total pressure 10 bars) in THF/H_2O 10/1 in the presence of sodium hydroxide (2 M) to afford sodium formate [71]. However, no reaction was observed when NEt_3 was used instead [62].

Exceptional TOF and TON in the decomposition of HCO_2H were reported by Hazari and coworkers when using the aliphatic pincer iron complex $[(^{iPr}P^{H}NP)Fe(H)(\eta^1\text{-OOCH})(CO)]$ **29** ($^{iPr}P^{H}NP$ = $HN\{CH_2CH_2(PiPr_2)\}_2$) prepared by the addition of formic acid across the nitrogen–iron bond in the amido complex **19** (Scheme 17.9) [63]. At 80 °C, 10^{-4} mol% of the catalyst and 10 mol% of $LiBF_4$ dissolved in 1,4-dioxane gave an initial TOF of 196 728 h^{-1} and a total TON of 983 642 over 9.5 hours. The same mechanism for HCO_2H decomposition as for Milstein's catalyst was proposed (Scheme 17.9). Key to success is the Lewis acid cocatalyst, which should stabilize the transition state **30**. This leads to the H-bound iron formate complex **31**, thus accelerating the decarboxylation to generate the dihydride complex **32**. No amine is required in this case. A proper base is instead necessary for the same catalytic system to promote the reverse reaction. Because of the poor stability of the catalyst in water and the necessity

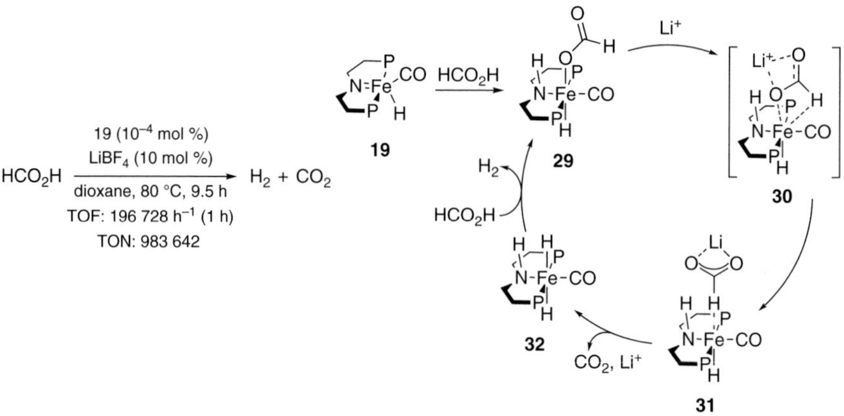

Scheme 17.9 Dehydrogenation of formic acid promoted by catalyst **29** and the proposed mechanism which highlights the role of the Lewis acid cocatalyst in easing decarboxylation of the coordinated formate anion (P = P*i*Pr_2).

to operate in an organic solvent, DBU (1,8-diazabicyclo(5.4.0)undec-7-ene) was selected as the base of choice to hydrogenate CO_2 to formate in THF under 69 bars total pressure and 80 °C [72].

17.2.2 Nickel

Enthaler, Junge and coworkers reported the first example of a nickel-based homogeneous catalyst able to promote the decomposition of HCO_2H in the presence of the high-boiling amine dimethyl-*n*-octylamine (HCO_2H/amine 11/10) [65a]. The catalyst is a nickel hydride complex **33** bearing a PCP pincer ligand (Scheme 17.10). At 80 °C in propylene carbonate, a maximum TOF of 240 h^{-1} and a maximum TON of 626 after three hours were achieved. Because of partial degradation of propylene carbonate in long-term experiments at this temperature, other solvents such as diethylene glycol dimethyl ether and 1,4-dioxane were used, in which, however, lower activities were recorded. The same catalyst was also used for the hydrogenation of sodium bicarbonate to sodium formate in MeOH at 150 °C and 55 bars of H_2, affording a TON of 3038 after 20 hours [65a].

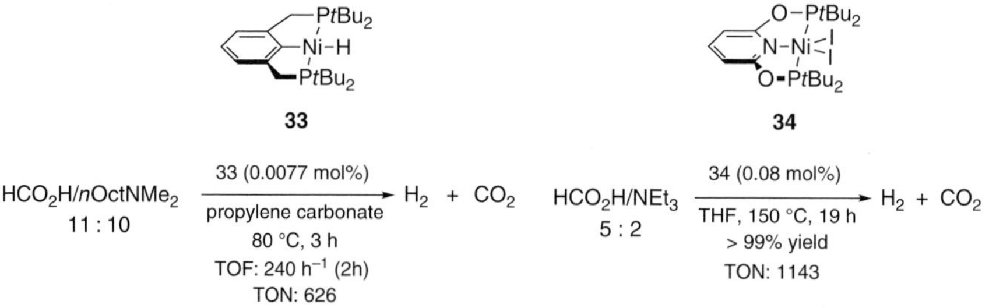

Scheme 17.10 Dehydrogenation of formic acid/amine adducts promoted by Ni-catalysts **33** and **34**.

The nickel complex **34** bearing a PNP pincer ligand built on 2,6-dihydroxy pyridine was only active above 140 °C and in the presence of TEAF. Full conversion of HCO_2H to H_2 and CO_2 was attained with 0.08 mol% of the catalyst at 150 °C in THF within eight hours corresponding to a TON of 1143 [65b]. $[Ni(PMe_3)_4]$ has also been shown to be able to decarboxylate formic acid in benzene at 80 °C with a TON of 70 and an initial TOF of 1.7 h^{-1}. Here, no base additive was required [65c].

17.2.3 Aluminum

The aluminum monohydride complex **35** (Scheme 17.11), supported by a pincer phenyl-substituted bis(imino)pyridine ligand, has proved to be competent for the selective dehydrogenation of HCO_2H as TEAF in THF. At 65 °C in the presence of 0.006% of the catalyst, an initial TOF of 5200 h^{-1} (after 15 minutes) and a TON of 2200 (after 1 hour) were achieved [67a]. The resting state **37**, isolated from a typical catalytic test solution, results from the reaction of the catalyst precursor **36** with 3 equiv. of formic acid. Two equivalents serve ligand protonation and one hydride protonation with ensuing release of dihydrogen. Catalytic turnover proceeds through β-hydride abstraction from one of the two bound formate anions in **37** to afford the aluminum–monohydride species **38** (Scheme 17.11). Protonation of the latter by formic acid releases hydrogen and regenerates **37**. The proposed mechanism has been recently corroborated by DFT calculations [67b].

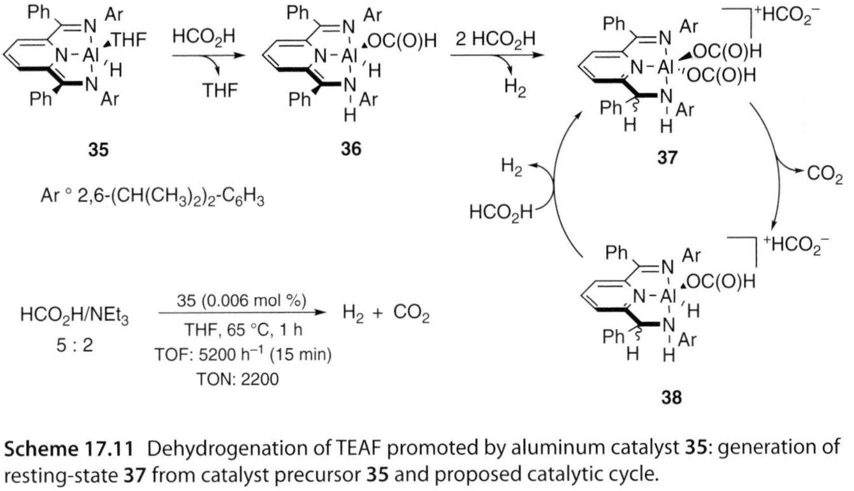

Scheme 17.11 Dehydrogenation of TEAF promoted by aluminum catalyst **35**: generation of resting-state **37** from catalyst precursor **35** and proposed catalytic cycle.

17.2.4 Miscellaneous

Evidence of HCO_2H acid decomposition to H_2 and CO_2 has also been reported when $Co(II)Cl_2$ [58], $\{Co(I)(H)[(PPh(OEt_2)_2]_4\}$ [64a], and the water-soluble Co(II)/4,4′,4″,4‴-tetrasulfophthalocyanine complex $\{[Co(II)(TSPc)]Na_4\}$ [64b]

have been used as catalysts. However, productivities were only scored using the *in situ* catalyst prepared by the combination of $Co(BF_4)_2{\cdot}(H_2O)_6$ and 2 equiv. of the PP_3 ligand. At 60 °C, the selective decomposition of HCO_2H acid in propylene carbonate yielded a TON of 197 after two hours [59e]. With the same catalytic system, it was possible to hydrogenate both sodium bicarbonate and CO_2 in high yields and TON [73].

H_2 and CO_2 were evolved from solutions of sodium formate in 95% aqueous MeOH (0.33 M) in the presence of 2.5 mol% of $Cr(CO)_6$ at 60 °C affording a maximum TON of 17 [55]. The reaction proceeds only under light irradiation, which is required to displace a CO molecule from $Cr(CO)_6$ and generate the catalytically active species $Cr(CO)_5$, which is coordinatively unsaturated and can therefore bind the formate anion. Following β-hydride elimination and the ensuing CO_2 release, protonation of the resulting monohydride species $[HCr(CO)_5]^-$ by water affords a dihydrogen complex from which hydrogen gas evolves and the catalytic cycle is completed.

The cyclopentadienyl molybdenum hydride compounds $[Cp^RMo(PMe_3)_{3-x}(CO)_xH]$ (Cp^R = Cp $(C_5H_5)^-$ or Cp* $(C_5(CH_3)_5)^-$; x = 0–3) are competent for the dehydrogenation of formic acid, although activities vary significantly within the series [56]. Activities were evaluated by NMR spectroscopy, showing $[Cp^*Mo(PMe_3)_2(CO)H]$ to be the most efficient catalyst. A solution of the catalyst (4 mol%) and the substrate in C_6D_6 afforded a TOF of 54 h^{-1} corresponding to 50% substrate conversion. Dehydrogenation proceeds through initial protonation of $[Cp^*Mo(PMe_3)_2(CO)H]$ at the metal, elimination of dihydrogen from the resulting dihydride, coordination of the formate anion, and decarboxylation of the latter via β-hydride elimination to regenerate the starting monohydride species. The composition of the best performing catalyst ensures the optimal balance between the σ-donating (PMe_3) and π-accepting (CO) properties of the ligands in order to favor the single elementary steps and allow for catalytic turnover.

The manganese(I) complex $[(^{iPr}P^HNP)Mn(CO)_2]$ ($^{iPr}P^HNP$ = $HN\{CH_2CH_2(PiPr_2)\}_2$) showed modest activity in the decomposition of formic acid. The highest TON, 190 after 14 hours, was achieved in the decomposition of a 0.87 M solution of formic acid in 1,4-dioxane at 65 °C using 0.3 mol% of the catalyst. Besides, the reaction was proven to be unselective as CO was released as a by-product. In contrast to the related iron(II) complex, the use of $LiBF_4$ as a cocatalyst had an inhibitory effect [57].

Simple copper salts also proved competent for the dehydrogenation of HCO_2H/amine adducts [66]. The activity depends on the properties of the amine. Although a linear correlation could not be established between basicity and activity, the latter turned out to be positively affected by more basic and more bulky amines. To prevent reduction to metallic copper and extend the catalyst lifetime, the presence of acetic acid to buffer the increase of pH due to formic acid conversion and the use of 1 equiv. to the metal of a chelating amine such as ethylene diamine have been proved beneficial. Under optimized conditions using ca. 1 mol% of CuI, hydrogen was evolved from a 1/1 mixture of HCO_2H/NEt_3 at 95 °C affording an overall TON of 72.

17.3 Hydrogen Generation from Alcohols

17.3.1 Hydrogen Generation with Respect to Energetic Application

The dehydrogenation of alcohols aiming for the generation of hydrogen has been in the focus of research for more than 40 years [74]. In the early 2000s, the interest in this field reached a new height because of the increasing concern about depletion of fossil fuels and the starting transition to a system based on renewable energy. In this paragraph, the focus lies on acceptorless alcohol dehydrogenation. Hereby, the amount of hydrogen produced per time and moderate reaction conditions are of first priority, whereas full conversion of the substrate plays only a secondary role. The use of non-noble metal-based catalysts for this reaction only started during recent years, and so far, applications have been limited to iron-based catalysts.

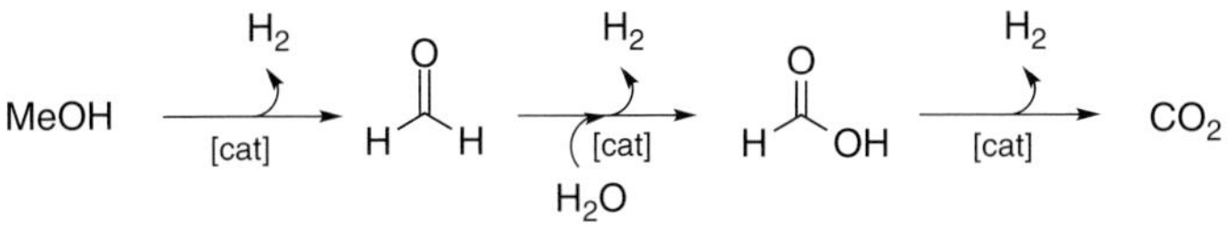

Scheme 17.12 Three-step aqueous-phase reforming of MeOH to hydrogen and carbon dioxide.

In 2013, Beller and coworkers reported the aqueous-phase dehydrogenation of methanol to hydrogen and carbon dioxide (Scheme 17.12), employing the aliphatic PNP iron catalysts **39** and **40** (Figure 17.3) [75]. For the first time, non-noble metal-catalyzed dehydrogenation of an alcohol and the application of an iron pincer catalyst were successfully performed. Notably, a very high base concentration of 8 M KOH significantly improved the catalyst activity. A TON of nearly 10 000 was achieved with 0.4 mol% of **40** under highly basic conditions in a 9 : 1 methanol/water solution at 91 °C. The highest activity of 702 h^{-1} for the first hour was obtained under similar conditions, although with an increased catalyst concentration of 1.9 mol%. Although this system showed a very high activity, rather low yields of up to 17.4% were accomplished. Subsequently, also other groups employed the structural motif of this kind of iron complexes. Notably, Bernskoetter, Hazari, Holthausen, and coworkers developed a base-free system for the dehydrogenation of aqueous methanol by using the amido complex **41** and the formate complex **29** in combination with Lewis acid cocatalysts (Figure 17.3) [76]. Employing 0.01 mol% **29** and 10 mol% $LiBF_4$ for the dehydrogenation of a 4 : 1 methanol/water mixture in refluxing ethyl acetate, a TON of 30 000 was obtained.

Reducing the catalyst loading to 0.006 mol% while increasing the reaction time to 94 hours, the so far highest TON of 51 000 for both a base-free and a first-row transition-metal-based system was achieved. However, the practical relevance for energy-related hydrogen generation is limited as the outstanding activities were obtained only for highly diluted solutions containing 160 μl of methanol and 80 μl of water in 10 ml of ethyl acetate. After a theoretical study on the mechanism of the dehydrogenation of ethanol involving the iron-based amido complex

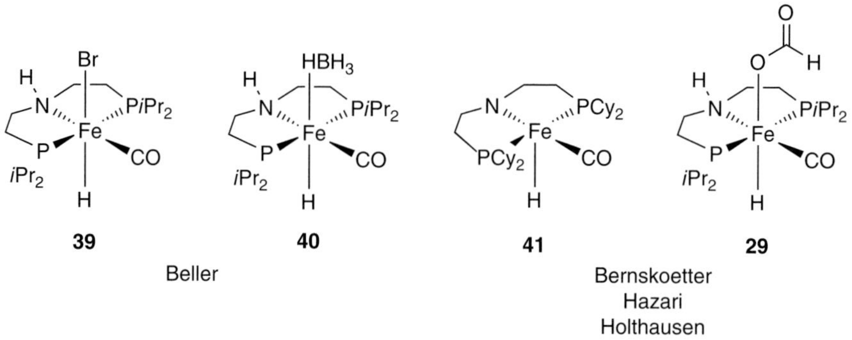

Figure 17.3 Fe-based catalysts applied in the dehydrogenation of MeOH water mixtures.

19 was published [77], catalysts **39** and **19** were employed for hydrogen generation from primary and secondary alcohols and diols without any base or further additives [78]. Exemplarily, 1,4-butanediol was dehydrogenatively coupled to γ-butyrolactone in 85% (isolated) yield, employing 0.1 mol% of **19** in refluxing toluene for 20 hours (Scheme 17.13).

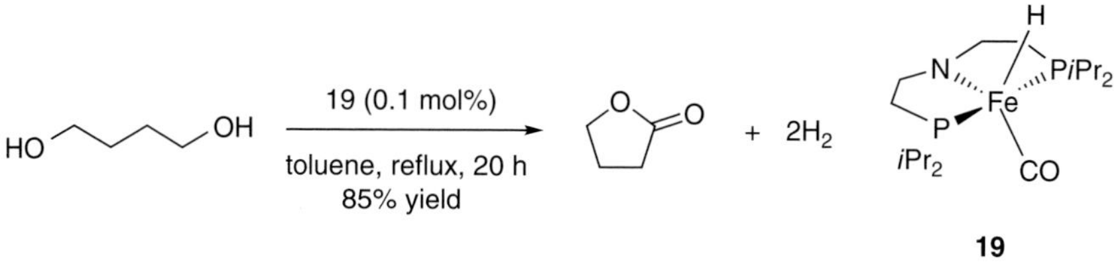

Scheme 17.13 Dehydrogenative coupling of 1,4-butanediol to γ-butyrolactone.

Following the same protocol, various primary alcohols were converted into esters at good yields. Additionally, hydrogen was generated from various secondary benzylic alcohols with catalyst loadings of 0.1–1 mol% **40** for 8–48 hours using refluxing toluene or THF as a solvent. Hong and coworkers reported an *in situ* system consisting of a 1/1/1 mixture of $Fe(acac)_2$ (acac, acetylacetonate), 1,10-phenanthroline and K_2CO_3 for the dehydrogenation of benzylic alcohols [79]. Good-to-quantitative yields were obtained for a range of substrates. However, the protocol failed to dehydrogenate primary and secondary aliphatic alcohols. Among the successfully dehydrogenated substrates, 2-phenyl ethanol released the highest hydrogen amount of 1.66 wt%.

17.3.2 Hydrogen Generation Coupled with the Synthesis of Organic Compounds

As shown in the previous paragraph especially for higher alcohols, only a small part of the theoretical hydrogen amount is released. Thus, the combination of hydrogen generation with the production of an additional valuable product appears favorable. In particular, the acceptorless alcohol dehydrogenation offers the possibility to synthesize important carbonyl compounds and their derivatives such as aldehydes, ketones, carboxylic acids, imines, lactones, lactams, Guerbet products, and others. Noteworthy, this type of reaction is the most atom efficient way to oxidize the starting material without using any oxidizing agents. Hereby, iron, cobalt, nickel, and manganese catalysts have been shown to feature the significant activity.

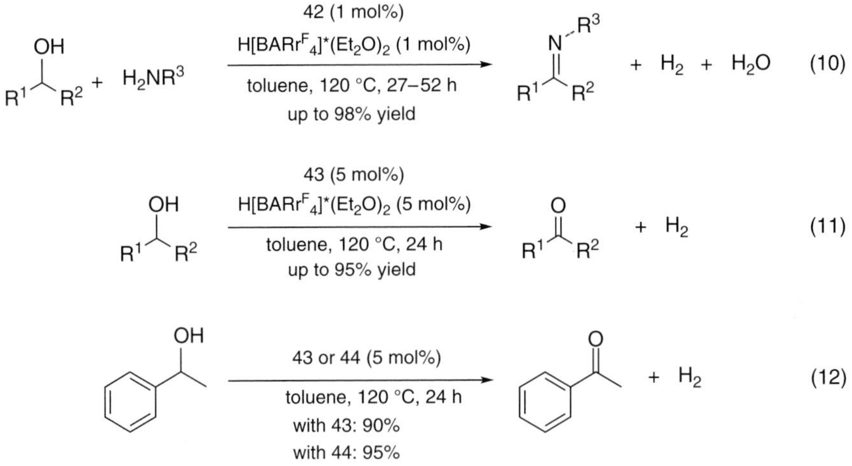

$$R^1CH(OH)R^2 + H_2NR^3 \xrightarrow[\text{toluene, 120 °C, 27–52 h; up to 98\% yield}]{\text{42 (1 mol\%), H[BArF}_4\text{]*(Et}_2\text{O)}_2\text{ (1 mol\%)}} R^1C(=NR^3)R^2 + H_2 + H_2O \quad (10)$$

$$R^1CH(OH)R^2 \xrightarrow[\text{toluene, 120 °C, 24 h; up to 95\% yield}]{\text{43 (5 mol\%), H[BArF}_4\text{]*(Et}_2\text{O)}_2\text{ (5 mol\%)}} R^1C(=O)R^2 + H_2 \quad (11)$$

$$PhCH(OH)CH_3 \xrightarrow[\text{toluene, 120 °C, 24 h; with 43: 90\%; with 44: 95\%}]{\text{43 or 44 (5 mol\%)}} PhC(=O)CH_3 + H_2 \quad (12)$$

Scheme 17.14 Cobalt-catalyzed dehydrogenative coupling of primary alcohols with amines and dehydrogenation of alcohols.

In 2013, Hanson and coworker reported the first cobalt-catalyzed dehydrogenation of secondary alcohols and the dehydrogenative coupling of alcohols with amines to give imines [80]. It is noteworthy that for the dehydrogenative coupling of benzylic alcohols with amines (Scheme 17.14, Eq. (10)), the neutral cobalt alkyl complex **42** (Figure 17.4) showed a comparable activity to ruthenium catalysts previously reported by Milstein and Madsen [81].

For the synthesis of ketones, a higher catalyst loading of 5 mol% of the *in situ*-generated cationic cobalt complex **43** was necessary (Scheme 17.14, Eq. (11)). In the same year, the group of Hanson investigated metal–ligand cooperativity by using the cationic cobalt catalyst **43** and the related N-methylated cobalt catalyst **44** for the hydrogenation of various olefins and ketones as well as the dehydrogenation of 1-phenylethanol (Scheme 17.14, Eq. (12)) [82].

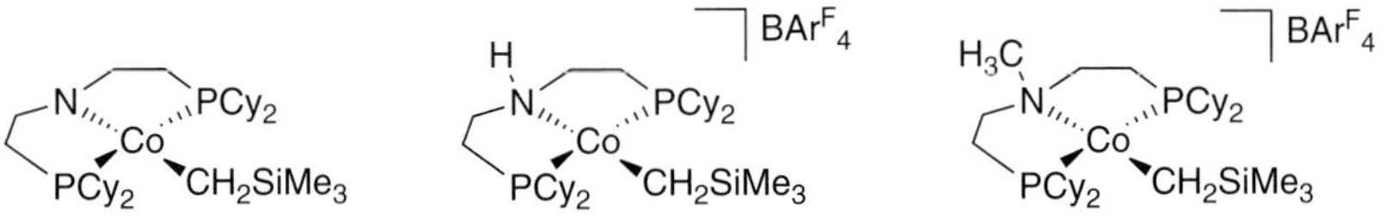

Figure 17.4 Co-based catalysts applied in the acceptorless dehydrogenation of alcohols.

Besides cobalt, iron pincer complexes were also applied for synthetic dehydrogenation reactions. The first iron-catalyzed acceptorless dehydrogenation of alcohols was reported by Nakazawa and coworkers, who used iron catalyst **45** for the dehydrogenation of 2-pyridylmethanol derivatives (Scheme 17.15) [83]. The addition of NaH proved to be essential for high activity. Thus, with 0.001 mol% **45** and 2 mol% NaH in refluxing toluene, a TON of 67 000 and a yield of 67% were reached. Interestingly, no other substrates could be dehydrogenated, which was explained by the essential chelating effect of the substrate's pyridine moiety.

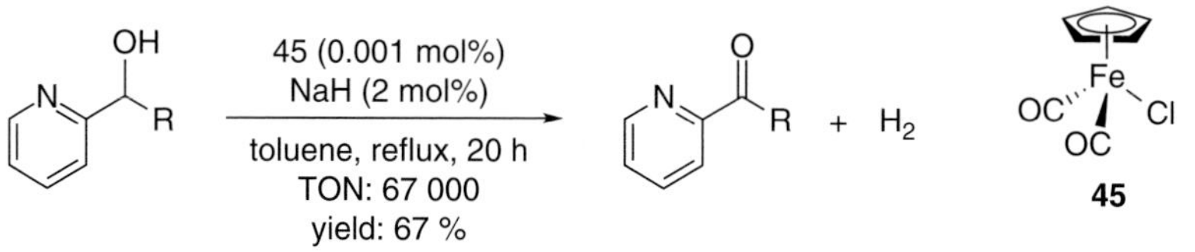

45 (0.001 mol%)
NaH (2 mol%)
toluene, reflux, 20 h
TON: 67 000
yield: 67 %
$+ H_2$
45

Scheme 17.15 Iron-catalyzed dehydrogenation of 2-pyridylmethanol derivatives.

Recently, Beller and coworkers successfully applied aliphatic iron pincer complex **40** for the synthesis of lactones and lactams from a broad scope of diols and hydroxyamines, respectively. Hereby, two sequential dehydrogenation reactions were catalyzed, leading in the first step to the corresponding ketone and in the second step to the lactone or lactam. Optimization of conditions gave good-to-excellent yields after five hours of reaction using 0.5 mol% (for the synthesis of lactams 1 mol%) **40**, 10 mol% K_2CO_3, and *t*-amyl alcohol as a solvent at 150 °C (Scheme 17.16) [84].

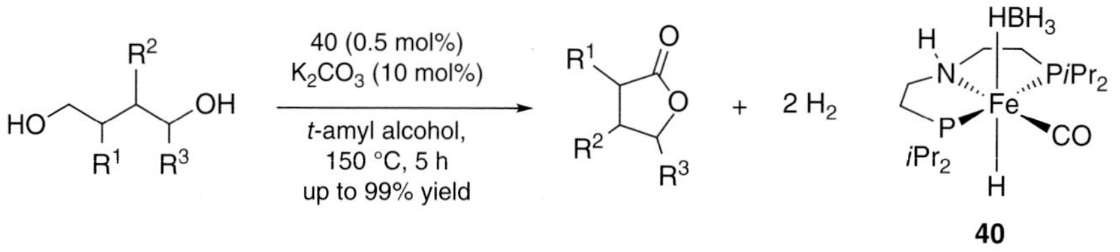

40 (0.5 mol%)
K_2CO_3 (10 mol%)
t-amyl alcohol,
150 °C, 5 h
up to 99% yield
$+ 2 H_2$
40

Scheme 17.16 Dehydrogenative coupling of diols and transfer hydrogenation of acetophenone with glycerol catalyzed by the aliphatic PNP pincer catalyst **40**.

Catalysts of the same family were used by Hazari, Crabtree, and coworkers for the synthesis of lactic acid from glycerol, a substrate, which is obtained in large amounts as a side product of biodiesel production (Scheme 17.17) [85].

Although formate complex **29** (Scheme 17.9) showed the highest activity (TON of 880), optimization experiments were performed with complex **40** (TON of 770) as it was the synthetically best accessible catalyst among those tested. Application of a 10-fold amount of catalyst leads to an only limited increase of conversion. This was explained by a bimolecular deactivation pathway. Consequently, when reducing the catalyst amount to 0.004 mol%, a TON of 1050 was obtained after six hours. However, the selectivity dropped to 55% as added NaOH catalyzed the competing glycerol etherification.

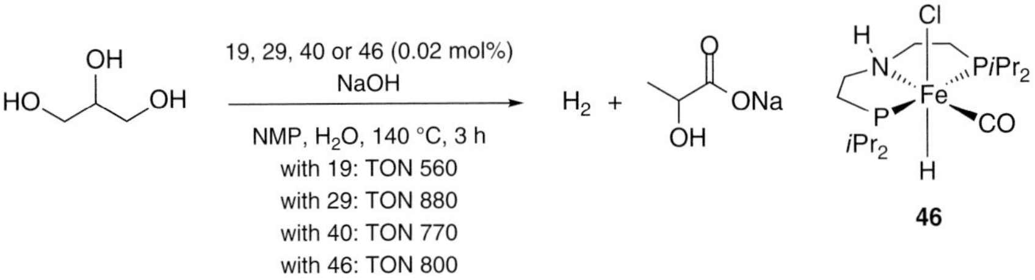

HO–CH₂–CH(OH)–CH₂–OH → (19, 29, 40 or 46 (0.02 mol%), NaOH; NMP, H_2O, 140 °C, 3 h) → H_2 + CH₃–CH(OH)–C(O)ONa

with 19: TON 560
with 29: TON 880
with 40: TON 770
with 46: TON 800

46: H–N, Cl, $PiPr_2$, Fe, P, CO, iPr_2, H

Scheme 17.17 Dehydrogenation of glycerol to lactic acid catalyzed by Fe-based aliphatic PNP pincer complexes.

17.4 Hydrogen Storage in Liquid Organic Hydrogen Carriers

LOHC are polycyclic aromatic hydrocarbons or heterocycles, which can be reversibly hydrogenated and dehydrogenated. Depending on the number of carbon atoms involved in the hydrogenation/dehydrogenation reaction, hydrogen storage capacities between approximately 3 and 7 wt% can be achieved practically. Figure 17.5 provides an overview of (poly)cyclic hydrocarbons and N-heterocycles that have gained increasing attention as alternative hydrogen storage materials.

A variety of heterogeneous catalysts, mainly based on Pd and Pt, but also on Ru, Rh, Ir, and Ni, have been employed for the dehydrogenation of LOHC [86]. Generally, dehydrogenation takes place at temperatures between 200 and 300 °C because of the unfavored thermodynamics. Currently, iridium is the noble metal of choice among homogeneous catalysts. Applying PCP pincer and other iridium catalysts, Jensen, Fujita, Yamaguchi, Xiao, and others reported successful

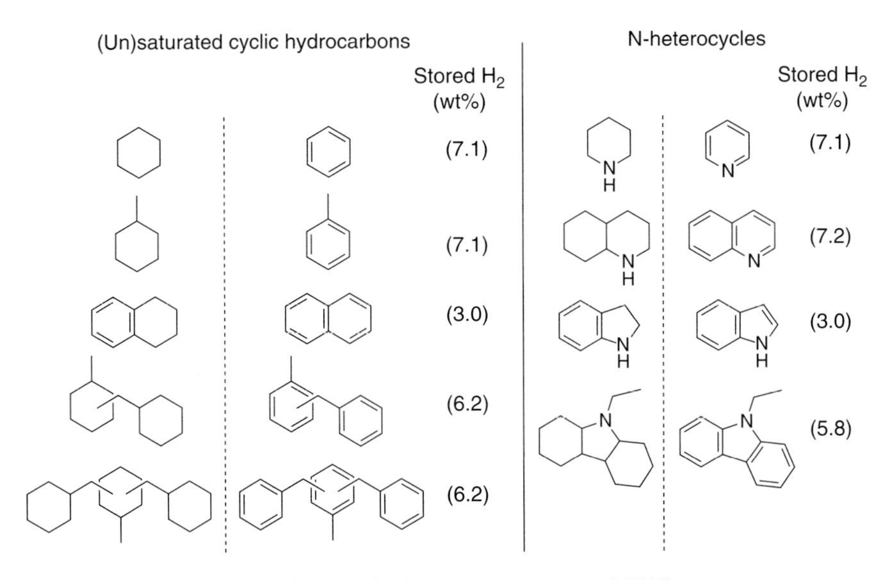

Figure 17.5 Exemplary liquid organic hydrogen carrier systems (LOHC).

hydrogen generation at temperatures between 100 and 200 °C [29a, 87]. Very recently, two non-noble metal-based catalyst systems were applied for the dehydrogenation of N-heterocycles. Because of the positive thermodynamic effect of introducing nitrogen atoms in the hydrocarbon ring, significantly less energy is needed for the hydrogen release compared to solely carbon-containing compounds [26, 88]. The first example of the non-noble metal-catalyzed hydrogenation and dehydrogenation of N-heterocycles was reported in 2014 by Jones and coworkers employing the molecular iron catalysts **19** and **40** (Scheme 17.18; structure of **19** is shown in Figure 17.2). Under optimized conditions using 3 mol% of catalyst **40** in xylene as a solvent, good-to-quantitative yields for the dehydrogenation of 1,2,3,4-tetrahydroquinaldines were reached within 30 hours. The position of the methyl group did not have a significant effect on the conversion. Noteworthy, for the hydrogenation of the unsaturated N-heterocycles, catalyst **39** (Figure 17.3) showed the best activities using 10 mol% of KO*t*Bu at 80 °C for 24 hours in THF at hydrogen pressures of 5 or 10 atm [89]. One year later, the cationic cobalt pincer complex **43** was applied for a similar class of N-heterocycles. However, for dehydrogenation, a significantly higher catalyst loading of 10 mol% was necessary to attain good conversions and for hydrogenation 5 mol% of the catalyst have been used (Scheme 17.18) [90]. Interestingly, it was not possible to either dehydrogenate 2,6-dimethylpiperidine or hydrogenate its unsaturated counterpart, which was successfully demonstrated by the aliphatic iron complexes **39** and **40**. In both cases, the benzene ring stayed unaffected because of the low temperatures applied, which reduces the exploitable hydrogen content.

17.5 Dehydrogenation of Ammonia Borane and Amine Boranes

The catalytic dehydrogenation of boron–nitrogen compounds gained first attention with the work of Roberts and coworkers (Pd/C) [91] and Manners and coworkers (Rh^{I}) [37a, 92]. Since then, much work has been done on this subject, extending the scope to a broad variety of boranes including ammonia borane $H_3N–BH_3$ (AB), amine boranes $R_2HN–BH_3$, hydrazine borane, and many others [93]. In this chapter, only a selection of compounds dehydrogenating AB and amine boranes using late non-noble metals will be presented. For a selection of other non-noble metals such as Ti or Zr [94], Sc or Y [95], Ca [96], and Cr or Mo [97], the reader is referred to the provided literature and the recent review by Rossin and Peruzzini [98].

17.5.1 Overview on Conditions for H_2 Liberation from Ammonia Borane and Amine Boranes

For H_2 liberation, there are several possibilities of dehydrogenation, namely, thermal dehydrogenation, thermal solvolysis, acid-catalyzed dehydrogenation, anionic dehydropolymerization, and metal-catalyzed solvolysis [20d]. In contrast

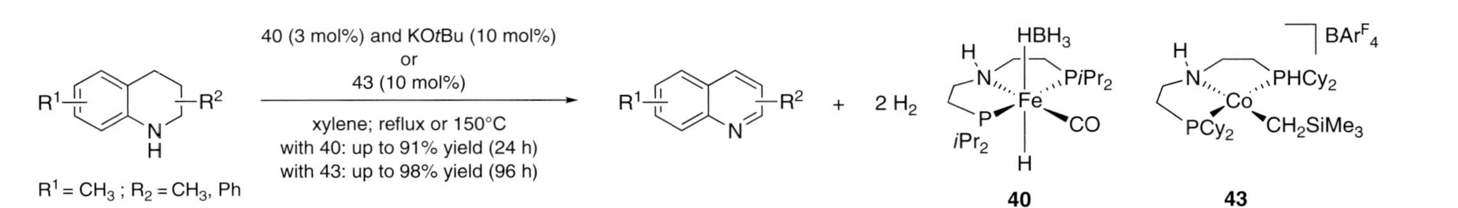

Scheme 17.18 Dehydrogenation of 1,2,3,4-tetrahydroquinaldines by iron catalyst **40** and cobalt catalyst **43**.

to previous substrates, the thermal dehydrogenation of AB and amine boranes is exothermic and yields a couple of different products. Ammonia borane itself decomposes in three steps, losing more than 1 equiv. of hydrogen between 107 and 117 °C, the 2nd equiv. at about 150 °C, and the rest at more than 500 °C [20d]. The other methods of dehydrogenation mentioned above focus on decreasing the dehydrogenation temperature as well as on directing the dehydrogenated products. Dehydrogenation in solvents such as alcohols or water can be done at 135 °C to release 3 equiv. of hydrogen from AB [99]. However, addition of a solvent lowers the overall gravimetric hydrogen storage capacity. Another option for H_2 liberation from AB is anionic dehydropolymerization starting with the generation of catalytic amounts of $H_2NBH_3^-$ upon addition of, e.g. $LiNH_2$ or LiH. Although the mechanistic details are unknown so far, an increased hydricity of the B—H bond in $H_2NBH_3^-$ is assumed to facilitate H_2 loss [20d]. Dehydrogenation of amine boranes and AB can also be catalyzed by Lewis or Brønsted acid, generating the more reactive boronium cation $[H_3NBH_2]^+$ by either hydride abstraction or B—H protonation with concomitant H_2 release [20d]. However, at acid concentrations higher than 0.5% and at elevated temperatures of 60 °C, undesired side products (aminodiborane $B_2H_5(\mu\text{-}NH_2)$, borazine) will be formed [20d]. Metal-catalyzed dehydrogenation has gained considerable attention during the last years mainly because of the broad variety of potentially available materials, which are assumed to meet the requirements of constant and reliable hydrogen release with respect to practical applications. With some materials based, e.g. on iridium or ruthenium, hydrogen liberation was achieved even at −30 °C and within times as short as four minutes.

17.5.2 Non-noble Metal-Catalyzed Dehydrogenation of Ammonia Borane and Amine Boranes

Although there are various catalysts based on heterogeneous materials and/or noble metals, an increasing number of homogeneously working catalysts based on non-noble metals have been developed during the past years [98]. Most catalysts comprise M—H or M—H_2 bonds, whose polarity (strong or weak) and nature (acidic or basic) determine their reactivity in AB dehydrogenation [98]. Concerning the number of released H_2 molecules, either polyborazanes $B_xN_xH_{4x+2}$ (linear) or polyaminoboranes $B_xN_xH_{4x}$ (cyclic, one molecule of H_2 released) or borazine $B_3N_3H_6$ (two molecules of H_2 released), or cross-linked borazine and polyborazylene BNH_x (more than two molecules of H_2 released) are formed during the course of reaction (see Figure 17.6) [98, 100]. Mechanistic investigations of the formed products are preferentially done with ^{11}B NMR spectroscopy using dimethylaminoborane (DMAB) as it releases exactly one molecule H_2 with concomitant formation of more stable products [98].

To date, the largest fraction of non-noble metal borane dehydrogenation catalysts are based on iron. A synopsis of these catalysts as well as of those based on cobalt, nickel, and copper is given in Figure 17.7.

The complexes **18**, **19** (for structures of **18** and **19** see Figure 17.2), and **47** belong to the class of complex metal hydrides (CMHs) because they bind at least one hydride ligand. The utilization of 5 mol% **47** in the dehydrogenation of AB

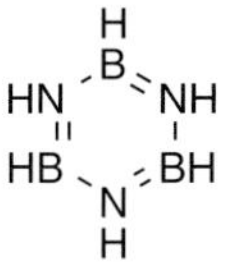

Borazine 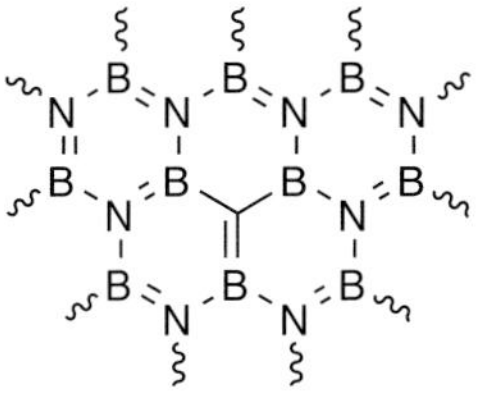 Polyborazylene 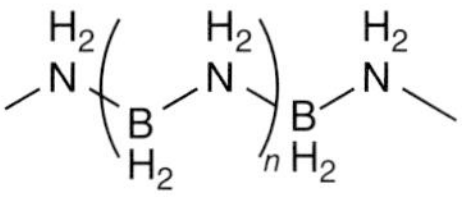 Polyaminoboranes

Figure 17.6 Selection of proposed products after dehydrogenation of ammonia borane. Source: Rossin and Peruzzini 2016 [98]. Copyright 2016. Adapted with permission of the American Chemical Society.

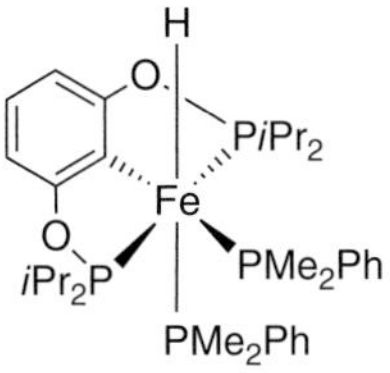

47 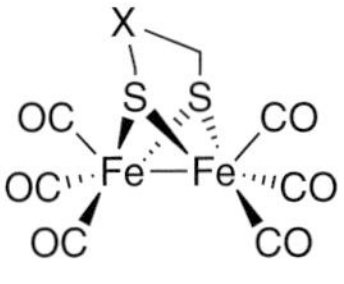 **48** 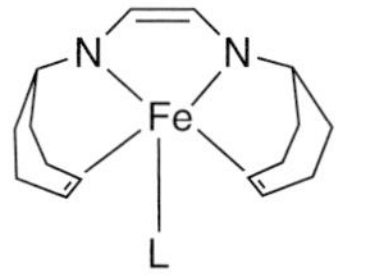 **49** 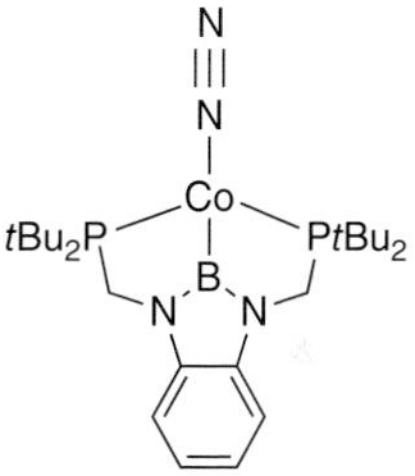 **50**

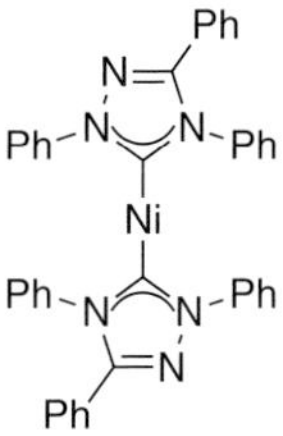

51 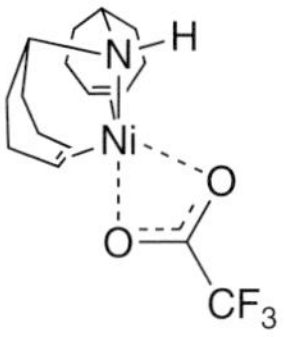 **52** 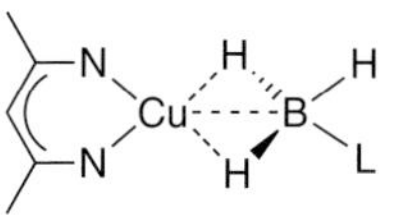 **53** 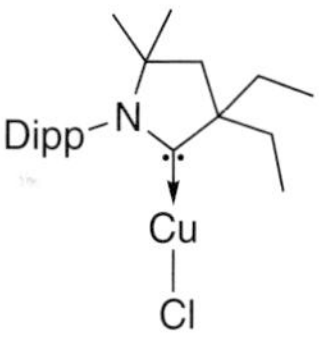 **54**

Figure 17.7 State-of-the-art non-noble metal catalysts applied in ammonia or amine borane dehydrogenation.

in THF yielded 2.3–2.5 equiv. of H_2 within 24 hours at a temperature of 60 °C [101]. Application of ^{31}P and ^{11}B NMR spectroscopy showed a substitution of one PMe_2Ph ligand trans to the hydride ligand by $H{-}BH_2{-}NH_3$ at room temperature or by $H{-}BH_2$ at 60 °C. The former species is assumed to transfer a hydride to the iron center and a proton to the *ipso*-carbon atom of the POCOP-derived pincer ligand MeO-^{iPr}POCOP (MeO-^{iPr}POCOP = 2,6-(OP^{i}Pr$_2$)$_2$-4-(OMe)-C_6H_2). From this stage, H_2 release occurs under the reformation of the initial species. The pincer ligand is also supposed to provide more stability.

As mentioned above, Schneider and coworkers also applied an iron pincer complex, **19**, in the catalytic dehydrogenation of AB in THF [102]. This system operates well at room temperature and releases 1 equiv. of H_2 within 10 hours. At the same time, AB is converted to mainly polyaminoborane and ^{11}B NMR studies as well as DFT calculations point toward the involvement of the iron center in both the hydrogen release and the B—N coupling. Similar to **47**, a hydride transfer from the BH_3 moiety to the iron center with concomitant proton transfer to the nitrogen atom at the PNP ligand occurs. Subsequently, $H_2N{=}BH_2$ release and dehydrogenation occur with simultaneous formation of an octahedral iron dihydride complex. Conversion of $H_2N{=}BH_2$ leads to either polyaminoborane or diaminoborane and BH_3, which deactivates the catalyst via the formation of the H—BH_3 ligand.

Complex **18** was also shown to be active in AB dehydrogenation as demonstrated recently by Baker and coworkers [103]. The focus of this work was, however, more put on the investigation of intermediates of the second dehydrogenation step of AB. It was found that besides a polyaminoborane material, another intermediate, namely, the tetrameric B-(cyclotriborazanyl)amine-borane instead of the trimeric B-(cyclodiborazanyl)amine-borane, is formed. This is relevant to catalyst design, which aims at fast dehydrogenation as well as at the avoidance of undesired products such as the volatile borazine, which is desired to efficiently being cross-linked to polyborazylene.

Inspired by the rich chemistry of hydrogenases, Darensbourg, Bengali, and coworkers applied hydrogenase mimics $[(\mu\text{-}SCH_2\text{-}X\text{-}CH_2)\{Fe(CO)_3\}_2]$ (X = CR_2, NR) **48** to the dehydrogenation of DMAB [104]. Substrate binding to the iron centers was reached via irradiation and subsequent CO loss. The bridging groups, which contain either CR_2 (R = H, Me, Et) or NR (R = Me, *t*Bu, Ph), determine the reactivity in the dehydrogenation reaction. It was shown that a CH function in the bridge positively influences the catalytic performance because of agostic interactions, and the amine functionality assisted the dehydrogenation process in a positive manner as well.

Iron complexes with the diazadiene ligand trop$_2$dad (dad, diazadiene; trop, 5*H*-dibenzo[*a,d*]cyclohepten-5-yl) have been proven to be active at room temperature in the dehydrogenation of AB, methylamine borane (MAB), and DMAB by Grützmacher and coworkers [105]. H_2 liberation was found to be dependent on the ligand L in **49**, of which THF resulted in complete conversion within 8.5 minutes when using MAB as a hydrogen source [105b]. The mechanism remained unclear so far, and selective poisoning with $P(OMe)_3$ as well as scanning electron microscopy (SEM) indicated homogeneous reaction conditions when using MAB. With AB and DMAB, the behavior of the catalyst is dependent on the solvent, influencing the formation of a heterogeneous species upon the usage of toluene [105b].

Baker et al. also reported a series of iron complexes (Fe(depe)$[N(SiMe_3)]_2$ (depe, 1,2-bis(diethylphosphino)ethane), Fe(depe)(N–N) (N–N = PhN–CH_2CH_2–NPh^{2-}), and $Fe(P\text{–}N)_2$ (P–N = $Cy_2P\text{–}CH_2CH_2\text{–}NPh^-$), which dehydrogenated AB at 60 °C in THF or diglyme [106]. Besides hydrogen, borazine and polyborazylene are formed. However, in all cases, a darkening of the reaction mixture was observed indicative of a rearrangement of the complexes. It is suggested that these complexes are just precursors of the catalytically competent species, whose identity is unclear so far.

A similar problem was observed by Manners and coworkers in 2011 using $[\{CpFe(CO)_2\}_2]$ as a catalyst [107]. Activation by light irradiation led to complete conversion of AB at room temperature within three hours forming polyborazanes and borazine. Using DMAB, a reaction time of four hours was required and *cyclo*-$(Me_2N–BH_2)_2$ was formed. However, in a later work, this catalyst was proven by poisoning experiments (PMe_3) to be converted to a heterogeneous species (iron nanoparticles) [108]. Moreover, using $[CpFe(CO)_2I]$ as a catalyst, no heterogeneous species were found. Using this catalyst for DMAB dehydrogenation, 55% of the activity compared with $[\{CpFe(CO)_2\}_2]$ was found [108].

Another system applying iron PNNP complexes also proved the formation of iron nanoparticles [109], which was evidenced by CO poisoning and addition of substoichiometric amounts (0.15 equiv.) of the ligand to a suitable iron precursor leading to the same activity. Depending on the solvent, different products were produced, e.g. $B(O^iPr)_3$ in isopropanol and B—N species in THF and diglyme. This is in contrast to presynthesized iron nanoparticles, where B—O species have been exclusively formed [110].

In the case of cobalt, only one example for homogeneous dehydrogenation of dimethylamine borane is known [111]. Reaction of cobalt PBP pincer complex **50** with DMAB in benzene at room temperature yielded hydrogen and cyclo-$(NMe_2–BH_2)_2$ within six hours. The complex was shown to reversibly add 2 equiv. of hydrogen under 1 atm of hydrogen with concomitant formation of a dihydridoborato–cobalt dihydride. Reaction with 2 equiv. of DMAB yielded a hydridoborane cobalt tetrahydroborate complex, which was proven to be inactive in DMAB dehydrogenation by theoretical studies. Moreover, theory revealed that the $LCo(H)_2$ species is the active one in DMAB dehydrogenation as well as styrene hydrogenation and transfer hydrogenation of styrene with amine borane [111b].

Dehydrogenation of AB was also achieved using nickel catalysts. Baker and coworkers used the biscyclooctadiene nickel (0) complex as a precursor with N-heterocyclic carbenes (NHCs) to form $Ni(NHC)_2$ **51** [112]. This complex performed AB dehydrogenation in benzene at 60 °C within four hours releasing more than 2 equiv. of H_2 with simultaneous formation of soluble cross-linked borazine. Kinetic isotope effect studies with (partially) deuterated AB revealed that both the N—H and the B—H bond are being broken in the rate-determining step. In the following years, several studies have been published that clarified different aspects of the underlying mechanism. Hall and coworkers theoretically investigated the role of the coordinated NHC ligand showing that a proton is initially transferred from a nitrogen atom of AB to the carbene carbon of the NHC [113]. This proton then moves to the nickel center to react with a hydridic H from the borane moiety. In other studies, hydrogenation of the free carbene [114] as well as the formation of a monocarbene intermediate $Ni(NHC)(H_2N{=}BH_2)$ were found [115].

Recently, Son, Hwang, Kang, and coworkers used $[Ni(MeCN)_6][BF_4]_2$ for AB dehydrogenation at 80 °C in nitromethane [116]. Under these conditions, ca. 1.7 equiv. of H_2 were released within 10 minutes. $B(OMe)_3$ and *N*-methyl-hydroxylamine were formed as decomposition products with the methyl groups originating from the solvent.

Hydrogen release from DMAB with amino olefin nickel(I) and nickel(0) complexes at room temperature was applied by Trincado, Grützmacher, and coworkers [117]. Bis(tropylidenyl)amine (trop$_2$NH) was used to synthesize Ni(trop$_2$NH)(OTf) (**52**) and Ni(trop$_2$NH)(PPh$_3$), of which the nickel(I) derivative was proven to be more active in DMAB dehydrogenation in THF. The H_2 release time is less than one minute, even at a low catalyst loading of 0.3 wt%. Mechanistic investigations using NMR spectroscopy revealed the existence of a bridged dinickel hydride and a nickel hydride complex, which, however, could not be isolated. Moreover, the involvement of nickel(0) species in the mechanism of catalytic H_2 release could not be excluded as the nickel(I) complex can be cleanly reduced to the nickel(0) complex under noncatalytic conditions.

There are also two groups applying [CpMn(CO)$_3$] in the dehydrogenation of various amine boranes [118]. The group of Kawano could show that after irradiation, loss of a CO ligand starts the catalytic cycle by providing a coordination site for the amine borane [118a]. The cycle proceeds via NH and BH activation and final release of H_2 from a dihydride species. The group of Bengali could show the importance of the central metal as no boron complex formation has been observed when using the similar rhenium complex [118b].

Very recently, two examples for the dehydrogenation of AB and DMAB using copper complexes were also reported. In the work of Crimmin, bis(σ-B–H) copper(II) complexes (e.g. **53**) were dissolved in C_6D_6 to liberate H_2 from DMAB at room temperature within 34 hours [119]. However, a copper mirror was formed during the course of reaction showing the instability of the original molecule.

Hu et al. used a cyclic(alkyl)(amino)carbene (CAAC) to obtain a stable homogeneous copper complex (**54**) [120]. AB dehydrogenation was performed at room temperature in 20 wt% water/acetone yielding NH_4BO_2 as a product. Depending on the second ligand at the copper center, 2.6 equiv. of H_2 were released within five minutes (L = Cl^-) and 2.8 equiv. within two minutes (L = BH_4^-). Remarkably, in the presence of potassium tetrakis(pentafluorophenyl)borate (KBAr^F), the solution containing **54**-Cl could be reused 15 times without any noticeable loss of activity.

17.6 Conclusion

The dehydrogenation reaction is of high importance for the realization of hydrogen storage as well as renewable energy storage. Various substrates such as formic acid, (bio)alcohols, LOHC, AB, and amine boranes are envisioned to play a key role in future hydrogen storage. Hydrogen liberation constitutes an essential part in the storage cycle and has to meet two main requirements with respect to practical applications: Firstly, H_2 should be provided instantaneously at a constant rate. Secondly, it should be liberated under appropriate conditions. Application of viable catalysts may significantly contribute to solve these issues. So far, the state-of-the-art systems are mainly based on precious metal-based catalysts. However, regarding better availability and cost efficiency, considerable efforts to develop base metal catalysts were undertaken.

More specifically, iron-, cobalt-, nickel-, aluminum-, molybdenum-, chromium-, manganese-, and copper-based complexes proved to be promising catalysts in the dehydrogenation of formic acid and AB. In contrast, efficient hydrogen release from LOHC and (bio)alcohols is limited to mostly Fe and Co catalysts so far [121]. Among the mentioned materials, especially iron and cobalt complexes seem to possess favorable catalytic properties. This might be partly due to their broader range of investigation and application. Applying iron pincer-type catalysts, outstanding activities (TOF) of up to 196 000 h^{-1} and productivities (TON) of 983 000 in the formic acid dehydrogenation and a TON of 51 000 in the aqueous phase methanol reforming have been observed. Additionally, iron and cobalt PNP pincer complexes proved to be able to promote the hydrogen release from LOHC with good to quantitative yields and with iron pincer complexes 2.5 equiv. of hydrogen were liberated from AB under ambient conditions (60 °C).

Key to present and further success seems to be the use of chelating, partly non-innocent ligands. These paved the way for the application of first-row transition metals, which would otherwise prefer one-electron redox changes. The pincer coordination and the presence of strong-field ligands such as phosphines, hydride, and CO serve to stabilize the metal in its low-spin state and avoid the formation of unpaired electron species. In this respect, mechanistic investigations are of special importance as it has been reported that, e.g. iron complexes tend to form iron nanoparticles, which might also contribute to the overall activity in dehydrogenation reactions.

Significant progress regarding activity and stability was achieved and a few non-noble metal-based catalysts have already been shown to be a viable alternative. However, among catalysts promoting the respective dehydrogenation reactions, noble metal-containing systems still clearly dominate, and it is obvious that no demonstration device for reversible hydrogen storage applying base metal catalysts has been realized so far. Further improvements in the activity and stability for iron, cobalt, as well as the other base metal catalysts, can be expected within the next decade. This progress will be the prerequisite for the substitution of the present noble metal-based state-of-the-art catalysts.

References

1 (a) Armaroli, N. and Balzani, V. (2007). *Angew. Chem. Int. Ed.* 46: 52; (2007). *Angew. Chem.* 119: 52. (b) Schiermeier, Q., Tollefson, J., Scully, T. et al. (2008). *Nature* 454: 816.

2 Eberle, U., Felderhoff, M., and Schüth, F. (2009). *Angew. Chem.* 121: 6732; (2009). *Angew. Chem. Int. Ed.* 48: 6608.

3 For excellent reviews see e.g.: (a) Enthaler, S., von Langermann, J., and Schmidt, T. (2010). *Energy Environ. Sci.* 3: 1207. (b) Boddien, A., Gärtner, F., Federsel, C. et al. (2012). *Organic Chemistry – Breakthroughs and Perspectives* (ed. K. Ding and L.-X. Dai), 685–724. Weinheim: Wiley-VCH. (c) Fujita, E., Muckerman, J.T., and Himeda, Y. (2013). *Biochim. Biophys. Acta* 1827: 1031. (d) Wang, W.-H., Himeda, Y., Muckerman, J.T. et al. (2015). *Chem. Rev.* 115: 12936. (e) Klankermayer, K., Wesselbaum, S., Beydoun, K., and Leitner, W. (2016). *Angew. Chem. Int. Ed.* 55: 2; (2016). *Angew. Chem.* 128: 7416.

4 Tanaka, R., Yamashita, M., Chung, L.W. et al. (2011). *Organometallics* 30: 6742.
5 Filonenko, G.A., van Putten, R., Schulpen, E.N. et al. (2014). *ChemCatChem* 6: 1526.
6 Kothandaraman, J., Goeppert, A., Czaun, M. et al. (2016). *J. Am. Chem. Soc.* 138: 778.
7 Olah, G. (2013). *Angew. Chem. Int. Ed.* 52: 104; (2013). *Angew. Chem.* 125: 112.
8 Nielsen, M., Alberico, E., Baumann, W. et al. (2013). *Nature* 495: 85.
9 (a) Sponholz, P., Mellmann, D., Cordes, C. et al. (2014). *ChemSusChem* 7: 2419. (b) Nielsen, M., Junge, H., Kammer, A., and Beller, M. (2012). *Angew. Chem. Int. Ed.* 51: 5711; (2012). *Angew. Chem.* 124: 5809. (c) Li, Y., Nielsen, M., Li, B. et al. (2015). *Green Chem.* 17: 193. (d) Nielsen, M., Kammer, A., Junge, H. et al. (2011). *Angew. Chem. Int. Ed.* 50: 9593; (2011). *Angew. Chem.* 123: 9767.
10 (a) Rodriguez-Lugo, R.E., Trincado, M., Vogt, M. et al. (2013). *Nat. Chem.* (5): 342. (b) Hu, P., Diskin-Posner, Y., Ben-David, Y., and Milstein, D. (2014). *ACS Catal.* 4: 2649. (c) Fujita, K.-I., Kawahara, R., Aikawa, T., and Yamaguchi, R. (2015). *Angew. Chem. Int. Ed.* 54: 9057; (2015). *Angew. Chem.* 127, 9185. (d) Campos, J., Sharninghausen, L.S., Manas, M.G., and Crabtree, R.H. (2015). *Inorg. Chem.* 54: 5079.
11 Wang, Z., Lu, S.-M., Li, J. et al. (2015). *Chem. Eur. J.* 21: 12592.
12 Onishi, N., Ertem, M.Z., Xu, S. et al. (2016). *Catal. Sci. Technol.* 6: 988.
13 (a) Mellmann, D., Sponholz, P., Junge, H., and Beller, M. (2016). *Chem. Soc. Rev.* 45: 3954. (b) Alsabeh, P.G., Mellmann, D., Junge, H., and Beller, M. (2014). *Top. Organomet. Chem.* 48: 45. (c) Boddien, A., Gärtner, F., Nielsen, M. et al. (2013). *Comprehensive Inorganic Chemistry II*, vol. 6 (ed. J. Reedijk and K. Poeppelmeier), 587–603. Oxford: Elsevier. (d) Grasemann, M. and Laurenczy, G. (2012). *Energy Environ. Sci.* 5: 8171.
14 Fellay, C., Yan, N., Dyson, P.J., and Laurenczy, G. (2009). *Chem. Eur. J.* 15: 3752.
15 Boddien, A., Gärtner, F., Federsel, C. et al. (2011). *Angew. Chem. Int. Ed.* 50: 6411; (2011). *Angew. Chem.* 123: 6535.
16 Fukuzumi, S. and Suenobu, T. (2013). *Dalton Trans.* 42: 18.
17 (a) Newson, E., Hayeter, T.H., Hottinger, P. et al. (1998). *Int. J. Hydrogen Energy* 23: 905. (b) Scherer, G.W.H., Newson, E., and Wokaun, A. (1999). *Int. J. Hydrogen Energy* 24: 1157.
18 Weitkamp, A.W. (1968). *Adv. Catal.* 18: 1.
19 (a) Teichmann, D., Stark, K., Müller, K. et al. (2012). *Energy Environ. Sci.* 5: 9044. (b) Brückner, N., Obesser, K., Bösmann, A. et al. (2014). *ChemSusChem* 7: 229.
20 (a) Staubitz, A., Robertson, A.P.M., and Manners, I. (2010). *Chem. Rev.* 110: 4079. (b) Wolf, G., Baumann, J., Baitalov, F., and Hoffmann, F.P. (2000). *Thermochim. Acta* 343: 19. (c) Baitalov, F., Baumann, J., Wolf, G. et al. (2002). *Thermochim. Acta* 391: 159. (d) Hamilton, C.W., Baker, R.T., Staubitz, A., and Manners, I. (2009). *Chem. Soc. Rev.* 38: 279.

21 Yang, J., Sudik, A., Wolverton, C., and Siegel, D.J. (2010). *Chem. Soc. Rev.* 39: 656.

22 See e.g. for formic acid/formate: (a) Jessop, P.G., Ikariya, T., and Noyori, R. (1995). *Chem. Rev.* 95: 259. (b) Jessop, P.G., Joó, F., and Tai, C.C. (2004). *Coord. Chem. Rev.* 248: 2425. (c) Jessop, P.G. (2007). *Homogeneous Hydrogenation of Carbon Dioxide, Handbook of Homogeneous Hydrogenation* (ed. J.G. deVries and C.J. Elsevier), 489–511. Weinheim: Wiley-VCH. (d) NIST webbook. webbook.nist.gov/chemistry/ (accessed 06 June 2011).

23 Alberico, E., Lennox, A.J.J., Neumann, L.K. et al. (2016). *J. Am. Chem. Soc.* 138: 14890–14904.

24 See e.g. for LOHC: (a) Lu, R.-F., Boëthius, G., Wen, S.-H. et al. (1751). *Chem. Commun.* 2009. (b) Hodoshima, S. and Saito, Y. (2004). *J. Chem. Eng. Jpn.* 37: 391. (c) Peters, W., Eypasch, M., Frank, T. et al. (2015). *Energy Environ. Sci.* 8: 641.

25 Makowski, P., Thomas, A., Kuhn, P., and Goettmann, F. (2009). *Energy Environ. Sci.* 2: 480.

26 Crabtree, R.H. (2008). *Energy Environ. Sci.* 1: 134.

27 Giustra, Z.X., Ishibashi, J.S.A., and Liu, S.-Y. (2016). *Coord. Chem. Rev.* 314: 134.

28 (a) Yang, M., Dong, Y., Fei, S. et al. (2014). *Int. J. Hydrogen Energy* 39: 18976. (b) Kariya, N., Fukuoka, A., and Ichikawa, M. (2002). *Appl. Catal., A* 233: 91.

29 See e.g.: (a) Fujita, K.-I., Tanaka, Y., Kobayashi, M., and Yamaguchi, R. (2014). *J. Am. Chem. Soc.* 136: 4829. (b) Chwodhury, A.D., Julis, J., Grabow, K. et al. (2015). *ChemSusChem* 8: 323. (c) Li, L., Mu, X., Liu, W. et al. (2015). *J. Am. Chem. Soc.* 137: 7576. (d) Chwodhury, A.D., Weding, N., Julis, J. et al. (2014). *Angew. Chem. Int. Ed.* 126: 6595; (2014). *Angew. Chem.* 53: 6477.

30 Markiewicz, M., Zhang, Y.Q., Bösmann, A. et al. (2015). *Energy Environ. Sci.* 8: 1035.

31 (a) Marder, T.B. (2007). *Angew. Chem. Int. Ed.* 46: 8116; (2007). *Angew. Chem.* 119, 8262. (b) Hélary, J., Salandre, N., Saillard, J. et al. (2009). *Int. J. Hydrogen Energy* 34: 169. (c) Demirci, U.B. and Miele, P. (2011). *Energy Environ. Sci.* 4: 3334.

32 Müller, K., Stark, K., Müller, B., and Arlt, W. (2012). *Energy Fuels* 26: 3691.

33 (a) Dixon, D.A. and Gutowski, M. (2005). *J. Phys. Chem. A* 109: 5129. (b) Autrey, T., Bowden, M., and Karkamkar, A. (2011). *Faraday Discuss.* 151: 157.

34 (a) Hausdorf, S., Baitalov, F., Wolf, G., and Mertens, F.O.R.L. (2008). *Int. J. Hydrogen Energy* 33: 608. (b) Reller, C. and Mertens, F.O.R.L. (2012). *Angew. Chem. Int. Ed.* 51: 11731; (2012). *Angew. Chem.* 124: 11901. (c) Smythe, N.C. and Gordon, J.C. (2010). *Eur. J. Inorg. Chem.* 509. (d) Sutton, A.D., Burrell, A.K., Dixon, D.A. et al. (2011). *Science* 331: 1426. (e) Summerscales, O.T. and Gordon, J.C. (2013). *Dalton Trans.* 42: 10075. (f) Hua, T.Q. and Ahluwalia, R.K. (2012). *Int. J. Hydrogen Energy* 37: 14382. (g) Tang, Z., Chen, H., Wu, L., and Yu, X. (2012). *J. Am. Chem. Soc.* 134: 5464. (h) Leitao, E.M. and Manners, I. (2015). *Eur. J. Inorg. Chem.* 2199.

35 Selected examples: (a) Chandra, M. and Xu, Q. (2007). *J. Power Sources* 168: 135. (b) Simagina, V.I., Storozhenko, P.A., Netskina, O.V. et al. (2008). *Catal. Today* 138: 253. (c) Lai, S.-W., Park, J.-W., Yoo, S.-H. et al. (2016). *Int. J. Hydrogen Energy* 41: 3428. (d) Rakap, M. (2015). *J. Alloys Compd.* 649: 1025. (e) Xin, G., Yang, J., Li, W. et al. (2012). *Eur. J. Inorg. Chem.* 5722.

36 Selected examples: (a) Xu, Q. and Chandra, M. (2006). *J. Power Sources* 163: 364. (b) Figen, A.K. (2013). *Int. J. Hydrogen Energy* 38: 9186. (c) Yao, Q., Lu, Z.-H., Huang, W. et al. (2016). *J. Mater. Chem. A* 4: 8579. (d) Mori, K., Taga, T., and Yamashita, H. (2015). *ChemCatChem* 7: 1285. (e) Wang, Y.-W., Lu, Z.-H., and Chen, X.-S. (2015). *Mater. Technol.* 30: A89. (f) Yamada, Y., Yano, K., Xu, Q., and Fukuzumi, S. (2010). *J. Phys. Chem. C* 114: 16456. (g) Hu, J., Chen, Z., Li, M. et al. (2014). *ACS Appl. Mater. Interfaces* 6: 13191.

37 (a) Jaska, C.A., Temple, K., Lough, A.J., and Manners, I. (2001). *Chem. Commun.* 962. (b) Jaska, C.A., Temple, K., Lough, A.J., and Manners, I. (2003). *J. Am. Chem. Soc.* 125: 9424. (c) Jaska, C.A. and Manners, I. (2004). *J. Am. Chem. Soc.* 126: 9776.

38 (a) Douglas, T.M., Chaplin, A.B., and Weller, A.S. (2008). *J. Am. Chem. Soc.* 130: 14432. (b) Sewell, L.J., Huertos, M.A., Dickinson, M.E. et al. (2013). *Inorg. Chem.* 52: 4509. (c) Johnson, H.C., Leitao, E.M., Whitell, G.R. et al. (2014). *J. Am. Chem. Soc.* 136: 9078.

39 Rossin, A., Caporali, M., Gonsalvi, L. et al. (2009). *Eur. J. Inorg. Chem.* 3055.

40 Roselló-Merino, M., López-Serrano, J., and Conejero, S. (2013). *J. Am. Chem. Soc.* 135: 10910.

41 (a) Denney, M.C., Pons, V., Hebden, T.J. et al. (2006). *J. Am. Chem. Soc.* 128: 12048. (b) Hebden, T.J., Denney, M.C., Pons, V. et al. (2008). *J. Am. Chem. Soc.* 130: 10812. (c) Dietrich, B.L., Goldberg, K.I., Heinekey, D.M. et al. (2008). *Inorg. Chem.* 47: 8583. (d) Paul, A. and Musgrave, C.B. (2007). *Angew. Chem. Int. Ed.* 46: 8153; (2007) *Angew. Chem.* 119: 8301. (e) Tang, C.Y., Phillips, N., Kelly, M.J., and Aldridge, S. (2012). *Chem. Commun.* 48: 11999.

42 (a) Käß, M., Friedrich, A., Drees, M., and Schneider, S. (2009). *Angew. Chem. Int. Ed.* 48: 905; (2009). *Angew. Chem.* 121: 922. (b) Friedrich, A., Drees, M., and Schneider, S. (2009). *Chem. Eur. J.* 15: 10339. (c) Marziale, A.N., Friedrich, A., Klopsch, I. et al. (2013). *J. Am. Chem. Soc.* 135: 13342.

43 Shvo's cat: (a) Conley, B.L. and Williams, T.J. (2010). *Chem. Commun.* 46: 4815. (b) Lu, Z., Conley, B.L., and Williams, T.J. (2012). *Organometallics* 31: 6705. (c) Zhang, X., Lu, Z., Foellmer, L.K., and Williams, T.J. (2015). *Organometallics* 34: 3732.

44 Blaquiere, N., Diallo-Garcia, S., Gorelsky, S.I. et al. (2008). *J. Am. Chem. Soc.* 130: 14034.

45 (a) Clapham, S.E., Hadzovic, A., and Morris, R.-H. (2201). *Coord. Chem. Rev.* 2004: 248. (b) Samec, J.S.M., Bäckvall, J.-E., Andersson, P.G., and Brandt, P. (2006). *Chem. Soc. Rev.* 35: 237: although the two reviews pertain to the reduction of multiple bonds by means of transfer hydrogenation from suitable hydrogen donors, the principles ruling the abstraction of hydrogen from the hydrogen donor by the metal catalyst (Scheme 1, step 1) and the mechanisms thereof apply as well to the reactions dealt with in this chapter.

46 (a) Noyori, R., Yamakawa, M., and Hashiguchi, S.J. (2002). *Org. Chem.* 24: 7933. (b) For a recent review on metal-ligand cooperativity see Khusnutdinova, J.R. and Milstein, D. (2015). *Angew. Chem. Int. Ed* 54: 12236; (2015). *Angew. Chem.* 127: 12406.

47 (a) Zhao, B., Han, Z., and Ding, K. (2013). *Angew. Chem. Int. Ed.* 52: 4744; (2013). *Angew. Chem.* 125: 4844. (b) Annibale, V.T. and Song, D. (2013). *RSC Adv.* 3: 11432. (c) Luca, O.R. and Crabtree, R.H. (2013). *Chem. Soc. Rev.* 42: 1440. (d) Schneider, S., Meiners, J., and Askevold, B. (2012). *Eur. J. Inorg. Chem.* 412. (e) Lyaskovsky, V. and de Bruin, B. (2012). *ACS Catal.* 2: 270. (f) Gunanathan, C. and Milstein, D. (2011). *Acc. Chem. Res.* 44: 588. (g) van der Vlugt, J.I. and Reek, J.N.H. (2009). *Angew. Chem. Int. Ed.* 48: 8832; (2009). *Angew. Chem.* 121: 8990.

48 (a) Friedrich, A., Drees, M., auf der Günne, J.S., and Schneider, S. (2009). *J. Am. Chem. Soc.* 131: 17552. (b) Casey, C.P., Johnson, J.B., Singer, S.W., and Cui, Q. (2005). *J. Am. Chem. Soc.* 127: 3100.

49 Zell, T. and Langer, R. (2015). *Recycl. Catal.* 2: 87.

50 Hietala, J., Vuori, A., Johnsson, P. et al. (2016). *Ullmann's Encyclopedia of Industrial Chemistry*. Wiley.

51 For a recent example see Reichert, J., Brunner, B., Jess, A. et al. (2015). *Energy Environ. Sci.* 8: 2985. and references therein.

52 Yuan, Z., Eden, M.R., and Gani, R. (2016). *Ind. Eng. Chem. Res.* 55: 3383.

53 Selected examples of noble metal based catalytic systems which have been shown to promote the reversible interconversion of formic acid/H_2–CO_2: (a) Himeda, Y., Miyazawa, S., and Hirose, T. (2011). *ChemSusChem* 4: 487. (b) Leitner, W., Dinjus, E., and Gassner, F. (1994). *J. Organomet. Chem.* 475: 257.

54 Aresta, M., Dibenedetto, A., and Angelini, A. (2014). *Chem. Rev.* 114: 1709.

55 Linn, D.E. Jr.,, King, R.B., and King, A.D. Jr., (1993). *J. Mol. Catal.* 80: 151.

56 (a) Neary, M.C. and Parkin, G. (2015). *Chem. Sci.* 6: 1859. (b) Shin, J.H., Churchill, D.G., and Parkin, G. (2002). *J. Organomet. Chem.* 642: 9.

57 Tondreau, M. and Boncella, J.M. (2016). *Organometallics* 35: 2049.

58 Morris, D.J., Clarkson, G.J., and Wills, M. (2009). *Organometallics* 28: 4133.

59 (a) Boddien, A., Loges, B., Gärtner, F. et al. (2010). *J. Am. Chem. Soc.* 132: 8924. (b) Boddien, A., Gärtner, F., Jackstell, R. et al. (2010). *Angew. Chem. Int. Ed.* 49: 8993; (2010). *Angew. Chem.* 122: 9177. (c) Boddien, A., Mellmann, D., Gärtner, F. et al. (2011). *Science* 333: 1733. (d) Ott, S. (2011). *Science* 333: 1714. (e) Mellmann, M., Barsch, E., Bauer, M. et al. (2014). *Chem. Eur. J.* 20: 13589.

60 Bertini, F., Mellone, I., Ienco, A. et al. (2015). *ACS Catal.* 5: 1254.

61 Montandon-Clerc, M., Dalebrook, A.F., and Laurenczy, G. (2016). *J. Catal.* 343: 62.

62 Zell, T., Butschke, B., Ben-David, Y., and Milstein, D. (2013). *Chem. Eur. J.* 19: 8068.

63 Bielinski, E.A., Lagaditis, P.O., Zhang, Y. et al. (2014). *J. Am. Chem. Soc.* 136: 10234.

64 (a) Onishi, M. (1993). *J. Mol. Catal.* 80: 145. (b) Kudrik, E.V., Makarov, S.V., Ageeva, E.S., and Dereven'kov, I.A. (2009). *Macroheterocycles* 2: 69.

65 (a) Enthaler, S., Brück, A., Kammer, A. et al. (2015). *ChemCatChem* 7: 65. (b) Lescot, C., Savourey, S., Thuéry, P. et al. (2016). *C. R. Chimie* 19: 57. (c) Neary, M.C. and Parkin, G. (2016). *Dalton Trans.* 45: 14645.

66 Scotti, N., Psaro, R., Ravasio, N., and Zaccheria, F. (2014). *RSC Adv.* 4: 61514.

67 (a) Myers, T.W. and Berben, L.A. (2014). *Chem. Sci.* 5: 2771. (b) Lu, Q.-Q., Yu, H.-Z., and Fu, Y. (2016). *Chem. Eur. J.* 22: 4584.

68 (a) Bianchini, C., Laschi, F., Peruzzini, M. et al. (1990). *Inorg. Chem.* 29: 3394. (b) Bianchini, C., Peruzzini, M., Polo, A. et al. (1991). *Gazz. Chim. Ital.* 121: 543. (c) Bianchini, C., Peruzzini, M., and Zanobini, F. (1988). *J. Organomet. Chem.* 354: C19. (d) Bianchini, C., Dante, M., and Peruzzini, M. (1997). *Inorg. Chem.* 36: 1061.

69 (a) Yang, X. (2013). *Dalton Trans.* 42: 11987. (b) Sánchez-de-Armas, R., Xue, L., and Ahlquist, M.S.G. (2013). *Chem. Eur. J.* 19: 11869.

70 Federsel, C., Boddien, A., Jackstell, R. et al. *Angew. Chem. Int. Ed.* 2010, 49: 9777; (2010). *Angew. Chem.* 122: 9971.

71 Langer, R., Diskin-Posner, Y., Leitus, G. et al. *Angew. Chem. Int. Ed.* 2011, 50: 9948; (2011). *Angew. Chem.* 123: 10122.

72 Zhang, Y., MacIntosh, A.D., Wong, J.L. et al. (2015). *Chem. Sci.* 6: 4291.

73 Federsel, C., Ziebart, C., Jackstell, R. et al. (2012). *Chem. Eur. J.* 18: 72.

74 (a) Charman, B. (1970). *J. Chem. Soc. B* 584. (b) Dobson, A. and Robinson, S.D. (1977). *Inorg. Chem.* 16: 137.

75 Alberico, E., Sponholz, P., Cordes, C. et al. *Angew. Chem. Int. Ed.* 2013, 52: 14162; (2013). *Angew. Chem.* 125: 14412.

76 Bielinski, E.A., Förster, M., Zhang, Y. et al. (2015). *ACS Catal.* 5: 2404.

77 Yang, X. (2013). *ACS Catal.* 3: 2684.

78 (a) Chakraborty, S., Lagaditis, P.O., Förster, M. et al. (2014). *ACS Catal.* 4: 3994. (b) Bonitatibus, P.J., Chakraborty, S., Doherty, M.D. et al. (2015). *Proc. Natl. Acad. Sci. U.S.A.* 112: 1687.

79 Song, H., Kang, B., and Hong, S.H. (2014). *ACS Catal.* 4: 2889.

80 Zhang, G. and Hanson, S.K. (2013). *Org. Lett.* 15: 650.

81 (a) Gnanaprakasam, B., Zhang, J., and Milstein, D. (1468). *Angew. Chem. Int. Ed.* 2010, 49; (2010). *Angew. Chem.* 122: 1510. (b) Maggi, A. and Madsen, R. (2012). *Organometallics* 31: 451.

82 Zhang, G., Vasudevan, K.V., Scott, B.L., and Hanson, S.K. (2013). *J. Am. Chem. Soc.* 135: 8668.

83 Kamitani, M., Ito, M., Itazaki, M., and Nakazawa, H. (2014). *Chem. Commun.* 50: 7941.

84 Peña-López, M., Neumann, H., and Beller, M. (2015). *ChemCatChem* 7: 865.

85 Sharninghausen, L.S., Mercado, B.Q., Crabtree, R.H., and Hazari, N. (2015). *Chem. Commun.* 51: 16201.

86 (a) Kustov, L.M., Tarasov, A.L., and Tarasov, B.P. (2013). *Int. J. Hydrogen Energy* 38: 5713. (b) Boufaden, N., Akkari, R., Pawelec, B. et al. (2016). *J. Mol. Catal. A: Chem.* 420: 96. (c) Sung, J.S., Choo, K.Y., Kim, T.H. et al. (2008). *Int. J. Hydrogen Energy* 33: 2721. (d) Jiang, Z., Pan, Q., Xu, J., and Fang, T. (2014). *Int. J. Hydrogen Energy* 39: 17442.

87 (a) Wang, Z., Tonks, I., Belli, J., and Jensen, C.M. (2009). *J. Organomet. Chem.* 694: 2854. (b) Yamaguchi, R., Ikeda, C., Takahashi, Y., and Fujita, K.-i. (2009). *J. Am. Chem. Soc.* 131: 8410. (c) Wu, J., Talwar, D., Johnston, S. et al. *Angew. Chem. Int. Ed.* 2013, 52: 6983; (2013). *Angew. Chem.* 125: 7121.

88 Moores, A., Poyatos, M., Luo, Y., and Crabtree, R.H. (2006). *New J. Chem.* 30: 1675.

89 Chakraborty, S., Brennessel, W.W., and Jones, W.D. (2014). *J. Am. Chem. Soc.* 136: 8564.

90 Xu, R., Chakraborty, S., Yuan, H., and Jones, W.D. (2015). *ACS Catal.* 5: 6350.

91 Green, I.G., Johnson, K.M., and Roberts, B.P. (1989). *J. Chem. Soc., Perkin Trans. 2* 1963.

92 Clark, T.J., Lee, K., and Manners, I. (2006). *Chem. Eur. J.* 12: 8634.

93 Moussa, G., Moury, R., Demirci, U.B. et al. (2013). *Int. J. Energy Res.* 37: 825.

94 (a) Pun, T., Lobkovsky, E., and Chirik, P.J. (2007). *Chem. Commun.* 3297. (b) Sloan, M.E., Staubitz, A., Clark, T.J. et al. (2010). *J. Am. Chem. Soc.* 132: 3831. (c) Beweries, T., Hansen, S., Kessler, M. et al. (2011). *Dalton Trans.* 40: 7689. (d) Klahn, M., Hollmann, D., Spannenberg, A. et al. (2015). *Dalton Trans.* 44: 12103. (e) Erickson, K.A., Stelmach, J.P.W., Mucha, N.T., and Waterman, R. (2015). *Organometallics* 34: 4693.

95 Hill, M.S., Kociok-Köhn, K., and Robinson, T.P. (2010). *Chem. Commun.* 46: 7587.

96 (a) Spielmann, J., Jansen, G., Bandmann, H., and Harder, S. (2008). *Angew. Chem. Int. Ed.* 47: 6292; (2008). *Angew. Chem.* 120: 6386. (b) Spielmann, J. and Harder, S. (2009). *J. Am. Chem. Soc.* 131: 5064. (c) Liptrot, D.J., Hill, M.S., Mahon, M.F., and MacDougall, D.J. (2010). *Chem. Eur. J.* 16: 8508. (d) Harder, S., Spielmann, J., and Tobey, B. (2012). *Chem. Eur. J.* 18: 1984.

97 (a) Kawano, Y., Uruichi, M., Shimoi, M. et al. (2009). *J. Am. Chem. Soc.* 131: 14946. (b) Buss, J.A., Edouard, G.A., Cheng, C. et al. (2014). *J. Am. Chem. Soc.* 136: 11272.

98 Rossin, A. and Peruzzini, M. (2016). *Chem. Rev.* 116: 8848.

99 Diwan, M., Diakov, V., Shafirovich, E., and Varma, A. (2008). *Int. J. Hydrogen Energy* 33: 1135.

100 (a) Pons, V. and Baker, R.T. (2008). *Angew. Chem. Int. Ed.* 47: 9600; (2008). *Angew. Chem.* 120, 9742. (b) Whittell, G.R. and Manners, I. (2011). *Angew. Chem. Int. Ed.* 50: 10288; (2011). *Angew. Chem.* 123: 10470. (c) Davis, B.L., Rekken, B.D., Michalczyk, R. et al. (2013). *Chem. Commun.* 49: 9095. (d) Stubbs, N.E., Jurca, T., Leitao, E.M. et al. (2013). *Chem. Commun.* 49: 9098. (e) Robertson, A.P.M., Leitao, E.M., Jurca, T. et al. (2013). *J. Am. Chem. Soc.* 135: 12670. (f) Metters, O.J., Chapman, A.M., Robertson, A.P.M. et al. (2014). *Chem. Commun.* 50: 12146. (g) Johnson, H.C., Cooper, T.N., and Weller, A.S. (2015). *Top. Organomet. Chem.* 49: 153.

101 Bhattacharya, P., Krause, J.A., and Guan, H. (2014). *J. Am. Chem. Soc.* 136: 11153.

102 Glüer, A., Förster, M., Celinski, V.R. et al. (2015). *ACS Catal.* 5: 7214.

103 Kalviri, H.A., Gärtner, F., Ye, G. et al. (2015). *Chem. Sci.* 6: 618.

104 Lunsford, A.M., Blank, J.H., Moncho, S. et al. (2016). *Inorg. Chem.* 55: 964.

105 (a) Lichtenberg, C., Viciu, L., Adelhardt, M. et al. (2015). *Angew. Chem. Int. Ed.* 54: 5766; (2015). *Angew. Chem.* 127: 5858. (b) Lichtenberg, C., Adelhardt, M., Gianetti, T.L. et al. (2015). *ACS Catal.* 5: 6230.

106 Baker, R.T., Gordon, J.C., Hamilton, C.W. et al. (2012). *J. Am. Chem. Soc.* 134: 5598.

107 Vance, J.R., Robertson, A.P.M., Lee, K., and Manners, I. (2011). *Chem. Eur. J.* 17: 4099.

108 Vance, J.R., Schäfer, A., Robertson, A.P.M. et al. (2014). *J. Am. Chem. Soc.* 136: 3048.

109 Sonnenberg, J.F. and Morris, R.H. (2013). *ACS Catal.* 3: 1092.

110 Yan, J.-M., Zhang, X.-B., Han, S. et al. (2008). *Angew. Chem. Int. Ed.* 47: 2287; (2008). *Angew. Chem.* 120: 2319.

111 (a) Lin, T.-P. and Peters, J.C. (2013). *J. Am. Chem. Soc.* 135: 15310. (b) Ganguli, G., Malakar, T., and Paul, A. (2015). *ACS Catal.* 5: 2754.

112 Keaton, R.J., Blacquiere, J.M., and Baker, R.T. (2007). *J. Am. Chem. Soc.* 129: 1844.

113 Yang, X. and Hall, M.B. (2008). *J. Am. Chem. Soc.* 130: 1798.

114 Zimmerman, P.M., Paul, A., Zhang, Z., and Musgrave, C.B. (2009). *Angew. Chem. Int. Ed.* 48: 2201; (2009). *Angew. Chem.* 121: 2235.

115 Zimmerman, P.M., Paul, A., and Musgrave, C.B. (2009). *Inorg. Chem.* 48: 5418.

116 Kim, S.-K., Hong, S.-A., Son, H.-J. et al. (2015). *Dalton Trans.* 44: 7373.

117 Vogt, M., de Bruin, B., Berke, H. et al. (2011). *Chem. Sci.* 2: 723.

118 (a) Kakizawa, T., Kawano, Y., Naganeyama, K., and Shimoi, M. (2011). *Chem. Lett.* 40: 171. (b) Muhammad, S., Moncho, S., Brothers, E.N., and Bengali, A.A. (2014). *Chem. Commun.* 50: 5874.

119 Nako, A.E., White, A.J.P., and Crimmin, M.R. (2015). *Dalton Trans.* 44: 12530.

120 Hu, X., Soleilhavoup, M., Melaimi, M. et al. *Angew. Chem. Int. Ed.* 2015, 54: 6008; (2015). *Angew. Chem.* 127: 6106.

121 During the editorial process of this book, a review article was published updating the state of the art for hydrogen generation from FA and alcohols. Noteworthy, recently especially manganese complexes came more into the focus due to their activity in dehydrogenation reactions. See: Sordakis, K., Tang, C., Vogt, L.K. et al. (2018). *Chem. Rev.* 118: 372–433.

18

Molecular Catalysts for Proton Reduction Based on Non-noble Metals

Catherine Elleouet, François Y. Pétillon, and Philippe Schollhammer

Université de Bretagne Occidentale, UMR CNRS 6521, 6 Avenue Victor Le Gorgeu, CS 93837, 29238 Brest Cedex 3, France

18.1 Introduction

During the past two decades, molecular catalysts for proton reduction based on non-noble metals have been extensively studied in the quest of finding alternative ways of H_2 production involving new molecular electrocatalysts that are able to replace platinum. With this goal in mind, numerous synthetic and bioinspired/biomimetic complexes have been designed [1–7]. This chapter will focus on recent advances in the production of H_2 through the reduction of protons. H^+/H_2 conversion is apparently a simple reaction, but, at a metal center, it may follow different pathways involving successive proton (C) and electron transfer (E) or concerted proton electron transfer (CPET). The first step of the mechanism depends on the possibility to protonate the catalytic species. This is controlled by the basicity of the metal site, the nature of the set of ligands, the presence of a base in the coordination sphere, and the strength of the acid used. If the catalyst is not protonable, the process starts through a reduction step, and in this case, the knowledge of its reductive properties in the absence of protons is required to know the identity of the effective catalytic species [8]. We would like to draw the attention of the readers to the fact that this chapter presents characteristics, such as overpotential, turnover number/turnover frequency (TON/TOF), and rate constant, that are generally used to evaluate the efficiency of the catalytic systems, as they have been reported. The reader should keep in mind that definitions, calculation methods, catalytic potential measurements, etc., are not always the same [9–14]. Moreover, the potentials are not systematically referenced to the standard potential of the $[Fe(\eta^5\text{-}C_5H_5)_2]^+/[Fe(\eta^5\text{-}C_5H_5)_2]$ (Fc^+/Fc) couple, measured in the solvent being used. Acids with different strengths are utilized, which also renders the comparison of the results found in the literature difficult.

18.2 Iron and Nickel Catalysts

The activity of iron- and nickel-based catalysts has been extensively examined during the past decades because of the structural X-ray determinations in

Non-Noble Metal Catalysis: Molecular Approaches and Reactions,
First Edition. Edited by Robertus J. M. Klein Gebbink and Marc-Etienne Moret.

the 1990s of the active sites of [FeFe] and [NiFe] hydrogenases, which are metalloenzymes that are able to catalyze the reversible H^+/H_2 conversion with an efficiency that is comparable to that of platinum [15–17]. Homobimetallic [FeFe] and heterobimetallic [NiFe] models of the active sites of these enzymes have been studied to have a better understanding of their functioning, with the hope to reproduce the high efficiency of these metalloenzymes toward H^+/H_2 conversion [18–25]. The design and activity of mononuclear and cluster compounds have also been widely examined.

18.2.1 Bioinspired Di-iron Molecules

General features for modeling [FeFe] hydrogenase activity have been highlighted. The functioning, at a molecular level, of the H-cluster (the active site) requires the combination of three functional moieties: (i) a redox functionality (Fe_4S_4 cluster); (ii) a pendant base in the dithiolate bridge, playing the role of a base relay in the proton shuttle; and (iii) an open coordination site where the H^+/H_2 conversion occurs. The proposed catalytic cycle involves proton and electron transfer steps at a di-iron center with relevant oxidation states of I and II, depending on the catalytic active stages of the related site (H_{red}, H_{ox}, H_{hyd}, H_{redH^+}, and H_{sred}) (Scheme 18.1) [15, 18]. Formation of active terminal hydride species is a key structural feature of the process.

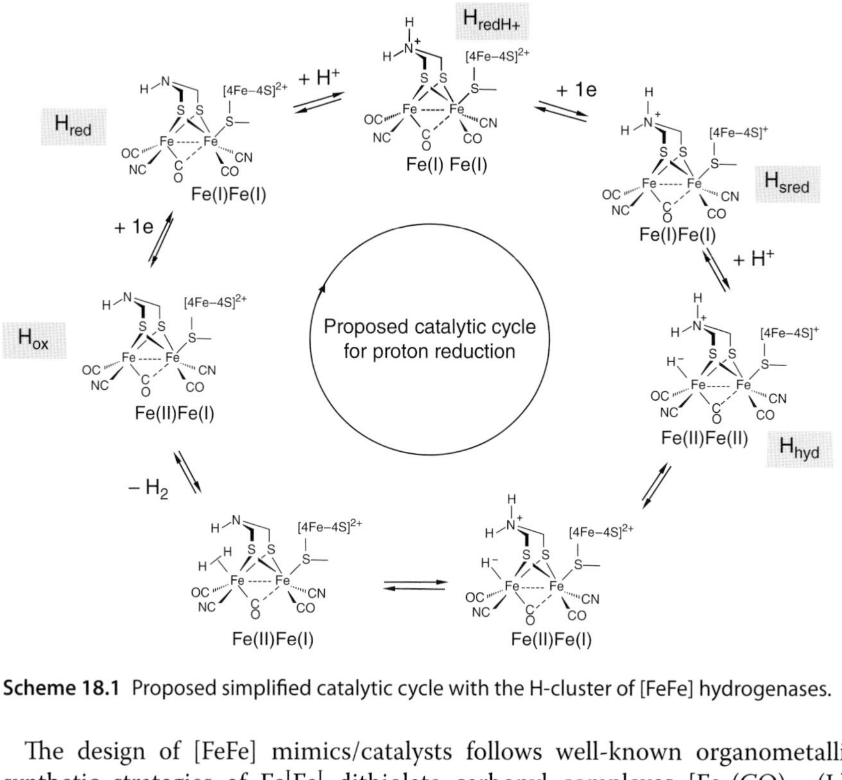

Scheme 18.1 Proposed simplified catalytic cycle with the H-cluster of [FeFe] hydrogenases.

The design of [FeFe] mimics/catalysts follows well-known organometallic synthetic strategies of Fe^IFe^I dithiolate carbonyl complexes $[Fe_2(CO)_{6-x}(L)_x$

(μ-dithiolate)] [26]. Several hundreds of complexes of this class have been synthesized over the past two decades. Most of them are able to electrocatalyze proton reduction in organic solvents, but they are generally less efficient than the enzyme (very low TONs or TOFs and large overpotentials). They induce different mechanisms depending on the strength of the acid and their ability to be protonated or not, before the reduction step. Efforts have been made for improving the catalytic efficiency of dithiolate di-iron complexes, with the following objectives: (i) to decrease the overpotential by using electron-withdrawing bridges, (ii) to introduce base functionalities in proximity to the di-iron site (on the dithiolate bridge or a terminal ligand) and to study their role as a proton relay that is crucial in the overall mechanism of proton reduction by the H-cluster, (iii) to shed light on the mechanisms of protonation of di-iron compounds and especially the role of bridging hydride species vs terminal ones and their use as electrocatalysts, (iv) to introduce redox ligands, (v) to use water as a solvent and source of protons instead of other acids, (vi) to graft di-iron systems on electrodes for designing new active materials, and (vii) to introduce a photosensitive system for developing renewable devices using solar fuels.

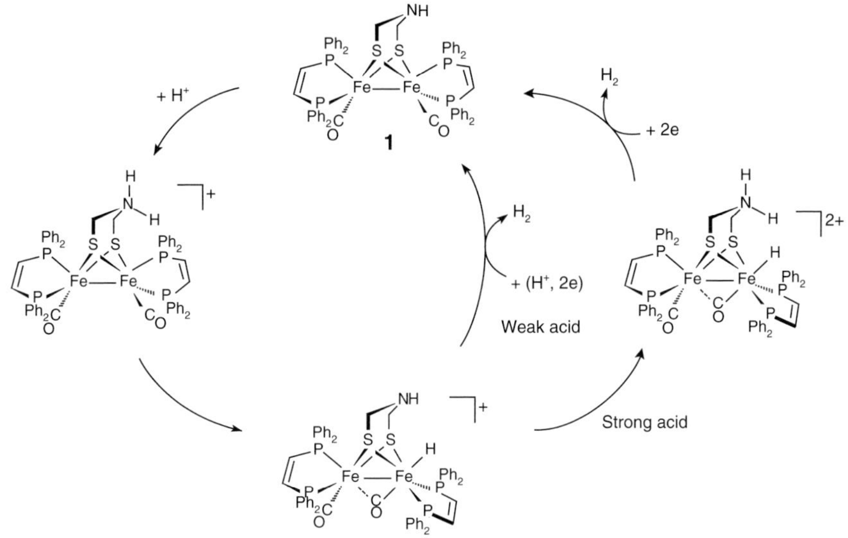

Scheme 18.2 Catalytic proton reduction mechanism with $[Fe_2(CO)_2(\kappa^2\text{-dppv})_2(\mu\text{-adt})]$ (**1**).

Obviously, this section cannot give an exhaustive survey of all the di-iron complexes with hydrogen evolution reaction (HER) activity. It will mainly focus on the more recent complexes having the highest efficiencies. In organic media, the best efficiency has been observed with the complex $[Fe_2(CO)_2(\kappa^2\text{-dppv})_2(\mu\text{-adt})]$ (dppv, 1,2-bis(diphenylphosphino)ethylene) having an azadithiolate bridge (**1**, adt, $(SCH_2)_2NH$) [27]. The electrocatalytic proton reduction proceeds via a mechanism that depends on the strength of the acid (pK_a) (Scheme 18.2). The first step is the protonation of the amine function. Tautomerization of the ammonium species affords a terminal hydride intermediate

$[Fe_2(t\text{-}H)(CO)_2(\kappa^2\text{-dppv})_2(\mu\text{-adt})]^+$, which can isomerize into its bridging form $[Fe_2(CO)_2(\kappa^2\text{-dppv})_2(\mu\text{-H})(\mu\text{-adt})]^+$; this latter transformation corresponds to a deactivation process. Indeed, in the presence of $ClCH_2CO_2H$ (pK_a^{MeCN} = 15.3) [28] in CH_2Cl_2, the complex $[Fe_2(t\text{-}H)(CO)_2(\kappa^2\text{-dppv})_2(\mu\text{-adt})]^+$ catalyzes proton reduction at −1.49 V vs (Fc^+/Fc) with a TOF of 5000 s^{-1} (overpotential = 0.71 V), whereas its bridging hydride isomer gives a HER response with a higher overpotential (0.90 V) and lower efficiency (TOF = 20 s^{-1}). When a stronger acid, such as CF_3CO_2H (pK_a^{MeCN} = 12.7) [29], is used, catalysis occurs at a less negative potential (−1.11 V; overpotential = 0.51 V), which corresponds to the reduction of the doubly protonated dication $[Fe_2(t\text{-}H)(CO)_2(\kappa^2\text{-dppv})_2(\mu\text{-adt}^H)]^{2+}$, with a remarkable TOF of 58 000 s^{-1} [27].

Recent advances in di-iron bioinspired chemistry show that the introduction of a redox ligand, such as a ferrocene, in the coordination sphere of di-iron models featuring an azadithiolate bridge, allows bidirectional H^+/H_2 catalysis. In the presence of sufficiently strong acids, such as $[H(OEt_2)_2][BAr^F_4]$ ($[BAr^F_4]^-$, $[B(3,5\text{-}C_6H_3(CF_3)_2)_4]^-$), the complex $[Fe_2(CO)_2(Fc')(\kappa^2\text{-dppv})(\mu\text{-adt}^{Bn})]$ (**2**, Fc′ = $[Fe(C_5Me_5)(C_5Me_4(CH_2PEt_2))]$; Bn = $CH_2C_6H_5$) undergoes protonation at the amine function and then protonates at one iron atom giving a terminal hydride (Scheme 18.3) [30]. An intramolecular electron transfer from the di-iron core to the ferrocene ligand may induce the release of H_2. In the absence of the ferrocene group, the terminal hydride isomerizes into a catalytically deactivated bridging hydride species. It should be noted that a comproportionation reaction, involving cationic and dicationic intermediates (Scheme 18.3), occurs during the catalytic cycle. Finally, the process is performed in CH_2Cl_2 with an overpotential of c. 0.54 V vs (Fc^{*+}/Fc^*) (Fc^* = $[Fe(C_5Me_5)_2]$) and a turnover rate of 6.6 h^{-1} when $[H(OEt_2)_2][BAr^F_4]$ is used.

Very recently, HER activity of complex **3** bearing a redox-active phosphole ligand has been investigated in CH_2Cl_2 media and in aqueous solution (Scheme 18.4) [31]. An ECCE mechanism at c. −1.40 V affords the active species, which is a mixed-valence bridging hydride form featuring a singly reduced, protonated ligand. The catalytic process arises at −2.00 V vs (Fc^+/Fc) in CH_2Cl_2 ($[Et_3NH][BF_4]$, pK_a^{MeCN} = 18.7) [28] according to an ECEC mechanism. Using H_2SO_4 (1 M), HER catalysis occurs in water at −0.66 V vs NHE (NHE, normal hydrogen electrode) with a TOF of 70 000 s^{-1}; the catalytic rate constant (k_{cat}) is 3.5×10^4 M^{-1} s^{-1}, close to that found in CH_2Cl_2 (10^5 M^{-1} s^{-1}).

Different strategies were used to obtain water-soluble di-iron complexes, which is a major goal for applicative devices. The use of ligands [32–34], known for their ability to improve the solubility of complexes in water, such as 1,3,5-triaza-7-phosphaadamantane (PTA) [33] or 3,7-diacetyl-1,7-triaza-5-phosphabicyclo[3.3.1]nonane (DAPTA) [34], has allowed the electrochemical study in water of the disubstituted complexes, $[Fe_2(CO)_4(PTA)(DAPTA)(\mu\text{-pdt})]$ and $[Fe_2(CO)_4(DAPTA)_2(\mu\text{-pdt})]$ (pdt, propanedithiolate), showing their ability to catalyze, in this medium, proton reduction in the presence of acetic acid (CH_3CO_2H, p$K_a^{H_2O}$ = 4.7) [35] at relatively low potential (c. −1.30 V vs NHE; −1.50 V vs Ag|AgCl). A water-soluble photocatalytic device, combining a

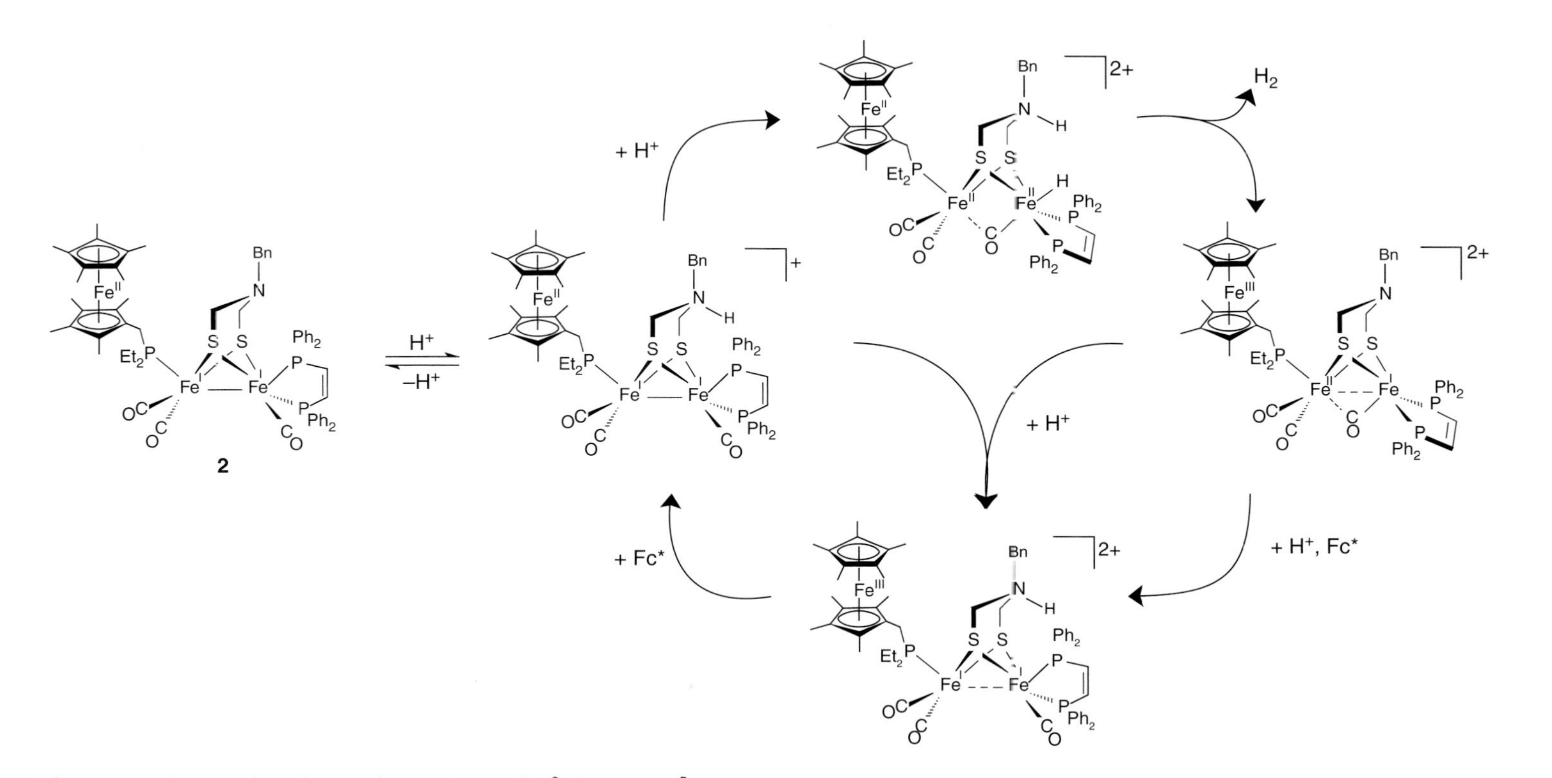

Scheme 18.3 Proposed mechanism for $[Fe_2(CO)_2(Fc')(\kappa^2\text{-dppv})(\mu\text{-adt}^{Bn})]$ (**2**) HER activity.

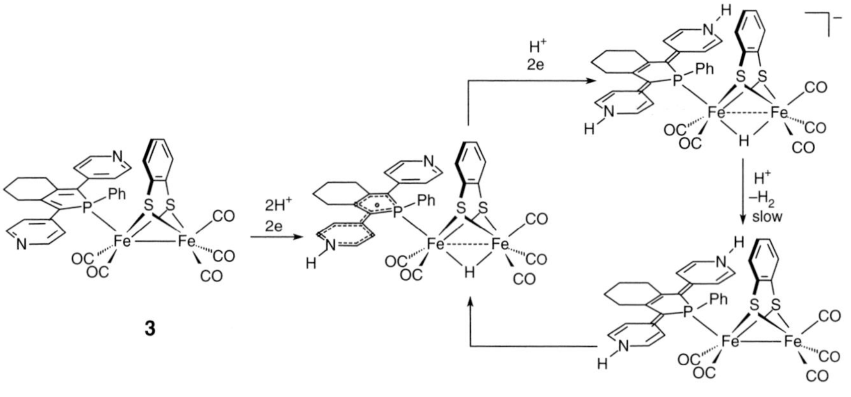

Scheme 18.4 Mechanism for HER activity of a redox-active phosphole-based complex.

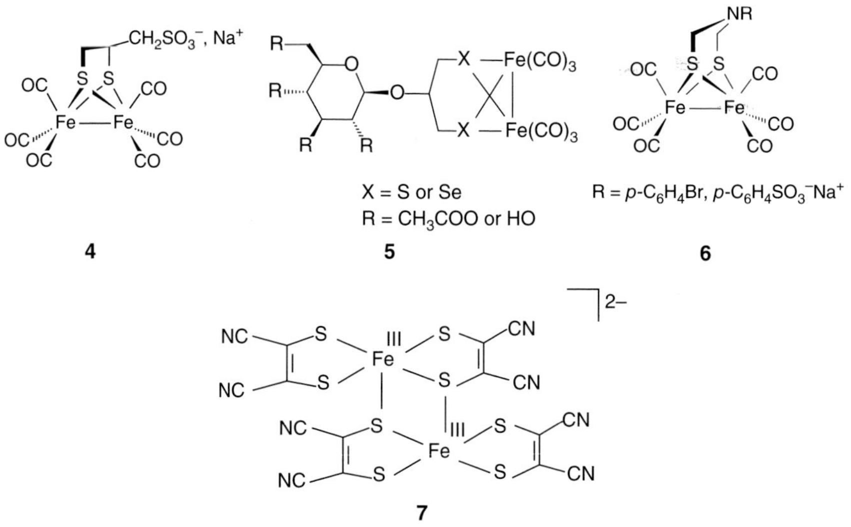

Figure 18.1 Selected examples of di-iron complexes having HER activity in water.

di-iron complex, in which a hydrophilic sulfonate group has been introduced, $Na[Fe_2(CO)_6(\mu\text{-edt}^{CH2SO3})]$ (**4**, $\text{edt}^{CH_2SO_3}$ = $–S–CH_2–CH(CH_2SO_3^-)–S–$) (Figure 18.1), $[Ru(\kappa^2\text{-bpy})_3]^{2+}$ (bpy, 2,2′-bipyridine) as a photosensitizer, and ascorbic acid produced more than 88 mol H_2 (mol cat)$^{-1}$ for four hours [36].

With this goal of designing water-soluble devices, di-iron complexes were also grafted to a protein [37] or a peptide [38, 39]. A water-soluble [FeFe] hydrogenases mimic was obtained by attaching a $\{Fe_2(CO)_6\}$ moiety to the S atom of two cysteinyl residues in the native CXXC sequence of cytochrome *c* giving a $[Fe_2(CO)_6(\mu\text{-}S\text{-Cys})_2]$ complex [37]. A TON of c. 80 over two hours at pH 4.7, in the presence of a Ru photosensitizer and ascorbate as sacrificial electron donor, was obtained.

Functionalization of the dithiolate/diselenolate bridge with a sugar residue, such as the tetra-*O*-acetyl-β-D-glucopyranoside, was realized to improve water solubility [40]. Catalytic properties for H_2 production were described for the selenolate complex (**5**) in H_2O/MeCN (5/1) mixtures (at −1.40 V vs Ag|Ag$^+$), whereas in similar conditions, the dithiolate analog decomposed. The use of β-cyclodextrins (β-CyD) to prepare aqueous solutions of di-iron mimics was also reported [41]. The sulfonated hexacarbonyl complex Na[$Fe_2(CO)_6$(μ-adt$^{p-C_6H_4SO_3}$)] (**6**) is soluble in a 10 mM aqueous NaCl solution after addition of β-CyD. In this medium, its reduction arises at c. −1.28 V vs Ag|AgCl|KCl (saturated solution), and a catalytic response, which increases upon addition of CH_3CO_2H, is observed at −1.40 V.

The catalytic activity of [$Fe_2(CO)_6$(μ-bdt)] (bdt, benzenedithiolate) for proton reduction in water was studied by preparing an aqueous micellar solution where the water-insoluble complex was dispersed by using sodium dodecyl sulfate (SDS) as the surfactant [42]. The efficiency and rate were higher than those determined in organic solvents. At pH < 6, catalytic current densities exceeding 0.5 mA cm^{-2} were recorded by DC polarography for an acid/catalyst ratio of 17. The catalytic process occurs at potential less negative than −0.70 V vs SHE (pH 3) (SHE, standard hydrogen electrode). TOF and TON were roughly estimated to be 2600 s^{-1} and 52 mol H_2 (mol cat)$^{-1}$, respectively. The disubstituted phosphite derivative [$Fe_2(CO)_4${P(OMe)$_3$}$_2$(μ-bdt)] was also described as a better electrocatalyst in SDS micellar solution than in MeCN (TOF = 4400 s^{-1} at pH 3; overpotential <0.50 V) [43]. Unlike the hexacarbonyl compound, catalysis is initiated by the protonation of the disubstituted complex. Visible light-driven H_2 evolution involving complexes [$Fe_2(CO)_{6-x}${P(OMe)$_3$}$_x$(μ-bdt)] ($x \leq 2$) was examined in the presence of eosin as a sensitizer and triethylamine as an electron donor. [$Fe_2(CO)_6$(μ-bdt)] appeared as a robust photocatalyst in aqueous SDS solution at pH 10. Upon irradiation at 455 nm, a TON of 117 mol H_2 (mol cat)$^{-1}$ was obtained in 4.5 hours for a complex/eosin ratio of 0.5. Mono- and disubstituted P(OMe)$_3$ derivatives are less efficient (TON of 2–4). The mechanism of the catalytic process at pH 10 is not yet definite.

Recently, a stable photoelectrochemical device for H_2 production was designed by adsorbing an indium phosphide (InP) nanocrystal on a gold electrode, in which the complex [$Fe_2(CO)_6$(μ-S_2)] was incorporated [44]. Such a modified electrode was used in aqueous Na[BF_4] solution at pH 7. Upon illumination, a photocurrent was observed at −0.40 V vs Ag|AgCl| 3 M Cl$^-$. However, the system suffered from low faradaic efficiency (60%) and low current density (375 mA cm^{-2}). More recently, the complex [$Fe_2(CO)_6$(μ-adt$^{p-C_6H_4Br}$)] (**6**) was immobilized on an edge plane graphite electrode [45]. A HER response was detected at an onset potential of −0.36 V vs NHE. A TOF of 6400 s^{-1} was determined at a potential of −0.50 V in a 0.25 M H_2SO_4 solution (TON ≫ 10^8 over c. eight hours, faradaic efficiency = 95%), which ranks this complex among the most efficient catalysts reported until now. It is worth noting that the non-bioinspired, air-stable catalyst [Fe(mnt)$_2$]$_2$$^{2-}$ (**7**; mnt, maleonitrile dithiolate) exhibits activity toward electrochemical H_2 evolution in aqueous acetate buffer solution at pH 5, with an overpotential of 0.56 V [46].

18.2.2 Mono- and Poly-iron Complexes

In comparison with the high number of di-iron systems, few mononuclear iron complexes have been described as HER catalysts. The first example was an iron–tetraphenylporphyrin complex which, in the Fe^0 oxidation state, efficiently reduces protons despite a substantial overpotential [47]. The introduction of a proton donor group/base relay in the second coordination sphere of the iron–porphyrin enhances the catalytic activity. Iron complexes with sterically demanding hanging dibenzofuran porphyrins (hangman porphyrin) (**8**, Figure 18.2) outperform non-hangman porphyrins [48]. The electrocatalytic process is observed at the potential of the Fe^{II}/Fe^{I} reduction in the presence of triphenylphosphine, the hanging group being a proton relay. Their HER activities were quantified in CH_3CN–$[Et_4N][TsO]$ (TsO^- = tosylate = *p*-toluenesulfonate). Very recently, an efficient iron–porphyrin complex featuring a triazole substituent as a base relay was reported [49].

Iron polypyridyl monophenolate complexes (**9**) are stable catalysts for proton reduction in organic solvents as well as in the presence of water. In organic solvents, the sulfinato Fe^{III} species (**10**) reduces protons from CF_3CO_2H at −1.57 V vs (Fc^+/Fc) with an overpotential of 0.80 V (k_{cat} = 3300 s^{-1}) [50]. Proton reduction occurs according to an ECEC or ECCE mechanism, whereas the first step is a protonation (CECE or CEEC mechanism) in the case of the iron polypyridyl monophenolate complexes (**8**). The catalytic activity of such a complex **9** with R = H is enhanced in the presence of water, achieving a TOF of 3000 s^{-1} with an overpotential of 0.80 V [51]. The functionalization of the phenolate ligand by an electron-withdrawing nitro group not only lowers the overpotential to a

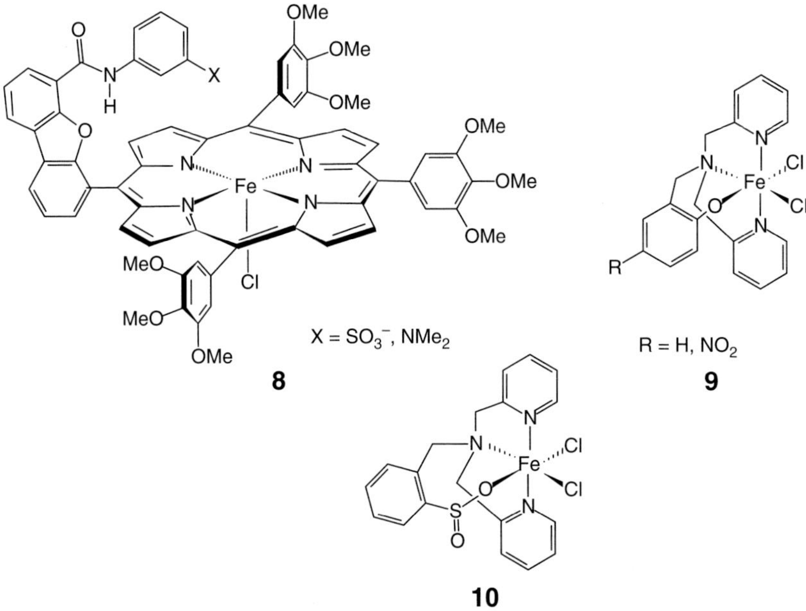

Figure 18.2 Monometallic hangman porphyrin and polypyridyl Fe^{III} complexes.

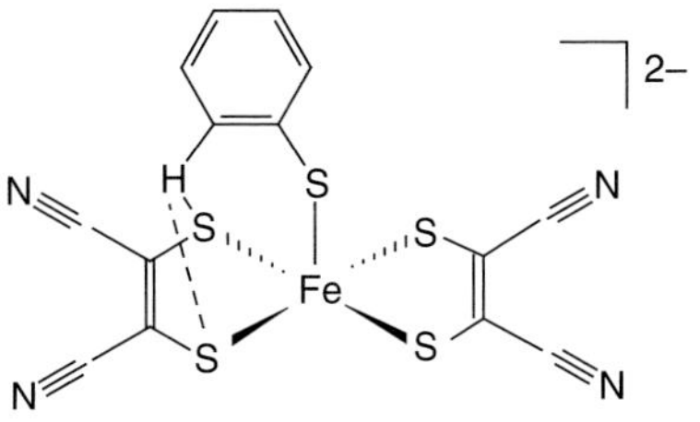

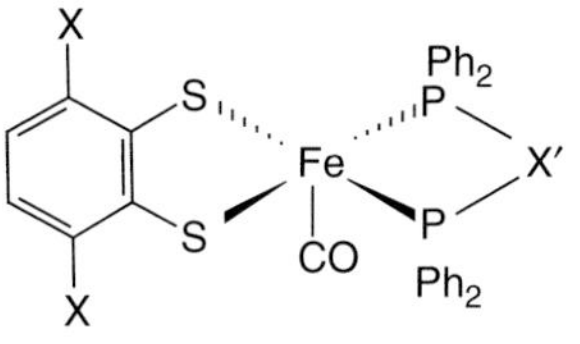

Figure 18.3 Examples of monoiron complexes with a dithiolate ligand.

value of 0.30 V but also its catalytic activity [52]. The photocatalytic activity of these compounds was studied in the presence of fluorescein (chromophore) and triethylamine (sacrificial electron source) in a 1 : 1 ethanol:water mixture showing TONs higher than 2100 after 24 hours [53]. Other Fe^{III}–Cl complexes, bearing pentadentate pyridylamino-bis(phenolate) ligands, have an electrocatalytic activity both in organic and phosphate-buffered (pH 7) solution [54, 55]. TOF values were determined in the range of 50–98 mol H_2 (mol cat)$^{-1}$ h^{-1} (overpotential = 0.94 V) in DMF (CH_3CO_2H) and 554–676 mol H_2 (mol cat)$^{-1}$ h^{-1} (overpotential = 0.84 V) in aqueous buffered solution. Other mononuclear iron(II) complexes containing bidentate-functionalized pyridine, considered as models of the active site of [Fe]-hydrogenases, were reported as poor HER catalysts and will not be described here [56].

S-coordinating ligands, such as maleonitrile dithiolate or thiophenolate, were also used for the design of HER catalysts (Figure 18.3). The complex $[Fe(mnt)_2(SPh)][PPh_4]_2$ (**11**) promotes the H_2 production at low potentials (E_p = −0.31 V vs Ag|AgCl in CH_3CN and E_p = −0.53 V in water) [57]. Penta-coordinated iron complexes with a $\{Fe^{II}(CO)P_2S_2\}$ core (**12**), featuring a functionalized benzene dithiolate and a chelating diphosphine, such as amine-functionalized diphosphines ($P^R{}_2N^{R'} = N^{R'}\{CH_2P^R{}_2\}_2$, $P^R{}_2N^{R'}{}_2 = (CH_2P^RCH_2N^{R'})_2$) or dppf (dppf, 1,1′-bis(diphenylphosphino) ferrocene), have been considered as functional biomimetic models of the distal iron in the active site of [FeFe]-hydrogenases [58–61]. Catalytic H_2 formation with a weak acid (CH_3CO_2H) ($pK_a{}^{MeCN}$ = 23.5) [62] at low overpotential was observed for $[Fe(CO)(\kappa^2\text{-bdt})(\kappa^2\text{-}\{(Ph_2PCH_2)_2N(X)\}]$ (X = 1,1-diethoxy-ethyl) in CH_3CN-$[Bu_4N][PF_6]$ 0.1 M (TOF > 500 s^{-1}, $[CH_3CO_2H] \approx$ 0.1–0.5 M, overpotential ≈ -1.65 V). The activity depends on geometrical constraints of the diphosphine ligands. In the case of dppf, which has a large bite angle (99°) [63], a TOF of 10 s^{-1} has been determined in 1.8 M CH_3CO_2H (0.1 M $[NBu_4][PF_6]$ in THF) with an overpotential of 0.17 V. When monodentate phosphines (PPh_3, PMe_3) are used instead of chelating diphosphine, octahedral Fe^{II} species are formed [64]. The complex $[Fe(CO)_2(PMe_3)_2(\kappa^2\text{-bdt})]$ catalyzes the proton reduction at −1.20 V according to a CECEC mechanism. It is proposed that the reduction of the protonated species results in the cleavage of one Fe—S bond giving a pentacoordinate iron complex.

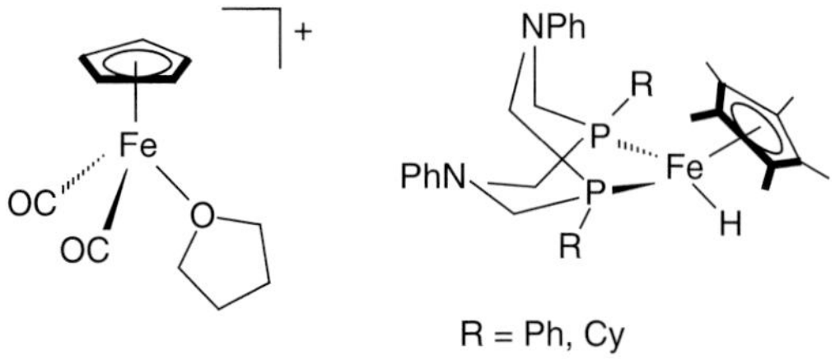

R = Ph, Cy

13 14

Figure 18.4 Selected examples of cyclopentadienyl monoiron complexes.

Cyclopentadienyl compounds have been considered as HER electrocatalysts (Figure 18.4). The complex $[FeCp(CO)_2(THF)][BF_4]$ (**13**) exhibits a catalytic wave assigned to H_2 production in the presence of CF_3CO_2H at −0.80 V vs Ag|AgCl, which is the potential corresponding to the reduced form $[FeCp(CO)_2]$ [65]. This catalyst becomes rapidly inactive because of dimerization into $[FeCp(CO)]_2$. Following a recent result on the activation of H_2 by the cyclopentadienyl complex $[FeHCp'(P^{tBu}{}_2N^{Bn}{}_2)]$ ($Cp' = C_5H_4C_6F_5$) [66], a series of Cp-iron compounds featuring pendant bases, $[FeXCp^*(P^R{}_2N^{Ph}{}_2)]$ ($Cp^* = C_5Me_5$; R = Ph, Cy; X = Cl, H), has been studied [67]. The hydride complexes **14** display catalytic currents at −1.86 V vs (Fc^+/Fc) (R = Ph) and −1.90 V (R = Cy) upon addition of $(CF_3SO_2)_2NH$ ($pK_a^{MeCN} = 1.0$) [68], but decomposition is observed in these strong acidic conditions, whereas no catalysis was detected with a weaker acid-like CF_3CO_2H because of the coordinating ability of the counter-anion $CF_3CO_2^-$ after protonation.

Fluorinated diglyoxime complexes $[Fe(dAr^Fg_2HBF_2)(Py)_2]$ (**15**, Figure 18.5) and $[Fe(dAr^FgBF_2)_2(Py)_2]$ (**16**) (Py, pyridine, dAr^FgH_{2-x} = 1,2-bis perfluorophenyl ethane-1,2-dionedioxime) catalyze H_2 production from CF_3CO_2H in CH_2Cl_2 or CH_3CN-$[Bu_4N][ClO_4]$ solution at −0.80 V and −0.90 V vs SCE, respectively. Although a TOF of 20 s^{-1} is obtained for the difluoroborated complex (**16**), a TOF of 200 s^{-1} is calculated for the monofluoroborated derivative (**15**) [69, 70]. The activity of the iron(II) chlatrochelate derivatives $[Fe(Nx_3BThioph)_2]$ (**17**, Nx = nioximate = cyclohexanedion-1,2-dioxime; BThioph = 2-thiophene boron) has also been studied in acetonitrile and in a mixed water–acetonitrile 1 : 1 solution [71]. When the acid is $[Et_3NH]Cl$ ($pK_a^{H_2O} = 10.7$) [35], the HER activity at the (Fe^{II}/Fe^{I}) reduction potential is higher than that of its cobalt analog, but in the presence of $HClO_4$, $[Fe(Nx_3BThioph)_2]$ is less efficient than the chlatrochelate cobalt complex. Finally, the iron–pincer complex $[Fe(H)_2(CO)(tBu\text{-}PNP)]$ (**18**, *t*Bu-PNP = 2,6-bis(di-*tert*-butylphosphinomethyl)pyridine) in the presence of trialkylamine has a TOF up to 836 h^{-1} and a TON up to 100 000 (after 10 days) at 40 °C in the presence of formic acid [72] ($pK_a^{H_2O} = 3.8$ [73]), whereas the complex $[FeH(PP_3)][BF_4]$ ($PP_3 = P(CH_2CH_2PPh_2)_3$) works without the need of an additional base [74].

Figures 18.6–18.8 depict selected polynuclear iron compounds with more than two iron centers used for HER catalysis. The linear tetrairon cluster $[Fe_4(CO)_8\{\mu_3\text{-}(SCH_2)_3CMe\}_2]$ (**19**), featuring a mixed valence $\{Fe^IFe^{II}Fe^{II}Fe^I\}$ core, promotes two electrocatalytic events at two different mild potentials [75, 76]. The former, which is associated with an ECEC process and arises at the first reduction potential of the complex (−1.22 V vs (Fc^+/Fc)), is slow. When 2,6-dimethylpyridinium tetrafluoroborate ($pK_a^{MeCN} = 14.0$) [29] is

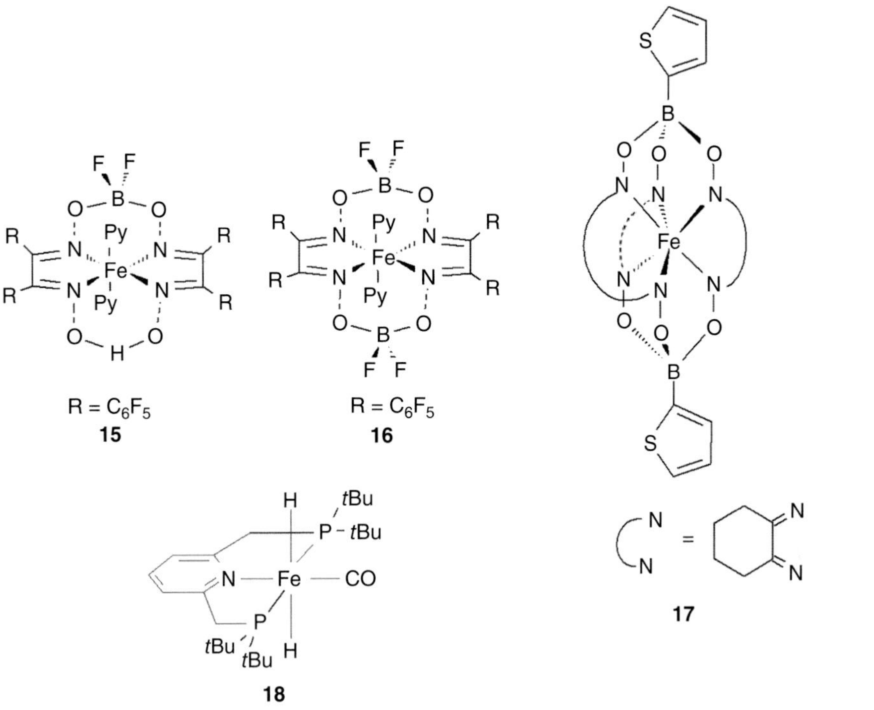

Figure 18.5 Diglyoxime and chlatrochelate mononuclear Fe^{II} systems.

used as acid, the rate of proton reduction increases at the second reduction potential (at −1.58 V vs (Fc^+/Fc)). DFT calculations suggested that the doubly reduced species is protonated at the terminal Fe position, whereas at the first potential, the ECE step may produce a μ-hydride species that would slow down the HER catalysis [77]. Replacement of the CMe– for a SiMe– group at the bridge head increases the basicity of the bridging sulfur atoms, making them easier to protonate during the catalytic process [78]. A series of original quasilinear tri-iron complexes $[Fe_3(CO)_{7-x}(PPh_3)_x(\mu\text{-edt})_2]$ (x = 0–2) (edt, ethanedithiolate) (**20**) and $[Fe_3(CO)_5(\kappa^2\text{-dppe})(\mu\text{-dithiolate})_2]$ (**21**, dppe, 1,2-bis(diphenylphosphino)ethane; dithiolate = propanedithiolate, azadithiolate) were also described, but they are not stable in acid media, in contrast to the complex **22** that is stable over a period of 24 hours, and they exhibit low HER catalytic activities [79–82].

The activity of tri-iron clusters having a nonlinear structure (Figure 18.7), such as $[Fe_3(CO)_{12}]$ (**23**), $[Fe_3(CO)_9(\mu_3\text{-S})_2]$ (**24**), $[Fe_3(CO)_7(\mu\text{-dppv})(\mu_3\text{-S})_2]$ (**25**, L = dppf and dppm; dppm: 1,1-bis(diphenylphosphino)methane)), and $[Fe_3(CO)_5(\kappa^2\text{-L})_2(\mu_3\text{-S})_2]$ (**26**, L = dppv), has also been studied [83–86]. A TON of 296 was determined after a bulk 10 hours electrolysis of $[Fe_3(CO)_9(\mu_3\text{-S})_2]$, in CH_2Cl_2-$[Bu_4N][PF_6]$, at −1.10 V vs (Fc^+/Fc) in the presence of $H(BF_4)\cdot OEt_2$ (pK_a^{MeCN} = 1.8) [68]. The mechanism of the process remains unclear. Substitution of carbonyl ligands by one or two diphosphines bridged between two iron atoms or chelated at one iron atom allowed the proton reduction at potentials more positive than that of $[Fe_3(CO)_9(\mu_3\text{-S})_2]$ [84, 85]. The positive shift

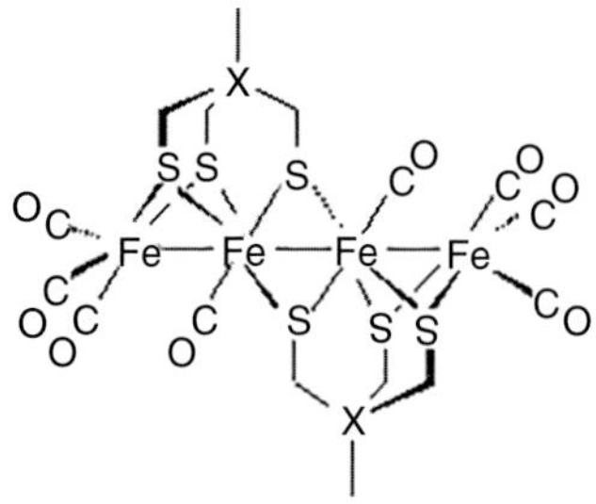

X = C, Si
19

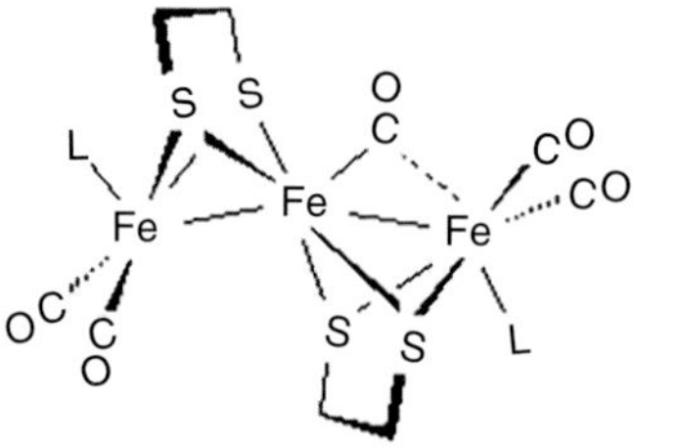

L = CO, PPh_3
20

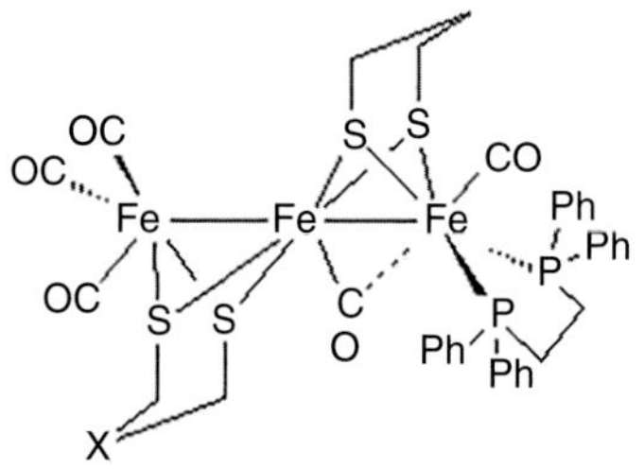

X = CH_2, NBn
21

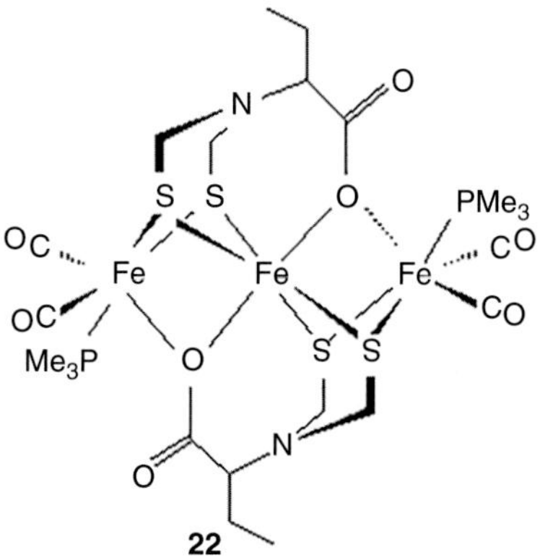

22

Figure 18.6 Quasi linear tetra- and tri-iron complexes.

23

24

P⁀P = dppm, dppf

25

P⁀P = dppv

26

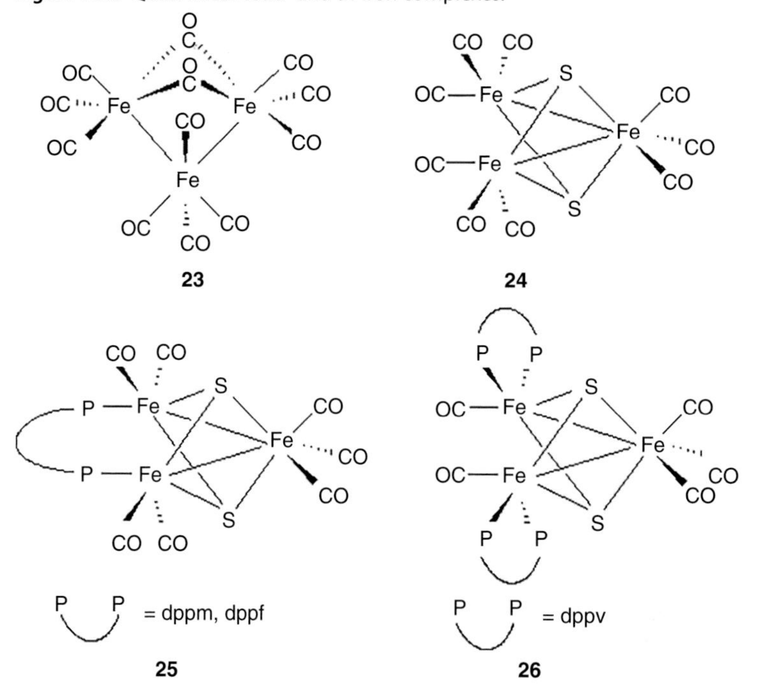

Figure 18.7 Bent tri-iron complexes.

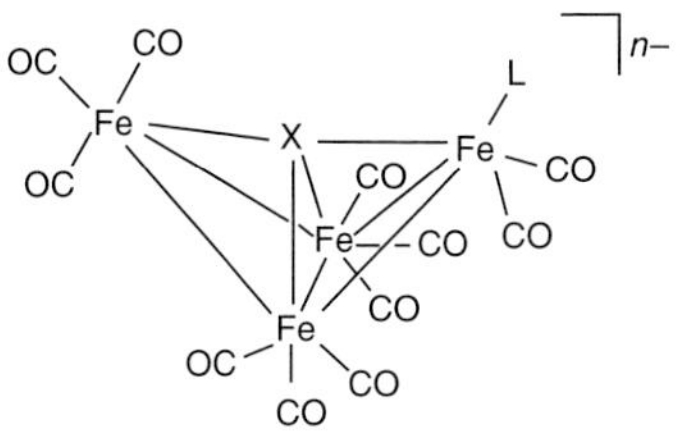

X = N, n = 1
X = C, n = 2
L = CO, PPh_3, $Ph_2P(CH_2)OH$

27

Figure 18.8 Nitride- and carbide-centered tetrairon clusters.

(1.01 V) in fact results from the protonation of the diphosphine-substituted clusters by acid (CF_3CO_2H) giving μ-hydride complexes, the catalytic activity being observed at the potential of reduction of the protonated species. $[Fe_3(CO)_{12}]$ has also been used as a water reduction catalyst in typical photocatalytic systems in which the active species would be $[HFe_3(CO)_{11}]^-$ [87, 88]. TONs up to 1330 within a one hour period were obtained.

Low-valent multinuclear compounds with more than three iron atoms (Figure 18.8), such as $[Fe_4N(CO)_{12}]^-$ and $[Fe_4C(CO)_{12}]^{2-}$ (**27**), also promote electrocatalytic hydrogen evolution from water [89–91]. The nitride-centered cluster $[Fe_4N(CO)_{12}]^-$ is reduced before successive protonation steps occur, affording H_2, whereas the dianionic carbide-centered complex $[Fe_4C(CO)_{12}]^{2-}$ undergoes two successive proton-coupled electron transfers. When a carbonyl is replaced by PPh_3 or $Ph_2P(CH_2)_2OH$ in the nitride-centered cluster $[Fe_4N(CO)_{12}]^-$, the hydroxyl group enables $[Fe_4N(CO)_{11}(Ph_2P(CH_2)_2OH]^-$ to catalyze proton reduction two times faster than $[Fe_4N(CO)_{11}(PPh_3)]^-$ because of the apparently high pK_a of $Ph_2P(CH_2)_2OH$.

18.2.3 Bioinspired [NiFe] Complexes and [NiMn] Analogs

Heteropolymetallic catalysts for proton reduction, and especially heterobimetallic [NiFe] systems, were investigated because of the structure of the active site of [NiFe] hydrogenases that features a {NiFe} core [15]. The functioning, at the molecular level, of such an active site is proposed to involve the formation of a bridging hydride species and the change in the redox state of the nickel atom from I to III, the iron atom retaining a redox state II (Scheme 18.5).

The electrochemical study of $[NiFeCp(CO)(xbsms)][BF_4]$ (H_2xbsms, 1,2-bis(4-mercapto-3,3-dimethyl-2-thiabutyl)benzene) (**28**, Figure 18.9) [92] in DMF, in the presence of CF_3CO_2H (pK_a^{DMF} = 4.1) [92], indicated that this complex catalyzes proton reduction with an overpotential of 0.73 V. A catalytic rate of 5 h^{-1} and a faradaic yield of 72% were determined after 4 h electrolysis at −1.83 V vs (Fc^+/Fc). In the presence of Et_3NH^+ (pK_a^{DMF} = 9.3) [35], no catalytic current was observed. The active species in the catalytic cycle was proposed to feature a bridging hydride. The effect of the replacement of the terminal ligand L (L = Cl^-, CO, MeCN) at the {FeCp*L} moiety was explored in analogous

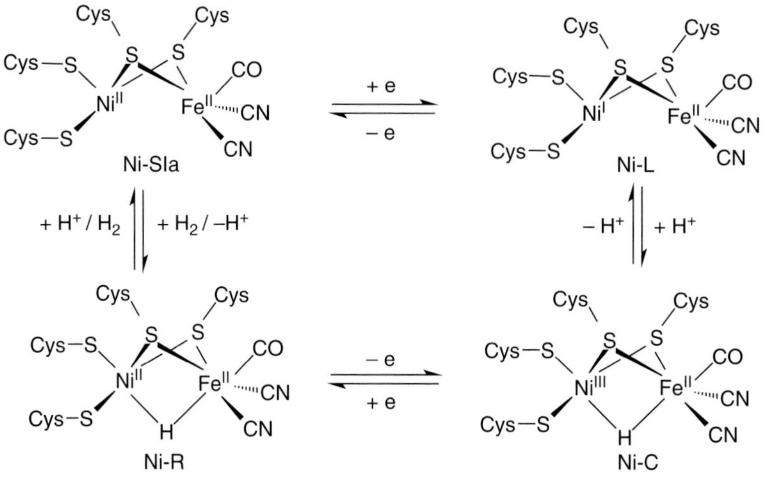

Scheme 18.5 Proposed H^+/H_2 conversion at the active site of [NiFe] hydrogenases (without global charge).

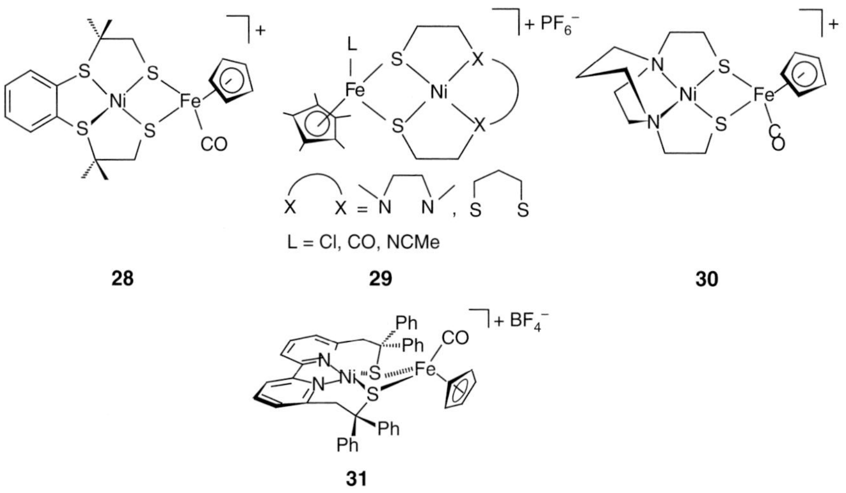

Figure 18.9 Bioinspired [NiFe] complexes with a {Cp′FeCO} moiety.

compounds **29** [93]. No catalytic activity was observed for MeCN compounds, which were converted, in the presence of protons, into trinuclear complexes. Cl-compounds were also unstable in acidic media. Only the CO-adduct complex $[FeNiS_4Cp^*CO]^+$ exhibits catalytic activity for proton reduction as indicated by electrochemical measurements. No activity was detected with a weak acid such as CH_3CO_2H, but with the strong acid $H(BF_4){\cdot}OEt_2$ in MeCN, a catalytic process was observed at a potential close to that of the primary reduction of this complex (c. −1.38 V vs (Fc^+/Fc)). Recent studies on complex **30**, featuring a heterocyclic diamine backbone, have shown the role of hemilabile bridging thiolates as efficient proton shuttles for the design of electrocatalysts for proton reduction [94]. The use of a tetradentate bipyridine bis-thiolate ligand (N_2S_2)

Figure 18.10 Bioinspired [NiFe] complexes with a $\{Fe(CO)_3\}$ moiety.

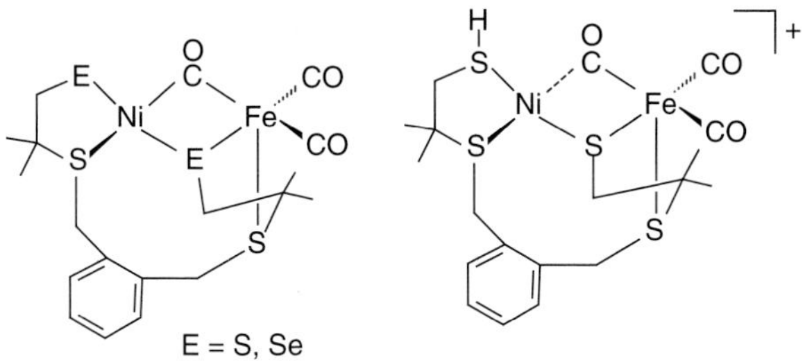

instead of the xbsms chelate afforded an analogous $[Ni^{II}Fe^{II}]$ compound (**31**) with electrocatalytic activity for proton reduction [95]. Upon addition of $[Et_3NH][BF_4]$ to this complex in MeCN solution, a catalytic wave was detected at −1.90 V vs (Fc^+/Fc) (measured at the half-wave), which corresponds to the second reduction process of this complex. An EECC mechanism, proposed on the basis of electrochemical and DFT studies, may involve the formation of a terminal hydride species at the nickel center, but no Ni^{III} species unlike the mechanism proposed for the [NiFe] hydrogenase active site. A TOF of 250 s^{-1} was estimated.

The precursor [Ni(xbsms)] has also been combined with a $\{Fe(CO)_3\}$ moiety, giving a new class of [NiFe] system (**32**, Figure 18.10) [96, 97]. A selenolate analog **33** was described as well [97]. Interestingly, S-protonated complex **32** could be prepared quantitatively by treatment of the complex [NiFe-xbsms] with 1 equiv. of $H(BF_4)\cdot OEt_2$ in CH_2Cl_2. CV (cyclic voltammetry) revealed the increase of the cathodic current upon addition of increasing amounts of CF_3CO_2H (up to 100 equiv.) in MeCN and CH_2Cl_2 solutions at a potential close to that of the reduction of the neutral complex (c. −1.75 V vs (Fc^+/Fc)). An overpotential of 0.54 V was estimated at the half-wave of the cathodic peak after addition of 100 equiv. of acid. S-protonated complex **32** also acts as a HER catalyst but could not be considered as an intermediate in the process involving the neutral precursor and CF_3CO_2H. Bulk electrolysis in MeCN, for 4 hours, showed the relative stability of the neutral complex, but the efficiency was modest (TOF ≈ 5–8 h^{-1}). Further results revealed that these compounds do not behave as molecular catalysts for proton reduction in organic solvents but are precursors of heterogeneous catalysts, electrodeposited at the electrode surface [97].

The complex $[NiFe(CO)_3(\kappa^2\text{-dppe})(\mu\text{-pdt})(\mu\text{-H})][BF_4]$ (**34**) is also an active HER catalyst (Figure 18.11) [98]. A catalytic current was observed at −1.37 V vs (Fc^+/Fc) after successive additions of CF_3CO_2H in CH_2Cl_2. The study was extended to a series of compounds of general formula $[NiFe(CO)_2L(\kappa^2\text{-dppe})(\mu\text{-pdt})(\mu\text{-H})][BF_4]$, obtained by replacement of one CO by L = $P(OPh)_3$, PPh_3, or PPh_2Py [99]. Catalytic events were detected in the presence of CF_3CO_2H in MeCN at a potential (potential at the half-height of the cathodic peak) of c. −1.30 V vs (Fc^+/Fc) (peak ≈ −1.20 V when L = CO). A TOF of 50 s^{-1} (20 s^{-1}, when L = CO) and an overpotential of c. 0.40 V (0.31 V, when L = CO) were estimated. Replacing dppe with dcpe

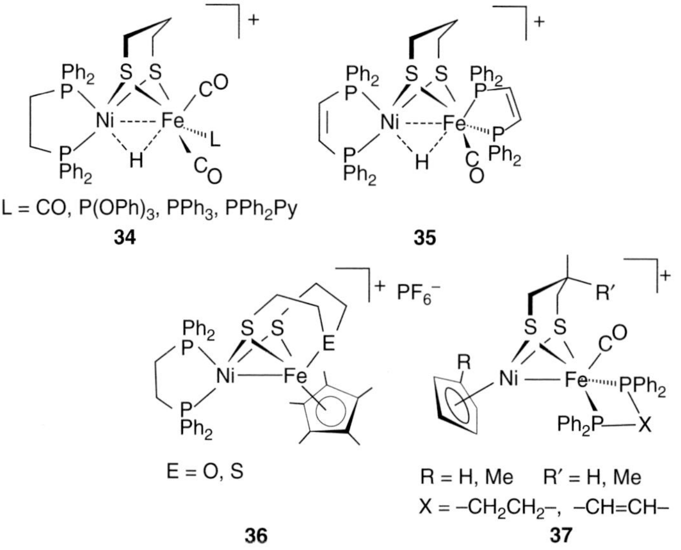

L = CO, $P(OPh)_3$, PPh_3, PPh_2Py

34 **35**

E = O, S

R = H, Me R′ = H, Me
X = $-CH_2CH_2-$, $-CH{=}CH-$

36 **37**

Figure 18.11 Selected examples of heterobimetallic [NiFe] complexes with a dithiolate bridge.

(1,2-bis(dicyclohexylphosphino)ethane) and pdt with edt on this [NiFe] platform resulted in catalysts that operate at overpotentials of c. 0.50 V with rate constants of c. 300 s^{-1} [100]. The edt complexes present faster turnover rates but with similar overpotentials with respect to pdt systems. As expected, the substitution of dppe for more basic dcpe increases the overpotential for proton reduction. Bis-chelated bridging hydride complex $[NiFe(CO)(\kappa^2\text{-dppv})_2(\mu\text{-pdt})(\mu\text{-H})][BF_4]$ (**35**) was also found to be a catalyst for dihydrogen evolution in MeCN in the presence of *p*-toluenesulfonic acid (HOTs) ($pK_a^{MeCN} = 8.7$) [28], CF_3CO_2H, or $ClCH_2CO_2H$ at potentials near the first reduction potential of this complex (c. −1.40 V vs (Fc^+/Fc)) [101]. A TOF of 480 s^{-1} was determined with HOTs. Compounds $[NiFeCp(\kappa^2\text{-dppe})\{\mu\text{-}\kappa^2\text{:}\kappa^3\text{-}S_2(CH_2CH_2)_2E\}][PF_6]$ (E = O, S) (**36**) are unstable in the presence of the strong acid $H(BF_4)\cdot OEt_2$, but electrochemical measurements showed their catalytic activity when the weaker acid $[Et_3NH][BF_4]$ was used in CH_2Cl_2 [102]. The potential at which catalysis arises is close to that of the second reduction process of these complexes, c. −1.50 V vs (Fc^+/Fc). The kinetics of the process seems to slow down when E = S. The combination of a {NiCp′} moiety with the iron precursor $[Fe(CO)_2(\kappa^2\text{-diphosphine})(\kappa^2\text{-dithiolate})]$ allows to obtain the new type of [NiFe] complexes $[NiFeCp'(CO)(\kappa^2\text{-diphosphine})(\mu\text{-dithiolate})]^+$ ($Cp' = C_5H_4R$, R = H, Me; diphosphine = dppe, dppv; dithiolate = $\{SCH_2\}_2CR'_2$, R′ = H, Me) (**37**) [103]. The treatment of the one-electron reduced form of complex **37**, where R = R′ = H and diphosphine, dppe, with various acids (HOTs; $H(BF_4)\cdot OEt_2$; [pyridinium]$[BF_4]$, $pK_a^{MeCN} = 12.3$; $[NH_4][PF_6]$, $pK_a^{MeCN} = 16.5$) [104] released H_2 with the formation of the starting cationic compound. This complex is a slow electrocatalyst at its first reduction potential (−1.16 V vs (Fc^+/Fc)), with an acid-independent rate of c. 4 s^{-1}. No hydride intermediate

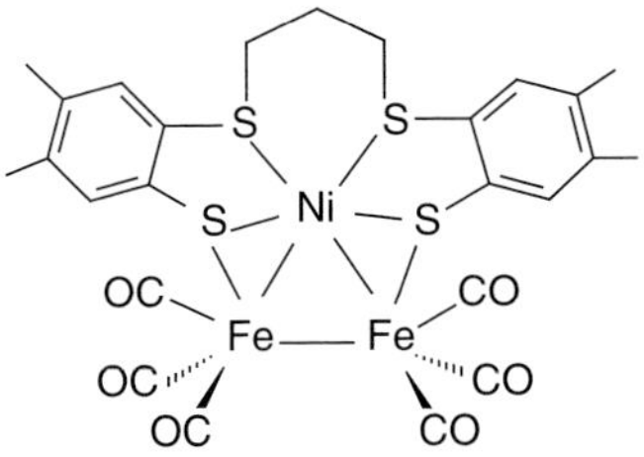

38

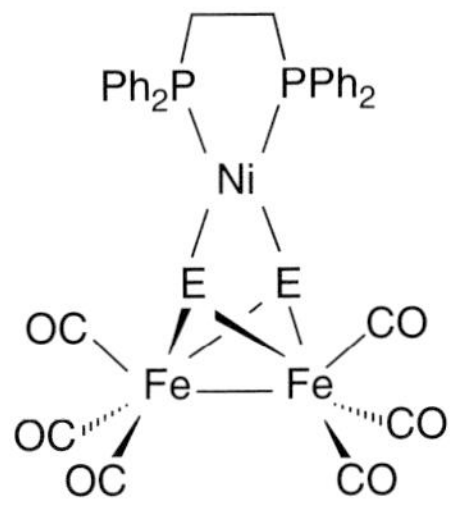

E = S, Se, Te

39

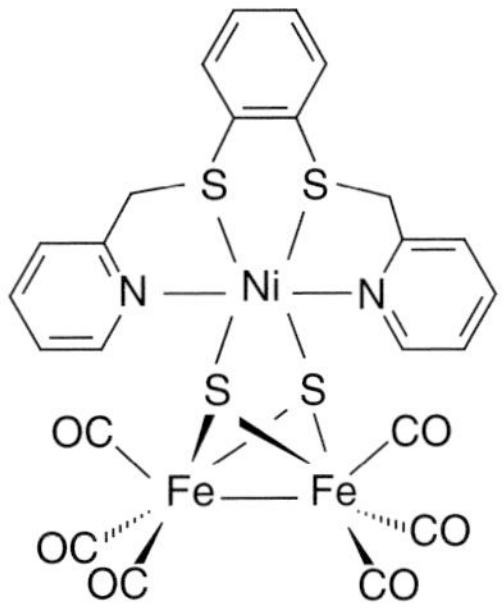

40

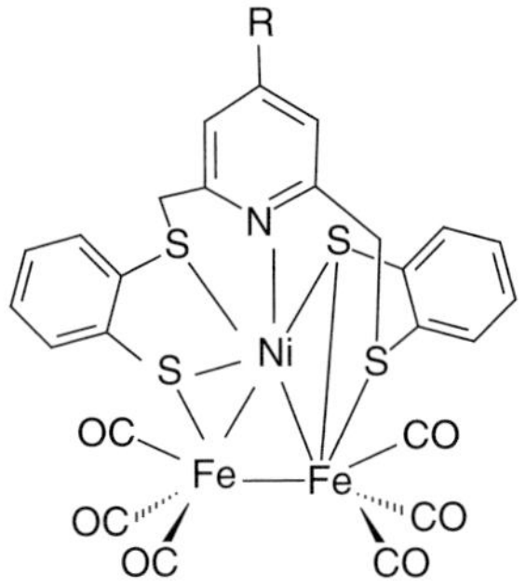

R = H, MeO, Cl, Br, *i*-Pr

41

Figure 18.12 Heterobimetallic [NiFe] clusters.

could be characterized, but DFT results suggested an ECEC sequence involving a bridging hydride $\{Ni^{III}(\mu H)Fe^{II}\}$ species.

Some $[NiFe_2]$ clusters were also reported as catalysts for proton reduction (Figure 18.12). A catalytic proton-reductive process was observed in CV upon addition of CF_3CO_2H (up to 50 mM) to a 1 mM CH_2Cl_2 solution of $[NiFe_2(CO)_6L]$ (**38**, L^{2-} = $\{(CH_3)_2C_6H_3S_2\}_2(CH_2)_3{}^{2-}$) [105]. A moderate activity, with a TOF of 6 h^{-1}, was observed after a bulk electrolysis at −1.64 V vs (Fc^+/Fc) in the presence of CF_3CO_2H (50 mM). Photochemical generation of H_2 was described with this compound using $[ReCl(CO)_3(\kappa^2\text{-bpy})]$ as a photosensitizer and TEOA (1 M)/[HTEOA][BF_4] (0.1 M) (TEOA, triethanolamine, pK_a^{MeCN} = 15.9) [35] as a combined sacrificial electron donor and proton source in MeCN [106]. A TON of 55 was measured after three hours photolysis ($\lambda < 420$ nm), but the precipitation of a black powder suggests decomposition of the catalyst. H_2 production was also observed in DMF but not in CH_2Cl_2. A slower rate of H_2 production was calculated (TOF = 23 h^{-1} over c. 17 minutes) in MeCN, when $[Ru(\kappa^2\text{-bpy})_3][PF_6]_2$ was used instead of $[ReCl(CO)_3(\kappa^2\text{-bpy})]$ (TOF = 52 h^{-1} over c. 22 minutes). A linked dyad [Re]-[$NiFe_2$] was designed, but an almost complete loss of activity toward proton

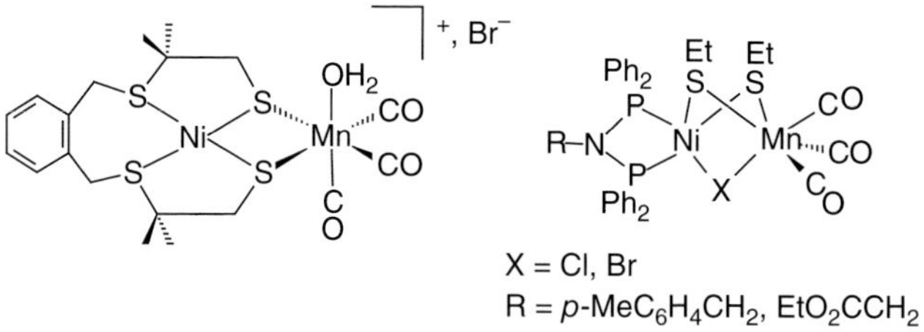

Figure 18.13 Heterobimetallic [NiMn] complexes.

reduction was observed because of the decomposition of this complex under catalytic conditions [107]. Other [$NiFe_2$] clusters based on a butterfly di-iron hexacarbonyl {$Fe_2(CO)_6(\mu\text{-}E)_2$} (E = S, Se, Te) core (**39–41**) were reported to have a moderate activity as electrocatalysts for proton reduction in MeCN in the presence of acids such as CH_3CO_2H, HOTs, and CF_3CO_2H [108–110].

Dinuclear [NiMn] complexes are also considered as bioinspired mimics of the active site of [NiFe] hydrogenases because of the isolobal analogy between the fragments {$Mn^I(CO)_3$} and {$Fe^{II}(CO)(CN)_2$} (Figure 18.13) [111]. The compound [$NiMn(xbsms)(CO)_3(H_2O)]^+$ (**42**) catalyzes dihydrogen evolution from CF_3CO_2H in DMF with an overpotential of 0.86 V (TOF ~ 4 h^{-1}) [112]. A heterolytic mechanism, involving the formation of a bridging hydride derivative, is proposed from electrochemical and theoretical studies. A series of complexes [$NiMn(CO)_3(\{PPh_2\}_2NR)(\mu\text{-}SEt)(\mu\text{-}X)$] (**43**) was prepared by reacting [$MnX(CO)_5$] with [$Ni(SEt)_2(\{PPh_2\}_2NR)$] (X = Cl, Br; R = *p*-$MeC_6H_4CH_2$) [113]. Electrochemical studies reveal their electrocatalytic behaviors toward proton reduction in the presence of CH_3CO_2H in MeCN, with overpotentials estimated to 0.47–0.48 V. Bulk electrolysis of MeCN solutions of these two complexes (0.5 mM) at −1.85 V vs (Fc^+/Fc) for 30 minutes allowed to determine a TON of c. 10 during this time. Other [MnRe] heterobimetallic complexes based on a {$MnRe(CO)_6S_2$} core were described as structural and functional models of the active site of hydrogenases [114]. They were modest electrocatalysts for H_2 production in the presence of CH_3CO_2H in acetonitrile.

18.2.4 Other Nickel-Based Catalysts

First evidences of the activity of mononuclear nickel-based complexes as HER catalysts were highlighted in the 1980s for Ni^{II} tetraazamacrocycle systems (**44**, Figure 18.14) [115]. [$Ni(cyclam)]^{2+}$ and [$Ni_2(bis\text{-}cyclam)]^{4+}$ cations (cyclam = 1,4,8,11-tetrazacyclotetradecane) display HER catalytic activity in water on a mercury electrode. Bulk one hour electrolysis of the bis-macrocyclic complex at −1.26 V vs SHE allowed to calculate a TON up to 100 [116]. More recently, Ni^{II} bis(diphosphine) complexes were widely studied using amine-functionalized diphosphines as biomimetic base relays (**45, 46**) [117–123]. The presence of amine in the second coordination sphere enhances the catalytic HER activity of these complexes because of the metal-based proximity and the stabilization of NH species [124]. The cationic complex

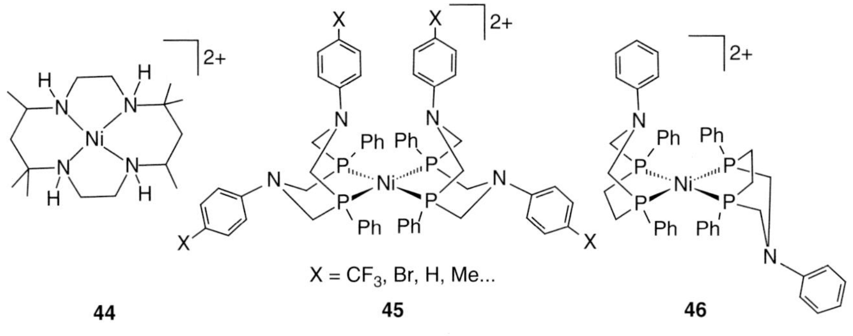

Figure 18.14 Tetraazamacrocycle and amine-functionalized diphosphine Ni^{II} complexes.

$[Ni(P^{Ph}{}_2N^{p\text{-}C_6H_4X}{}_2)_2]^{2+}$ (**45**, $P^{Ph}{}_2N^{p\text{-}C_6H_4X}{}_2$ = $\{CH_2PPhCH_2N^{p\text{-}C_6H_4X}\}_2$) exhibits high activity toward proton reduction in acetonitrile. When X = $CH_2P(O)(OEt)_2$, a TOF of 1850 s^{-1} and an overpotential of 0.37 V were determined at a glassy carbon electrode in acidic MeCN solution ($[(DMF)H][CF_3SO_3]$, $pK_a{}^{MeCN}$ = 6.1) to which water is added [125]. A TOF > 4.10^4 s^{-1} and an overpotential of 0.40 V were found in acidic ionic liquid–water medium when X = *n*-hexyl [126]. A similar range of efficiency was reported for $[Ni(\{CH_2P^{Ph}CH_2\}_2N^{Ph})_2]^{2+}$ (**46**) containing two seven-membered cyclic diphosphine ligands ($P^{Ph}{}_2N^{Ph}$ = 1,3,6-triphenyl-1-aza-3,6-diphosphacycloheptane). TOFs of 33 000 and 106 000 s^{-1} were found in pure MeCN and in the presence of 1.2 M of water in MeCN solution, respectively, at −1.13 V vs (Fc^+/Fc) [127]. The efficiency of this class of Ni-based catalysts prompted the use of such complexes for the design of electrode materials. Grafting of Ni^{II} bis(diphosphine) complex to multiwalled carbon nanotubes and using glassy carbon with Nafion gave electrodes that are able to catalyze HER in water with a TON of c. 10^5 (10 hours experiment/−0.30 V vs NHE/0.5 M H_2SO_4) [128, 129]. Several further developments of the DuBois systems have been reported [130–137].

Numerous Ni^{II} systems based on sulfur donor ligands have also been reported (Figure 18.15) [138–148]. In 2017, the nickel pyrazinedithiolate complex $[Ni(dcpdt)_2]^{2-}$ (**47**, dcpdt, 5,6-dicyanopyrazine-2,3-dithiolate) with a $\{NiS_4\}$ core has been described as an efficient HER catalyst, showing low overpotentials (0.33–0.40 V at pH 4–6) and a TOF up to 20 000 day^{-1}

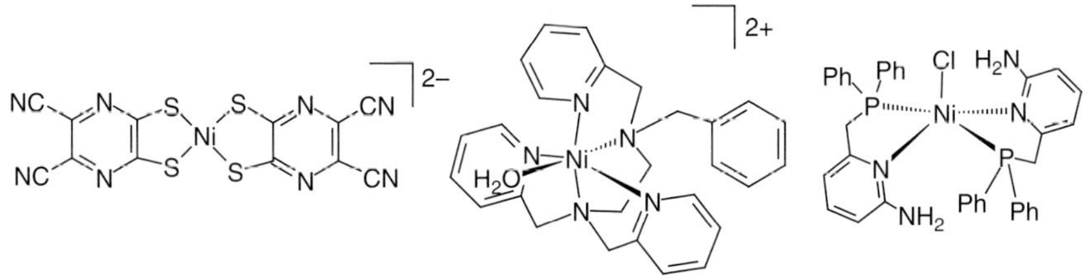

Figure 18.15 Selected examples of mononuclear Ni-based complexes with various chelating ligands.

with high faradaic efficiency (92–100%) [149]. Several other examples of complexes featuring cyclopentadienyl [150], porphyrin [151–154], tetradentate macrocyclic [155–157], and pyridine/polypyridyl [158–163] ligands catalyze the reduction of protons. Recently, the high efficiency of the N_5-pentadentate nickel complex $[Ni(bztpen)(H_2O)]^{2+}$ (bztpen, *N*-benzyl-*N,N*′, *N*′-tris(2-pyridylmethylethylenediamine)) (**48**) was reported [164]. A CPE experiment (CPE, controlled potential electrolysis) was performed for 60 hours in neutral phosphate buffer at −1.25 V on a mercury pool electrode, and a TON of 308 000 mol H_2 (mol cat)$^{-1}$ was thus calculated (TOF = 1650 mol H_2 (mol cat)$^{-1}$ h^{-1} cm^{-2}). A novel nickel complex $[NiL_2Cl]Cl$ was prepared with a bidentate amine-functionalized phosphinopyridyl ligand (L) (**49**). By electrochemical measurements, a TOF of 8400 s^{-1} was found, in MeCN, in the presence of acetic acid, with an overpotential of 0.59 V [165]. To date, only scarce examples of dinuclear nickel complexes having a catalytic HER activity are known [166–170].

18.3 Other Non-noble Metal-Based Catalysts: Co, Mn, Cu, Mo, and W

18.3.1 Cobalt

Numerous reviews have been dedicated to mononuclear cobalt-based HER catalysts [171–173]. The scope of this field being very broad, this section will focus only on recent results concerning this class of HER catalysts, for which a huge choice of ligand platforms is available: N_4-macrocyclic, hexaamino, porphyrin, phthalocyanine, polypyridine, cyclopentadienyl, glyoxime, amine-containing diphosphine, and thiolate systems (Figure 18.16). Several mechanisms have been

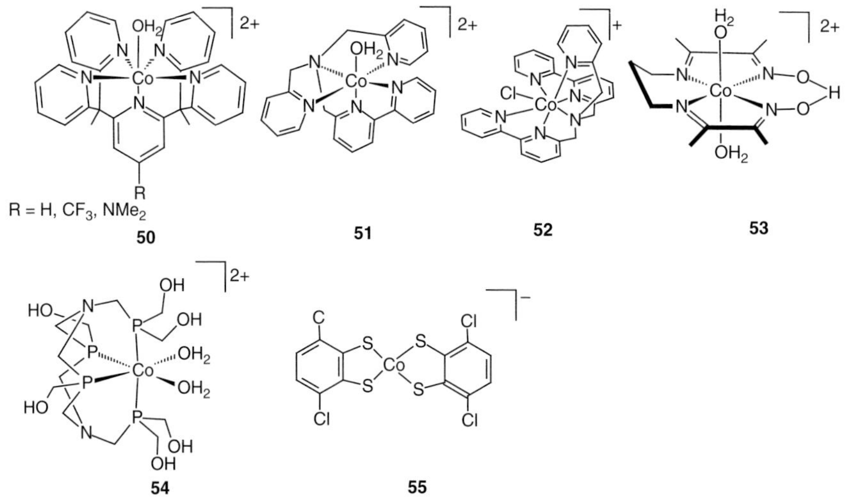

R = H, CF_3, NMe_2

50 **51** **52** **53**

54 **55**

Figure 18.16 Selected examples of Co-based mononuclear complexes.

proposed for proton reduction catalyzed by Co complexes. In these systems, the formation of a $[Co^{III}–H]^+$ intermediate via the protonation of a Co^I species is proposed as a key step of the process. Then, mononuclear heterolytic pathways ($[Co^{III} - H]^+ + H^+ \rightarrow [Co^{III}]^{2+} + H_2$, or $[Co^{III} - H]^+ + e^- \rightarrow [Co^{II} - H]$ followed by $[Co^{II} - H] + H^+ \rightarrow [Co^{II}]^+ + H_2$) or a bimolecular homolytic pathway ($2[Co^{III} - H]^+ \rightarrow 2[Co^{II}]^+ + H_2$) have been formally proposed for supporting the H_2 release by cobalt compounds.

Stable polypyridine-based catalysts display an interesting efficiency. A TON of 5.5×10^4 mol H_2 (mol cat)$^{-1}$ was measured for the pentapyridine complex $[Co(PY_5Me_2)(H_2O)]^{2+}$ (**50**, PY_5Me_2, 2,6-bis{1,1-bis(2-pyridyl)ethyl}pyridine) with high faradaic efficiency (99%), after CPE at −1.30 V vs SHE on a Hg pool electrode in phosphate buffer (pH 7). This catalyst is stable over 60 hours. It is worth noting that the introduction of an electron-withdrawing group (CF_3) on the central pyridine strongly reduces its activity and that of an amine functionality (NMe_2) increases the overpotential [174]. CPE for one hour of an aqueous solution of the pentacoordinate pyridyl-amine cobalt(II) complex $[Co(DPA\text{-}Bpy)(H_2O)]^{2+}$ (**51**, DPA-Bpy, *N*,*N*-bis(2-pyridinylmethyl)-2,2′-bipyridine-6-methanamine) in pH 7 phosphate buffer at −1.40 V vs SHE using a Hg pool electrode gave a TON > 300 mol H_2 (mol cat)$^{-1}$ [175].

The catalytic behaviors of other cobalt complexes with tripyridine–diamine pentadentate ligands [176] and bis(terpyridyl) [177] were also examined. Recently, a first heptacoordinate polypyridine cobalt complex **52** was described as an efficient catalyst in water for light-driven H_2 production. A TON of 16 300 mol H_2 (mol cat)$^{-1}$ was achieved in two hours of irradiation (λ = 475 nm) using $[Ru(\kappa^2\text{-bpy})_3]^{2+}$ as the photosensitizer [178]. Cobaloxime derivatives **53** are efficient homogeneous photocatalysts in water. For example, they were used successfully as supported catalysts to prepare highly efficient electrode materials, reaching TONs up to 5.5×10^4 in seven hours at pH 4.5, at a potential of −0.59 V vs RHE (reversible hydrogen electrode) [179]. The use of a multihydroxy-functionalized tetraphosphine ($P_4N_2 = (CH_2P^RCH_2N^{CHCH_2PR_2})_2(R = CH_2OH)$) afforded $[Co(P_4N_2)(H_2O)_2]^{2+}$ catalyst **54**, which is able to produce 9×10^4 mol H_2 (mol cat)$^{-1}$ over 20 hours in neutral phosphate buffer solution at −1.00 V vs NHE, on a mercury electrode with a faradaic efficiency close to 100% [180]. Bis-dithiolate complexes have also been described as efficient photocatalysts. A TON of 6000 mol H_2 (mol cat)$^{-1}$ and a TOF of 1400 h^{-1} were determined for **55** after a 12 hours photocatalysis experiment in a MeCN:H_2O (1 : 1) mixture (λ = 520 nm/$[Ru(\kappa^2\text{-bpy})_3]^{2+}$/ascorbic acid) [181]. It should be noted that in several cases, the molecular complex is not the HER catalyst, but this latter generates, through reductive decomposition, Co nanoparticles at the electrode as the active species [182–187]. This fact points out the risk of misidentification of the real catalytic species in the case of lack of stability in acidic media.

Few catalysts based on di-, multinuclear or mixed-cobalt cores are reported in the literature (Figure 18.17). The protonation of the complex $[Co_2Cp^*_2(\mu\text{-bdt})]$ (**56**) with $H(BF_4)\cdot OEt_2$ in CH_2Cl_2 afforded the bridging hydride species $[Co_2Cp^*_2(\mu\text{-bdt})(\mu\text{-H})][BF_4]$ [188]. CV of this cationic complex revealed a catalytic response for proton reduction in CH_2Cl_2 in the presence of CF_3CO_2H with

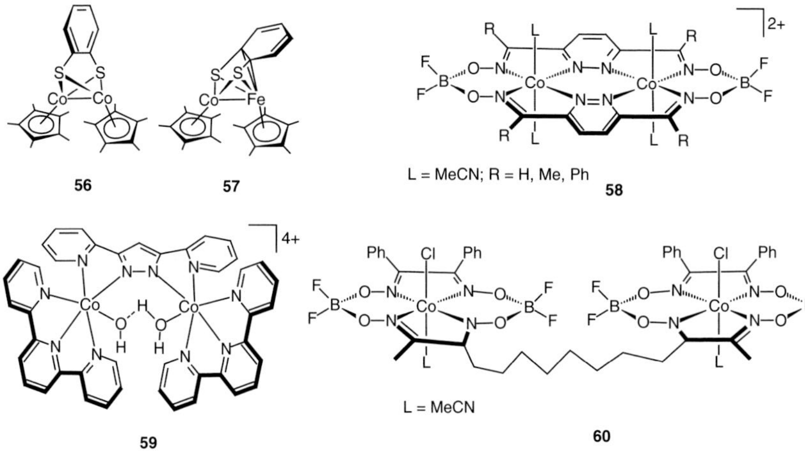

Figure 18.17 Examples of dicobalt complexes.

an estimated overpotential of 0.42 V (half-wave potential). A bulk electrolysis at −1.60 V vs (Fc^+/Fc) of 1 mM of this hydride complex, in the presence of 30 mM of CF_3CO_2H, generates 0.28 mmol of H_2 (faradaic efficiency ≈ 90%) over a period of 400 seconds. The use of a weaker acid, such as CH_3CO_2H, does not induce an observable catalytic event. The treatment of the heterobimetallic cobalt–iron complex [$CoFeCp^*_2$(μ-η^2:η^4-bdt)] (**57**) with H(BF_4)·OEt_2 in CH_2Cl_2 resulted in the formation of the cationic complex [$CoFeCp^*_2$(μ-η^2:η^4-bdt)][BF_4] along with H_2, through a one-electron oxidation. Dicobalt(II) macrocycles were designed by using pyridazine dioxime building blocks (**58**) [189]. Study of their reductive behaviors in the presence of protons by electrochemical methods suggested that they act as electrocatalysts for proton reduction in the presence of 2,6-dichloroanilinium tetrafluoroborate (pK_a^{MeCN} = 5.1) [190] in acetonitrile at the ($Co^{II}Co^{I}$/$Co^{I}Co^{I}$) redox potential. More detailed studies on the electrocatalytic properties were not reported. Other ways to prepare dicobalt systems were described by using combinations of well-known ligand-building blocks such as bis(pyridyl)pyrazolato (bpp) and terpyridine (trpy) (**59**) [191], difluoroboryl-bridged diglyoxime assembly (**60**) [192], and bis(phenolate) N_6O_2 tetrakis-Schiff base macrocycles as templates (**61**, **62**) [193].

Addition of successive amounts of CF_3CO_2H in MeCN solution of the bis(pyridyl)pyrazolato and terpyridine dicobalt(III) complex [Co_2(OH)(H_2O)$(trpy)_2$(μ-bpp)]$^{4+}$ (**59**) triggers catalytic currents at potentials near that of the couple ($Co^{II}Co^{II}$/$Co^{II}Co^{I}$) (−0.78 V vs SCE) [191]. H_2 was produced by electrolysis at −1.10 V vs SCE for 10 minutes with a faradaic yield of 79%. Kinetics and the mechanism of proton (CF_3CO_2H) reduction to H_2 were explored by using [$CoCp_2$] as the reductant and a homolytic pathway was ruled out (Scheme 18.6).

The properties of the dicobaloxime compound **60**, featuring two monomeric Co^{III} units linked by an octamethylene bridge, are close to those of mononuclear units, without noticeable enhancement of the efficiency of H_2 production [192]. An electrocatalytic event was observed at −0.40 V vs SCE in acetonitrile in the presence of HOTs and a rate constant of 1100 M^{-1} s^{-1} was determined.

$$[Co^{III}Co^{III}]^{4+} + 3\,[CoCp_2] \longrightarrow [Co^{II}Co^{I}]^{+} + 3\,[CoCp_2]^{+}$$
$$[Co^{II}Co^{I}]^{+} + H^{+} \rightleftharpoons [Co^{II}Co^{III}\text{–}H]^{2+}$$
$$[Co^{II}Co^{III}\text{–}H]^{2+} + H^{+} \longrightarrow [Co^{II}Co^{III}]^{3+} + H_2$$
$$[Co^{II}Co^{I}]^{+} + [Co^{II}Co^{III}]^{3+} \longrightarrow 2\,[Co^{II}Co^{II}]^{2+}$$

Scheme 18.6 Proposed mechanism of H^+/H_2 conversion for dicobalt(III) catalysts.

A heterolytic mechanism, through the protonation of a $[Co^{II}\text{–}H]$ intermediate, was proposed to occur rather than a homolytic pathway involving two $[Co^{III}\text{–}H]^+$ species. Photocatalytic activity of analogous dicobaloxime compounds $[\{Co(dmgH)(dmgH_2)\}_2L]$ ($dmgH_2$, dimethylglyoxime, dmgH, dimethylglyoximate; L = 1,3-bis(4-pyridyl)propane or 1,3-bis(imidazol-1-ylmethyl)benzene) has been described [194]. Depending on the nature of the bridging group, TONs of c. 1000–1100 were obtained, over two hours with irradiation at $\lambda > 420$ nm, in $MeCN{:}H_2O$ (3 : 1) mixtures (pH 8) in the presence of eosin Y dye (EY^{2-}), as photosensitizer and TEOA as a sacrificial electron donor.

A novel class of Co-based compounds was obtained by using tetrakis-Schiff base macrocycles **61** and **62** (Figure 18.18) [193]. Addition of increasing amounts

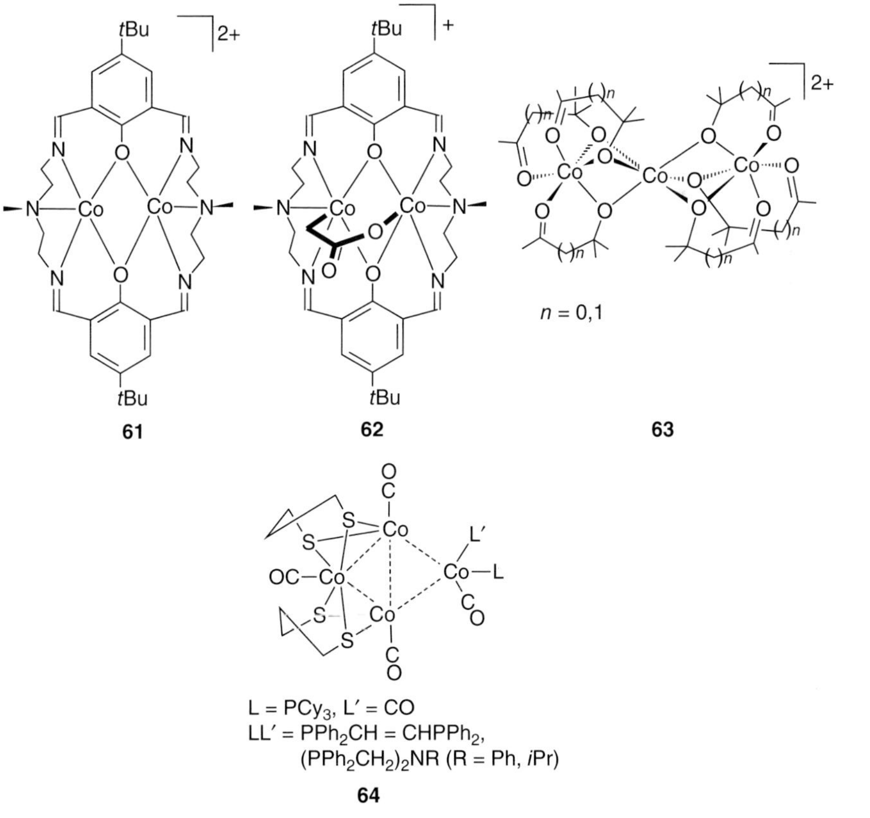

61 **62** **63** (n = 0,1)

L = PCy_3, L′ = CO
LL′ = $PPh_2CH{=}CHPPh_2$, $(PPh_2CH_2)_2NR$ (R = Ph, *i*Pr)

64

Figure 18.18 Other bi- and polymetallic cobalt complexes.

of CF_3CO_2H or CH_3CO_2H to MeCN solutions of **61** resulted in the observation of typical electrocatalytic responses corresponding to an irreversible catalytic reduction, proportional to the concentration of acid, starting at c. −1.28 V vs (Fc^+/Fc), close to the potential of the ($Co^{II}Co^{I}$/$Co^{I}Co^{I}$) redox couple. Relatively high overpotentials were noted. CPE of a solution of **61** was carried out at −1.88 V vs (Fc^+/Fc) in the presence of an excess of CH_3CO_2H, allowing the detection of H_2 (faradaic efficiency of 72–94%). Similar properties were observed with complex **62**.

Efficient proton reduction was reported for other tri- and tetranuclear complexes of cobalt **63** and **64** (Figure 18.18) [195, 196]. Linear trimetallic [$Co^{III}Co^{II}Co^{III}$] complexes **63** ($[Co_3(C_xH_yO)_6][BF_4]_2$, $x = 5$, $y = 9$ or $x = 6$, $y = 11$), with bridging acyl–alkoxy groups, behave as electrocatalysts in MeCN for the reduction of HOTs with relatively low-onset overpotential (c. 0.17 V) and TOF estimated at 80 s^{-1}. A homogeneous mechanism was proposed to involve sequential steps of one-electron transfer and protonation at the oxygen atom of a bridging alkoxy and at the metal center [195]. Tetracobalt clusters **64**, having some structural analogies with dithiolate di-iron models, were recently prepared [196]. Preliminary electrochemical investigations revealed their activity as electrocatalysts for proton reduction in the presence of increasing concentrations of CF_3CO_2H in THF. Bulk electrolysis at a potential of −2.15 V vs (Fc^+/Fc) in the presence of **64** (30 μmol) and CF_3CO_2H (160 μL) produced 800 μL of H_2 for c. 1.5 hours.

18.3.2 Manganese

The use of manganese complexes in homogeneous HER catalysis remains scarce. To date, a limited number of mononuclear Mn-based catalysts have been identified (Figure 18.19) [197]. Organometallic allenylidene and vinylidene manganese complexes **65** moderately catalyze the reduction of $H(BF_4)\cdot OEt_2$ in organic media (MeCN, CH_2Cl_2) [198]. The mechanism of proton reduction involves the protonation of the metallacumulene ligand giving a carbyne complex that is reduced into a 19-electron radical intermediate, which then releases H_2 through a homolytic C—H bond cleavage. Recent electrochemical studies show that the complex $[Mn(mesbpy)(CO)_3(MeCN)][CF_3SO_3]$ (**66**, mesbpy, 6,6′-dimesithylene-2,2′-bipyridine) electrocatalyzes proton reduction by using CF_3CO_2H, with a TOF of 5500 s^{-1} and an overpotential of 0.90 V in acetonitrile [199]. The complex [Mn(^{Ph_2PPr}PDI)(CO)]Br (**67**), featuring a pentadentate pyridine di-imine (^{Ph_2PPr}PDI), has a modest activity toward proton reduction (MeOH, H_2O) in MeCN only in the presence of CO_2 [200]. A bulk electrolysis of a CO_2-saturated solution of **67** (5 mM) and MeOH (1.05 M) in MeCN at −2.20 V vs (Fc^+/Fc) over 47 minutes produced H_2 as the sole product. A faradaic efficiency of c. 97% and a TOF of 0.176 h^{-1} were calculated. The activity of another Mn^{I} polypyridyl complex was very recently reported [201].

The electrocatalytic proton reduction behavior of various thiolato- and selenato-bridged dimanganese complexes was evidenced (Figure 18.20). The tris-thiolato dimanganese(I) hexacarbonyl anion $[Mn_2(CO)_6(\mu\text{-}SPh)_3]^-$ (**68**) catalyzes electrochemical proton reduction in CH_3CN in the presence of

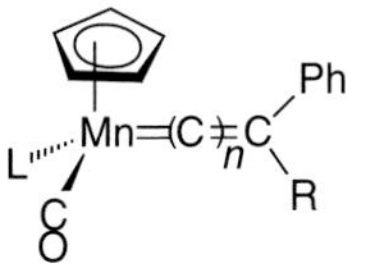

$n = 1, 2$
L = CO, PPh_3
R = H, Ph
65

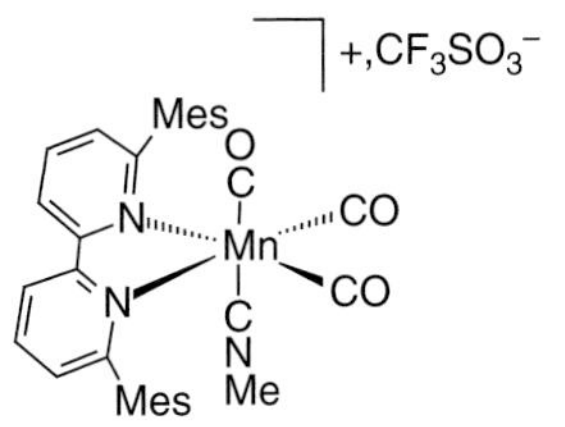

66

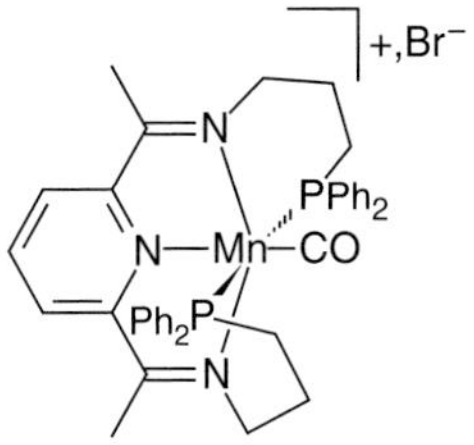

67

Figure 18.19 Mn-based mononuclear catalysts.

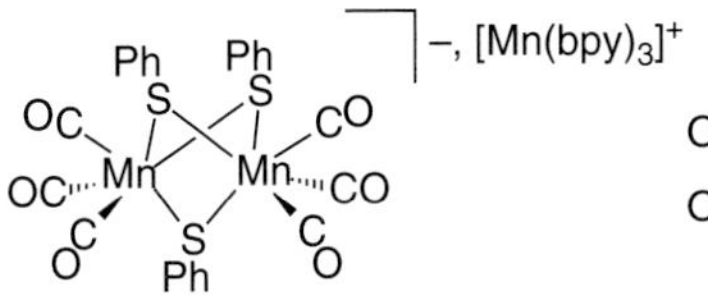

68

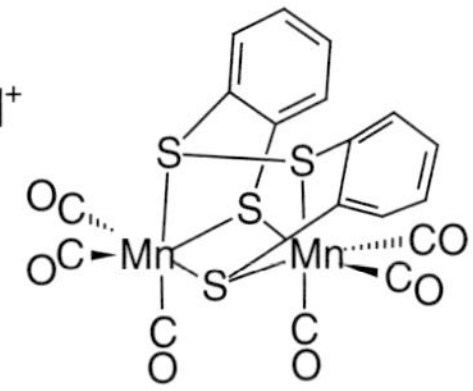

69

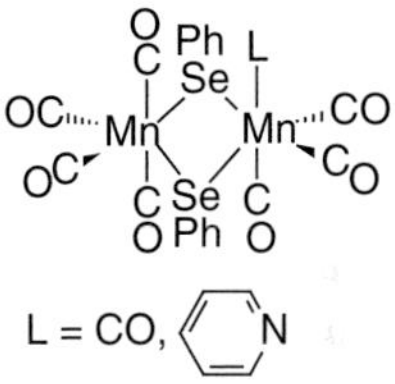

L = CO,

70

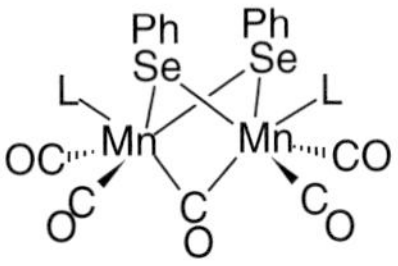

L = $P(C_4H_9)_3$
71

Figure 18.20 Examples of dimanganese complexes.

CF_3CO_2H, with a low overpotential (0.61 V) and a very high TOF (up to 44 600 s^{-1}) [202]. A CECE mechanism, involving transient S-protonation, was proposed for the proton reduction process based on experimental results. The competitive formation of a tetramer, $[Mn_4(CO)_{12}(\mu\text{-SPh})_4]$, was observed in the presence of a large excess of acid (50 equiv.). The neutral dinuclear hexacarbonyl manganese(I) complex $[Mn_2(CO)_6(\mu\text{-}S_4(C_6H_4)_2)]$ (**69**), featuring a disulfide bond, catalyzes proton reduction (CH_3CO_2H in DMF) at −2.10 V vs (Fc^+/Fc) with an overpotential near 0.60 V [203]. Electrochemical observations combined with DFT calculations suggest that an initial two-electron reduction may induce the cleavage of the dimer into the dianionic species $[Mn(CO)_3(\kappa^2\text{-bdt})]^{2-}$ that would be successively protonated at sulfur and manganese atoms before releasing H_2. Electrocatalytic efficiencies of selenolato-manganese carbonyl complexes **70** and **71** in the presence of CF_3CO_2H in CH_2Cl_2 have also been studied [204]. It is worth noting that the photolysis of $[Mn_2(CO)_{10}]$ in a cyclohexane/water biphasic system resulted in the formation of 1.80 (±0.16) mol of H_2/ mol of $[Mn_2(CO)_{10}]$. The formation of $[MnH(CO)_5]$ as a key intermediate in this process was proposed [205].

18.3.3 Copper

Very recently, HER catalytic activity has been observed with molecular copper complexes (Figure 18.21). The pentacoordinated copper(II) complex [Cu(bztpen)][BF_4]$_2$ (**72**) is an efficient electrocatalyst for proton reduction (apparent rate constant >10 000 s^{-1}, onset overpotential of 0.42 V, pH 2.5) [206]. In acidic aqueous solutions, an efficiency of 7000 mol H_2 (mol cat)$^{-1}$ h^{-1} cm^{-2} was determined from CPE at −0.90 V vs SHE for two hours (faradaic efficiency of 96%). The mechanism for H_2 evolution may involve two sequential PCET steps, affording N-protonated Cu^I-pyridinium and Cu^{II}-hydride key species, respectively [207]. The electrocatalytic activity toward proton reduction of a stable water-soluble polypyridine copper(II) complex [Cu(dpp)(ClO_4)$_2$] (**73**, dpp, 2,9-di(pyridin-2-yl)-1,10-phenanthroline) was reported [208]. Bulk two hours electrolysis in neutral phosphate buffer solution at a potential of −1.40 V vs SHE generated H_2 with a rate of 734 mol H_2 (mol cat)$^{-1}$ cm^{-2} (faradaic efficiency of 95%). The relatively low overpotential of 0.52 V allowed light-driven production in water (6 mol H_2 (mol cat)$^{-1}$). The activity of Cu–corrole complexes **74** was also studied [209, 210]. The influence of substituents bound to the macrocycle on the catalytic properties was evidenced. Their efficiencies were evaluated by CV in acetonitrile in the presence of CF_3CO_2H. Stopped flow and spectroelectrochemistry gave some insights into the mechanism, showing that [Cu(cor)]$^{2-}$ and [CuH(cor)]$^-$ are key intermediates. Recently, electrochemical experiments showed that di- and trinuclear triazenido Cu^I and Cu^{II} compounds **75** and **76** catalyze H_2 evolution [211–214]. A series of mono- and dinuclear copper(II) compounds **77** and **78** based on 1,1-bis(6-{1*H* imidazol-2-yl}pyridin-2-yl)ethan-1-ol and 2-(1*H* imidazol-2-yl)-6-methylpyridin were prepared [215, 216]. CPE at −0.81 V vs SHE in water under acidic conditions (phosphate buffer) led to the formation of Cu^0 and Cu_2O deposits on the electrode, which are the active catalysts in the HER process.

18.3.4 Group 6 Metals (Mo, W)

Several molecular Mo–S HER catalysts have been described as mimics of efficient catalytic Mo–S materials (Figure 18.22). Water-soluble Mo^{VI} anions such as [MoO(S_2)$_2$(L_2)](Et_4N) (**79**, L_2 = picolinate or pyrimidine-2-carboxylate) electrocatalyze the HER process. CPE for one hour at pH 3 (citric acid buffer:MeCN (6 : 4)) at an overpotential of 0.54 V generates H_2 with a TOF in the range of 27–48 s^{-1} [217]. Calculations on the complex [Mo(O)(S_2)$_2$(bpy)] reveal that, in acidic operating conditions, the {Mo = O} moiety may act as a proton relay. A two-electron reduction of a sulfide would be followed by two successive protonation steps at the reduced sulfide and at the {Mo = O} moiety. The interaction of these hydride- and proton-like hydrogen atoms may lead to the release of H_2 [218]. The electrocatalytic HER efficiency of [Mo(O)(S_2)$_2$(bpy^R)] (R = H, *t*-Bu, OMe) is affected by the substitution on the bpy^R ligand. Acid-independent rate constants of 95, 246, and 227 s^{-1} were found for R = H, *t*-Bu, OMe, respectively, from CV experiments in DMF (ν = 0.2 V s^{-1}) with CF_3CO_2H as the source of proton. Overpotentials in the range of 0.62 V (R = H) to 0.86 V (R = *t*-Bu) were calculated [219].

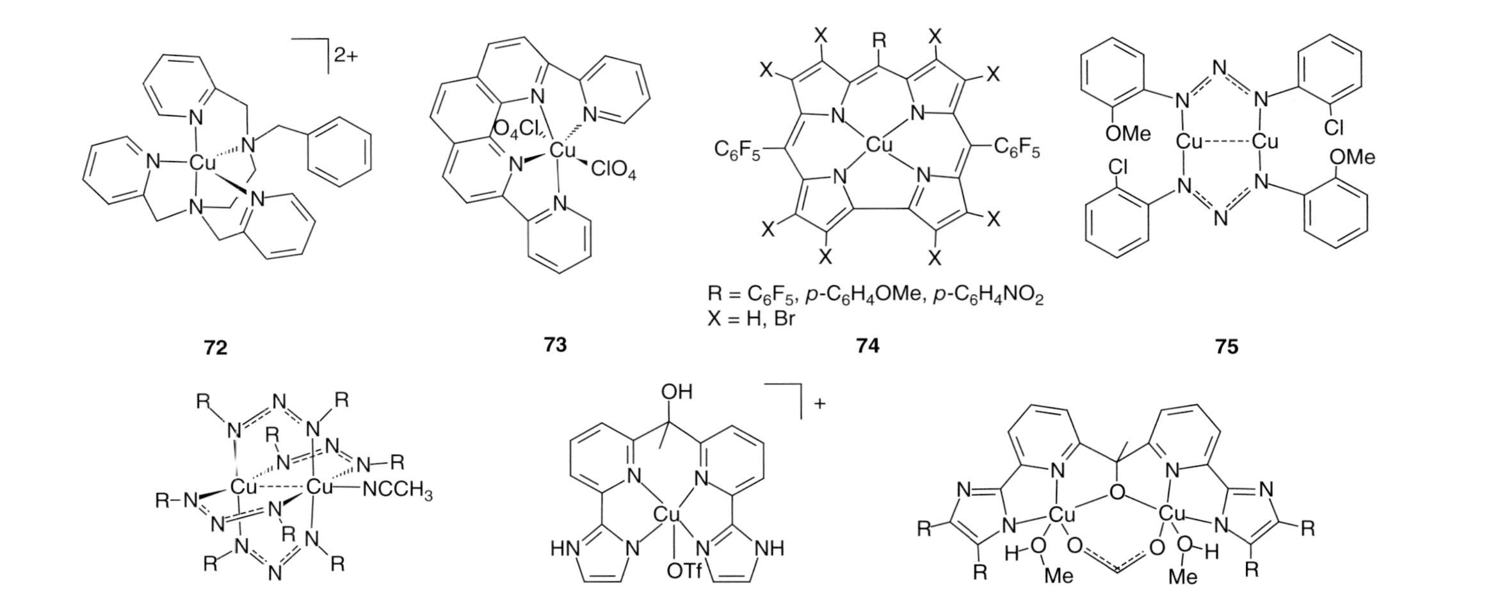

Figure 18.21 Examples of recent Cu-based HER catalysts.

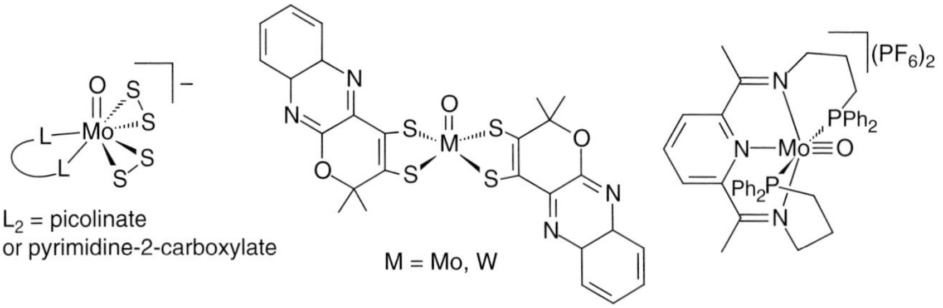

L_2 = picolinate or pyrimidine-2-carboxylate

M = Mo, W

79 **80** **81**

Figure 18.22 Selected examples of Mo-based HER catalysts.

The bioinspired molybdenum-dithiolene–oxo complex $[Bu_4N]_2[MoO(qpdt)_2]$ (**80**, $qpdt^{2-}$, quinoxaline-pyrane-fused dithiolene) displays HER activity in MeCN with a low-onset potential and a TOF of $1030\,s^{-1}$ at −1.30 V vs Ag|AgCl. DFT calculations suggested that the oxo ligand plays a key role, and the transfer of two electrons and two protons may generate the reactive hydroxo hydrido intermediate $[Mo^{IV}(OH)(H)(qdptH)(qdpt)]$. This species is able to photocatalyze the reduction of protons efficiently with a TON of 500 and TOF of $203\,h^{-1}$, showing a good stability in aqueous/organic media [220]. Similarly, the bis(dithiolene) tungsten(VI) complex $[PPh_4]_2[W(O)_2(S_2C_2(CO_2Me)_2)_2]$ catalyzes proton reduction in acidic organic media (CH_3CO_2H/MeCN) with good faradaic yields (75–95%), good activity (rate constant of $100\,s^{-1}$), and an overpotential of 0.70 V. DFT calculations suggest that the catalytic HER mechanism involves a W-hydroxo–hydride intermediate [221]. A HER catalytic activity is observed in 5.0 M H_2O/acetonitrile solution in the presence of the pentadentate bis(imino)pyridine-based complex $[MoO(^{Ph2PPr}PDI)][PF_6]_2$ (**81**). CPE at −2.5 V vs (Fc^+/Fc) produces H_2 with a faradaic efficiency of 96%. A rate constant of $55\,s^{-1}$ was determined by CV. The dicationic complex is proposed to act as a precatalyst generated through a two-electron reduction step in the reduced neutral Mo^{II} catalytic species [222].

Molybdenum–polypyridyl complexes with a pentadentate ligand such as $PY5Me_2$ display high efficiency and robustness (Figure 18.23) [223]. The oxo Mo^{IV} complex $[MoO(PY_5Me_2)](CF_3SO_3)_2$ (**82**) is active for more than 71 hours in phosphate buffer on a Hg pool electrode at −1.4 V vs SHE. Values of $8500\,h^{-1}$ and 6.1×10^5 mol H_2 (mol cat)$^{-1}$ have been reported for the TOF and TON,

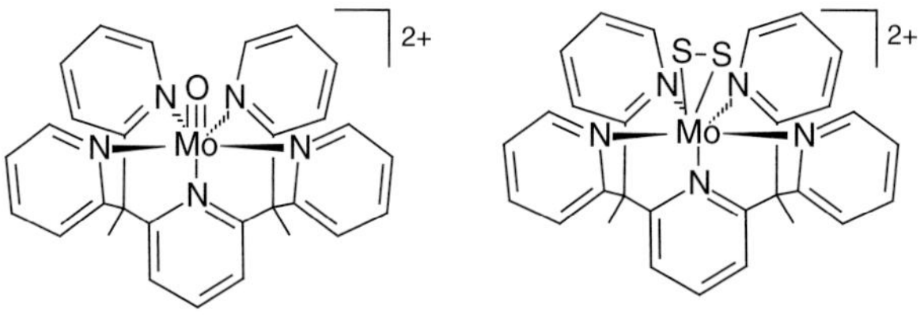

82 **83**

Figure 18.23 Molybdenum–polypyridyl complexes.

respectively. It is worth noting that the efficiency of this complex is preserved in seawater [224, 225]. Similar activity is observed for the related compound $[Mo(\kappa^2\text{-}S_2)(PY_5Me_2)](CF_3SO_3)_2$ (**83**), mimicking the edge sites of active MoS_2 materials. TON values of 3.5×10^3 and 1.9×10^7 mol H_2 (mol cat)$^{-1}$ were estimated as lower and upper limits, after 23 hours bulk electrolysis in pH 3 acetate buffer on a Hg pool electrode applying an overpotential of 0.78 V [226]. The robust dihydride cations $[MH_2(\kappa^2\text{-}RCO_2)(\kappa^2\text{-dppe})_2]^+$ (M = Mo or W; R = H, Me, CF_3, aryl) are capable of photoelectrocatalyzing proton reduction. The reduction potentials of these catalysts can be turned by varying R and the metal M. Electrocatalysis of dihydrogen production from CH_3CO_2H in a CH_3CN electrolyte by the $[MoH_2(\eta^2\text{-}CH_3CO_2)(\eta^2\text{-dppe})_2]^+$ cation has been undertaken at −1.4 V vs Ag|AgCl. A TON of 125 was reached after CPE for 40 hours. The hemilabile carboxylate ligand $CH_3CO_2^-$ is proposed to act as a proton relay [227].

The dimeric cluster $[Mo_2S_{12}]^{2-}$ is the smallest unit possessing both terminal and bridging disulfide ligands. Electrochemical studies have shown that this anionic cluster is a promising and efficient heterogeneous HER catalyst when supported on FTO (fluorine-doped tin oxide) under acidic conditions (0.5 M H_2SO_4). An overpotential of only 0.16 V is required to reach a current density of 10 mA cm^{-2}, and a TOF of c. 3 s^{-1} (normalized to per Mo atom in the catalyst) has been calculated [228]. Other dinuclear group 6 clusters, such as the neutral complex $[Mo_2Cp_2(\mu\text{-}S)_2(\mu\text{-}S_2CH_2)]$ (**84**, Figure 18.24) and related derivatives are also active as electrocatalysts for proton reduction. Their efficiency depends on the strength of the acid being used, the choice of the solvent, and the Cp and S substituents. In the presence of **84**, the catalytic reduction of *p*-cyanoanilinium tetrafluoroborate ($pK_a^{MeCN} = 7.6$) [229] in 0.3 M $[Et_4N][BF_4]$/acetonitrile buffered at pH 7.6 is observed at −0.64 V vs (Fc^+/Fc), with an overpotential of 0.12 V. The catalytic process is initiated by the protonation of a sulfide ligand. H_2 release, the rate-determining step, is proposed to occur through either a bis-hydrosulfido intermediate or a hydrosulfide–molybdenum hydride species [230].

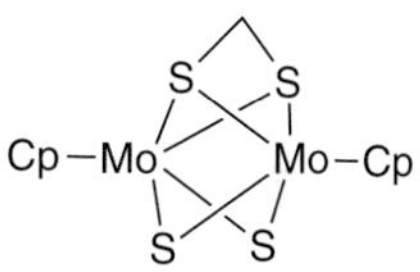

84

Figure 18.24 A dimolybdenum-based HER catalyst.

Heterobimetallic models for hydrogenases, involving group 6 metals, have been synthesized in view to achieve efficient catalysts via a cooperative activation by adjacent metals (Figure 18.25). Only scarce examples of such systems have been studied. Proton reduction can be electrocatalyzed by the heterobimetallic [NiW] cluster $[NiW(SNS)_2(dppe)]$ (**85**, $(SNS)H_3$, bis(2-mercapto-4-methylphenyl)amine)), which contains a redox-active {W(SNS)} metalloligand. A CPE experiment in MeCN in the presence of *p*-cyanoanilinium tetrafluoborate confirmed the production of H_2 at a potential of −1.25 V vs (Fc^+/Fc) with a faradaic efficiency of c. 80%. Despite a high efficiency at early electrolysis times, a complete deactivation, because of the decomposition of the catalyst, was observed after 2000 seconds [231]. Iron–molybdenum hybrid models $[Fe^{II}Mo^0]$ of the type $[FeMo(CO)_5(\kappa^2\text{-L})(\mu\text{-pdt})]$ (L = diphosphine) were also prepared and tested as HER catalysts [232]. Finally, heterobimetallic

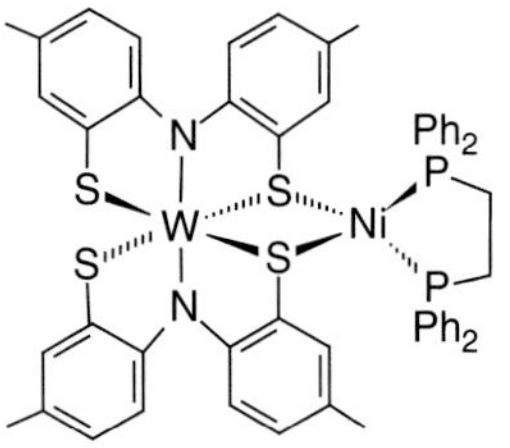

Figure 18.25 Examples of [NiMo] and [NiW] heterobimetallic complexes.

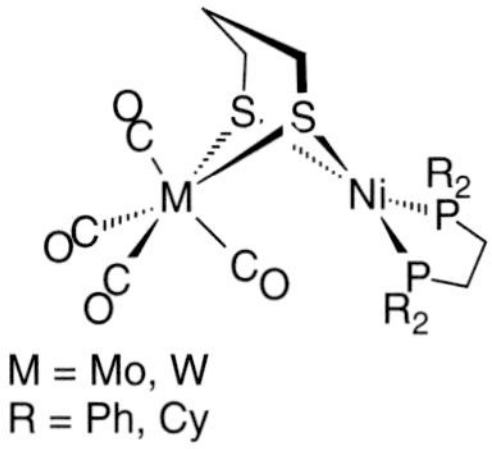

[NiMo] and [NiW] dithiolate complexes [Ni(κ^2-diphosphine)(μ-pdt)M(CO)$_4$] (**86**, diphosphine, dppe, dcpe, 1,2-bis(dicyclohexylphosphine)ethane; M = Mo, W) have also been prepared [233]. With a TOF of 47 s^{-1}, the parent hydride cation (diphosphine, dppe, M = W) exhibits modest activity for proton reduction ($H(BF_4)\cdot OEt_2$), which arises at −1.21 V ($E_{cat/2}$ = potential at which half of the catalytic current is reached) vs (Fc^+/Fc). The readily tunable redox properties and pK_a of this series of complexes encourages further investigations of their activity.

18.4 Conclusion

The design of homogeneous HER non-noble metal catalysts is an intensive field of research. Complexes of the first-row transition metals, such as Mn, Fe, Ni, Co, and Cu, as well as group 6 metals (Mo, W), have been widely developed either because of the presence of some of these metallic elements in the active sites of hydrogenases or in efficient catalytic HER solid-state materials. The combination of these metal centers with various ligand platforms has allowed important progresses in the efficiency of such HER catalysts. The improvement of the long-term stability of these systems is crucial, as well as their compatibility and efficiency in aqueous solution. Their use for the design of electrode materials and efficient photochemical devices is still in progress and appears as an attainable goal.

References

1 Weigand, W. and Schollhammer, P. (eds.) (2015). *Bioinspired Catalysis: Metal–Sulfur Complexes*. Weinheim: Wiley-VCH.
2 Thoi, V.S., Sun, Y., Long, J.R., and Chang, C.J. (2013). *Chem. Soc. Rev.* 42: 2388–2400.
3 McKone, J.R., Marinescu, S.C., Brunschwig, B.S. et al. (2014). *Chem. Sci.* 5: 865–878.
4 Wang, M., Chen, L., and Sun, L. (2012). *Energy Environ. Sci.* 5: 6763–6778.
5 Berardi, S., Drouet, S., Francàs, L. et al. (2014). *Chem. Soc. Rev.* 43: 7501–7519.

6 Gan, L., Jennings, D., Laureanti, J., and Jones, A.K. (2015). *Homo- and Heterobimetallic Complexes in Catalysis* (ed. P. Kalck), 233–272. Cham: Springer International Publishing.
7 Tran, P.D. and Barber, J. (2012). *Phys. Chem. Chem. Phys.* 14: 13772–13784.
8 Capon, J.-F., Gloaguen, F., Pétillon, F.Y. et al. (2009). *Coord. Chem. Rev.* 253: 1476–1494.
9 Felton, G.A.N., Mebi, C.A., Petro, B.J. et al. (2009). *J. Organomet. Chem.* 694: 2681–2699.
10 Fourmond, V., Jacques, P.-A., Fontecave, M., and Artero, V. (2010). *Inorg. Chem.* 49: 10338–10347.
11 Costentin, C., Drouet, S., Robert, M., and Savéant, J.-M. (2012). *J. Am. Chem. Soc.* 134: 11235–11242.
12 Costentin, C. and Savéant, J.-M. (2014). *ChemElectroChem* 1: 1226–1236.
13 Costentin, C., Passard, G., and Savéant, J.-M. (2015). *J. Am. Chem. Soc.* 137: 5461–5467.
14 Gloaguen, F. (2016). *Inorg. Chem.* 55: 390–398.
15 Lubitz, W., Ogata, H., Rüdiger, O., and Reijerse, E. (2014). *Chem. Rev.* 114: 4081–4148.
16 Lubitz, W., Reijerse, E., and van Gastel, M. (2007). *Chem. Rev.* 107: 4331–4365.
17 Frey, M. (2002). *ChemBioChem* 3: 153–160.
18 Schilter, D., Camara, J.M., Huynh, M.T. et al. (2016). *Chem. Rev.* 116: 8693–8749.
19 Tard, C. and Pickett, C.J. (2009). *Chem. Rev.* 109: 2245–2274.
20 Simmons, T.R., Berggren, G., Bacchi, M. et al. (2014). *Coord. Chem. Rev.* 270–271: 127–150.
21 Capon, J.-F., Gloaguen, F., Pétillon, F.Y. et al. (2008). *Eur. J. Inorg. Chem.* 4671–4681.
22 Harb, M.K., Apfel, U.-P., Sakamoto, T. et al. (2011). *Eur. J. Inorg. Chem.* 986–993.
23 Behnke, S.L. and Shafaat, H.S. (2016). *Comments Inorg. Chem.* 36: 123–140.
24 Ohki, Y. and Tatsumi, K. (2011). *Eur. J. Inorg. Chem.* 973–985.
25 Canaguier, S., Artero, V., and Fontecave, M. (2008). *Dalton Trans.* 315–325.
26 Li, Y. and Rauchfuss, T.B. (2016). *Chem. Rev.* 116: 7043–7077.
27 Rauchfuss, T.B. (2015). *Acc. Chem. Res.* 48: 2107–2116.
28 Felton, G.A.N., Glass, R.S., Lichtenberger, D.L., and Evans, D.H. (2006). *Inorg. Chem.* 45: 9181–9184.
29 Izutsu, K. (1990). *Acid–Base Dissociation Constants in Dipolar Aprotic Solvents*. Oxford, Boston, MA, Brookline Village, MA: Blackwell Scientific Publications; Distributors, USA, Publishers' Business Services.
30 Lansing, J.C., Camara, J.M., Gray, D.E., and Rauchfuss, T.B. (2014). *Organometallics* 33: 5897–5906.
31 Becker, R., Amirjalayer, S., Li, P. et al. (2016). *Sci. Adv.* 2: e1501014.
32 Wang, F., Wang, W.-G., Wang, X.-J. et al. (2011). *Angew. Chem. Int. Ed.* 50: 3193–3197.
33 Mejia-Rodriguez, R., Chong, D., Reibenspies, J.H. et al. (2004). *J. Am. Chem. Soc.* 126: 12004–12014.

34 Na, Y., Wang, M., Jin, K. et al. (2006). *J. Organomet. Chem.* 691: 5045–5051.

35 Izutsu, K. (2002). *Electrochemistry in Nonaqueous Solutions.* Weinheim: Wiley-VCH.

36 Cao, W.-N., Wang, F., Wang, H.-Y. et al. (2012). *Chem. Commun.* 48: 8081–8083.

37 Sano, Y., Onoda, A., and Hayashi, T. (2011). *Chem. Commun.* 47: 8229–8231.

38 Roy, A., Madden, C., and Ghirlanda, G. (2012). *Chem. Commun.* 48: 9816–9818.

39 Jones, A.K., Lichtenstein, B.R., Dutta, A. et al. (2007). *J. Am. Chem. Soc.* 129: 14844–14845.

40 Apfel, U.-P., Halpin, Y., Gottschaldt, M. et al. (2008). *Eur. J. Inorg. Chem.* 5112–5118.

41 Singleton, M.L., Reibenspies, J.H., and Darensbourg, M.Y. (2010). *J. Am. Chem. Soc.* 132: 8870–8871.

42 Quentel, F., Passard, G., and Gloaguen, F. (2012). *Energy Environ. Sci.* 5: 7757–7761.

43 Quentel, F., Passard, G., and Gloaguen, F. (2012). *Chem. Eur. J.* 18: 13473–13479.

44 Nann, T., Ibrahim, S.K., Woi, P.-M. et al. (2010). *Angew. Chem. Int. Ed.* 49: 1574–1577.

45 Dey, S., Rana, A., Dey, S.G., and Dey, A. (2013). *ACS Catal.* 3: 429–436.

46 Yamaguchi, T., Masaoka, S., and Sakai, K. (2009). *Chem. Lett.* 38: 434–435.

47 Bhugun, I., Lexa, D., and Savéant, J.-M. (1996). *J. Am. Chem. Soc.* 118: 3982–3983.

48 Graham, D.J. and Nocera, D.G. (2014). *Organometallics* 33: 4994–5001.

49 Rana, A., Mondal, B., Sen, P. et al. (2017). *Inorg. Chem.* 56: 1783–1793.

50 Cavell, A.C., Hartley, C.L., Liu, D. et al. (2015). *Inorg. Chem.* 54: 3325–3330.

51 Connor, G.P., Mayer, K.J., Tribble, C.S., and McNamara, W.R. (2014). *Inorg. Chem.* 53: 5408–5410.

52 Hartley, C.L., DiRisio, R.J., Chang, T.Y. et al. (2016). *Polyhedron* 114: 133–137.

53 Hartley, C.L., DiRisio, R.J., Screen, M.E. et al. (2016). *Inorg. Chem.* 55: 8865–8870.

54 Tang, L.-Z., Lin, C.-N., and Zhan, S.-Z. (2016). *Polyhedron* 110: 247–253.

55 Zhou, L.-L., Tang, L.-Z., Zhang, Y.-X., and Zhan, S.-Z. (2015). *Polyhedron* 92: 124–129.

56 Song, L.-C., Hu, F.-Q., Wang, M.-M. et al. (2014). *Dalton Trans.* 43: 8062–8071.

57 Begum, A. and Sarkar, S. (2012). *Eur. J. Inorg. Chem.* 40–43.

58 Beyler, M., Ezzaher, S., Karnahl, M. et al. (2011). *Chem. Commun.* 47: 11662–11664.

59 Orthaber, A., Karnahl, M., Tschierlei, S. et al. (2014). *Dalton Trans.* 43: 4537–4549.

60 Eady, S.C., Breault, T., Thompson, L., and Lehnert, N. (2016). *Dalton Trans.* 45: 1138–1151.

61 Roy, S., Mazinani, S.K.S., Groy, T.L. et al. (2014). *Inorg. Chem.* 53: 8919–8929.

62 Kütt, A., Leito, I., Kaljurand, I. et al. (2006). *J. Org. Chem.* 71: 2829–2838.
63 Birkholz, M.-N. (née Gensow), Freixa, Z., and van Leeuwen, P.W.N.M. (2009). *Chem. Soc. Rev.*, 2009, 38: 1099–1118.
64 Kaur-Ghumaan, S., Schwartz, L., Lomoth, R. et al. (2010). *Angew. Chem. Int. Ed.* 49: 8033–8036.
65 Artero, V. and Fontecave, M. (2008). *C.R. Chim.* 11: 926–931.
66 Liu, T., DuBois, D.L., and Bullock, R.M. (2013). *Nat. Chem.* 5: 228–233.
67 Weber, K., Weyhermuller, T., Bill, E. et al. (2015). *Inorg. Chem.* 54: 6928–6937.
68 Kütt, A., Rodima, T., Saame, J. et al. (2011). *J. Org. Chem.* 76: 391–395.
69 Rose, M.J., Gray, H.B., and Winkler, J.R. (2012). *J. Am. Chem. Soc.* 134: 8310–8313.
70 Harshan, A.K., Solis, B.H., Winkler, J.R. et al. (2016). *Inorg. Chem.* 55: 2934–2940.
71 Dolganov, A.V., Belov, A.S., Novikov, V.V. et al. (2013). *Dalton Trans.* 42: 4373–4376.
72 Zell, T., Butschke, B., Ben-David, Y., and Milstein, D. (2013). *Chem. Eur. J.* 19: 8068–8072.
73 Tonneau, J. (1991). *Tables de chimie: un mémento pour le laboratoire.* Bruxelles: De Boeck Université.
74 Mellmann, D., Barsch, E., Bauer, M. et al. (2014). *Chem. Eur. J.* 20: 13589–13602.
75 Tard, C., Liu, X., Hughes, D.L., and Pickett, C.J. (2005). *Chem. Commun.* 133–135.
76 Cheah, M.H., Tard, C., Borg, S.J. et al. (2007). *J. Am. Chem. Soc.* 129: 11085–11092.
77 Surawatanawong, P. and Hall, M.B. (2010). *Inorg. Chem.* 49: 5737–5747.
78 Apfel, U.-P., Troegel, D., Halpin, Y. et al. (2010). *Inorg. Chem.* 49: 10117–10132.
79 Ghosh, S., Hogarth, G., Holt, K.B. et al. (2011). *Chem. Commun.* 47: 11222–11224.
80 Rahaman, A., Ghosh, S., Unwin, D.G. et al. (2014). *Organometallics* 33: 1356–1366.
81 Beaume, L., Clémancey, M., Blondin, G. et al. (2014). *Organometallics* 33: 6290–6293.
82 Huang, Y., Gao, W., Åkermark, T. et al. (2012). *Eur. J. Inorg. Chem.* 4259–4263.
83 Li, Z., Zeng, X., Niu, Z., and Liu, X. (2009). *Electrochim. Acta* 54: 3638–3644.
84 Gao, W., Sun, J., Li, M. et al. (2011). *Eur. J. Inorg. Chem.* 1100–1105.
85 Kaiser, M. and Knör, G. (2015). *Eur. J. Inorg. Chem.* 4199–4206.
86 Mebi, C.A., Brigance, K.E., and Bowman, R.B. (2012). *J. Braz. Chem. Soc.* 23: 186–189.
87 Mejía, E., Luo, S.-P., Karnahl, M. et al. (2013). *Chem. Eur. J.* 19: 15972–15978.
88 Fischer, S., Bokareva, O.S., Barsch, E. et al. (2016). *ChemCatChem* 8: 404–411.

89 Rail, M.D. and Berben, L.A. (2011). *J. Am. Chem. Soc.* 133: 18577–18579.
90 Nguyen, A.D., Rail, M.D., Shanmugam, M. et al. (2013). *Inorg. Chem.* 52: 12847–12854.
91 Loewen, N.D., Thompson, E.J., Kagan, M. et al. (2016). *Chem. Sci.* 7: 2728–2735.
92 Canaguier, S., Field, M., Oudart, Y. et al. (2010). *Chem. Commun.* 46: 5876–5878.
93 Yang, D., Li, Y., Su, L. et al. (2015). *Eur. J. Inorg. Chem.* 2965–2973.
94 Ding, S., Ghosh, P., Lunsford, A.M. et al. (2016). *J. Am. Chem. Soc.* 138: 12920–12927.
95 Brazzolotto, D., Gennari, M., Queyriaux, N. et al. (2016). *Nat. Chem.* 8: 1054–1060.
96 Weber, K., Krämer, T., Shafaat, H.S. et al. (2012). *J. Am. Chem. Soc.* 134: 20745–20755.
97 Wombwell, C. and Reisner, E. (2015). *Chem. Eur. J.* 21: 8096–8104.
98 Barton, B.E., Whaley, C.M., Rauchfuss, T.B., and Gray, D.L. (2009). *J. Am. Chem. Soc.* 131: 6942–6943.
99 Barton, B.E. and Rauchfuss, T.B. (2010). *J. Am. Chem. Soc.* 132: 14877–14885.
100 Carroll, M.E., Barton, B.E., Gray, D.L. et al. (2011). *Inorg. Chem.* 50: 9554–9563.
101 Ulloa, O.A., Huynh, M.T., Richers, C.P. et al. (2016). *J. Am. Chem. Soc.* 138: 9234–9245.
102 Sun, P., Yang, D., Li, Y. et al. (2016). *Organometallics* 35: 751–757.
103 Chambers, G.M., Huynh, M.T., Li, Y. et al. (2016). *Inorg. Chem.* 55: 419–431.
104 Sarmini, K. and Kenndler, E. (1999). *J. Biochem. Biophys. Methods* 38: 123–137.
105 Perra, A., Davies, E.S., Hyde, J.R. et al. (2006). *Chem. Commun.* 1103–1105.
106 Summers, P.A., Dawson, J., Ghiotto, F. et al. (2014). *Inorg. Chem.* 53: 4430–4439.
107 Summers, P.A., Calladine, J.A., Ghiotto, F. et al. (2016). *Inorg. Chem.* 55: 527–536.
108 Song, L.-C., Sun, X.-J., Zhao, P.-H. et al. (2012). *Dalton Trans.* 41: 8941–8950.
109 Song, L.-C., Sun, X.-J., Jia, G.-J. et al. (2014). *J. Organomet. Chem.* 761: 10–19.
110 Song, L.-C., Cao, M., and Wang, Y.-X. (2015). *Dalton Trans.* 44: 6797–6808.
111 Fontecilla-Camps, J.C., Volbeda, A., Cavazza, C., and Nicolet, Y. (2007). *Chem. Rev.* 107: 4273–4303.
112 Fourmond, V., Canaguier, S., Golly, B. et al. (2011). *Energy Environ. Sci.* 4: 2417–2427.
113 Song, L.-C., Li, J.-P., Xie, Z.-J., and Song, H.-B. (2013). *Inorg. Chem.* 52: 11618–11626.
114 Zhao, J., Ma, Y., Bai, Z. et al. (2012). *J. Organomet. Chem.* 716: 230–236.
115 Fisher, B.J. and Eisenberg, R. (1980). *J. Am. Chem. Soc.* 102: 7361–7363.
116 Collin, J.P., Jouaiti, A., and Sauvage, J.P. (1988). *Inorg. Chem.* 27: 1986–1990.

117 Ginovska-Pangovska, B., Dutta, A., Reback, M.L. et al. (2014). *Acc. Chem. Res.* 47: 2621–2630.
118 DuBois, D.L. (2014). *Inorg. Chem.* 53: 3935–3960.
119 Shaw, W.J., Helm, M.L., and DuBois, D.L. (2013). *Biochim. Biophys. Acta* 1827: 1123–1139.
120 Small, Y.A., DuBois, D.L., Fujita, E., and Muckerman, J.T. (2011). *Energy Environ. Sci.* 4: 3008–3020.
121 Rakowski, DuBois, M. and DuBois, D.L. (2009). *Chem. Soc. Rev.* 38: 62–72.
122 Rakowski, Dubois, M. and Dubois, D.L. (2009). *Acc. Chem. Res.* 42: 1974–1982.
123 DuBois, M.R. and DuBois, D.L. (2008). *C.R. Chim.* 11: 805–817.
124 Wilson, A.D., Shoemaker, R.K., Miedaner, A. et al. (2007). *Proc. Natl. Acad. Sci. U.S.A.* 104: 6951–6956.
125 Kilgore, U.J., Roberts, J.A.S., Pool, D.H. et al. (2011). *J. Am. Chem. Soc.* 133: 5861–5872.
126 Pool, D.H., Stewart, M.P., O'Hagan, M. et al. (2012). *Proc. Natl. Acad. Sci. U.S.A.* 109: 15634–15639.
127 Helm, M.L., Stewart, M.P., Bullock, R.M. et al. (2011). *Science* 333: 863–866.
128 Le Goff, A., Artero, V., Jousselme, B. et al. (2009). *Science* 326: 1384–1387.
129 Tran, P.D., Le Goff, A., Heidkamp, J. et al. (2011). *Angew. Chem. Int. Ed.* 50: 1371–1374.
130 Gentil, S., Lalaoui, N., Dutta, A. et al. (2017). *Angew. Chem. Int. Ed.* 56: 1845–1849.
131 Zhou, Y., Yang, S., and Huang, J. (2017). *Phys. Chem. Chem. Phys.* 19: 7471–7475.
132 Huan, T.N., Jane, R.T., Benayad, A. et al. (2016). *Energy Environ. Sci.* 9: 940–947.
133 Meyer, K., Bashir, S., Llorca, J. et al. (2016). *Chem. Eur. J.* 22: 13894–13899.
134 Gross, M.A., Creissen, C.E., Orchard, K.L., and Reisner, E. (2016). *Chem. Sci.* 7: 5537–5546.
135 Tran, P.D., Morozan, A., Archambault, S. et al. (2015). *Chem. Sci.* 6: 2050–2053.
136 Weingarten, A.S., Kazantsev, R.V., Palmer, L.C. et al. (2014). *Nat. Chem.* 6: 964–970.
137 Martindale, B.C.M., Joliat, E., Bachmann, C. et al. (2016). *Angew. Chem. Int. Ed.* 55: 9402–9406.
138 Zarkadoulas, A., Field, M.J., Papatriantafyllopoulou, C. et al. (2016). *Inorg. Chem.* 55: 432–444.
139 Gan, L., Groy, T.L., Tarakeshwar, P. et al. (2015). *J. Am. Chem. Soc.* 137: 1109–1115.
140 Wang, D., Zhang, Y., and Chen, W. (2014). *Chem. Commun.* 50: 1754–1756.
141 Downes, C.A. and Marinescu, S.C. (2016). *Dalton Trans.* 45: 19311–19321.
142 Han, Z., McNamara, W.R., Eum, M.-S. et al. (2012). *Angew. Chem. Int. Ed.* 51: 1667–1670.
143 Begum, A., Moula, G., and Sarkar, S. (2010). *Chem. Eur. J.* 16: 12324–12327.
144 Zhang, W., Hong, J., Zheng, J. et al. (2011). *J. Am. Chem. Soc.* 133: 20680–20683.

145 Khrizanforova, V.V., Knyazeva, I.R., Matveeva Sokolova, V.I. et al. (2015). *Electrocatalysis* 6: 357–364.

146 Wise, C.F., Liu, D., Mayer, K.J. et al. (2015). *Dalton Trans.* 44: 14265–14271.

147 Lin, C.-N., Xue, D., Zhou, Y.-H. et al. (2017). *J. Electroanal. Chem.* 785: 58–64.

148 Zarkadoulas, A., Field, M.J., Artero, V., and Mitsopoulou, C.A. (2017). *ChemCatChem* 9: 2308–2317.

149 Koshiba, K., Yamauchi, K., and Sakai, K. (2017). *Angew. Chem. Int. Ed.* 56: 4247–4251.

150 Sondermann, C. and Ringenberg, M.R. (2017). *Dalton Trans.* 46: 5143–5146.

151 Han, Y., Wu, Y., Lai, W., and Cao, R. (2015). *Inorg. Chem.* 54: 5604–5613.

152 Han, Y., Fang, H., Jing, H. et al. (2016). *Angew. Chem. Int. Ed.* 55: 5457–5462.

153 Yuan, Y.-J., Tu, J.-R., Lu, H.-W. et al. (2016). *Dalton Trans.* 45: 1359–1363.

154 Bediako, D.K., Solis, B.H., Dogutan, D.K. et al. (2014). *Proc. Natl. Acad. Sci. U.S.A.* 111: 15001–15006.

155 Cao, J.-P., Fang, T., Fu, L.-Z. et al. (2014). *Int. J. Hydrogen Energy* 39: 10980–10986.

156 Chen, L., Chen, G., Leung, C.-F. et al. (2015). *ACS Catal.* 5: 356–364.

157 Xue, D., Lv, Q.-Y., Lin, C.-N., and Zhan, S.-Z. (2016). *Polyhedron* 117: 300–308.

158 Kankanamalage, P.H.A., Mazumder, S., Tiwari, V. et al. (2016). *Chem. Commun.* 52: 13357–13360.

159 Yuan, Y.-J., Lu, H.-W., Tu, J.-R. et al. (2015). *ChemPhysChem* 16: 2925–2930.

160 Powers, D.C., Anderson, B.L., and Nocera, D.G. (2013). *J. Am. Chem. Soc.* 135: 18876–18883.

161 Han, Z., Shen, L., Brennessel, W.W. et al. (2013). *J. Am. Chem. Soc.* 135: 14659–14669.

162 Luo, S., Siegler, M.A., and Bouwman, E. (2016). *Eur. J. Inorg. Chem.* 4693–4700.

163 Yang, Y., Wang, M., Xue, L. et al. (2014). *ChemSusChem* 7: 2889–2897.

164 Zhang, P., Wang, M., Yang, Y. et al. (2014). *Chem. Commun.* 50: 14153–14156.

165 Tatematsu, R., Inomata, T., Ozawa, T., and Masuda, H. (2016). *Angew. Chem. Int. Ed.* 55: 5247–5250.

166 Moraes Braga Martin, M.d.G., Vidotti, M., and Nunes, F.S. (2012). *Int. J. Hydrogen Energy* 37: 14094–14102.

167 Cui, H., Wang, J., Hu, M. et al. (2013). *Dalton Trans.* 42: 8684–8691.

168 Tsang, C.-S., Chen, L., Li, L.-W. et al. (2015). *Dalton Trans.* 44: 13087–13092.

169 Lin, C.-N., Ren, S.-T., Ye, L.-P. et al. (2016). *Inorg. Chem. Commun.* 69: 24–27.

170 Mondragon, A., Flores-Alamo, M., Martinez-Alanis, P.R. et al. (2015). *Inorg. Chem.* 54: 619–627.

171 Artero, V., Chavarot-Kerlidou, M., and Fontecave, M. (2011). *Angew. Chem. Int. Ed.* 50: 7238–7266.

172 Queyriaux, N., Jane, R.T., Massin, J. et al. (2015). *Coord. Chem. Rev.* 304–305: 3–19.
173 Kaeffer, N., Chavarot-Kerlidou, M., and Artero, V. (2015). *Acc. Chem. Res.* 48: 1286–1295.
174 Sun, Y., Bigi, J.P., Piro, N.A. et al. (2011). *J. Am. Chem. Soc.* 133: 9212–9215.
175 Singh, W.M., Baine, T., Kudo, S. et al. (2012). *Angew. Chem. Int. Ed.* 51: 5941–5944.
176 Zhang, P., Wang, M., Gloaguen, F. et al. (2013). *Chem. Commun.* 49: 9455–9457.
177 Aroua, S., Todorova, T.K., Mougel, V. et al. (2017). *ChemCatChem* 9: 2099–2105.
178 Lucarini, F., Pastore, M., Vasylevskyi, S. et al. (2017). *Chem. Eur. J.* 23: 6768–6771.
179 Andreiadis, E.S., Jacques, P.-A., Tran, P.D. et al. (2013). *Nat. Chem.* 5: 48–53.
180 Chen, L., Wang, M., Han, K. et al. (2014). *Energy Environ. Sci.* 7: 329–334.
181 McNamara, W.R., Han, Z., Yin, C.-J. et al. (2012). *Proc. Natl. Acad. Sci. U.S.A.* 109: 15594–15599.
182 Anxolabéhère-Mallart, E., Costentin, C., Fournier, M. et al. (2012). *J. Am. Chem. Soc.* 134: 6104–6107.
183 Anxolabéhère-Mallart, E., Costentin, C., Fournier, M., and Robert, M. (2014). *J. Phys. Chem. C* 118: 13377–13381.
184 El Ghachtouli, S., Guillot, R., Brisset, F., and Aukauloo, A. (2013). *ChemSusChem* 6: 2226–2230.
185 Sconyers, D.J. and Blakemore, J.D. (2017). *Chem. Commun.* 53: 7286–7289.
186 El Ghachtouli, S., Fournier, M., Cherdo, S. et al. (2013). *J. Phys. Chem. C* 117: 17073–17077.
187 Kaeffer, N., Morozan, A., Fize, J. et al. (2016). *ACS Catal.* 6: 3727–3737.
188 Tong, P., Xie, W., Yang, D. et al. (2016). *Dalton Trans.* 45: 18559–18565.
189 Szymczak, N.K., Berben, L.A., and Peters, J.C. (2009). *Chem. Commun.* 6729–6731.
190 Greb, L., Tussing, S., Schirmer, B. et al. (2013). *Chem. Sci.* 4: 2788–2796.
191 Mandal, S., Shikano, S., Yamada, Y. et al. (2013). *J. Am. Chem. Soc.* 135: 15294–15297.
192 Valdez, C.N., Dempsey, J.L., Brunschwig, B.S. et al. (2012). *Proc. Natl. Acad. Sci. U.S.A.* 109: 15589–15593.
193 Kal, S., Filatov, A.S., and Dinolfo, P.H. (2014). *Inorg. Chem.* 53: 7137–7145.
194 Wang, Z.-Y., Rao, H., Deng, M.-F. et al. (2013). *Phys. Chem. Chem. Phys.* 15: 16665–16671.
195 Ahn, H.S., Davenport, T.C., and Tilley, T.D. (2014). *Chem. Commun.* 50: 3834–3837.
196 Li, P., Zaffaroni, R., de Bruin, B., and Reek, J.N.H. (2015). *Chem. Eur. J.* 21: 4027–4038.
197 Valyaev, D.A., Lavigne, G., and Lugan, N. (2016). *Coord. Chem. Rev.* 308: 191–235.
198 Valyaev, D.A., Peterleitner, M.G., Semeikin, O.V. et al. (2007). *J. Organomet. Chem.* 692: 3207–3211.
199 Sampson, M.D. and Kubiak, C.P. (2015). *Inorg. Chem.* 54: 6674–6676.

200 Mukhopadhyay, T.K., MacLean, N.L., Gan, L. et al. (2015). *Inorg. Chem.* 54: 4475–4482.

201 Rao, G.K., Jamshidi, M.P., Dawkins, J.I.G. et al. (2017). *Dalton Trans.* 46: 6518–6522.

202 Hou, K., Poh, H.T., and Fan, W.Y. (2014). *Chem. Commun.* 50: 6630–6632.

203 Hou, K. and Fan, W.Y. (2014). *Dalton Trans.* 43: 16977–16980.

204 Hou, K., Lauw, S.J.L., Webster, R.D., and Fan, W.Y. (2015). *RSC Adv.* 5: 39303–39309.

205 Kee, J.W., Chong, C.C., Toh, C.K. et al. (2013). *J. Organomet. Chem.* 724: 1–6.

206 Zhang, P., Wang, M., Yang, Y. et al. (2014). *Angew. Chem. Int. Ed.* 53: 13803–13807.

207 Liao, R.-Z., Wang, M., Sun, L., and Siegbahn, P.E.M. (2015). *Dalton Trans.* 44: 9736–9739.

208 Xin, Z.-J., Liu, S., Li, C.-B. et al. (2017). *Int. J. Hydrogen Energy* 42: 4202–4207.

209 Lei, H., Fang, H., Han, Y. et al. (2015). *ACS Catal.* 5: 5145–5153.

210 Liang, X., Niu, Y., Zhang, Q. et al. (2017). *Dalton Trans.* 46: 6912–6920.

211 Cao, J.-P., Fang, T., Wang, Z.-Q. et al. (2014). *J. Mol. Catal. Chem.* 391: 191–197.

212 Fang, T., Zhou, L.-L., Fu, L.-Z. et al. (2015). *Polyhedron* 85: 355–360.

213 Fang, T., Lu, H.-X., Zhao, J.-X. et al. (2015). *J. Mol. Catal. Chem.* 396: 304–309.

214 Xue, D., Luo, S.-P., Chen, Y.-Y. et al. (2017). *Polyhedron* 132: 105–111.

215 Kügler, M., Scholz, J., Kronz, A., and Siewert, I. (2016). *Dalton Trans.* 45: 6974–6982.

216 Nestke, S., Kügler, M., Scholz, J. et al. (2017). *Eur. J. Inorg. Chem.* doi: 10.1002/ejic.201700154.

217 Garrett, B.R., Click, K.A., Durr, C.B. et al. (2016). *J. Am. Chem. Soc.* 138: 13726–13731.

218 Garrett, B.R., Polen, S.M., Pimplikar, M. et al. (2017). *J. Am. Chem. Soc.* 139: 4342–4345.

219 Garrett, B.R., Polen, S.M., Click, K.A. et al. (2016). *Inorg. Chem.* 55: 3960–3966.

220 Porcher, J.-P., Fogeron, T., Gomez-Mingot, M. et al. (2015). *Angew. Chem. Int. Ed.* 54: 14090–14093.

221 Gomez-Mingot, M., Porcher, J.-P., Todorova, T.K. et al. (2015). *J. Phys. Chem. B* 119: 13524–13533.

222 Pal, R., Laureanti, J.A., Groy, T.L. et al. (2016). *Chem. Commun.* 52: 11555–11558.

223 Zee, D.Z., Chantarojsiri, T., Long, J.R., and Chang, C.J. (2015). *Acc. Chem. Res.* 48: 2027–2036.

224 Karunadasa, H.I., Chang, C.J., and Long, J.R. (2010). *Nature* 464: 1329–1333.

225 Thoi, V.S., Karunadasa, H.I., Surendranath, Y. et al. (2012). *Energy Environ. Sci.* 5: 7762–7770.

226 Karunadasa, H.I., Montalvo, E., Sun, Y. et al. (2012). *Science* 335: 698–702.

227 Webster, L.R., Ibrahim, S.K., Wright, J.A., and Pickett, C.J. (2012). *Chem. Eur. J.* 18: 11798–11803.

228 Huang, Z., Luo, W., Ma, L. et al. (2015). *Angew. Chem. Int. Ed.* 54: 15181–15185.

229 Edidin, R.T., Sullivan, J.M., and Norton, J.R. (1987). *J. Am. Chem. Soc.* 109: 3945–3953.

230 Appel, A.M., DuBois, D.L., and Rakowski, DuBois, M. (2005). *J. Am. Chem. Soc.* 127: 12717–12726.

231 Rosenkoetter, K.E., Ziller, J.W., and Heyduk, A.F. (2016). *Inorg. Chem.* 55: 6794–6798.

232 Bouchard, S., Clémancey, M., Blondin, G. et al. (2014). *Inorg. Chem.* 53: 11345–11347.

233 Schilter, D., Fuller, A.L., and Gray, D.L. (2015). *Eur. J. Inorg. Chem.* 4638–4642.

19

Nonreductive Reactions of CO_2 Mediated by Cobalt Catalysts: Cyclic and Polycarbonates

Thomas A. Zevaco[1] *and Arjan W. Kleij*[2,3]

[1] *Institute of Catalysis Research and Technology (KIT), Hermann-von-Helmholtz-Platz 1, 76344 Eggenstein-Leopoldshafen, Germany*
[2] *Catalan Institution for Research and Advanced Studies (ICREA), Passeig Lluis Companys 23, 08010 - Barcelone, Spain*
[3] *Institute of Chemical Research of Catalonia (ICIQ), The Barcelona Institute of Science and Technology, Avinguda Països Catalans 16, 43007 Tarragona, Spain*

19.1 Introduction

Catalysts based on cobalt, representing an attractive, relatively cheap, and abundant metal, have been progressively used by the synthetic community, giving new potential in important areas such as cross-coupling reactions, enantioselective transformations, hydroformylation, and activation of π-bonds such as those encountered in alkynes. [1] Their privileged use in the recently emerged area of CO_2 activation and conversion is probably less known , where the typical characteristics of cobalt compounds (rich coordination chemistry, access to different, relatively stable oxidation states, and chemical stability) have been exploited to design among the best reported catalysts to date for the prominent category of nonreductive coupling reactions of epoxides and CO_2 into organic carbonates. The vast majority of the catalytic efforts has focused on the formation of aliphatic polycarbonates with less attention on cyclic carbonates, and this chapter aims to highlight the advances made in catalyst design and development, amplification of the synthetic potential, and the mechanistic lessons learned from experimental and theoretical studies. Through examination of the different cobalt-catalyzed coupling reactions of CO_2 with epoxides to produce organic carbonates (cyclic or polymeric, see Figure 19.1), three main classes of complexes are noteworthy. These complexes involve salens (probably the most emblematic class of ligands in this context), porphyrins, and related amino-phenoxide chelating ligands. This chapter will focus on homogeneous systems, leaving aside the heterogeneous catalyst systems whose exact mode of action remains the subject of discussion [3].

Non-Noble Metal Catalysis: Molecular Approaches and Reactions,
First Edition. Edited by Robertus J. M. Klein Gebbink and Marc-Etienne Moret.

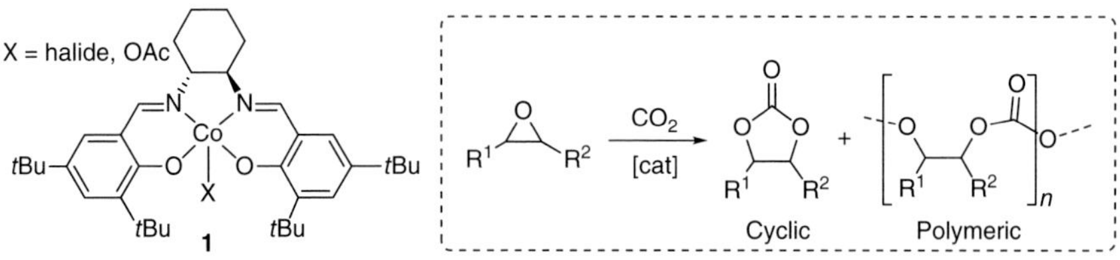

Figure 19.1 The cobalt–salen catalyst reported by Coates and coworkers [2] and the formation of cyclic and polymeric carbonates from CO_2 and epoxides (on the right).

19.2 Cocatalysts for CO_2/Epoxide Couplings: Salen-Based Systems

Among the efficient homogeneous catalytic systems used, the cobalt–salen complexes (Figure 19.1, **1**) stand out as highly efficient catalytic systems for CO_2 conversion, combining a "cheap" metal element with a versatile and modular ligand architecture. The research group of Coates had a major impact in this area, building their pioneering work on a catalytic system first reported by Jacobsen and coworkers [4], who used chiral (salcy)Co(III) carboxylates in the hydrolytic kinetic resolution of epoxides with remarkable efficiencies (salcy = *N*,*N*′-bis(3,5-di-*tert*-butylsalicylidene)-1,2-diaminocyclohexane). Coates et al. extended the utilization of these cobalt complexes to the copolymerization of propylene oxide (PO) and CO_2 and reported in 2003 an – at that time – outstanding selectivity in the formation of highly regioregular polypropylene carbonate (PPC) with head-to-tail (H–T) linkages around 80% at ambient temperature and under 55 bar of CO_2 pressure [2]. The addition of a cocatalyst (i.e. quaternary ammonium halides, NBu_4X) substantially enhanced the catalytic activity of the cobalt–salen complexes at higher temperatures and lower CO_2 pressures and led to an excellent selectivity of >99% PPC vs propylene carbonate (PC) with a very low amount of ether linkages (<1%, usually obtained via ring-opening polymerization (ROP) of the epoxide substrate) [5]. One of the proposed explanations for this improvement was a stabilization of the active Co(III) salen catalyst by the ionic cocatalyst, preventing the reduction to inactive Co(II) salen complexes.

The cobalt–salen complexes revealed a considerable copolymerization potential in combination with a wide range of neutral Lewis bases (*N*-methyl imidazole, pyridine, 4-dimethylaminopyridine, 1,5,7-triazabicyclo-[4,4,0]dec-5-ene, 1,8-diazabicyclo[5.4.0] undec-7-ene, and 7-methyl-1,5,7-triazabicyclo[4.4.0]dec-5-ene) or with "onium" salts (mostly ammonium or phosphonium halides/azides) as cocatalysts. This significant catalytic potential is largely due to the modular and easy-to-optimize salen ligand architecture, giving straightforward formation of a broad ligand library. Additionally, the overall high stability of salen-based systems against oxidation and moisture also adds to their catalytic application potential. The possibility to influence the stereochemistry of the copolymerization and chemoselectivity of the reaction toward polycarbonate or cyclic carbonate using an appropriate salen backbone and cocatalyst is also a major advantage in favor of the use of this class of catalysts. The term "catalyst" should be, however, used

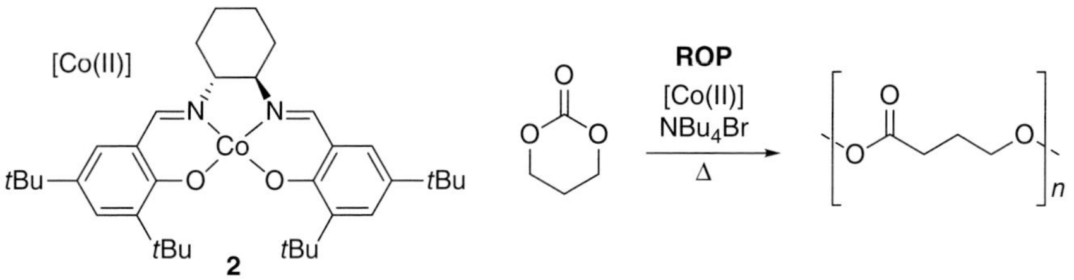

Figure 19.2 A cobalt–salen catalyst (2) reported by Darensbourg and Moncada [6] and the ring-opening polymerization (ROP) of a six-membered cyclic carbonate.

with some precaution if one considers that not only the structure of the ligand and the nature of the metal but also the initiating group present in the metal complex (Figure 19.1; X = Cl, Br, NO_3, OAc, and 2,4-dinitro-phenoxy) plays a significant role. Actually, the Co(salen) complexes should be rather regarded as catalyst precursors, with the actual catalytic active alkoxo species being generated *in situ*.

Interestingly, the research group of Darensbourg widened at an early stage the cyclic ether substrate scope and reported the successful copolymerization of oxetane with CO_2 in the presence of a Co(salen) complex (**2**) with tetrabutyl ammonium bromide (TBAB) as the cocatalyst (Figure 19.2) [6]. *In situ* infrared spectroscopy was used to study the mechanism of the reaction and revealed that the copolymerization is actually better described as an anion-mediated ROP of the *in situ*-formed monomeric cyclic trimethylene carbonate. These first successful attempts led to many subsequent studies aiming at a better understanding of the mechanism at work in epoxide/CO_2 coupling chemistry and at optimizing yields and selectivities obtained with these cobalt–salens. Darensbourg and Lu nicely summarized the different approaches and monitoring efforts performed with the cobalt–salens in tutorial reviews published in 2012 with a focus on the mechanistic pathways [7]. Interestingly, many other research groups complemented these overviews with comprehensive reviews focusing (more) on stereoselective (co)polymerization [8] and with wide-ranging inventories of the homogeneous catalysts used in (co)polymerization [9] or related cyclization reactions [10] involving epoxides, with the cobalt catalysts being prominently discussed in these latter accounts.

Depending on the complexity of the investigated system and the nature of the cocatalyst, both monometallic and bimetallic mechanisms are plausible with Co(salen)-based systems. However, the monometallic mechanism seems to be more likely in the case of binary catalysts, i.e. the combination of a Co(salen) complex and a cocatalytic additive. A prerequisite for a high catalytic activity and a successful control of the coupling reaction is the presence of a chiral salen ligand with sterically demanding substituents in the ortho- and para-positions of the aromatic side groups of the salen scaffold. On the other hand, for good activity, an ionic cocatalyst should have an anion with a rather low nucleophilicity and good leaving ability (Br^-, I^-) and a rather bulky cation (n-Bu_4N or bis(triphenylphosphine)iminium, PPN). The use of a neutral base instead of an onium salt seems to improve the activity of the catalyst system only when the base is strong enough and has low competing coordination ability for the cobalt center.

Many research groups have backed up their experimental findings with theoretical studies using different computational methodologies. Some of the more recent reports allow for a practical comparison between Cr-, Al-, Zn-, and Co-based salen catalysts confirming in many cases the catalytic activities found during the screening experiments. For instance, the group of Deng compared Zn–, Al–, and Co–salens in the formation of cyclic carbonates using density functional theory (DFT) methods based on the B3LYP functional and a polarized 6-311G-(3df,3pd) basis set/PCM solvation model (Gaussian 09 package). On the basis of these theoretical studies, the cobalt complex is expected to have the best catalytic activity among the three complexes, with the Zn–salen being slightly more active than the Al complex [11]. Recently, Darensbourg investigated related Cr– and Co–salens in copolymerization reactions using DFT-based methods. These complexes with a simplified, non-substituted salen ligand were investigated using the M06-L functional and a diffused BS2+ basis set/SMD solvation model (Gaussian 09 package). The free energy barriers found for the reactions involving Co–salen and a range of epoxides, including PO and cyclohexene oxide (CHO), were systematically lower (0.5–3 $kcal\,mol^{-1}$) than those found for the Cr-analog, in agreement with the better catalytic activity observed at lower temperatures for the Co–salen derivatives [12].

As mentioned earlier, the high structural versatility of the salen ligand architecture provides many possible variations, whereas a wide range of substitution patterns of the salicylidene fragments have been reported to date, the bridging unit is in many cases based on either a simple ethylenediamine (salen) or a chiral 1,2-diamino-cyclohexane (salcy) group. Based on a former study of Coates [13], Merna and coworkers replaced the 1,2-diamino-cyclohexane fragment by a nonsymmetrically substituted 1,2-diamino-benzene [14]. These cobalt(III) sal*phen* derivatives displayed high catalytic activity for CHO/CO_2 copolymerization (chemoselectivity >99%) to pure poly(cyclohexene)carbonate (PCHC) polymers with high molar masses of 15–30 $kg\,mol^{-1}$ and narrow polydispersities ($Đ = 1.07–1.23$), whereas the chemoselectivity of PO/CO_2 copolymerization was lower (<89%). As documented earlier, the nature of the initiating group played a key role in the PO/CO_2 copolymerization reaction with respect to catalytic activity and selectivity, with the best results being obtained using Co–salphen complexes having a trichloroacetate axial ligand incorporated.

The pivotal role of the cocatalyst(s) led many research groups to optimize its function and notably its separation from the products for potential recycling. The research group of Nozaki further developed the basic concept of integrating the catalyst and cocatalyst into one, bifunctional system. The obtained one-component Co–salen complex with tethered piperidinyl/piperidinium groups (Figure 19.3, **3**) exhibited high selectivity toward aliphatic PPC with high conversion and improved catalytic activity. This system allowed for the first time the terpolymerization of two epoxides (PO and 1-hexene oxide) with CO_2 [15]. The interplay between the axial acetate ligand that initiates the copolymerization and the piperidinium-based salen framework that influences the nucleophilicity of the propagating copolymer species was proposed to lead to

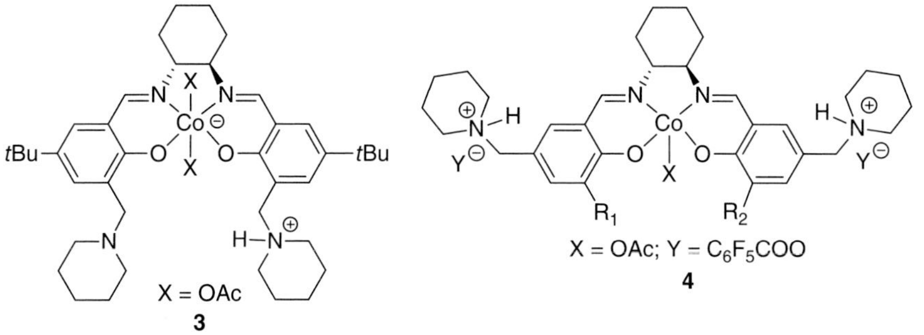

Figure 19.3 Bifunctional Co–salens **3** and **4** reported by Nozaki and coworkers [15, 16].

the observed high chemoselectivity. The same group further optimized the salen ligand and confirmed that substituents exerting a high steric demand near the cobalt center are important to influence both the regio- and enantioselectivities. As a result of these ligand optimization studies, they reported the preparation of a so-called stereogradient copolymer using a modified one-component, bifunctional Co–salen complex (Figure 19.3, **4**), that favors the incorporation of (*S*)-PO by a selectivity factor *s* of 3.5. The stereogradient copolymer produced from racemic PO exhibits regions enriched in (*S*)-PO inserted at the beginning of the reaction and regions enriched in (*R*)-PO formed after full consumption of (*S*)-PO. In an intermediary state, when the amount of (*R*)-PO is 3.5 times higher than that of the (*S*)-enantiomer, the copolymer displays a region with alternating (*R*)/(*S*)-PO incorporation [16].

Other examples of bifunctional catalysts have also been reported, thereby unifying a Lewis acidic center and a functional group acting as a nucleophile. For instance, the group of Lee developed highly active bifunctional Co–salen catalysts containing sophisticated linear [17] or branched [18] quaternary ammonium salts at the para-position of the salicylidene fragments (Figure 19.4, **5** and **6**). Apart from these bifunctional cobalt complexes, an unusual cobalt catalyst (Figure 19.4, 7) was also reported by Lee that displayed very high catalytic activity with turnover frequencies (TOFs) of up to 20 000 h^{-1} and generating up to 2.0 kg of polymer per gram of catalyst (57 kg per g Co: turnover number (TON) = 32 900) in just two hours. The reported polycarbonate/cyclic carbonate selectivities are also very high, as well as the stereoselectivity of the monomer insertion with head-to-tail coupling efficiencies of up to 94%. The high activity of **7** was correlated with the unusual binding mode of the cobalt ion to the substituted salen ligand; the imine nitrogen donors of the salen ligand do not directly stabilize the metal via coordination, but instead the cobalt ion forms a complex with the counter anions (2,4-dinitrophenolates) of the tethered quaternary ammonium groups, being concurrently "side-on" *O*,*O*′-coordinated to the N_2O_2 plane [19].

Interestingly, Lee and coworkers also reported another unusual coordination of the salen ligand in the well-known Jacobsen catalyst, i.e. (salcy)Co(III)OAc, with the tetradentate salen ligand being "wrapped" in a cis-β configuration around the cobalt center with a *O*,*O*′-chelating acetate (Figure 19.4, **8**) [20]. This

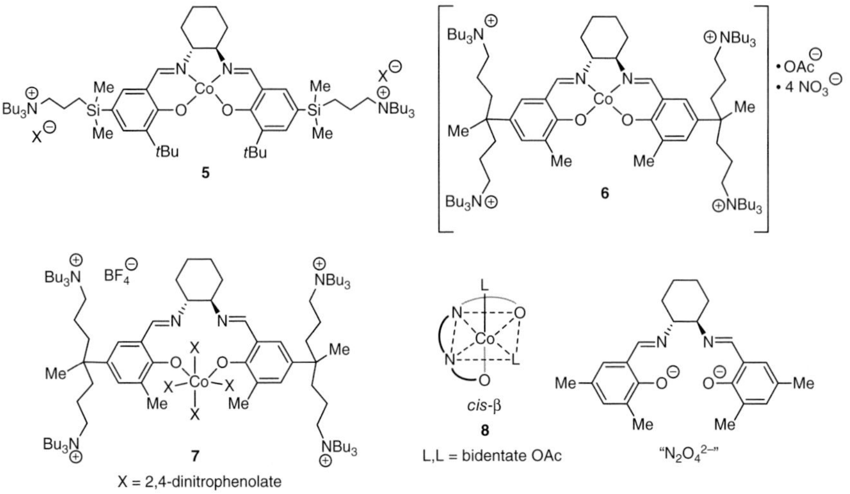

Figure 19.4 Bifunctional Co(salen) complexes reported by Lee and coworkers [17–20].

unusual coordination mode was proposed as a reason for the unusual catalytic activity in the epoxide/CO_2 copolymerization reactions reported for some of the Co–salen systems. The quality of the copolymers obtained with this system is high, as demonstrated by the high molecular weights (M_n up to 300 000 g mol^{-1}) and narrow $Đ$'s.

These bifunctional Co–salens have also been used to form terpolymers via addition of a second epoxide or other chain extenders [21]. Notably, the molecular weight could be easily controlled over a broad range (M_n = 2000–600 000 g mol^{-1}) by adding chain transfer agents such as alcohols, thereby influencing the chain length of the copolymer growing at the cobalt center. The synthesis of the original catalysts was further optimized toward tetra-nitrato derivatives (Figure 19.4; introducing X = NO_3^- in **7**), whose handling on a large scale is safer as compared to the potentially explosive 2,4-dinitrophenoxo variants, retaining rather similar catalytic performances. The bifunctional cobalt-based catalytic systems of Lee et al. belong to a category of rare systems to fulfill the prerequisites for commercialization owing to their high activity, stability, and easy separation and were accordingly patented in South Korea by SK Innovation Co., Ltd. [22].

Similar types of bifunctional Co–salens were developed by Lu, who grafted strong Lewis bases such as 1,5,7-triazabicyclo-[4,4,0]dec-5-ene (TBD) or more commonly used alkyl-triethyl ammonium units onto the salen ligand (Figure 19.5). The resulting complexes are among the most active reported to date for the copolymerization of CO_2 and epoxides, producing, for instance, PPC with a TOF around 10 880 h^{-1} under relatively mild conditions (100 °C, 25 bar CO_2) with a catalyst concentration of only 0.01 mol%. The methyldiethyl ammonium-substituted Co–salen **9** was highly efficient for CHO/CO_2 coupling to give PCHC (120 °C, 25 bar CO_2) with high TOFs of up to 6105 h^{-1} [25].

The same catalysts are also very promising for the controlled terpolymerization of CHO, PO, and CO_2 to selectively give polymeric materials with higher

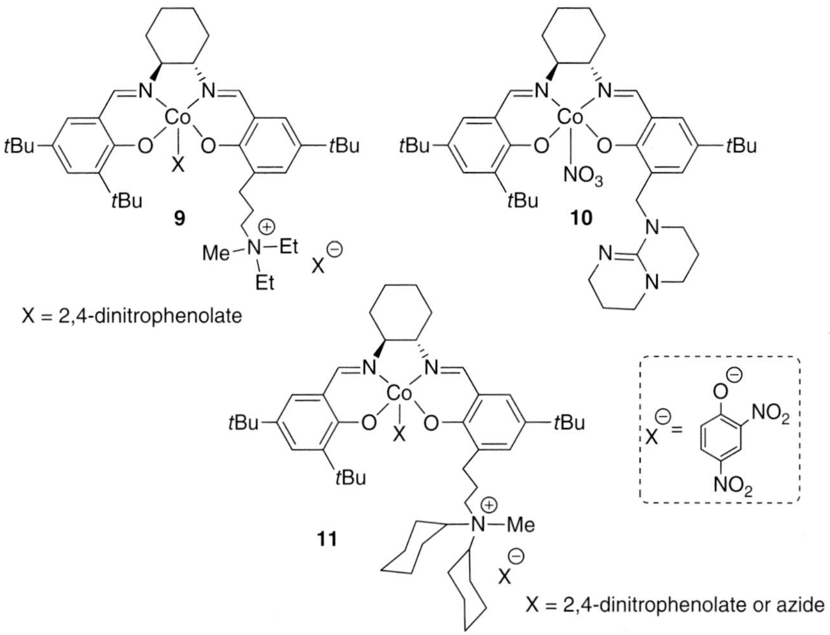

Figure 19.5 Bifunctional Co–salens developed by Lu, Darensbourg et al. [23, 24].

glass transition temperatures (T_g's) than typically found for pure PPC, demonstrating the trend toward an increasing control of the copolymerization and the generation of new attractive polymeric materials [26].

Further to this, the Lu group also focused on other high-quality polymeric materials obtained via copolymerization of CO_2 with other epoxides and investigated in detail monomers such as epichlorohydrin [27] and styrene oxide [23] to give isotactic poly(chloropropylene carbonate) and poly(styrene carbonate), respectively, with high chemo/regioselectivity and in good yields. For instance, the complex with a pendant TBD fragment (Figure 19.5, **10**) delivered an alternating copolymer with 82% head-to-tail linkages and around 87% ee (after hydrolysis) using an optically pure (*S*)-styrene oxide. A slightly modified complex bearing an adamantane group on one side of the salen and a bulky methyl-dicyclohexyl ammonium group on the other side (not shown in Figure 19.5) led to the formation of poly(chloropropylene carbonate) with an enantioselectivity of 94% ((*R*)-configured) using enantiomeric pure (*R*)-epichlorohydrin as substrate, indicative of retaining 97% of the original stereochemistry at the methine carbon of the epoxide incorporated into the polycarbonate [28].

Along these lines, Darensbourg and Chung used a similar bifunctional, one-component Co–salen catalyst with a functional pendant group comprising a methyldi-cyclohexyl ammonium moiety and incorporating a 2,4-dinitro-phenoxide or azide as the initiating group (Figure 19.5, **11**). These complexes were tested in the coupling of CO_2 with *cis*-2-butene oxide, *trans*-2-butene oxide, *iso*-butene oxide, and 2,3-epoxy-2-methylbutane. The

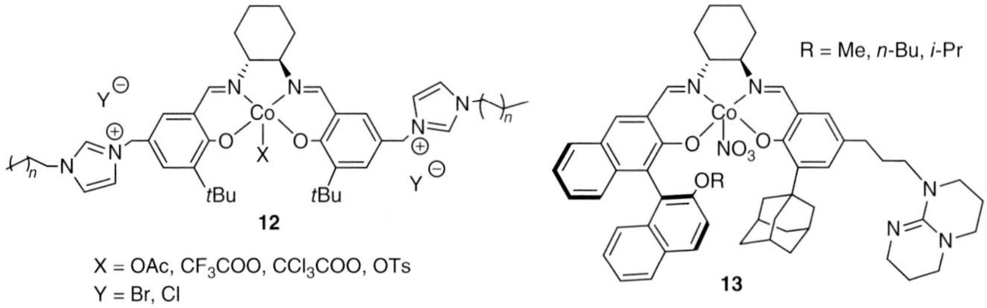

X = OAc, CF_3COO, CCl_3COO, OTs
Y = Br, Cl

R = Me, *n*-Bu, *i*-Pr

Figure 19.6 Bifunctional Co–salen complexes **12** with pendant imidazolium halide groups reported by Jing and coworkers [29] and multichiral Co–salens **13** developed by Lu and coworkers [30].

reaction with the most reactive substrate of this latter series, i.e. *cis*-2-butene oxide, produced highly selectively copolymers (>99% at 40 °C) in 75% yield, whereas increasing the temperature to 70 °C led to significant amounts of *trans*-cyclic carbonate. The cyclic *trans*-butene carbonate results from a "back-biting" process within the copolymer that requires a temperature onset. Comparatively, the other epoxides in this series exclusively delivered the cyclic carbonates as products [24].

The research group of Jing reported on a bifunctional, one-component chiral Co–salen comprising imidazolium fragments that are usually part of ionic liquids (Figure 19.6, **12**) [29]. The catalytic system was examined with a range of racemic epoxides with the aim to induce kinetic resolution and isolate enantiomerically pure cyclic carbonates. The Co–salen complex equipped with imidazolium chloride groups and acetate as the leaving group gave rise to enantiomerically enriched PC with ee's ranging from 43% to 57%. According to the results gathered in this study, the reactivity trend followed the order OAc > CF_3CO_2 > CCl_3CO_2 > OTs when varying the axial ligand, whereas the observed enantiomeric excess showed a different trend: OTs > OAc > CCl_3CO_2 > CF_3CO_2. Unfortunately, no mechanistic postulation was provided to explain these trends.

The pathway to high-quality polymeric materials embraces not only the utilization of a broader scope of epoxide substrates but also a strict control of the regio- and stereoselectivity of the CO_2/epoxide coupling reaction and thus generally of the polymer microstructure. Lu and coworkers investigated in detail Co–salens with chiral features within different regions of the salen framework. Fascinating "multichiral" Co–salen complexes were prepared comprising a chiral BINOL (1,1′-binaphthol) fragment on one side of the salen moiety and on the other side a 3-adamantyl substituent and a tethered TBD group acting as a sterically hindered organic base (Figure 19.6, **13**) [30].

This design aimed at influencing the chiral environment around the central metal ion via the BINOL fragment, whereas the tethered TBD fragment should enable high activity of the catalyst in different deployments at low temperatures and low catalyst loadings. These Co–salen complexes are able to perform the asymmetric, regio- and stereoselective alternating copolymerization of various racemic aliphatic terminal epoxides. The results obtained with these catalysts

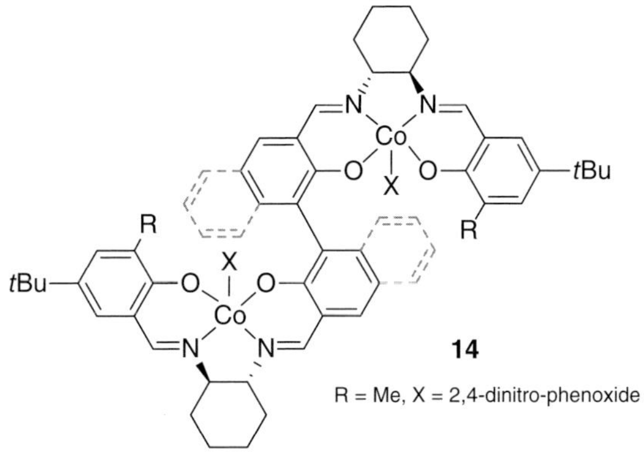

Figure 19.7 Dinuclear Co–salen complexes **14** investigated by Lu and coworkers [32].

validated the general design concept, and the Co–salens facilitated preferential (*R*)- over (*S*)-PO insertion into the polymer backbone with an observed K_{rel} of up to 24 for the (*R*) enantiomer with the ring-opening occurring preferentially at the methylene carbon to give copolymers with >99% head-to-tail contents. Interestingly, the combination of this type of "multichiral" Co–salen complexes with a specific cocatalyst (i.e. the 2,4-dinitrophenolate salt of bis-(triphenylphosphine)iminium; PPN-DNP) generates chiral cyclic carbonates in low-to-moderate yields (12–43%) depending on the nature of the epoxide substituents but with appreciable ee's ranging from 68% to 89% at optimized temperatures and catalyst loadings. To favor cyclic carbonate formation, the use of an excess of a nucleophilic cocatalyst (anion) with good leaving group ability is necessary, whereas the use of a bulky cation leads to high asymmetric induction [31]. Interestingly, Lu's group further developed these Co–salen structures toward dinuclear (*S*,*S*,*S*,*S*)-configured multi-chiral cobalt complexes with two Co–salens bridged through a biphenyl/binaphthyl linker (Figure 19.7, **14**). These complexes exhibited similar enantioselectivities as their former variants with, however, an improved catalytic activity observed for CHO and the less reactive cyclopentene oxide copolymerization reactions in the presence of PPNCl as the cocatalyst, producing high-molecular-weight aliphatic polycarbonates (M_n up to 107.9 kg mol^{-1}, $Đ = 1.4$, ee > 99%) with a perfectly isotactic structure [32].

19.3 Co–Porphyrins as Catalysts for Epoxide/CO_2 Coupling

The most basic structural feature encountered in the transition metal salens is the planar coordination geometry of the ligand to the metal, the metal occupying either the center of this N_2O_2 chelate or being "side-on" *O*,*O*′-coordinated to the N_2O_2 plane as in the case of the Lee's complexes [19]. A comparable planar structure is found in transition metal porphyrin complexes with the only difference that an N_4 chelation is provided, while retaining a dianionic nature. The well-defined coordination geometry around the metal ion in porphyrins allowed them to act as ideal systems for mechanistic studies in

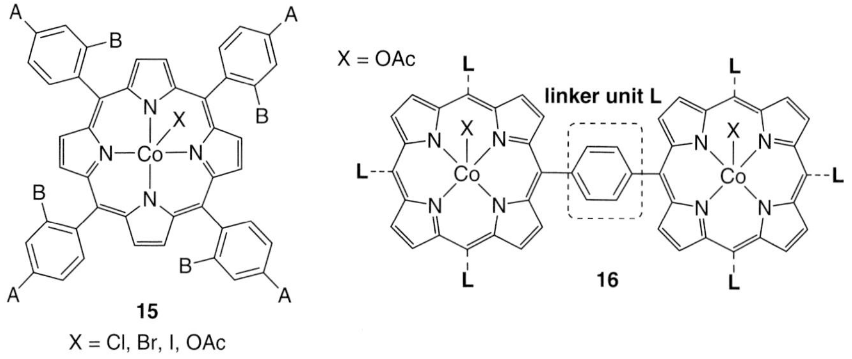

Figure 19.8 General structure of (substituted) Co–TPPs with ortho- and para-substitution [34, 35]. On the right, the metalloporphyrin-based organic polymers reported by Chen [36].

the catalytic coupling of epoxides with CO_2, with an early example provided by aluminum-(5,10,15,20-tetraphenylporphyrin) (Al(TPP)) reported by the Inoue group [33]. Hence, it is perhaps not surprising that also some Co-based porphyrin compounds are among the most effective catalysts for epoxide/CO_2 couplings. Compared to their salen analogs, the Co–porphyrins (Figure 19.8) do not exhibit the same versatile/modular chemistry, likely because of more complex synthetic procedures required for their preparation and fewer possible modifications of the ligand backbone; therefore, most of the studies deal with the parent TPP system. The use of Co–porphyrins (Figure 19.8, **15**) as catalysts in the formation of organic carbonates was reported for the first time by the research group of Nguyen using Co(TPP) in combination with 4-dimethylamino-pyridine (DMAP) as cocatalyst to produce PC at 120 °C [34]. The system was further optimized by Jing and coworkers, who successfully used phenyltrimethyl ammonium tribromide (PTAT) as a highly efficient cocatalyst and investigated the role of ortho- (B = –OMe, $-NO_2$, and $-NH_2$) and para-substitution (A = $-CO_2H$) of the aromatics in Co–porphyrins, reporting high yields of cyclic carbonate only in the case of the ortho-substituted methoxy derivative (and the plain Co-TPP) [35]. Interestingly, this group reported the positive effect of a subsequent addition of trifluoroacetic acid on the course of the reaction using the para-carboxy-substituted Co–porphyrin/PTAT catalytic system [35b].

The separation of a homogeneous catalyst from a reaction mixture enriched with cyclic carbonate(s) can be quite tedious because of the relatively high polarities and boiling points of cyclic carbonates. This complication led many groups over the years to support the catalysts onto different matrices or to integrate them as part of an "organized framework." This aspect was recently investigated in more detail by the group of Chen [36], who developed a convenient and simple strategy for the construction of porphyrin-based organic polymers (designated as (Por)OP) and metalloporphyrin-based organic polymers (M(Por)OP: M = Zn and Co; Figure 19.8, **16**), which display a high density of catalytic sites that are covalently linked. The (Por)OP starting material was obtained via condensation of pyrrole with terephthaldehyde (1,4-diformylbenzene) in propanoic acid and treated with Zn and Co acetate to yield the M(Por)OP networks. The catalytic tests performed with the heterogeneous catalyst Co(Por)OP using PO and CO_2 as substrates led to PC in high yields (86%). The catalyst is essentially insoluble in almost all organic solvents, allowing thus for an easy catalyst recycling. Interestingly, other related "conjugated microporous polymers" with incorporated aluminum and cobalt sites were investigated by the group of Deng [37] and Son and coworkers [38] in the formation of cyclic carbonates derived from propylene oxide and epichlorohydrin, and excellent recycling potential (up to 22 cycles with virtually no loss in activity) was described.

In the context of copolymerization of epoxide and CO_2, two research groups simultaneously reported on binary catalytic systems based on Co(TPP) and nucleophilic additives such as PPNCl [39] and 4-dimethylaminopyridine [40]. These Co(TPP)-based catalysts delivered copolymers under mild conditions (i.e. room temperature and 1 atm CO_2) with 99% carbonate linkages when using CHO or PO as substrates. Upon using PO as the substrate, after 24 hours, a PPC polymer was obtained with a high molecular weight ($M_n \sim 115\,kg\,mol^{-1}$) and 90% head-to-tail linkages. The lower Lewis acidity of the Co–porphyrin compared to its Al and Cr analogs is partly responsible for the higher polymer selectivity obtained with the Co system. This reinitiated the interest in this class of copolymerization catalysts, exemplified in the work carried out by the group of Rieger [41]. Rieger's group investigated a range of para-substituted cobalt(III) tetra-(alkoxyphenyl)- and tetra-nitrophenylporphyrins in the coupling of PO with CO_2 in the presence of a suitable cocatalyst. The use of para-alkoxy-substituted Co–porphyrins selectively led to alternating PPC at low temperatures with high head-to-tail epoxide incorporation. The polymer selectivity and conversion were higher than noted for reactions catalyzed by the parent Co(TPP) catalyst and can be electronically correlated with the nature of the alkoxy chains present at the meso-positions of the porphyrin framework. On the contrary, the para-nitrophenyl derivative only gave PC as product in moderate yield, thus showing a tool to fine-tune the chemoselectivity of the PO/CO_2 coupling process by variation of the Lewis acidity of the Co center. The same research group further investigated the use of dinuclear Co–porphyrin complexes having alkoxy linkers and compared their activities with a mononuclear Co–porphyrin compound [42]. The catalytic activities were evaluated in the presence of PPNCl as cocatalyst for PO/CO_2 copolymerization and revealed to be rather similar, thus indicating that no cooperative effect is achieved using bimetallic Co–porphyrin complexes. These results are in sharp contrast with those achieved with Co–salen complexes, where a bimetallic mechanism plays a key role and bimetallic catalysts are essentially preferred over monometallic ones [32].

Generally speaking, the Lewis acidity of the metal center in a metalloporphyrin determines the fate of the epoxide/CO_2 coupling reaction. Although Al(TPP) and Cr(TPP) are efficient systems to produce cyclic carbonates, their high Lewis acidity prevents selective copolymerization/terpolymerization reactions, leading instead to the formation of polyethers via ROP of the epoxide or cyclic carbonate formation as side or polymer degradation product. The group of Chisholm clearly demonstrated this correlation in a comprehensive study dealing with the relative catalytic activities of different Al–, Cr–, and Co–porphyrins in the ROP of PO and the PO/CO_2 coupling reaction [43]. Among the porphyrin ligands studied, the use of simple TPP provided the best results in terms of activity (TOF) and molecular weight (M_n) compared with the use of other, substituted porphyrin ligands.

19.4 Cocatalysts Based on Other N_4-Ligated and Related Systems

If one generally considers the salen ligands with an acyclic N_2O_2 coordination geometry, porphyrins having a cyclic N_4 coordination, and the success of their Co complexes in the epoxide/CO_2 coupling, it is obvious that the use of other tetradentate ligands with an acyclic N_4-chelating coordination was also considered to create new opportunities in this area of research using dianionic multidentate donors. In this context, the group of Zevaco investigated the use of N_4 bpb (bpb = *N*,*N*-bis(2-pyridinecarboxamide)-1,2-benzene; Figure 19.9, **17** : A = C) [44] and *N*,*N*-bis(2-pyrazine-carboxamide)-1,2-benzene ligands (Figure 19.9, **17** : A = N) [45], easily obtained from the condensation of 1,2-diaminobenzenes with picolinic acid and 2-pyrazine-carboxylic acid, respectively.

This general class of N_4 ligands displays advanced synthetic versatility as well as planar coordination geometries and can be seen as a possible ligand alternative to porphyrins, providing a more "open" N_4 coordination mode to metal ions. The metal complexes are air stable in the solid state and can be stored for months without observable degradation. Although the compounds obtained with zinc are neutral, the complexes based on iron(III), chromium(III), and cobalt(III) are ionic in nature, which may be advantageous when their axial ligand site is occupied by an (ammonium) halide. This actually allows their use as a catalyst in the coupling of CO_2 with epoxides without the presence of any supplementary cocatalytic additive.

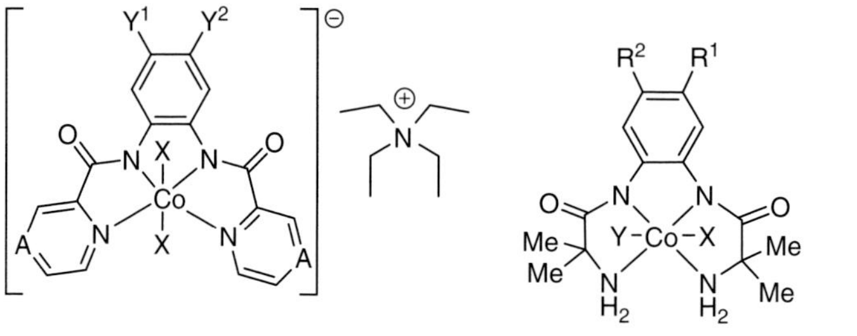

17: Y^1 = H, Cl, NO_2, Me; Y^2 = H, Cl, Me; X = OAc, Br, Cl; A = C, N

18: X = Y = Cl, Br or: X = Cl, Br; Y = solvent; $R^1 = R^2$ = H, Cl, Me or: R^1 = H; R^2 = NO_2

Figure 19.9 N_4-ligated Co complexes **17** reported by Zevaco and coworkers [44, 45] and the systems developed by Ghosh and coworkers (**18**) [46].

In the case of the Co acetate complexes (Figure 19.9, **17**: X = OAc, A = C or N), catalytic tests with CHO led to alternating polycarbonates with conversions ranging from 35% to 83% depending on the substitution pattern on the ligand. Using PO as a substrate selectively afforded PC with yields up to 70% (35 bar CO_2, 80 °C). Considering the role of the substituents, a positive influence was found when the aromatic linker was substituted with electron-donating groups (such as Me) and a negative effect if an electron-withdrawing group (such as NO_2) was present. The cobalt chloride and bromide bpb complexes, as observed for the acetate-derived complexes, selectively mediate the formation of alternating polycarbonates with molecular weights (M_n) ranging from 6500 to 8800 g mol^{-1} with narrow Đ's. Again, the use of PO only gave rise to the formation of the cyclic carbonate product. A wider catalytic screening demonstrated that terminal epoxides with electron-withdrawing functional groups are easily converted to the corresponding cyclic carbonates with high chemoselectivity and in good yields. In addition, recycling tests indicated that bpb catalysts can be reused after distillation of the carbonate product without significant loss of activity.

Generally, PC was produced in high yields using the different metal complexes based on the bpb ligands with a reactivity order [Cr] > [Co] ~ [Fe], whereas for the coupling of CHO and CO_2, the difference between the different metal complexes was more pronounced with a reactivity trend being [Co] ≫ [Fe] > [Cr]. As far as the axial ligands/leaving groups in the Co bpb complexes are concerned, the reactivity trend found in PO conversion to PC is Br > Cl > OAc. Unlike for the cobalt acetate complexes, for the cobalt halide complexes, a positive influence of electron-withdrawing groups in the ligand backbone was clearly observed in agreement with the related work of Ghosh and coworkers, who used Co–amidoamine complexes. This is likely explained by a more efficient activation of the epoxide as a result of a higher electrophilic character of the metal center. The systems developed by the Ghosh group are based on Co(III) complexes that involve N_4-chelating bis-amido-bisamine ligands (Figure 19.9, **18**) [46]. These stable Co(III) complexes display good activity and high selectivity for the formation of cyclic carbonates using an external cocatalyst (DMAP). PO was converted to PC in three hours with an appreciable average TOF of 662 h^{-1} at 20 bar of CO_2 and 130 °C. This catalytic system works with a relatively broad range of epoxide substrates, including internal epoxides such as cyclo-octene oxide (T = 150 °C) to selectively give the cyclic carbonates without traces of the related polyether/polycarbonates.

In the context of (a)cyclic N_4 ligand architectures, the group of Kilic demonstrated that a N_4-chelating ligand scaffold is not necessarily required to obtain a planar coordination geometry around the metal to give suitable cobalt catalysts that are effective in the formation of cyclic carbonates. Mono- and multinuclear cobaloxime and related organocobaloxime complexes bearing bidentate dimethylglyoxime ligands were evaluated in their studies. The monometallic cobaloxime and organocobaloxime complexes delivered cyclic carbonates in good yields (up to 72%; 100 °C, 16 bar, two hours) from coupling of various epoxides and CO_2, and particularly, the use of epichlorohydrin showed relative good activity (TOF·h^{-1} up to 358) in the presence of DMAP as a cocatalyst [47].

19.5 Aminophenoxide-Based Co Complexes

The activity of the most efficient transition metal-based salen catalysts can be optimized via subtle modification of the salen backbone and generally monometallic complexes are isolated. Taking into account the success in the coupling of CO_2 with different types of epoxides mediated by dinuclear zinc complexes (cf., the zinc β-diiminates reported by the Coates group) [48] and the work of North and coworkers using μ-oxo-bridged Al–salen catalysts [49], research groups have become increasingly interested in the development of new bimetallic structures. In this respect, "oxo"-bridged dinuclear cobalt complexes were designed and optimized for this chemistry by the Williams group, focusing on dinuclear and trinuclear systems supported by macrocyclic amino-diphenoxide ligands that can be seen as a sophisticated evolution of the salen ligand. These ligands, which may be regarded as reduced Robson-type ligands [50], comprise two phenolic fragments and two C_3-diamino bridges (instead of a single C_2-bridge which is the most standard one in salen chemistry) to form a macrocyclic, tetraamine diphenoxide N_4O_2 ligand with two available metal coordination sites and with the phenolic oxygen atoms acting as bridging ligands. Notably, one of the bridging acetates in these complexes functions as the initiator of the copolymerization reaction (Figure 19.10).

The dinuclear zinc complex **19** was the first (bifunctional) catalyst tested [51] and showed promising activity under just 1 bar pressure of CO_2 to produce alternating PCHC from CHO; soon after, active Fe(III) and Mg(II) catalysts were also developed. The dinuclear macrocyclic di- and triacetato cobalt complexes **20** and **21** were the most active of the series in this type of chemistry, with a maximum TOF of 480 h^{-1} under 1 atm CO_2 at 100 °C, the activity trend being $[Co_2] > [Mg_2] > [Fe_2] > [Zn_2]$ using the same ligand framework [52].

Bimetallic cobalt(II) diacetato complexes **20** as well as a mixed valence cobalt(II)/(III) triacetato complex **21** were developed in this context and selectively delivered PCHC in high yield and appreciable molecular weights (M_n = 5000–9000 g mol^{-1}), with the di-Co(II) complexes being less active at higher temperatures than the mixed valence systems. The high catalytic activity of these binuclear cobalt complexes was attributed to a cooperative mechanism

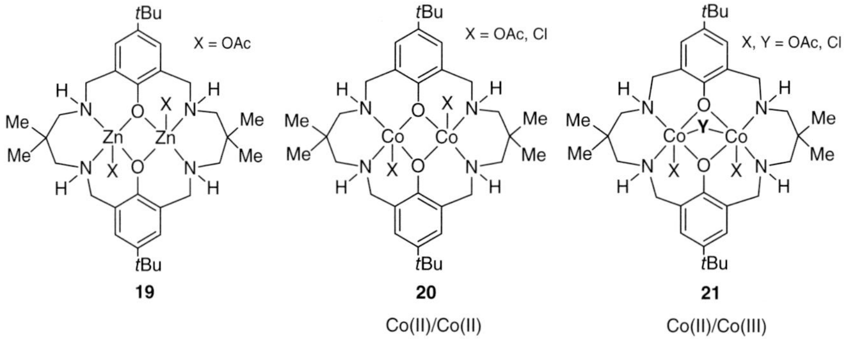

Figure 19.10 Macrocyclic, bifunctional bimetallic complexes **19–21** developed by Williams and coworkers [51, 52].

implying both metallic centers to be involved in substrate activation and polymer propagation, enabled by the rigid ligand scaffold that allows intermetallic distances between 3.02 and 3.06 Å.

Two years later, the Williams group investigated a related series of di-Co(II) halide complexes using the same macrocyclic ligand [53]. This new generation of dinuclear dichloro Co complexes displayed an unusual bowl-shaped structure, with a halide acting as a bridge between the two cobalt centers at the “concave” face of the molecule (intermetallic distances around 2.92 Å) and two nonbridging halides residing at the convex side of the complex. Catalytic evaluation of these latter di-cobalt halide species was carried out with CHO as the benchmark substrate and applying various neutral cocatalysts, including pyridine, 1-methyl imidazole, and DMAP. These experiments did give rise to the formation of the corresponding PCHC in high selectivity, but with yields slightly lower than those obtained with the acetate analogs. Pyridine, the weakest donor within the tested series, turned out to be the most efficient cocatalyst [53]. The catalytic screening study together with comprehensive X-ray diffraction data, *in situ* ATR-IR for monitoring intermediate species formed during the reaction, kinetic evaluations, and DFT investigations led to a copolymerization mechanistic proposal involving only one of the nonbridging chloride groups of the convex side initiating the reaction and implicating copolymer growth on this side of the complex. The first-order dependence of the initial rates on both (bimetallic) catalyst and CHO concentration suggests that both metals are involved and that copolymer growth occurs through alternating coordination of a subsequent epoxide monomer to one of the Co centers, following migratory insertion into the carbonate-ended polymer species, thus interacting with the other Co ion.

A couple of other studies were inspired by the success of the reduced Robson-type macrocyclic ligands of Williams, among others the contributions of Kozak, Kerton, and coworkers [54] as well as that of Rieger and coworkers [55]. The Kerton group focused particularly on cobalt(II) complexes bearing tripodal tetradentate, N_2O_2-chelating bisamino–bisphenoxide ligands with pendant nitrogen-containing donor groups (Figure 19.11). These complexes are efficient as catalysts for PO/CO_2 coupling under mild reaction conditions (room temperature and 34 bar CO_2), used in conjunction with TBAB and $PPNN_3$ (N_3 = azide). Surprisingly, the cobalt(II) complexes led to higher TONs than

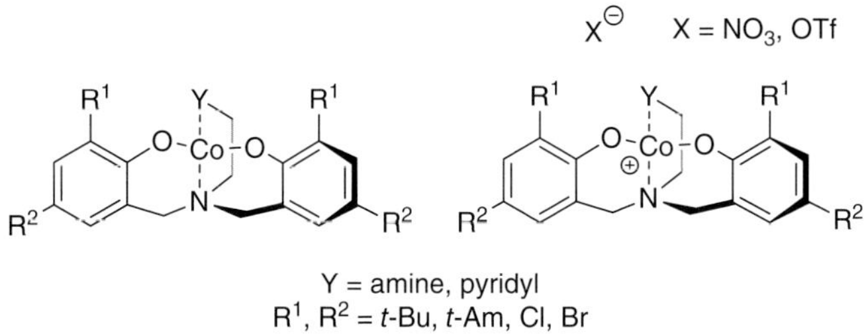

$X^{\ominus}$ X = NO_3, OTf

Y = amine, pyridyl

R^1, R^2 = *t*-Bu, *t*-Am, Cl, Br

22: Co(II) **23**: Co(III)

Figure 19.11 Co(II)/Co(III) complexes **22** and **23** used by Kerton and coworkers [54] and Rieger and coworkers [55] for PO/CO_2 and CHO/CO_2 couplings.

the analogous cobalt(III) complexes. The use of ligands with a pendant pyridyl group afforded more active catalysts than those containing other groups such as dimethylamine, and giving TONs of up to 2025 in the presence of TBAB.

The Rieger group followed a similar approach and developed other versions of these tripodal ligands with a dimethylamino donor group and different substituents localized on the phenolic fragments of the ligands (R^1, R^2 = $-C(Me)_2Ph$, Ph, Cl, and Br in **22**). The corresponding monometallic cobalt(II) complexes display appreciable catalytic activity (TON = 480) in the formation of PC in combination with TBAB as cocatalyst at 80 °C and 50 bar CO_2 pressure. The copolymerization screening tests were performed with CHO and PO (the latter not giving any observable polymer formation) and revealed that the cobalt(II) complex derived from a tetrachloro-substituted amino–bisphenoxide ligand was effective in the formation of PCHC carbonate if used with DMAP as the cocatalyst, providing reasonable polycarbonate quality product (M_n = 6700 g mol^{-1}, $Đ$ = 1.35; TON = 60).

19.6 Conclusion and Outlook

The content of this chapter clearly demonstrates the high versatility in the preparation of Co(II)- or Co(III)-based catalysts for the coupling reaction of epoxides and CO_2 toward cyclic and polycarbonates. Generally, these ligands are able to combine advantageous features including accessibility, modularity, and ease of synthesis. Structurally well-characterized cobalt catalysts can thus be prepared in a straightforward manner, and the field of CO_2/epoxide couplings has long been dominated by salen- and porphyrin-based systems. Recent activities in this area, however, have focused on the introduction of competitive and complementary systems that are based on chiral ligands, dinucleating ligand frameworks, and ligands that enforce more flexible and/or different coordination geometries (cf., aminotriphenoxides). The availability of new and more flexible ligands systems, for instance, allows for a wider variety of substrates/monomers to be converted in the presence of CO_2. Consequently, more functional carbonates can be accessed for use in areas such as polymer chemistry and fine chemical synthesis. As far as dinuclear complexes are concerned, their utility is highly useful in the area of copolymerization of epoxides and CO_2 to afford polycarbonates. New dinuclear complex designs offer potentially new reactivity and selectivity profiles that can alter polymer microstructures and eventually their properties including glass transition behavior and mechanical strength. Despite the fact that less attention has been paid to the use of chiral Co complexes, significant advances have been made to direct the copolymerization process of (*rac*)-epoxides toward chiral polymers and cyclic carbonate synthesis.

Based on the most recent developments in catalyst design, it can thus be foreseen that further developments may be expected in material science applications of the aforementioned polycarbonate and related polymers. Additionally, recent work based on the use of vinyl-substituted cyclic carbonates has also opened up new possibilities for cyclic carbonates to become key synthetic intermediates in (transition) metal-catalyzed fine chemical and pharmaceutical synthesis

[56]. Another development that may be of interest is the use and conversion of renewable feed stocks such as terpenes [57], as coupling reactions between terpene scaffolds and CO_2 may offer new academic and commercial opportunities. The long and important impact of Co-based catalysis in CO_2 transformations to various organic products will definitively continue to attract the synthetic communities to provide new opportunities for the above-mentioned application areas.

Acknowledgments

The authors wish to thank ICIQ, ICREA, and the Spanish Ministerio de Economía y Competitividad (MINECO) through project CTQ2017-88920-P and the Severo Ochoa Excellence Accreditation 2014–2018 through project SEV-2013-0319 and the Helmholtz Research School "Energy-Related Catalysis" of the Karlsruhe Institute of Technology.

References

1 (a) Cahiez, G. and Moyeux, A. (2010). *Chem. Rev.* 110: 1435. (b) Omae, I. (2007). *Appl. Organomet. Chem.* 21: 318. (c) Pellissier, H. and Clavier, H. (2014). *Chem. Rev.* 114: 2775.
2 (a) Qin, Z.Q., Thomas, C.M., Lee, S., and Coates, G.W. (2003). *Angew. Chem. Int. Ed.* 42: 5484. (b) For a review see: Coates, G.W. and Moore, D.R. (2004). *Angew. Chem. Int. Ed.* 43: 6618.
3 For a recent review see: Luo, M., Li, Y., Zhang, Y.-Y., and Zhang, X.-H. (2016). *Polymer* 82: 406.
4 Tokunaga, M., Larrow, J.F., Kakiuchi, F., and Jacobsen, E.N. (1997). *Science* 277: 936.
5 Lu, X.-B. and Wang, Y. (2004). *Angew. Chem. Int. Ed.* 43: 3574.
6 Darensbourg, D.J. and Moncada, A.I. (2009). *Macromolecules* 42: 4063.
7 (a) Darensbourg, D.J. and Lu, X.-B. (2012). *Chem. Soc. Rev.* 41: 1462. (b) Lu, X.-B., Ren, W.-M., and Wu, G.-P. (2012). *Acc. Chem. Res.* 45: 1721.
8 (a) Taherimehr, M. and Pescarmona, P.P. (2014). *J. Appl. Polym. Sci.* 131: 41141. (b) Kielland, N., Whiteoak, C.J., and Kleij, A.W. (2013). *Adv. Synth. Catal.* 355: 2115.
9 (a) Trott, G., Saini, P.K., and Williams, C.K. (2016). *Phil. Trans. R. Soc. A.* 374: 20150085. (b) Childers, M.I., Longo, J.M., Van Zee, N.J. et al. (2014). *Chem. Rev.* 114: 8129.
10 Martín, C., Fiorani, G., and Kleij, A.W. (2015). *ACS Catal.* 5: 1353.
11 Wang, T.-T., Xie, Y., and Deng, W.-Q. (2014). *J. Phys. Chem. A* 118: 9239.
12 Darensbourg, D.J. and Yeung, A.D. (2015). *Polym. Chem.* 6: 1103.
13 Cohen, C.T. and Coates, G.W. (2006). *J. Polym. Sci. A: Polym. Chem.* 44: 5182.
14 Hošt'álek, Z., Mundil, R., Císařová, I. et al. (2015). *Polymer* 63: 52.
15 Nakano, K., Kamada, T., and Nozaki, K. (2006). *Angew. Chem. Int. Ed.* 45: 7274.

16 Nakano, K., Hashimoto, S., Nakamura, M. et al. (2011). *Angew. Chem. Int. Ed.* 50: 4868.
17 Noh, E.K., Na, S.J., Sujith, S. et al. (2007). *J. Am. Chem. Soc.* 129: 8082.
18 Sujith, S., Min, J., Seong, J. et al. (2008). *Angew. Chem. Int. Ed.* 47: 7306.
19 Na, S.J., Sujith, S., Cyriac, A. et al. (2009). *Inorg. Chem.* 48: 10455.
20 Cyriac, A., Jeon, J.Y., Varghese, J.K. et al. (2012). *Dalton Trans.* 41: 1444.
21 (a) Seong, J.E., Na, S.J., Cyriac, A. et al. (2010). *Macromolecules* 43: 903. (b) Cyriac, A., Lee, S.H., and Lee, B.Y. (2011). *Polym. Chem.* 2: 950.
22 Lee, B.Y., Sudevan, S., Noh, E.K., and Min J.K. (2014). Process for producing polycarbonates and a coordination complex used therefor. US Patent 0,249,323 (A1).
23 Wu, G.-P., Wei, S.-H., Ren, W.-M. et al. (2011). *Energy Environ. Sci.* 4: 5084.
24 Darensbourg, D.J. and Chung, W.-C. (2014). *Macromolecules* 47: 4943.
25 Ren, W.-M., Liu, Z.-W., Wen, Y.Q. et al. (2009). *J. Am. Chem. Soc.* 131: 11509.
26 Ren, W.-M., Zhang, X., Liu, Y. et al. (2010). *Macromolecules* 43: 1396.
27 Wu, G.-P., Wei, S.-H., Ren, W.-M. et al. (2011). *J. Am. Chem. Soc.* 133: 15191.
28 Wu, G.-P., Xu, P.-X., Lu, X.-B. et al. (2013). *Macromolecules* 46: 2128.
29 Duan, S., Jing, X., Li, D., and Jing, H. (2016). *J. Mol. Catal. A: Chem.* 411: 34.
30 Ren, W.M., Liu, Y., Wu, G.-P. et al. (2011). *J. Polym. Sci., Part A: Polym. Chem.* 49: 4894.
31 Ren, W.-M., Wu, G.-P., Lin, F. et al. (2012). *Chem. Sci.* 3: 2094.
32 Liu, Y., Ren, W.-M., Liu, J., and Lu, X.-B. (2013). *Angew. Chem. Int. Ed.* 52: 11594.
33 (a) Takeda, N. and Inoue, S. (1978). *Bull. Chem. Soc. Jpn.* 51: 3564. (b) Aida, T. and Inoue, S. (1983). *J. Am. Chem. Soc.* 105: 1304.
34 Paddock, R.L., Hiyama, Y., McKay, J.M., and Nguyen, S.T. (2004). *Tetrahedron Lett.* 45: 2023.
35 (a) Ortho-substitution: Jin, L., Jing, H., Chang, T. et al. (2007). *J. Mol. Cat. A: Chem.* 261: 262. (b) Para-substitution: Li, B., Zhang, L., Song, Y. et al. (2012). *J. Mol. Cat. A: Chem.* 26: 363–364.
36 Chen, A., Zhang, Y., Chen, J. et al. (2015). *J. Mater. Chem. A* 3: 9807.
37 Xie, Y., Wang, T.-T., Liu, X.-H. et al. (2013). *Nat. Commun.* 4: 1960.
38 Chun, J., Kang, S., Kang, N. et al. (2013). *J. Mater. Chem. A* 1: 5517.
39 Qin, Y., Wang, X., Zhang, S. et al. (2008). *J. Polym. Sci., Part A: Polym. Chem.* 46: 5959.
40 Sugimoto, H. and Kuroda, K. (2008). *Macromolecules* 41: 312.
41 Anderson, C.E., Vagin, S.I., Xia, W. et al. (2012). *Macromolecules* 45: 6840.
42 Anderson, C.E., Vagin, S.I., Hammann, M. et al. (2013). *ChemCatChem* 5: 3269.
43 Chatterjee, C., Chisholm, M.H., El-Khaldy, A. et al. (2013). *Inorg. Chem.* 52: 4547.
44 (a) Adolph, M., Zevaco, T.A., Walter, O. et al. (2012). *Polyhedron* 48: 92. (b) Adolph, M., Zevaco, T.A., Altesleben, C. et al. (2014). *Dalton Trans.* 43: 3285.
45 Adolph, M., Zevaco, T.A., Altesleben, C. et al. (2015). *New J. Chem.* 39: 9700.
46 Ramidi, P., Gerasimchuk, N., Gartia, Y. et al. (2013). *Dalton Trans.* 42: 13151.
47 Kilic, A., Ulusoy, M., Aytar, E., and Durgun, M. (2015). *J. Ind. Eng. Chem.* 24: 98.

48 Moore, D.R., Cheng, M., Lobkovsky, E.B., and Coates, G.W. (2003). *J. Am. Chem. Soc.* 125: 11911.

49 Clegg, W., Harrington, R.W., North, M., and Pasquale, R. (2010). *Chem. Eur. J.* 16: 6828.

50 Pilkington, N.H. and Robson, R. (1970). *Aust. J. Chem.* 23: 2225.

51 Kember, M.R., Knight, P.D., Reung, P.T.R., and Williams, C.K. (2009). *Angew. Chem. Int. Ed.* 48: 931.

52 Kember, M.R., White, A.J.P., and Williams, C.K. (2010). *Macromolecules* 43: 2291.

53 Kember, M.R., Jutz, F., Buchard, A. et al. (2012). *Chem. Sci.* 3: 1245.

54 Saunders, L.N., Ikpo, N., Petten, C.F. et al. (2012). *Catal. Commun.* 18: 165.

55 Reiter, M., Altenbuchner, P.T., Kissling, S. et al. (2015). *Eur. J. Inorg. Chem.* 10: 1766.

56 (a) Khan, A., Yang, L., Xu, J. et al. (2014). *Angew. Chem. Int. Ed.* 53: 11257. (b) Guo, W., Martínez-Rodríguez, L., Martín, E. et al. (2016). *Angew. Chem. Int. Ed.* 55: 11037.

57 (a) Byrne, C.M., Allen, S.D., Lobkovsky, E.B., and Coates, G.W. (2004). *J. Am. Chem. Soc.* 126: 11404. (b) Hauenstein, O., Reiter, M., Agarwal, S. et al. (2016). *Green Chem.* 18: 760. (c) Fiorani, G., Stuck, M., Martín, C. et al. (2016). *ChemSusChem* 9: 1304.

20

Dinitrogen Reduction

Fenna F. van de Watering and Wojciech I. Dzik

University of Amsterdam, Van't Hoff Institute for Molecular Sciences (HIMS), Department of Homogeneous, Supramolecular and Bio-Inspired Catalysis, Science Park 904, 1098 XH, Amsterdam, The Netherlands

20.1 Introduction

Dinitrogen has a well-deserved reputation of being one of the most inert molecules. This feature has become the origin of the name of the element with atomic number 7 in many languages. In some, it is described as "life less" (as coined by Lavoisier from Greek αζωω), e.g. azote (French), azoto (Italian), azot (Polish), and азот (Russian), as it does not support respiration, and in some as a substance that causes suffocation, e.g. Stickstoff (German), stikstof (Dutch), dusík (Czech), and kväve (Swedish) [1]. Ironically, life that we know is not possible without this element, as all living organisms are based on nitrogen-containing molecules (i.e. amino and nucleic acids). However, the biosynthesis of these adducts can only be accomplished once nitrogen is fixed as ammonia [2]. The two most prominent dinitrogen reduction pathways are the industrial Haber–Bosch process [3–6] and the biological reduction by cyanobacteria [7], and each accounts roughly for 50% of the annual ammonia production [8–10]. The industrial process uses N_2 and H_2, which are reacted over a heterogeneous iron or ruthenium catalyst, forming ammonia. The overall process, which includes methane reforming for H_2 production, requires high temperatures and pressures and consumes more than 1% of the annual production of world energy [11, 12]. Nitrogenases, the metalloenzymes responsible for ammonia production in the biological process, can perform the reaction at ambient conditions from N_2, protons, and electrons and use energy in the form of MgATP [13]. The active site of the most common enzyme is composed of an iron–molybdenum cluster [14]. Other, less common nitrogenases are vanadium–iron or iron only [15, 16].

Despite its chemical inertness, N_2 can be reduced by nitrogenases already at ambient conditions. This fact triggered much effort in the development of synthetic catalysts capable of ammonia synthesis under mild conditions. Finding systems that would enable reduction of dinitrogen in a manner that is less energy intensive than the Haber–Bosch process could lead to a more sustainable

Non-Noble Metal Catalysis: Molecular Approaches and Reactions,
First Edition. Edited by Robertus J. M. Klein Gebbink and Marc-Etienne Moret.

economy. In addition, development of synthetic models of nitrogenase enzymes may provide more fundamental understanding in how this small inert molecule can be transformed into useful nitrogen-containing compounds. In recent years, a great progress has been made in this field, and each year new, catalytic systems for dinitrogen reduction are being reported.

The purpose of this chapter is to give an overview of the approaches used to tackle the challenging problem of N_2 reduction at ambient conditions and of the catalytic systems that are capable of this transformation. To avoid detouring from the main focus of this book, systems that contain molybdenum will not be discussed in detail. Comprehensive reviews that give more insight into the models and mechanisms of these Mo-based catalysts are available [10, 17–19]. In this chapter, we will focus on (catalytic) reactions involving either iron or cobalt because to date these are the only base metals that are catalytically active in dinitrogen reduction.

First, the binding of dinitrogen to metal complexes is discussed (Section 20.2). In the following sections, we will discuss various approaches toward (catalytic) reduction of dinitrogen using base metals. Based on the known reactivity of dinitrogen complexes, two different pathways can be envisioned. The first approach involves initial homolytic cleavage of the N_2 molecule into nitrides and can be regarded as a Haber–Bosch-inspired approach (Section 20.3.1). The thus formed nitrides can react, e.g., with protons under reductive conditions to form ammonia. Addition of other electrophiles, e.g. carbocations, can also be envisioned, which could lead to the formation of alkylamines or nitriles; however, to date, no such systems were reported for base metals [20]. The second approach is inspired by the action mode of nitrogenase in which the coordinated N_2 molecule undergoes reductive protonation in a stoichiometric (Section 20.3.2) or catalytic (Section 20.3.3) manner. Section 20.4 covers the catalytic formation of silylamines from molecular dinitrogen. The chapter ends with conclusions and an outlook (Section 20.5).

20.2 Activation of N_2

For most synthetic (in)organic chemists, the inertness of dinitrogen is a very advantageous property because nitrogen atmosphere can be used to prevent the presence of moisture and oxygen, thus enabling many advanced chemical transformations. This inertness, however, renders fixation of dinitrogen as one of the more challenging chemical reactions for the reasons described below.

Dinitrogen complexes have been prepared with the majority of transition metals [21, 22]. Binding of N_2 to a transition metal center involves both σ-donation to the metal and back donation from the metal to the empty π^*-orbital(s). However, because of a large electronegativity of dinitrogen, N_2 is a weak σ-donor. Additionally, N_2 is a poor π-acceptor because of a weak overlap of the contracted π^*-orbitals with the metal d-orbitals [23]. Overall, the donation of electron density from the metal to the dinitrogen molecule is the most important interaction [24]. Therefore, N_2 will preferentially bind to electron-rich metals with a strong capacity of π-back-bonding. When choosing a system for the activation of N_2, one has to take into account the following factors: the energy of d-orbitals of transition metals decreases from left to right in the row and thus the amount of charge transfer from the metal to the N_2 ligand is higher for early and lower for late transition metals. This means that, for example, iron forms a stronger bond with dinitrogen than cobalt, and copper dinitrogen complexes are very unusual. The addition of electrons (e.g. reduction of the metal) follows the same trend: lowering the oxidation state will lead to a stronger metal–dinitrogen bond. The use of strongly donating ligands will increase the electron richness of the metal. Although multidentate alkyl phosphines seem to be the privileged ligands in this case, strong binding was also observed for hydride complexes [25, 26]. Needless to say, such electron-rich platforms for N_2 activation are extremely air sensitive. In addition, dinitrogen coordination is less favorable to occur with high-spin complexes, and therefore, the use of weak field sulfide ligands leads to limited success, as these ligands generally form high-spin complexes [27]. This is in big contrast to the nitrogenase, where multiple sulfides per active site are present.

The amount of electron transfer to the coordinated N_2 moiety can be probed by measuring the N≡N stretch frequency using infrared or Raman spectroscopy. However, the extent of activation of the triple bond is not a direct measure for performance in N_2 reduction [28] because a more electron-rich metal center can also be more prone to side reactions that lead to deactivation of the catalyst. As can be seen from Figure 20.1, there is no direct correlation between the vibrational frequencies of N_2 bound to the NH_3-forming catalysts and their efficiency. Moreover, many (pre)catalysts that are capable of catalyzing the formation of tris(trimethylsilyl)amine, often do not even feature a bound N_2 ligand. This demonstrates that the extent of spectroscopic activation of dinitrogen ligand does not predict the overall performance of the complex; it displays if the dinitrogen molecule is bound to the metal center and the electron richness of the complex.

Further aspects to be considered in the design of systems for catalytic reduction of N_2 are factors such as the ease of release of the products; nitrogen-containing products bind very strongly to the early transition metals, and therefore, to date, catalytic N_2 reduction to ammonia has been disclosed only for molybdenum, iron, and cobalt. Another factor is the potential poisoning of the catalyst with side products formed during catalysis (e.g. H_2) or impurities present in the reactants.

20.3 Reduction of N_2 to Ammonia

20.3.1 Haber–Bosch-Inspired Systems

In the Haber–Bosch process, ammonia is formed by the reaction of dinitrogen and dihydrogen over an iron catalyst. The reaction proceeds through the stages depicted in Figure 20.2: dissociative chemisorption of both gasses followed by stepwise formation of N—H bonds between chemisorbed atoms and subsequent release of gaseous ammonia [29, 30]. The homolytic splitting of the N_2 molecule forming a surface-bound nitride is the rate-determining step [11, 30–34].

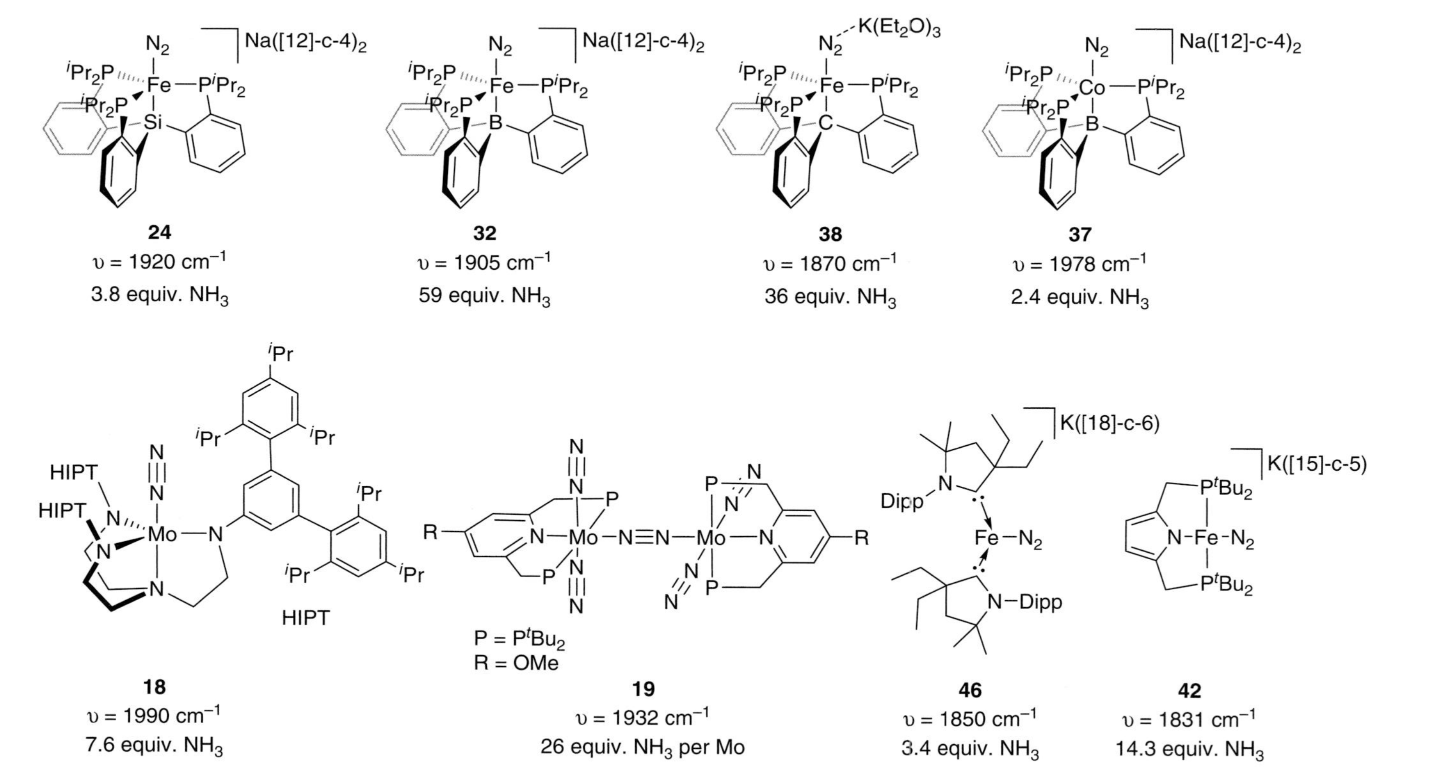

Figure 20.1 Selected dinitrogen reduction catalysts, their N_2 ligand IR frequencies, and the maximum yield of ammonia they provide.

$$N_2(g) + 3\,H_2(g) \xrightarrow[200\ \text{atm},\ 400\ ^\circ\text{C}]{\text{Fe/K/Al}_2\text{O}_3} 2\,NH_3(g)$$

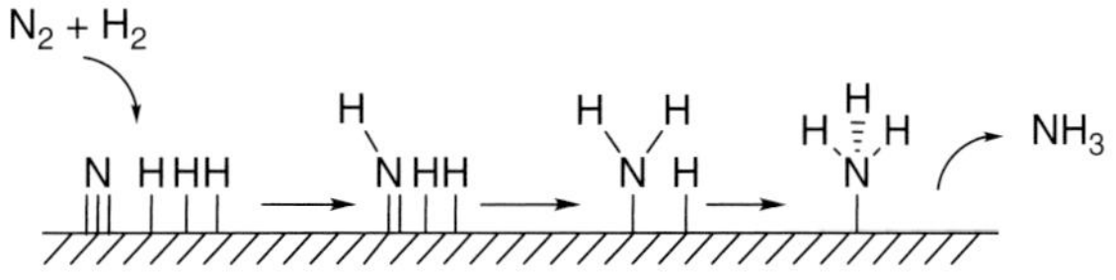

Figure 20.2 The Haber–Bosch process (top) and its mechanism (bottom).

The rate of chemisorption is enhanced by the addition of potassium oxide, which alters the electronic properties of the catalyst surface [11, 33, 35], whereas a framework of Al_2O_3 and CaO stabilizes the catalyst, preventing sintering of the active iron particles [8, 31, 32]. In the quest for processes that can work under lower pressures, ruthenium on graphite (Kellogg Advanced Ammonia Process [KAAP]) [36] and promoted cobalt [12, 37] catalysts were developed.

Homolytic cleavage of N_2 is possible at low temperatures and pressures, and one can envision that the thus formed nitrido species can undergo follow-up reactions similar to the surface-bound nitrides formed during the Haber–Bosch process. The first well-defined complex that was able to perform the homolytic splitting of the dinitrogen molecule forming a molybdenum–nitrido complex was reported in 1995 [38]. Since then, many well-defined metal complexes (Ti, Hf, Mo, Nb, Ta, V, Cr, U, and Re) were shown to split the N_2 molecule [17, 39], and some of them allowed for further reactions of the thus formed nitrides [39, 40].

In 2001, an attempt to homolytically cleave the dinitrogen triple bond using iron was made by the Holland group. The dimeric highly reduced (β-diket^{tBu})(K)Fe(μ-N_2)Fe(K)(β-diket^{tBu}) (β-diket^{tBu} = [HC(C(^{t}Bu)NC$_6$H$_3$(^{i}Pr)$_2$)$_2$]$^-$) complex (**1**) showed extreme weakening of the dinitrogen triple bond ($\upsilon(N_2) = 1589\ cm^{-1}$) (Figure 20.3) [41]. Quite remarkably, coordination of a potassium cation side-on to the dinitrogen ligand further increases the back-bonding of iron into the π*-orbitals of N_2. This slightly resembles the operation of the Haber–Bosch process, where promoters as potassium are used to facilitate N_2 binding to the iron centers [35]. Analogous cobalt ($\upsilon\ (N_2) = 1599\ cm^{-1}$) [42] (**2**) and nickel adducts (**3**) ($\upsilon\ (N_2) = 1696\ cm^{-1}$) [43] were also reported; however, no further reactivity of these complexes was disclosed.

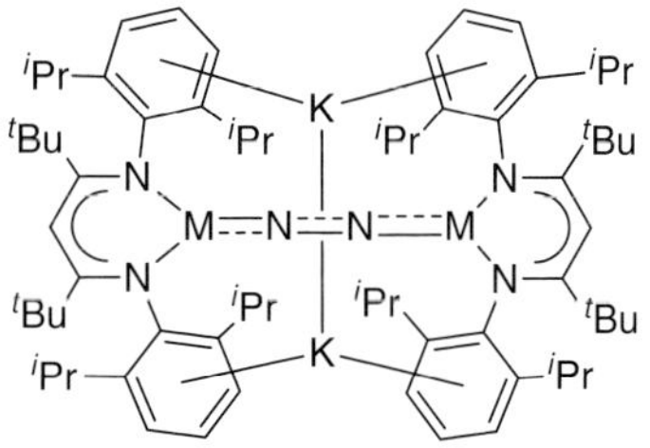

Figure 20.3 Dimeric β-diketiminate N_2 complexes of iron (1), cobalt (2), and nickel (3).

1, M = Fe $\upsilon = 1589\ cm^{-1}$
2, M = Co $\upsilon = 1599\ cm^{-1}$
3, M = Ni $\upsilon = 1696\ cm^{-1}$

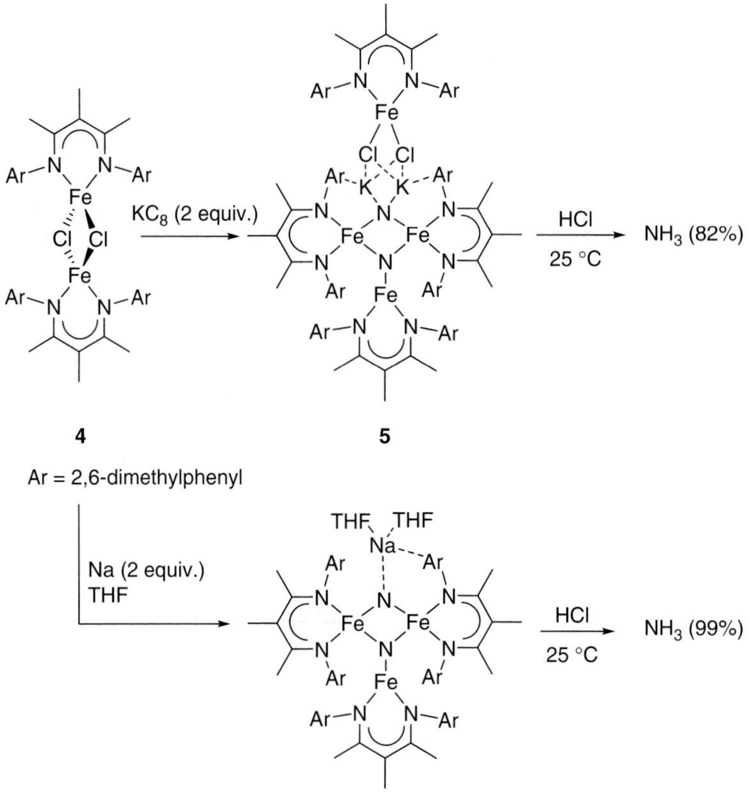

Figure 20.4 Triiron β-diketiminate N_2 complexes that split the dinitrogen bond and subsequent release of ammonia upon addition of hydrogen chloride.

The use of a less sterically demanding β-diketiminate ligand (β-diket^{Me} = MeC (C(Me)N$C_6H_3(Me)_2)_2]^-$) coordinated with iron allowed for splitting of the N_2 triple bond to form the tetrairon dinitrido complex **5** (Figure 20.4) [44]. The core of complex **5** consists of two nitrides surrounded by three iron centers and two potassium cations to which a fourth iron center is bound via bridging chloride anions. Addition of HCl to complex **5** resulted in the release of 82% of ammonia. Further studies showed that if the reduction of the parent complex (β-diket^{Me})(K)Fe(μ-Cl)$_2$Fe(K)(β-diket^{Me}) (**4**) is conducted with 2 equiv. of sodium, the triiron sodium complex (**6**) is formed, which upon reacting with HCl releases ammonia in 99% yield [45]. In the presence of excess reductant (4 equiv.), no cleavage of N_2 is observed, and triiron complexes with bridging N_2 ligands are formed, which precludes any catalytic turnover.

The group of Murray used a rigid β-diketiminate ligand that could be used as a template in the formation of a triiron complex (**7**) (Figure 20.5) [46]. Upon reduction of the triiron(I) complex **7** with KC_8, homolytic splitting of the N≡N bond occurred, forming complex **8**. Surprisingly, the complex incorporates three NH_x fragments, which suggest that (at least one of) the NH fragments is formed via

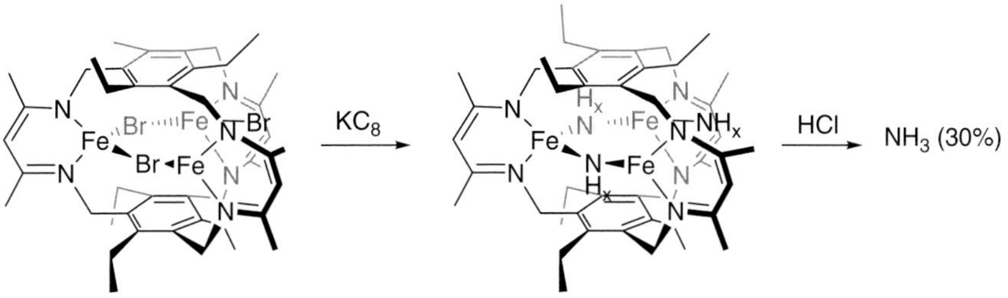

$$\mathbf{7} \xrightarrow{KC_8} \mathbf{8} \xrightarrow{HCl} NH_3\ (30\%)$$

Figure 20.5 Rigid triiron β-diketiminate complex and its subsequent reduction resulting in homolytic splitting of the N_2 molecule.

reaction between two triiron complexes. The source of protons for complex **8** could not be established. Protonation of **8** with hydrochloric acid resulted in the release of 30% of ammonia.

The above examples show that homolytic cleavage of the N≡N bond followed by protonation can yield ammonia already at room temperature and ambient pressure. The difference compared with the Haber–Bosch reaction is, however, that the hydrogen atoms are introduced in the form of protons and electrons and not as dihydrogen. Although not yet catalytic, this approach demonstrates that it is possible to prepare synthetic models that show a similar mechanism as the Haber–Bosch process. Hopefully, related systems will be further explored in the future to yield some Haber–Bosch-inspired catalysts.

20.3.2 Nitrogenase-Inspired Systems

The active site of the most common nitrogenase enzyme consists of a protein-embedded [Mo:7Fe:9S:C]:homocitrate cofactor (FeMo-co) [47–50] (Figure 20.6), which catalyzes the formation of ammonia from dinitrogen, protons, and electrons. The electrons are delivered by a [4Fe-4S] cluster [13], and although the detailed mechanistic picture remains under debate, it is clear that the reduction takes place at the iron center [13, 51–56]. The reduction follows a mechanism in which protons are added to the dinitrogen core in an alternate manner to the terminal and proximal nitrogen atoms (Figure 20.7).

$$N_2 + 8\ H^+ + 8\ e^- + 16\ MgATP \xrightarrow{\text{FeMo-nitrogenase}} 2\ NH_3 + H_2 + 16\ MgADP + P_i$$

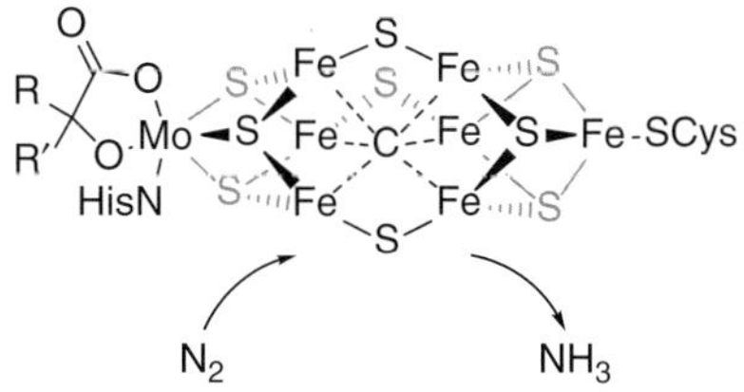

Figure 20.6 Overall reaction scheme for nitrogen reduction by nitrogenase (top). FeMoco, the active site of the FeMo based nitrogenase (bottom).

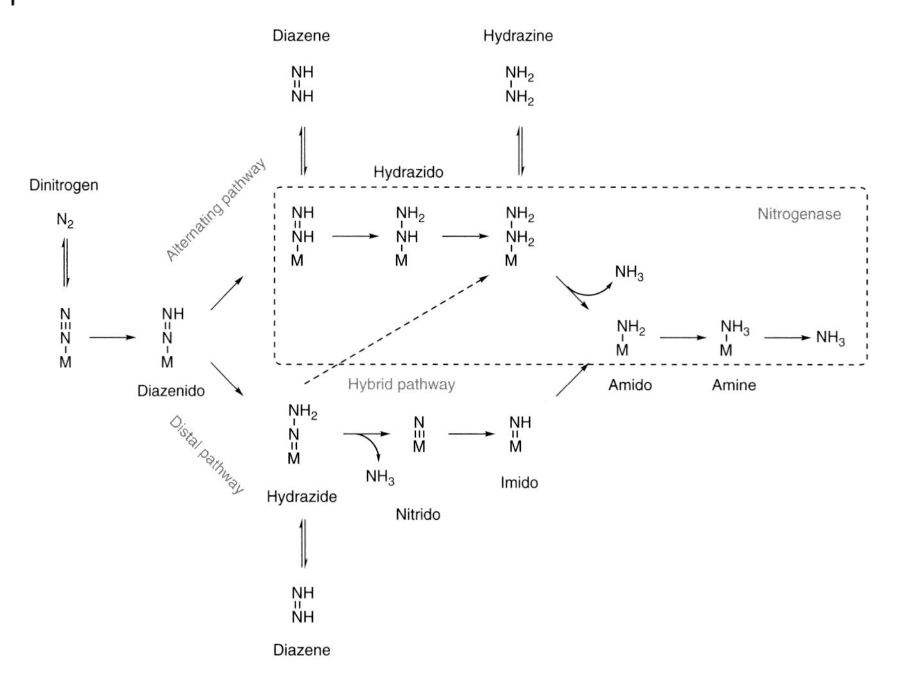

Figure 20.7 Two possible pathways of dinitrogen reduction to ammonia: alternating (top) and distal (bottom).

During the reduction, bridging iron hydrides are formed, which reductively eliminate H_2 leaving a reduced vacant iron atom to which N_2 can bind [57].

The reduction of dinitrogen on transition metal centers can proceed through two limiting pathways: alternating and distal (Figure 20.7). In the alternating pathway, the protons are transferred to the terminal and proximal dinitrogen atoms producing coordinated diazene and hydrazine. The release of the first equivalent of ammonia leads to the formation of an amido complex that undergoes subsequent protonation to yield the second equivalent of ammonia. This mechanism has been shown to be operational for FeMo-co [13, 58] and for iron-phosphine-based [59] model systems. In the distal pathway, the protonation occurs initially on the terminal nitrogen atom, and after release of ammonia, a nitride complex is formed, which after three subsequent protonation and electron transfer steps releases the second equivalent of ammonia [57]. This pathway was shown to operate for molybdenum systems of Schrock and Nishibayashi [60–62]. Interconversion between the two pathways is also possible as recently showed by Rittle and Peters, who reported that the hydrazide intermediate $Fe{=}N{-}NH_2^+$ can transform to $Fe{-}NH_2NH_2^+$ through double reduction and protonation (Figure 20.14) [60]. This brings about the possibility of a hybrid pathway in which initial two proton additions occur on a distal pathway, whereas after interconversion, the subsequent steps follow the alternating pathway.

20.3.2.1 Early Mechanistic Studies on N_2 Reduction by Metal Complexes

As the most common nitrogenase enzyme exists as a cluster of molybdenum and iron atoms [47–50], the main research on room temperature N_2 reduction was

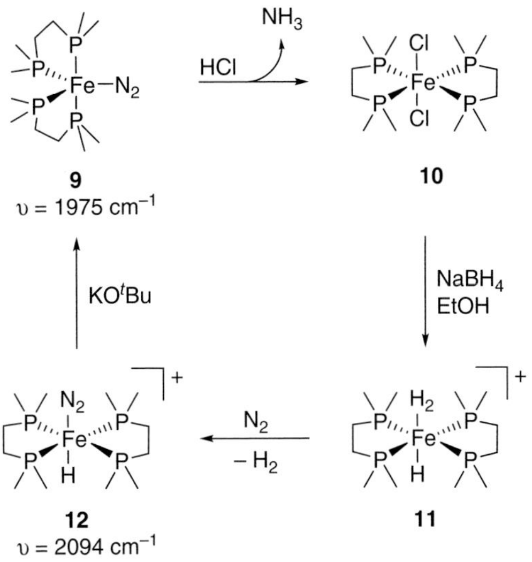

Figure 20.8 The Leigh cycle.

focused on these two metals. The first examples of substoichiometric ammonia formation from dinitrogen were reported in 1964 with molybdenum [63], and in 1966 with iron [64]. These two systems involved mixing of the "naked" metal salts MoO_4^{2-} or $FeCl_3$ with a reductant, in the presence of acid and dinitrogen, but their mechanisms were not investigated further. Further studies on iron as the active metal in the nitrogenase enzyme followed roughly 30 years later. Leigh studied a well-defined $[Fe^0(dmpe)_2N_2]$ (**9**) (dmpe = $Me_2P–CH_2CH_2–PMe_2$) ($\upsilon(N_2)$ = 1975 cm^{-1}) complex (Figure 20.8) that formed ammonia (20%) and $Fe^{II}(dmpe)_2Cl_2$ (**10**) upon addition of HCl [65]. The iron atom is the only electron source for the reduction, so a maximum yield of 33.3% of ammonia could theoretically be reached if the dinitrogen reaction was paired to the oxidation of Fe^0 to Fe^{II}. Iron(0) dinitrogen complex **9** could be re-formed by reaction with $NaBH_4$/EtOH to yield a hydrido dihydrogen complex (**11**) [28], followed by substitution of H_2 with N_2 (**12**) ($\upsilon(N_2)$ = 2094 cm^{-1}) and reductive deprotonation by base, which closed the so-called Leigh cycle. Tyler and coworkers reported that for related systems, the reduction step could be performed by using molecular hydrogen in the presence of proton and chloride scavengers instead of $NaBH_4$. In this way, H_2 was used as a source of electrons for N_2 reduction [66].

Theoretical and experimental mechanistic investigations on the $Fe(dmpe)_2$ and similar iron diphosphine complexes [67–75] showed that an alternating pathway (Figure 20.7) was operational, similar to the way the natural system is suggested to proceed [13, 58]. Most importantly, these studies confirmed that a well-defined iron system is able to mediate the formation of ammonia from dinitrogen. A related system in which the dmpe ligand was replaced by the depe ligand (depe = $Et_2P—CH_2CH_2—PEt_2$) proved to be active in catalytic reduction of dinitrogen (see Section 20.3.3, Figure 20.22).

20.3.2.2 Iron–Sulfur Systems

Although the nitrogenase enzyme consists of multiple sulfur atoms in its core, the number of synthetic models that contain sulfur atoms are limited. In fact, sulfides and thiolates are weak-field ligands, which favor formation of high-spin complexes, which is generally unfavorable for N_2 binding [27]. Some groups prepared mono- or dinuclear iron-containing sulfide/thiolate complexes, but although these complexes could form ammonia from hydrazine, none of them were shown to bind and subsequently reduce the π-acidic N_2 ligand [76–82]. Sellmann reported various N_2 thiolate-containing ruthenium complexes and hypothetical intermediates (N_2H_2, N_2H_4 NH_3) [81, 83] of a dinitrogen reduction cycle. However, this cycle could not be made catalytic [84]. The Peters group investigated sulfur-containing iron complexes with thioethers. Thioethers, have better σ-donating and π-accepting properties than thiolates and sulfides [27], and thus, it was hypothesized that the use of thioether donors may favor dinitrogen binding. Tripodal mono-, bis-, and tris-thioether ligands with a central silicon donor atom were prepared, from which only the iron complex with the ligand containing one thioether and two electron-donating phosphines coordinated a weakly activated ($\upsilon(N_2) = 2156\,cm^{-1}$) dinitrogen ligand (**13**) (Figure 20.9) [27]. Addition of a hydride [25, 26] resulted in a more activated dinitrogen bond (**14**) ($\upsilon(N_2) = 2055\,cm^{-1}$) and also allowed the bis-thioether complex to coordinate dinitrogen (**15**) ($\upsilon(N_2) = 2060\,cm^{-1}$). Although these complexes show some resemblance with the nitrogenase active site, they could not be applied in dinitrogen reduction.

Some success was reached with the tripodal diiron sulfide-bridged dinitrogen complex (**16**) ($\upsilon(N_2) = 2017$ and $1979\,cm^{-1}$) (Figure 20.10), which not only coordinates with dinitrogen but is also able to reduce it. Stoichiometric amounts of ammonia (1.8 equiv.) are formed upon addition of a reductant and an acid to **16** [85]. The reactivity is attributed to the tripodal structure of the ligand. In

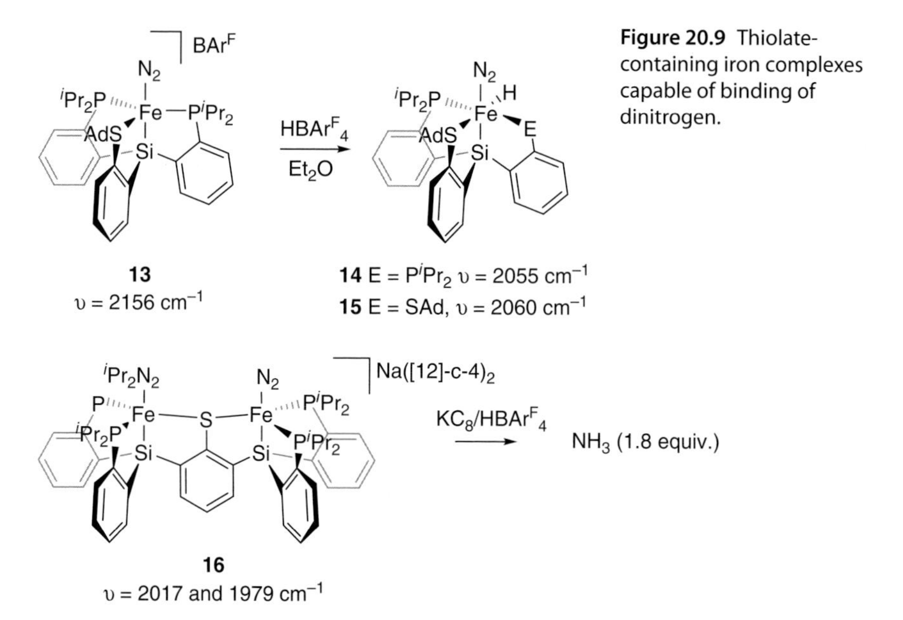

Figure 20.9 Thiolate-containing iron complexes capable of binding of dinitrogen.

Figure 20.10 Sulfur-containing diiron complex that coordinates dinitrogen and forms up to 1.8 equiv. of ammonia when subjected to protons and electrons.

Figure 20.11 Iron dinitrogen complex with a ligand featuring only thiolato and carbon donors.

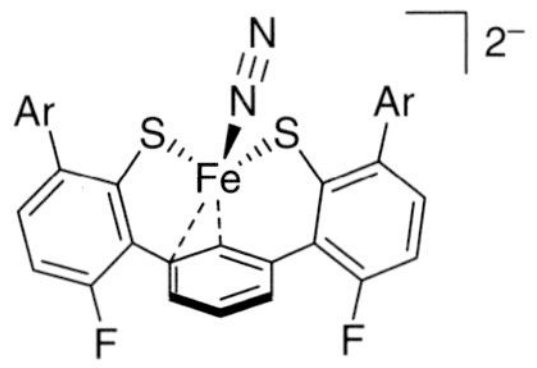

17
$\upsilon = 1880\ cm^{-1}$

the following sections, more successful examples of systems featuring tripodal ligands are shown. Longer reaction times or different reagent concentrations did not lead to higher yields of ammonia; probably, the catalyst is unstable under the reaction conditions applied.

Recently, the group of Holland and coworkers reported a high-spin ($S = 1$) iron(0) dinitrogen complex coordinated with a tridentate ligand featuring two thiolato and a carbon-based coordination site (Figure 20.11) [86]. The remarkably low frequency of the N—N bond stretch ($\upsilon(N_2) = 1880\ cm^{-1}$) shows that thiolates can enable substantial back donation to the coordinated N_2 moiety. Complex **17** is a synthetic model with closest resemblance to the nitrogenase site that is able to coordinate to dinitrogen; however, no reactivity studies were disclosed.

Synthetic Fe–S clusters [87, 88] that show much resemblance with the nitrogenase-active site are known; however, these clusters do not bind N_2 [57]. Interestingly, recently, FeMoS–SnS and FeS–SnS clusters embedded in chalcogels were shown to reduce dinitrogen under white light irradiation and in the presence of aqueous pyridinium hydrochloride and sodium ascorbate in solutions to ammonia (up to 17 equiv.) [89, 90]. Their mechanism has, however, not been investigated.

20.3.3 Catalytic Ammonia Formation

The first molecular catalyst that was able to reduce dinitrogen to ammonia (7.6 equiv. per Mo) under ambient conditions was found in 2003 by Yandulov and Schrock [62]. The catalyst (**18**) ($\upsilon(N_2) = 1990\ cm^{-1}$) contained a molybdenum atom surrounded by a bulky tripodal $HIPTN_3N$ ligand ($(HIPTN_3N)^{3-} = (\{(HIPT)NCH_2CH_2\}_3N)^{3-}$, HIPT = $\{3,5\text{-}(2,4,6\text{-}^iPr_3C_6H_2)_2C_6H_3\}$) Figure 20.12) [91]. The ligand was specifically chosen for its large steric bulk, which prevents the formation of inactive μ-N_2 dimers. Based on the isolation of most of the intermediates of the catalytic cycle [62, 92–95] and density functional theory (DFT) calculations [96–98], a distal pathway (Figure 20.7) was proposed. Molybdenum systems supported with PNP-type ligands (**19** [99] and **20** [100]) described by the group of Nishibayashi and coworkers were found to have even higher activity, which reached up to 63 equiv. of NH_3 per metal center. As will be shown below, both tripodal and PNP ligand scaffolds could be successfully used for the development of base metal systems for catalytic ammonia synthesis.

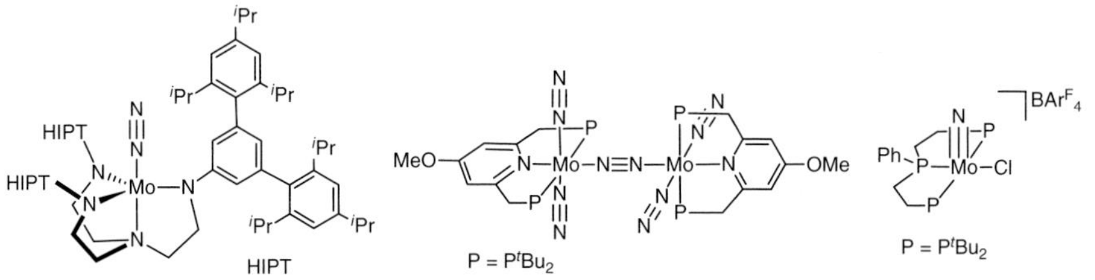

18
υ = 1990 cm^{-1}
7.6 equiv. NH_3

19
υ = 1932 cm^{-1}
26 equiv. NH_3 per Mo

20
63 equiv. NH_3

Figure 20.12 Selected nitrogen reduction molybdenum catalysts.

20.3.3.1 Tripodal Systems

The activity of iron and cobalt systems supported by tripodal tetradentate ligands was investigated by the group of Peters. These ligands feature three phosphine donors and a coordinating central X-type (Si, C) or Z-type (B) atom. Similar to Schrock's molybdenum system, these complexes feature the ability to accommodate both a π-acidic N_2 ligand and π-basic N_2-derived intermediates (e.g. N^{3-} and NH^{2-}) [101].

Initial investigations were performed for iron complexes featuring a silyl donor atom, which allowed for the isolation of the $(SiP^{Ph}{}_3)Fe(N_2)$ (**21**) ($SiP^{Ph}{}_3{}^-$ = $(2\text{-}Ph_2PC_6H_4)_3Si^-$) complex with a moderately activated N_2 moiety ($\upsilon(N_2)$ = 2041 cm^{-1}) (Figure 20.13) [102].

Complex **21** reacts with HBF_4 in tetrahydrofuran (THF) at room temperature to form hydrazine in 17% yield, which could be improved to 47% if the one-electron donor $CrCl_2$ is added. Substitution of the phosphine phenyl

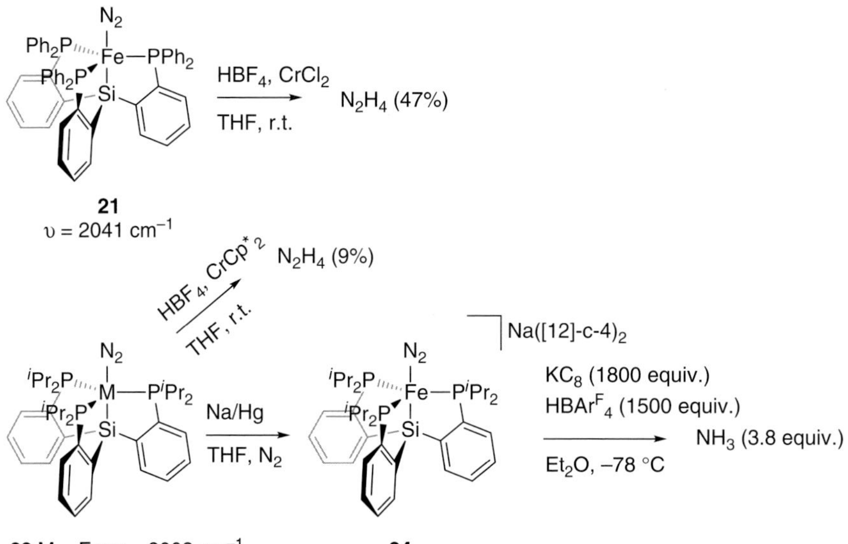

21
υ = 2041 cm^{-1}

22 M = Fe; υ = 2003 cm^{-1}
23 M = Co; υ = 2063 cm^{-1}

24
υ = 1920 cm^{-1}

Figure 20.13 Reactivity of $(SiP^R{}_3)Fe(N_2)$ complexes in the reduction of N_2.

groups for the more electron-donating isopropyl groups led to a more activated dinitrogen complex $(SiP^{i\text{-}Pr}{}_3)Fe(N_2)$ (**22**) ($\upsilon(N_2) = 2003\,cm^{-1}$). The cobalt analog $(SiP^{i\text{-}Pr}{}_3)Co(N_2)$ (**23**) was also prepared and showed the expected weaker activation of dinitrogen $2063\,cm^{-1}$ [103]. The reaction of complex **22** with HBF_4 in the presence of $CrCp^*{}_2$ led to the formation of only 9% of hydrazine. It was proposed that the more reducing nature of **22** caused preferential reduction of protons over dinitrogen reduction, giving rise to lower hydrazine yields [102]. One-electron reduction of **22** to $[(SiP^{i\text{-}Pr}{}_3)Fe(N_2)][Na([12]\text{-}c\text{-}4)_2]$ (**24**, [12]-c-4 = 12-crown-4) led to an even stronger charge transfer to the N_2 ligand ($\upsilon(N_2) = 1891\,cm^{-1}$). Interestingly, when **24** was subjected to excess KC_8 (50 equiv.) and $HBAr^F{}_4$ (46 equiv.) in diethyl ether at −78 °C, substoichiometric amounts of ammonia (0.7 equiv.) were observed, which was later optimized to give 3.8 ± 0.8 equiv. of ammonia (see below) [104].

Although iron complexes supported with the tris(phosphino)silyl ligands did not show true catalytic activity, they proved to be good scaffolds for mechanistic investigations. The iron(II) hydrazine and ammonia complexes with $SiP^{i\text{-}Pr}{}_3$ ligands (**25** and **26**, respectively) were prepared, which support their possible intermediacy in the reduction of N_2 (Figure 20.14) [105]. More importantly,

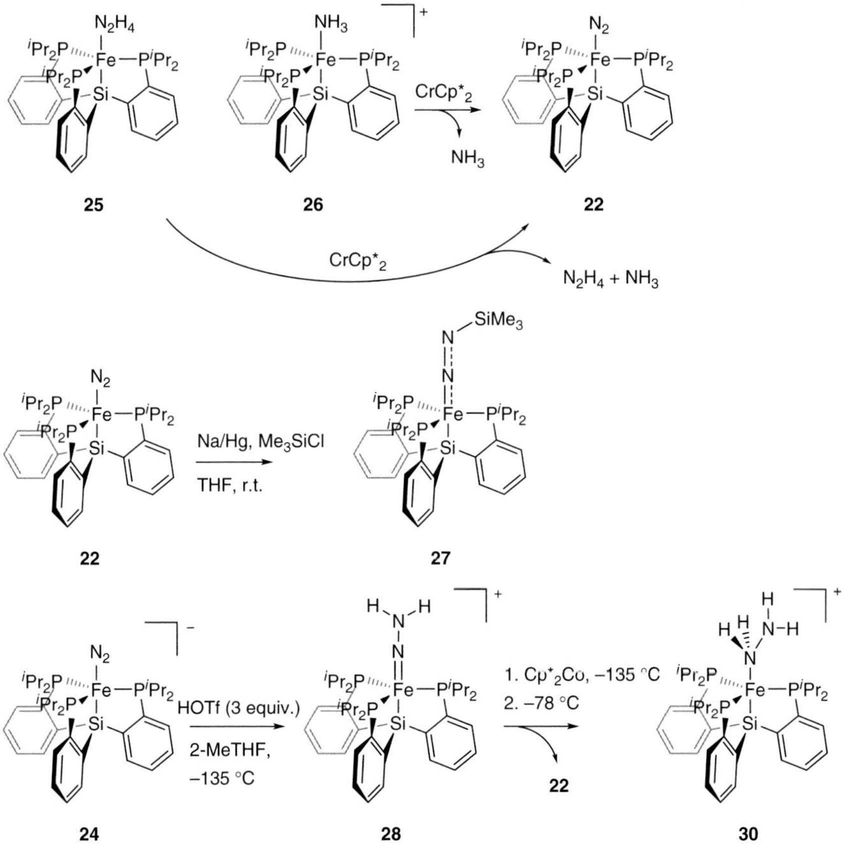

Figure 20.14 Mechanistic investigations on the $(SiP^{i\text{-}Pr}{}_3)Fe(N_2)$ system.

one-electron reduction of complexes **25** or **26** using $CrCp^*_2$ in THF under dinitrogen at room temperature resulted in the formation of the iron(I)-N_2 adduct **22**, accompanied by the liberation of NH_3 and/or N_2H_4, which showed the potential to make the system catalytic. The addition of 1 equiv. of Me_3SiCl and Na/Hg to **22** in THF resulted in the formation of the stable diazenido complex $(SiP^{i\text{-}Pr}{}_3)Fe(N_2SiMe_3)$ (**27**), which shows that the terminal nitrogen atom is prone to reaction with electrophiles [105]. Stoichiometric reaction of **24** with protons could be studied at low temperatures. Addition of 3 equiv. of HOTf in 2-MeTHF to $[(SiP^{i\text{-}Pr}{}_3)FeN_2]^-$ (**24**) at −135 °C led to the formation of the hydrazide complex $[(SiP^{i\text{-}Pr}{}_3)Fe{=}NNH_2]^+$ (**28**), one of the intermediates of ammonia synthesis along the distal pathway (Figure 20.7) [60]. Interestingly, compound **28** reacts with its one-electron reduced analog $[(SiP^{i\text{-}Pr}{}_3)Fe{=}NNH_2]$ (**29**) to form $[(SiP^{i\text{-}Pr}{}_3)FeNH_2NH_2]^+$ (**30**) and the neutral dinitrogen complex **22** through disproportionation of the diazenido ligand. Complex **30** can be viewed as one of the intermediates along the alternating pathway. These results show that N_2 bound to iron can be doubly protonated to generate a distal intermediate and subsequently disproportionate to an alternating intermediate. As such, a hybrid distal/alternating pathway is viable for the iron system.

A real breakthrough in catalytic dinitrogen fixation with iron was achieved with the use of the tripodal, neutral Lewis acidic borane ($BP^{i\text{-}Pr}{}_3$ = tris[2-(diisopropyl-phosphino)-phenyl]borane) scaffold [106]. The use of this borane ligand was expected to stabilize intermediates with multiple iron–nitrogen bonds, which would be formed along the distal pathway. The dinitrogen ligand in $(BP^{i\text{-}Pr}{}_3)FeN_2$ (**31**) is rather weakly activated ($\upsilon(N_2)$ = 2011 cm^{-1}) (Figure 20.15); however, one-electron reduction with sodium amalgam in the presence of a crown ether results in the formation of $[(BP^{i\text{-}Pr}{}_3)Fe(N_2)][Na([12]\text{-}c\text{-}4)]$ complex **32**, which not only reveals strong spectroscopic activation of N_2 ($\upsilon(N_2)$ = 1905 cm^{-1}) but also is catalytically active in N_2 reduction. Complex **32** catalyzed the formation of 7 equiv. of ammonia when reacted with $[H(Et_2O)_2][BAr^F{}_4]$ (46 equiv.) and KC_8 (50 equiv.) at −78 °C in Et_2O under 1 atm of N_2 (Figure 20.15) [104]. The choice of reagents was of crucial importance as the use of other reductants ($CoCp^*_2$, $CrCp^*_2$, K) or acids (HOTf, LutH$[BAr^F{}_4]$, HCl) decreased the yield to less than 0.6 equiv. of NH_3. Even more crucial was the purity of the acid. In an optimized procedure using ~30 times higher loading of KC_8 (1600–1800 equiv.) and highly purified $HBAr^F{}_4$ (1500 equiv.) in diethyl ether at −78 °C, 59 ± 6 equiv. of NH_3 was formed. Under these conditions, $SiP^{iPr}{}_3{}^-$ complex **24** catalyzed the formation of 3.8 ± 0.8 equiv. of NH_3, and the $CP^{iPr}{}_3{}^-$ complex **38** (Figure 20.19) 36 ± 7 equiv. of NH_3 [107]. Electrochemical studies revealed that reduction of N_2 is possible already at −2.2 V (vs Fc/Fc^+), implying that chemical reductants that are weaker than KC_8 can also be used. However, in comparison with sodium amalgam, the use of KC_8 led to higher yields of NH_3 because of a lower rate of the competing background proton reduction. Electrochemical reduction of N_2 could be performed; however, only 2.2 equiv. of ammonia per iron atom was produced.

Reactivity studies of the iron $BP^{i\text{-}Pr}{}_3$ complexes shed light on the possible mechanism of the catalytic reaction. Reaction of **32**$[Na([12]\text{-}c\text{-}4)]$ with 10 equiv. of $[H(Et_2O)_2][BAr^F{}_4]$ allowed for the characterization of the transient

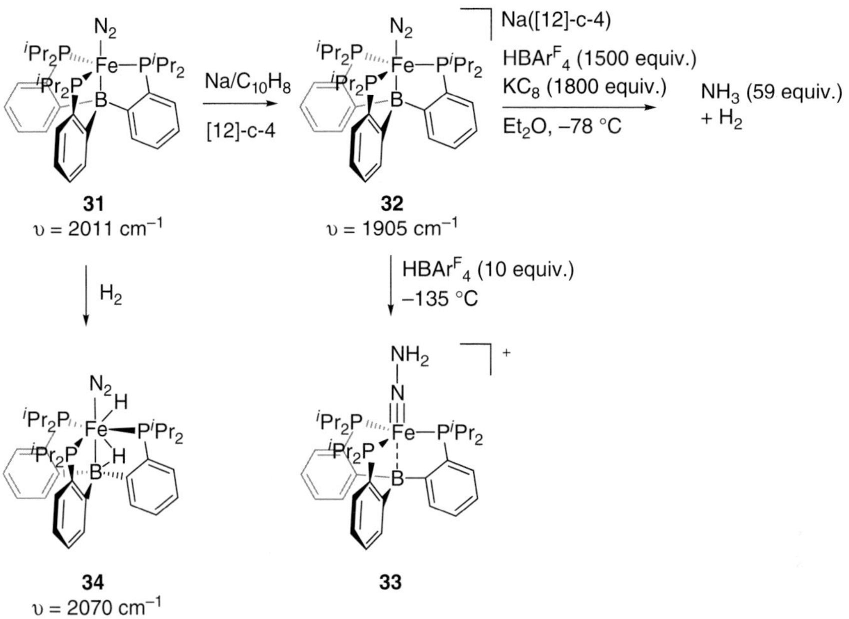

Figure 20.15 Reactivity of the $(BP^{i\text{-}Pr}{}_3)FeN_2$ (31) complex with acid and reductant and with molecular hydrogen.

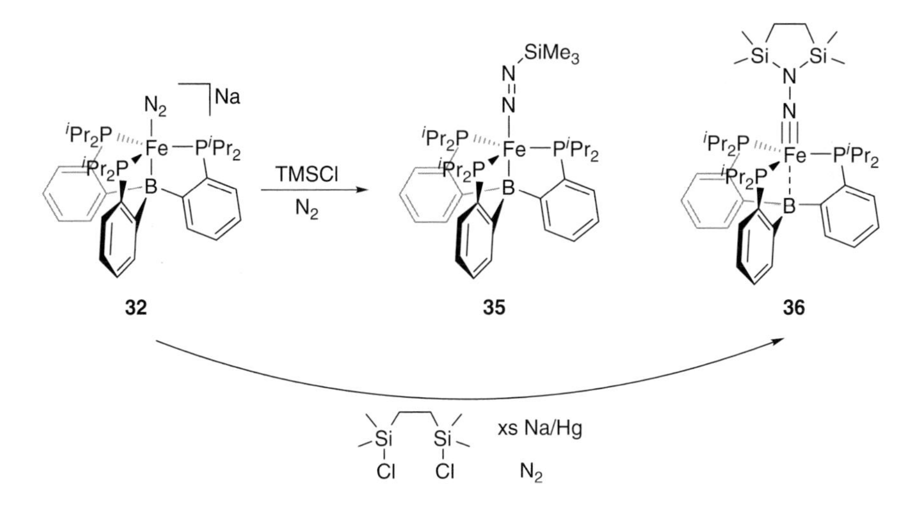

Figure 20.16 Stepwise addition of silicon electrophiles to $[(BP^{i\text{-}Pr}{}_3)FeN_2]^-$ (**32**).

iron–hydrazido complex $[(BP^{i\text{-}Pr}{}_3)Fe{\equiv}N{-}NH_2]^+$ (**33**) [108]. The formation of this species indicates that the two initial protonation steps during the catalytic cycle proceed through a distal pathway. Isolation of mono- and disilylated intermediates **35** and **36** upon reaction with silylchlorides further supports the distal mechanism in the initial steps of the catalytic reaction (Figure 20.16) [109].

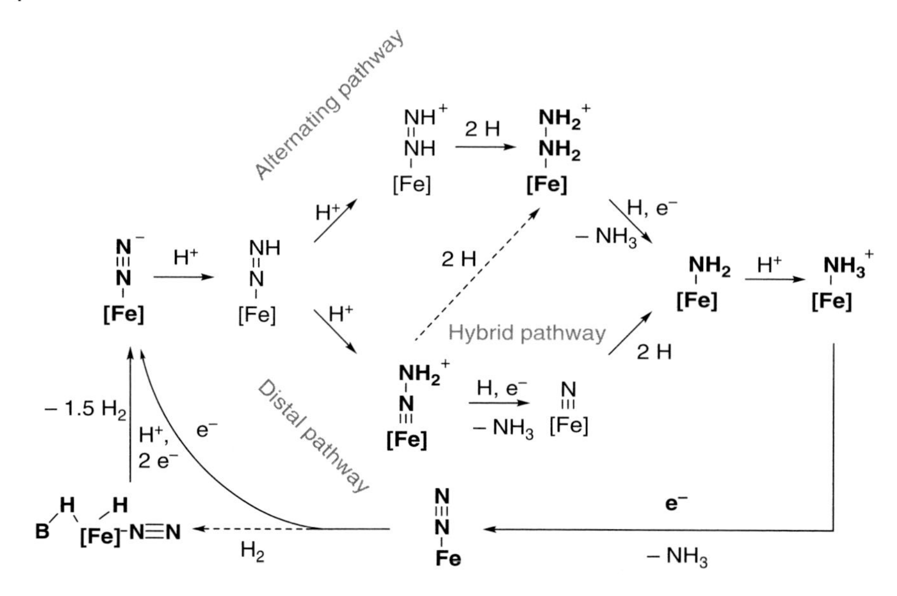

Figure 20.17 Possible mechanisms of reduction of N_2 to NH_3 using $[(BP^{i\text{-}Pr}_3)FeN_2]^-$ (**32**) as the catalyst. Characterized intermediates are shown in bold.

The mechanism for the catalysis with complex **32** may proceed similarly to the mechanism proposed for $(SiP^{i\text{-}Pr}_3)Fe(N_2)$ complex **24** (Figure 20.14): the addition of two protons along the distal pathway going to the hybrid pathway, after which an intermediate with coordinated hydrazine is formed and subsequent addition proceeds through the alternating pathway (Figure 20.17). This hypothesis is supported by the isolation of nitrogen fixation intermediates featuring N—NH_2, N_2H_4, NH_2, and NH_3 ligands and the release of ammonia upon reduction of these intermediates [108, 110]. Kinetic studies revealed first order in catalyst concentration and zeroth order in acid concentration. The initial turnover frequency (TOF) is $1.2 \pm 0.1\ \text{min}^{-1}$, which is the highest reported for any synthetic system. This activity at −78 °C is remarkably high considering the fact that the FeMo nitrogenase from *Klebsiella pneumoniae* exhibits a TOF of approximately $80\ \text{min}^{-1}$ at room temperature [111]. Further studies revealed no significant decomposition of the catalyst during the course of the reaction, and the decreasing efficiency at higher turnover was assigned partly because of the buildup of NH_3, which acts as an inhibitor.

Proton reduction to form H_2 is the main side reaction of nitrogen reduction under acidic conditions, which (apart from the background reaction between protons and KC_8) is catalyzed by complex **32**. The hydrogen evolution activity of **32** is, however, much lower than that of the silicon analog $[(SiP^{iPr}_3)Fe(N_2)]^-$ **24**, which, in turn, is a worse nitrogen reduction catalyst. Because both dinitrogen and proton reduction are competing reactions that occur simultaneously during the catalytic runs, it seems that increasing the selectivity toward nitrogen reduction is an important goal for improved efficiency of ammonia formation. During catalysis, the buildup of a dihydrido complex $(BP^{i\text{-}Pr}_3)(\mu\text{-H})Fe(H)(N_2)$ ($\upsilon(N_2) = 2070\ \text{cm}^{-1}$) [112] (**34**) (Figure 20.15) was observed with Mössbauer

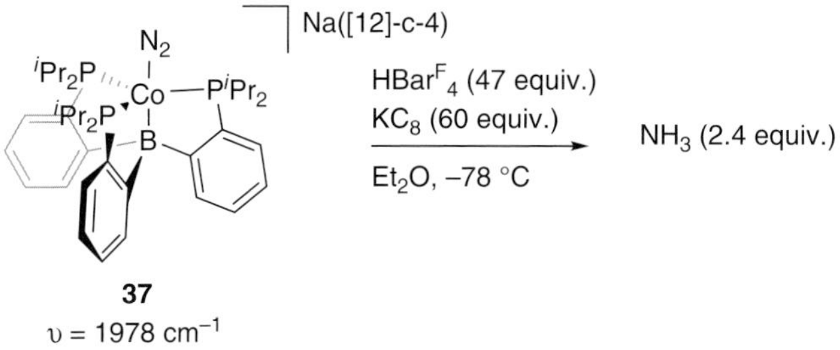

Figure 20.18 Ammonia formation catalyzed by $[(BP^{i\text{-}Pr}_3)Co(N_2)][Na([12]\text{-}c\text{-}4)_2]$.

spectroscopy. This hydride complex shows low reactivity toward N_2 reduction but can be converted back to **32** by addition of $[H(Et_2O)_2][BAr^F_4]$ and KC_8. Thus, complex **34** is likely a dormant state that temporarily exits the catalytic cycle and eventually converts back to the active catalyst **32** in time, releasing H_2 (Figure 20.17).

The cobalt analog $[(BP^{i\text{-}Pr}_3)Co(N_2)][Na([12]\text{-}c\text{-}4)_2]$ (**37**) ($\upsilon(N_2) = 1978\,cm^{-1}$) was also tested for its catalytic activity toward dinitrogen reduction (Figure 20.18) [113]. The dinitrogen ligand in the cobalt complex is less activated than in iron complex **32** because of the lower π-basicity of Co vs Fe. When **37** was subjected to $[H(Et_2O)_2][BAr^F_4]$ and KC_8 at −78 °C in Et_2O, super-stoichiometric amounts of NH_3 (2.4 equiv.) were observed. The $(SiP^{i\text{-}Pr}_3)Co$ and $(CP^{i\text{-}Pr}_3)Co$ analogs were also prepared and tested, but these did not produce any ammonia [114].

The Peters group also investigated tripodal systems with the $CP^{iPr}{}_3{}^-$ ($CP^{iPr}{}_{3-}$ = tris(o-diisopropylphosphinophenyl)methyl) ligand that features a carbon atom as the central donor [115]. The rationale behind the use of this ligand was that if the N_2 ligand binds terminally to the iron centers of nitrogenase, it would bind trans to the interstitial carbon atom [47–50, 53]. By stabilizing the negative charge on the carbon atom, the phenyl rings bound to the $CP^{iPr}{}_3{}^-$ ligand should allow for a flexible C—Fe bond, which should lead to the stabilization of various oxidation states of iron. The negatively charged iron(0) complex $[(CP^{iPr}{}_3)Fe(N_2)][K(Et_2O)_{0.5}]$ ($\upsilon(N_2) = 1870\,cm^{-1}$) (**38**$[K(Et_2O)_{0.5}]$) (Figure 20.19) showed a stronger activation of N_2 compared to silicon complex **23** or boron complex **32**. Reacting complex **38** with KC_8 and $[H(Et_2O)_2][BAr^F_4]$ at −78 °C under conditions optimized for $[(BP^{i\text{-}Pr}_3)Fe(N_2)]^-$ (**32**) in Et_2O led to

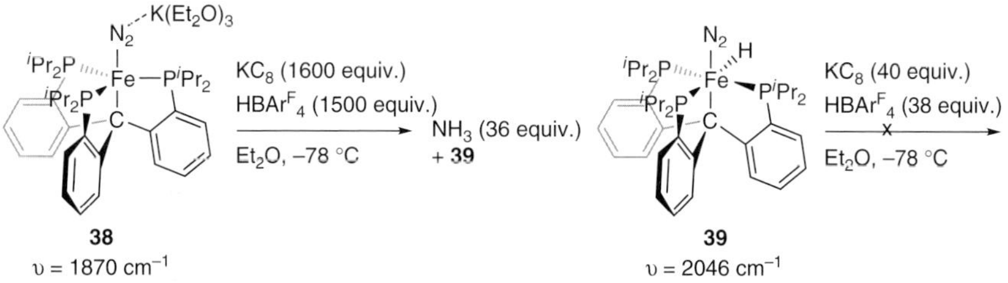

Figure 20.19 Catalysis with $(CP^{iPr}{}_3)Fe(N_2)^-$ (**38**) forming up to 36 equiv. of NH_3. The coformed $(CP^{iPr}{}_3)Fe(N_2)H$ complex (**39**) was found to be an inactive decomposition product.

the formation of 36 ± 7 equiv. of NH_3. Thus, despite stronger activation of the N_2 triple bond, the efficiency of complex **38** in ammonia production is slightly lower than for $BP^{i\text{-}Pr}{}_3$ complex **32**, which produced 59 ± 6 equiv. of NH_3.

To understand the limiting factors of the system, the reaction mixture was analyzed for catalyst decomposition products. The reaction mixture showed the formation of the dinitrogen mono-hydride complex $(CP^{i\text{-}Pr}{}_3)Fe(H)(N_2)$ (**39**) ($\upsilon(N_2) = 2046\,cm^{-1}$) in time reaching 70%. In contrast to the boron dihydrido complex $(BP^{i\text{-}Pr}{}_3)(\mu\text{-}H)Fe(H)(N_2)$ (**34**), **39** is inactive toward protons and reductant and does not regenerate to **38**. The initial rate of **38** has not been measured; however, it seems likely that formation of **39** upon protonation can be one of the reasons for the lower overall efficiency of **38** as compared to **34**.

20.3.3.2 Iron and Cobalt PNP Systems

The most efficient molybdenum systems for the reduction of dinitrogen to ammonia are based on PNP and PPP pincer complexes, which catalyze the formation of up to 63 equiv. of ammonia (Figure 20.12) [61, 99, 100, 116, 117]. Consequently, Nishibayashi and coworkers investigated the activity of iron and cobalt dinitrogen complexes supported with anionic PNP ligands. Treatment of the $(PNP^{tBu})Fe(N_2)$ complex (PNP^{tBu-} = 2,5-bis(di-*tert*-butylphosphinomethyl) pyrrolide) (**40**) with KC_8 (200 equiv.), $[H(OEt_2)_2]BAr^F{}_4$ (184 equiv.) in Et_2O at −78 °C under 1 atm of N_2 for one hour, led to the formation of 14.3 ± 0.4 equiv. of ammonia, 1.8 ± 0.2 equiv. of hydrazine, and 12.3 equiv. of H_2. (Figure 20.20) [118]. The nature of the solvent and reductant had a large influence on the performance of this system. Although in MeO^tBu comparable yields were observed, the use of coordinating THF resulted in the formation of 2.9 ± 0.2 equiv. of ammonia and 2.4 ± 0.1 equiv. of hydrazine. No activity was observed when toluene was used as a solvent. Using $CoCp^*{}_2$ ($Cp^* = C_5Me_5{}^-$) instead of KC_8 as the reductant decreased the yield by 50%. The low temperature was essential

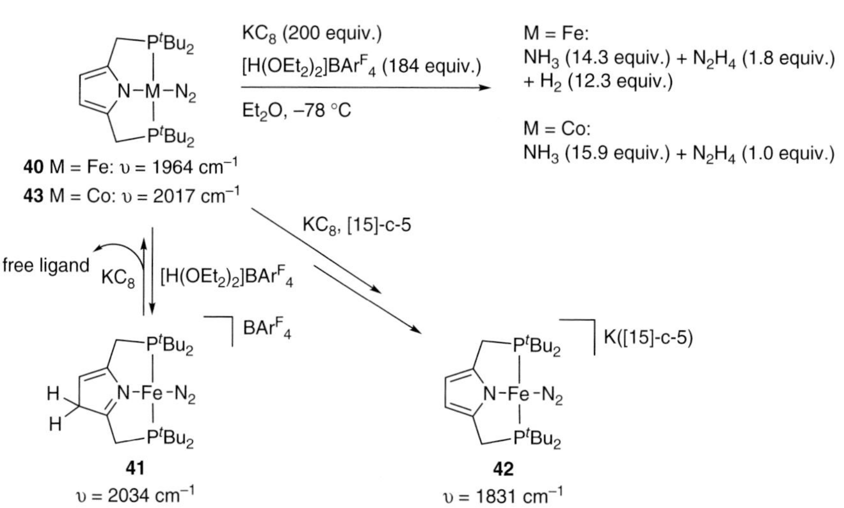

Figure 20.20 Formation of ammonia with $(PNP)MN_2$ complex **40** (M = Fe) and **43** (M = Co) and formation of reaction intermediates.

for the reduction of N_2 to occur. When the reaction was carried out at room temperature, 5.2 equiv. of H_2 was generated, and no dinitrogen reduction was observed. This clearly demonstrates that the competing proton reduction reaction is disfavored at low temperatures. The evolution of hydrazine during the catalytic ammonia formation indicates that the iron–hydrazine complex may be involved as a key intermediate during the reaction. Mechanistic studies were performed to shed light on the possible mechanism. Reaction of complex **40** with $[H(OEt_2)_2]BAr^F{}_4$ at room temperature does not lead to the protonation of the coordinated N_2. Instead, protonation occurs on the pyrrole ligand backbone forming complex **41**. The subsequent addition of reductant to this complex partially recovers **40** but also shows the formation of free PNP-H. This decomposition pathway accounts for a lower yield of ammonia (2.6 ± 0.2 equiv.) when the protonated complex is used as a catalyst. Thus, the formation of **41** is a likely deactivation pathway of the catalyst. DFT studies show that protonation of the dinitrogen ligand in complex **40** is thermodynamically unfavorable; however, for the one-electron reduced $[(PNP^{tBu})Fe^0(N_2)]^-$ (**42**) complex, protonation on the terminal nitrogen atom should be kinetically favorable over the protonation of the ligand. Complex **42** is isolable and shows strong activation of the N_2 ligand ($\upsilon(N_2) = 1832\,cm^{-1}$), which renders it a likely intermediate in the catalytic reaction.

The use of the anionic PNP^{tBu} ligand allowed for the first truly catalytic reduction of N_2 to ammonia using cobalt [119]. Under identical conditions to the ones applied for iron complex **40**, the cobalt complex $PNP^{tBu}CoN_2$ (**43**) catalyzed the formation of 15.9 ± 0.2 equiv. of ammonia and 1.0 ± 0.4 equiv. of hydrazine. A cobalt complex in which the *tert*-butyl groups of the PNP ligand were substituted with cyclohexyl groups showed slightly lower activity. The use of THF as a solvent led to a decrease in activity. Also, the choice of reductant is crucial as the use of metallic potassium instead of KC_8 resulted in the formation of no ammonia nor hydrazine. Compared to iron complex **40**, cobalt catalyst **43** seems to be more stable under catalytic conditions, as no protonation of the ligand could be detected. However, the overall performance in N_2 reduction is very similar.

20.3.3.3 The Cyclic Aminocarbene Iron System

Ung and Peters investigated whether catalytic reduction of N_2 can be performed using iron systems that do not contain phosphine ligands. A two-coordinate iron complex $(CAAC)_2Fe$ (**44**) (CAAC = cyclic(alkyl)(amino)carbene) supported with a CAAC π-accepting ligand reversibly binds N_2 at low temperatures (Figure 20.21) [120]. One-electron reduction of the transient dinitrogen complex $(CAAC)_2Fe(N_2)$ (**45**) with KC_8 in the presence of crown ether at −95 °C led to a clean formation of the isolatable negatively charged, moderately activated dinitrogen complex $[(CAAC)_2Fe(N_2)][K([18]\text{-c-}6)]$ (**46**) ($\upsilon(N_2)$ = $1850\,cm^{-1}$). Attempts to perform the reduction of complex **45** at room temperature resulted in decomposition, and at −78 °C, only traces of **46** were observed, indicating that significant amounts of the N_2-bound complex **45** are present only at extremely low temperatures. Therefore, catalytic runs were performed at −95 °C. When **44** was reacted with reductant (50 equiv.) and $HBAr^F{}_4\cdot 2Et_2O$ (50 equiv.) in Et_2O, 3.3 ± 1.1 equiv. of ammonia was formed. The use of **46**[K([18]-c-6)] at similar

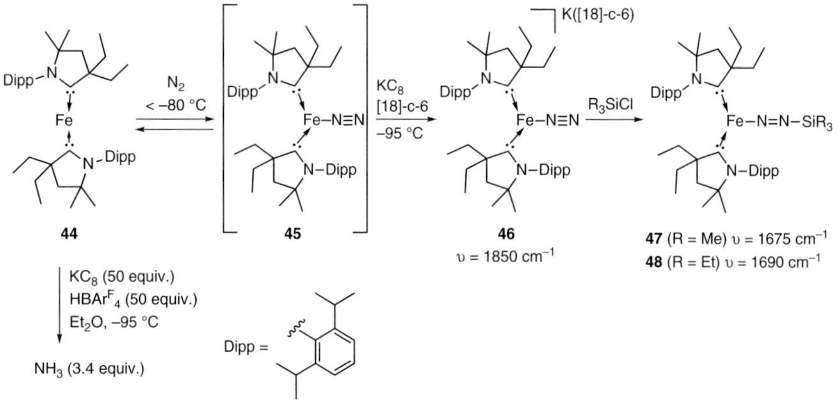

Figure 20.21 Ammonia formation by $(CAAC)_2Fe$ (44) and isolation of some reaction intermediates.

conditions led to a not significantly different ammonia yield (2.6 ± 0.6 equiv.), which suggests that the dinitrogen complex is indeed capable of catalyzing reduction of dinitrogen. Increasing the temperature to −78 °C or higher led to the formation of minor amounts of ammonia (0.9 ± 0.3 equiv. at −78 °C and 0.4 ± 0.2 equiv. at 23 °C), which is in accord with the weak binding of N_2 at higher temperatures. Additionally, the capability of activation of N_2 coordinated with the $(CAAC)_2Fe$ center toward addition of nucleophiles was probed by reacting **46** with chlorosilanes forming silyldiazenido iron complexes $(CAAC)_2Fe(N_2SiMe_3)$ (**47**) and $(CAAC)_2Fe(N_2SiEt_3)$ (**48**), of which the latter could be isolated. $(CAAC)_2Fe$ complex **44** was also an active catalyst for silylation of dinitrogen (see below).

20.3.3.4 The Diphosphine Iron System

Recently, the group of Ashley reinvestigated the Leigh system (Figure 20.22), and by careful choice of reductant and proton source, they were able to reach catalytic formation of hydrazine using $[Fe^0(depe)_2(N_2)]$ (**49**) ($\upsilon(N_2)1985 = cm^{-1}$) as the catalyst. During the reductive protonation of N_2 coordinated to **49**, the one-electron oxidized $[Fe^I(depe)_2(N_2)]^+$ (**50**) species is formed as a side product. The use of $CoCp^*_2$ as the reductant allowed for the reduction of complex **50** back to **49**, thus making a catalytic turnover possible. When **49** was reacted with $CoCp^*_2$ (270 equiv.) and $[PhNH_2]OTf$ (360 equiv.) at −78 °C in Et_2O, 24.5 ± 0.2 equiv. of hydrazine and 24.5 ± 0.2 equiv. of ammonia were formed. The choice of Et_2O as

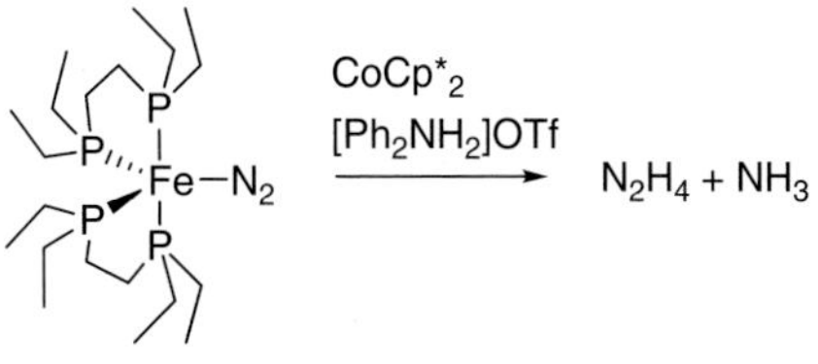

Figure 20.22 Catalytic formation of hydrazine by $[Fe^0(depe)_2(N_2)]$ (**49**).

a solvent was crucial, as in THF, the competing reaction between $CoCp^*_2$ and $[PhNH_2]OTf$ led to only substoichiometric formation of hydrazine (0.6 equiv) and ammonia (0.4 equiv.). The advantage of using $[PhNH_2]OTf$ as an acid in Et_2O stems from its sparing solubility in Et_2O and, as a consequence, low proton concentration. The selectivity toward hydrazine was also likely caused by solubility factors: the hydrazine salt $[N_2H_5]OTf$ is insoluble in diethyl ether and thus removed from the reaction by precipitation. The remaining bottleneck is the low stability of the catalyst, which leads to the decomposition under acidic condition, which is accompanied with protonation of the depe ligand. This system is remarkable as it is the first one that is selective toward the formation of hydrazine over ammonia.

20.4 Reduction of N_2 to Silylamines

One of the reasons of a still low performance of molecular catalysts for ammonia synthesis is poisoning with the product and dihydrogen formed as the side product. This is not the case in catalytic silylation reactions in which silylamines are formed from molecular dinitrogen and halosilanes in the presence of a suitable reductant. Therefore, this reaction generally yields more turnovers of N_2 fixation. The ability of transition metals to catalyze the formation of silylamines from dinitrogen was serendipitously discovered by Shiina in 1972, who observed consumption of N_2 during reductive silylation of benzene derivatives when using a nichrome wire stirrer [121]. Various transition metal chlorides were tested on their reactivity among which $CrCl_3$ performed best, producing up to 5.4 equiv. of the silylamine. Several other metal chlorides were also active, which included manganese, iron, cobalt, and nickel (Table 20.1).

Although single electrophilic addition of halosilanes to a transition-metal-coordinated N_2 molecule is well known [101, 122, 123], there was only one report on *catalytic* silylation before 2011. In 1989, Hidai and coworkers disclosed that a well-defined molybdenum tetraphosphine-based dinitrogen complex *cis*-$[Mo(N_2)_2(PMe_2Ph)_4]$ (**51**) ($\upsilon(N_2)$ = 1991 and 1913 cm^{-1}) (Figure 20.23) produced up to 24 equiv. of $N(SiMe_3)_3$ upon reacting $SiMe_3Cl$ and sodium with N_2 in THF [124]. This system has been further improved by the group

Table 20.1 Activity of transition metal chlorides in silylation of dinitrogen.

Catalyst	TON	Catalyst	TON
$TiCl_4$	0.8	$CoCl_2$	1.2
VCl_3	0.9	$NiCl_2$	0.2
$CrCl_3$	5.4	CuCl	0.0
$MnCl_2$	1.2	$MoCl_5$	1.0
$FeCl_3$	2.3	WCl_6	0.2

Conditions: 0.01 mol catalyst, 0.5 mol trimethylchlorosilane, 0.5 mol lithium wire, 150 ml THF, room temperature, 30 hours. TON = turnover number.

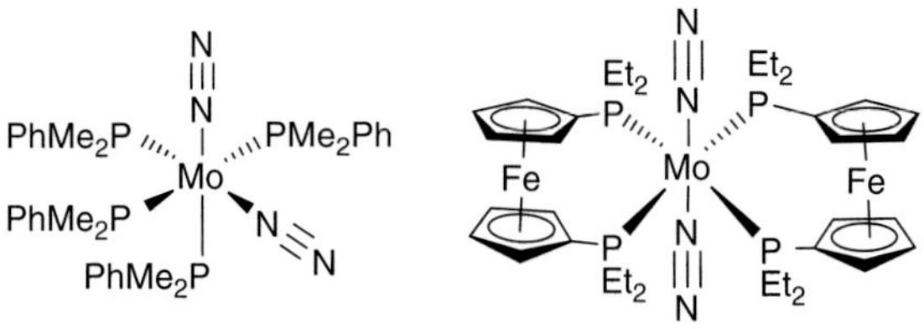

51

υ = 1991 and 1913 cm^{-1}

24 equiv. $N(SiMe_3)_3$

52

υ = 2093 cm^{-1}

226 equiv. $N(SiMe_3)_3$

Figure 20.23 Selected molybdenum catalysts that are active in the reduction of N_2 to silylamines.

of Nishibayashi, which employed bidentate ferrocenyldiphosphine ligands for stabilization of the molybdenum complex. The complex *trans*-$[Mo(N_2)_2(depf)_2]$ (depf = 1,10-bis(diethylphosphino)ferrocene) (**52**) ($\upsilon(N_2)$ = 2093 cm^{-1}) catalyzed the formation of 90 equiv. of tris(trimethylsilyl)amine in 20 hours. The turnover number (TON) could be pushed to 226 equiv. by running the reaction for 200 hours and adding a second batch of reactants halfway, which set the record for catalytic silylation of N_2. The very high activity of the ferrocenyldiphosphine-containing system could, in part, be caused by the activity of ferrocenes in nitrogen silylation reaction (vide infra).

20.4.1 Iron

In a subsequent study, Nishibayashi and coworkers showed that simple organoiron complexes are effective precatalysts for the reduction of N_2 to silylamines (Figure 20.24) [125]. Treatment of $Fe(CO)_5$ (**53**) or $FeCp_2$ (**54**) with 600 equiv. of trimethylsilyl chloride and sodium in THF under 1 atm. of N_2 for 20 hours at room temperature yielded 25 and 13 equiv. of $N(SiMe_3)_3$, respectively. Ferrocenes with trimethylsilyl-substituted cyclopentadienyl rings were even more effective, and for $[\{\eta^5\text{-}C_5H_2(SiMe_3)_3\}_2Fe]$ (**55**), the TON reached 34. Significant activity was also observed for $Fe(CO)_3(SiMe_3)_3$ (**56**), $Fe(CpSiMe_3)_2$ (**57**), and $[FeCp(CO)_2]_2$ (**58**). The use of THF as the solvent for **53** was crucial, as in benzene, hexane, or diethyl ether, no desired product was formed, whereas in 1,2-dimethoxyethane (TON = 2) and 1,4-dioxane (TON = 0.2), the yield was greatly diminished. Also, when lithium was used instead of sodium, the yield dropped to only 5 equiv. of $N(SiMe_3)_3$. An incubation period of approximately one hour during which the active catalyst is formed was observed when the reaction was followed in time. The fact that all precatalysts showed similar behavior in terms of activity, regardless of the initial catalyst structure, led to the conclusion that the same species is likely responsible for catalysis. Unfortunately, no dinitrogen-containing active species could be observed during catalysis. Based on the mercury test, the formation of active Fe-nanoparticles was also disregarded.

In the proposed mechanism (Figure 20.25), the catalytic cycle starts with the formation of $Fe^{II}(SiMe_3)_2(THF)$ (**59**), which coordinates dinitrogen, forming $Fe^{II}(N_2)(SiMe_3)_2(THF)$ (**60**). The dinitrogen reduction proceeds via a distal pathway (Figure 20.7) in which the addition of the first 3 equiv. of trimethylsilyl

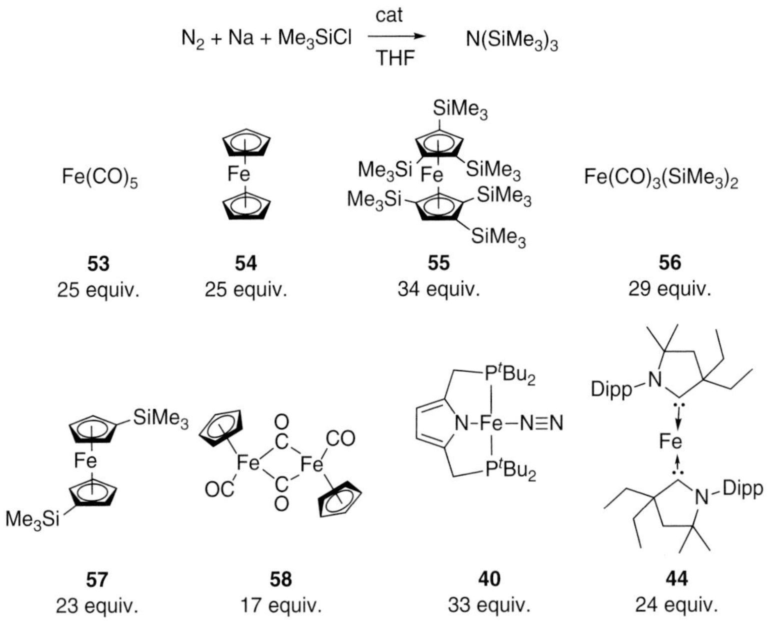

$$N_2 + Na + Me_3SiCl \xrightarrow[\text{THF}]{\text{cat}} N(SiMe_3)_3$$

$Fe(CO)_5$ **53** 25 equiv.; **54** 25 equiv.; **55** 34 equiv.; $Fe(CO)_3(SiMe_3)_2$ **56** 29 equiv.; **57** 23 equiv.; **58** 17 equiv.; **40** 33 equiv.; **44** 24 equiv.

Figure 20.24 Iron based precatalysts for silylation of dinitrogen.

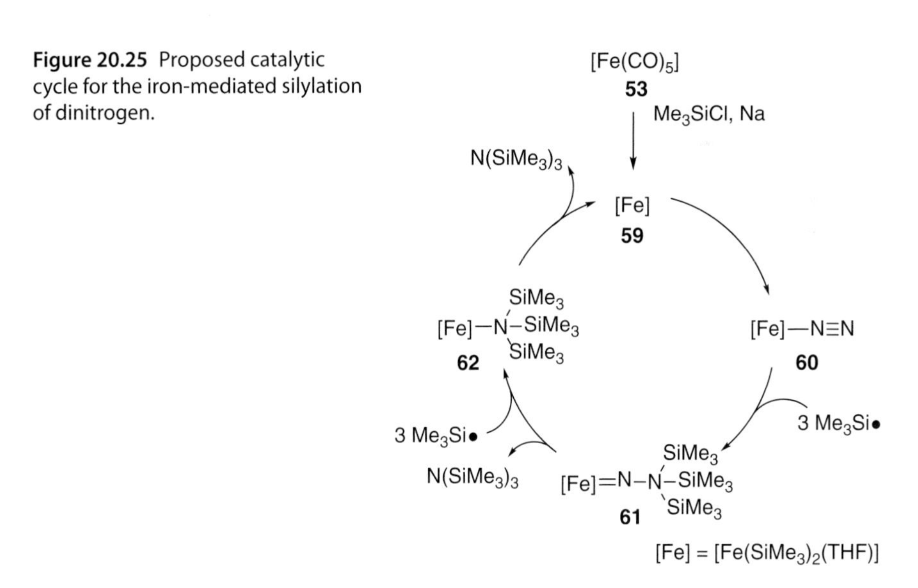

Figure 20.25 Proposed catalytic cycle for the iron-mediated silylation of dinitrogen.

radicals, which are formed by reduction of Me_3SiCl with sodium, leads to the formation of the first equivalent of $N(SiMe_3)_3$ and thus cleavage of the N—N bond. This results in the formation of an iron nitrido species that subsequently reacts with trimethylsilyl radicals to release the second equivalent of $N(SiMe_3)_3$. DFT calculations show the feasibility of such mechanism. This mechanism is substantially different than the proposed mechanisms by

Nishibayashi's molybdenum catalyst [126] and Lu's Co—Co dimer [82] (see below, Figure 20.27).

Ung and Peters' carbene-supported iron complex **44** (Figure 20.24) was also active in catalyzing silylation of N_2 [120]. When **44** was treated at room temperature with 600 equiv. of Me_3SiCl and KC_8, 24 ± 2.7 equiv. of $(Me_3Si)_3N$ were formed. At −78 °C, only 7 ± 1.0 equiv. is formed presumably because of a slower generation of the trimethylsilyl radical at low temperatures. Nishibayashi's iron dinitrogen complex **40**, which is catalytically active in ammonia formation at −78 °C, also revealed catalytic activity reaching the formation of 33 equiv. of tris(trimethylsilyl)amine at room temperature [118].

20.4.2 Cobalt

Catalytic silylation of N_2 using cobalt catalysts was independently reported by the groups of Lu and Nishibayashi. Following their discovery that simple organometallic iron complexes catalyze the formation of $N(SiMe_3)_3$, the group of Nishibayashi and coworkers evaluated the activity of $Co_2(CO)_8$ (**63**) (Figures 20.26 and 20.27) [127]. Initial screening revealed that the use of dimethoxyethyl ether (DME) as a solvent led to higher yields compared to THF (36 vs 25 equiv.), whereas in non-coordinating solvents (diethyl ether, dioxane, benzene, and hexane), the reduction of N_2 did not proceed at all. The hypothesis that the active species is stabilized by bidentate coordination of DME led to the development of the $Co_2(CO)_8 + 2$ bpy (bpy = 2,2′-bipyridine) system,

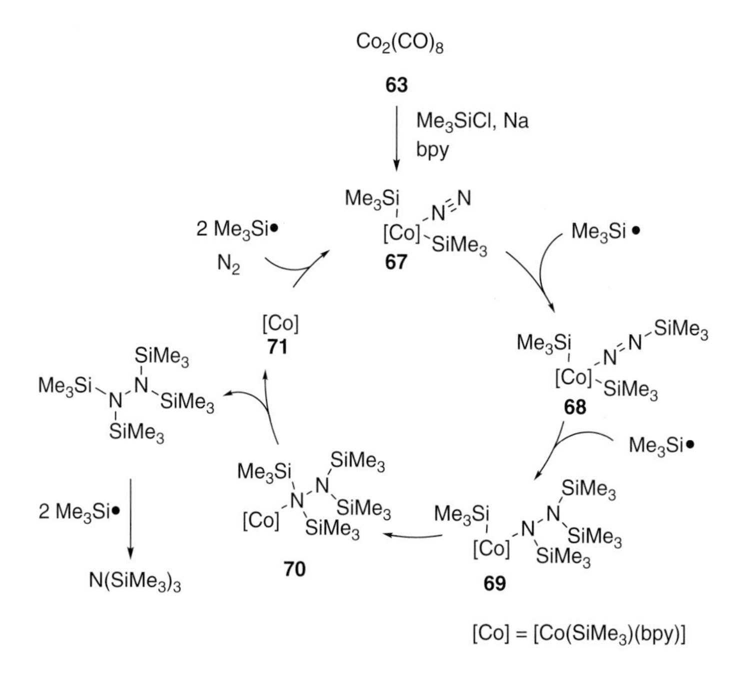

Figure 20.26 Proposed catalytic cycle for silylation of dinitrogen using the $Co_2(CO)_8$/bpy system.

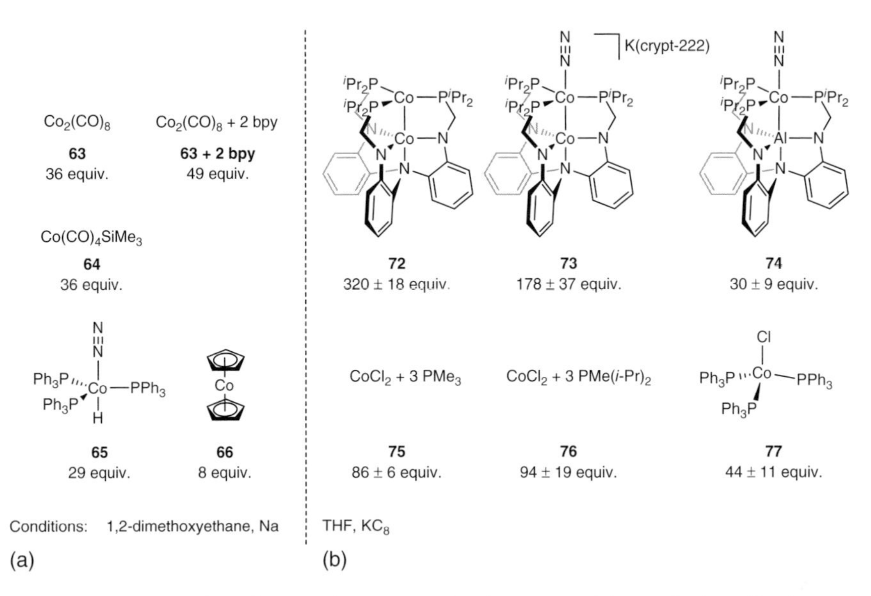

Figure 20.27 Cobalt-based precatalyst systems for dinitrogen silylation disclosed by the groups of Nishibayashi (a) and Lu (b).

which produced up to 49 equiv. of $N(SiMe_3)_3$ per cobalt atom. $Co(CO)_4SiMe_3$ (**64**), $CoH(PPh_3)_3N_2$ (**65**), and $CoCp_2$ (**66**) were less active (Figure 20.27). Similar to the iron systems mentioned before (Figure 20.24), the time profile of this reaction showed an incubation period of one hour, which pointed to a slow formation of the catalytically active species. The incubation period was proposed to involve the formation of a tris-trimethylsilyl cobalt(III) species (**67**); however, no experimental evidence for the formation of such species under the strongly reducing reaction conditions was provided. DFT calculations showed that this hypothetical species could promote radical addition to N_2 to form tetra(trimethylsilyl)hydrazine cobalt (**69**). Dissociated tetra(trimethylsilyl)hydrazine would undergo further reduction outside of the metal coordination sphere (Figure 20.26).

Carbonyls of other transition metals were also tested: $Mn_2(CO)_{10}$, $CpMn(CO)_3$, $Ru_3(CO)_{12}$, and $Mo(CO)_6$ allowed for the formation of 4, 3, 6, and 6 equiv. of $N(SiMe_3)_3$ per metal atom, respectively, whereas $NiCp_2$ and $Ni(cod)_2$ showed no activity.

A different approach was chosen by the group of Lu and coworkers who used a well-defined dicobalt complex as the precatalyst [128]. This complex consists of a dicobalt core supported by a trianionic ligand that features three phosphine donors that bind to a Co^I center and three amido and one amine donor that bind to a Co^{II} center (**72**) (Figure 20.27). In the presence of N_2 and 2.2.2-cryptand, the reaction with KC_8 results in one-electron reduction of Co^I to Co^0 and coordination of N_2, which results in a rather moderate activation of the N≡N bond (**73**) ($\upsilon(N_2) = 1994\,cm^{-1}$). In the presence of 2000 equiv. of Me_3SiCl and potassium graphite in THF, complex **72** provides 195 ± 25 turnovers of silylamine formation in 12 hours corresponding to 30% yield in Me_3SiCl. When an additional

2000 equiv. of Me_3SiCl and KC_8 was added and the reaction was performed for an additional 12 hours, the TON reached 320 ± 18. The complex remains bimetallic throughout at least the first 10 silyl radical additions. Measurement of the initial rates reveals a pseudo-first-order kinetics with a TOF of $1\,min^{-1}$. This, together with the lack of an incubation period, is consistent with a well-defined homogeneous catalyst. The one-electron-reduced anionic complex **73** reveals no significantly different activity. The use of KC_8 as a reductant for catalyst **72** proved to be crucial as the use of alkali metals led to a large decrease of activity, which was explained by the larger contact area of finely dispersed potassium graphite compared to metallic particles of K, Na, or Li. Bimetallic aluminum cobalt complex **74** ($\upsilon(N_2) = 2081\,cm^{-1}$) [129] was still active in the dinitrogen reduction reaction but reached a much lower activity yielding 30 ± 9 equiv of $N(SiMe_3)_3$.

The proposed mechanism (Figure 20.28) [128] for the dicobalt tripodal complex, supported by DFT, is similar to the mechanism calculated for the molybdenum-catalyzed formation of tris(trimethylsilyl)amine [124, 126]. Starting from the anionic dinitrogen complex **73**, it follows a double (distal) addition at the terminal nitrogen (**78** and **79**). Dissociation of one of the phosphine arms (**80**) allows addition on the proximal nitrogen (**81**). Association of the phosphine arm (**72**) leads to a release of $[N_2(SiMe_3)_3]^-$, which in the presence of Me_3SiCl and reductants spontaneously converts to $N(SiMe_3)_3$ in a noncatalyzed pathway. Subsequent reduction of **72** and coordination of dinitrogen (**73**) closes

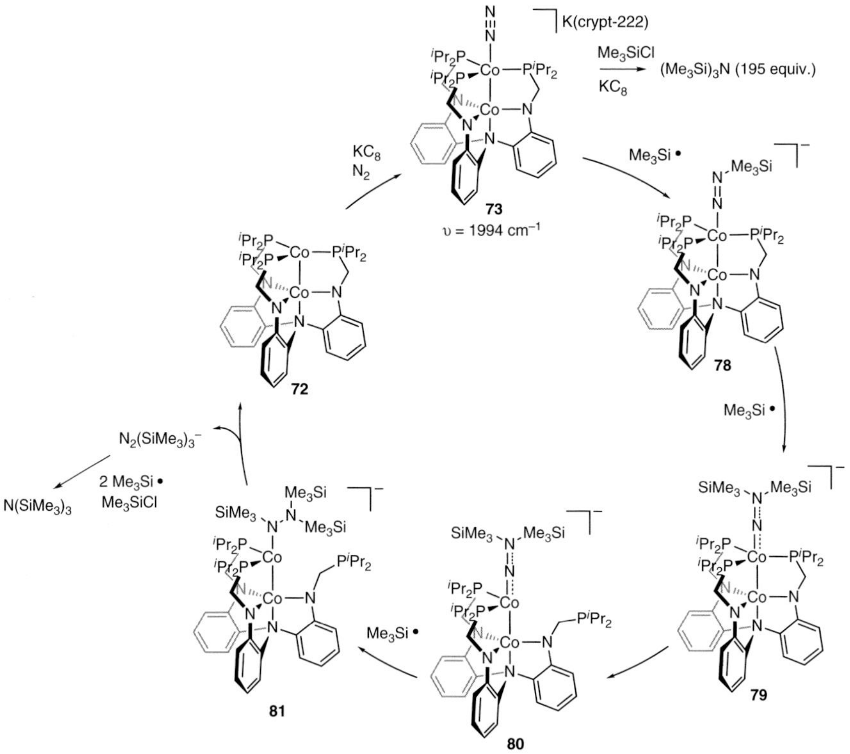

Figure 20.28 Proposed catalytic cycle for silylation of dinitrogen using dicobalt complex **72**.

the cycle. The dissociation of the phosphine arm in combination with the bimetallic character is thought to be the key for high turnovers and stability of the catalyst [130].

Other cobalt complexes were also found active under these conditions. For instance, in the presence of electron-donating phosphines (**75–77**), $CoCl_2$ reached a TON of up to 94 ± 19. Studies from our group on the silylation of N_2 using $CoH(PPh_3)_3N_2$ (**65**) as the precatalyst in THF showed that despite having coordinated dinitrogen, this complex is not the active catalyst and no direct N—Si bond formation is observed on the cobalt-bound N_2. The induction period before any $N(SiMe_3)_3$ is formed suggests that the sole presence of activated N_2 bound to a metal is not sufficient for its activation, and the putative active species has to be first formed by reacting with chlorosilane and a reductant [131].

20.5 Conclusions and Outlook

Because of the inherent inertness of dinitrogen, fixation of N_2 is a reaction that requires overcoming of a very high activation barrier. As a consequence, the industrial processes for ammonia production require the use of very high temperatures and pressures. Natural enzymes, however, can catalyze the formation of ammonia already at room temperature and under atmospheric pressure. This fact has been an inspiration for the development of catalysts that can reduce dinitrogen under similarly mild conditions. In recent years, the understanding of factors that influence the reactivity of the N_2 molecule reached a level that allowed for the development of the first catalyst systems that use base metals.

The current state-of-the-art iron and cobalt systems reach the performance of the best molybdenum-based catalysts yielding 59 and 16 equiv. of ammonia per metal center, respectively, at −78 °C. In comparison, the best molybdenum system reaches up to 63 equiv. of ammonia at room temperature. The mechanistic aspects of dinitrogen reduction to form ammonia using iron and cobalt systems are relatively well understood, and the possible catalyst deactivation pathways were in most cases investigated. The mechanistic studies performed on the catalytically active iron complexes show that the first two proton additions to the coordinated N_2 moiety occur at the distal nitrogen atom to form a hydrazido species, which subsequently undergoes protonation to form hydrazine, which is an intermediate during ammonia synthesis by nitrogenase. This is in stark contrast to the molybdenum systems, which operate via a distal pathway via a metal nitrido intermediate. An alternative approach in which the formation of ammonia would start with the homolytic splitting of the N_2 molecule has not yet led to catalytic turnover.

Several factors constitute to the current low overall performance of these catalysts. Impurities present in the reagents have detrimental effects on activity, and rigorous purification of reactants is necessary. Background reaction between the reductant and protons results in the formation of sufficient amounts of dihydrogen to slow down or even totally shut off the catalytic ammonia formation reaction. For this reason, reduction of dinitrogen has to be performed at very low temperatures under which the competing proton reduction reaction is retarded.

Thus, key to improvement of these catalysts would be preventing the rate of proton reduction. Lastly, product inhibition is another limiting factor.

Even though the exact nature of the catalytically active species is rather poorly understood, much more efficient catalysts are known for the dinitrogen silylation reaction. Cobalt is particularly effective in this transformation, and the reaction is believed to proceed via addition of silyl radicals to the metal bound N_2. As in many cases the structure of the catalyst is not known, rational design of the catalytic system is difficult. Many of the precatalysts used undergo transformations that lead to the formation of active species as evidenced by long induction periods of the catalytic reaction.

Despite increased understanding of the chemistry behind dinitrogen reduction at ambient conditions, much more research is needed before synthetically applicable catalysts are developed. However, especially if these catalysts could be coupled with electrochemical electron sources, they could lead to the development of greener routes toward amines. Overcoming these challenges could lead to the development of homogeneous systems that could be applied in the synthesis of nitrogen-containing compounds.

Acknowledgments

We thank the National Research School Combination Catalysis (NRSC-C) (F.F.W.) and the Netherlands Organization for Scientific Research (NWO-CW, VENI grant 722.013.002 for W.I.D.) for funding.

References

1 Nitrogen. http://www.vanderkrogt.net/elements/element.php?sym=N (accessed 23 November 2016).
2 van der Ham, C.J.M., Koper, M.T.M., and Hetterscheid, D.G.H. (2014). *Chem. Soc. Rev.* 43: 5183–5191.
3 Aftalion, F. (2001). *A History of the International Chemical Industry*. Chemical Heritage Foundation.
4 Logsdon, S. (2008). *Encyclopedia of Soil Science*. The Netherlands: Springer.
5 Smith, B.E., Richards, R.L., and Newton, W.E. (eds.) (2004). *Catalysts for Nitrogen Fixation: Nitrogenases, Relevant Chemical Models and Commercial Processes*. The Netherlands: Springer.
6 Hazari, N. (2010). *Chem. Soc. Rev.* 39: 4044–4056.
7 Howard, J.B. and Rees, D.C. (1996). *Chem. Rev.* 96: 2965–2982.
8 Cherkasov, N., Ibhadon, A.O., and Fitzpatrick, P. (2015). *Chem. Eng. Process.* 90: 24–33.
9 Khoenkhoen, N., de Bruin, B., Reek, J.N.H., and Dzik, W.I. (2015). *Eur. J. Inorg. Chem.* 2015: 567–598.
10 Walter, M.D. (2016). *Adv. Organomet. Chem.* 65: 261–377.
11 Schlögl, R. (2003). *Angew. Chem. Int. Ed.* 42: 2004–2008.

12 Zybert, M., Karasińska, M., Truszkiewicz, E. et al. (2015). *Pol. J. Chem. Technol.* 17: 138–143.
13 Hoffman, B.M., Dean, D.R., and Seefeldt, L.C. (2009). *Acc. Chem. Res.* 42: 609–619.
14 Seefeldt, L.C., Hoffman, B.M., and Dean, D.R. (2009). *Annu. Rev. Biochem.* 78: 701–722.
15 Rees, J.A., Bjornsson, R., Schlesier, J. et al. (2015). *Angew. Chem. Int. Ed.* 54: 13249–13252.
16 Zhao, Y., Bian, S.M., Zhou, H.N., and Huang, J.F. (2006). *J. Integr. Plant Biol.* 48: 745–755.
17 Tanabe, Y. and Nishibayashi, Y. (2016). *Chem. Rec.* 16: 1549–1577.
18 Tanabe, Y. and Nishibayashi, Y. (2013). *Coord. Chem. Rev.* 257: 2551–2564.
19 Miyazaki, T., Tanabe, Y., Yuki, M. et al. (2013). *Chem. Eur. J.* 19: 11874–11877.
20 Klopsch, I., Kinauer, M., Finger, M. et al. (2016). *Angew. Chem. Int. Ed.* 55: 4786–4789.
21 Fryzuk, M.D. and Johnson, S.A. (2000). *Coord. Chem. Rev.* 200–202: 379–409.
22 Holland, P.L. (2010). *Dalton Trans.* 39: 5415–5425.
23 Crabtree, R. (2009). *The Organometallic Chemistry of the Transition Metals.* Hoboken, NJ: Wiley.
24 McWilliams, S.F., Rodgers, K.R., Lukat-Rodgers, G. et al. (2016). *Inorg. Chem.* 55: 2960–2968.
25 Rittle, J., McCrory, C.C.L., and Peters, J.C. (2014). *J. Am. Chem. Soc.* 136: 13853–13862.
26 Sacco, A. and Aresta, M. (1968). *Chem. Commun.* 1223–1224.
27 Takaoka, A., Mankad, N.P., and Peters, J.C. (2011). *J. Am. Chem. Soc.* 133: 8440–8443.
28 Hills, A., Hughes, D.L., Jimenez-Tenorio, M. et al. (1993). *J. Chem. Soc., Dalton Trans.* 3041–3049.
29 Hellman, A., Baerends, E.J., Biczysko, M. et al. (2006). *J. Phys. Chem. B* 110: 17719–17735.
30 Ertl, G. (2001). *Chem. Rec.* 1: 33–45.
31 Ertl, G. (1980). *Catal. Rev. Sci. Eng.* 21: 201–223.
32 Rayment, T., Schloegl, R., Thomas, J.M., and Ertl, G. (1985). *Nature* 315: 311–313.
33 Ertl, G. (2008). *Angew. Chem. Int. Ed.* 47: 3524–3535.
34 Grunze, M., Golze, M., Hirschwald, W. et al. (1984). *Phys. Rev. Lett.* 53: 850–853.
35 Whitman, L.J., Bartosch, C.E., Ho, W. et al. (1986). *Phys. Rev. Lett.* 56: 1984–1987.
36 Maxwell, G. (2006). *Synthetic Nitrogen Products: A Practical Guide to the Products and Processes.* Springer US.
37 Lloyd, L. (2011). *Handbook of Industrial Catalysts.* New York: Springer US.
38 Laplaza, C.E. and Cummins, C.C. (1995). *Science* 268: 861–863.
39 Klopsch, I., Finger, M., Würtele, C. et al. (2014). *J. Am. Chem. Soc.* 136: 6881–6883.

40 Curley, J.J., Sceats, E.L., and Cummins, C.C. (2006). *J. Am. Chem. Soc.* 128: 14036–14037.
41 Smith, J.M., Lachicotte, R.J., Pittard, K.A. et al. (2001). *J. Am. Chem. Soc.* 123: 9222–9223.
42 Ding, K., Pierpont, A.W., Brennessel, W.W. et al. (2009). *J. Am. Chem. Soc.* 131: 9471–9472.
43 Pfirrmann, S., Limberg, C., Herwig, C. et al. (2009). *Angew. Chem. Int. Ed.* 48: 3357–3361.
44 Rodriguez, M.M., Bill, E., Brennessel, W.W., and Holland, P.L. (2011). *Science* 334: 780–783.
45 Grubel, K., Brennessel, W.W., Mercado, B.Q., and Holland, P.L. (2014). *J. Am. Chem. Soc.* 136: 16807–16816.
46 Lee, Y., Sloane, F.T., Blondin, G. et al. (2015). *Angew. Chem. Int. Ed.* 54: 1499–1503.
47 Spatzal, T., Aksoyoglu, M., Zhang, L. et al. (2011). *Science* 334: 940–940.
48 Lancaster, K.M., Roemelt, M., Ettenhuber, P. et al. (2011). *Science* 334: 974–977.
49 Lancaster, K.M., Hu, Y., Bergmann, U. et al. (2013). *J. Am. Chem. Soc.* 135: 610–612.
50 Wiig, J.A., Hu, Y., Lee, C.C., and Ribbe, M.W. (2012). *Science* 337: 1672–1675.
51 Holland, P.L. (2005). *Can. J. Chem.* 83: 296–301.
52 Kästner, J. and Blöchl, P.E. (2007). *J. Am. Chem. Soc.* 129: 2998–3006.
53 Hinnemann, B. and Nørskov, J.K. (2004). *J. Am. Chem. Soc.* 126: 3920–3927.
54 Peters, J.C. and Mehn, M.P. (2006). Bio-organometallic approaches to nitrogen fixation chemistry. In: *Activation of Small Molecules: Organometallic and Bioinorganic Perspectives*. Weinheim: Wiley-VCH.
55 Crossland, J.L. and Tyler, D.R. (2010). *Coord. Chem. Rev.* 254: 1883–1894.
56 Siegbahn, P.E.M. (2016). *J. Am. Chem. Soc.* 138: 10485–10495.
57 Čorić, I. and Holland, P.L. (2016). *J. Am. Chem. Soc.* 138: 7200–7211.
58 Hoffman, B.M., Lukoyanov, D., Dean, D.R., and Seefeldt, L.C. (2013). *Acc. Chem. Res.* 46: 587–595.
59 Yelle, R.B., Crossland, J.L., Szymczak, N.K., and Tyler, D.R. (2009). *Inorg. Chem.* 48: 861–871.
60 Rittle, J. and Peters, J.C. (2016). *J. Am. Chem. Soc.* 138: 4243–4248.
61 Arashiba, K., Miyake, Y., and Nishibayashi, Y. (2011). *Nat. Chem.* 3: 120–125.
62 Yandulov, D.V. and Schrock, R.R. (2003). *Science* 301: 76–78.
63 Haight, G.P. and Scott, R. (1964). *J. Am. Chem. Soc.* 86: 743–744.
64 Vol'Pin, M.E. and Shur, V.B. (1966). *Nature* 209: 1236–1236.
65 Leigh, J. and Jimenez-Tenorio, M. (1991). *J. Am. Chem. Soc.* 113: 5862–5863.
66 Gilbertson, J.D., Szymczak, N.K., and Tyler, D.R. (2004). *Inorg. Chem.* 43: 3341–3343.
67 Balesdent, C.G., Crossland, J.L., Regan, D.T. et al. (2013). *Inorg. Chem.* 52: 14178–14187.
68 Doyle, L.R., Hill, P.J., Wildgoose, G.G., and Ashley, A.E. (2016). *Dalton Trans.* 45: 7550–7554.
69 Tyler, D.R. (2015). *Z. Anorg. Allg. Chem.* 641: 31–39.

70 Field, L.D., Li, H.L., and Magill, A.M. (2009). *Inorg. Chem.* 48: 5–7.

71 Crossland, J.L., Balesdent, C.G., and Tyler, D.R. (2009). *Dalton Trans.* 3: 4420–4422.

72 Crossland, J.L., Balesdent, C.G., and Tyler, D.R. (2012). *Inorg. Chem.* 51: 439–445.

73 Field, L.D., Li, H.L., Dalgarno, S.J., and Turner, P. (2008). *Chem. Commun.* 33: 1680–1682.

74 Crossland, J.L., Zakharov, L.N., Tyler, D.R., and Uni, V. (2007). *Inorg. Chem.* 46: 10476–10478.

75 Field, L.D., Li, H.L., Dalgarno, S.J. et al. (2011). *Inorg. Chem.* 50: 5468–5476.

76 Chen, Y., Zhou, Y., Chen, P. et al. (2008). *J. Am. Chem. Soc.* 130: 15250–15251.

77 Chen, Y., Liu, L., Peng, Y. et al. (2011). *J. Am. Chem. Soc.* 133: 1147–1149.

78 Yuki, M., Miyake, Y., and Nishibayashi, Y. (2012). *Organometallics* 31: 2953–2956.

79 Vela, J., Stoian, S., Flaschenriem, C.J. et al. (2004). *J. Am. Chem. Soc.* 126: 4522–4523.

80 Stubbert, B.D., Vela, J., Brennessel, W.W., and Holland, P.L. (2013). *Z. Anorg. Allg. Chem.* 639: 1351–1355.

81 Sellmann, D. and Sutter, J. (1997). *Acc. Chem. Res.* 30: 460–469.

82 Chang, Y.H., Chan, P.M., Tsai, Y.F. et al. (2014). *Inorg. Chem.* 53: 664–666.

83 Sellmann, D., Böhlen, E., Waeber, M. et al. (1985). *Angew. Chem. Int. Ed. Engl.* 24: 981–982.

84 Sellmann, D., Hille, A., Rösler, A. et al. (2004). *Chem. Eur. J.* 10: 819–830.

85 Creutz, S.E. and Peters, J.C. (2015). *J. Am. Chem. Soc.* 137: 7310–7313.

86 Čorić, I., Mercado, B.Q., Bill, E. et al. (2015). *Nature* 526: 96–99.

87 Ohki, Y., Ikagawa, Y., and Tatsumi, K. (2007). *J. Am. Chem. Soc.* 129: 10457–10465.

88 Hashimoto, T., Ohki, Y., and Tatsumi, K. (2010). *Inorg. Chem.* 49: 6102–6109.

89 Liu, J., Kelley, M.S., Wu, W. et al. (2016). *Proc. Natl. Acad. Sci. U.S.A.* 113: 5530–5535.

90 Banerjee, A., Yuhas, B.D., Margulies, E.A. et al. (2015). *J. Am. Chem. Soc.* 137: 2030–2034.

91 Yandulov, D.V., Schrock, R.R., Rheingold, A.L. et al. (2003). *Inorg. Chem.* 42: 796–813.

92 Schrock, R.R. (2005). *Acc. Chem. Res.* 10338: 955–962.

93 Kinney, R.A., McNaughton, R.L., Chin, J.M. et al. (2011). *Inorg. Chem.* 50: 418–420.

94 Schrock, R.R. (2008). *Angew. Chem. Int. Ed.* 47: 5512–5522.

95 Yandulov, D.V. and Schrock, R.R. *Proc. Natl. Acad. Sci. U.S.A.* 2006, 103: 17099–17106.

96 Schenk, S., Le Guennic, B., Kirchner, B., and Reiher, M. (2008). *Inorg. Chem.* 47: 3634–3650.

97 Reiher, M., Le Guennic, B., and Kirchner, B. (2005). *Inorg. Chem.* 44: 9640–9642.

98 Schenk, S., Kirchner, B., and Reiher, M. (2009). *Chem. Eur. J.* 15: 5073–5082.

99 Kuriyama, S., Arashiba, K., Nakajima, K. et al. (2014). *J. Am. Chem. Soc.* 136: 9719–9731.
100 Arashiba, K., Kinoshita, E., Kuriyama, S. et al. (2015). *J. Am. Chem. Soc.* 137: 5666–5669.
101 Betley, T.A. and Peters, J.C. (2003). *J. Am. Chem. Soc.* 125: 10782–10783.
102 Mankad, N.P., Whited, M.T., and Peters, J.C. (2007). *Angew. Chem. Int. Ed.* 46: 5768–5771.
103 Whited, M.T., Mankad, N.P., Lee, Y. et al. (2009). *Inorg. Chem.* 48: 2507–2517.
104 Anderson, J.S., Rittle, J., and Peters, J.C. (2013). *Nature* 501: 84–87.
105 Lee, Y., Mankad, N.P., and Peters, J.C. (2010). *Nat. Chem.* 2: 558–565.
106 Moret, M.E. and Peters, J.C. (2011). *Angew. Chem. Int. Ed.* 50: 2063–2067.
107 Del Castillo, T.J., Thompson, N.B., and Peters, J.C. (2016). *J. Am. Chem. Soc.* 138: 5341–5350.
108 Anderson, J.S., Cutsail, G.E., Rittle, J. et al. (2015). *J. Am. Chem. Soc.* 137: 7803–7809.
109 Moret, M.-E. and Peters, J.C. (2011). *J. Am. Chem. Soc.* 133: 18118–18121.
110 Anderson, J.S., Moret, M.-E., and Peters, J.C. (2013). *J. Am. Chem. Soc.* 135: 534–537.
111 Thorneley, R.N. and Lowe, D.J. (1984). *Biochem. J.* 224: 887–894.
112 Fong, H., Moret, M.E., Lee, Y., and Peters, J.C. (2013). *Organometallics* 32: 3053–3062.
113 Suess, D.L.M., Tsay, C., and Peters, J.C. (2012). *J. Am. Chem. Soc.* 134: 14158–14164.
114 Del Castillo, T.J., Thompson, N.B., Suess, D.L.M. et al. (2015). *Inorg. Chem.* 54: 9256–9262.
115 Creutz, S.E. and Peters, J.C. (2014). *J. Am. Chem. Soc.* 136: 1105–1115.
116 Tanaka, H., Arashiba, K., Kuriyama, S. et al. (2014). *Nat. Commun.* 5: 1–11.
117 Kuriyama, S., Arashiba, K., Nakajima, K. et al. (2015). *Chem. Sci.* 6: 3940–3951.
118 Kuriyama, S., Arashiba, K., Nakajima, K. et al. (2016). *Nat. Commun.* 7: 12181.
119 Kuriyama, S., Arashiba, K., Tanaka, H. et al. (2016). *Angew. Chem. Int. Ed.* 55: 14291–14295.
120 Ung, G. and Peters, J.C. (2015). *Angew. Chem. Int. Ed.* 54: 532–535.
121 Shiina, K. (1972). *J. Am. Chem. Soc.* 94: 9266–9267.
122 Hidai, M. and Mizobe, Y. (1995). *Chem. Rev.* 95: 1115–1133.
123 Oshita, H., Mizobe, Y., and Hidai, M. (1993). *J. Organomet. Chem.* 456: 213–220.
124 Komori, K., Oshita, H., Mizobe, Y., and Hidai, M. (1989). *J. Am. Chem. Soc.* 111: 1940–1941.
125 Yuki, M., Tanaka, H., Sasaki, K. et al. (2012). *Nat. Commun.* 3: 1–6.
126 Tanaka, H., Sasada, A., Kouno, T. et al. (2011). *J. Am. Chem. Soc.* 133: 3498–3506.
127 Imayoshi, R., Tanaka, H., Matsuo, Y. et al. (2015). *Chem. Eur. J.* 21: 8905–8909.

128 Siedschlag, R.B., Bernales, V., Vogiatzis, K.D. et al. (2015). *J. Am. Chem. Soc.* 137: 4638–4641.

129 Alex Rudd, P., Planas, N., Bill, E. et al. (2013). *Eur. J. Inorg. Chem.* 2: 3898–3906.

130 Cammarota, R.C., Clouston, L.J., and Lu, C.C. (2016). *Coord. Chem. Rev.* 334: 100–111.

131 Dzik, W.I. (2016). *Inorganics* 4: 21.

Index

Non-Noble Metal Catalysis: Molecular Approaches and Reactions,
First Edition. Edited by Robertus J. M. Klein Gebbink and Marc-Etienne Moret.